AF539574

Structural Styles in Petroleum Exploration

C R JONES

Structural Styles in Petroleum Exploration

James D. Lowell

OGCI Publications
Oil & Gas Consultants International Inc.
Tulsa

4554 South Harvard Avenue
Tulsa, Oklahoma 74135

Printed in the United States of America

Library of Congress Catalog Card Number: 84-62622
International Standard Book Number: 0-930972-08-2

First printing—May, 1985
Second printing—July, 1987

To
Suzanne and Pearl

Contents

PREFACE

The intent of this book is to analyze the structural styles that we explore for oil and gas in terms of mechanics of deformation, profile and plan expression—both surface and subsurface, seismic expression, producing examples, and plate tectonic processes and habitats. In this attempt the book differs from other structural geology texts and is directed specifically to the petroleum explorationist.

The book is divided into chapters according to structural styles (Chapter 1), Basement involved—wrench assemblage (Chapter 2), compressive fault blocks and basement thrusts (Chapter 3), extensional fault blocks (Chapter 4), warps—arches, domes, and sags (Chapter 5), detached—thrust-fold belt assemblage (Chapter 6), normal-fault assemblage (Chapter 7), salt structures (Chapter 8), and shale structures (Chapter 9). Finally, Chapter 10 deals with superimposed styles, particularly structural inversion. References are given at the end of each chapter.

The various chapters represent the conversion of sections of a manual used in teaching courses in Structural Geology on a worldwide basis since 1976 for Oil and Gas Consultants International, Inc. The reader will appreciate that the literature of structural geology is voluminous; examples were chosen which seemed best suited to illustrate the structural styles concept. A large number of illustrations has been incorporated in the book; text has been kept short.

Chapter 1 (Structural styles, their plate-tectonic habitats, and hydrocarbon traps in petroleum provinces), reprinted from the July 1979 bulletin of The American Association of Petroleum Geologists, is for the most part an expanded abstract or summary of the book. Since publication of that article, concepts of some styles have been modified. For example, it is now known that detachment structures are an important part of the basement-involved compressive fault block style and probably the wrench assemblage as well. However, the concept of styles divided according to involvement or detachment of basement still serves as a convenient means of classifying structures (see, for example, Bally, 1983).

Clay models have frequently been used to illustrate in profile and plan view the structural patterns of many styles. Clay is an excellent experimental medium to demonstrate fault and fracture patterns for two reasons. First, clay fractures readily, and second, predictable patterns can be consistently reproduced for any given stress condition. Deficiencies of clay are that it is homogeneous and isotropic, whereas rocks are heterogeneous and anisotropic.

For over 30 years, I have benefited from associations and conversations with so many geologists and geophysicists that it would be impossible to name them all. However, I particularly wish to acknowledge my connections with Columbia, Nebraska, and Washington & Lee universities, American Overseas Petroleum Limited, Esso Production Research Company and Exxon Company U.S.A., and Northwest Exploration Company. In recent years most beneficial have been my contacts with participants in more than 60 training courses in Structural Geology, sponsored by Oil and Gas Consultants International, and my associations with colleagues and explorationists of the Rocky Mountain West.

REFERENCE

Bally, A. W., 1983, Seismic expression of structural styles: American Association Petroleum Geologists Studies in Geology Series #15, V.1, p. viii–ix.

1 STRUCTURAL STYLES—INTRODUCTION

The following article by T. P. Harding and myself is essentially an expanded abstract or summary of this book. A table of structural styles and an illustration of plate tectonic habitats of styles follows the article; both illustration and table are considered to be more comprehensive than their counterparts in the article.

STRUCTURAL STYLES, THEIR PLATE-TECTONIC HABITATS, AND HYDROCARBON TRAPS IN PETROLEUM PROVINCES[1]

Abstract Broadly interrelated assemblages of geologic structures constitute the fundamental structural styles of petroleum provinces. These assemblages generally are repeated in regions of similar deformation, and their associated hydrocarbon traps can be anticipated prior to exploration. Styles are differentiated on the basis of basement involvement or detachment of sedimentary cover. Basement-involved styles include wrench-fault structural assemblages, compressive fault blocks and basement thrusts, extensional fault blocks, and warps. Detached styles are decollement thrust-fold assemblages, detached normal faults (''growth faults'' and others), salt structures, and shale structures.

These basic styles are related to the larger kinematics of plate tectonics and, in some situations, to particular depositional histories. Most styles have preferred plate-tectonic habitats: (1) wrench faults at transform and convergent plate boundaries; (2) compressive fault blocks and basement thrusts at convergent boundaries, particularly in forelands and orogenic belts; (3) extensional fault blocks at divergent boundaries in all stages of completion and certain parts of convergent boundaries; (4) basement warps in a variety of plate-interior and boundary settings; (5) decollement thrust-fold belts in trench inner walls and foreland zones of convergent boundaries; (6) detached normal faults, usually in unstable, thick clastic wedges (mostly deltas); (7) salt structures primarily in interior grabens that may evolve to completed divergent boundaries; and (8) shale structures in regions with thick overpressured shale sequences.

Important differences in trend arrangements and structural morphologies provide criteria for differentiation of styles. These differences also result in different kinds of hydrocarbon traps. Wrench-related structural assemblages are concentrated along throughgoing zones and many have en echelon arrangements. The basic hydrocarbon trap is the en echelon anticline, in places assisted by closure directly against the wrench fault itself. Compressive and extensional fault styles typically have multiple, repeated trends, which combine to form zigzag, dogleg, or other grid patterns. Their main trap types are fault closures and drape folds above the block boundaries. Basement warps (domes, arches, etc.) are mostly solitary features and commonly provide long-lived positive areas for hydrocarbon concentration in broadly flexed closures.

Most decollement thrust-fold structures are arranged in long, sinuous belts and are repeated in closely spaced, wavelike bands. Effective closures include slightly to moderately disrupted compressive anticlines and lead

[1]Article by T. P. Harding and J. D. Lowell. Reprinted with permission from The American Association of Petroleum Geologists Bulletin, V. 63, n. 7, (July, 1979), p. 1016–1058.

edges of thrust sheets. Most detached normal faults are listric faults that occur in coalescing, cuspate bands parallel with the strike of contemporaneous sedimentation. Their basic hydrocarbon traps are associated rollover anticlines which are uniquely concentrated along the downthrown sides of major faults. Salt and shale structures are present both as buoyantly rising pillows, domes, ridges, etc., and as highly complex injected features caused by tectonic forces. Stratigraphic factors, such as truncation, wedging, onlap, and unconformity, add to the variety of traps in all styles.

In many places the structures of a petroleum province are either, or both, a gradation between the described fundamental styles and a mix of several styles. These structures can be further complicated by superimposition of fundamentally different tectonic environments. Additional modification of structures can result from still other factors inherent in the deformed region or in the particular tectonic event.

INTRODUCTION

The intent of this paper is to provide an initial framework for better understanding and predictability of structural trends and types of traps in petroleum provinces. Almost all geologic structures, if viewed in enough detail, have unique geometries and histories. On a more regional scale, however, certain general characteristics define broad categories of structures. These characteristics are the substance of this report.

Our structural classification employs the concepts of structural styles and comparative tectonics. The style of a particular region describes its dominant structural geometry. Basic styles are defined by an assemblage of

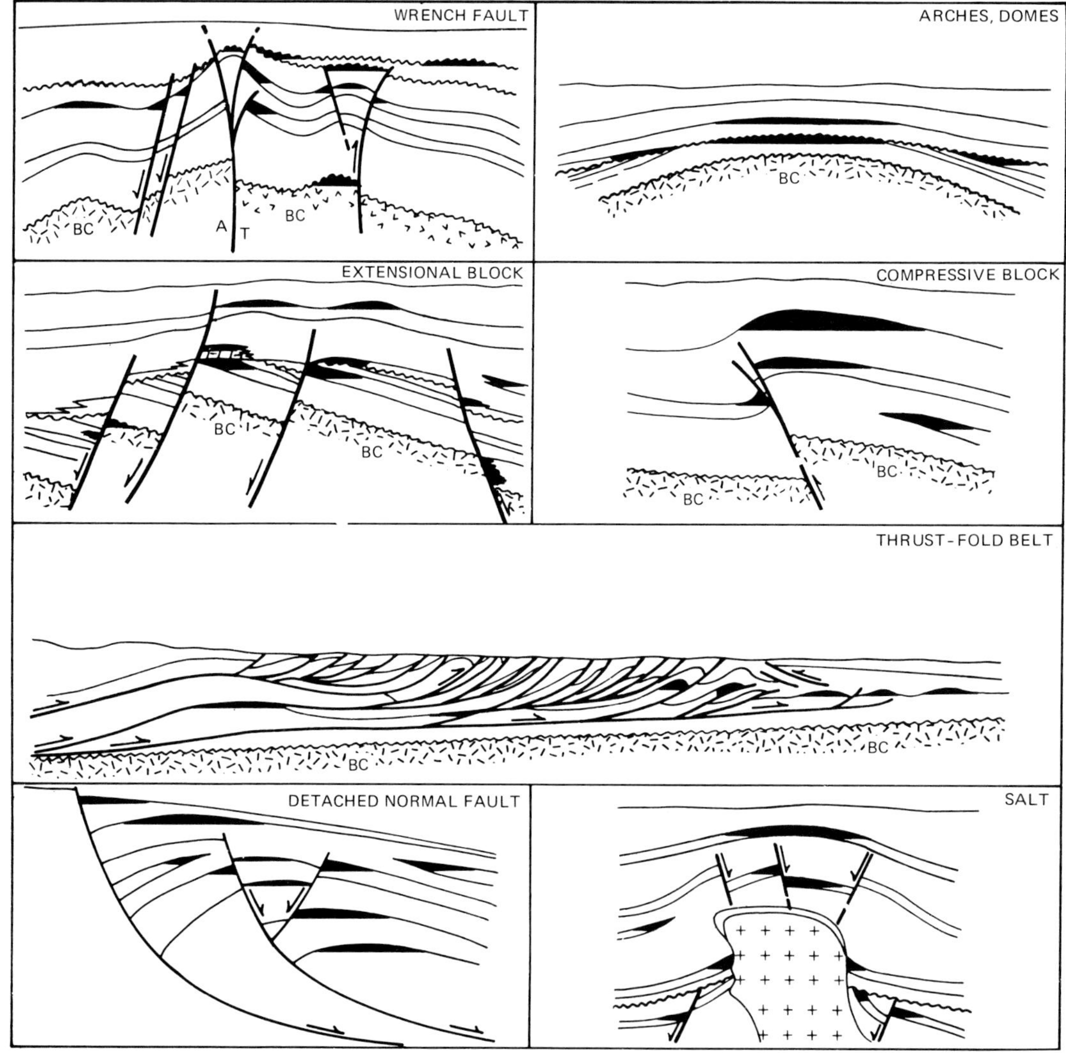

Fig. 1-1—Schematic diagrams of hydrocarbon traps (black areas) most commonly associated with structural styles of sedimentary basins. Purely stratigraphic type traps and traps associated with basement thrusts are omitted. Salt-related closures modified after Halbouty (1967, fig. 6). BC, basement complex; T, displacement toward viewer; A, away from viewer.

tectonically related elements and their spatial arrangement. Although styles are often modified by local rock differences, such as ductility contrasts and preexisting fabric, and tectonic events, including intensity, duration, and timing, the principal style or styles are usually discernible. For basic style definitions, morphologic types, essential repeatable characteristics, and general trend arrangements are emphasized rather than number of structures present in a specific region, their geographic localities, or particular histories. The concept of comparative tectonics involves the use of basic styles, well documented in one area, as guides for structural interpretation of a lesser known but similarly deformed area.

Several structural classifications have been made (Badgley, 1965, p. 50–97). None, however, have dealt directly with subsurface structures and associated hydrocarbon traps in sedimentary basins. Moreover, recent data in support of large-scale lithospheric plate movement show some earlier classifications that featured only a limited number of deformational mechanisms (e.g., the "verticalist" approach of Beloussov, 1959) are incomplete. Plate tectonics has also provided a unified concept of deformational processes and tectonic habitats to which structures can be related. Finally, deep-penetration reflection seismic surveys processed with modern techniques have revealed subsurface structures, such as detached normal faults ("growth faults"), that were poorly known at the time of past classifications. Modern seismic data have also resolved critical older tectonic controversies, demonstrating conclusively, for example, the validity of regional detachment or decollement in thrust-fold belts. (Bally et al., 1966).

In this paper we categorize structural styles in petroleum provinces and discuss key characteristics for their differentiation. Each style is treated as to common plate-tectonic settings, typical structural patterns and mor-

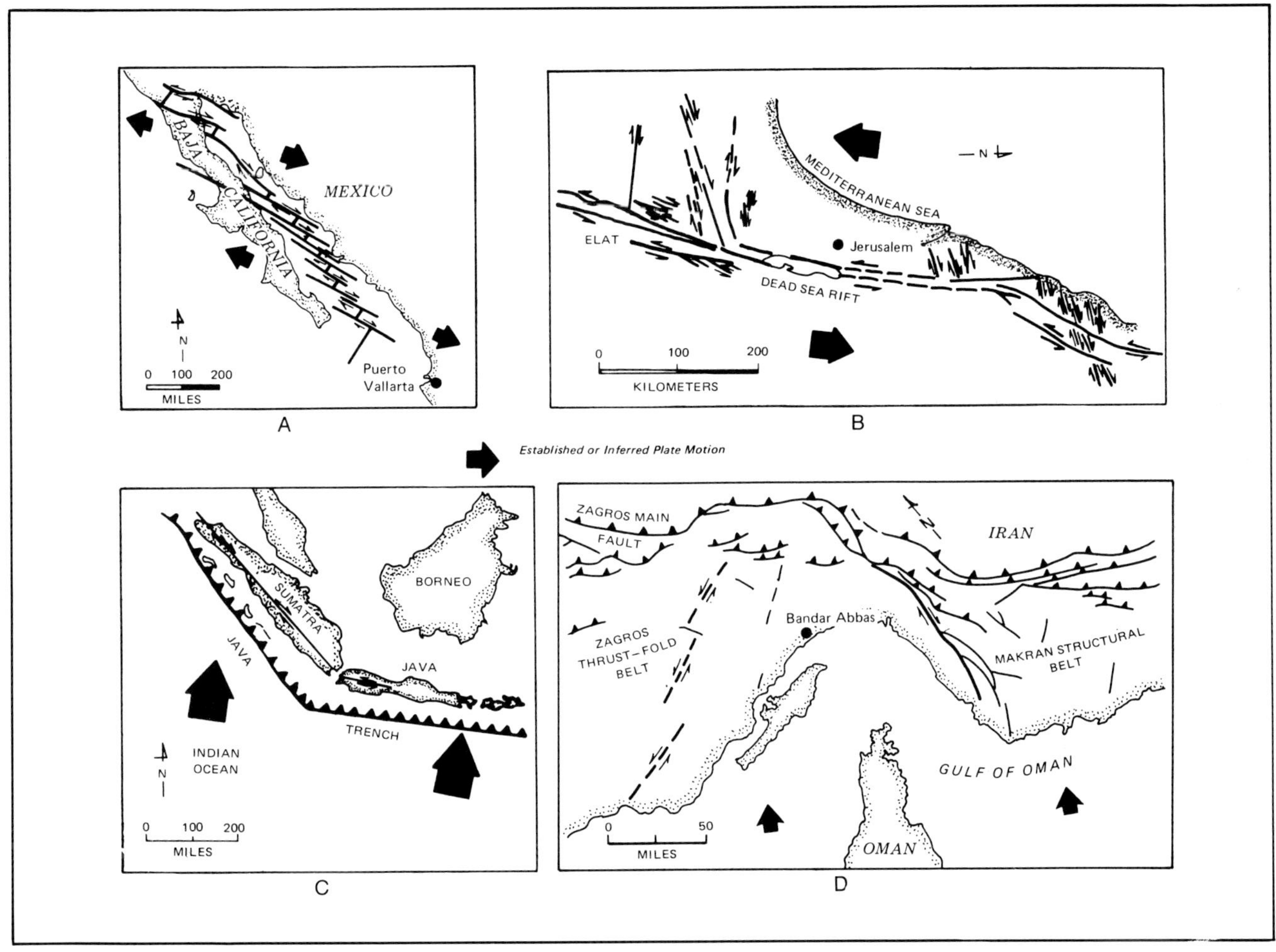

Fig. 1-2—Wrench "set" examples. A, from transform plate boundary; strike slip occurs on parallel fault set. B, example from transform plate boundary; short, right-lateral antithetic strike-slip faults intersect single, main displacement zone at various oblique angles. C, example from convergent plate boundary; aligned left- and right-lateral faults presumably owe their opposing displacements to differences in sense of obliquity of plate encroachment that results from bend to plate boundary. D, example from convergent plate boundary; strike-slip displacement may have developed analogous to conjugate shears. Adjoins right margin of figure 1-3.

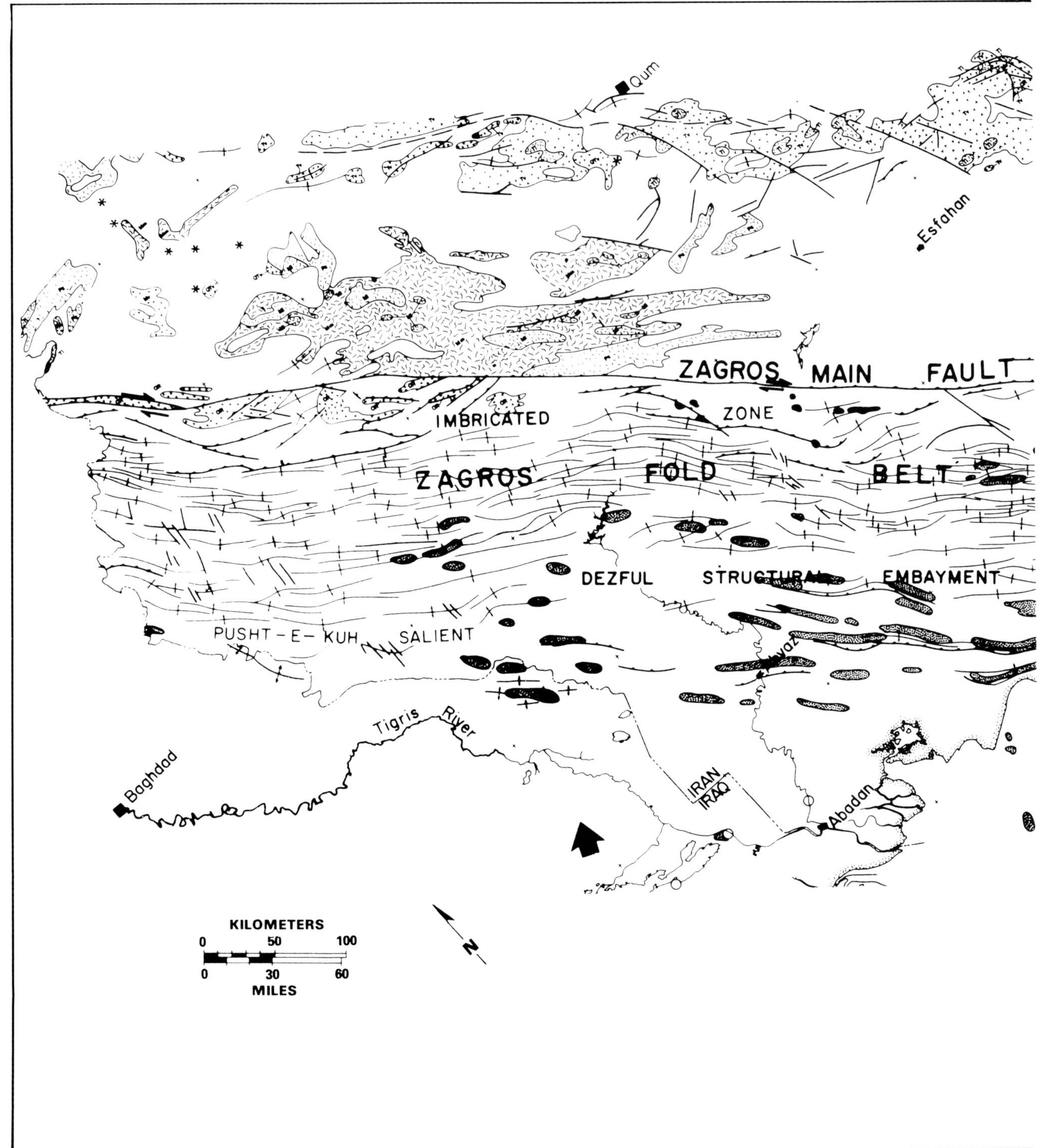

Fig. 1-3—Zagros orogenic belt, Iran. Frontal or external folds are considered to be detached and probably thrust in part. Basement-involved thrusts are present adjacent to Zagros Main fault, which has a late-stage component of right-lateral strike-slip. Buoyantly and tectonically injected salt structures also contribute to mix of structural styles. Generalized from Stocklin and Nabavi (1973).

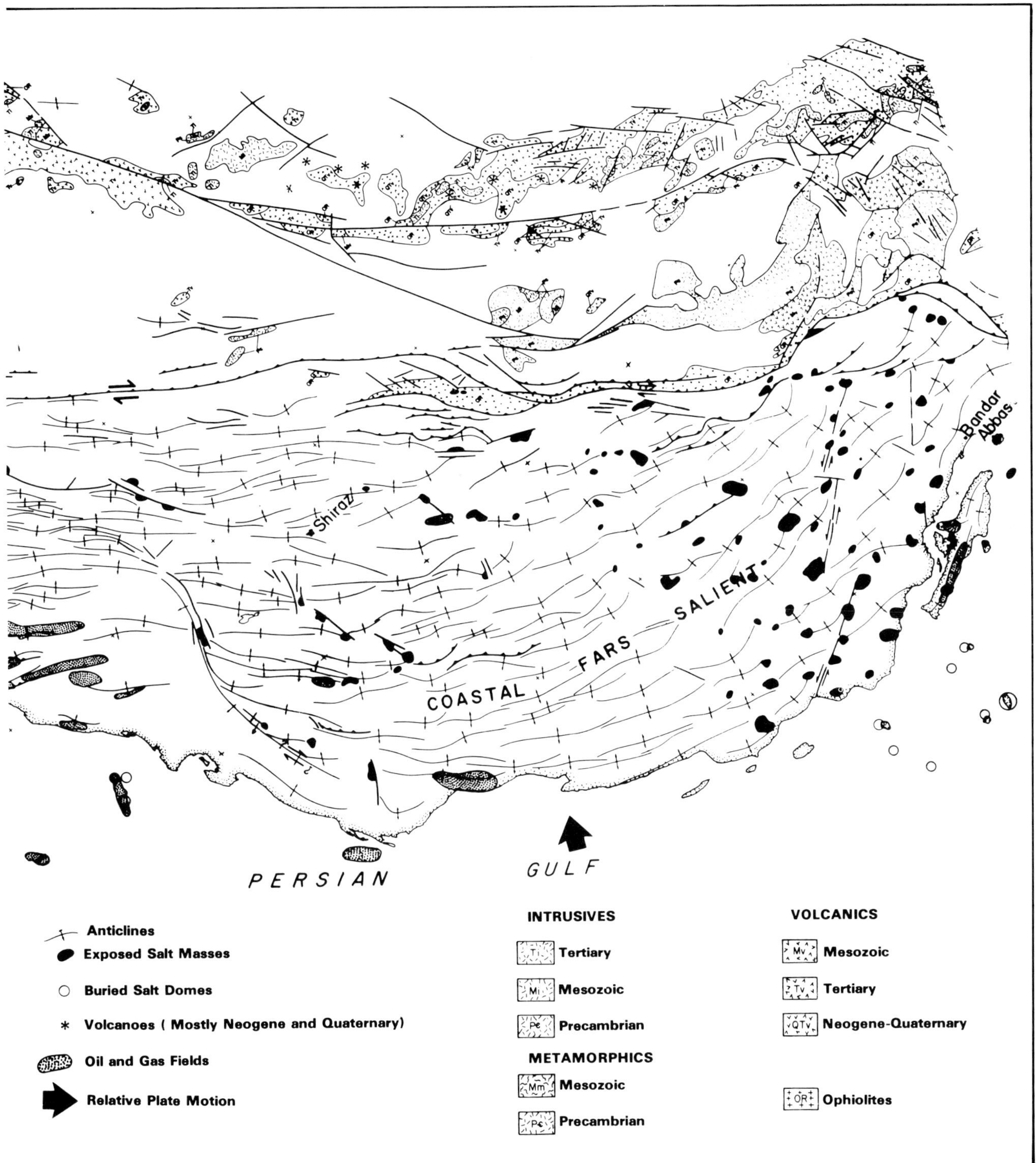
Shiraz
Bandar Abbas
COASTAL FARS SALIENT
PERSIAN GULF
Anticlines
Exposed Salt Masses
Buried Salt Domes
Volcanoes (Mostly Neogene and Quaternary)
Oil and Gas Fields
Relative Plate Motion
INTRUSIVES
Tertiary
Mesozoic
Precambrian
METAMORPHICS
Mesozoic
Precambrian
VOLCANICS
Mesozoic
Tertiary
Neogene-Quaternary
Ophiolites

TABLE 1-1
Structural Styles and Habitats*

Structural Style	Dominant Deformational Force	Typical Transport Mode	Plate-Tectonic Habitats: Primary	Plate-Tectonic Habitats: Secondary
		BASEMENT INVOLVED		
Wrench-fault assemblages	Couple	Strike slip of subregional to regional plates	Transform boundaries	Convergent boundaries: 1. Foreland basins 2. Orogenic belts 3. Arc massif Divergent boundaries: 1. Offset spreading centers
Compressive fault blocks and basement thrusts	Compression	High to low-angle convergent dip slip of blocks, slabs, and sheets	Convergent boundaries: 1. Foreland basins 2. Orogenic belt cores 3. Trench inner slopes and outer highs	Transform boundaries (with component of convergence)
Extensional fault blocks	Extension	High to low-angle divergent dip slip of blocks and slabs	Divergent boundaries: 1. Completed rifts 2. Aborted rifts aulacogens Intraplate rifts	Convergent boundaries: 1. Trench outer slope 2. Arc massif 3. Stable flack of foreland and fore-arc basins 4. Back-arc marginal seas (with spreading) Transform boundaries: 1. With component of divergence 2. Stable flank of wrench basins
Basement warps: arches, domes, sags	Multiple deep-seated processes (thermal events, flowage, isostasy, etc.)	Subvertical uplift and subsidence of solitary undulations	Plate interiors	Divergent, convergent, and transform boundaries Passive boundaries
		DETACHED		
Decollement thrust-fold assemblages	Compression	Subhorizontal to high-angle convergent dip slip of sedimentary cover in sheets and slabs	Convergent boundaries: 1. Mobile flank (orogenic belt) of forelands 2. Trench inner slopes and outer highs	Transform boundaries (with component of convergence)
Detached normal-fault assemblages ("growth faults" and others)	Extension	Subhorizontal to angle divergent dip slip of sedimentary cover in sheets, wedges, and lobes	Passive boundaries (delta)	
Salt structures	Density contract Differential loading	Vertical and horizontal flow of mobile evaporites with arching and/or piercement of sedimentary cover	Divergent boundaries: 1. Completed rifts and their passive margin sags 2. Aborted rifts; aulacogens	Regions of intense deformation containing mobile evaporite sequence
Shale structures	Density contrast Differential loading	Dominantly vertical flow of mobile shales with arching and/or piercement of sedimentary cover	Passive boundaries (deltas)	Regions of intense deformation containing mobile shale sequence

*Styles in regions underlain by continental crust are emphasized; see Dickinson (1974) for discussion of most kinds of settings. Nontectonic types of drape folds, such as supratenuous or compaction folds, and structures resulting from igneous or metamorphic activity, are excluded.

phologies, critical differences from other styles, and associated hydrocarbon traps. Last, factors influencing variations in style expression and occurrence are discussed.

STRUCTURAL STYLES IN PETROLEUM PROVINCES

Classification of Structural Styles

Our classification of structural styles is based primarily on the involvement or noninvolvement of basement in the observed structures. Additional criteria are inferred deformational force and mode of tectonic transport inferred from strain features of the structures (Table 1-1). Although basement can exhibit widely different competence and mechanical behavior, in petroleum provinces it is usually rigid crystalline igneous or metamorphic rock. The degree of basement involvement is critical to petroleum exploration, for it indicates not only how structures propagated, but, qualitatively, how much of the sedimentary section is in a trap configuration (Fig. 1-1).

Structural styles are initially treated separately as discrete "end members" to facilitate description and to emphasize their distinctive aspects. Styles apply in scale to trends of tectonically interrelated, contemporaneous structures, but consideration is also given to the components common to single features. Each style has provided hydrocarbon traps (Fig. 1-1).

Identification of Styles

Identifying structural styles is one of the critical tasks in petroleum exploration and is often difficult because a single characteristic seldom is unique to one structural assemblage, and identification must be made early with few data. Identification depends mostly on (1) recognition of key structural elements such as en echelon folds and faults, trap-door blocks, rollover anticlines, etc.; (2) critical differences in local trend arrangements; and (3) gross regional patterns of structures. Following is a resume of the most common distinguishing characteristics:

Drag fold—Fold formed by drag of sedimentary rocks along a fault.

Drape (forced) fold—Fold formed in sedimentary layers by a forcing member from below, usually a basement fault block.

En echelon—Consistently overlapped structures aligned parallel with one another but oblique to the zone of deformation in which they occur.

Intersecting or *grid*—Structures repeated in multiple alignments crisscrossing over broad areas (Fig. 1-12); these can combine to form *zigzag* or *dogleg* (Fig. 1-11, index map) and clustered features.

Irregularly clustered—Concentrations of structures that lack consistent spatial arrangements.

Parallel—Similar elements in parallel alignment. In some places parallel trends may be repeated in closely spaced, wavelike bands, constituting *belts* whose sinuosity delimits *salients* and *reentrants* (Fig. 1-3).

Relay—Inconsistently overlapped elements aligned parallel with one another and with the zone of deformation in which they occur.

Solitary—Isolated, singular features generally not aligned with other structures of similar characteristics.

Trap-door blocks—Blocks formed by intersection of two faults with maximum relative uplift at or near the point of intersection.

Zonal—Structures in a discrete, elongate, linear trend. Local trends can be longitudinal (parallel), oblique, or transverse to the larger or dominant structural features of an area.

BASEMENT-INVOLVED STRUCTURES

Wrench-Fault Assemblages

Transform plate boundaries are the primary environment of wrench-fault structural assemblages, but divergent and convergent boundaries are also very important habitats (Table 1-1; Fig. 1-2). On transform boundaries the movement between lithospheric plates results in a side-by-side motion in concert with the strike-slip kinematics of the style itself. Strike slip in this example is commonly distributed throughout a set of parallel faults (Fig. 1-2A), or it can be concentrated on a single master fault (Fig. 1-2B). Parallel transform faults are also present at divergent boundaries, where they offset spreading axes. Where transforms intersect divergent continental margins, they can delimit subbasins with different depositional histories, as along the West African and East Greenland shelves (Lehner and DeRuiter, 1977; Surlyk, 1977).

Master wrench faults also develop subparallel with convergent plate boundaries (Fig. 1-2C) and are attributed to obliquity of plate encroachment (Fitch, 1972). These are longitudinal wrench faults (a term suggested by D. R. Seely, pers. com., 1971) and typically occupy axial positions in orogenic belts or magmatic arcs (e.g., Zagros Main fault, figs. 1-3, 1-4).

Strike-slip faults trending at an angle to the alignment of the convergent plate boundary are termed "oblique wrench faults." They are present in orogenic belts (Fig. 1-2D) and forelands, and many seem to have patterns and senses of displacement that conform to conjugate shear sets. Foreland wrench faults generally have smaller displacements and less widely distributed associated structures than those of either the transform or longitudinal type. The region north of the Himalayas may contain notable exceptions (Tapponnier and Molnar, 1976).

Continental lithospheric segments of divergent margins not near oceanic crustal joins and intraplate graben systems incorporate wrench faulting less commonly. Intraplate regions appear to be the least common habitat. In North America, the known midplate wrench faults

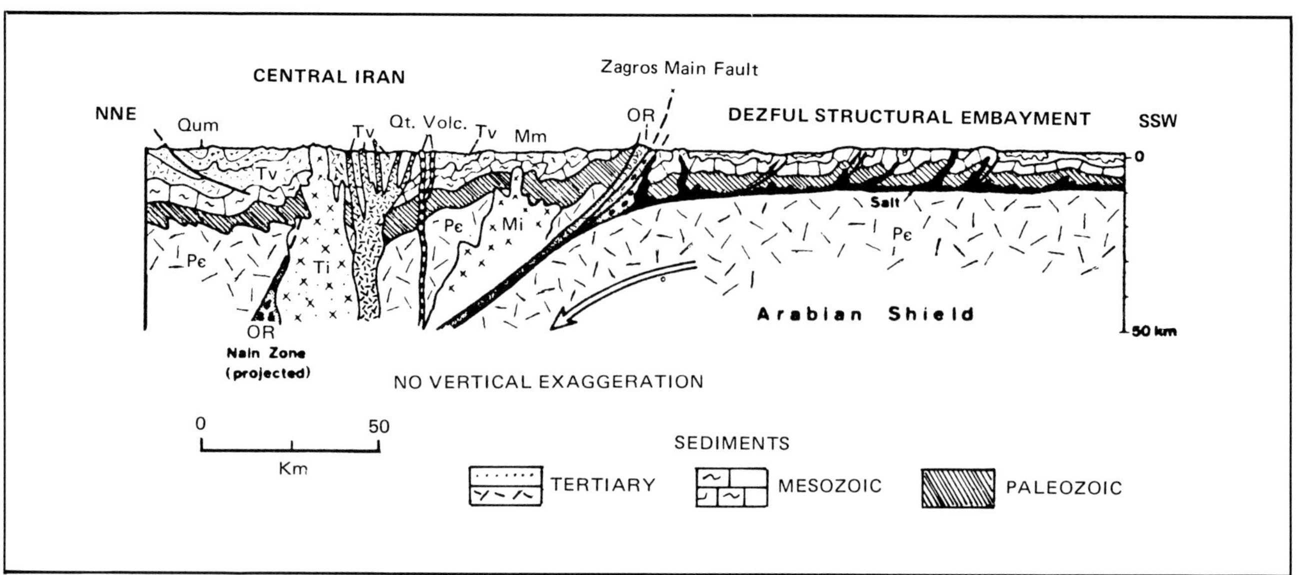

Fig. 1-4—Generalized cross section of Zagros orogenic belt; extends southwest from city of Qum (see Fig. 1-3). See figure 1-3 for explanation of symbols. (Modified after Gansser, 1974).

mostly have small displacements, plus limited associated structures which commonly have a fault (rather than fold) response to the wrench tectonics. The Cottage Grove fault of the southern Illinois basin appears to be a characteristic example (Wilcox et al., 1973). Not all midplate wrench faults developed this way, however. The Rough Creek fault zone, also of the Illinois basin area, has associated en echelon folds and other compressive features suggesting a moderately large right-lateral displacement.

The structures associated with wrench faults (Fig. 1-5) are more diverse than those of any other style and include most elements that are fundamental to other styles. Wrench assemblages have both compressional and extensional features (Harding, 1973), or are dominated by compressional (Harland, 1971; Lowell, 1972; Harding, 1974), or extensional structures (Harland, 1971; Harding, 1974). These three substyles, side-by-side, convergent, and divergent wrenching, result from the configuration of the laterally moving blocks, or from the orientations of their boundaries relative to regional plate motion, or both (Wilcox et al., 1973).

The structural patterns of a single wrench zone show much repetition. Tectonic relations between several wrench faults, however, are much less consistent. Wrenches develop as singular, major displacement zones more commonly than other kinds of faults, such as thrusts, block faults, or detached normal faults, which tend to occur in sets. The diversity of several documented wrench zones (Fig. 1-2) demonstrates that their occurrence cannot be adequately explained by any single, unifying scheme. We thus differ, for example, with Moody and Hill's (1956) and Moody's (1973) universal conjugate wrench sets. They considered such sets to be the result of pervasive meridional and equatorial compression, with faults formed at 30° angles to maximum principal compressive stress. Their approach disregards the kinematic and structural differences between different types of plate boundaries and tectonic settings. Their views are inconsistent with relations demonstrated by earthquake focal mechanisms; moreover, plate rotations would disrupt the orientations of any such global fault system.

The wide variety of features included in wrench-fault assemblages has resulted in confusion with other styles. Where direct offset data are lacking, two characteristics provide a basis for initial distinction of wrench faults. The oblique resolution of stresses along a finite, throughgoing tectonic boundary (Fig. 1-5, center) commonly causes structures (1) to be arranged en echelon and (2) to be confined to a relatively narrow, persistent, linear zone.

In profile view, some wrench faults have a characteristic seismic signature termed a "flower structure" (R. F. Gregory, pers. com., 1970). It is expressed as an upward-spreading fault zone, whose elements usually have reverse separations (Fig. 1-6); the spreading fault system need not be symmetrical; half-flowers are also known (Lowell, 1972). Development of flower structures is enhanced where strike slip is accompanied by components of convergence (Lowell, 1972) and where the rocks are highly mobile. Sylvester and Smith (1976) have documented surface examples along the San Andreas fault in Mecca Hills, California, and have described their genesis.

Flower-structure reversals can be differentiated from those in other structural styles by recognizing fault displacements directly below the reversal that indicate a high-angle fault stem involving basement (Fig. 1-6). "Negative" flower structures have also been observed and consist of shallow sags overlying upward-spreading strike-slip faults with normal separation.

A great variety of hydrocarbon traps (Fig. 1-1) occurs with the wrench-fault substyles. By far the most prolific

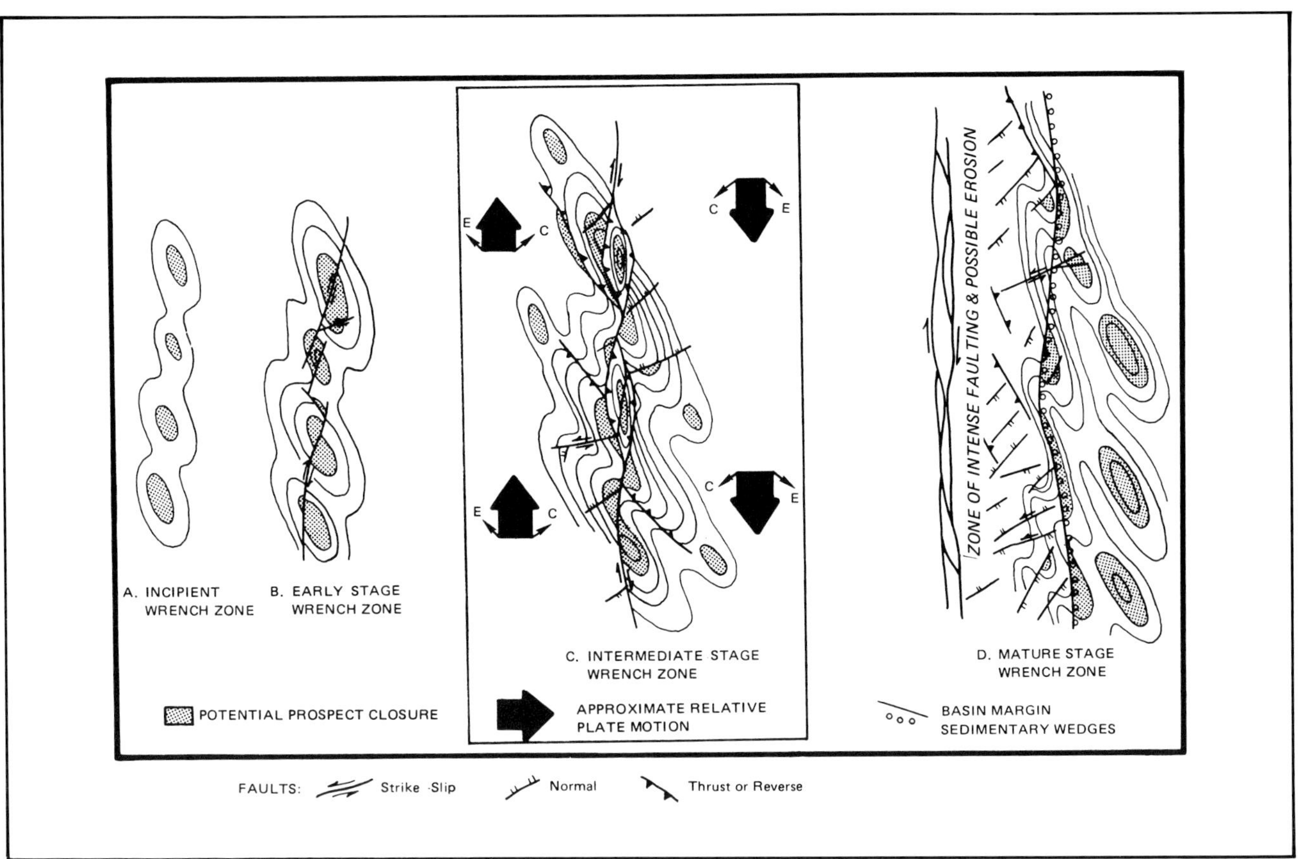

Fig. 1-5—Schematic diagram of structural assemblage associated with major wrench fault and stages in evolution that cause changes in hydrocarbon trapping; modeled after California examples (discussed in Harding, 1974). Arrows, C, E represent compressive and extensional vectors, respectively, that arise from coupling motion between adjacent plates.

have been those associated with en echelon folds. En echelon normal fault blocks, subthrust bed terminations, and flower structures have also been effective traps. Closure types and changes in the prospective fairway that develop with increased strike slip have been described previously (Fig. 1-5; Harding, 1974, 1976).

Compressive Fault Blocks and Basement Thrusts

Compressive fault blocks (Fig. 1-7) and basement thrusts preferentially occur on convergent-plate boundaries. The former are typically more areally restricted, being confined mostly to forelands, whereas the latter are found in forelands, orogenic belts, and landward walls of oceanic trenches (Table 1-1). Compressive blocks are discussed first, then basement thrusts with which compressive blocks seem to be transitional in foreland settings, and finally, basement thrusts in settings other than forelands.

Forelands, are developed in two convergent-margin positions, back-arc and peripheral (adapted from Dickinson, 1974). Back-arc forelands lie between the magmatic-volcanic arc and the craton and commonly have thrust-fold belts directed toward the plate interior or craton (''Andean'' or ''Cordilleran'' type, Dewey and Bird, 1970). Peripheral forelands develop with continental collision and have thrust-fold belts lying between the magmatic-volcanic arc and the former trench (''collision'' or ''Himalayan'' type, Figs. 1-3, 1-4). Folds and thrusts in this example are directed toward the plate boundary or position of the earlier trench.

Compressive blocks are best known from the Laramide back-arc foreland of Wyoming. In Wyoming and elsewhere, although orogenic belts are laterally extensive adjacent forelands with significant compressive block faulting are seemingly rare. Lowell (1974) and Dickinson and Snyder (1978) have postulated that some of the reverse displacement on fault blocks in Wyoming is caused, respectively, by either buoyancy from or physical contact with an underlying subducted lithospheric slab. The limits of the deep-seated slab would have controlled the regional distribution of foreland structures. Burchfiel and Davis (1975) have attributed structures of the Wyoming province to a thermally weakened, brittle crust.

Prucha et al. (1965) used the term ''upthrust'' for reverse faults that bound the Wyoming uplifts and considered that the uplifts were caused by differential vertical

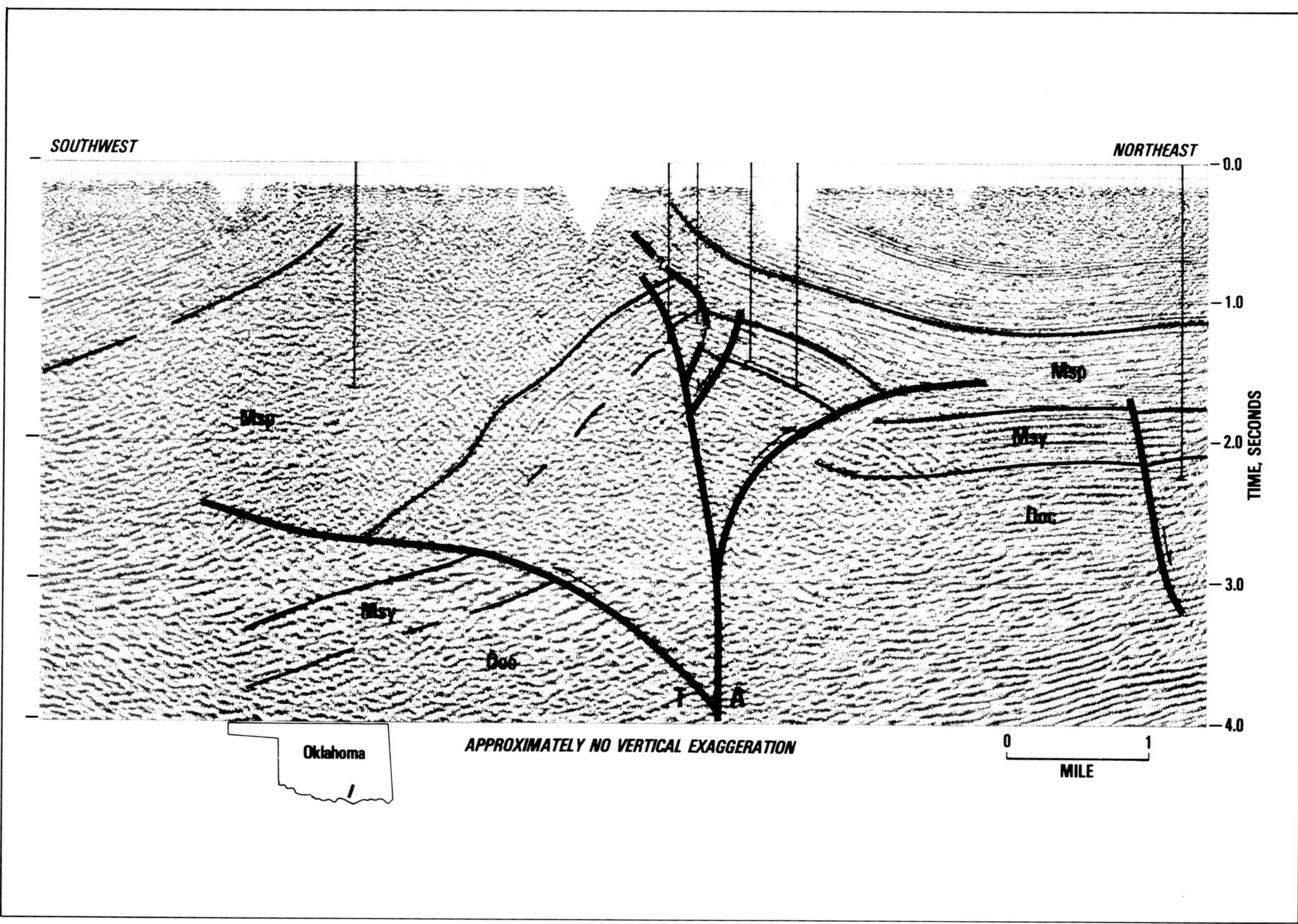

Fig. 1-6—Migrated seismic profile across wrench-fault zone in Ardmore basin of Oklahoma demonstrating flower-structure geometry (adapted from unmigrated interpretation by R. F. Gregory and E. C. Lookabaugh, 1973). Msp, Mississippian Springer, Msy, Mississippian Sycamore, and Ooc, Ordovician Oil Creek reflectors. T, displacement toward viewer, A away from viewer.

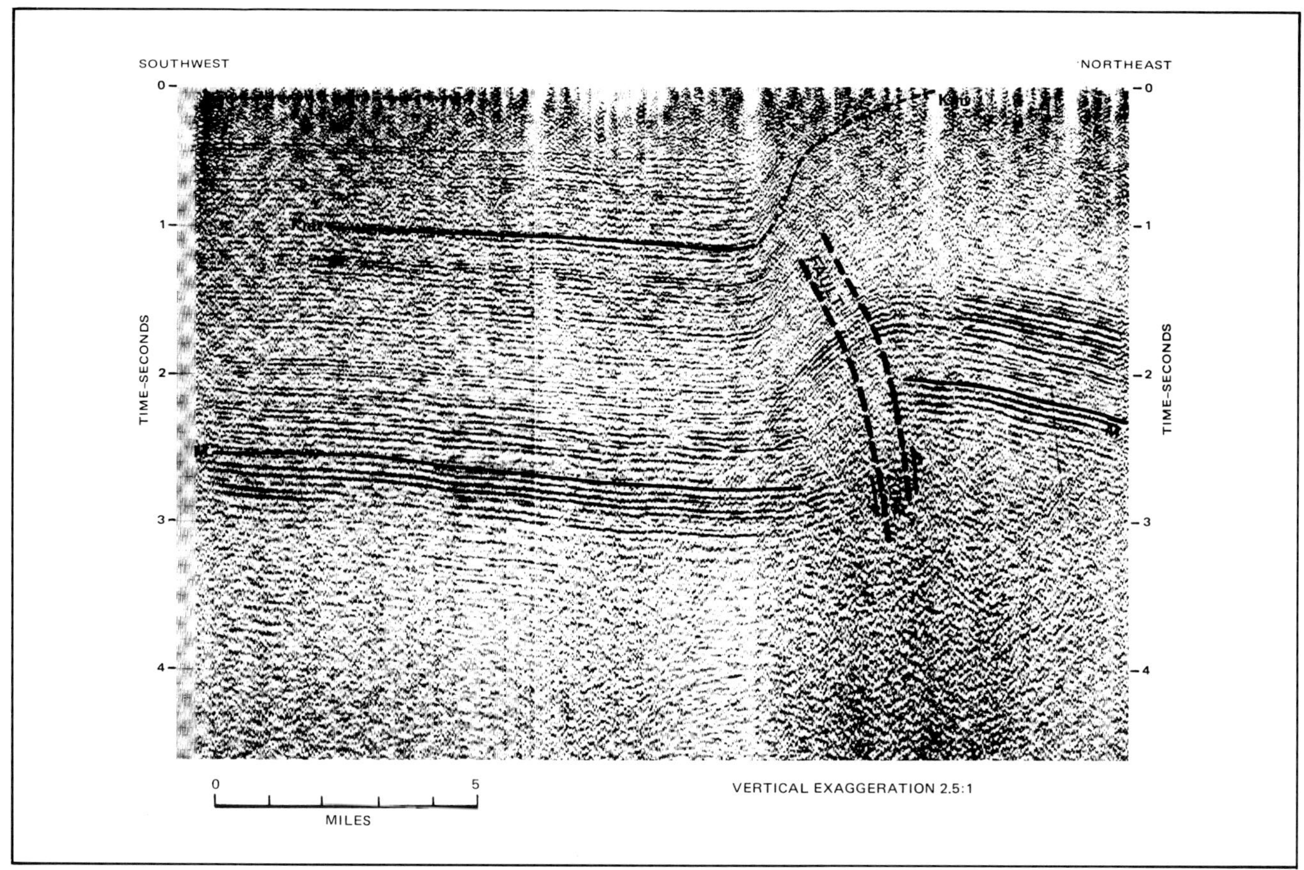

Fig. 1-7—Southwest-northeast seismic profile across Rangely oil field structure, northwestern Colorado. Inclination of fault is approximated by syncline's axial plane which dips steeply under upthrown block. Flexure is caused dominantly by fault drag and drape over edge of tilted, compressive fault block. Anticlines formed in this way parallel uplifted block edges; their traces reflect pattern of primary block-bounding faults. Compressive effects, down to deepest limit of this control (3 or 4 sec), are concentrated in fault zone (with no apparent deformation in reflectors on either side). M, Mississippian; Kmv, Cretaceous Mesaverde reflectors.

movement. The upthrusts characteristically flatten upward from nearly vertical fault surfaces through reverse faults into less steeply dipping thrusts (Fig. 1-7). Upthrust or reverse faults, however, can also be important components of the convergent wrench style (Lowell, 1972), both as wrench faults themselves (Fig. 1-6; Sylvester and Smith, 1976) and as integral elements of the associated fold set that is away from the wrench fault.

We do not use the term "upthrust" for compressive blocks because the proposed fault profile may not be consistent with deeper intrabasement relations revealed by recent seismic data (Smithson et al., 1978) and because the genetic implications seem incompatible with coexisting basement thrusts and wrench faults that require at least some basement compression and lateral movement. Furthermore, Reches (1978a) has recently demonstrated layer-parallel shortening within the sedimentary cover in foreland monoclinal structures in the Colorado Plateau.

In our treatment of compressive blocks we are concerned with the structural geometry of the uppermost basement and sedimentary cover where trap closures develop. At these levels compressive blocks can have bounding fault surfaces that range from subvertical, with reverse (Fig. 1-7) or, rarely, normal segments (e.g., Rattlesnake Mountain anticline of Bighorn basin, Wyoming; Pierce and Nelson, 1968) to lower dipping thrusts, particularly on zones with greater structural relief. Block faults can also have subordinate amounts of strike slip.

In the Bighorn basin of the Wyoming foreland the variable trend of block-bounding faults and their overlying and paralleling drape folds or monoclines is the most distinctive style characteristic. Trends are dominantly northwest-southeast longitudinal structures and subordinately north-south and east-west oblique structures (Fig. 1-8). The variously aligned elements can combine to form rhombic or cross-trended patterns, or outline individual discrete blocks. Blocks include large rectilinear features (e.g., Beartooth uplift), uplifts with dogleg boundaries (northwest terminus of Bighorn uplift, Fig. 1-8), clusters of trapdoor blocks (Pryor Mountains, Fig. 1-8), and solitary trap doors or drape anticlines of

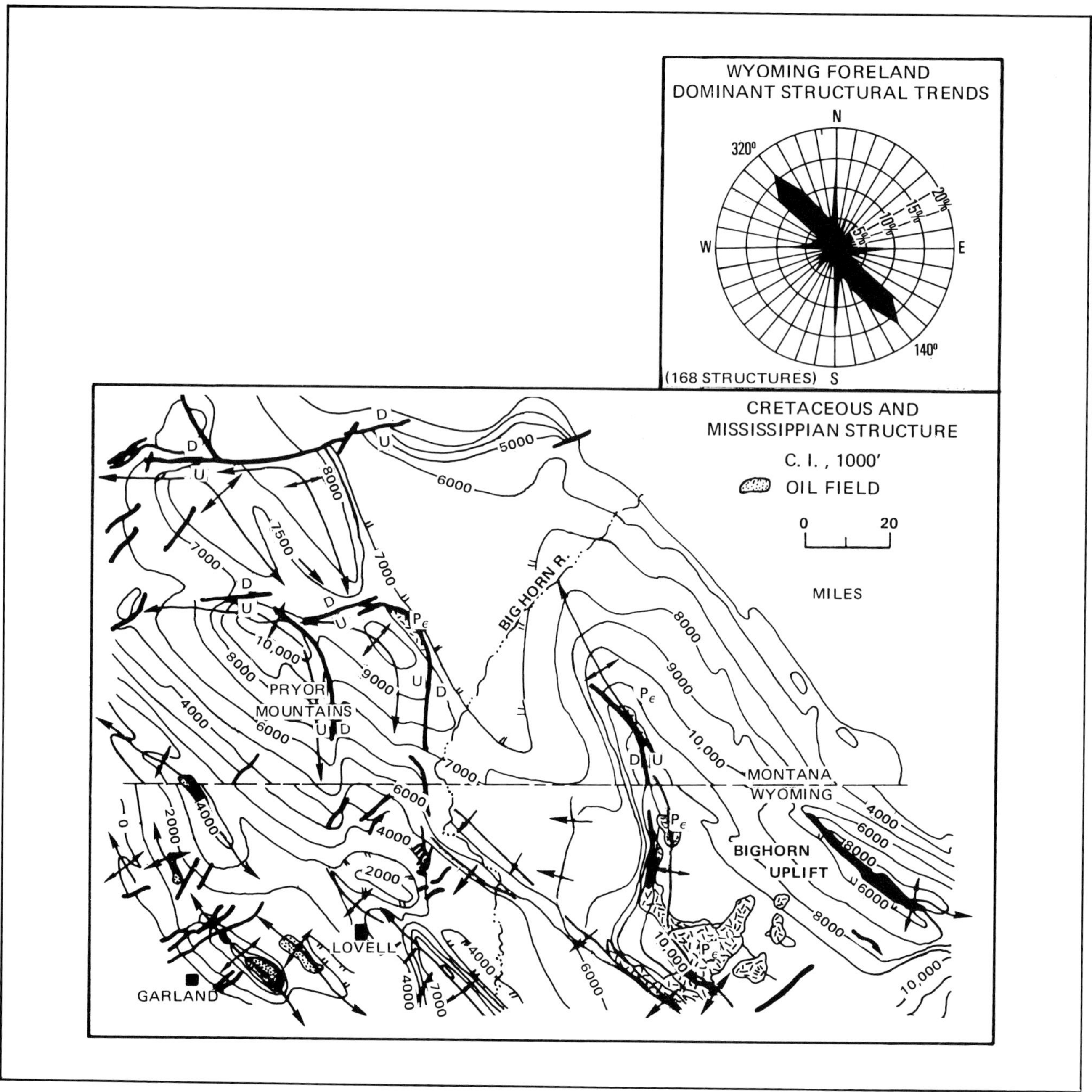

Fig. 1-8—Structures at northeast corner of Bighorn basin, Wyoming, demonstrate three dominant trend orientations of general Wyoming foreland province (insert plot includes orientations of major faults, folds, uplifts, and lineaments). Fold axes parallel dominant block-bounding faults where latter are exposed at surface or are demonstrated by seismic (not shown). Similar faults may be inferred in structures lacking this control on basis of consistent, fold-fault relations observed elsewhere. (Adapted from Pierce et al., 1947; Dobbin and Erdman, 1955; Darton, 1906).

several repeated alignments. These distinctive block features are present within or disrupt a more general relay pattern formed by the northwest-southeast folds (Fig. 1-8, southwest quadrant).

In profile view (Fig. 1-7), a threefold vertical zonation is common in higher structural levels and further demonstrates the fault-block geometry of the style. The near top of basement level is a tilted fault block. A steep drag fold develops at intermediate levels, and a gentle drape fold, or tilted monocline, lies at shallow levels where preserved from erosion. Blocks appear as rotated slabs in cross sections, and associated flexures typically are markedly asymmetric (Fig. 1-7). Symmetric, curvilinear flexures are rare.

Internally, individual structures range from simple to complex; large normal faults that parallel the structural axis but lie well downdip from the backlimb (i.e., flank of flexure away from the block boundary) have been observed (Howard, 1966). Secondary crestal normal faults, both longitudinal and transverse, are common on some drape flexures (Wise, 1963). Some transverse faults have strike-slip components and offset flexure axes, and terminate mostly at high angles at the block boundary. They can also act as basement-involved tear faults that displace block edges (Foose et al., 1961).

Compressive blocks appear to grade into low-angle, large-displacement basement thrusts on some foreland structures. Deep reflection profiling across the Wind River Mountains (see Fig. 1-20, right side) has shown a thrust surface that can be traced to a depth of a least 24 km (15 mi) at an average dip of 30 to 35°. A minimum horizontal displacement of 13 to 21 km (8 to 13 mi) exceeds the minimum vertical displacement of 8 mi (Smithson et al., 1978).

The multidirectional trends of foreland block structures present problems in interpretation that we will see are similar to those of extensional block faulting. General contemporaneity of the block-associated flexures is demonstrated at block corners along the west flank of the Bighorn uplift, where flanking monoclines have equal relief on either edge of the corner and are not offset at the point of junction (Fig. 1-8). Similar relations are apparent at the Pryor Mountains trap-door faults and at the corners of the Beartooth uplift. Elsewhere we have not examined in detail the relative timing of the differently aligned structures; tectonic overprint with different orientation remains a possibility.

Buried, older zones of basement weakness control development of several structures (Foose et al., 1961; Reches, 1978a) but we do not know whether this is a universal prerequisite for the generation of grid patterns (Hoppin and Palmquist, 1965). Marked similarities in trend patterns with extensional block-faulted terranes (e.g., clustered trap doors, platforms with dogleg boundaries) suggest that some structures may result from compression and inversion of a preexisting normal-fault fabric imparted to the Wyoming province during earlier phases of rifting (Stewart, 1972). The deep listric basement-involved normal faults could provide listric surfaces for the Laramide thrust displacements and at the same time would impart the fault-block patterns observed at the surface (see Lowell, 1974, fig. 2).

Because of the diversity of styles and the lack of control for fault attitudes within the basement, Wyoming uplifts have been given many different interpretations. Stone (1969) thought that wrenching controlled the basic tectonics. He explained fault blocks, such as the Pryor Mountains, as bounded by a thrust on one edge and a strike-slip tear fault on the other edge. Structural relations demonstrate that both boundaries have identical drape-flexure style and lack evidence of strike slip. Foreland structures have also been interpreted as due to differential vertical uplift (Stearns, 1975), a mechanism that fails to explain the large thrust overlap of the Wind River Mountains, and the co-occurrence in some areas of wrench faults. Deep, intrabasement thrusting (Bally, 1975) has received increased attention as the primary tectonic control. We believe that the Wyoming foreland contains a mix of Laramide structural styles, compressive fault blocks, basement thrusts, wrench faults, and basement warps, and should not be attributed to a single type of deformation.

Differentiation of block faulting from detached thrusting or convergent wrenching is critical in regions of compressive deformation. With sparse data (e.g., a single seismic profile) differentiation can be difficult, especially where upthrust or reverse faults are a potential element of more than one style. Regionally, the grid pattern of block structures differs distinctly from the wave-like salient and reentrant patterns of conventional fold-thrust belts (cf. Figs. 1-8, 1-3). The grid pattern is also distinctive from the en echelon structures and through-going, straight master fault of many wrench zones (cf. Figs. 1-8, 1-5). Trap-door clusters and doglegs are thought to be especially diagnostic of block faulting. Upthrust faults at transform plate boundaries can be distinguished from shallow, steep fault profiles of the compressive block style by the former's more nearly unidirectional orien-

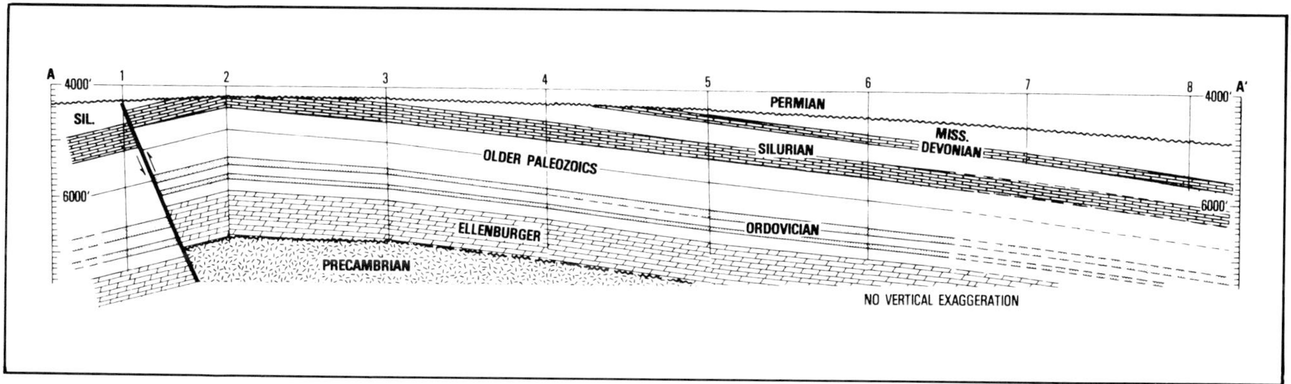

Fig. 1-9—Cross-section AA' across Keystone field, Permian basin, Texas. Steep dip of fault at top of basement and in sedimentary cover is indicated by well control. Seismic and subsurface control demonstrate similar steep faults at these levels elsewhere in basin (Elam, 1969). See figure 1-10 for location of section. (After Osborne, 1957).

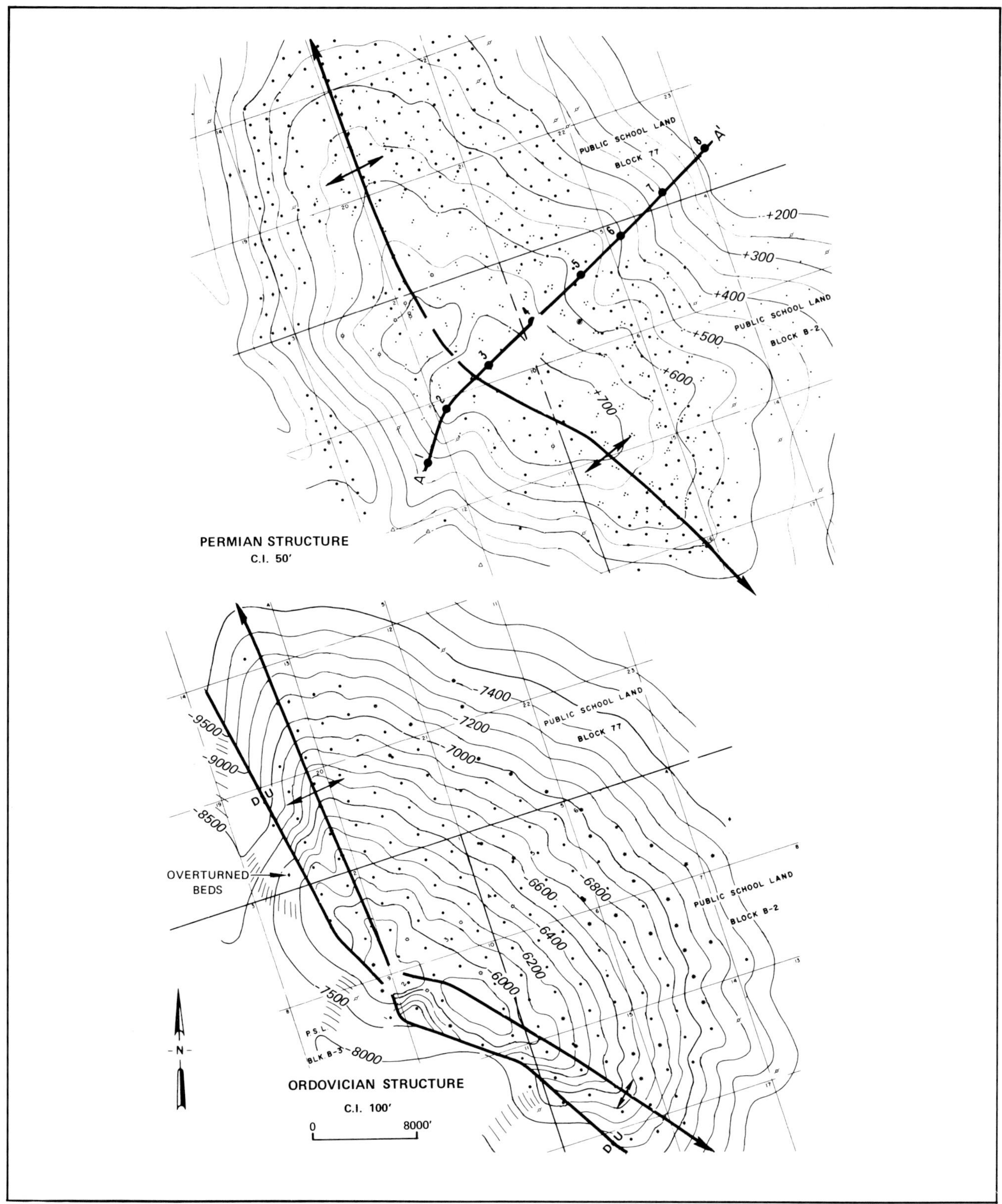

Fig. 1-10—Structure at Keystone field, west edge of Central basin platform, Permian basin, Texas (from Osborne, 1957). Shallow flexure (Permian, upper diagram) reflects drape over buried Ordovician trap-door block (lower) and has characteristic culmination opposite obtuse-angle junction of bounding faults. Other angles of fault junctions in area range from obtuse to acute. Trap-door flexures associated with latter appear as triangular faulted domes.

tation and association with en echelon folds.

In cross section, the curvilinear fold profiles of wrench assemblages and thrust-fold belts are commonly quite different from the rotated slab and monoclinal-step appearance of many block structures. The latter are markedly asymmetric, whereas wrench-associated folds are often symmetric. Both wrench flower structures and compressive fault blocks, however, can have similar-appearing reverse dislocations. Identification of a relatively shallow rollover and branching faults that dip inward toward a narrow, common stem is important in recognizing the wrench-associated structure (cf. Figs. 1-6, 1-7); half-flower structures, however, are harder to distinguish.

The two block-fault styles, compressive and extensional, are differentiated by the character of the block-bounding faults and by the steeper flexures and compressive features obviously associated with the former style (cf. Figs. 1-7, 1-11).

Compressive block faulting has created prolific traps

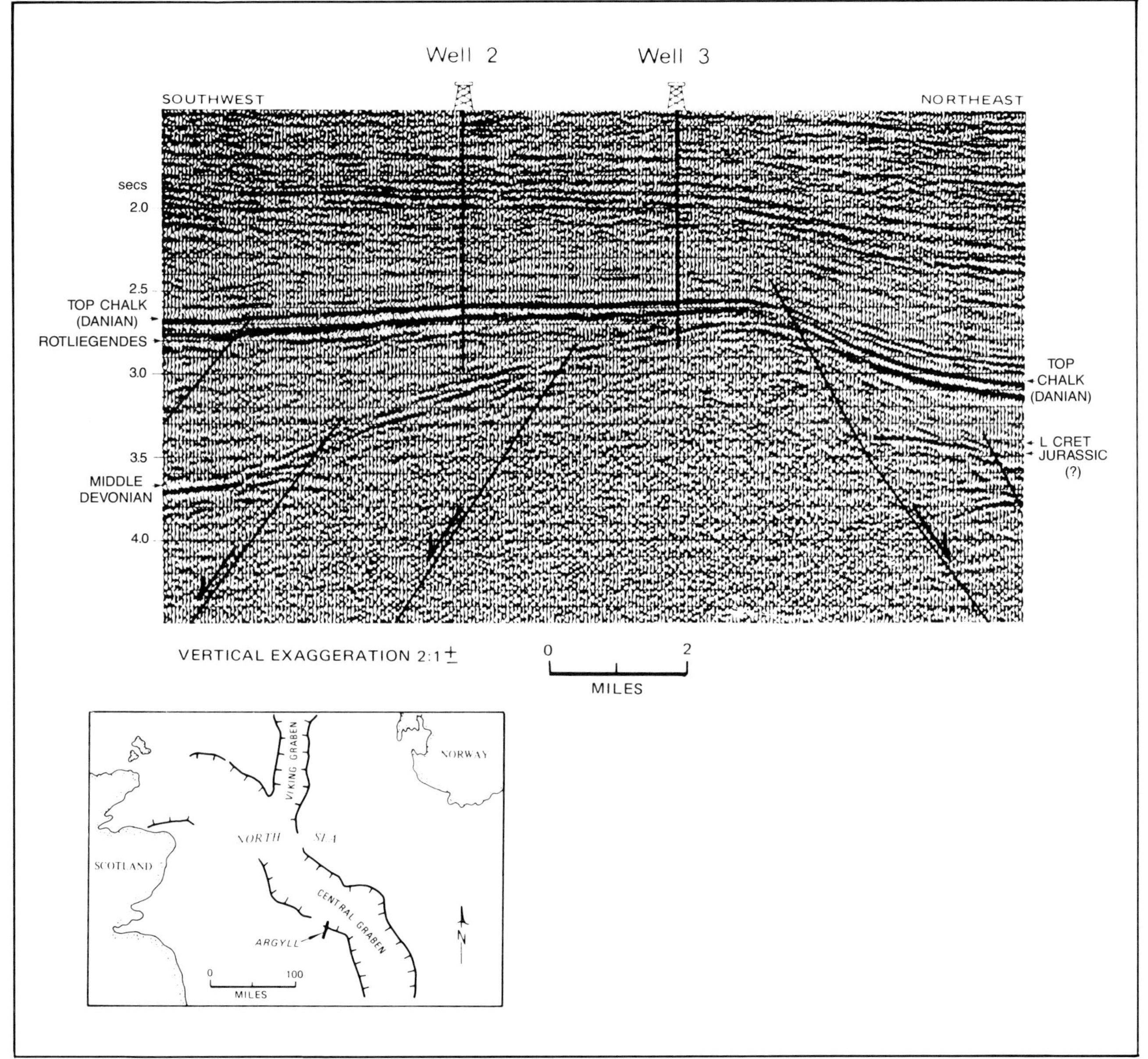

Fig. 1-11—Seismic profile across subunconformity trap (Rotliegendes sandstone) at Argyll field, western margin of Central graben, North Sea. Three interrelated structural levels are present; slablike, unflexed, rotated normal-fault block at depth (Middle Devonian), fault block with dip-slip fault drag flexure at intermediate level (Danian chalk), and shallow, intact drape flexure (above 2.4 sec). In index map, note dogleg configuration of Central graben (i.e., proceeding from far north end, graben trend is first south, then southeast, and then back to original south trend) and bifurcation at junction with Viking graben. (From Pennington, 1975).

(Fig. 1-1) in the Rocky Mountain foreland and the Permian basin of west Texas (Figs. 1-9, 1-10; Elam, 1969). Specific closures include culminations on drape anticlines (Fig. 1-8, southwest quadrant), trap doors (Fig. 1-10), cross-faulted noses, and backlimb subsidiary flexures and faults. Additional production has come from subthrust warps closed against block-bounding faults and from various associated stratigraphic traps.

Basement thrusts in several other convergent-plate settings have not been found prospective for petroleum because of their extremely complex deformation under conditions of ultra-high pressure or temperature, or both. Such basement thrusts include orogenic belts that flank back-arc and peripheral forelands (Fig. 1-3) and faults beneath the outer high in the landward slope of oceanic trenches. At the latter, the downgoing or underthrust oceanic lithosphere is involved in deep thrust slices together with the detached thrusts (Seely et al., 1974).

Extensional Fault Blocks

Normal faults occur subordinately in all style assemblages, but certain suites of structures (Fig. 1-11) and tectonic settings (Table 1-1) are dominated by regional, deep-seated normal faults which constitute a discrete fault-block style. Such normal faulting is perhaps the most widespread of all styles. It dominates divergent margins in early stages of development, the oceanic crust formed at spreading centers, and some intraplate regions. In cross section, normal faulting is one of the least complex styles. In plan view, its patterns are highly variable and difficult to predict.

The Gulf of Suez (Fig. 1-12) typifies the basic style observed at divergent margins in various stages of development and in intraplate grabens. In the Suez graben, regional faults are distinctly multidirectional and create a grid or intersecting system (Robson, 1971). Blocks with oblique, zigzag edges and longitudinal fault blocks with relay patterns are both present; either can occur alone or in series. Fault plots (Fig. 1-12) show a perferred longitudinal orientation that parallels the overall graben trend. Faults oblique in mostly two directions about this regional trend are also important and about equally developed. Transverse regional faults are rare. Alignment plots are similar for both synthetic (i.e., downthrown toward the basin axis) and antithetic (i.e. upthrown toward the basin) faults with either major or minor displacements (Robson, 1971). Overprints of several different episodes of normal faulting result in still more complex patterns.

Individual fault blocks are internally complex (Fig. 1-13, 1-14). Second-order faults can repeat the regional fault trends or can have inconsistent transverse orientations, but they generally terminate at block boundaries.

On a regional scale, the normal faults form intraplate rifts that range in complexity from simple fault troughs with persistent straightaways to grabens with many junctions, multiple bifurcations, and doglegs (Fig. 1-15; in-

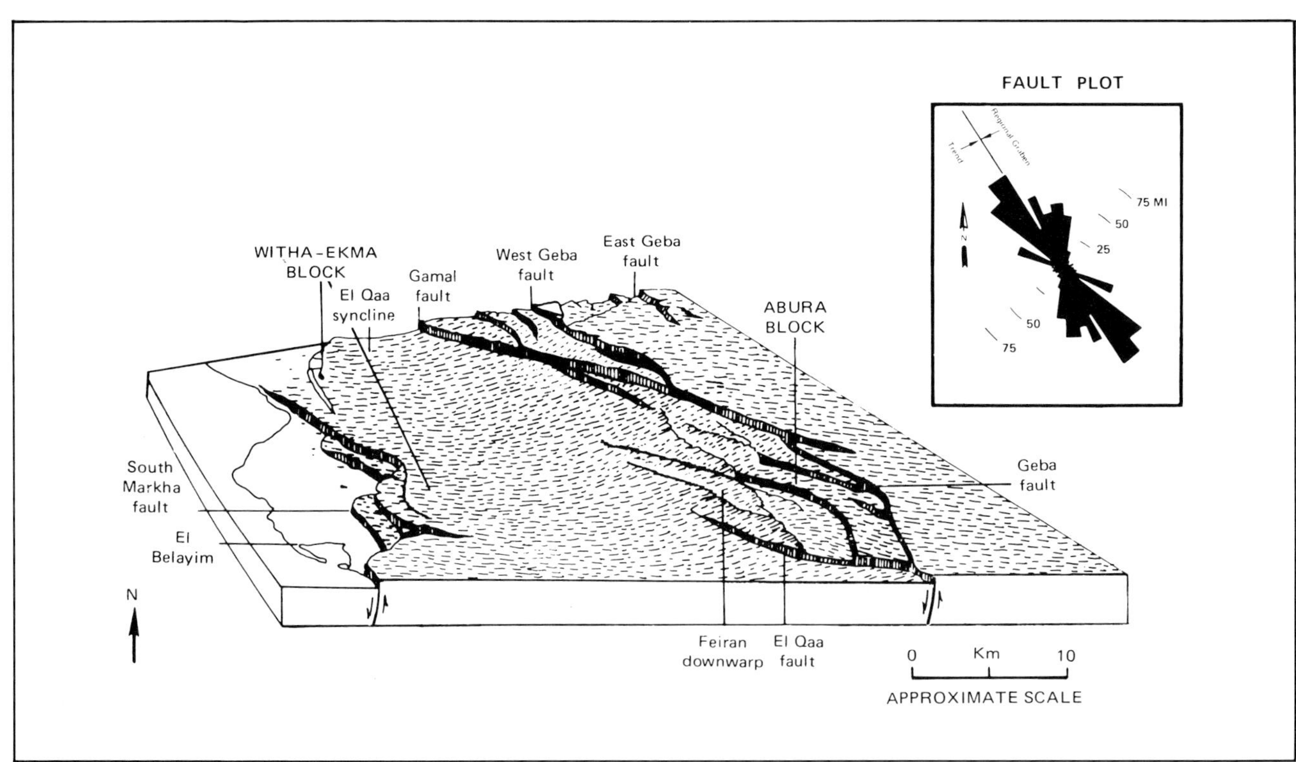

Fig. 1-12—Perspective block diagram of normal faults at east-central border of Gulf of Suez graben, Egypt. Fault frequency (insert) is plotted by cumulative lengths for each 5° quadrant of fault strike. Structure is controlled by extensive surface exposures (see Robson, 1971, pl. 1).

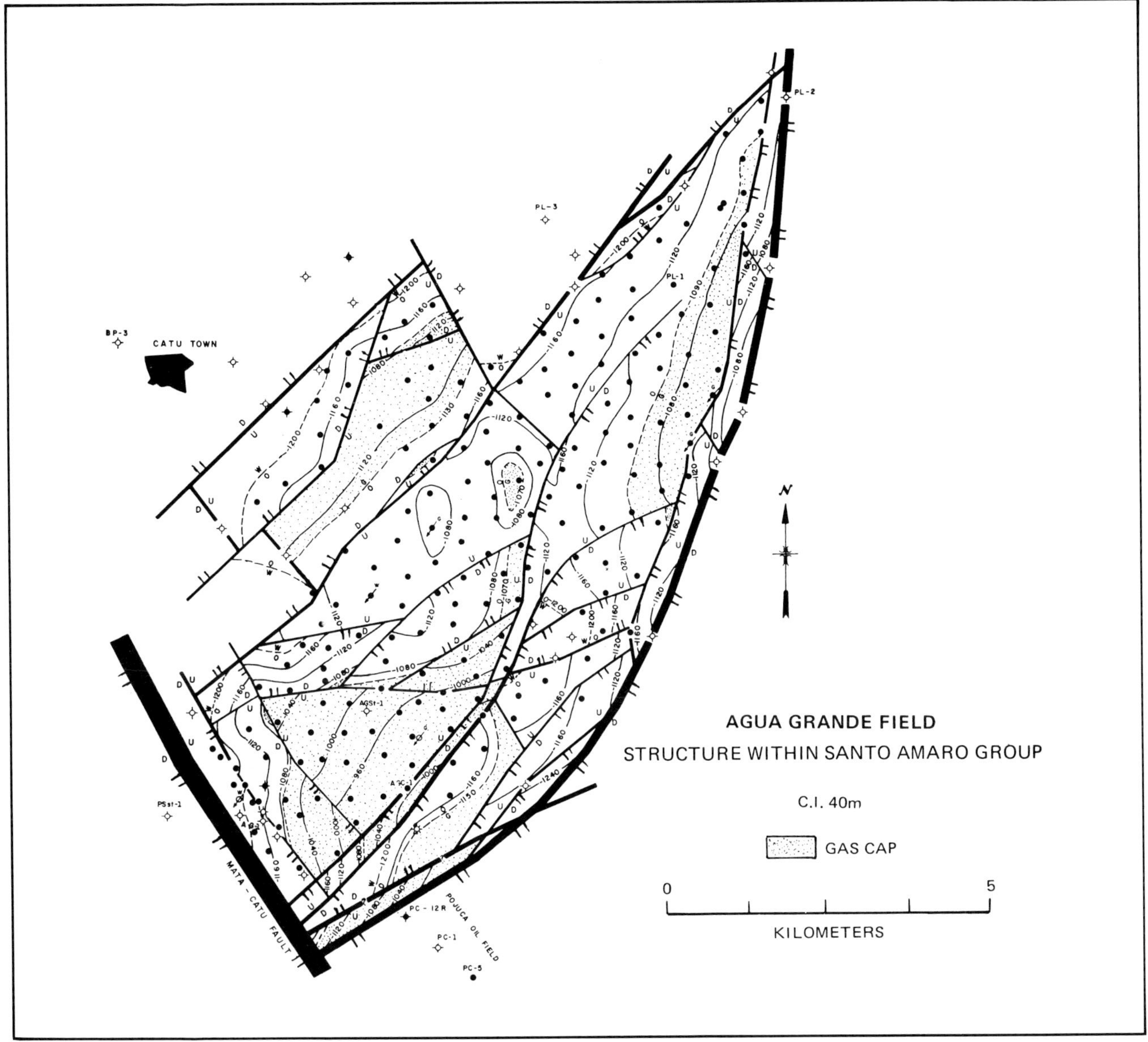

Fig. 1-13—Trap-door structure at Agua Grande oil field, Recôncavo basin, Brazil. Longitudinal faults trend northeast, parallel with graben axis, and oblique faults trend generally east-northeast and nearly north-south to north-northwest. (From Ghignone and de Andrade, 1970).

dex map, Fig. 1-11). Divergent plate boundaries inherit their outline mainly from the trends of earlier intraplate rifts.

The high frequency of longtitudinal faults appears to satisfy the common concept that normal faults trend perpendicular to regional extension (Anderson, 1942). The oblique faults do not. They interact at block corners, uplift intersecting sides equally, and appear to be unified with the relay elements into one system to achieve a well-integrated graben subsidence. The various fault directions thus appear to be contemporaneous in some places, which precludes explanation of the pattern by multiple periods of extension.

The oblique faults, because of their roughly 60° intersections in map view, have been considered to be conjugate strike-slip faults. Surface mapping in the Gulf of Suez area (Robson, 1971; El-Tarabili and Adawy, 1972) and elsewhere (King, 1965) and subsurface control demonstrate, however, that the faults are notable for their absence of significant strike slip. Other investigators (El-Tarabili and Adawy, 1972) have assumed that preexisting zones of crustal weakness influenced the later position of oblique normal faults, and this has occurred in some areas (King, 1965). In others, the typical normal-fault system continues to develop without the aid of older zones of basement weakness (Illies, 1970). The persistent recurrence of such faults in many different areas and tectonic settings, and at many scales, suggests that mul-

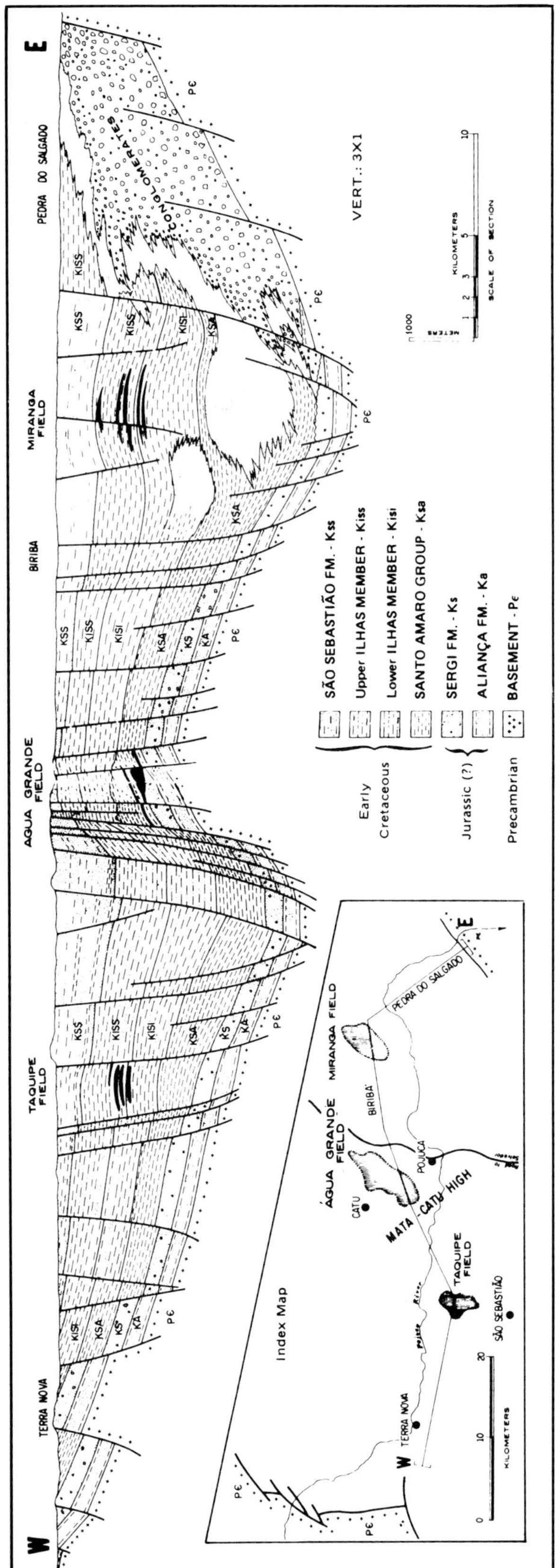

Fig. 1-14—Regional cross section of Recôncavo basin, Brazil (from Ghignone and de Andrade, 1970, fig. 4).

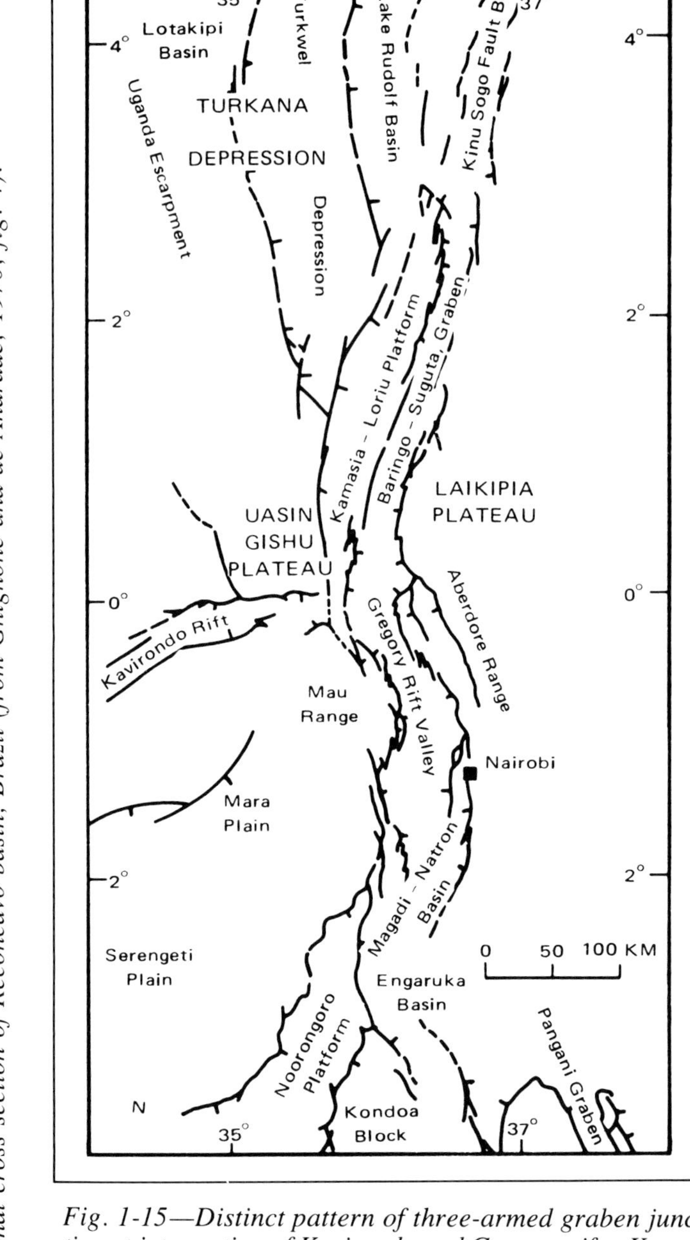

Fig. 1-15—Distinct pattern of three-armed graben junction at intersection of Kavirondo and Gregory rifts, Kenya. Northern and southern arms splay or spread out from junction, and each arm has multidirectional sets of normal faults. Main elements of southern arm between Gregory rift valley and eastern border of Kondoa block delineate dogleg. (From Baker et al., 1972).

tidirectional fault sets may be the fundamental result of extensional rupture. Reches (1978b) has reached a similar conclusion on theoretical grounds and the multidirectional dip-slip pattern has been reproduced with single-stage tectonic models (Freund and Merzer, 1976).

Asymmetric blocks with steep sides bounded by listric normal faults and gentle sides comprised of constant dips characterize the style in cross section (Fig. 1-11; Lowell and Genik, 1972). Some graben profiles show rotation and downstepping of blocks consistently toward the trough axis (Fig. 1-14, right side) or toward one side of a half-graben. Others show rotation on both synthetic and antithetic faults which causes the basement surface to be at various inclinations and depths. Some graben segments can inherit a consistent regional slope from prior structural events or can have it imposed during a rift-related regional arching phase. Along strike the position of the graben axis relative to these earlier or imposed slopes can change, causing abrupt variation in profile appearance.

Listric normal faults cause individual block rotations that can result in reverse drag on the downthrown side of faults. When reverse drag is opposite regional tilt a low-side rollover can occur (e.g., apparently at Taquipe field; fig. 1-14) comparable to those that form on some detached normal faults. Extreme block rotations result from multiple stages of listric faulting (Proffett, 1977). Rotational effects attributed to listric faults are not consistently apparent, however, even on adjacent blocks. Presumably this is due to differences in amount of displacement and to different levels at which faults flatten, the smaller the displacement and the deeper the level of flattening, the less obvious the rotation of dip into the downthrown side of the fault.

Cross sections are further complicated by the complex patterns of oblique and longitudinal faults that can dip either toward or away from the graben axis. Intersection of different fault sets brings different block rotations into the line of section and causes a loss of lateral structural continuity. Also, block rotation can be modified by post-faulting, en masse subsidence of the rift system.

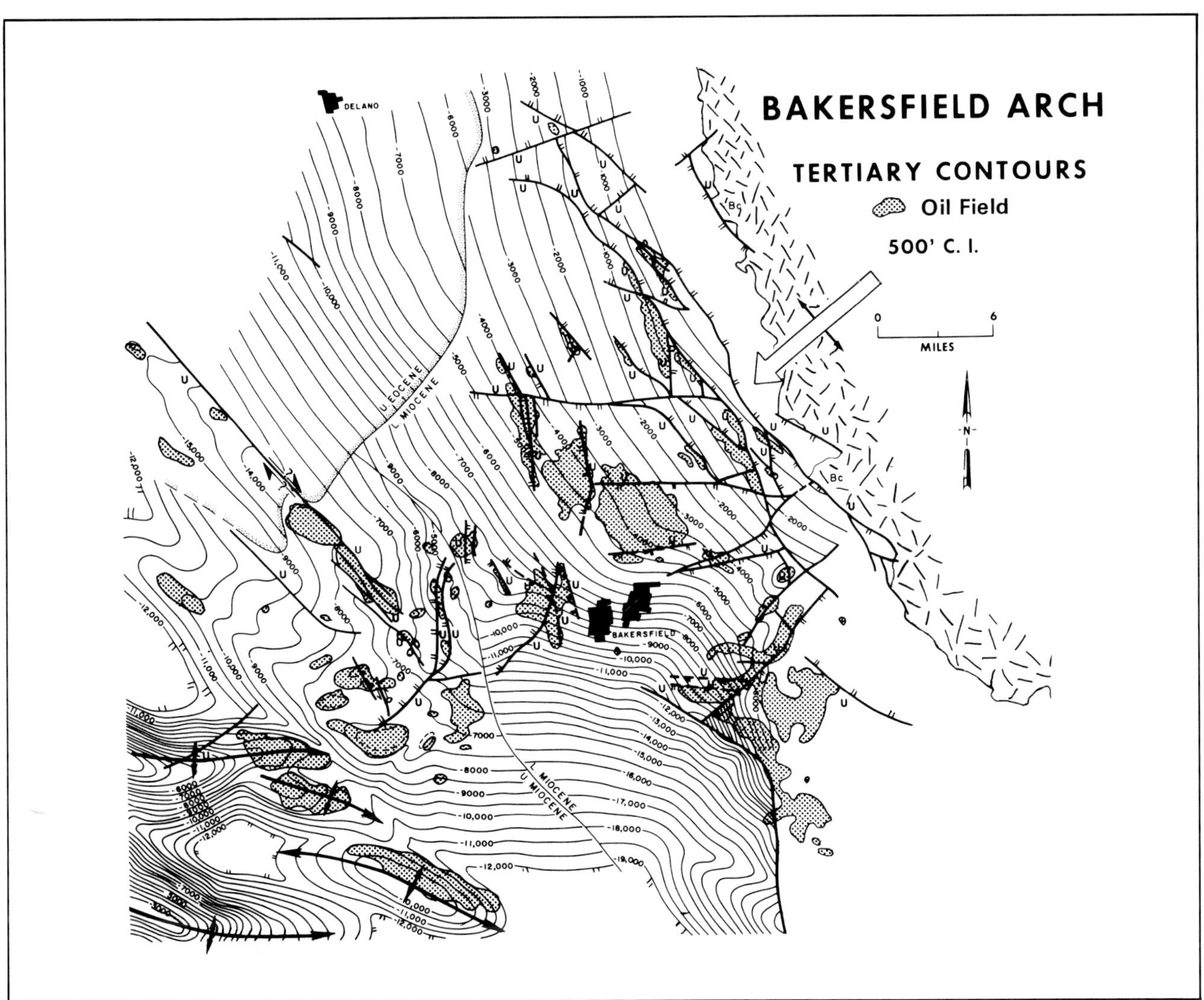

Fig. 1-16—Bakersfield arch, on stable east flank of San Joaquin Valley basin, California. Individual hydrocarbon traps are provided by secondary normal faulting, apparently concentrated by transverse arching, and by small basement domes at lower part of arch. Interference from en echelon fold set associated with San Andreas fault (west of map) is evident at southwest plunge of arch. Contours are of top of Tertiary. C.I. = 500 ft.

In many places the sedimentary cover is draped over block edges (Fig. 1-11). Where faults are downthrown in a direction opposite the direction of block tilt, drape forms reversals on the upthrown sides of and parallel with the faults. The pattern of drape is multidirectional, reflecting the grid patterns of the controlling faults. Where low-side reverse drag is also present graben structures can include a mixture of low- and high-side dip reversals.

Convergent plate boundaries are also an important normal-fault habitat (Table 1-1). Convergent settings that can be dominated by this style, progressing from plate boundary to plate interior, are:

1– the outer trench slope of downgoing plates in front (seaward) of the trench and arc complex,
2– the inner or arc-massif flank of fore-arc basins,
3– intra-arc basins within the magmatic-volcanic arc proper,
4– back-arc marginal seas produced by back-arc spreading, and
5– the stable or craton flank of back-arc and peripheral foreland basins (see Dickinson, 1974, for elaboration of settings).

On some transform margins, basins oriented parallel with a bounding major strike-slip zone have had significant normal faulting on their opposite, stable flanks. Examples include the Greater Oficina-Temblador region of eastern Venezuela opposite the El Pilar fault, the Bakersfield arch of the San Joaquin Valley basin opposite the San Andreas fault (Fig. 1-16), and the Magallanes basin of the Andean Scotia arc.

In the San Joaquin Valley, normal faulting was coeval with the period of greatest displacement and deformation along the San Andreas fault (Harding, 1976). At that time thick basin fill was deposited in sags flanking the developing Bakersfield arch (Fig. 1-16, bottom and left margins). Here and in Greater Oficina, the family of normal faults is similar to orientation plots for the Suez graben (Fig. 1-12). Fault-trend maxima are parallel with both regional strike of the shelf and the bounding strike-slip zone on the opposite basin flank. Drape flexing of beds is rare and hydrocarbon traps owe their closure principally to the bends, splays, and updip intersections of the oblique and longitudinal faults (Fig. 1-16). Similar fault closures have been found on the stable flank of some foreland basins (e.g., German Molasse basin).

The plan-view pattern permits differentiation of this style from others, except for compressive fault blocks (see previous section). In profile, however, compressive effects are notably absent; flexing, if present, is more limited than with compressive blocks. The steps and rotated-slab appearance of normal fault blocks contrast with the curvilinear fold profiles common to most other styles.

Rift systems are a prolific hydrocarbon setting. Here oil and gas traps depend mostly on the multidirectional faults and their associated drape flexures (Fig. 1-1, 1-13). Important trap-door closures occur at fault junctions, where oblique faults cause rollovers to intersect, and regional dip or block rotation maximizes structural relief at the point of fault intersection. Horsts and, less commonly, the downthrown sides of faults provide additional structural closures. The high edges of blocks are ideally situated to localize development of reefs and other reservoir facies. Subunconformity truncation traps in rotated blocks (Fig. 1-11) and other kinds of buried topographic closures are common. Their morphologies also reflect an ultimate control by the multidirectional fault pattern.

Basement Warps—Arches, Domes, and Sags

Basement warps are generally large simple flexures with gently dipping broad flanks. Regional positive elements include arches (Fig. 1-16), domes, massifs, anteclises, upwarps, swells, platforms, broad noses, and structural terraces. Negative features are downwarps, depressions, syneclises, sags, and troughs. Basement warps develop at both local and regional scales and typically are solitary. More intensely deformed basement folds also occur as elements of well-defined structural trends. These have been treated previously under other styles, but distinctions from the basement-warp category are not always obvious.

Basement warps are present in all habitats and are the dominant style of many plate-interior or craton regions (Table 1-1). They are present on the stable flanks of some convergent margin basins (e.g., Thornton arch of Sacramento Valley fore-arc basin), stable flanks of transform margin basins (e.g., Bakersfield arch of San Joaquin Valley, Fig. 1-16) and in divergent margin basins (e.g., Cape Fear arch of Atlantic seaboard).

The Williston basin and adjoining shelf on the west (Fig. 1-17) demonstrate the style of basement warps and other structures common to intracratonic settings. Structures are of two general types: warps of various shapes and sizes, and narrow linear-faulted or fault-related features. Structures of the first type include the Miles City arch; Porcupine dome (Fig. 1-17, no. 7); Nesson anticline (Fig. 1-17, no. 1), a broad southward-plunging nose or arch with several culminations; the Bowdoin (Fig. 1-17, no. 2) and Poplar domes (Fig. 1-17, no. 3) west of the basin proper; and the central-basin deep (Fig. 1-17, no. 4), which is an irregularly shaped sag or downwarp. The second type of structure, linear fault-associated features, is exemplified by the Cedar Creek anticline (Fig. 1-17, no. 5) which is essentially a monocline facing up the basin flank. Most of these Williston basin structures are solitary and do not share common geographic alignments or obvious trend associations; instead, their orientations include all quadrants of the compass.

Additional styles occurring in intracratonic regions are normal faults, either randomly dispersed or in well-defined zones and a few solitary wrench faults.

The structural genesis of basement warps is not well understood and they probably have multiple origins. Warps lack diagnostic features, such as drag and drape

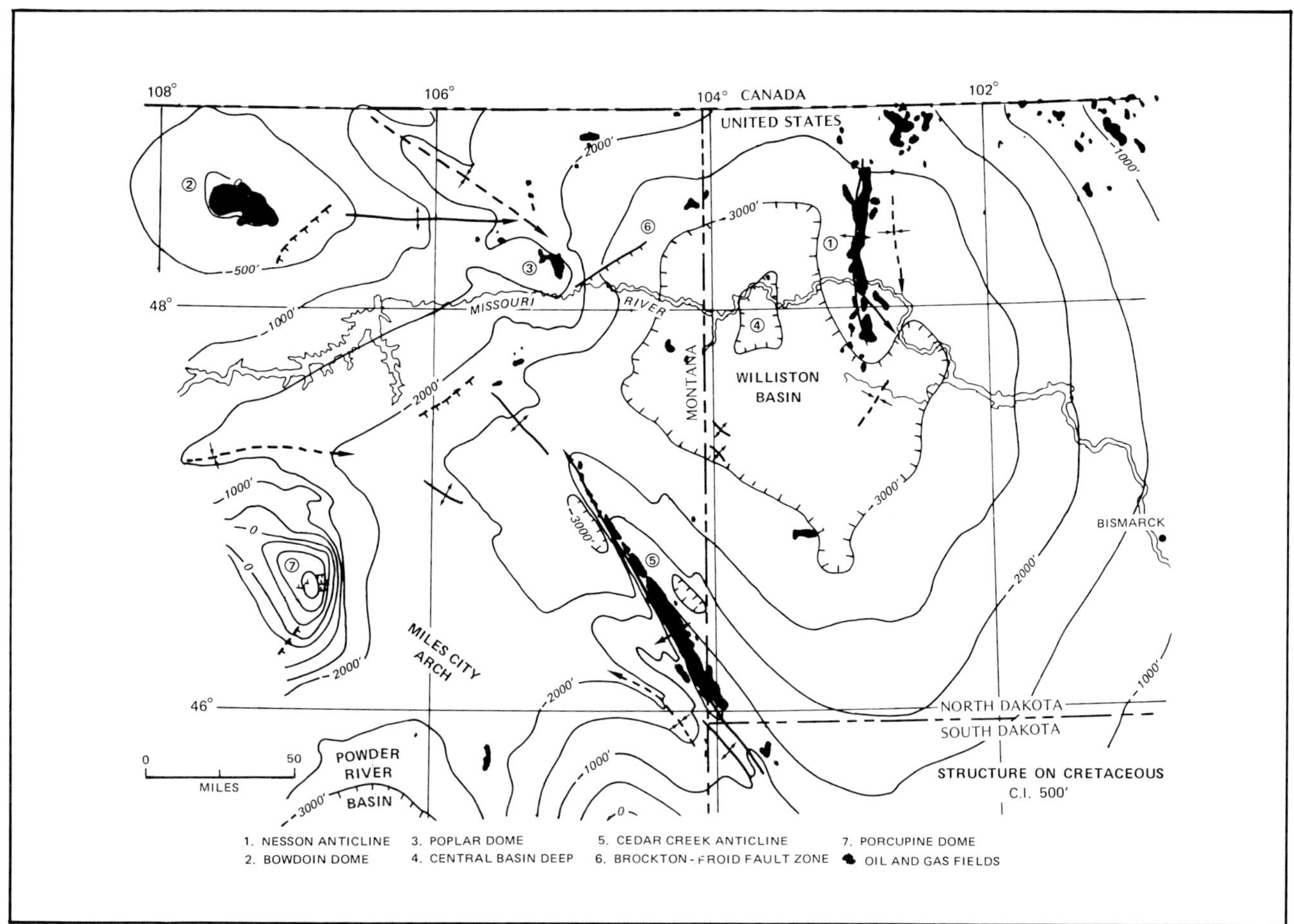

Fig. 1-17—Regional structure and main elements of Williston basin area. Trend orientations are diverse and continuity from one structure to another is absent (except in Bowdoin and Poplar domes). (Cohee, 1962).

folds or trend patterns, to indicate directions of tectonic transport. Also lacking is an obvious plate-boundary association that could suggest a particular type of deformation. Differential regional subsidence or uplift has intuitive appeal as a cause of many intraplate warps. For example, the Miles City arch appears to be a residual high separating the Williston and younger Powder River basins (Fig. 1-17). The main Williston basin structures had their early growth in Ordovician to Mississippian time (Hansen, 1972), synchronous with the first major phase of basin subsidence (Mallory, 1972). Differential subsidence or uplift could have occurred because of (1) irregularities in fundamental subcrustal processes and (2) large inhomogeneities within the crust.

Green (1977), in a review of crustal processes, stated that the following have all been considered mechanisms of continental lithospheric thinning and subsidence: differential lithospheric cooling, phase changes, ductile flowage, subcrustal erosion, injection and stoping of dense material and related magma-chamber collapse, and surface erosion following thermal uplift. In addition, deep-seated listric normal faulting can also cause lithospheric thinning and associated subsidence.

Crustal inhomogeneities can modify structural responses to those deep-seated processes. Steep linear zones of weakness such as preexisting wrench faults may be reactivated in a dip-slip mode. The reactivation could develop basement steps or linear monoclines such as the Paleozoic precursor of the Cedar Creek anticline (Fig 1-17, no. 5). Some basement warps may be influenced by earlier buried fault blocks to the extent that they merge with the shallow expressions of fault-block styles (perhaps Porcupine dome, Fig. 1-17, no. 7).

The origin of domal features in contrast to linear or blocklike structures is much more speculative. Irregularly shaped intrabasement masses with differing physical properties perhaps could subtly and locally retard or enhance regional subsidence.

Tectonic loading and sediment loading of fore-deeps have been considered mechanisms of warping in foreland settings. Crustal compression may be an additional agent for basement warping in such regions. The Williston basin structures were rejuvenated during the Laramide (Fig. 1-17) when compression was prevalent elsewhere in the Rocky Mountains province. Using earthquake focal plane solutions, Sykes and Sbar (1974) have demonstrated present compression within the North American plate.

The style and occurrence of basement warps pose problems for the simplistic plate-tectonic approach that restricts structural development to the edges of so-called rigid plates and that implies mainly horizontal plate movement. Many basement warps in intraplate settings suggest first-order vertical movement in their development. Other features suggest midplate compression. All argue against simple plate tectonics as the exclusive cause of deformation. Clearly, vertical movements must occur within a larger scope of horizontal plate translations.

Basement warps can be differentiated from other basement-involved folds by their solitary occurrence, inconsistent orientations, lack of distinct trend patterns, and general lack of dependency on faulting for their development. Basement folds associated with block faulting reflect block-fault patterns as a genetic control. Basement folds associated with thrust or wrench faults demonstrate the characteristic trend patterns of these styles.

Positive basement warps, particularly where they intervene in areas of thick sedimentary cover, are the focal points for oil migration and accumulation (Fig. 1-16). They commonly persist over long periods of time and are particularly effective in localizing truncation, unconformity, convergence, and onlap in the sedimentary section. All of these factors can enhance entrapment of oil and gas (Fig. 1-1). Associated closures commonly are considerably larger and less segmented than the more complicated structures of other styles.

DETACHED STRUCTURES

Decollement Thrust-Fold Assemblages

Thrust-fold assemblages are an essential element of many convergent plate boundaries (Table 1-1). They may be present along the mobile flanks of back-arc and peripheral forelands, and at trench inner slopes and outer highs (Dickinson, 1974). They occur as wide zones of deformed sedimentary cover on the external side, or outer fringe, of many orogenic belts (Fig. 1-3). Gradations to basement thrusting are common toward the internal, or core, regions (Fig. 1-4). The detached structures regionally are festooned into externally gently convex salients and more sharply concave reentrants (Fig. 1-3). Both salients and reentrants are comprised of parallel bands of thrusts and associated folds having relay trend patterns. Locally, thrusts overlap in cuspate arrangement convex in the direction of tectonic transport. Anticlines, where present, lie in the hanging walls of the thrust faults and have axes parallel with the trends of the thrusts.

In profile, complex listric thrust faults that often merge with bedding at depth are characteristic (Fig. 1-19). Lithostratigraphic changes and attendant ductility contrasts are responsible for localization of the fault surfaces so that thrusts are generally parallel or subparallel with bedding in incompetent rocks and oblique to bedding in competent rocks. Thus, the thrusts alternately follow

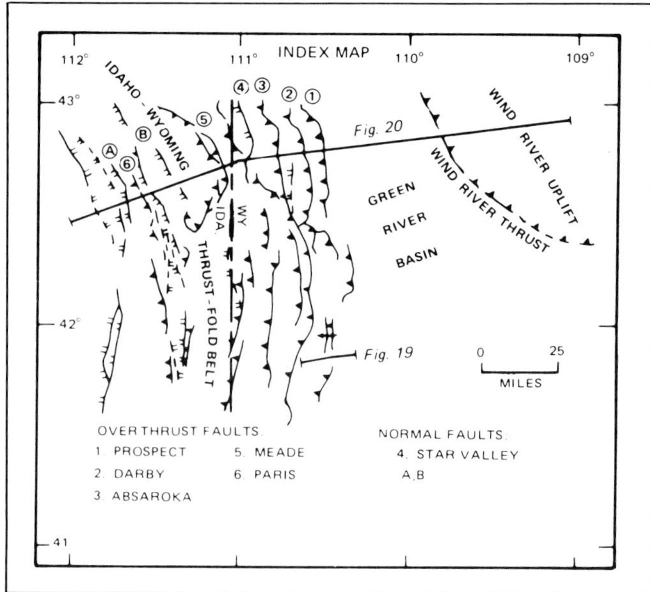

Fig. 1-18—Index map across Idaho-Wyoming thrust-fold belt. Seismic section and cross sections are shown in figures 1-19 and 1-20. (Adapted from Cohee, 1962).

bedding, then ramp upward as step faults (Fig. 1-20). The faults almost invariably cut upsection in the direction of relative tectonic transport of the hanging wall.

Where massive carbonate beds are a significant part of the deformed section, thrust sheets typically consist of repeated slabs. Where more ductile lithologies prevail, hanging-wall folds dominate. Such folds are characteristically asymmetric in the direction of relative transport of the upper plate. Structures generally are younger from the internal to the external side of the deformed belt (Fig. 1-20).

Anticlines with imbricate thrusts and broad folded thrust-fault structures (Fig. 1-19) occur across the breadth of a thrust-fold belt. The latter seem to be more common on the internal side and are usually located where faults step up in the sedimentary cover. Some thrust faults, backthrusted or directed against the direction of relative transport of the upper plate, bound belts at their fronts and are termed ''delta structures.''

The process of thrusting areally extensive but thin sheets of sedimentary rocks for long distances has always been an enigma of structural geology. Overthrusting by compression (or ''pushing'' at the rear of thrust sheets) does not resolve yield-strength and stress-transmission problems in moving relatively weak materials. Gravity sliding, which invokes body forces to overcome these problems, also is not a viable universal mechanism of thrust emplacement because:

1– necessary regions of tectonic denudation are rarely observed,

2– penetrative deformation of sedimentary cover in the internal parts of orogenic belts is not explained,

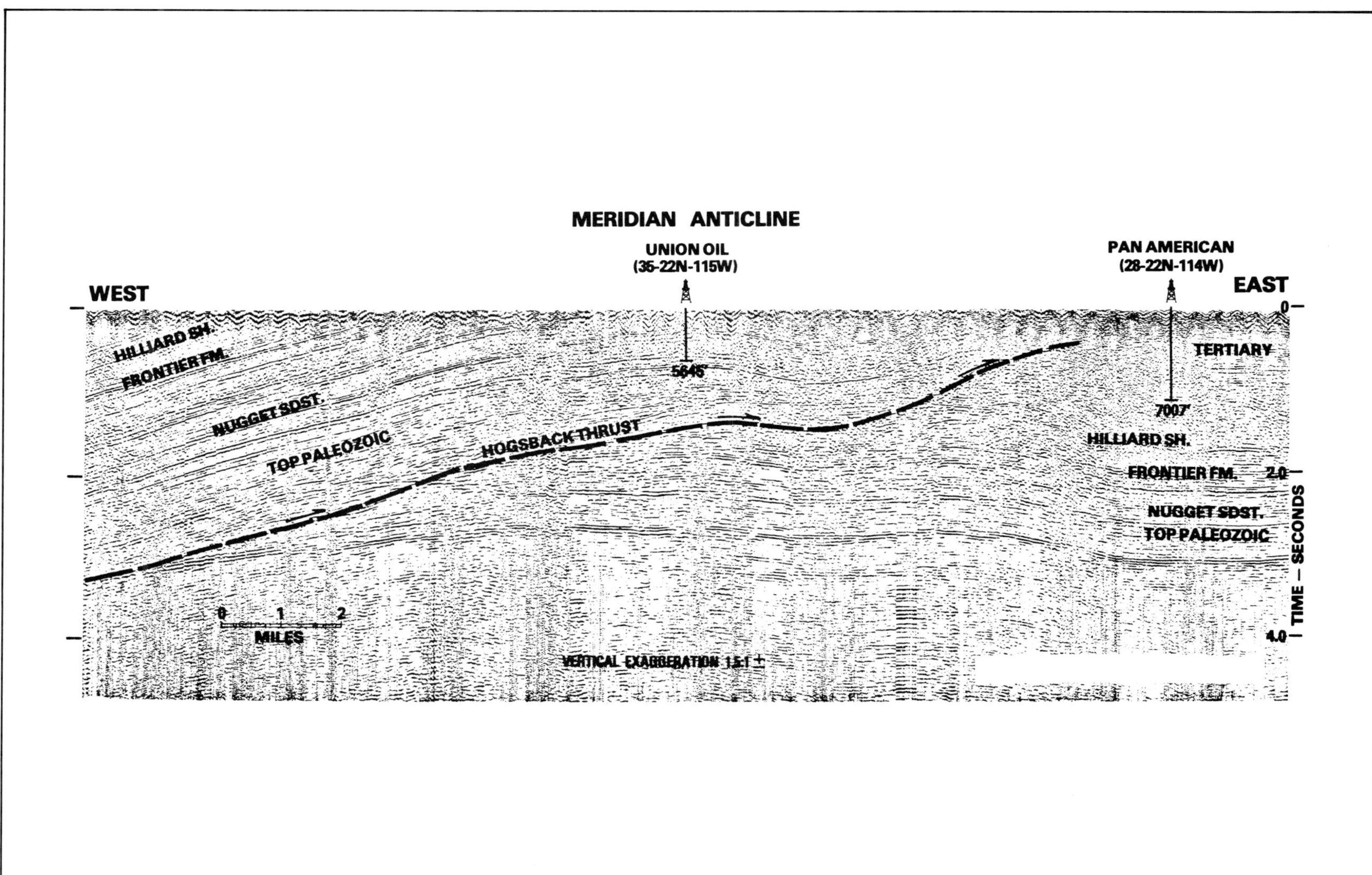

Fig. 1-19—Seismic line across Meridian anticline and Hogsback thrust on east side of Idaho-Wyoming thrust-fold belt near latitude of Kemmerer, Wyoming (see Fig. 1-18). Fault alternately parallels bedding and steps upsection. Velocity pull-up on time section lies beneath leading edge of thrust, which has introduced relatively high-velocity carbonate rocks. Paleozoic rocks in hanging wall in part of structure near label "Hogsback thrust" are repeated by imbricate faults that splay from main detachment. (From Royse et al., 1975).

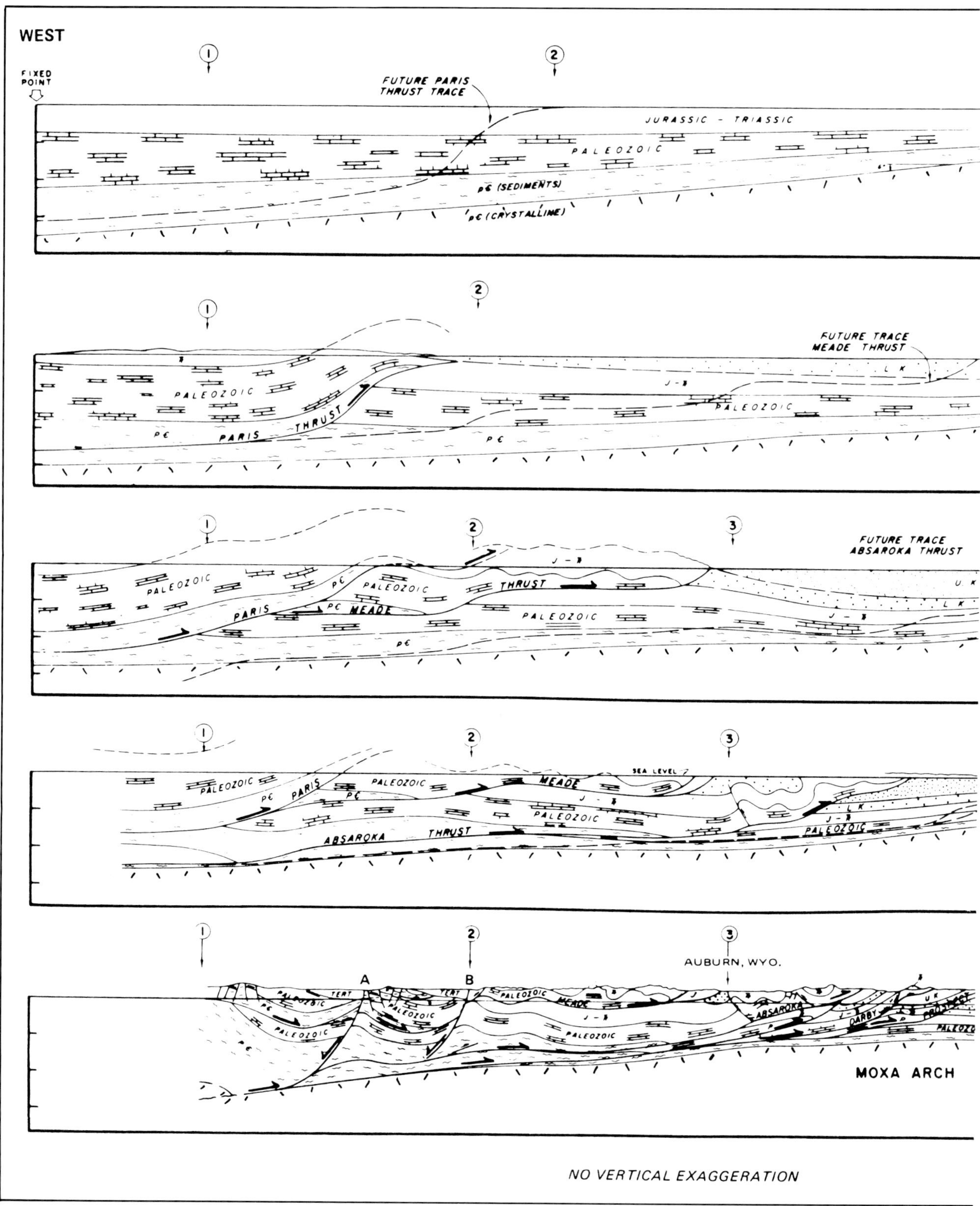

Fig. 1-20—Sequential evolution of Idaho-Wyoming thrust-fold belt at latitude of Auburn and Pinedale, Wyoming (see Fig. 1-18). Detachments form progressively from west to east, as rigid basement slab is presumably underthrust from east to west (Lowell, 1977). Wedging action of underthrusting lifts earlier formed western thrust sheets (e.g., Paris thrust) so that older rocks are exposed. Reference numbers 1 through 7, which ultimately become present-day 30' longitude positions, demonstrate cumulation of shortening in sedimentary cover. A, B in bottom cross section are later listric normal faults. (Generalized from Royse et al., 1975).

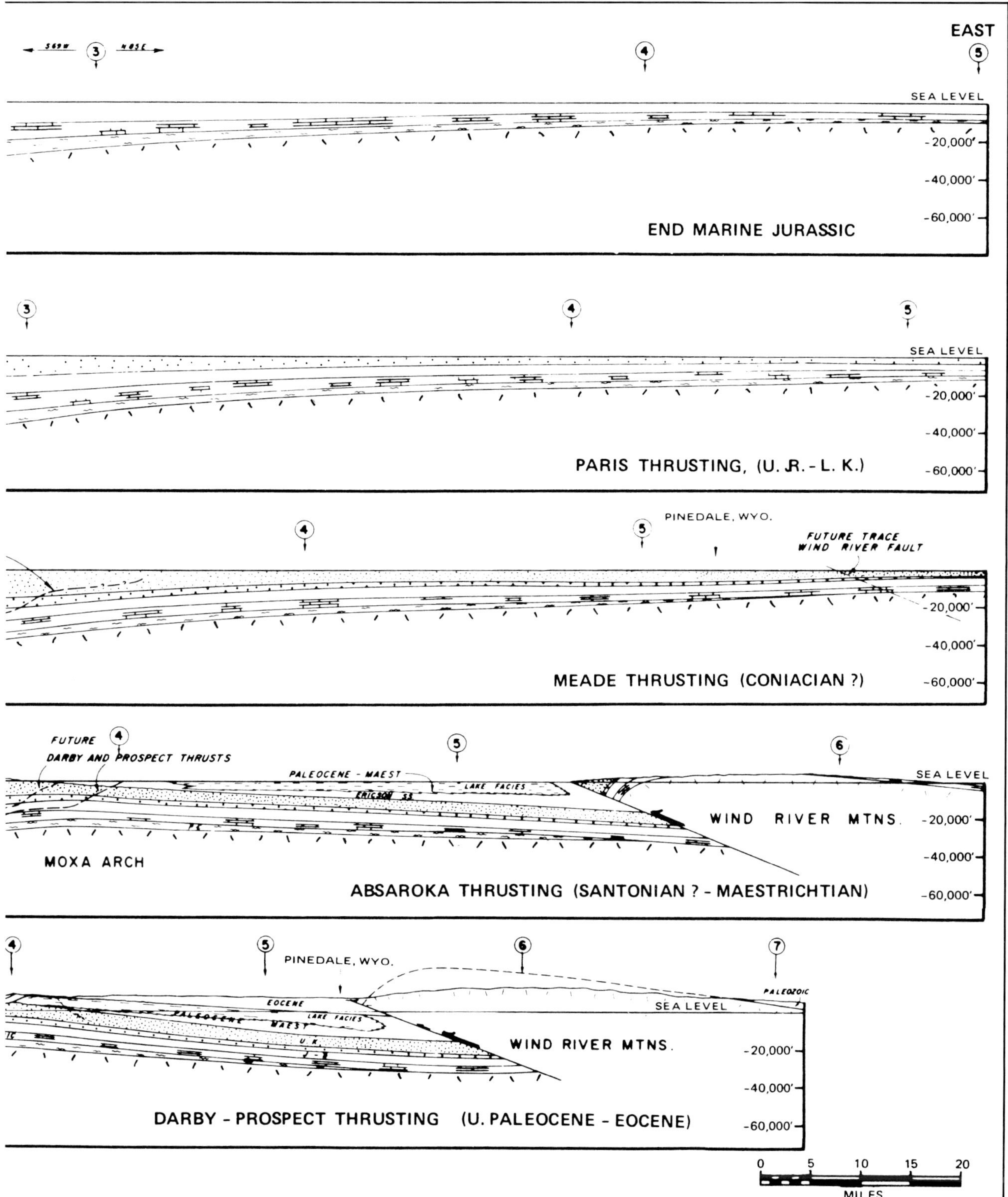
EAST
SEA LEVEL
-20,000'
-40,000'
-60,000'
END MARINE JURASSIC
SEA LEVEL
-20,000'
-40,000'
PARIS THRUSTING, (U. JR. - L. K.)
-60,000'
PINEDALE, WYO.
FUTURE TRACE WIND RIVER FAULT
-20,000'
-40,000'
MEADE THRUSTING (CONIACIAN ?)
-60,000'
FUTURE DARBY AND PROSPECT THRUSTS
PALEOCENE - MAEST
LAKE FACIES
ERICSON SS
SEA LEVEL
WIND RIVER MTNS.
-20,000'
MOXA ARCH
-40,000'
ABSAROKA THRUSTING (SANTONIAN ? - MAESTRICHTIAN)
-60,000'
PINEDALE, WYO.
PALEOZOIC
EOCENE
PALEOCENE
LAKE FACIES
MAEST
U. K.
SEA LEVEL
WIND RIVER MTNS.
-20,000'
-40,000'
DARBY - PROSPECT THRUSTING (U. PALEOCENE - EOCENE)
-60,000'
0 5 10 15 20
MILES

3– basement slopes required for sliding are usually in the wrong direction,

4– the chronologic sequence of thrusting is reversed, and

5– the laterally extensive nature of thrust-fold belts that follow plate boundaries for thousands of miles is not compatible with the local aprons that would be expected from simple sliding of sedimentary cover off of uplifted regions.

Although gravity sliding is not a suitable mechanism for emplacement of thrust sheets on a regional scale, it has created smaller deformed belts where a slope on the base of the detached sedimentary section exists over a limited area, such as the Bearpaw Mountains of north-central Montana.

As an alternative to regional gravity sliding, a model of gravity spreading has been proposed (Bucher, 1956; Price, 1969, 1973; Price and Mountjoy, 1970; Elliott, 1976) in which only a slope on top of the sedimentary prism, and not on the actual thrust surfaces, is required to facilitate thrust-sheet emplacement. The structural results of gravity spreading can also be accommodated by the model of underthrusting described in the next paragraph and the two need not be mutually exclusive. However, at least in the Idaho-Wyoming thrust belt, thrusting began before any significant surface slope was attained (Fig. 1-20).

Large-scale plate movement provides a rationale for underthrusting between oceanic and continental lithospheric plates and within continental lithosphere. The model of stripping and stacking of sedimentary cover by crustal underthrusting in the landward slopes of oceanic trenches (Seely et al., 1974) appears to satisfy many of the critical relations in thrust-fold belts generally (Lowell, 1977). Underthrusting obviates problems of rock strength and stress transmission in that there need be no deformation of layered sedimentary rocks lying on rigid sialic basement until these rocks reach a zone of uncoupling (Fig. 1-20). The wedging action provided by the insertion of successive thrust sheets from below is also a mechanism for rotation and uplift of earlier formed thrusts and thereby accounts for the exposure of older rocks in the more internal parts of most orogenic belts. Finally, the age progression of detached thrusting from older to younger toward the underthrust plate is compatible with and even predicted by underthrusting (Fig. 1-20).

Thrust structures can be identified on seismic profiles by two main criteria: (1) shallow compressive folding above essentially undeformed reflectors and (2) stretches of backlimb dips that ultimately converge with underlying reflectors, generally in a direction opposite to tectonic transport (Fig. 1-19, left margin). In the latter way, thrusts do not demonstrate the persistent regional uplift present in many compressive fault blocks. The typical convergence of beds on one flank only reflects the asymmetry of the detachment. The bed geometry contrasts with the more symmetric convergence of reflectors apparent down both flanks of other detachments, such as flower structures (cf. Figs. 1-6, 1-19) and salt pillows.

The decollement thrust-fold assemblage is further distinguishable from other thrust styles on the basis of distribution patterns. On a very large scale, many thrust-fold assemblages have great lateral persistence. Wrench-fault-associated assemblages have more finite, limited distributions. The distribution of compressive fault blocks is more random.

The sinuosity of most thrust-fold belts is also distinctive. On an intermediate scale, individual relay trends curve with this sinuosity, thus conforming to the outline of salients or reentrants. Thrusts and folds associated with wrench faults ideally trend obliquely to the outline of the deformed zone (cf. Figs. 1-3, 1-5). On a more local scale, the relay patterns of thrust belts differ from both the en echelon trends of wrench-associated structures and the rhombic or cross-trended patterns of fault blocks (cf. Figs. 1-3, 1-8).

Most petroleum in thrust-fold belts has been trapped in asymmetric, hanging-wall folds and the lead edges of thrust sheets (Fig. 1-1). Effective closures are commonly located at a relatively external part of a belt and at generally shallow to moderate (10,000 ft; 3,000 m) structural levels. Closures at these positions have been only slightly to moderately disrupted by faulting allowing reservoir continuity and migration pathways to remain essentially intact. Structures in this setting along the Zagros fold-belt of Iran account for more than three-fourths of the production from the world's thrust-fold belts. Enormous volumes of hydrocarbons are contained in fractured carbonate rocks on very high-amplitude, structurally simple, detached anticlines (Fig. 1-4). Significant production occurs also in the Canadian Rockies where traps are mainly at the lead edges of thrust sheets. Recently, important discoveries have been made in the Idaho-Wyoming thrust-fold belt; all traps found so far are hanging-wall folds.

Detached Normal-Fault Assemblages

Normal faults detached from basement occur in a wide variety of tectonic terranes but many are secondary elements of other structural styles. Most normal faults on the crests of folds and above diapirs (Fig. 1-1) are secondary faults. In other places, for example the Gulf Coast basin of Texas and Louisiana, detached normal faults occur in regional zones and are a dominant mode of deformation. Such zones commonly have unique structural assemblages and comprise a distinct structural style (Fig. 1-21).

The most common types of primary detached normal faults are down-to-basin faults that develop in ductile lithologies and have a close relation between displacement history and sedimentation. Interval thicknesses increase abruptly along the downthrown sides of these faults and there is a generally progressive downward increase in throw through much of the fault profile. These syndepositional faults also have been termed "re-

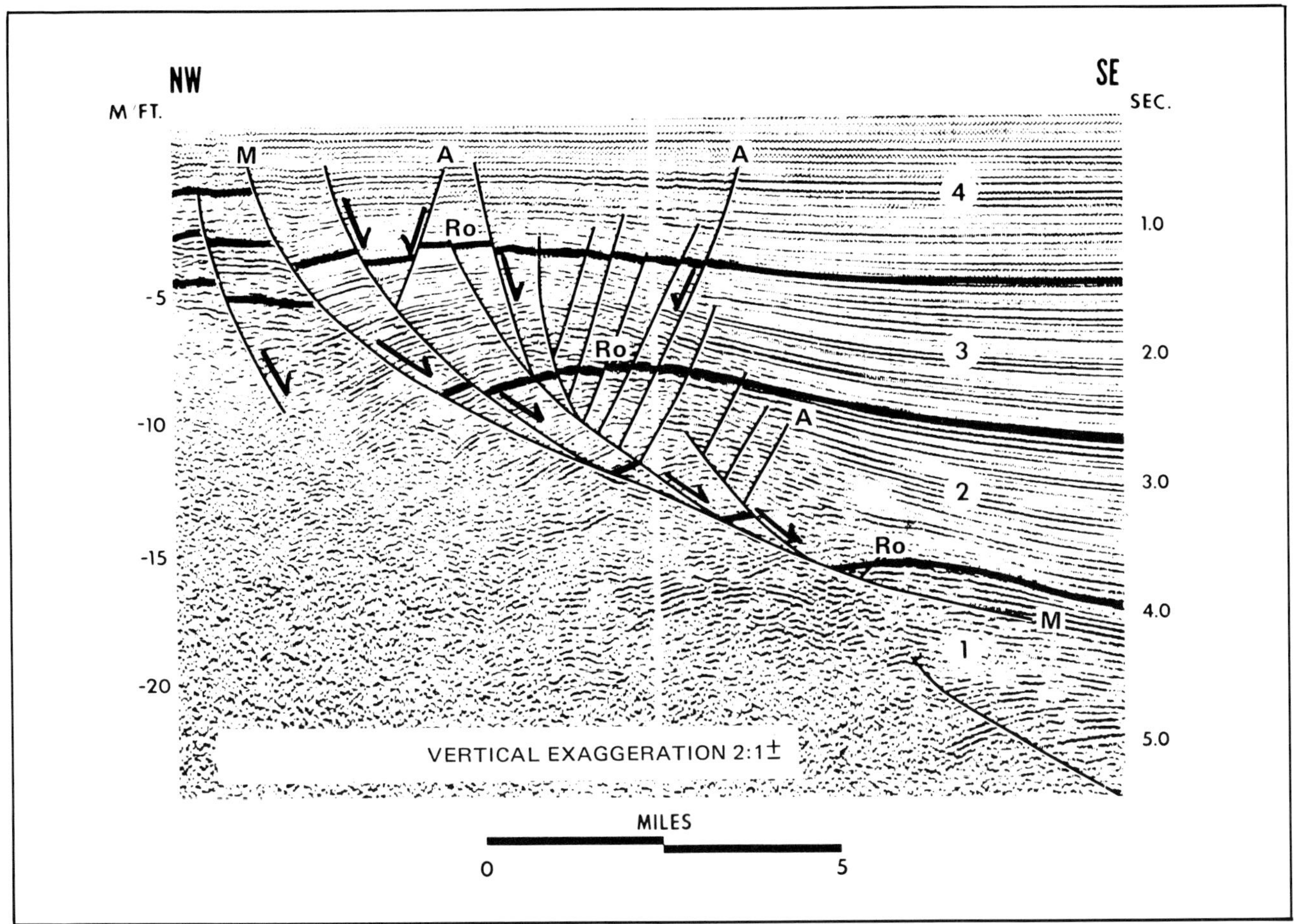

Fig. 1-21—Syndepositional, detached normal fault assemblage in Tertiary sediments of south Texas part of Gulf Coast basin. M, master down-to-basin synthetic fault. A, antithetic faults. Ro, rollover anticline with basinward migration of crest at depth. Northwest-dipping events beneath M fault would migrate to right to form part of rollover anticline. Some assemblages have only synthetic faults. (From Bruce, 1973).

gional faults," "growth faults" (Ocamb, 1961), "contemporaneous faults" (Hardin and Hardin, 1961), and "down-to-basin faults" (Cloos, 1968). These attributes, however, are obviously not restricted to just one type of fault and such terminology does not describe a discrete style.

Postdepositional detached normal faults, whose structural geometries are similar to the syndepositional types, have developed where a gradient has been present for sliding, for example, the topographically highest end of the Heart Mountain fault (Pierce, 1957). Detached normal faults, some of which flatten on the ramped portion of earlier thrust faults, are also present in thrust-fold belts (Fig. 1-20, faults A, B). Such faulting occurred in the Idaho-Wyoming thrust belt during a later episode of regional extension that was unrelated to thrusting but did affect deposition and preservation of section on the downthrown side of faults. Others apparently developed at the conclusion of thrusting and do not show contemporaneous expansion of sedimentary section. All of these postdepositional detached normal faults developed in relatively competent rocks. Their importance in petroleum provinces is minor compared to the Gulf Coast variety.

Syndepositional detached normal faults develop a dominant style in thick, generally regressive clastic sequences that have been built out into unconfined depositional sites, such as large deltas and the edges of continental shelves on passive continental margins (Todd and Mitchum, 1977, fig. 9). There, rapidly deposited, semiconsolidated sediment is free to move basinward, In addition to the process of creep, faulting is often closely associated with the growth of salt (Lehner, 1969; Spindler, 1977) and shale structures (Bruce, 1973; Dailly, 1976). The largest displacement faults, termed "synthetic," are downthrown toward the basin. Some secondary faults on individual structures have up-to-basin displacements ("antithetic") and generally terminate against the synthetic faults (Fig. 1-21). Other antithetic faults are primary, and have major displacements and structural characteristics similar to the regional synthetic faults (Spindler, 1977).

In profile view, many syndepositional faults dip subparallel with the flank of an underlying shale ridge or salt diapir, and may use the contacts with the overlying sediments for their detachment zones (Fig. 1-22). The faults are distinctively listric and can flatten downdip to merge parallel with bedding (Fig. 1-21) or dissipate within a chaotic slide-flow zone (J. S. Shelton pers. com., 1978; Rider, 1978). In some places the depositional top of salt is a detachment surface (W. F. Bishop, 1973). Syndepositional faults can also terminate in an adjacent shale ridge (Fig. 1-22) or die out in the bottom of "depopods" between shale or salt diapirs (McGookey, 1975, fig. 13). Compressional features have been reported at the toes or downdip ends which are sites of crumpling and telescoping of mobile section within the detached masses (Amery, 1969; Humphris, 1976; Dailly, 1976). Upward the displacement can decrease to nil owing to greater sediment filling of the downthrown side of the fault. Continued basin subsidence and late rejuvenation of some buried faults can reextend them to the surface, which has produced faults with both earlier growth and later postdepositional characteristics. The shallower, rejuvenated segments have displacement magnitudes that are constant throughout their profile, and formation thicknesses are the same in upthrown and downthrown blocks.

Detached normal-fault assemblages are complex (Fig. 1-21). The listric nature of many detached faults requires a backward rotation of beds into the steeper part of the fault (Cloos, 1968). In some places where rotation reverses basinward dip, a detached "rollover" anticline is formed on the downthrown side of and parallel with the master fault (top of Fig. 1-23). Downthrown beds are thickest adjacent to the fault and generally thin away from it in this example. Structural warping and consequent closure parallel with fault strike are thought to be the result of fault curvature and displacement variations along the trend of the fault that also cause changes in stratigraphic thickness within the downthrown block (Roux, 1977). The broad shallow rollover is commonly complicated at depth by increased faulting which segments the reverse dip into separated rotated fault blocks and increases the breadth of the counter-regional dip (Fig. 1-23, lower map). The overall rollover structure shifts basinward with depth but dip reversal can occur within different fault blocks at different locations and depths, and some blocks have no internal flexing (Fig. 1-24, at level of Kings Bayou sand).

The extensional tectonics in the Gulf Coast basin are further complicated by the presence of other kinds of

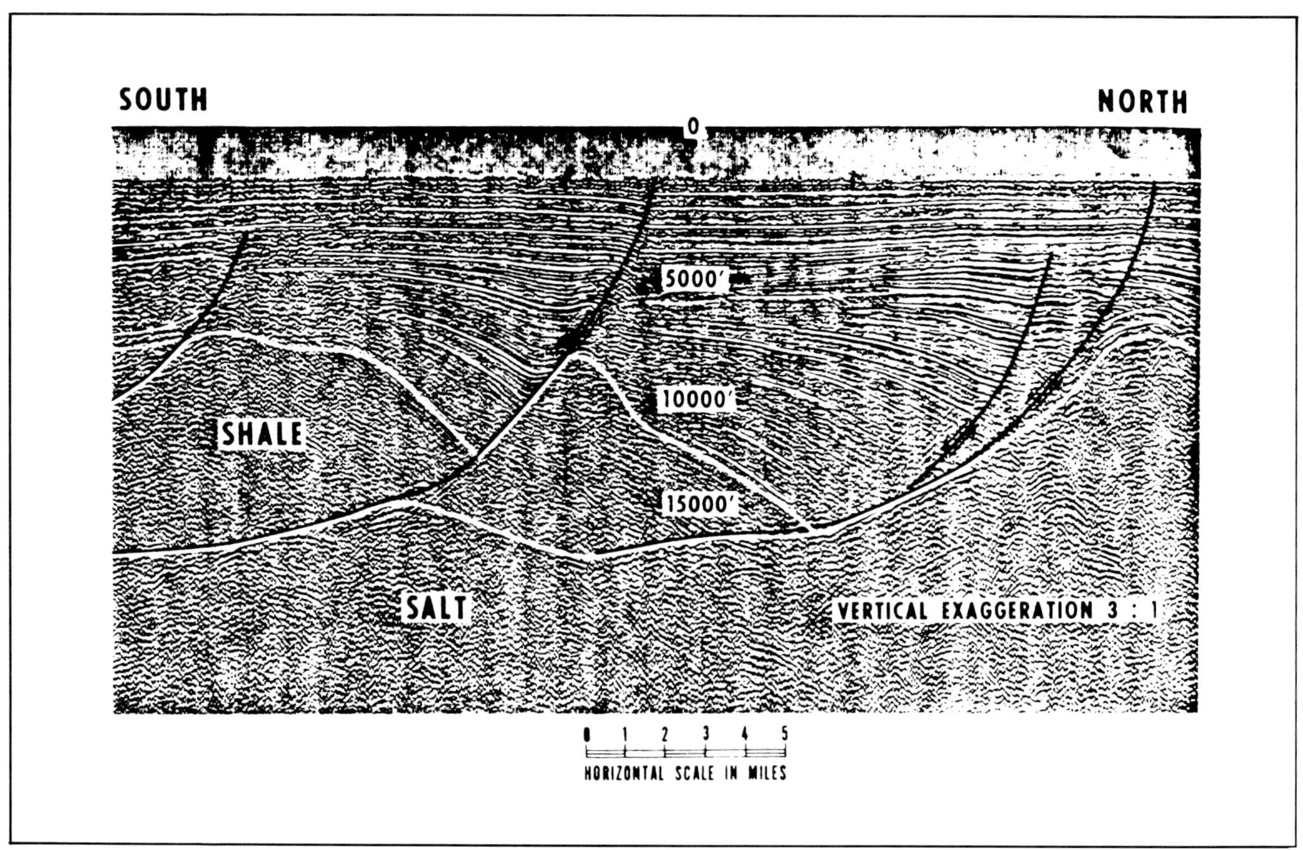

Fig. 1-22—Syndepositional detached normal faults in outer continental shelf of Gulf Coast basin, Louisiana and Texas. Faults sole out at salt-overburden contact identified by strong reflector and associated diffractions below 4,500 m (15,000 ft) in center of cross section. Presence of diapiric shale core (center) or salt diapir (right) below upthrown side of large growth fault is reported to be very common in this province (Woodbury et al., 1973).

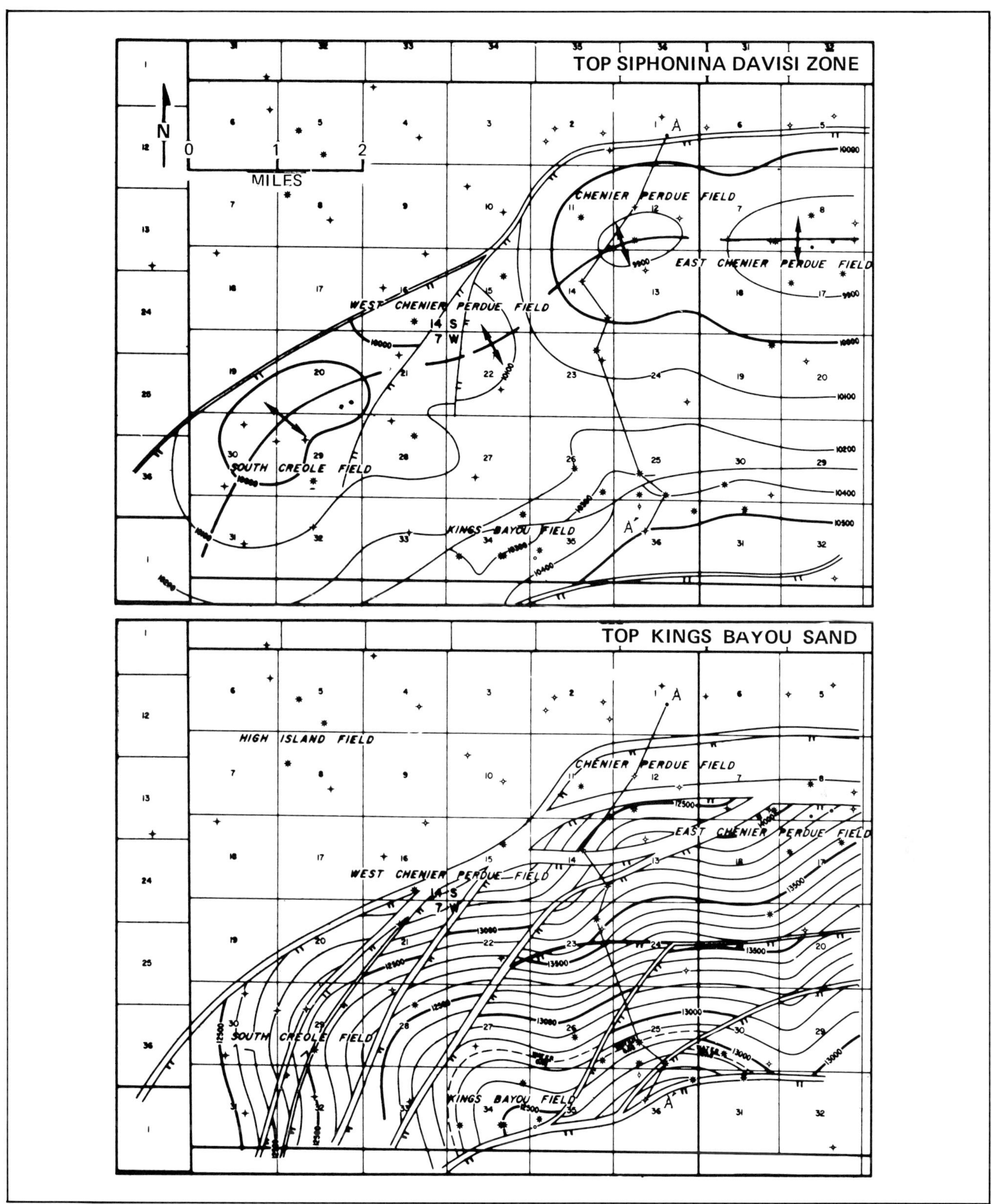

Fig. 1-23—Detail of detached normal fault pattern, southern Louisiana Gulf Coast. Parallelism of axis of rollover anticline on downthrown side of associated down-to-basin fault is apparent in top map. Rotation of regional dip into deeper faults increases landward-dipping segments that form individual fault traps in bottom map. (From Sloane, 1971).

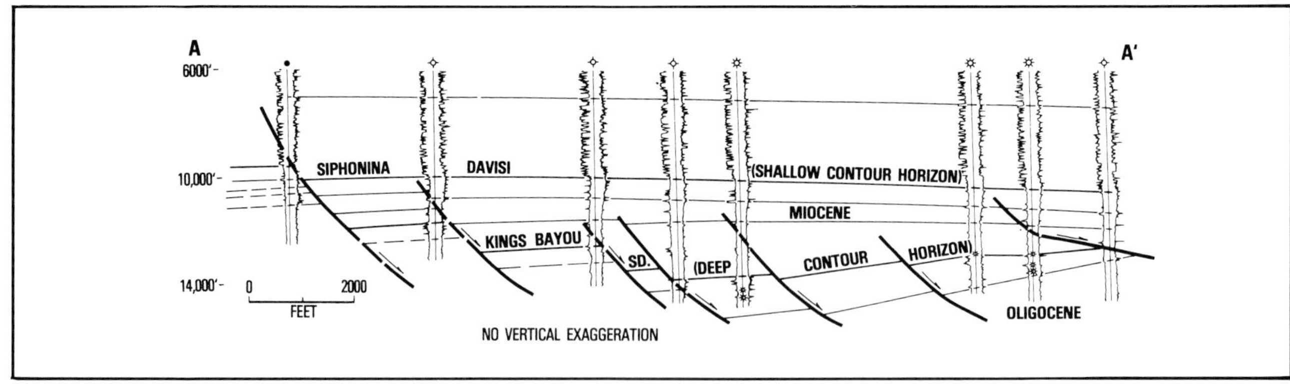

Fig. 1-24—North-south cross-section AA', across East Cameron area, United States Gulf Coast, showing normal fault pattern. Rollover has been removed by rotation of regional dip (Fig. 1-23). Location of cross section is shown on figure 1-23. (Adapted from Sloane, 1971).

normal faults. These include numerous local keystone faults above salt or shale domes and turtle structures, and more areally extensive detached "collapse faults" (Seglund, 1974) which circumscribe salt or shale withdrawal basins. Older basement-involved normal faults, thought to have facilitated opening of the Gulf of Mexico (Freeland and Dietz, 1971), may have been rejuvenated during prolonged subsidence of the basin. If so, they could cut upward into the regional detached structures and could even trigger the shallower detached faulting.

Regional elements of the detached normal fault style in the United States Gulf Coast (Cohee, 1962) and in the

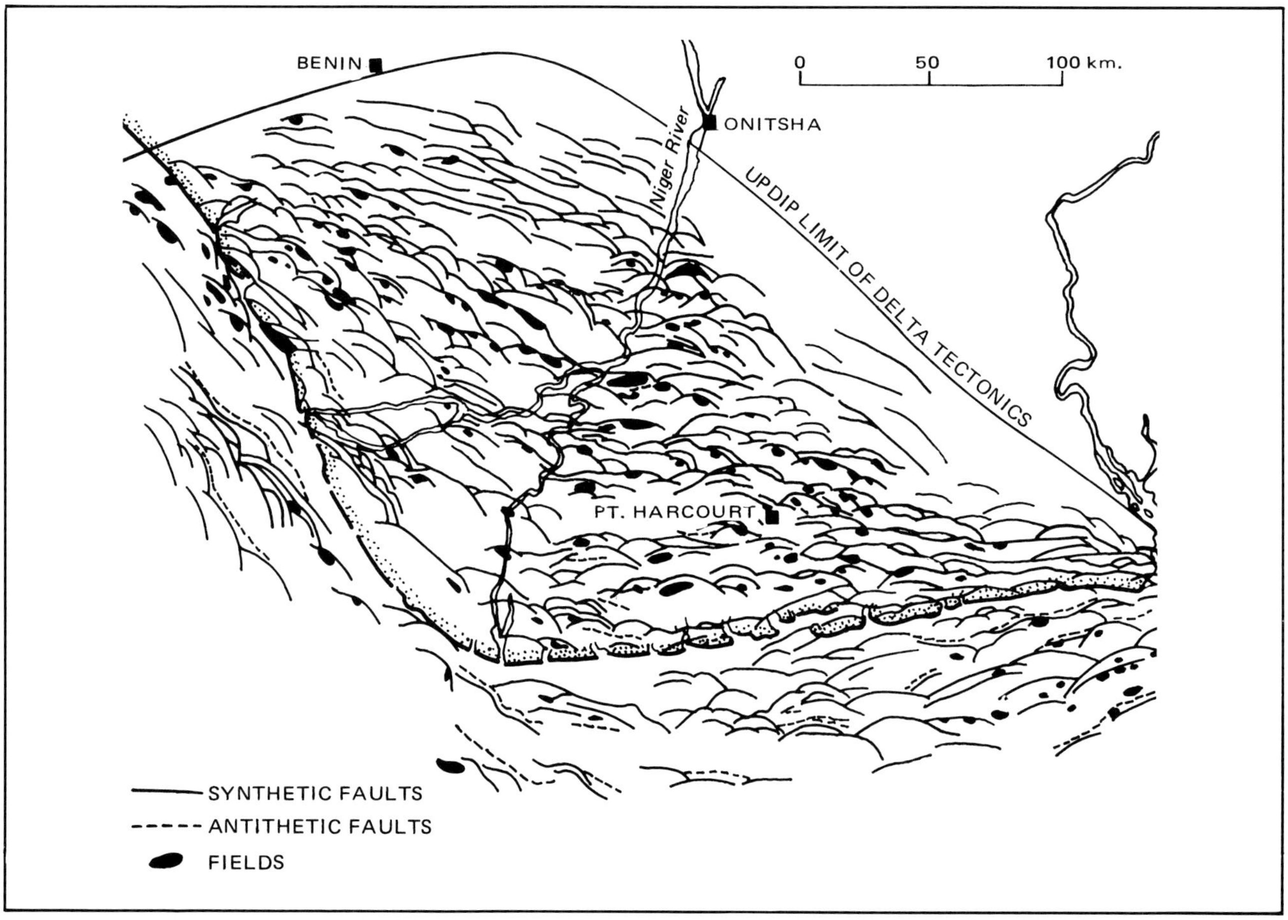

Fig. 1-25—Syndepositional detached normal faults in Niger delta demonstrate parallel crescentic trends and complex linkage of individual faults that are concave toward direction of transport. (After Weber and Daukoru, 1976).

Niger delta (Fig. 1-25) are present in wide, arcuate, parallel bands that trend subparallel with basin outline and/or depositional strike. Individual faults in map view are usually cuspate, concave in the direction of tectonic transport, and link to form an internally complex, anastomosing pattern. Fault zones become progressively younger in the direction of sedimentary progradation. Within a particular zone, however, the local timing for individual faults may reverse this sequence.

The detached normal-fault style in the Gulf Coast basin has been attributed to several causes. Large basement subsidence is required to provide a site for relatively rapid, thick deposition and to maintain a regional slope for basinward transport. Positive uplifts, other than those caused by diapirism, are lacking. Yorston and Weisser (1976) and W. F. Bishop (1973) have emphasized gravity sliding in several areas. Lehner (1969), W. F. Bishop (1973), and Spindler (1977) have cited salt withdrawal in areas with significant salt. Differential sedimentary loading and compaction of shales have been emphasized by Bruce (1973); Roux (1977) associated the style with differential subsidence and compaction. Cloos (1968) has reproduced the basinwide fault pattern and characteristics of individual structural assemblages with regional gravity creep and local sliding models, respectively. Many authors (e.g., Bruce, 1973) have noted the importance of overpressured shales in the faulting process.

Yorston and Weisser (1976) suggested that two different detached normal-fault assemblages can result from downslope gravity sliding: peripheral breakaway faults, and local slope transport faults. The first occurs near the updip edge of the regional fault system. In the northeastern Gulf Coast, they utilize salt for their detachment. Movement in this region apparently was facilitated by sediment loading of downdropped blocks and consequent flowage of salt from beneath the loaded blocks (W. F. Bishop, 1973). The development of local slope transport faults is thought to involve successive detachment and basinward creep of deltaic lobes deposited on depositional slopes closer to the delta front.

In south Louisiana and the adjacent Gulf of Mexico, where salt is thick and extensive, salt mobility is an important factor in regional detached normal faulting. McGookey (1975, figs. 12, 13) has documented syndepositional detached normal faults that occur as boundaries between rising salt diapirs and depopods that subside deeply into underlying salt. Regional detached normal-fault systems develop primarily on the seaward side of salt ridges or stocks. Synchroneity of salt withdrawal and displacements on a major up-to-basin antithetic fault has been demonstrated in this region (Spindler, 1977).

Salt is not usually present in south Texas, where Bruce (1973, fig. 2) has attributed the regional detached faulting mostly to local differential loading of basin fill by sand depocenters. Faults have formed at the landward flanks of these depocenters and flatten with depth. The result is a basinward-dipping normal fault that utilizes the flank of a residual shale mass as its detachment zone. Basinward-dipping faults on the landward side of the shale ridges in south Texas are usually postdepositional and have relatively small displacements and no associated rollover anticline (Bruce, 1973, figs. 3, 9). These faults mostly dip into the adjacent shale masses at relatively steep angles and do not have an associated glide plane.

Roux (1977) differed from other authors and proposed that the listric fault profile is in part at least a result of postfaulting compaction of shales and, to a lesser degree, sands; the fault profile is flattened appreciably as the original bed thicknesses are compacted during subsidence of the fault system. Roux (1977) stated that because fault segments exhibiting growth must have formed at surfaces of sedimentation these faults initially had steep dips throughout large segments of their profiles similar to faults present today at modern depositional surfaces. Rotation of beds into the listric faults during continued displacement may also rotate the adjoining fault segments and further decrease basinward fault dip.

The localization of anticlinal rollovers consistently within the downthrown block differs from the typical expression of all other structural styles in sedimentary basins. This characteristic and the direct evidence of detachment accompanying listric extensional faulting are key recognition criteria. Style similarities with basement-involved extensional block faulting do occur, particularly where low-side rotation of beds into the fault accompanies the block faulting, but this is far from universal in the latter style (Fig. 1-14). Furthermore, the rollover anticline has a flexed curvilinear profile that is different from the slablike profiles of most extensional fault blocks (cf. Figs. 1-11, 1-21). In map view, syndepositional detached normal faults lack the grid aspects of fault blocks, and trap-door structures are absent. Tectonic settings and correlations across the fault are helpful in establishing the downthrown block for differentiation from overthrust anticlines, which also lie above listric fault surfaces and have similarly asymmetric profiles on seismic lines (cf. Figs. 1-19, 1-21).

Syndepositional detached normal-fault assemblages of the type present in the Gulf Coast basin and Niger delta are by far the most prolific hydrocarbon-producing features of this style. Closures are mainly rollover anticlines and, less commonly, the upthrown sides of faults (Figs. 1-1, 1-24). Exploitation can be complex because basinward shift of structure and increased trap segmentation by faulting both occur with depth. A vertical stacking of quite different closures (fold versus fault) can result (Fig. 1-24).

Salt Structures

Salt has a capacity for mobile rise that is independent of tectonic forces and it is in this context that salt structures are considered a separate style (Fig. 1-26). Salt also can be mobilized by tectonic forces to form stocks, sills, dikelike masses, and other irregularly shaped bodies which are commonly injected along faults, folds, and fault-fold intersections. The Romanian Carpathians provide ex-

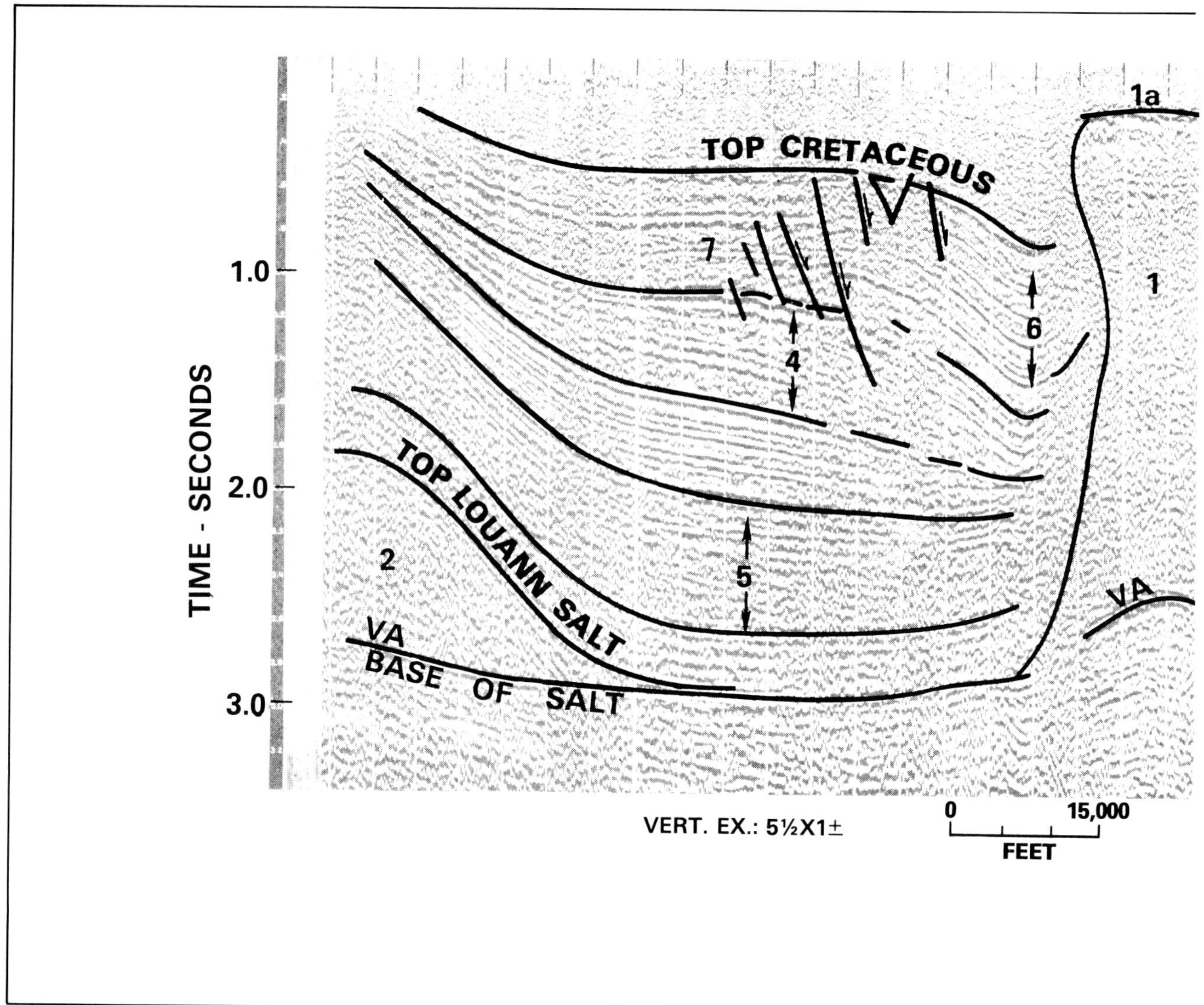

Fig. 1-26—Seismic section across Hainesville dome (center), east Texas, with structures as indicated. Higher velocity of salt relative to surrounding rocks at velocity anomalies at left and center results in "pull-up" of time section. Velocity anomaly at right shows as depression, because salt here has lower velocity than laterally adjacent rocks. (Adapted from Tucker and Yorston, 1973).

amples of salt features that have been clearly modified and reshaped by tectonic forces that formed the thrust-fold belt (Fig. 1-27). Differentiation between buoyant and tectonic salt structures is sometimes difficult. Distinctions can be arbitrary, for virtually every transition exists between the two types (O'Brien, 1968). Those demonstrably of tectonic origin are here considered as modifers of the particular structural assemblage with which they have developed.

The preferred habitats of buoyant salt structures are divergent continental margins, their superimposed passive margin sags, and aborted rift systems (Table 1-1). The narrow restricted troughs, which form at the initiation of rifting, provide ideal settings for thick evaporite accumulations, particularly in hot, arid, low latitudes. Tectonic salt structures occur mostly where original salt basins (typically grabens) have been transported by large-scale plate movement into other, commonly compressive, deformational environments (H. R. Hopkins, pers. com., 1977).

Basin history and configuration control original limits, thicknesses, and composition of salt and overburden. In this way, they influence the general distribution (Fig. 1-28) and types of buoyant salt features. Basinward increase in original salt thickness is ideally expressed by the following progression of structures (Trusheim, 1960; Hughes, 1968): (1) downbends in sedimentary rocks above areas of salt withdrawal along the original depositional edge of the salt; (2) pillows of salt that can coalesce along strike to form low ridges; (3) anticlines and low domes

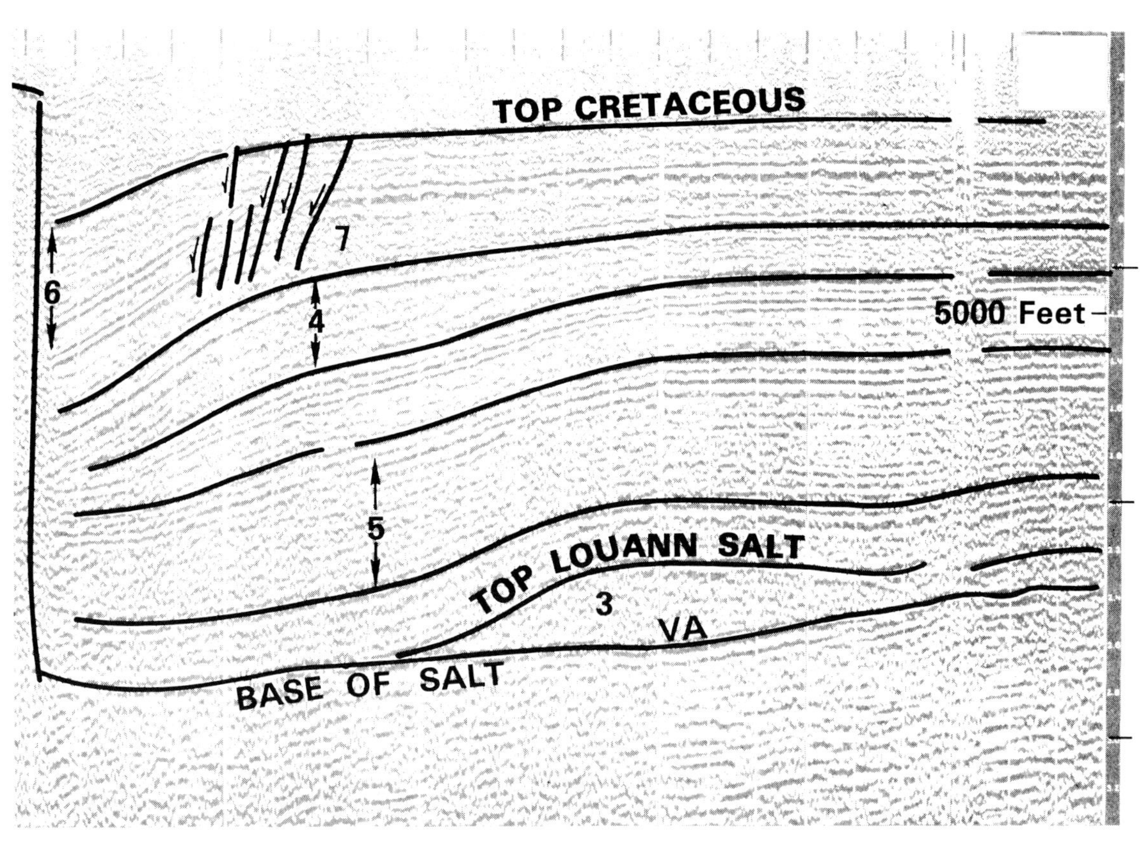

STRUCTURES: SALT RELATED

1. PIERCEMENT SALT DOME (1a. CAP ROCK)
2. NON - PIERCEMENT SALT ANTICLINE
3. RESIDUAL SALT HIGH
4. TILTED TURTLE STRUCTURE
5. PRIMARY SALT WITHDRAWAL SYNCLINE
6. RIM SYNCLINE
7. NORMAL FAULTS

VA - VELOCITY ANOMALY

of salt that can increase in relief to piercement features; and (4) walls of piercement salt.

In the coastal plain of the Gulf Coast and southward to the lower continental shelf the progression of structure is from (1) small, elongate, isolated spines to (2) progressively closer spaced, larger spines, and (3) almost continuous ridges and irregularly shaped large massifs (Stude, 1978). Growth sequences of offshore Gulf Coast structures are from initial salt swells to narrower ridges or stocks and finally to narrow salt chimneys or spines 2 to 4 n. mi in diameter (Lehner, 1969). Both the form of the salt features themselves and the effect on the adjacent sedimentary section of simultaneous lateral and vertical salt movement are important in understanding and exploring salt basins.

Distribution of salt domes in the absence of tectonic influence is typically in irregular clusters (Fig. 1-28, upper). In some places salt ridges or walls tend to be subparallel (Fig. 1-28, middle) and may be aligned with basin margins. Extensive massifs and ridges are closely spaced in a large area of the lower continental shelf of the Gulf Coast basin, but mostly lack discernible trend patterns (Martin, 1976).

A widely accepted mechanism for diapirism has been the movement of salt by buoyancy (Nettleton, 1934). Alternatives, which explain certain aspects of diapirism not resolved by the buoyancy model, have been proposed by Barton (1933), Bornhauser (1969), and R. S. Bishop (1978). According to R. S. Bishop (1978) initial salt movement occurs during progradation of sedimentary overburden across essentially exposed salt deposits on the basin floor (McGookey, 1975). The mobile evaporites flow both horizontally and vertically as waves or pillows in front of the advancing overburden. Flowage can occur before overburden is sufficiently thick to cause a density inversion between salt and sedimentary cover. The sedimentation rate, progradation rate, regional dip, sediment density, and thickness of both the mobile sub-

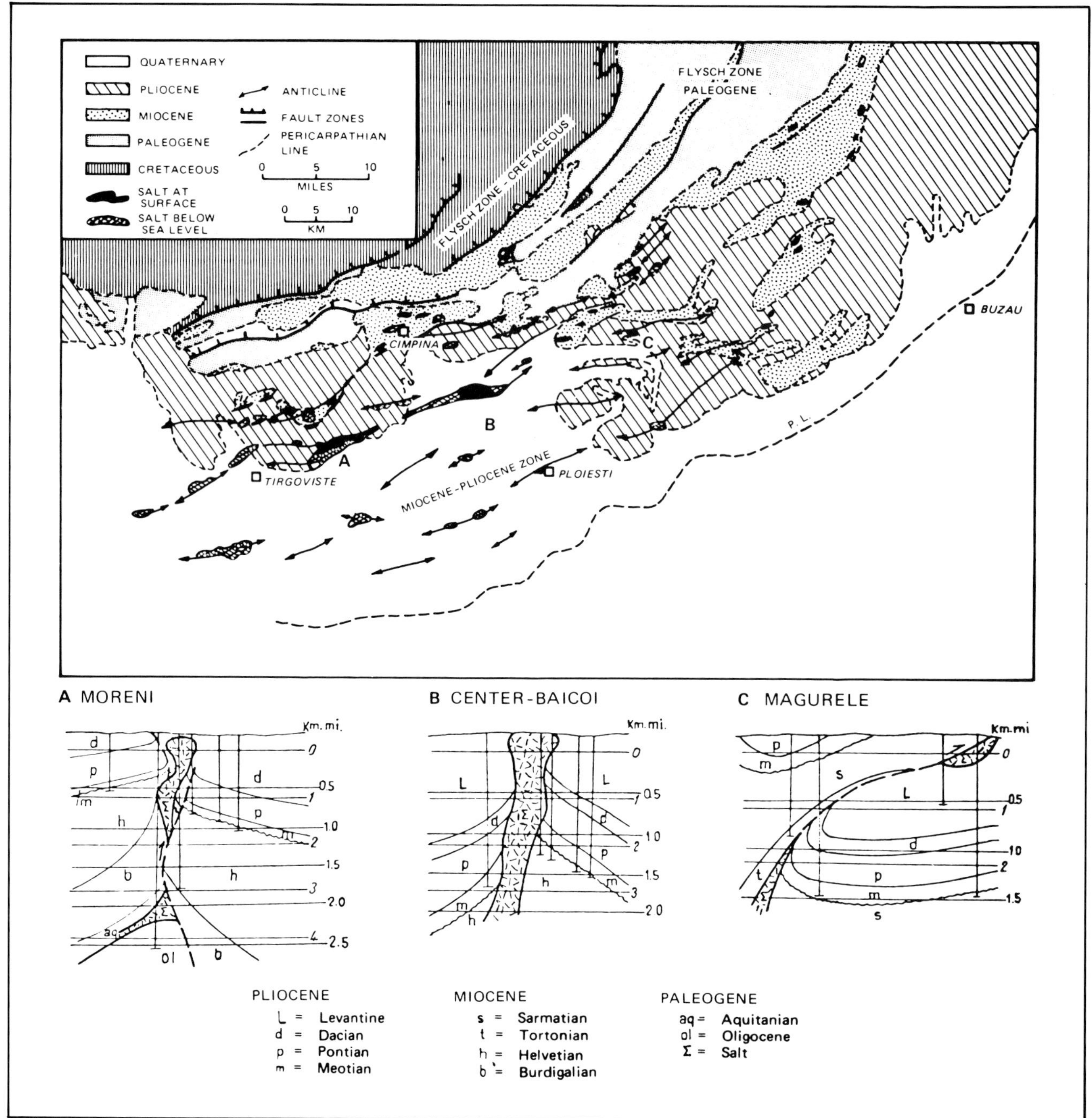

Fig. 1-27—Salt features in Ploesti area of Romanian Carpathian thrust-fold belt are influenced by tectonic forces, as attested by diapir, B and by diapirs localized along fault surface, A, C. (From Paraschiv and Olteanu, 1970).

strate and overburden all affect the size and shape of incipient salt structures. Slow, even accumulation of overburden results in salt structures with random distribution. Focused loading concentrates structures in specific sites and can align them parallel with thick areas of overburden (R. S. Bishop, 1978, fig. 11).

As overburden buries the initial salt structure, horizontal flow is restricted but salt buoyancy may develop and vertical growth may continue as an unrestrained extrusion to the seafloor. Extrusion alleviates the space problems that arise where the rising diapir is required to force aside its overburden.

If the density inversion (buoyancy) disappears, extrusion will cease and diapirism continues only as a buried intrusion. The cause of diapirism at this stage may be selective dissolution and removal of salt by migrating subsurface fluids (upsetting mass equilibrium) or differential loading (e.g., salt flowing out from under a pref-

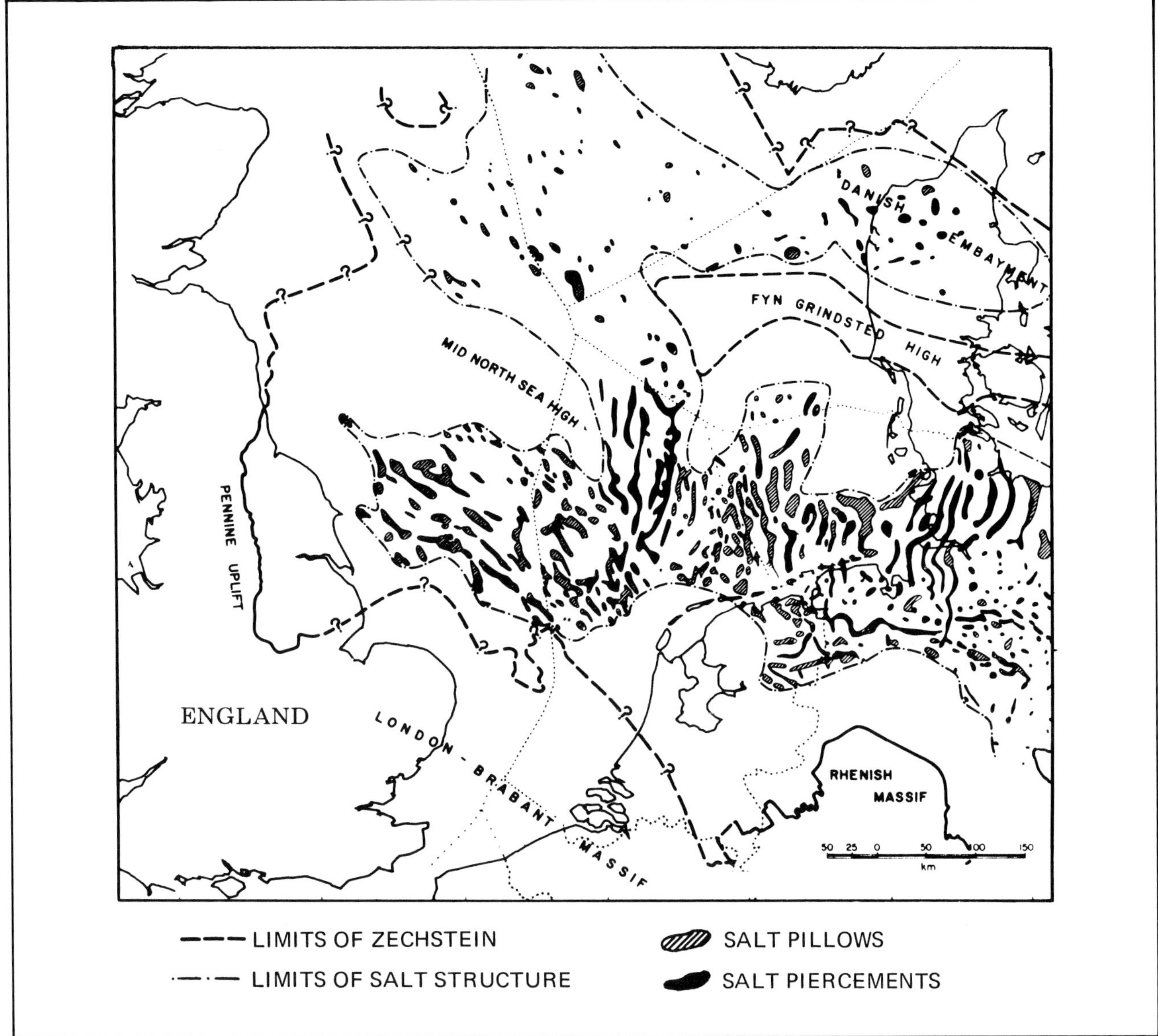

Fig. 1-28—Distribution of primarily buoyant salt features in southern North Sea and adjacent parts of Netherlands, northwest Germany, and northern Denmark as known in 1968. Randomly trended, clustered structures prevail in north part; trends are more linear and locally parallel in south. (From Heybroek, 1968).

erentially loaded rim syncline). According to R. S. Bishop (1978) growth of many Gulf Coast diapirs alternated between extrusion and intrusion.

Continuous vertical diapirism can be restrained by significant increases in overburden thickness or strength. Growth then becomes sporadic and occurs only by fracturing and forceful injection of salt into the overburden, which occur only when the pressure within the diapir, arising from sediment loading, becomes sufficient to rupture the overburden (R. S. Bishop, 1978). Deeply buried diapirs may expand laterally prior to vertically fracturing their overburden. When fracturing does not occur, the diapir ceases to grow, even if evaporite supply is sufficient.

The vertical rise of salt cores causes localized stretching and detached normal keystone faulting of the sedimentary overburden. Faulting can be highly complex and patterns are mostly radial and transverse. Stude (1978) has documented the faulting sequence at the South Timbalier Block 54 salt dome in offshore Louisiana, "The older graben faults moved farther from the crest of the salt and became inactive (as the dome continued to pierce through the overburden and widen). These faults were replaced by younger faults over the salt crest." Currie (1956) has modeled salt-dome growth and demonstrated similar timing but other sequences are also possible (Stude, 1978).

Distinctive profile characteristics facilitate identification of salt structures on seismic lines. On figure 1-26, residual salt highs are considered remnants of the original salt bed left by incomplete evacuation of salt. Turtle structures are inverted sediment-filled lows that formed above areas of early salt withdrawal. Rim synclines reflect the growth history of the associated salt dome.

The most prolific salt-related traps are those that are not pierced by evaporites at reservoir levels (Fig. 1-1). These include reservoirs concordant above salt pillows and anticlines. Much less common are turtle structures and passive drape closures over residual highs. Other significant traps are formed by reservoirs turned up against and sealed by salt, in places beneath salt overhangs. The two basic trap types can be further modified by stratigraphic pinchouts and unconformities. Moreover, the crestal normal faulting of the sedimentary section in response to arching above salt features can disrupt and segment closures (Gussow, 1968; Lafayette Geological Society, 1970).

Shale Structures

Shale structures (Fig. 1-29) range from nondiapiric residual shale masses such as those that underlie and can help motivate detached normal faults (Fig. 1-22), to highly mobile, diapiric mud lumps. They are not limited to any particular structural or plate-tectonic environment, as attested by their widespread occurrence. Shale structures, however, are perhaps best known in thick, modern clastic deltas (Table 1-1; R. S. Bishop, pers. com. 1977) where high depositional rates create thick sequences that retard water expulsion from fine mud fractions. Muds are consequently undercompacted, have abnormally high pore-fluid pressures, and thus can react in a highly mobile buoyant manner.

In south Texas, shale ridges develop by differential compaction resulting from differential loading and shale diagenesis. Ridges here are aligned subparallel with regional strike, and are commonly tens of miles long, up to 25 mi (40 km) wide, and 10,000 ft (3,000 m) high (Bruce, 1973). In Trinidad, shale diapirism appears to be tectonically motivated, with diapirs present along major strike-slip faults. Shale structures can also result from crowding and shortening of the sedimentary cover during downslope creep as at the toes of syndepositional detached normal-fault assemblages.

Morphologic characteristics of salt structures are also shared by most abnormally pressured shale structures formed in passive tectonic environments. Indeed, the two are known to occur together to form compound domal masses (Atwater and Forman, 1959; Musgrave and Hicks, 1968). On reflection seismic sections, both salt and shale flowage usually are manifested by reflection cutouts (Musgrave and Hicks, 1968).

Differentiation of salt and shale is possible when acoustic velocity information is available inasmuch as shale flowage masses typically have only about half the velocity of salt. Moreover, shale features seem to lack the well-defined rim synclines frequently associated with salt bodies, probably because salt flowage can occur at

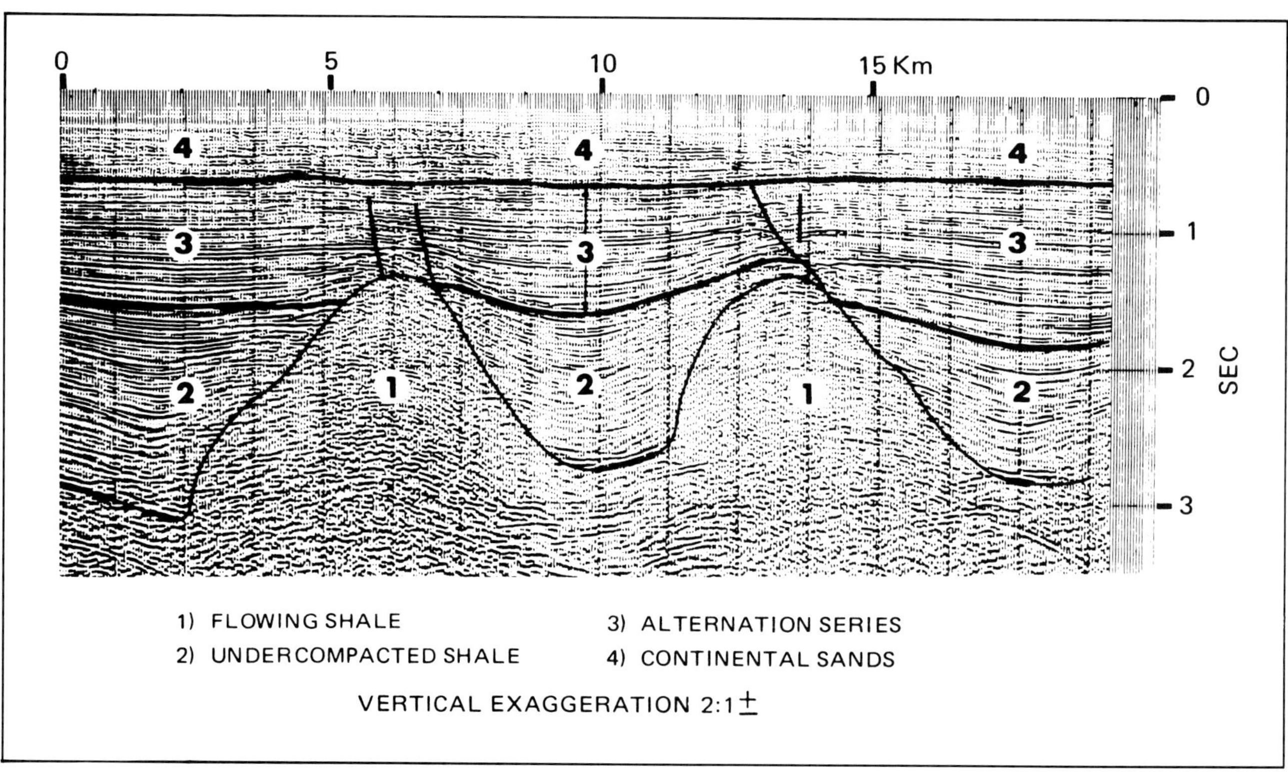

Fig. 1-29—Seismic expression of shale diapirs within deltaic sequence. (From Dailly, 1976).

substantial distances from a developing dome, whereas shale flowage involves less lateral movement. Slower flow rates over longer periods in salt-structure growth may also be important factors.

Greater shale compactibility may be a main cause of differences in the development of shale and salt structures according to R. S. Bishop (1978). Salt diapirs can pierce thicker overburden and rise to greater heights. Bishop (1978) wrote:

> During burial, the ratio of salt to the overburden density increases, and the salt diapir may stay near the surface and remain vertically unconstrained throughout sedimentation. In contrast, the ratio of shale density to its overburden may decrease during burial (due to shale compaction), thereby causing intrusion at a slower rate than sedimentation. This results in deeper burial of the shale diapirs; the consequence is vertical constraint and perhaps cessation of the diapirism.

Closures related to shale masses have not trapped nearly so much petroleum as that attributable to most other structural styles. O'Brien (1968), however, has cited numerous shale domes and features associated with mud volcanoes as significant sources of hydrocarbons.

CONCLUSIONS AND APPLICATIONS

In general terms, the common geologic structures of sedimentary basins are distributed in definable, interrelated suites that constitute styles. Their preferred tectonic habitats result broadly from lithospheric plate movements but are also controlled by other factors. The type of deformation and tectonic history of specific regions can only be determined by correct differentiation of styles. The style approach can then provide a first approximation of the kinds of hydrocarbon traps and the general distribution patterns to be expected during exploration.

Variations in Structural Style

Many factors contribute to the development of individual closures and can influence the expression of structural style or control its areal limits. Of prime importance are the physical properties of the deformed region, such as thickness and lithology of the sedimentary cover, and mobility and preexisting structural fabric of the basement. Mobile lithologies in the basin fill, such as evaporites and undercompacted shales, directly control the occurrence of salt and shale structures and are critical in the development of syndepositional detached normal faults. Greater basement ductility can enhance or even control the presence of folds in basement-involved styles (e.g., west side of San Joaquin Valley, Harding, 1976) although a lack of ductility may result in a brittle structural response (e.g., Wyoming foreland). Older basement fabric can localize fault-block closures.

Style expression is also influenced by tectonic variables such as intensity, rate, duration, and stage of deformation (Fig. 1-5), and differences of applied stress. In addition, sedimentation contemporaneous with deformation and the structural level of available control can determine the structural geometry observed. Gentle rejuvenation of dip-slip faulting concurrent with sedimentation, for example, results in drape flexures of greater amplitude that become the shallow expression of block faulting. In each of the described situations, however, the basic style category is still discernible.

Diversity of Styles Occurrence

For convenience, we have described each style individually, but actual occurrence of structures frequently indicates transition between styles. Detached thrust systems, for example, can ultimately root downdip in basement thrusts and in places merge along strike with block like compressive features. In other settings, syndepositional detached normal faults change downdip into detached thrust features. Transitions in the wrench style from structures that are dominantly compressive to those that are extensional can occur along a single zone.

The structures of a sedimentary basin commonly include several styles and transitions between styles and these can have either separate or overlapped distributions. In the San Joaquin Valley (Harding, 1976, fig. 1) a wrench-fault assemblage dominates the mobile southwest flank, a salient of locally detached thrust faults predominates at the south end where right- and left-lateral strike-slip faults converge, and basement warps with secondary normal faults dominate the stable northeast flank.

Some extensional settings can combine basement fault blocks with salt and/or shale structures and syndepositional detached normal faults, often superimposed at different levels (Fig. 1-22, in part). Compressive settings can mix detached thrusts and folds, basement thrusts or compressive block faults, wrench-related features, and, with the presence of mobile substrata, tectonic salt and/or shale structures (Fig. 1-3). Each basin has a unique history and styles can be further complicated by the overprints of fundamentally different stress systems through time.

No style is universally present within its preferred habitat, nor is a style always consistently developed where it is present. Foreland settings present the greatest extremes. In contrast with the Wyoming foreland, the Alberta basement foreland of the Canadian Rockies on strike has very few structures (King, 1969). Late Paleozoic structures along the foreland of the approximately coeval Ouachita-Llanoria fold-thrust belt are widespread and highly varied (King, 1969). Included are (1) normal block faults on the stable north flank of the Arkoma basin, (2) wrench faults along the trend of the Ardmore-Anadarko basin, (3) broad basement arch at the Llano uplift, and (4) dominantly compressive fault blocks in the Permian basin. Obviously, forces of plate convergence can be manifested in different ways. Earlier structures apparently controlled the style and/or distribution of some of this foreland deformation, but still other factors may have been important.

Identification of Structural Style

Two major factors—variations in style expression and complications in style occurrence—can make differentiation of styles exceedingly difficult; this may be especially true of convergent margin settings and their orogenic belts.

Many structural patterns can be developed in several ways, and "false" patterns arising from unique circumstances must be ruled out. For example, an overprint of faults trending obliquely to an earlier thrust-fold belt could result in a misleading en echelon appearance. Linear strike-slip fault sets superimposed on each other could result in a fault-block appearance. A pattern usually typical of a particular tectonic regime can be simply the result of unique physical properties in the deformed terrane such as a strong basement grain or anisotropy within the sedimentary cover. Determination of several recognition criteria is often essential. Consideration of alternatives is always necessary and requires equal familiarization with all styles.

REFERENCES

Amery, G. B., 1969, Structure of Sigsbee Scarp, Gulf of Mexico: *American Association Petroleum Geologists Bull.*, V. 58, p. 2480–2482.

Anderson, E. M., 1942, The dynamics of faulting and dyke formation with application to Britain: Edinburgh and London, Oliver Boyd, 206 p. (rev. ed. 1951).

Atwater, G. I., and Forman, M. J., 1959, Nature of growth of southern Louisiana salt domes and its effect on petroleum accumulation: *American Association Petroleum Geologists Bull.*, V. 43, p. 2592–2622.

Badgley, P. C., 1965, Structural and tectonic principles: New York, Harper and Row, 521 p.

Baker, B. H., Mohr, P. A., and Williams, L. A. J., 1972, Geology of the eastern rift system of Africa: *Geological Society America Spec. Paper 136,* 67 p.

Bally, A. W., 1975, A geodynamic scenario for hydrocarbon occurrences: *9th World Petroleum Cong. Proc.*, V. 2, p. 33–44.

———, Gordy, D. L., and Stewart, G. A., 1966, Structure, seismic data, and orogenic evolution of southern Canadian Rocky Mountains: *Bull. Canadian Petroleum Geology,* V 14, p. 337–381.

Barton, D. C., 1933, Mechanics of formation of salt domes, with special reference to Gulf Coast salt domes of Texas and Louisiana: *American Association Petroleum Geologists Bull.*, V. 17, p. 1025–1083.

Beloussov, V. V., 1959, Types of folding and their formation: *International Geological Review,* V. 1, p. 1–21.

Bishop, R. S., 1978, Mechanism for emplacement of piercement diapirs: *American Association Petroleum Geologists Bull.*, V. 62, p. 1561–1583.

Bishop, W. F., 1973, Late Jurassic contemporaneous faults in north Louisiana and south Arkansas: *American Association Petroleum Geologists Bull.*, V. 57, p. 858–877.

Bornhauser, M., 1969, Geology of Day dome (Madison County, Texas)—a study of salt emplacement: *American Association Petroleum Geologists Bull.*, V. 53, p. 1411–1420.

Bruce, C. H., 1973, Pressure shale and related sediment deformation—mechanisms for development of regional contemporaneous faults: *American Association Petroleum Geologists Bull.*, V. 57, p. 878–886.

Bucher, W. H., 1956, Role of gravity in orogenesis: *Geological Society America Bull.*, V. 67, p. 1295–1318.

Burchfiel, B. C., and Davis, G. A., 1975, Nature and controls of Cordilleran orogenesis, western United States—extensions of an earlier synthesis: *American Journal Science,* V. 275-A, p. 363–396.

Cloos, E., 1968, Experimental analysis of Gulf Coast fracture patterns: *American Association Petroleum Geologists Bull.*, V. 52, p. 420–444.

Cohee, G. V., 1962, Tectonic map of the United States: *United States Geological Survey and American Association Petroleum Geologists,* scale, 1:2,500,000.

Currie, J. B., 1956, Role of concurrent deposition and deformation of sediments in development of salt-dome graben structures: *American Association Petroleum Geologists Bull.*, V. 40, p. 1–16.

Dailly, G. C., 1976, A possible mechanism relating progradation, growth faulting, clay diapirism and overthrusting in a regressive sequence of sediments: *Bulletin Canadian Society Petroleum Geologists,* V. 24, p. 92–116.

Darton, N. H., 1906, Cloud Peak-Fort McKinney folio, Wyoming: *United States Geological Survey Geological Atlas,* Folio 142, 16 p.

Dewey, J. F., and Bird, J. M., 1970, Mountain belts and the new global tectonics: *Journal Geophysical Research,* V. 75, p. 2625–2647.

Dickinson, W. R., 1974, Plate tectonics and sedimentation, in Tectonics and sedimentation: *Society Economic Paleontologists Mineralogists Spec. Pub. 22,* p. 1–27.

———, and Snyder, W. S., 1978, Plate tectonics of the Laramide orogeny, *in* Laramide folding associated with basement block faulting in western United States: *Geological Society America Mem. 151,* p. 355–366.

Dobbin, C. E., and Erdman, C. E., 1955, Structure contour map of the Montana plains: *United States Geological Survey Oil and Gas Inv. Map OM 178A,* scale, 1:500,000.

Elam, J. G., 1969, The tectonic style in the Permian basin and its relationship to cyclicity, *in* Cyclic sedimentation in the Permain basin: *West Texas Geological Society Pub. 69-56,* p. 55–79.

Elliott, D., 1976, The motion of thrust sheets: *Journal Geo-*

physical Research, V. 81, p. 949–963.

El-Tarabili, E., and Adawy, N., 1972, Geologic history of Nukhul-Baba area, Gulf of Suez, Sinai, Egypt: *American Association Petroleum Geologists Bull.,* V. 56, p.882–902.

Fitch, T. J., 1972, Plate convergence, transcurrent faults, and internal deformation adjacent to southeast Asia and the western Pacific: *Journal Geophysical Research,* V. 77, p. 4432–4460.

Foose, R. M., Wise, D. U., and Garbarini, G. S., 1961, Structural geology of the Beartooth Mountains, Montana and Wyoming: *Geological Society America Bull.,* V. 72, p. 1143–1172.

Freeland, G. L., and Dietz, R. S., 1971, Plate tectonic evolution of Caribbean-Gulf of Mexico region: *Nature,* V. 232, p. 20–23.

Freund, R., and Merzer, A. M., 1976, The formation of rift valleys and their zigzag fault patterns: *Geological Magazine,* V. 113, p. 561–568.

Gansser, A., 1974, Ophiolitic melange, a world-wide problem on Tethyan examples: *Eclogae Geol. Helvetiae,* V. 67, p. 479–507.

Ghignone, J. I., and de Andrade, G., 1970, General geology and major oil fields of Recôncavo basin, Brazil, *in* Geology of giant petroleum fields: *American Association Petroleum Geologists Mem. 14,* p. 337–358.

Green, A. R., 1977, The evolution of the earth's crust, *in* The earth's crust—its nature and physical properties: *American Geophysical Union Geophysical Mon. 20,* p. 1–17.

Gussow, W. C., 1968, Salt diapirism: importance of temperature, and energy source of emplacement, *in* Diapirism and diapirs: *American Association Petroleum Geologists Mem. 8,* p. 16–52.

Halbouty, M. T., 1967, Salt domes—Gulf region, United States and Mexico: Houston, Gulf Publishing Company, 425 p.

Hansen, A. R., 1972, The Williston basin, *in* Geologic atlas of the Rocky Mountain region, United States of America: Rocky Mountain Association Geologists, p. 265–269.

Hardin, F. R., and Hardin, G. C., Jr., 1961, Contemporaneous normal faults of Gulf Coast and their relation to flexures: *American Association Petroleum Geologists Bull.,* V. 45, p. 238–248.

Harding, T. P., 1973, Newport-Inglewood trend, California—an example of wrenching style of deformation: *American Association Petroleum Geologists Bull.,* V. 57, p. 97–116.

———, 1974, Petroleum traps associated with wrench faults: *American Association Petroleum Geologists Bull.,* V. 58, p. 1290–1304.

———, 1976, Tectonic significance and hydrocarbon trapping consequences of sequential folding synchronous with San Andreas faulting, San Joaquin Valley, California: *American Association Petroleum Geologists Bull.,* V. 60, p. 356–378.

Harland, W. B., 1971, Tectonic transpression in Caledonian Spitzbergen: *Geological Magazine,* V. 108, p. 27–42.

Heybroek, P., 1968, Geologische Waarnemingen op do Noordzee: *Geologie en Mijnbouw,* V. 47, no. 3, p. 209–210 (illustrations with no. 4).

Hoppin, R. A., and Palmquist, J. C., 1965, Basement influence on later deformations, the problem, techniques of investigation, and examples from Bighorn Mountains, Wyoming: *American Association Petroleum Geologists Bull.,* V. 49, p. 993–1003.

Howard, J. H., 1966, Structural development of the Williams Range thrust, Colorado: *Geological Society America Bull.,* V. 77, p. 1247–1264.

Hughes, D. J., 1968, Salt tectonics as related to several Smackover fields along the northeast rim of the Gulf of Mexico basin: *Gulf Coast Association Geological Societies Trans.,* V. 18, p. 320–330.

Humphris, C. C., Jr., 1976, Salt movement on continental slope, northern Gulf of Mexico (abs.).: *American Association Petroleum Geologists Bull.,* V. 60, p. 683.

Illies, J. H., 1970, Graben tectonics as related to crust mantle interaction, *in* Illies, J. H., and St. Muellar, eds., Graben problems: Stuttgart, Schweizerbart, p. 4–27.

King, P. B., 1965, Geology of the Sierra Diablo region, Texas: *United States Geological Survey Prof. Paper 480,* 185 p.

———, 1969, Tectonic map of North America, scale, 1:5,000,000: United States Geological Survey.

Lafayette Geological Society, 1970, Typical oil and gas fields of southwest Louisiana, V. 2, variously paged.

Lehner, P., 1969, Salt tectonics and Pleistocene stratigraphy on continental slope of northern Gulf of Mexico: *American Association Petroleum Geologists Bull.,* V. 53, p. 2431–2479.

———, and DeRuiter, P. A. C., 1977, Structural history of Atlantic margin of Africa: *American Association Petroleum Geologists Bull.,* V. 61, p. 961–981.

Lowell, J. D., 1972, Spitsbergen Tertiary orogenic belt and the Spitsbergen fracture zone: *Geological Society America Bull.,* V. 83, p. 3091–3102.

———, 1974, Plate tectonics and foreland basement deformation: *Geology,* V. 2, p. 275–278.

———, 1977, Underthrusting origin for thrust-fold belts with applications to the Idaho-Wyoming belt: *Wyoming Geological Association Guidebook,* 29th Annual Field Conference, p. 449–455.

———, and Genik, G. J., 1972, Sea-floor spreading and structural evolution of southern Red Sea: *American Association Petroleum Geologists Bull.,* V. 56, p. 247–259.

Lowell, J. D., Genik, G. J., Nelson, T. H., and Tucker, P. M., 1975, Petroleum and plate tectonics of the southern Red Sea, *in* Fisher, A. G., and Judson, S., eds., Petroleum and global tectonics: Princeton Univ. Press, p. 129–153.

Mallory, W. W., ed., 1972, Geologic atlas of the Rocky Mountain region, United States of America: Rocky Mountain Association Geologists, 331 p.

Martin, R. G., 1976, Geological framework of northern and eastern continental margins, Gulf of Mexico, *in* Beyond the shelf break: American Association Petroleum Geologists Marine Geology Comm. Short Course, V. 2, p. A-1-A-28; also in *Amerian Association Petroleum Geologists Studies in Geology,* no. 7, 1978, p. 21–42.

McGookey, D. P., 1975, Gulf Coast Cenozoic sediments and structure, an excellent example of extra-continental sedimentation: *Gulf Coast Association Geological Societies Trans.,* V. 25, p. 104–120.

Moody, J. D., 1973, Petroleum exploration aspects of wrench-fault tectonics: *American Association Petroleum Geologists Bull.,* V. 57, p. 449–476.

———, and Hill, M. J., 1956, Wrench-fault tectonics: *Geological Society America Bull.,* V. 67, p. 1207–1246.

Moore, D. G., and Buffington, E. C., 1968, Transform faulting and growth of the Gulf of California since the late Pliocene: *Science,* V. 161, p. 1238–1241.

Musgrave, A. W., and Hicks, W. G., 1968, Outlining shale masses by geophysical methods, *in* Diapirism and diapirs: *American Association Petroleum Geologists Mem. 8,* p. 122–136.

Nettleton, L. L., 1934, Fluid mechanics of salt domes: *American Association Petroleum Geologists Bull.,* V. 18, p. 1125–1204.

O'Brien, G. D., 1968, Survey of diapirs and diapirism, *in* Diapirism and diapirs: *American Association Petroleum Mem. 8,* p. 1–9.

Ocamb, R. D., 1961, Growth faults of south Louisiana: *Gulf Coast Association Geological Societies Trans.,* V. 11, p. 139–173.

Osborne, W. C., 1957, Keystone field, *in* Occurrence of oil and gas in west Texas: *Texas Univ. Pub. 5716,* p. 156–171.

Paraschiv, D., and Olteanu, Gh., 1970, Oil fields in Mio-Pliocene zone of eastern Carpathians (District of Ploiesti), *in* Geology of giant petroleum fields: *American Association Petroleum Geologists Mem. 14,* p. 399–427.

Pennington, J. J., 1975, Geology of Argyll field, *in* Woodland, A. W., ed., Petroleum and the continental shelf of northwest Europe: New York, John Wiley & Sons, p. 285–297.

Pierce, W. G., 1957, Heart Mountain and South Fork detachment thrusts of Wyoming: *American Association Petroleum Geologists Bull.,* V. 41, p. 591–626.

———, and Nelson, W. H., 1968, Geologic map of the Pat O'Hara Mountain quadrangle, Park County, Wyoming: *United States Geological Survey Quad. Map GQ-755,* scale, 1:62,500.

———, Andrews, D. A., and Keroher, J. K., 1947, Structure contour map of the Big Horn basin, Wyoming and Montana: *United States Geological Survey Oil and Gas Inv. Prelim. Map 74,* scale approx. 1 in. = 3 mi.

Ponte, F. C., Fonsesca, J. D. R., and Morales, R. G., 1977, Petroleum geology of eastern Brazil continental margin: *American Association Petroleum Geologists Bull.,* V. 61, p. 1470–1482.

Price, R. A., 1969, The southern Canadian Rockies and the role of gravity in low-angle thrusting, foreland folding, and the evolution of migrating foredeeps (abs.): *Geological Society America Abs.* with Programs, pt. 7, p. 284–286.

———, 1973, Large-scale gravitational flow of supracrustal rocks, southern Canadian Rockies, *in* deJong, K. A. and Scholten, R., eds., Gravity and tectonics: New York, *Wiley-Interscience Pub.,* p. 491–502.

———, and Mountjoy, E. W., 1970, Geologic structure of the Canadian Rocky Mountains between Bow and Athabasca Rivers—a progress report: *Geological Association Canada Spec. Paper 6,* 25 p.

Proffett, J. M., Jr., 1977, Cenozoic geology of the Yerington district, Nevada, and implications for the nature and origin of Basin and Range faulting: *Geological Society America Bull.,* V. 88, p. 247–266.

Prucha, J. J., Graham, J. A., and Nickelsen, R. P., 1965, Basement controlled deformation in Wyoming province of Rocky Mountains foreland: *American Association Petroleum Geologists Bull.,* V. 49, p. 966–992.

Reches, Z., 1978a, Development of monoclines, pt. 1, Structure of the Palisades Creek branch of the East Kaibab monocline, Grand Canyon, Arizona, *in* Laramide folding associated with basement block faulting in western United States: *Geological Society America Mem. 151,* p. 235–271.

———, 1978b, Analysis of faulting in three-dimensional strain field: *Tectonophysics,* V. 47, p. 109–129.

Rider, M. H., 1978, Growth faults in western Ireland: *American Association Petroleum Geologists Bull.,* V. 62, p. 2191–2213.

Robson, D. A., 1971, The structure of the Gulf of Suez (clysmic) rift, with special reference to the eastern side: *Journal Geological Society,* V. 127, p. 247–276.

Roux, W. F., Jr., 1977, The development of growth fault structures: American Association Petroleum Geologists Structural Geology School Course Notes, 33 p.

Royse, F., Jr., Warner, M. A., and Reese, D. C., 1975, Thrust belt structural geometry and related stratigraphic problems, Wyoming-Idaho-northern Utah, *in* Symposium on deep drilling frontiers in the central Rocky Mountains: Rocky Mountain Association Geologists, p. 41–54.

Seely, D. R., Vail, P. R., and Walton, G. G., 1974, Trench slope model, *in* Burk, C. A., and Drake, C. C., eds., The geology of continental margins: New York, Springer-Verlag, p. 249–260.

Seglund, J. A., 1974, Collapse-fault systems of Louisiana Gulf Coast: *American Association Petroleum Geologists Bull.,* V. 58, p. 2389–2397.

Sloane, B. J., 1971, Recent developments in the Miocene Planulina gas trend of south Louisiana: *Gulf Coast Association Geological Societies Trans.,* V. 21, p. 199–210.

Smith, R. B., Mabey, D. R., and Eaton, G. P., 1976, Penrose Conference report—Regional geophysics and tectonics of the intermountain west: *Geology,* V. 4, p. 437–438.

Smithson, S. B., et al., 1978, Nature of the Wind River thrust, Wyoming, from COCORP deep-reflection data and from gravity data: *Geology,* V. 6, p. 648–652.

Spindler, W. M., 1977, Structure and stratigraphy of a small Plio-Pleistocene depocenter, Louisiana continental shelf: *Gulf Coast Association Geological Societies Trans.,* V. 27, p. 180–196.

Stearns, D. W., 1975, Laramide basement deformation in the Bighorn basin—the controlling factor for structures in the layered rocks: *Wyoming Geological Association Guidebook,* 27th Annual Field Conference, p. 149–158.

Stewart, J. H., 1972, Initial deposits in the Cordilleran geosyncline: evidence of a Late Precambrian (>850 m.y.) continental separation: *Geological Society America Bull.,* V. 83, p. 1345–1360.

Stocklin, J., and Nabavi, M. H., 1973, Tectonic map of Iran: Iran Geological Survey, scale 1:2,500,000.

Stone, D. S., 1969, Wrench faulting and Rocky Mountains tectonics: *The Mountain Geologist,* V. 6, no. 2, p. 67–69.

Stude, G. R., 1978, Depositional environments of the Gulf of Mexico South Timbalier Block 54—salt dome and salt dome growth models: *Gulf Coast Association Geological Societies Trans.,* V. 28, p. 627–646.

Surlyk, F., 1977, Stratigraphy, tectonics, and palaeogeography of the Jurassic sediments of the areas north of Kong Oscars Fjord, East Greenland: *Grønlands Geologiske Undersogelse Bull. 123,* 56 p.

Sykes, L. R., and Sbar, M. L., 1974, Focal mechanism solutions of intraplate earthquakes and stresses in the lithosphere, *in* Kristjansson, L., ed., Geodynamics of Iceland and the North Atlantic area: Dordrecht, Holland, D. Reidel Pub. Co., p. 207–224.

Sylvester, A. G., and Smith, R. R., 1976, Tectonic transpression and basement-controlled deformation in San Andreas fault zone, Salton trough, California: *American Association Petroleum Geologists Bull.,* V. 60, p. 2081–2102.

Tapponnier, P., and Molnar, P., 1976, Slip-line theory and large-scale plate tectonics: *Nature,* V. 264, p. 319–324.

Todd, R. G., and Mitchum, R. M., Jr., 1977, Seismic stratigraphy and global changes of sea level, part 8: identification of Upper Triassic and Lower Cretaceous seismic sequences in Gulf of Mexico and offshore West Africa, *in* Seismic stratigraphy—applications to hydrocarbon exploration: *American Association Petroleum Geologists Mem. 26,* p. 145–163.

Trusheim, F., 1960, Mechanism of salt migration in northern Germany: *American Association Petroleum Geologists Bull.,* V. 44, p. 1519–1540.

Tucker, P. M., and Yorston, H. J., 1973, Pitfalls in seismic interpretation: *Society Exploration Geophysicists Mon.* Ser. 2, 50 p.

Vroman, A. J., 1967, On the fold pattern of Israel and the Levant: *Israel Geological Survey Bull.,* V. 43, p. 23–32.

Weber, K. J., and Daukoru, E., 1976, Petroleum geology of the Niger delta: *9th World Petroleum Cong. Proc.,* V. 2, p. 209–221.

Wilcox, R. E., Harding, T. P., and Seely, D. R., 1973, Basic wrench tectonics: *American Association Petroleum Geologists Bull.,* V. 57, p. 74–96.

Wise, D. U., 1963, Keystone faulting and gravity sliding driven by basement uplift of Owl Creek Mountains, Wyoming: *American Association Petroleum Geologists Bull.,* V. 47, p. 586–598.

Woodbury, H. O., et al., 1973, Pliocene and Pleistocene depocenters, outer continental shelf, Louisiana and Texas: *American Association Petroleum Geologists Bull.,* V. 51, p. 2428–2439.

Yorston, H. J., and Weisser, G. H., 1976, The seismic analysis of glide plane fault structures; styles, mechanisms, and relationship to sedimentation (abs.): Society Exploration Geophysicists, 46th Annual Internat. Mtg. Abs., p. 85.

BASEMENT–INVOLVED

STRUCTURAL STYLE	WRENCH ASSEMBLAGE	COMPRESSIVE BLOCKS AND THRUSTS	EXTENSIONAL BLOCKS	WARPS ARCHES, DOMES, SAGS
DOMINANT DEFORMATIONAL FORCE	Couple	Compression and uplift	Extension and uplift	Uplift
DOMINANT TRANSPORT EXPRESSION	Strike-slip of sub-regional to regional plates	High to low angle convergent dip-slip of blocks, slabs, and sheets	High to low angle divergent dip-slip of blocks and slabs	Differential movement of solitary features
KEY STRUCTURAL ELEMENTS	• **En echelon** folds (± symmetrical) and associated reverse or thrust faults • Synthetic, antithetic, and P shears • **En echelon** and other normal faults • Through-going wrench • Flower (and half-flower) structures with up-thrusts parallel to wrench zone	Foreland — • Asymmetric blocks - trapdoors - steps • Drape flexures - high to low relief • Keystone normal faults • Tears • Gravity features • Thrust overlap Backarc - Rotation fans, polyphase closures, cleavage Forearc - Thrust slivers of oceanic crust	• Asymmetric blocks - trapdoors - steps - titled horst and half graben • Drape flexures - moderate to low relief • Fault gap	Gentle bends in absence of significant faulting
DOMINANT TRENDS REGIONAL LOCAL	Zonal, parallel **En echelon** with fold oblique to fault	Relay and dog-leg	Dog-leg and relay	Solitary
PRINCIPAL PETROLEUM TRAPS	**En echelon** folds low-side closures	Asymmetric blocks	Tilted blocks	Arches, domes
PLATE TECTONIC ENVIRONMENTS				
PRIMARY	Transform boundaries	Convergent boundaries - Forelands - Orogenic belt cores - Trench inner slopes and outer highs	Divergent boundaries (including aulacogens) Intra-plate rifts	Plate interiors
SECONDARY	Convergent boundaries - Longitudinal faults - Foreland conjugate shears Divergent boundaries - Offsetting spread centers and - Intersecting continental margins		Convergent boundaries - Arc massifs - Backarc spreading marginal basins - Oceanward walls of trenches - Stable flanks of foreland basins Transform boundaries - With divergence - Stable flanks of wrench basins	

STRUCTURAL STYLES AND THEIR PLATE TECTONIC ENVIRONMENTS

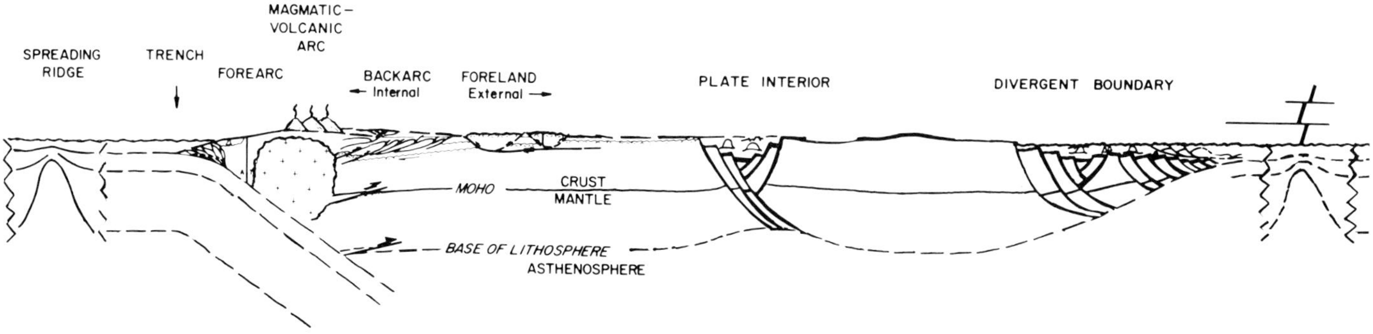

DETACHED FROM BASEMENT				
THRUST-FOLD ASSEMBLAGE	NORMAL FAULT ASSEMBLAGE	SALT STRUCTURES	SHALE STRUCTURES	STRUCTURAL STYLE
Compression	Extension	Density contrast Differential loading	Density contrast Differential loading	DOMINANT DEFORMATIONAL FORCE
Sub-horizontal to high angle convergent dip-slip of sedimentary sheets and slabs	Sub-horizontal to high angle divergent dip-slip of sedimentary sheets, wedges and lobes	Vertical and horizontal flow of mobil evaporites with arching and/or piercement of sedimentary cover	Vertical flow of mobile shales with arching and/or piercement of sedimentary cover	DOMINANT TRANSPORT EXPRESSION
• Listric imbricate thrusts – folded thrusts – back thrusts • Asymmetric concentric folds • Minor to major listric normal faults • Minor tears	• Listric normal faults – antithetic faults • Rollover anticlines on downthrown sides of faults • Expanded sections	Salt – • Pillows • Ridges • Domes • Walls • Residual highs Section – • Downbends above areas of salt withdrawal • Rim synclines • Turtle structures Faults – • Radial • Concentric	• Flowage • Mud lumps and volcanoes (wth gas emissons) • Erratic blocks	KEY STRUCTURAL ELEMENTS
Salient-reentrant Relay	Curvilinear – bifurcate and coalesce	Solitary (ridges/walls parallel basin edge)	Solitary	DOMINANT TRENDS REGIONAL LOCAL
Asymmetric, thrusted hanging-wall folds Lead Edges	Rollover anticlines	Salt-cored anticlines Beds upturned against salt	Shale-cored folds	PRINCIPAL PETROLEUM TRAPS
PLATE TECTONIC ENVIRONMENTS				
Convergent boundaries – Backarc – Landward walls of trenches	Passive boundaries – Deltas	Divergent boundaries Plate interiors – Aborted rift systems	Passive boundaries – Deltas	PRIMARY
Transform boundaries – with convergence	Convergent boundaries – Trench landward walls		Transform boundaries	SECONDARY

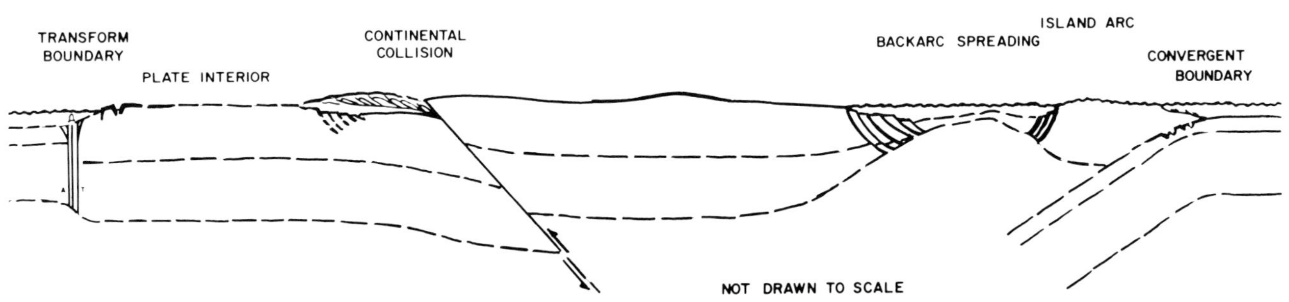

2 WRENCH ASSEMBLAGE

INTRODUCTION

Though wrench or strike-slip faults are among the most important components of wrenching, numerous other structures are developed so that a great variety of features is collectively grouped as the wrench assemblage. Structures of the wrench assemblage deserve considerable attention for two major reasons. First, the wrench assemblage provides an excellent vehicle for analyzing mechanics of faulting, which can in turn be applied to other structural styles. Second, patterns of other styles are sometimes mistaken for those of the wrench assemblage. In the discussion that follows the close interrelationship of these two points is evident.

Wrench or strike-slip faults are more or less vertical, generally straight, and throughgoing (all as seen from considerations of shear failure criteria—Fig. 2-1). They sometimes have a braided pattern in both cross-sectional and plan views and may have both normal and reverse senses of displacement. When the block across the fault opposite the observer has moved to the right, displacement is termed right lateral (left lateral for the opposite situation).

Because wrenching is a deep-seated process, basement is almost invariably involved. Faults propagate from the basement upward and outward through the sedimentary section so that a cross section resembles the longitudinal view of a cone (Fig. 2-2). There is usually greater displacement at lower structural levels (including a greater chance for a throughgoing wrench) than in structurally higher sedimentary strata. However, the zone of deformation is wider at higher levels than at lower ones where the zone can eventually become a discrete break (Fig. 2-2).

SLIP AND PIERCING POINTS

Piercing points (the intersection at a point of a geologic line with a fault surface) are necessary for determining slip. Slip is the distance between two formerly adjacent points on either side of a fault, measured on the fault surface or parallel to it. The matching of piercing points does not necessarily show the exact path of slip, but it does show the end result. The distinction of slip and separation is essential for any fault, but perhaps especially so for wrench faults where large horizontal movement was long viewed with suspicion.

Figure 2-3 represents a subsurface application of piercing points. The zero edge of the basal Oil Creek sand provides a piercing point that demonstrates 64 km (40 mi) of left-lateral slip on the Washita Valley fault zone (WVFZ). By "unfolding" observable folds along the WVFZ and restoring reverse faults to a pre-fault position, Brown (1981) has stated that very little strike-slip movement is needed along the WVFZ. Although the amount may well be less than 40 miles, some lateral slip is required because of en echelon structures associated with the WVFZ. Moreover, some of the aforementioned reverse faults may be wrench-related flower structures (Fig. 1-6).

GEOMETRY

The strain ellipse-conjugate shear rationale for fault planes or surfaces to be at some angle (θ) to the direction of maximum principal compressive stress (σ_1) is as follows:

Envision the drawing below as a side view of a cylindrical bar with unit cross-sectional plane represented by dashed line AB and oblique plane represented by line AC.

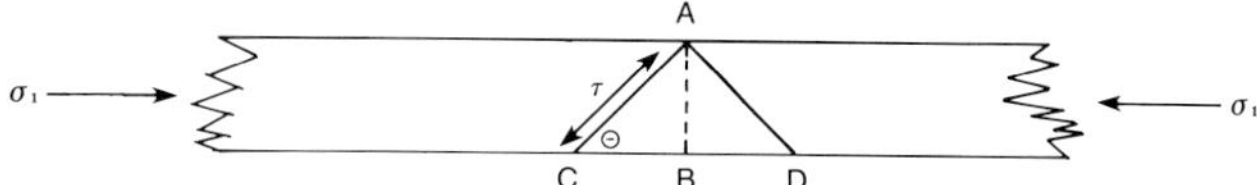

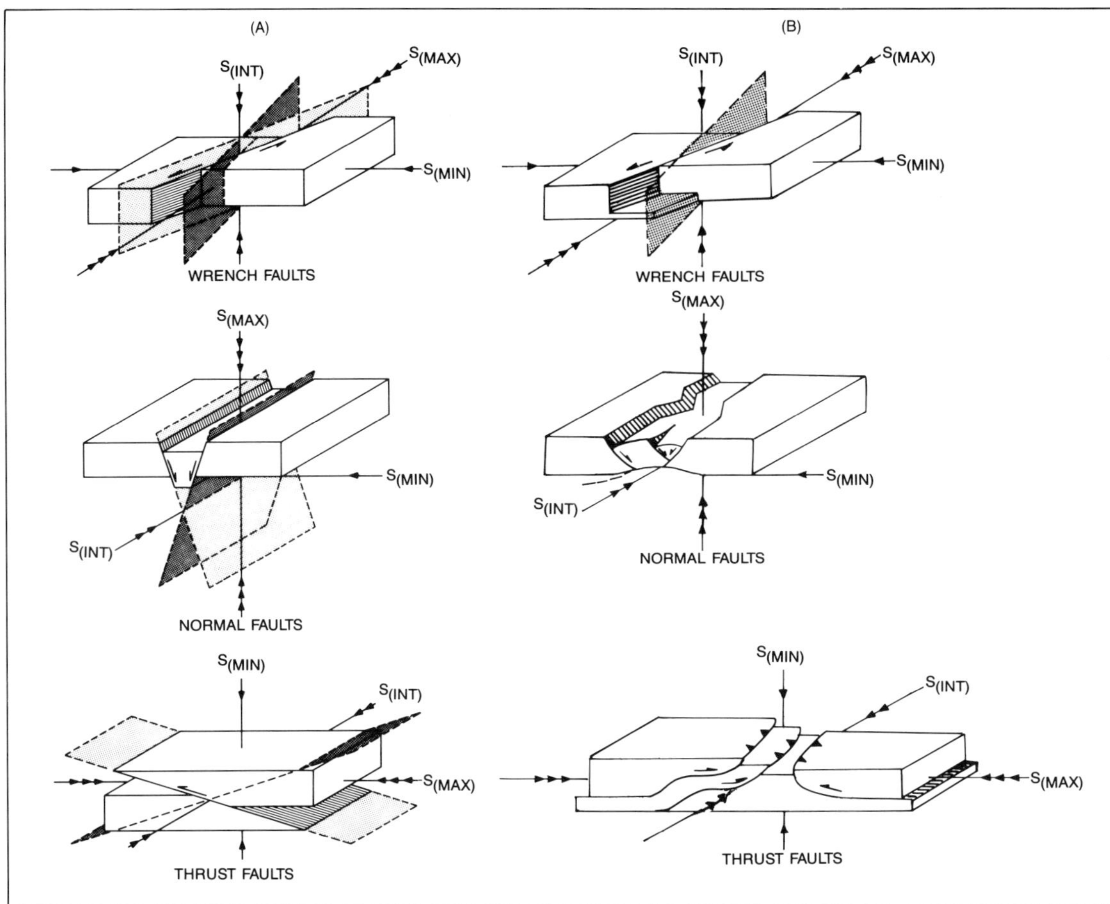

Fig. 2-1—Comparison of relation of conjugate shear faults to orientations of maximum $S_{(max)}$, intermediate $S_{(int)}$, and least $S_{(min)}$ principal compressive stress from conventional shear failure criteria (A) after Anderson (1951), and from empirical observations (B). In A all faults are shown as planar vertical or inclined, whereas in B, faults can alternate as vertical, listric, ramped, and flat. Wrench faults generally are straight and vertical or near vertical, but locally, may flatten (2-1B top), a characteristic that facilitates detachment in this style. Normal faults are typically listric sometimes with ramps and flats (Gibbs, 1984) and show both relay and dog-leg map patterns (2-1B middle) in contrast to relay only (2-1A middle). Similarly, thrust faults are listric with ramps and flats but have a relay pattern in plan view (2-1B bottom). Permission to publish 2-1A by Oliver & Boyd Publishers.

The force acting along AC is σ_1 (or CBD) cos θ. Since we are interested in shear stress τ, which is force acting on some area, the area of AC is

$$\frac{AB}{\sin\theta} \text{ or } \frac{1}{\sin\theta}$$

and $$\tau = \frac{\sigma_1 \cos\theta}{\frac{1}{\sin\theta}} = \sigma_1 \cos\theta \sin\theta$$

Substituting values for θ (after Lovering, 1928):

θ (angle between shear plane and direction of maximum principal compressive stress σ_1)	Shear stress ($\tau = \sigma_1 \cos\theta \sin\theta$)
0°	0.0000
10°	0.1710
20°	0.3214
30°	0.4330
40°	0.4925
45°	0.5000
50°	0.4925
60°	0.4330
70°	0.3214
80°	0.1710
90°	0.0000

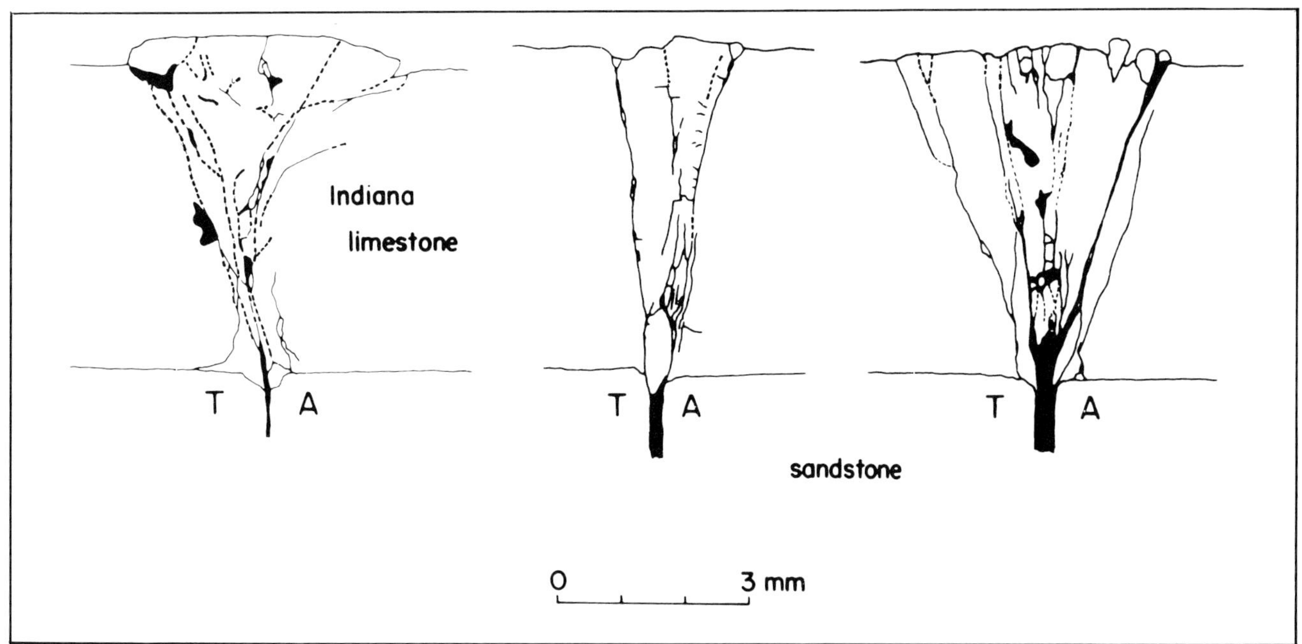

Fig. 2-2 (Bartlett et al., 1981)—Cross-sectional views of limestone experimentally deformed by strike-slip and showing cone or wedge-shaped zone of deformation. Solid lines are fractures having separation; dashed lines are traces of gouge; solid black indicates missing material (lost in thin section preparation). Permission to publish by Elsevier Science Publishers B.V. Amsterdam.

From simple trigonometry, greatest shear takes place when maximum principal compressive stress (σ_1) acts on a surface or plane at 45° to it. If trigonometry was the only factor, conjugate shears would invariably form at 90° (2θ) to one another. However, the internal coefficient of friction of the particular deforming material must be overcome for failure of that material by shear. Taking friction into consideration, θ is usually less than 45°, often 30° ($2\theta = 60°$), though this is merely a rule of thumb with many exceptions. For example, increased confining pressure increases the value of 2θ, as does decreased competence of rocks such that conjugate shears in soft shales have large dihedral angles (2θ).

It should be noted that θ values of 0° and 90° theoretically define surfaces of no shear. The 0° condition is that of tension fractures (in reality minor shear adjustments can occur on these features), whereas straight-on compression takes place across fractures where θ is 90°.

Many other factors can influence shear failure. Shear failure analyses assume elastic behavior of the deforming media. Yet all geologic materials do not deform elastically. Plastic and other forms of yielding are also important. A combination of ductility and bedding anisotropy can cause failure to be by bedding-plane shear, as exemplified in thrust-fold belt assemblages.

Nonrotational (pure) shear is termed nonrotational because no external rotation occurs; however, internal rotation does occur. In the clay model experiment of figures 2-4 and 2-5, conjugate shears from nonrotational shear are created by lengthening (σ_3) parallel to the long axis of the pictured ellipse and shortening (σ_1) parallel to the short axis of the ellipse. The angle between shears (2θ) increases approximately seven degrees during deformation as internal rotation causes clockwise rotation of NE-SW trending shears and counter-clockwise rotation of NW-SE trending shears. Tension fractures in nonrotational shear experiments are caused by lengthening (σ_3) as shown in figure 2-6. The angle 2θ is 0° and is parallel to the direction of shortening (σ_1) or the short axis of the ellipse. Nonrotational shear accounts for a minority of wrench structures.

Most wrench assemblages are related to rotational (simple) shear (Fig. 2-7) where external rotation occurs in addition to internal rotation, and a predictable geometric array of structures is developed. Figure 2-8 is a diagram of a couple (right-lateral) that effects rotational shear. It has a compressional component (C) and an extensional component (E). The ellipse was originally a circle. En echelon folds, which are usually the earliest of the structural assemblage to develop, approximate in trend the long axis of the ellipse. The theoretical angle or orientation of the fold to the wrenching couple is 45°; in practice the angle is usually about 30°, but can be even less. Parallel to the fold axis (normal to the compressional direction) thrust or reverse faults can occur during later stages of fold deformation.

Immediately after inception of the folds, and sometimes

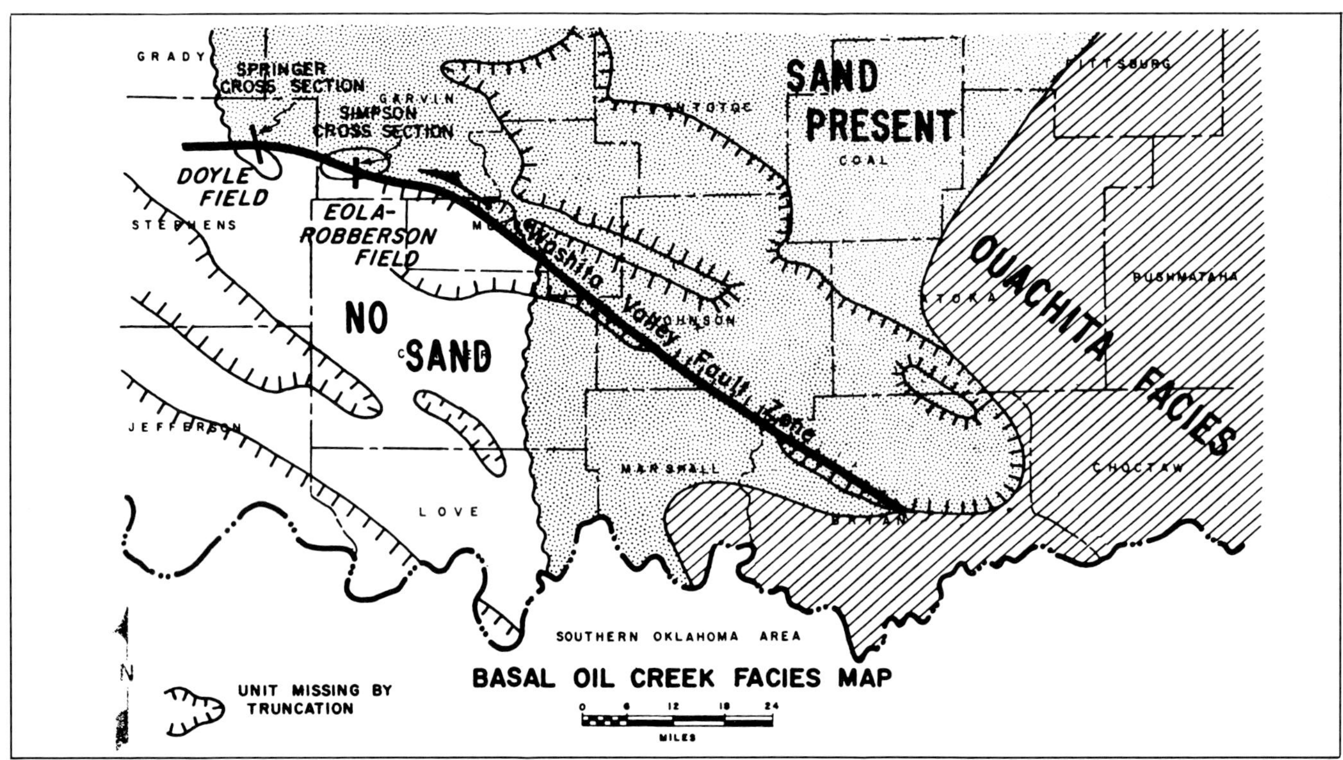

Fig. 2-3 (Tanner, 1967)—Basal Oil Creek facies map showing about 65 km (40 mi) of left-lateral slip on basis of offset zero-sand line. Permission to publish by American Association of Petroleum Geologists.

in the absence of fold development, synthetic strike-slip faults with the same displacement as the originating couple (right-lateral) develop together with a system of left-lateral faults (labeled antithetic). Faults are defined as synthetic where internal rotation (IR) and external rotation (the couple itself) are opposite in sense. Opposing internal and external rotation keeps the trend of these faults close to that of the causative couple and insures that the sense of movement on faults and couple is the same (right lateral). Faults are antithetic where internal and external rotation are in the same sense. Additive internal and external rotation rotates these faults at a high angle to the couple and the motion sense is different (left lateral for the faults vs. right lateral for the couple). Although antithetic faults can be numerous, their importance and amount of displacement are clearly subordinate to synthetic faults. Examples are known where synthetic faults offset axes of earlier formed folds. Synthetic and antithetic faults are sometimes referred to as Riedel shears. (In *nonrotational* deformation shears are equally developed so it is not possible to determine synthetic and antithetic faults).

An additional fracture that forms in clay model experiments is the "P" shear. "P" shears approximate the trend of the originating couple and appear to be the attempt of that fundamental trend to express itself at higher structural levels as a throughgoing fault system is developed. Many large strike-slip faults (e.g., San Andreas fault) may be, in essence, "P" shears. "P" faults undergo synthetic movement and tend to form immediately after inception of the conjugate synthetic and antithetic faults.

Perpendicular to the vector of extension and to the trend of fold axes are tension fractures, which often show small shear displacements. In nature these features can occur to the exclusion of other wrench-related structures, e.g., the Lake Basin fault zone in south-central Montana and a north-south trending zone in eastern Oklahoma.

In theory the entire array of wrench assemblage structures can occur at once. In fact this rarely happens and usually only some of the structures are evident in any one locality. Angles of various structural elements can also vary considerably from the theoretical directions (Fig. 2-8). Finally, theoretically all of the faults shown on figure 2-8 are

Fig. 2-4 (Cloos, 1955)—Experimental setup for nonrotational (pure) shear. Lengthening (σ_3) as shown necessitates shortening (σ_1) at 90°. From Geological Society of America Bulletin, Used with permission.

vertical, save those on the flanks of folds. In reality they are only statistically vertical, often varying a few degrees either way, and, depending on stratigraphic displacements, can appear as very high angle reverse or normal faults.

All of the structures presented on figure 2-8 occur in a distinct en echelon arrangement, meaning that structures of each type (i.e., folds, synthetic faults, etc.) are consistently overlapped and parallel to one another but oblique to the trend of the overall zone of deformation in which they occur. *A true en echelon pattern seems to require wrenching, or at least some component of wrenching, and thus may be unique to the wrench assemblage style. No other structural styles appear to have consistent en echelon patterns.*

In addition to purely geometric criteria of wrenching just cited, Chapin and Cather (1981) have summarized some associated characteristics of sedimentation:

1– syntectonic sedimentary basins that tend to be unusually narrow, deep, asymmetric, and structurally complex and contain sediments that exhibit rapid facies changes, thick facies units of limited lateral extent, and discordance between composition of clasts and possible adjacent source areas;

2– large and rapid dip-slip displacement during wrenching, which may result in the presence of significant volumes of scarp-derived breccias that grade abruptly basinward into finer-grained sediments;

3– a migrating locus of sedimentation as basins open so that syntectonic sediments, young in one direction and build unusually thick, rapidly deposited, sedimentary prisms;

4– local unconformities of the same age as thick sedimentary units nearby.

SUBSTYLES

Parallel, convergent, and divergent substyles (Fig. 2-9) of the wrench assemblage occur depending on:

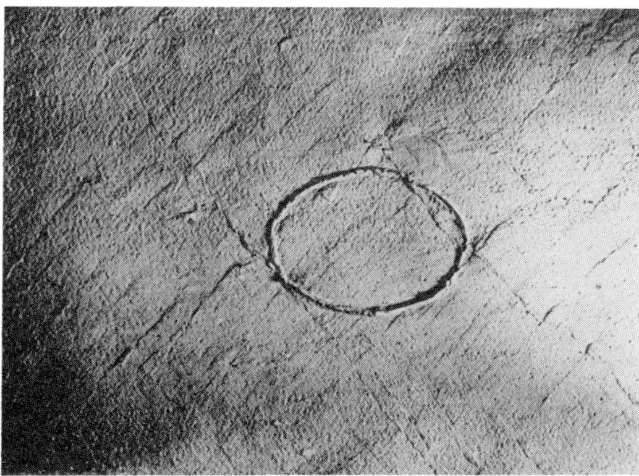

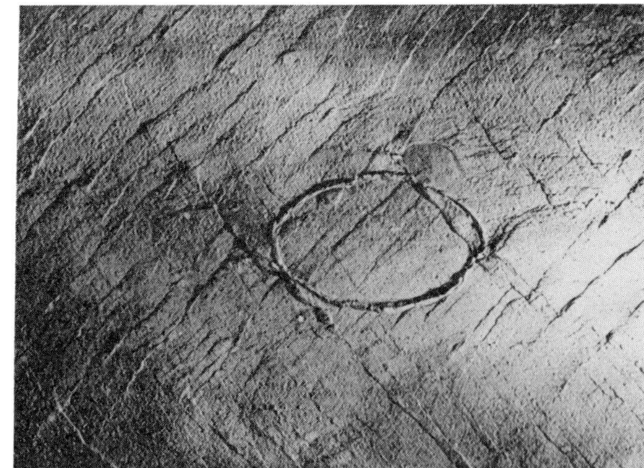

Fig. 2-5 (Cloos, 1955)—Clay experiment showing conjugate shears from nonrotational (pure) shear. From Geological Society of America Bulletin. Used with permission.

1– difference in line of attack of adjacent plates and

2– changes in fault trend relative to movement of plates (Fig. 2-10).

In parallel side-by-side movement the blocks on either side of the fault or fault zone move parallel to one another with the resultant partial structural array illustrated by figure 2-11. Relatively little lateral displacement is required to create this structural geometry—e.g., less than 610 m (2,000 ft) along parts of the Newport-Inglewood trend in the Los Angeles Basin.

Convergent wrenching is treated in detail in the article on the "Spitsbergen Tertiary orogenic belt and the Spitsbergen fracture zone" (Lowell, 1972).

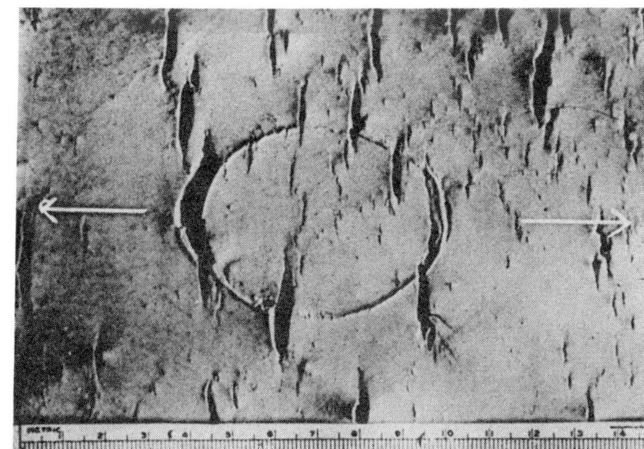

Fig. 2-6 (Cloos, 1955)—Clay experiment showing tension fractures in a nonrotational shear experiment. From Geological Society of America Bulletin. Used with permission.

Fig. 2-7 (Cloos, 1955)—Experimental setup for rotational (simple) shear by left-lateral couple. From Geological Society of America Bulletin. Used with permission.

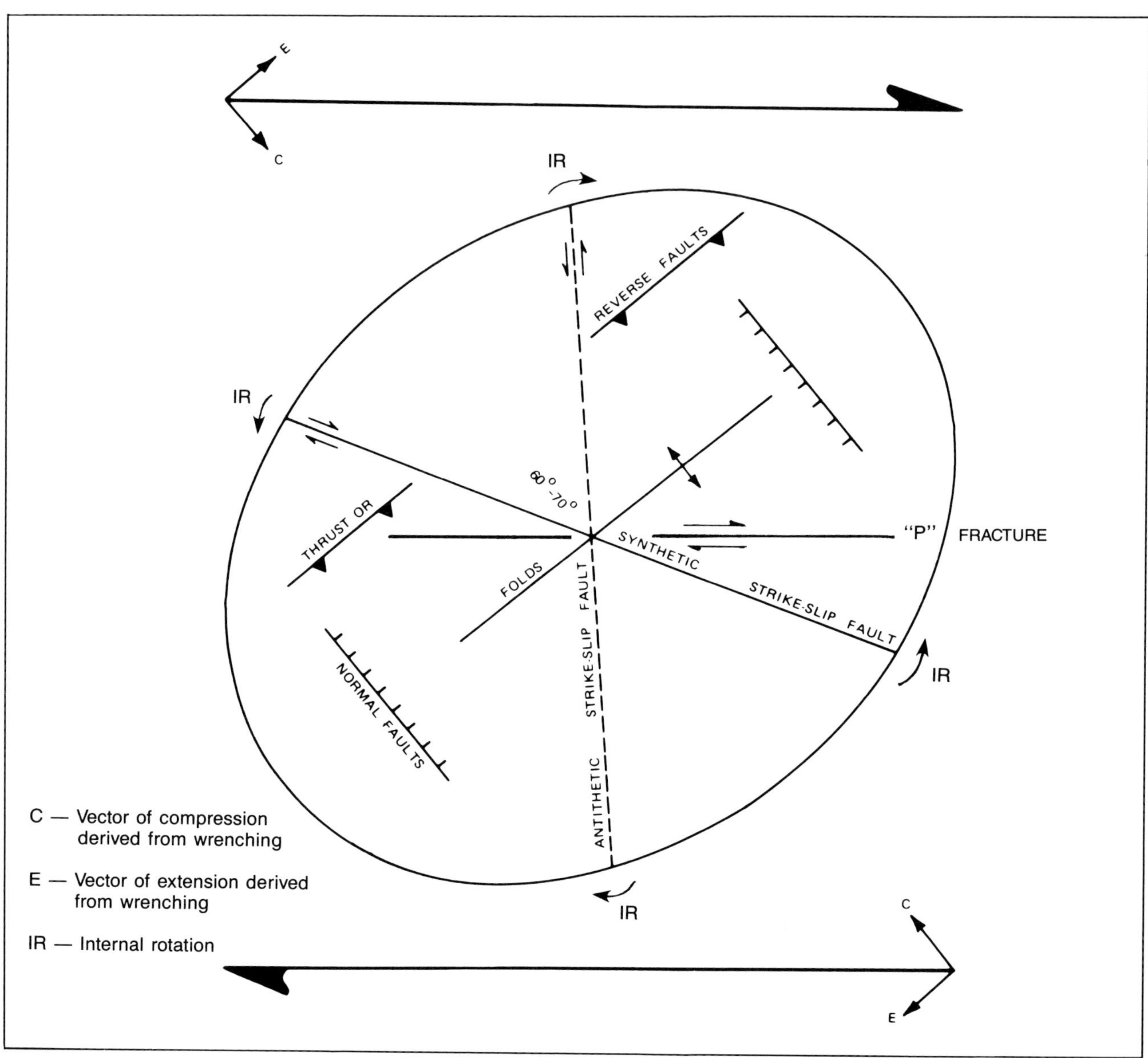

Fig. 2-8 (modified from Harding, 1974a)—Wrench assemblage from a right-lateral couple. Note that a left-lateral couple produces a reverse geometry (view diagram from the back). Permission to publish by American Association of Petroleum Geologists.

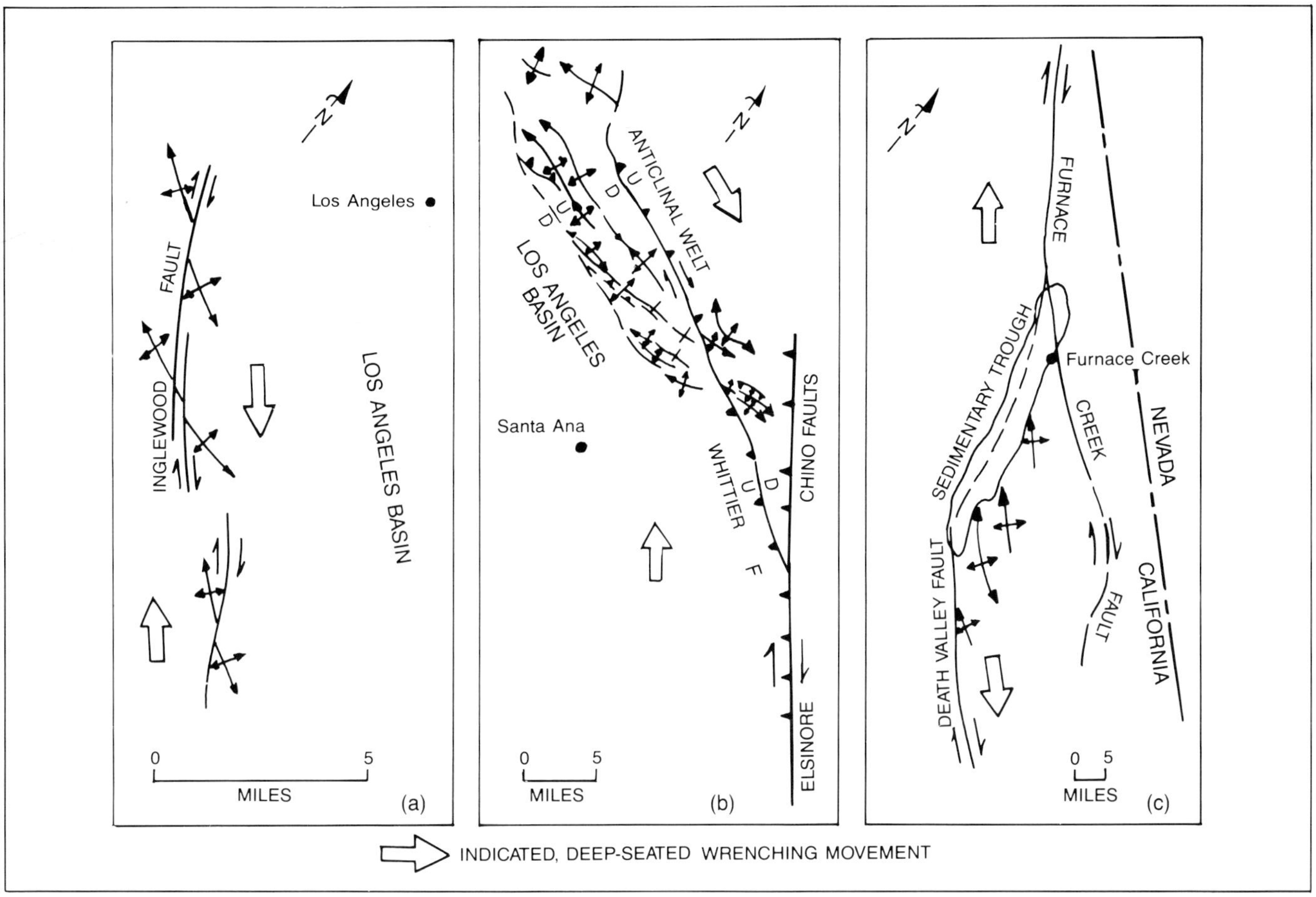

Fig. 2-9 (Harding, 1974b)—Parallel (a), convergent (b), and divergent (c) wrench examples. Reprinted by permission.

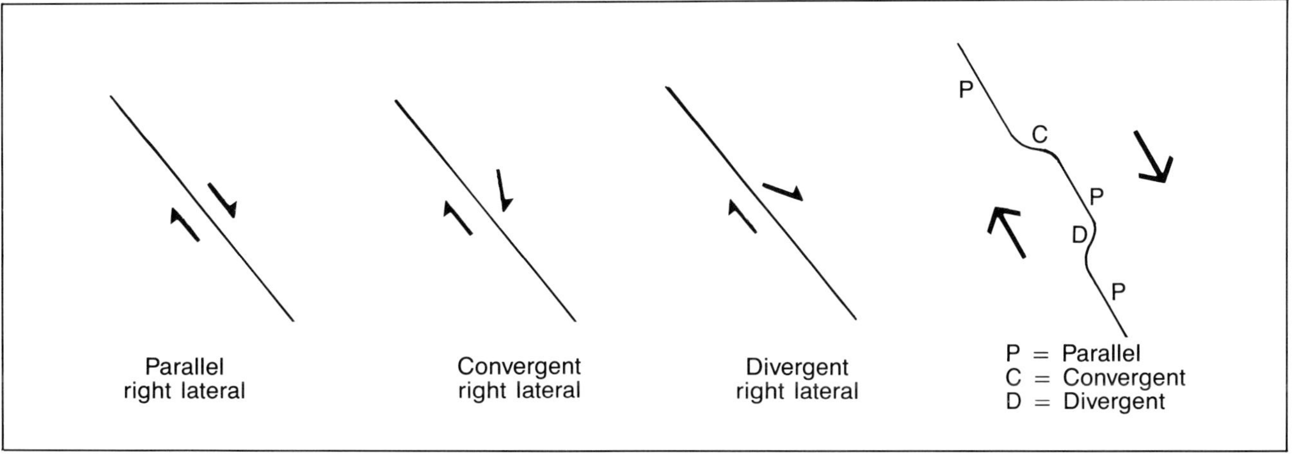

Fig. 2-10—Parallel, convergent, and divergent wrenching as a result of changes in plate motion and changes in fault trends.

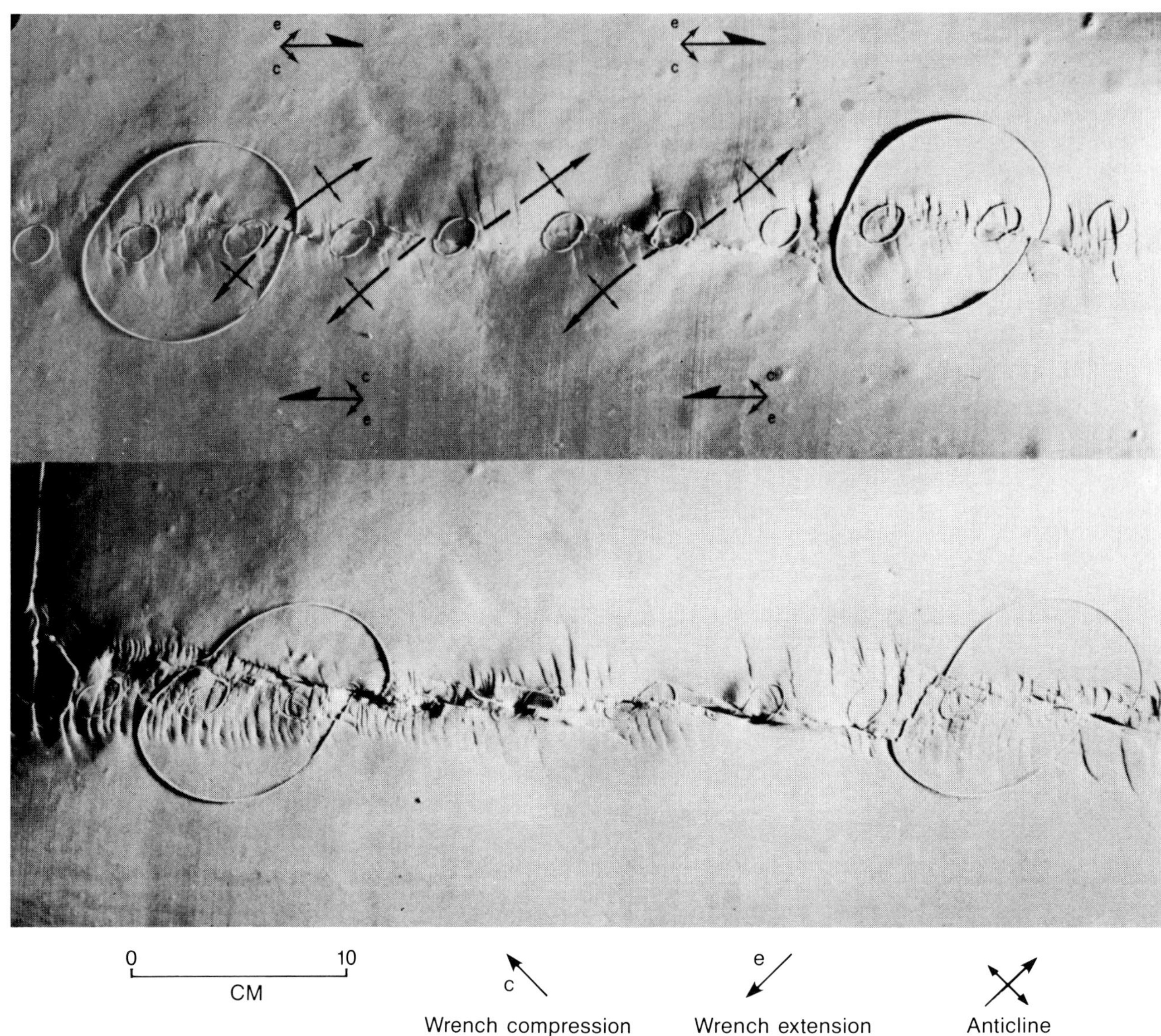

Fig. 2-11 (Harding, 1973)—Clay model of right-lateral parallel wrenching (A, B—two stages in plan views after Wilcox et al., 1973). Structural pattern consists of right-handed en echelon folds and left-handed en echelon, right-lateral synthetic, and left-lateral antithetic faults. Horst and graben slices are formed along youthful wrench-fault zone by anastomosing and interbraiding of synthetic faults (stage B). This process leads to development of mature, through-going wrench faults. Permission to publish by American Association of Petroleum Geologists.

SPITSBERGEN TERTIARY OROGENIC BELT AND THE SPITSBERGEN FRACTURE ZONE[1]

Abstract The Tertiary fold and thrust belt of Spitsbergen was created by compressional dextral strike-slip movement on the Spitsbergen fracture zone system as Spitsbergen moved past northern Greenland during the opening of the Norwegian and Greenland seas. This interpretation thus follows that of Harland (1965, 1969), but gives primary emphasis to strike-slip as a motor for deformation. The Spitsbergen orogenic belt as a ''strike-slip orogenic belt'' differs from a ''subduction orogenic belt'' in having a discrete pattern of en echelon folds, in having a narrow zone of deformed sedimentary cover with a much greater degree of basement involvement, in having presumably different cross-sectional profiles of thrusts, upthrust versus downward flattening, in being shorter in length, and probably in lacking alpine ophiolites and lacking metamorphism. Deformation is associated with a laterally moving and rising, lithospheric welt, rather than a downgoing lithospheric slab.

INTRODUCTION

Severe deformation affecting early Tertiary and older rocks along the west side of Spitsbergen[2] has been known for more than a century (Nordenskiold, 1866; see Orvin, 1940, for excellent discussion). A solution was recently suggested by Harland (1965, 1969), who related the deformation to the opening of the Norwegian and Greenland Seas and termed it the ''Western Spitsbergen Orogeny.'' Harland interpreted the western Spitsbergen deformation as a result of compression from the northward movement of Greenland against Spitsbergen that accompanied the general opening of the northernmost Atlantic. He further recognized that ''the deformation structures in Spitsbergen. . .show a marked dextral pattern of curvature in the folds and thrusts and dextral movement along some oblique strike-slip faults,'' and he therefore speculated that ''the compression was coupled to strike slip. . .and that a wrench regime obtained while Greenland was pressing northward'' (Harland, 1969, p. 839, 841). The purpose of this article is to emphasize the wrench regime—that is, that the western Spitsbergen orogeny was caused mainly by strike-slip motion and that compression was subordinate. This may be termed transpressive movement in Harland's (1971) recent terminology.

[1]Geological Society of America Bull., V. 83, 1972, p. 3091–3102.

[2]The Spitsbergen archipelago has been officially designated ''Svalbard.'' The main island, Vestpitsbergen, has been renamed ''Spitsbergen.''

CONTINENTAL DRIFT AND THE SPITSBERGEN FRACTURE ZONE

A predrift restoration closes the northernmost Atlantic and Arctic Eurasian oceans by placing Svalbard immediately adjacent to the Lomonosov Ridge, northern Greenland, and Ellesmere Island (Fig. 2-12). The remarkable late Paleozoic, Mesozoic, and early Cenozoic stratigraphic similarities between Svalbard and the Sverdrup Basin in Ellesmere Island have been previously recounted (Harland, 1965, 1969). According to present seismicity (Horsfield and Maton, 1970), the opening of the northernmost Atlantic and Arctic Eurasian oceans has been effected by spreading on segments of the Mid-Atlantic Ridge which has been transformed to the mid-ocean ridge of the Eurasian Arctic basin by right-lateral slip along the Spitsbergen fracture zone (Fig. 2-13). The Spitsbergen orogenic belt is separated by 100 km (62 mi) of submerged continental shelf from the Spitsbergen fracture zone which bounds the edge of the continental lithosphere. The orogenic belt is interpreted, as Harland (1969) speculated, to be a result of right-lateral slip (with a compressional component additional to that inherent in any strike-slip fault) along the Spitsbergen fracture zone and other submerged parallel faults to the east as Spitsbergen moved past northern Greenland.

SPITSBERGEN TERTIARY OROGENIC BELT

Structure

The analyses of folds and faults of Tertiary age in the belt along the west (more accurately the southwest) side of Spitsbergen (Fig. 2-14) have led to conflicting views as to the dominant type of deformation. Orvin (1940, p. 48) considered that the Tertiary deformation in Spitsbergen ''comes within the category of 'Faltengebirge' (fold mountains) of Stille.'' Harland (1969, p. 839) stated that the account by Orvin of the fold belt ''as a somewhat faulted limb of a monocline bounding the main Tertiary basin''. . .is oversimplified. . .''recent investigations showing that a complex fold and thrust structure with many tectonic units is involved''; Challinor (in Harland, 1969, p. 839) reports ''structures analogous to the eastern front of the Canadian Rocky Mountains.'' I concur that the Spitsbergen Tertiary orogenic belt is not the result of simple uplift. However, although there may be local similarities in structural form, neither is it analogous to the Canadian Rockies in the sense that the latter

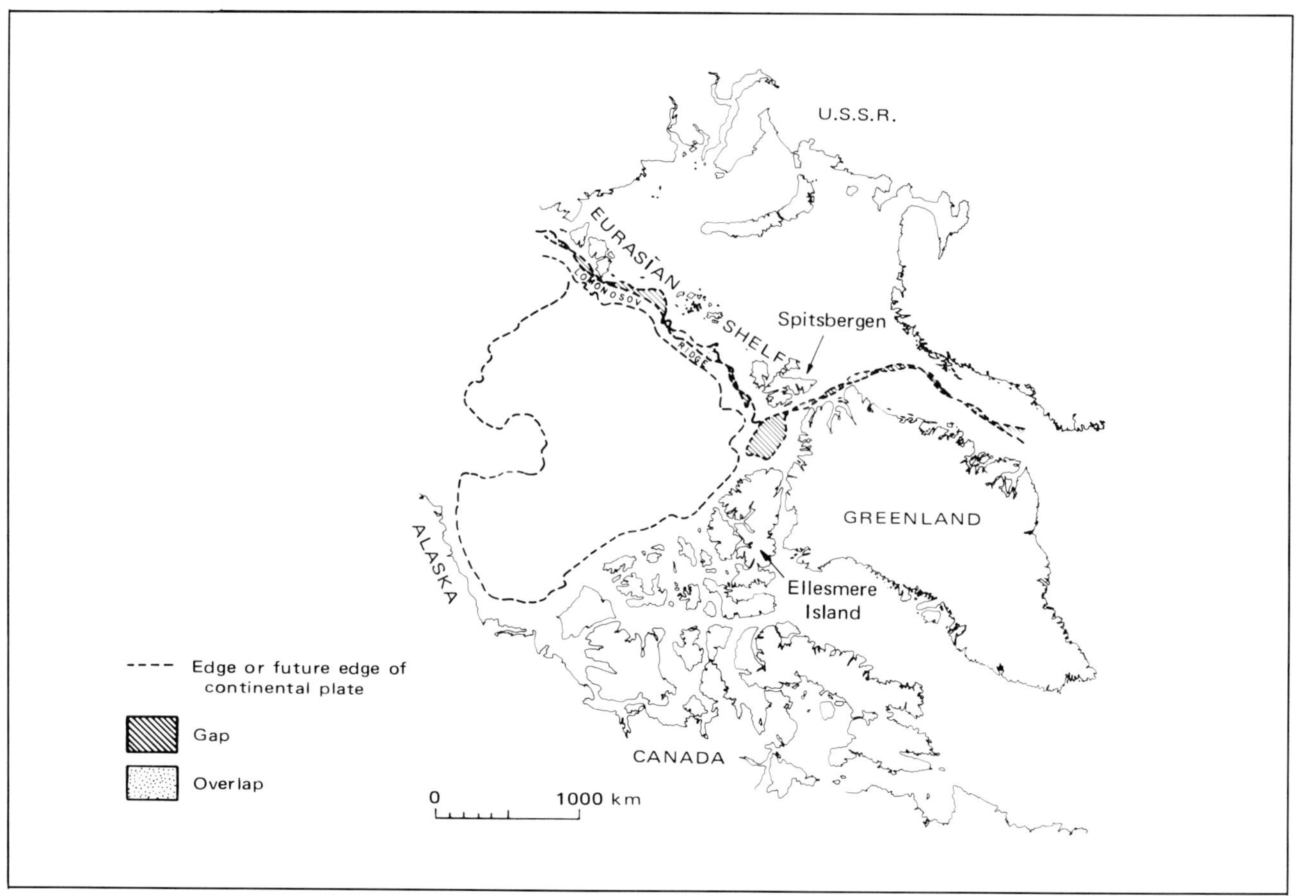

Fig. 2-12—Position of Spitsbergen in a pre-Tertiary drift restoration. Edge of continental plate was chosen to follow the steepest gradient along the continental slopes. Some movement between Ellesmere Island and Greenland probably preceded movement between Spitsbergen and Greenland.

is dominantly compressional. Rather, the Spitsbergen orogenic motor has been compressional dextral strike-slip faulting.

Figures 2-14 through 2-17 show in western Spitsbergen a folded and substantially thrusted belt which clearly precludes simple uplift as the driving force of the western Spitsbergen orogeny. Moreover, en echelon folds are known along the Tertiary deformation front (Fig. 2-14) in an orientation consistent with right-lateral slip. Faults of upthrust profile are well known along strike-slip faults (Fig. 2-18) and are often diagnostic of them (Allen, 1965; Lowell, 1969). With the knowledge that strike-slip must have occurred in the Spitsbergen Tertiary orogenic belt, the thrust faults shown on Figures 2-14 through 2-17 have been interpreted as upthrusts and shown to steepen with depth. The general structural character of the Tertiary fold belt suggests the tectonic architecture of a welt with a complexly structured, upthrusted edge. A welt of this type has been produced by compressional strike-slip in clay model experiments (Fig. 2-19) and is illustrated conceptually (Fig. 2-20). By analogy with the clay and conceptual models, the Spitsbergen Tertiary orogenic belt would lie along one edge of the welt and also show elements of the en echelon fold pattern.

Width

Orvin (1940, p. 43, 46) postulated that "the folding zone is in fact broader than it would seem from the younger (Tertiary) beds today, as the zone has also had a westerly extension where the younger beds later have been removed" and that "one cannot ignore the possibility of the folding in Spitsbergen being only an eastern, and perhaps smaller part, of an alpine folding out at sea west of Spitsbergen." These comments are very much to the point of the present interpretation because dextral movement only on the Spitsbergen fracture zone, located some 100 km (62 mi) from the exposed fold belt, could hardly account for deformation in the belt. Rather, dextral movement probably occurred east of the Spitsbergen fracture zone on several parallel faults now submerged west of the Spitsbergen orogenic belt. As a precedent, a zone of approximately 100 km (62 mi) width of strike-slip faults is known in southwesternmost California (U.S. Geological Survey, 1962). Harland (1969, p. 841) also recognized that "the possibility must be considered that the (Spitsbergen) fracture zone comprised several parallel faults. . . and that some of the north-south faults in Spitsbergen are part of that movement pattern."

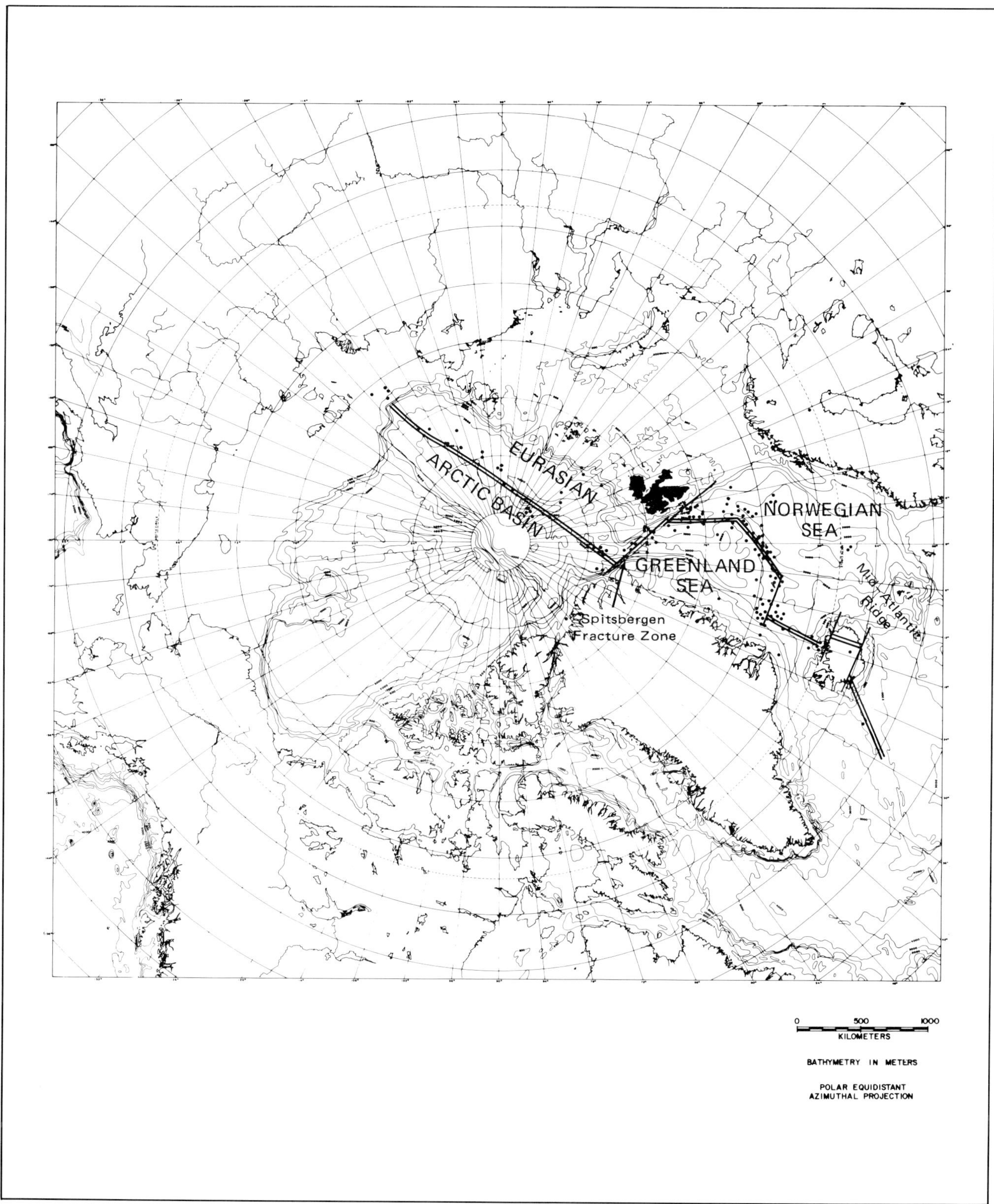

Fig. 2-13—Position of mid-ocean ridges and Spitsbergen fracture zone outlined by earthquake epicenters and bathymetry.

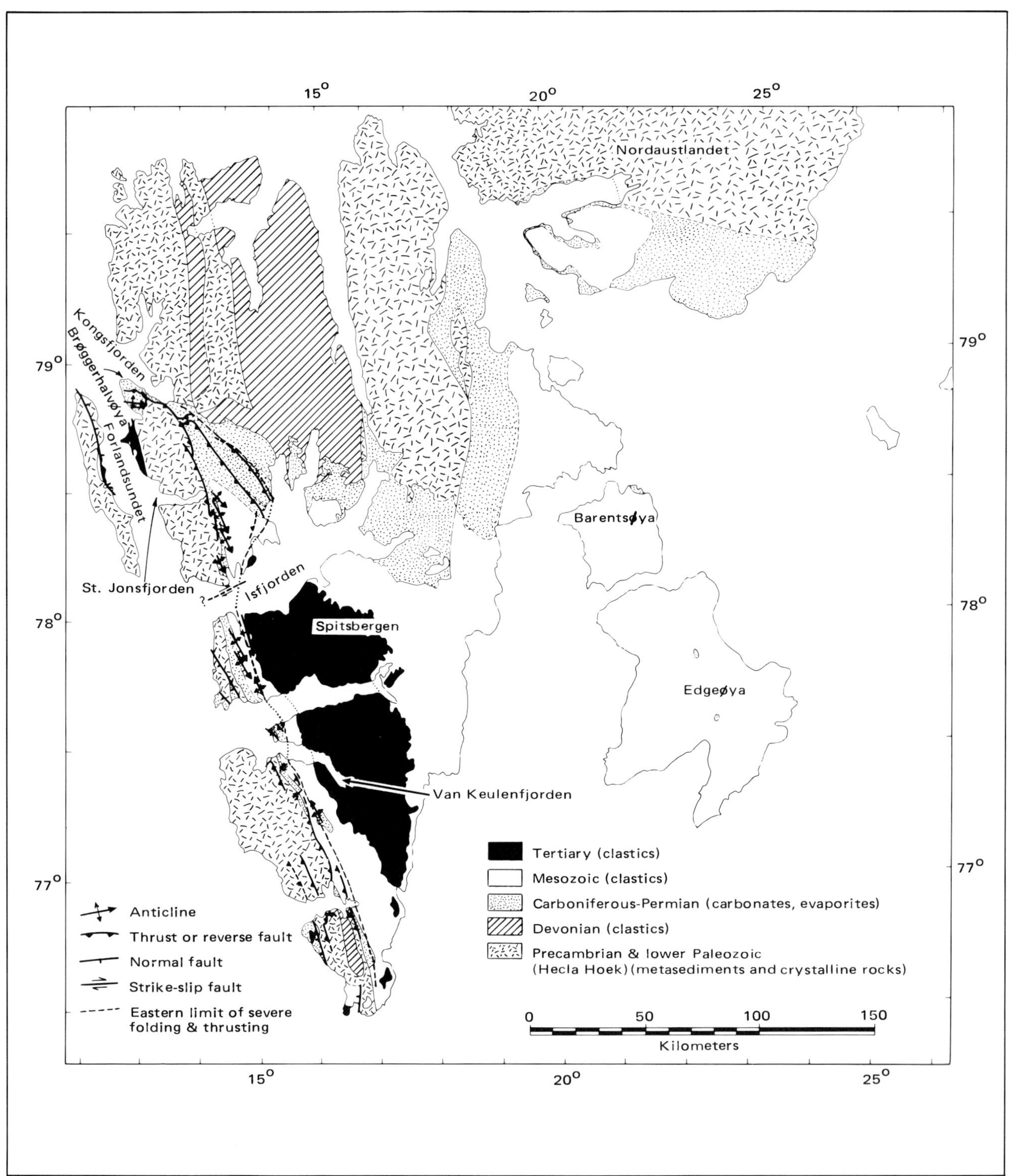

Fig. 2-14—Generalized geologic and tectonic map of Spitsbergen showing mainly the Tertiary structures of the Spitsbergen orogenic belt. The excursion of the limit of severe folding and thrusting to the east between Isfjorden and Kongsfjorden coincides with the presence of gypsum at the top of the Lower Permian that is not present to the south; the several hundred meters of gypsum may have served as a mobile layer. (Map originally prepared at larger scale; contacts and faults are accurate.)

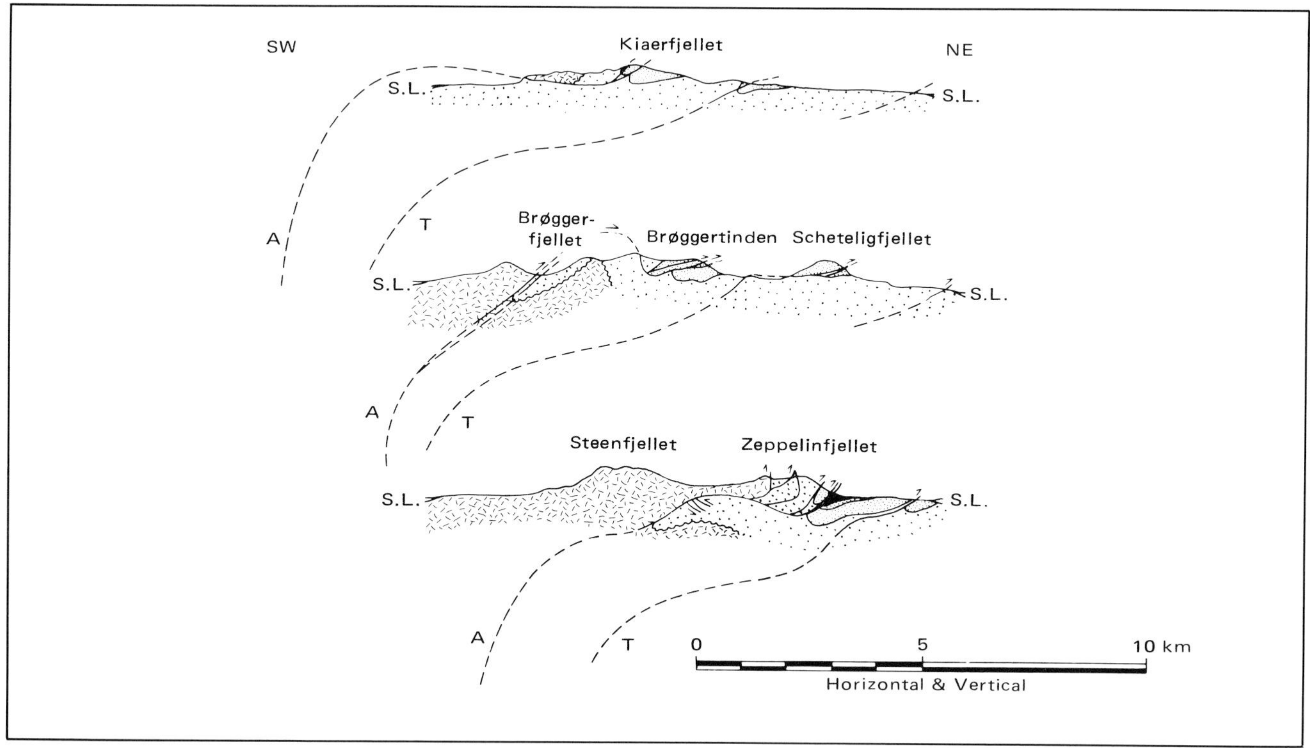

Fig. 2-15—Structure sections through Broggerhalvoya (modified from Challinor, 1967). In these and succeeding sections, the thrusts are interpreted to steepen with depth and be upthrusts related to strike-slip faulting (see Fig. 2-16 for legend).

Age

Evidence from marine surveys aids in dating the terrestial deformation, and the ages from the two different sources are in satisfactory accord. The youngest rocks cropping out in Spitsbergen are continental clastic rocks of Oligocene to questionably Miocene age (Flood et al., 1971). They are involved in the deformation so that the western Spitsbergen orogeny can only be dated as post-Oligocene-Miocene(?). The youngest marine beds on Spitsbergen are Eocene-Oligocene. Correlation of magnetic anomalies in the North Atlantic Ocean indicates emplacement of new sea floor began 63 my ago (late Paleocene) in the northernmost Atlantic and Artic regions (Pitman and Talwani, 1972); considerable uplift and faulting in these regions no doubt preceded measurable sea-floor spreading (compare Lowell and Genik, 1972). From these data, I interpret that uplift by the compressional strike-slip mechanism (Fig. 2-19) on the Spitsbergen fracture zone system ended marine sedimentation in the Eocene-Oligocene (Battfjellet Formation) and that the rising welt was the sediment source of the subsequent continental sequence. The continental beds, present only in the center of the Tertiary syncline (Fig. 2-14), would have been tilted while folding and thrusting occurred along the orogenic front to the west.

Prior to transform faulting along the west side of Spitsbergen and rifting by normal faulting along the north side, earlier uplift is indicated by the absence of Upper Cretaceous rocks in the whole of Svalbard and the presence of continental beds in the lower paleocene. Extensive intrusions of dolerite dikes and sills in latest Jurassic earliest Cretaceous times (Parker, 1967) appear to have been an even earlier precursor to continental breakup.

Harland (1969, p. 841) has stated that along the Spitsbergen fracture zone system, "there probably was a change. . . from transcurrent faulting with compression to transcurrent faulting with extension," (transtension of his later terminology in 1971) and that the Forlandsundet graben seems to be a late extension feature (see also Atkinson, 1963). I agree and further suggest that the change from compression to extension accompanying strike-slip may have occurred when the continental lithospheric plates containing Svalbard and Greenland ceased to scrape against each other as new sea floor was emplaced between them (about 15 my ago, using an average sea-floor spreading half-rate of 1 cm/yr; see Vogt et al., 1970). Normal faults creating Forlandsundet and continuing farther south along the west side of Spitsbergen were the products of this change in lithospheric regimen and attending stresses. Subsidence by late normal faulting also logically explains the submergence of the 100-km-wide strip between the west coast of Spitsbergen and the Spitsbergen fracture zone; otherwise, this area should be positive, according to its postulated compressional strike-slip origin.

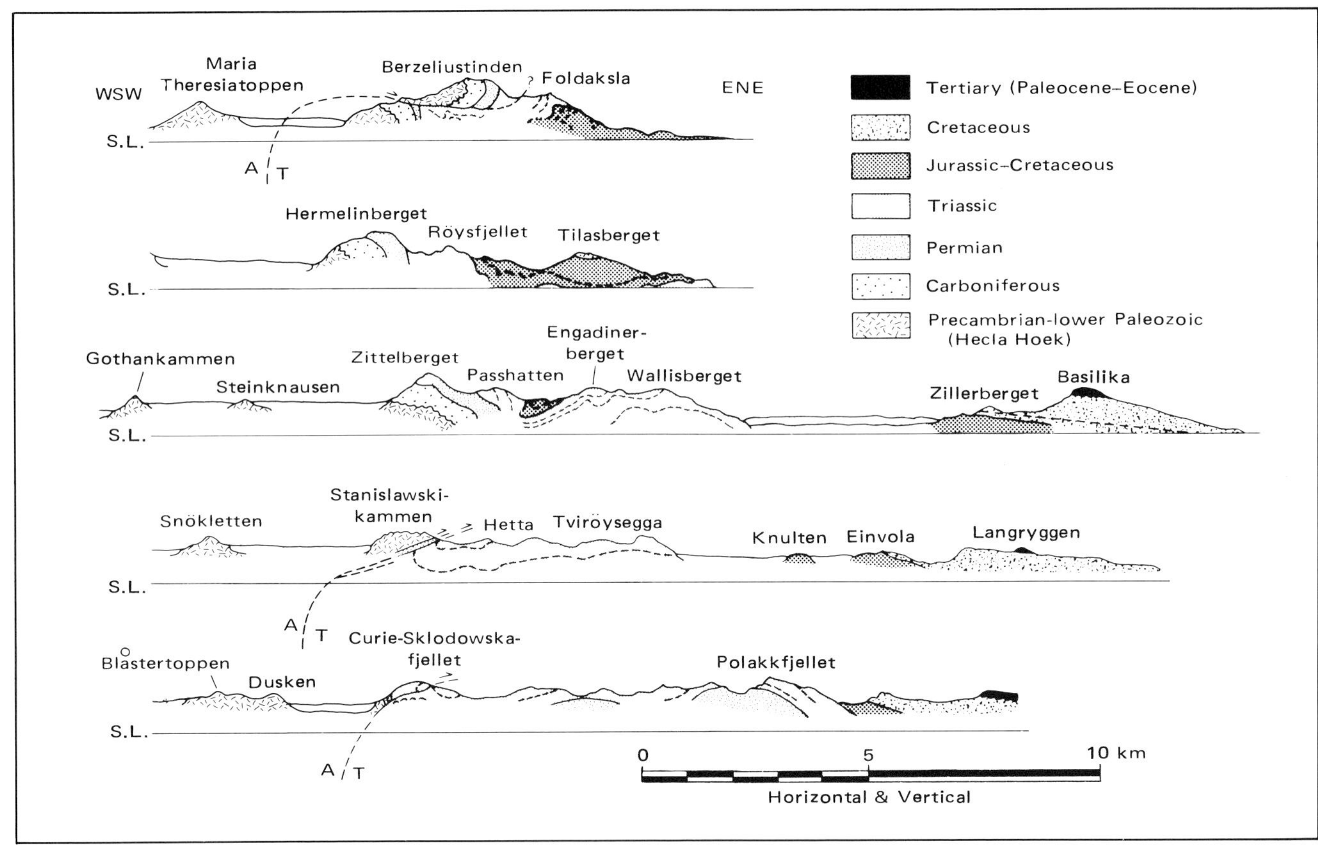

Fig. 2-16—Structure sections south of Van Keulenfjorden (modified from Rozycki, 1959).

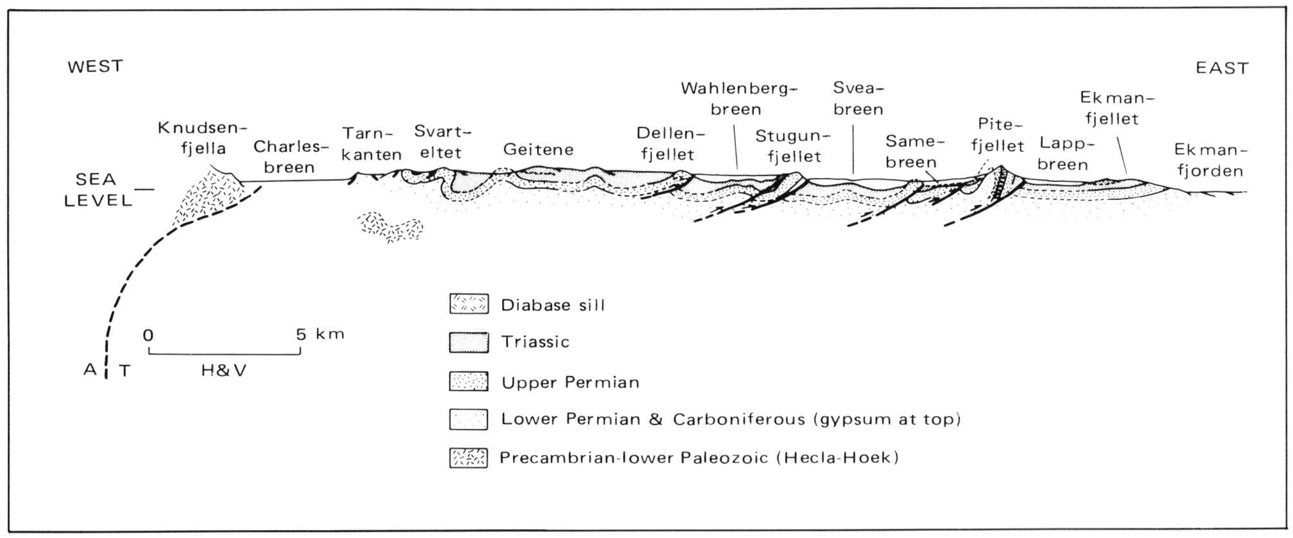

Fig. 2-17—Structure section east of St. Jonsfjorden. Note greater width of folding and thrusting interpreted as caused by mobile gypsum at top of Lower Permian.

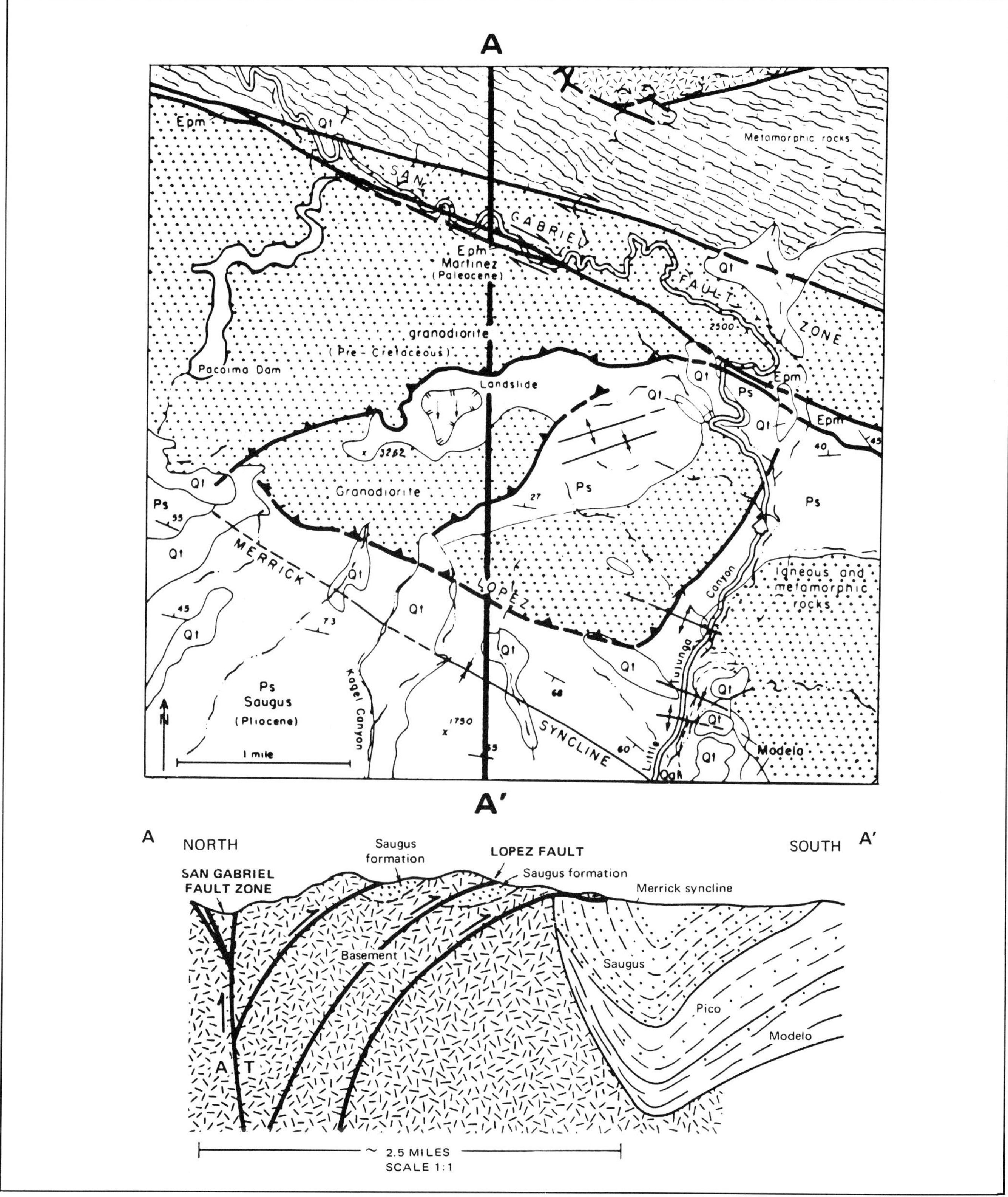

Fig. 2-18—Map and cross section in Little Tujunga area, California, showing curving traces of thrust faults which become vertical with depth in the San Gabriel fault zone. The San Gabriel fault has been reported to have 24 to 40 km (15–25 mi) of right-lateral movement on the basis of offset source areas and depositional sites of anorthosite and norite conglomerates (after Jennings and Troxel, 1954; Howell, 1954).

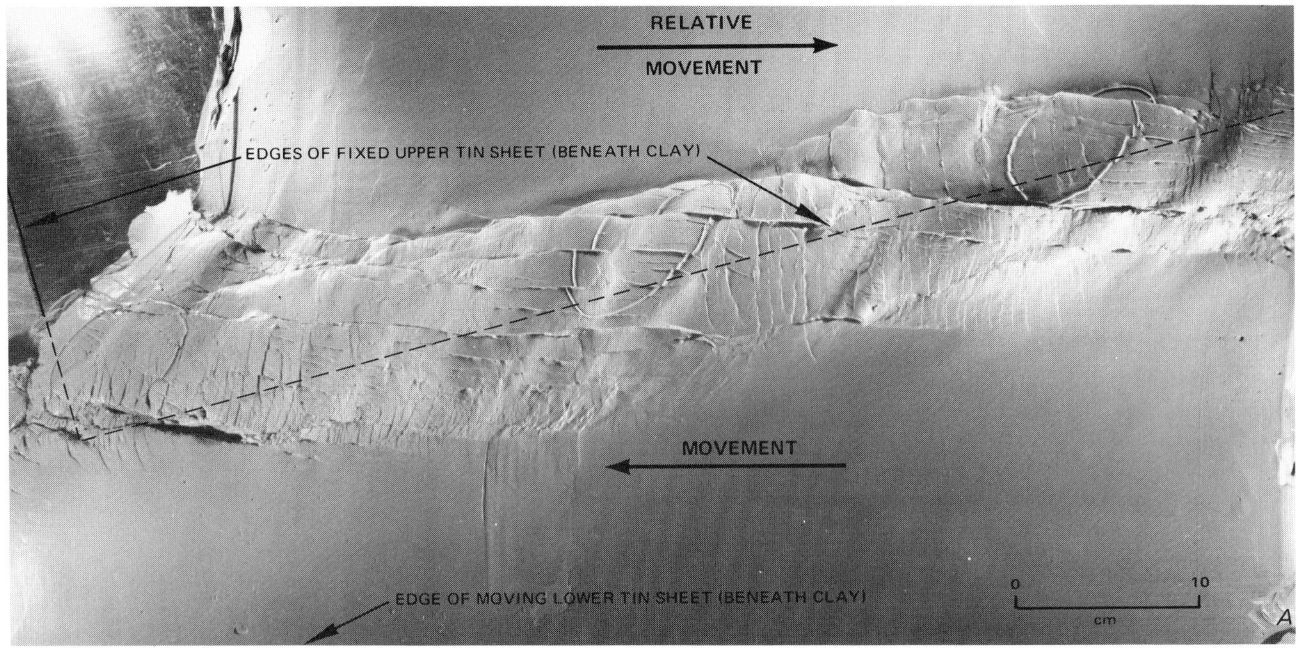

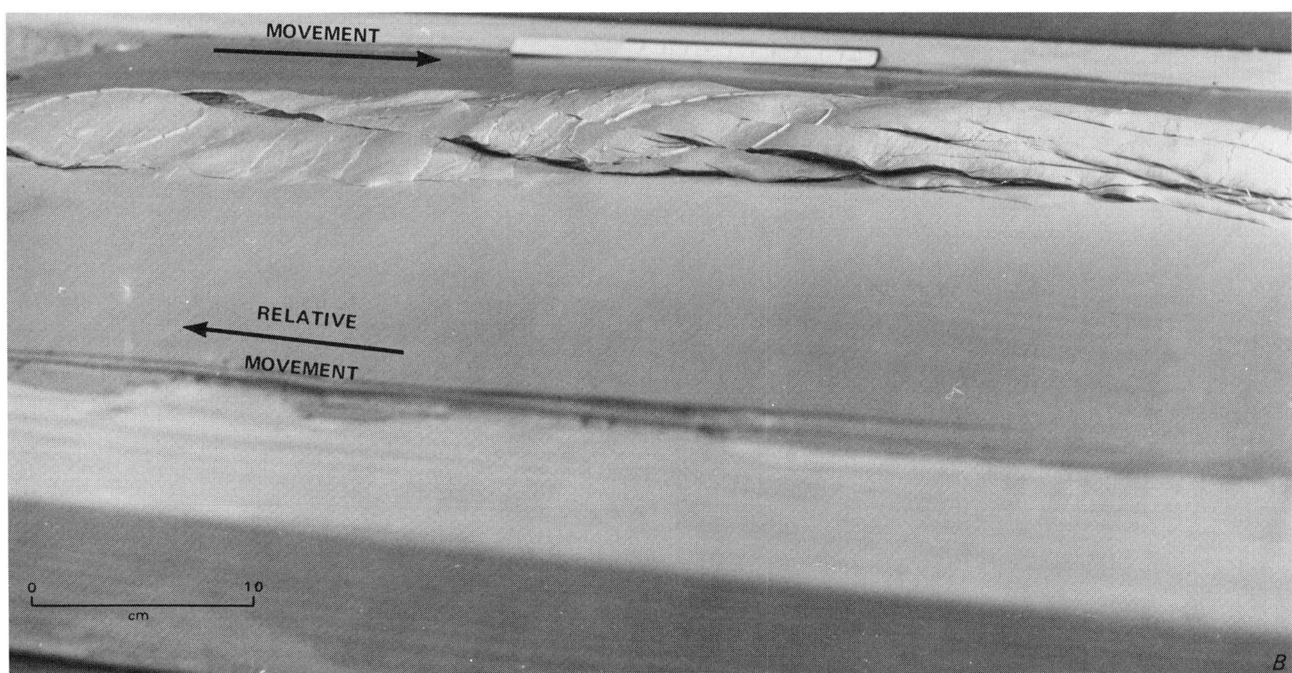

Fig. 2-19—A. Plan view of clay model of a thrust-bounded welt caused by and paralleling the trend of an underlying strike-slip zone. Strike-slip structures in the clay were created by convergent right-lateral motion between two underlying tin sheets; the lower tin was pulled to the left beneath the stationary upper tin, which was canted at an angle of 15°. A compressional component additional to that inherent in any strike-slip was thus introduced.

First formed on the top of the originally smooth clay surface were en echelon anticlines approximately parallel to the long axes of the deformed ellipses. Shortly thereafter, and as the anticlines continued to grow, synthetic right-lateral shears trending at low angles to the underlying wrench zone developed along with antithetic left-lateral shears trending at high angles to the underlying wrench. In the more advanced stages of lateral movement, an uplift (or welt) formed parallel to the underlying wrench zone. The edges of the uplift are clearly thrusted. Unfortunately, it is not possible to view the internal structure within the clay, since structure is smeared and destroyed by cutting, nor can structure be preserved by drying the clay, then cutting. However, by analogy with sand models and also experiments with limestone (Fig. 2-2), which can be cut to view their interiors, most of the thrusts must almost certainly be continuous with deeper vertical faults generated by the lateral movement at the base of the clay, and, if so, they are upthrusts. B. Side view of clay model of figure 2-19A showing the welt with thrusted edges evidenced by overhangs and lobate thrust fronts.

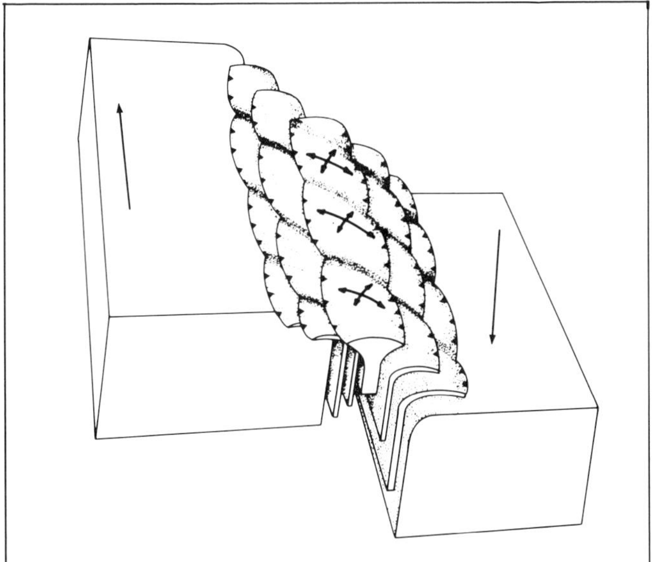

Fig. 2-20—Conceptual diagram of an upthrust-bounded welt created by convergent strike-slip or transform motion. Portions of two plates or blocks are shown, the one moving at a low convergent angle into the other in the same plane, causing each segment to move a small distance successively past the other in a right-slip direction. A space problem is created, and the easiest direction of relief for the crowded material is upward. A welt with downward-tapering wedges and upthrust margins is created. The upthrusts are not necessarily so symmetrically disposed as shown and the deeper faults may tend to coalesce and braid rather than be parallel.

Folding precedes faulting in this type of deformation, which explains the observations by Orvin (1940) that the faults are younger than the folds in the Spitsbergen Tertiary orogenic belt.

COMPARISON WITH OTHER OROGENIC BELTS

The exposed width of the Spitsbergen Tertiary orogenic belt, varying between 12 and 45 km (7.5–28 mi), is sufficient to permit comparison with other orogenic belts; an additional 100 km (62 mi) wide submerged strip, here interpreted as a subsided portion of the orogenic belt, unfortunately cannot be used for comparison. The outcropping part of the Spitsbergen belt differs markedly from mountains, such as the Alps, for example, which traditionally have been attributed mainly to compression, now often related to subduction (see White et al., 1970, for summary). First, the Spitsbergen Tertiary orogenic belt has a discrete pattern of en echelon folds, whereas the Alps do not (Beck-Mannagetta, 1964; Geological maps of Switzerland, 1:200,000, 8 sheets). Secondly, the width of the zone of deformed sediments between the most external surface thrust(s) and the more internal position where autochothonous or parautochthonous basement crops out and is involved in the deformation is much narrower in Spitsbergen than in the Alps, or also in the southern Apalachians (U.S. Geological Survey, 1962). The width of this zone in Spitsbergen averages 5 km (3 mi), increasing to 27 km (17 mi) only where gypsum is present; the zone (from the most external Helvetic surface thrusts to the north edge of the Aar Massif) ranges from 20 to 50 km (12.5–31 mi) in width in the Swiss Alps; and the zone (from the northwesternmost surface thrusts to the west edge of the Blue Ridge) is from 45 to more than 100 km (28–62 mi) wide in the southern Appalachians. This points up the fact that basement (Hecla Hoek mostly metamorphosed sedimentary rocks) is much more directly involved in the external portion of the Spitsbergen belt than in that of the other two in which there is a good deal more detached deformation of imbricate thrust character. The interpreted upthrust profile in Spitsbergen contrasts with the flatter thrust profiles of the Alps, the southern Appalachians, and also the Canadian Rockies (Bally et al., 1966).

Finally, the exposed portion of the Spitsbergen Tertiary orogenic belt differs from the Alps and numerous other orogenic belts in lacking metamorphism (Orvin, 1940) and in having no alpine ophiolites. The latter have been interpreted as having been ultimately emplaced in orogenic belts by subduction processes (Temple and Zimmerman, 1969; see Dewey and Bird, 1971, for summary). The more internal portion of the Spitsbergen belt is, of course, submerged, but it would be a prediction of the present interpretation that no evidence of either appreciable metamorphism or alpine ophiolites will be found there.

In conclusion, I believe there is sufficient contrast between the Tertiary orogenic belts of Spitsbergen and, for example, the Alps, that the former can be considered as a "strike-slip orogenic belt" caused by some compressive but mainly lateral plate movement, and the latter a "subduction orogenic belt" caused by undoubtedly some lateral but mainly compressive plate movement. In Spitsbergen, deformation is associated with a laterally moving and rising, lithospheric welt; in the Alps, much of the deformation can be related to downgoing lithospheric slabs.

In the absence of subduction, a strike-slip orogenic belt should lack igneous activity of the large-scale batholithic intrusive and extensive volcanic belt type, and deep-water turbidite-flysch sediments of trench origin should also be lacking; none of these features is known in Spitsbergen. The Spitsbergen Tertiary orogenic belt also differs from many others, including the Alps, Appalachians, and Canadian Rockies, in not having had a prior geosynclinal history; late Paleozoic, Mesozoic, and early Cenozoic sedimentary rocks of Spitsbergen are of shelf origin. Lack of geosynclinal conditions in Spitsbergen is related to sedimentation having taken place entirely on continental lithosphere. However, lack of a geosynclinal history is not necessarily true for all strike-slip orogenic belts. Areas which were structured by convergent strike-slip along a boundary between continental and oceanic lithosphere would no doubt have been geosynclinal.

The length of a strike-slip orogenic belt is less than that of a subduction orogenic belt, since, as for Spitsbergen, the length of the structured belt cannot exceed the length of the associated zone of transform faulting; subduction orogenic belts on the other hand may extend the length of one side of a continent. Where subduction zones were offset by transform faults (arc-arc type), however, subduction orogenic belts may include segments, trending at approximately right angles to the overall trend of the belt, which were structured by convergent strike-slip.

Many of the above distinctions probably can be made in other orogenic belts. If convergent strike-slip and head-on compression are seen as end members of a continuous series, then a full spectrum of structuring is possible, depending on angles of plate approach.

In summary oblique convergent wrenching (transpression) introduces a compressional component greater than that inherent in parallel wrenching. Numerous en echelon folds and an anticlinal welt(s), often bounded by faults of upthrust profile, are the result. Sigmoidal-shaped folded faults appear to be another characteristic (Fig. 2-21).

In addition to the Tertiary orogenic belt of Spitsbergen numerous other of the world's orogenic belts could be caused by convergent wrenching. These include parts of the Alborz Mountains in Iran (Stocklin, 1968), portions of the Andes in Venezuela (Deratmiroff, 1971) and Colombia (Julivert, 1970), and the Caucasus (Beloussov, 1962) and Dnepr-Donetz (Alverson et al., 1967) in the U.S.S.R.

Types of plate interaction resulting in convergent wrench (strike-slip) and compressional-subduction orogenic belts are suggested in figure 2-22, and criteria for differentiating the two are given in table 2-1. No single criterion is definitive, but several criteria occurring together are strongly suggestive of origin.

Oblique divergent wrenching (transtension) emphasizes an extensional component greater than that inherent in parallel wrenching. In direct contrast to the en echelon folds and upthrust-bounded welts associated with convergent wrenching, en echelon folds are subdued and sags or sedimentary troughs (Fig. 2-23) bounded by very high-angle normal faults (Fig. 2-24) dominate. Moreover, folds that are present seem as much related to movement on the high angle normal faults as to the lateral movement and hence tend to parallel the wrench zone (as in extensional deformation—Chapter 4) rather than be oblique to it (Harding, 1983, fig. 2; Harding et al., 1984).

The Scipio-Albion trend (Figs. 2-25, 2-26) has been interpreted as a divergent wrench feature with the axial traces of sags (Fig. 2-26b) defining a pattern of synthetic faults arising from divergent left-lateral movement. Divergence also helps to explain secondary dolomitization of dolomitic limestone by facilitating fluid percolation.

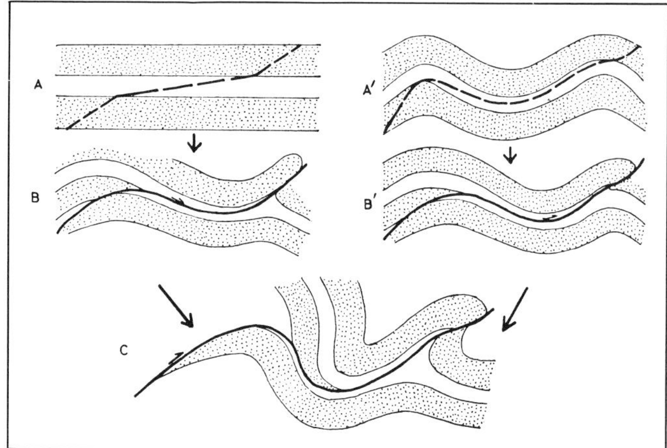

Fig. 2-21 (Challinor, 1967)—Evolution of folded fault surface from convergent wrenching in the Brøggerhalvøya area of Spitsbergen. Any dislocation along a potential surface of displacement in a previously folded sequence (A') immediately creates a folded fault surface (B'). Additional deformation is shown in C. Contrast with folded fault structures in thrust-fold belt settings (Figs. 6-13, 6-14). Permission to publish by Geological Magazine.

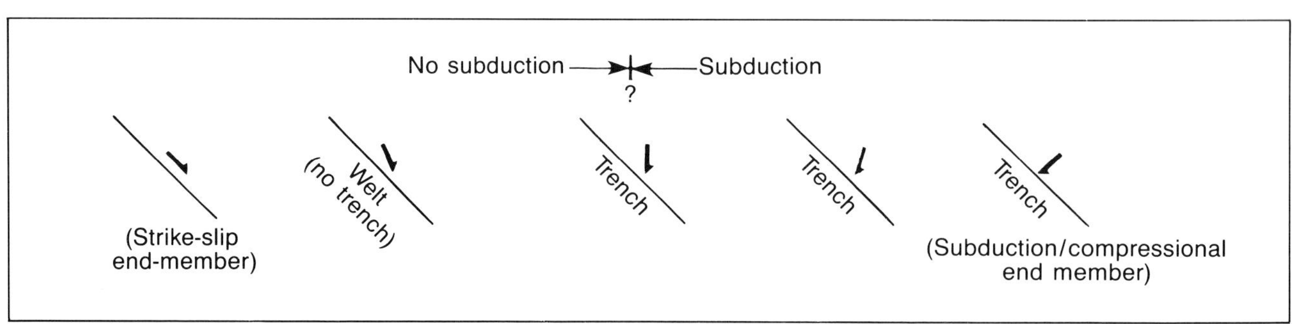

Fig. 2-22—Thrust-fold belt end members dependent on plate motion.

TABLE 2-1

Characteristics of convergent strike-slip and compressional orogenic belts

CONVERGENT STRIKE-SLIP OROGENIC BELT	COMPRESSIONAL/SUBDUCTION OROGENIC BELT
· En echelon folds · Great degree of basement involvement proximal to thrust belt · Mostly upthrusts · Narrow zone of deformed sedimentary cover lacking foreland detached structures · Crustal shortening minimal · No alpine ophiolites · No to little metamorphism · No batholithic intrusions · Volcanic belts not commonly associated · Infrequent island arc association · Deep-water sediments (turbidite, flysch) or trench (?) origin not common · Melange not known · Prior eugeosynclinal-miogeosynclinal association not common · Lesser sediment thicknesses deformed · No particular earlier deformation · Molasse sedimentation modest · Maximum length on order of 1,000 km · Generally straight	· Parallel-subparallel, wavelike folds; no systematic en echelon folds · Lesser degree of basement involvement proximal to thrust belt · Mostly sled-runner thrusts · Wide zone of deformed sedimentary cover including foreland detached or foothills structures · Crustal shortening can be extreme · Alpine ophiolites frequently associated · Metamorphism · Batholithic intrusions well known · Volcanic belts frequent · Island arcs frequently associated · Deep-water sediments common · Melange well known · Prior eugeosynclinal-miogeosynclinal association common · Great sediment thicknesses deformed · Earlier deformation may be normal faulting which effected pull-apart · Molasse sedimentation may be extensive · Maximum length on order of 10,000 km · Straight or curved.

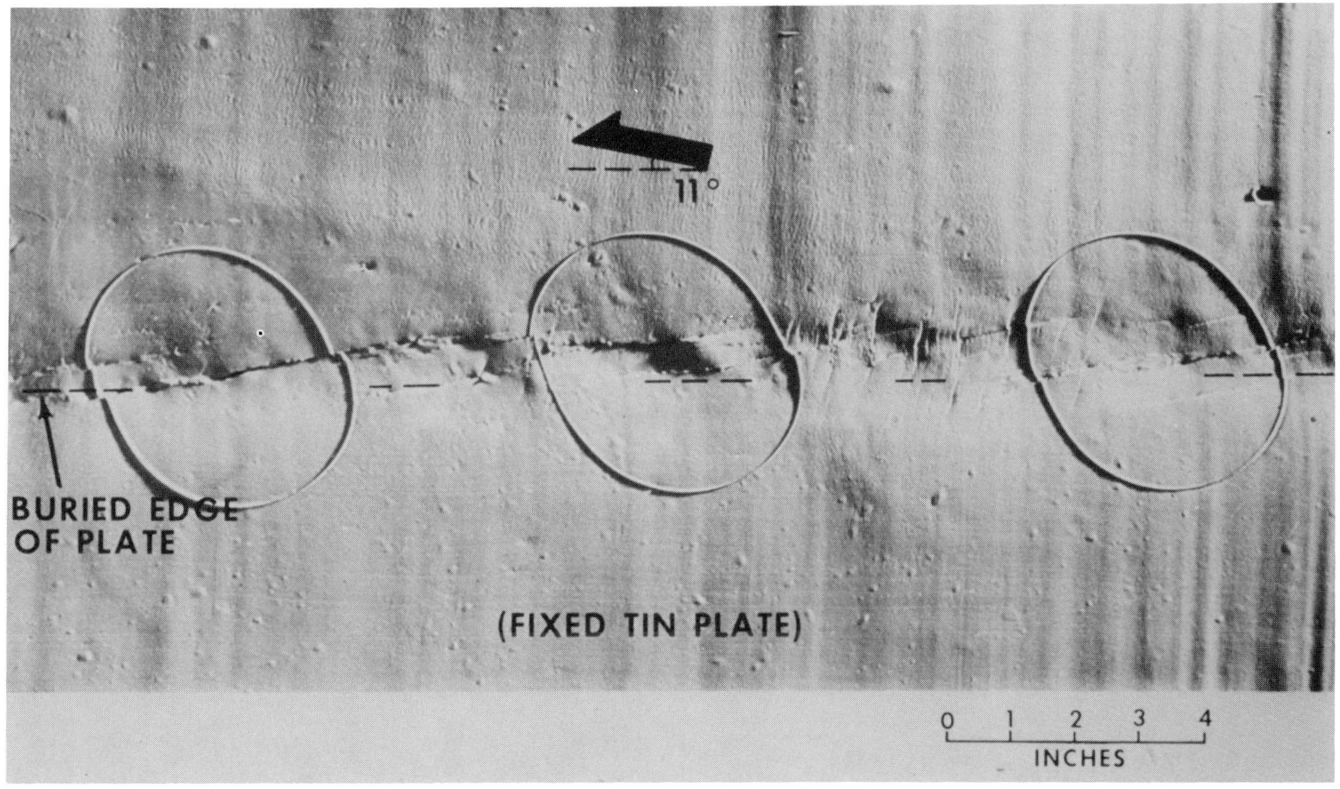

Fig. 2-23 (Wilcox et al., 1973)—Clay model of 11° divergent left-lateral wrench fault. Permission to publish by American Association of Petroleum Geologists.

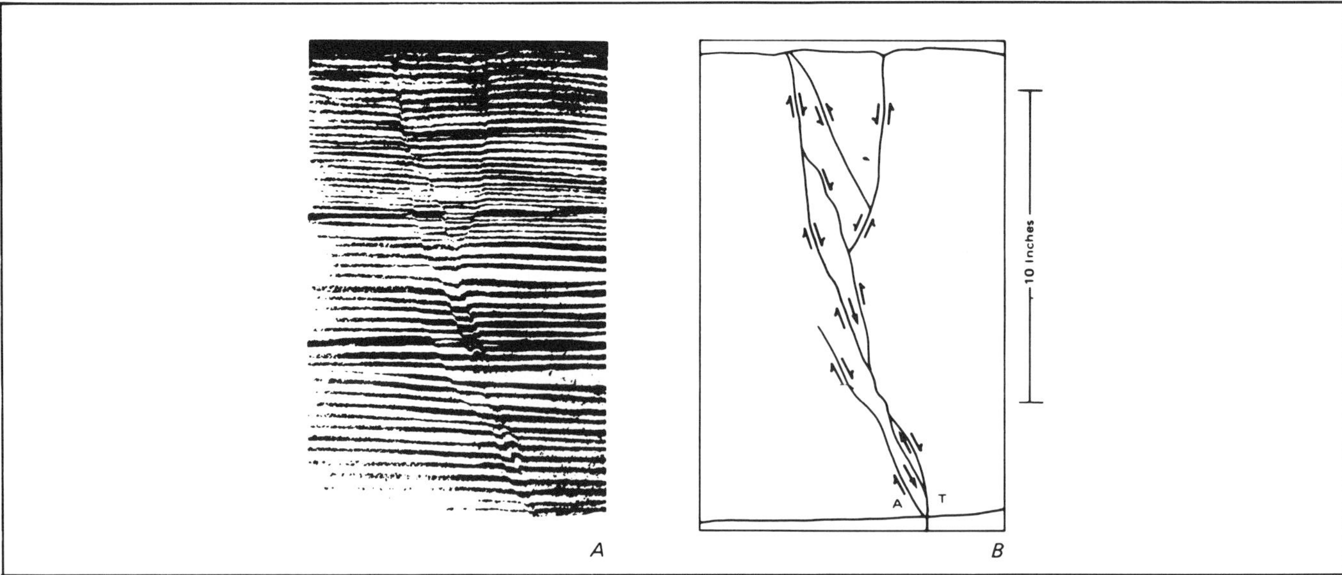

Fig. 2-24 (Emmons, 1969)—Cross-section of divergent wrench experiment showing that faults propagate as very high angle normal faults from a wrench displacement at depth. Most motion sense is normal, but a few reverse offsets also occur. Permission to publish by Elsevier Science Publishers B.V. Amsterdam.

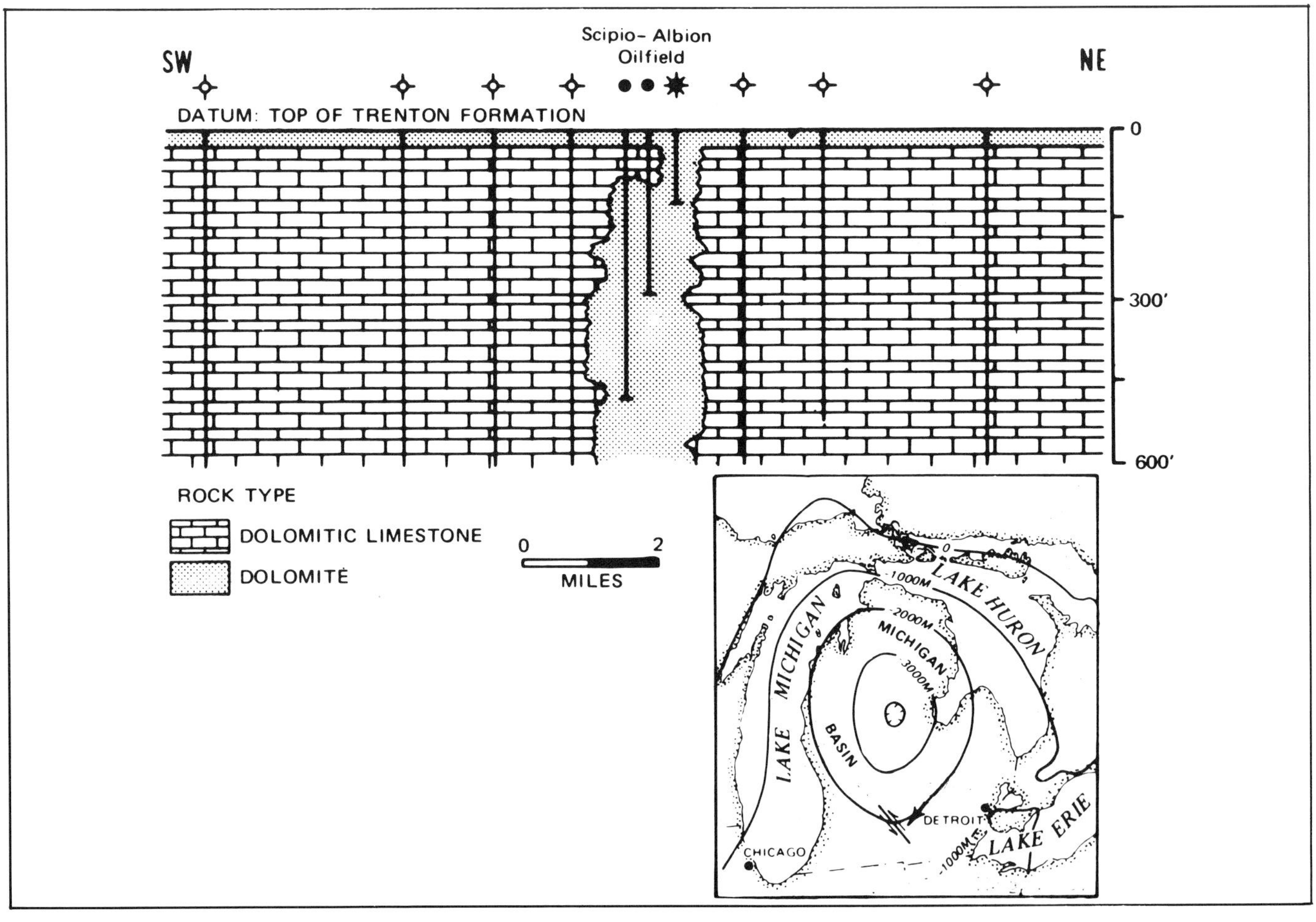

SW
Scipio- Albion
Oilfield
NE
DATUM: TOP OF TRENTON FORMATION
0
300′
600′
ROCK TYPE
DOLOMITIC LIMESTONE
DOLOMITE
0
2
MILES
LAKE HURON
1000M
2000M
MICHIGAN
3000M
LAKE MICHIGAN
BASIN
DETROIT
LAKE ERIE
1000M
CHICAGO

Fig. 2-25 (Harding, 1974a)—Index map and cross section of Scipio-Albion trend, Michigan. Permission to publish by American Association of Petroleum Geologists.

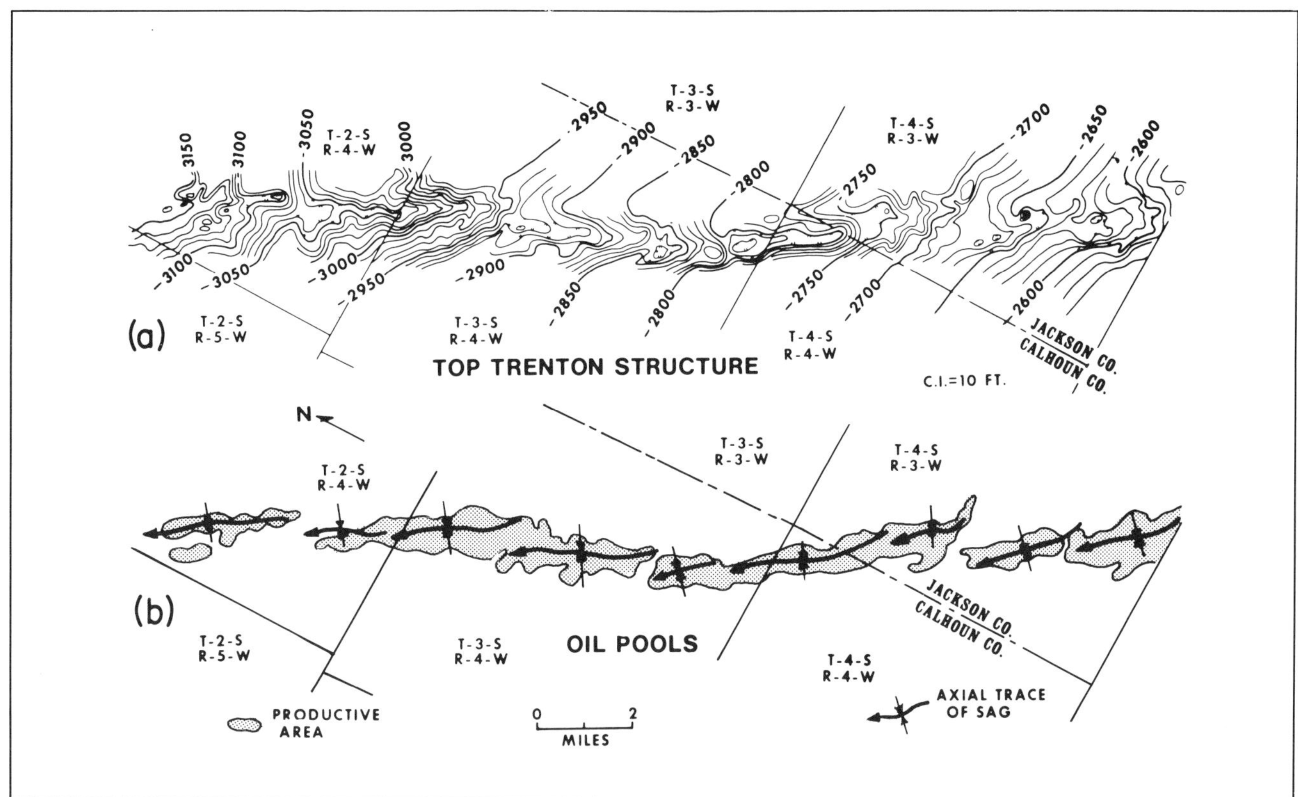

Fig. 2-26 (Harding, 1974a)—Structure contour map and oil pools of Scipio-Albion trend, southern Michigan Basin. Permission to publish by American Association of Petroleum Geologists.

SEISMIC EXPRESSION

Seismic record quality often deteriorates in the vicinity of wrench faults. Part of the reason for this deterioration is the mismatch of sedimentary units and attendant difficulty in correlation across the fault zone. Moreover, deformation within the zone can be complex with associated cataclasis, steep dips, and discontinuities, all of which lead to complex ray-path geometries and make recognition of wrenching difficult. Figure 2-27 shows a fault zone suspected of wrench motion on the basis of mismatched seismic reflections and velocities, but data in the zone are poor.

An upthrust-bounded welt or flower structure created by convergent wrenching is interpreted from seismic in southern Oklahoma, and some of its characteristics are discussed in Chapter 1 (Fig. 1-6). Similar structures have also been interpreted from seismic in western offshore Sabah by Bol and Van Hoorn (1978; Figs. 2-28 through 2-31). In many examples of this substyle the bedding attitudes and normal and reverse faults in the shallower part of the welt can be readily determined, but the deep structure is highly interpretive owing to loss of record quality. The manner in which the shallower faults root into a vertical system is usually conjectural.

Space problems at depth are also created by convergent wrenching. The more the convergence and the more the movement involved, the more intense and wider must be the deep zones of cataclasis. Possibly a slight overlap of plates may take place (incipient subduction). It is also probable that the process involves more detachment than heretofore suspected. In this regard Lehner et al. (1983) have suggested that the deformation of an accretionary wedge of folded and imbricated sediments (appearing as a forearc thrust-fold belt) in the Caribbean offshore of Colombia (Fig. 2-32) is related to a far-removed convergent wrench system associated with the Andes and the Caribbean-South American plate boundary. Furthermore, wrenching associated with the San Andreas system in California may be expressed as low-angle thrusts detached in the upper crust along the central California margin (Crouch et al., 1984).

For the ridges bounded by upthrusts in the Sabah offshore, Bol and Van Hoorn (1978) made the following comments:

> The formation of these ridges has been originally ascribed to flow and diapirism of undercompacted clay as a result of excessive loading by the overlying sediments. Various compelling arguments counter this hypothesis:
>
> *1– The absence of slow continuous growth which would be expected by the viscous flow of clay;
>
> *2– The absence of clay withdrawal features in the synclines;

*These two points do not necessarily support the re-interpretation for they are not required of clay flowage. See Chapter 9—Author.

3– The synchronous origin of these ridges;

4– The presence of continuous trends associated with positive gravity anomalies, suggesting basement involvement;

5– The tightly folded, indurated state of the lower Miocene clays in some of the ridge cores (i.e. Labuan, Klias) suggesting absence of ductile flow;

6– The apparent absence of a major detachment horizon underneath the ridges.

Modern seismic data disprove conclusively the clay diapiric origin of the ridges and suggest that the cause of the structures is compressional stress in the overburden. Because of deep erosion the upper, more concentric, part of these folds has been removed and today only the steeply compressed core of the structures remains. Seismic interpretation shows that the crestal part of the ridges is characterised by steeply dipping upthrusts. The rapid changes in orientation and dip direction of the fault planes and evidence of reversals in throw along faults, suggest basement-induced faulting. The precise nature of this faulting is unknown but is thought to be related to strike-slip movements. Since the resulting principal compressive stress may vary from horizontal to vertical, a wide range of unpredictable structural styles can therefore be anticipated in the sedimentary cover.

A divergent wrench fault has been interpreted by Harding (1983) on seismic lines (Figs. 2-33, 2-34) from the Andaman Sea, an area of known transform and associated strike-slip faulting. Criteria discussed in the previous section, such as few en echelon folds (but some folds paralleling the wrench zone), sags, and numerous normal faults, suggest the divergent wrench style.

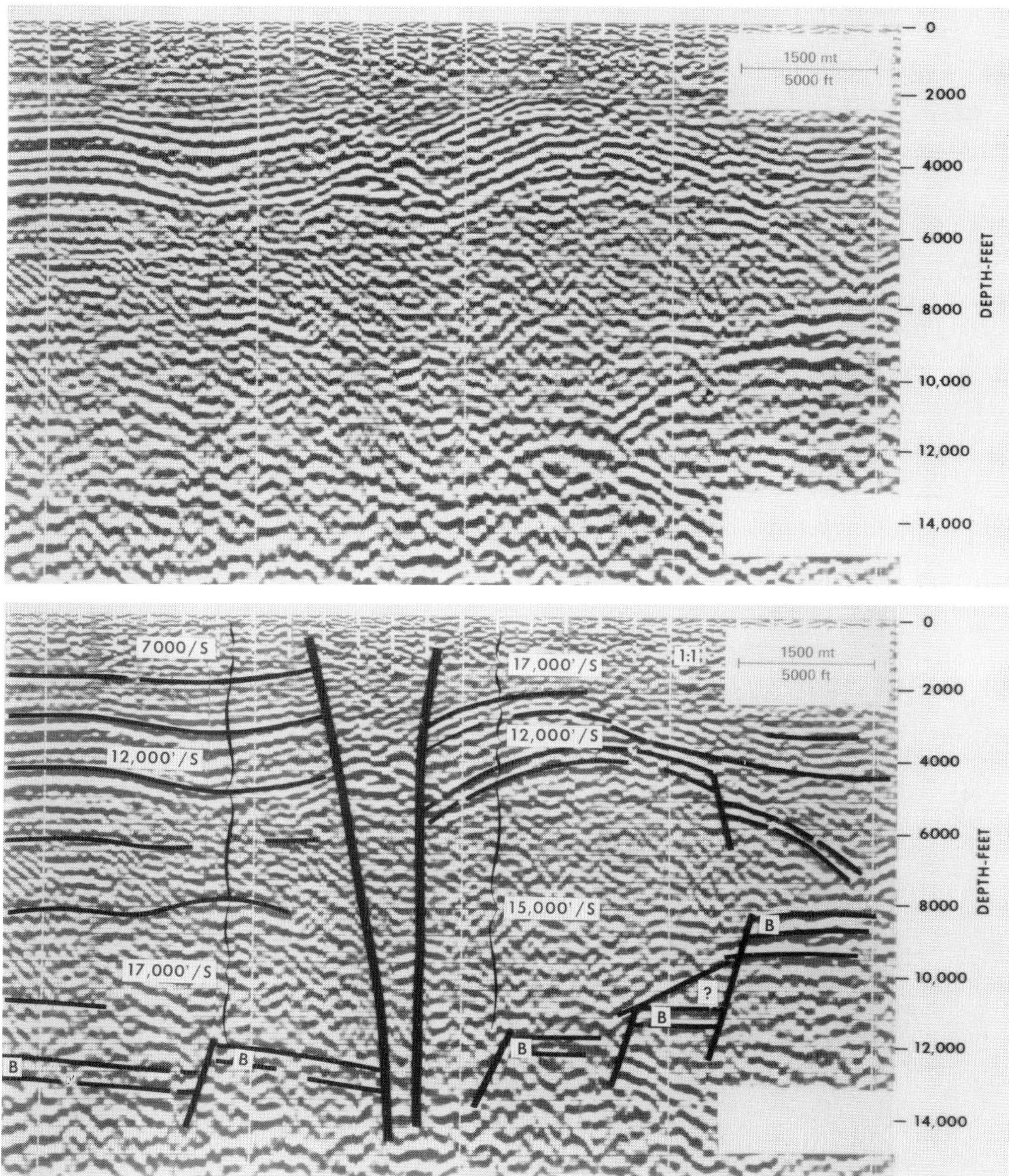

Fig. 2-27 (Lowell et al., 1975)—Reflection seismic section from former Esso-Mobil concession area in Ethiopian waters of the southern Red Sea showing fault interpreted as having had strike slip on the basis of mismatched velocities and seismic reflections. (B-basement). From Petroleum and Global Tectonics, copyright © Princeton University Press. Used with permission.

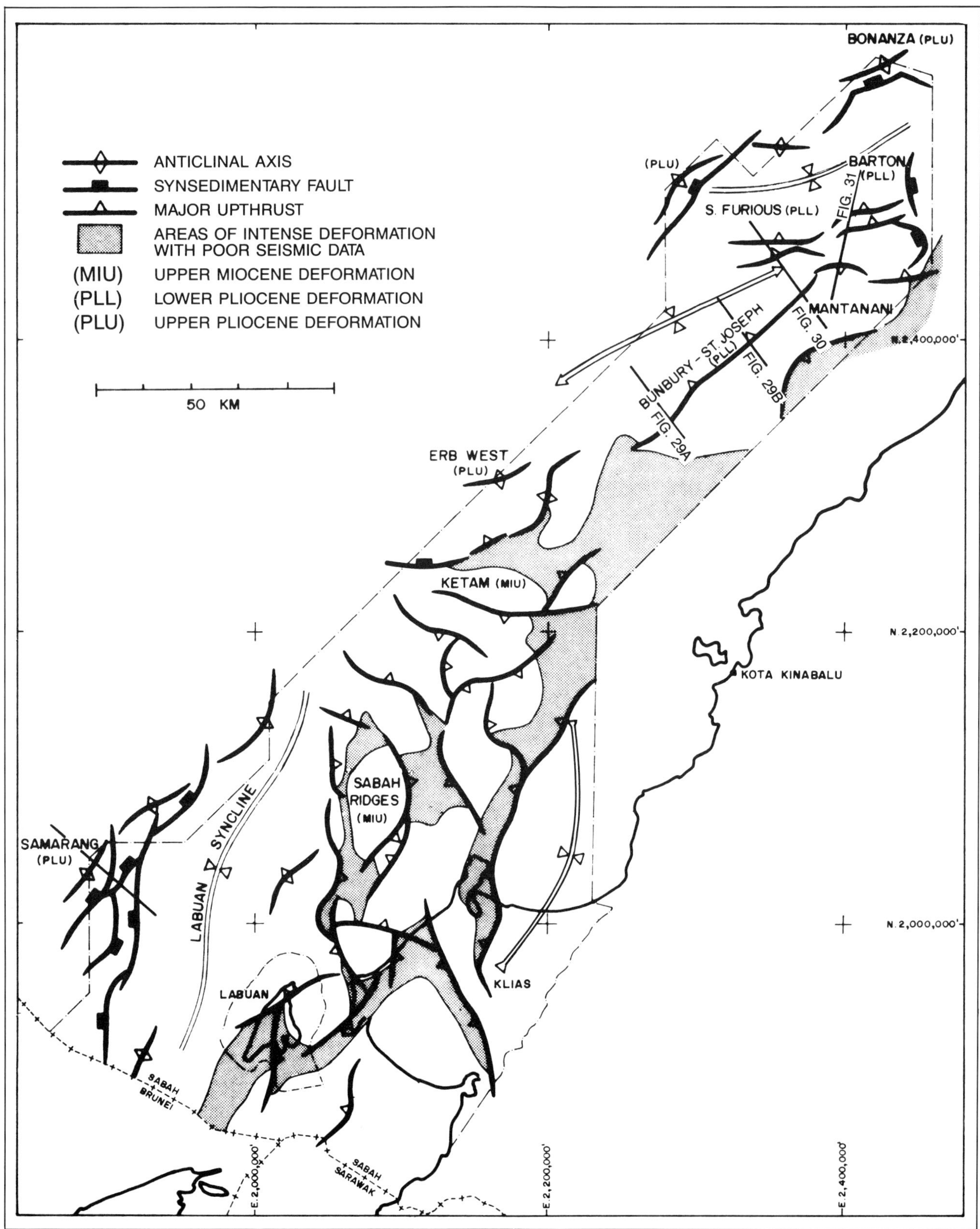

Fig. 2-28 (Bol and Van Hoorn, 1978)—Index map of offshore Sabah. Permission to publish by Geological Society of Malaysia.

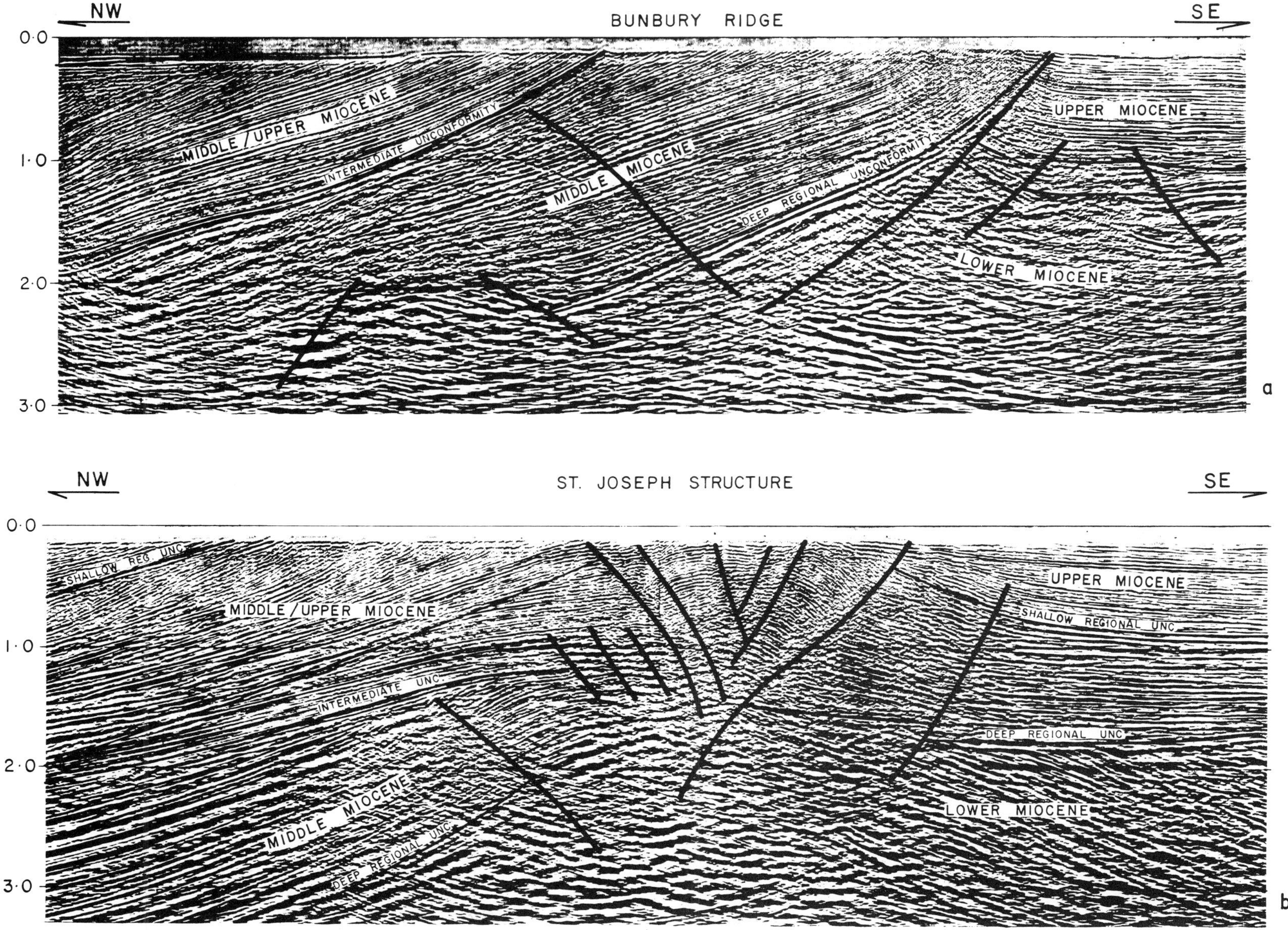

Fig. 2-29 (Bol and Van Hoorn, 1978)—On Bunbury-St. Joseph trend northwest-dipping faults appear to be aligned en echelon at shallow depths, but converge and steepen at depth. Section migrated. See figure 2-28 for lines of section. Permission to publish by Geological Society of Malaysia.

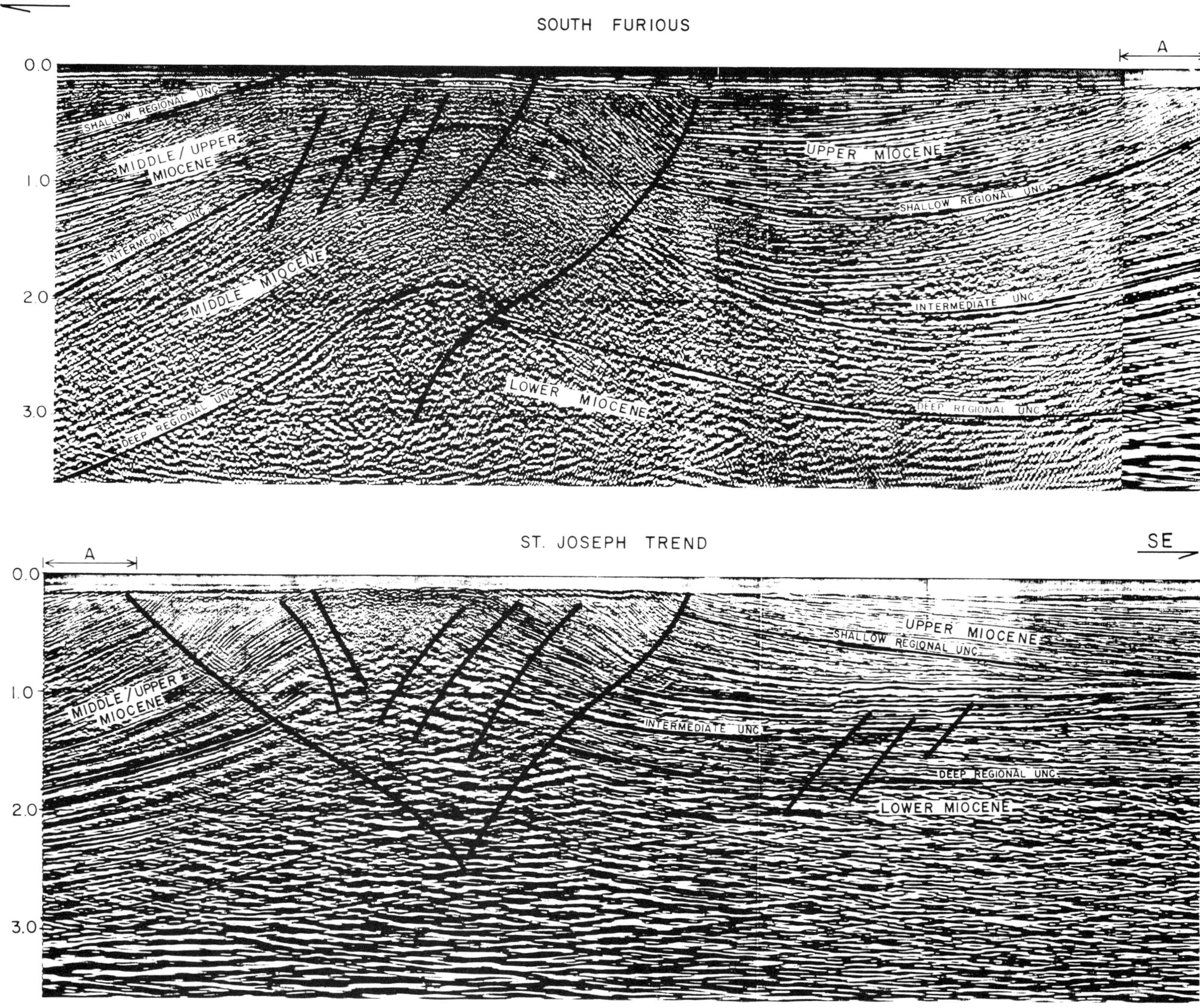

Fig. 2-30 (Bol and Van Hoorn, 1978)—Seismic line (migrated) through South Furious-St. Joseph trend. See figure 2-28 for line of section. Permission to publish by Geological Society of Malaysia.

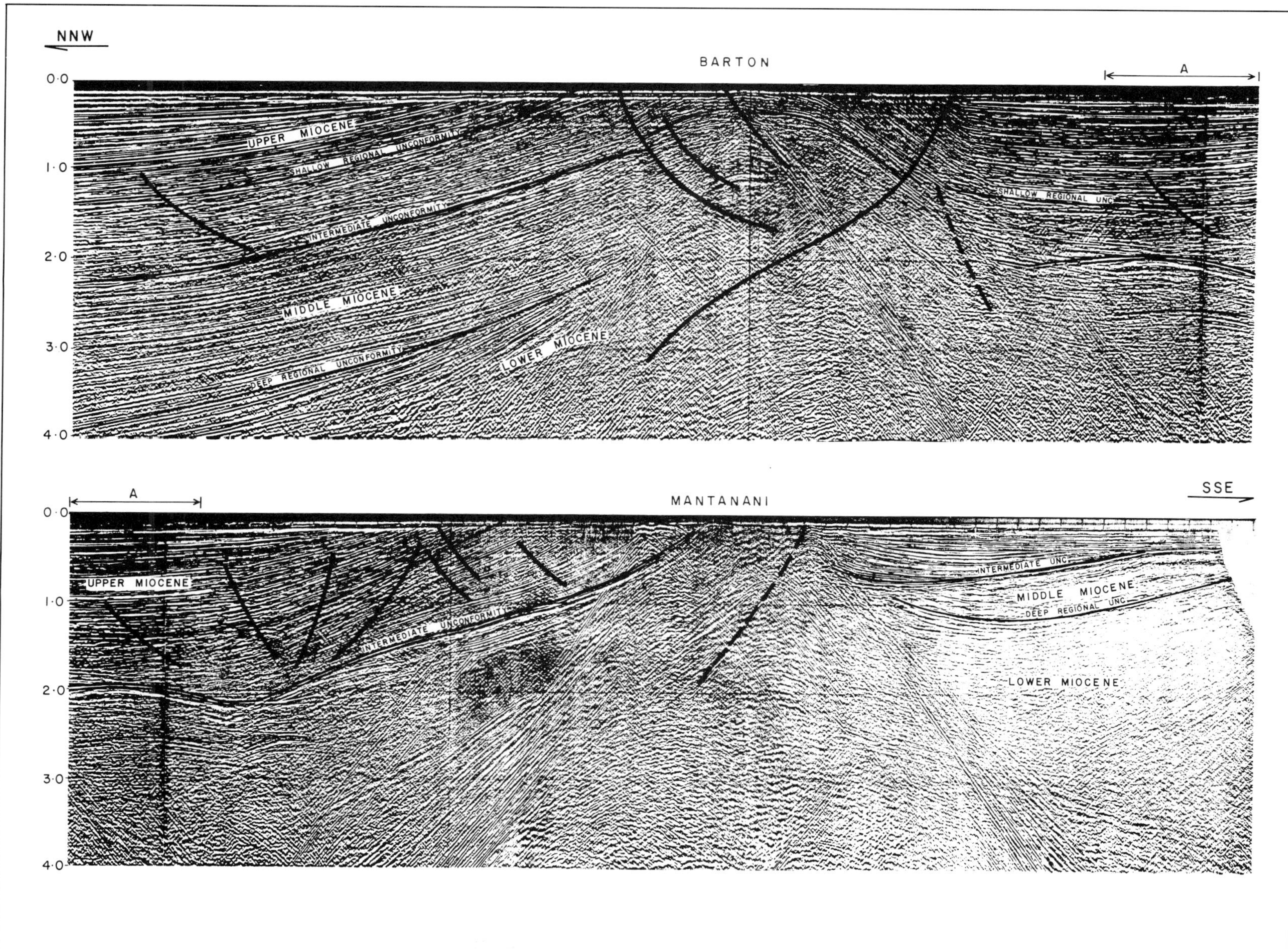

Fig. 2-31 (Bol and Van Hoorn, 1978)—Seismic line through Barton trend showing half-flower structure. Section migrated. See figure 2-28 for line of section. Permission to publish by Geological Society of Malaysia.

NNW

SSE

1225 1300 1400 1500 1539

0 SEC. 1 2 3 4 5 6 7

0 1 2 3KM

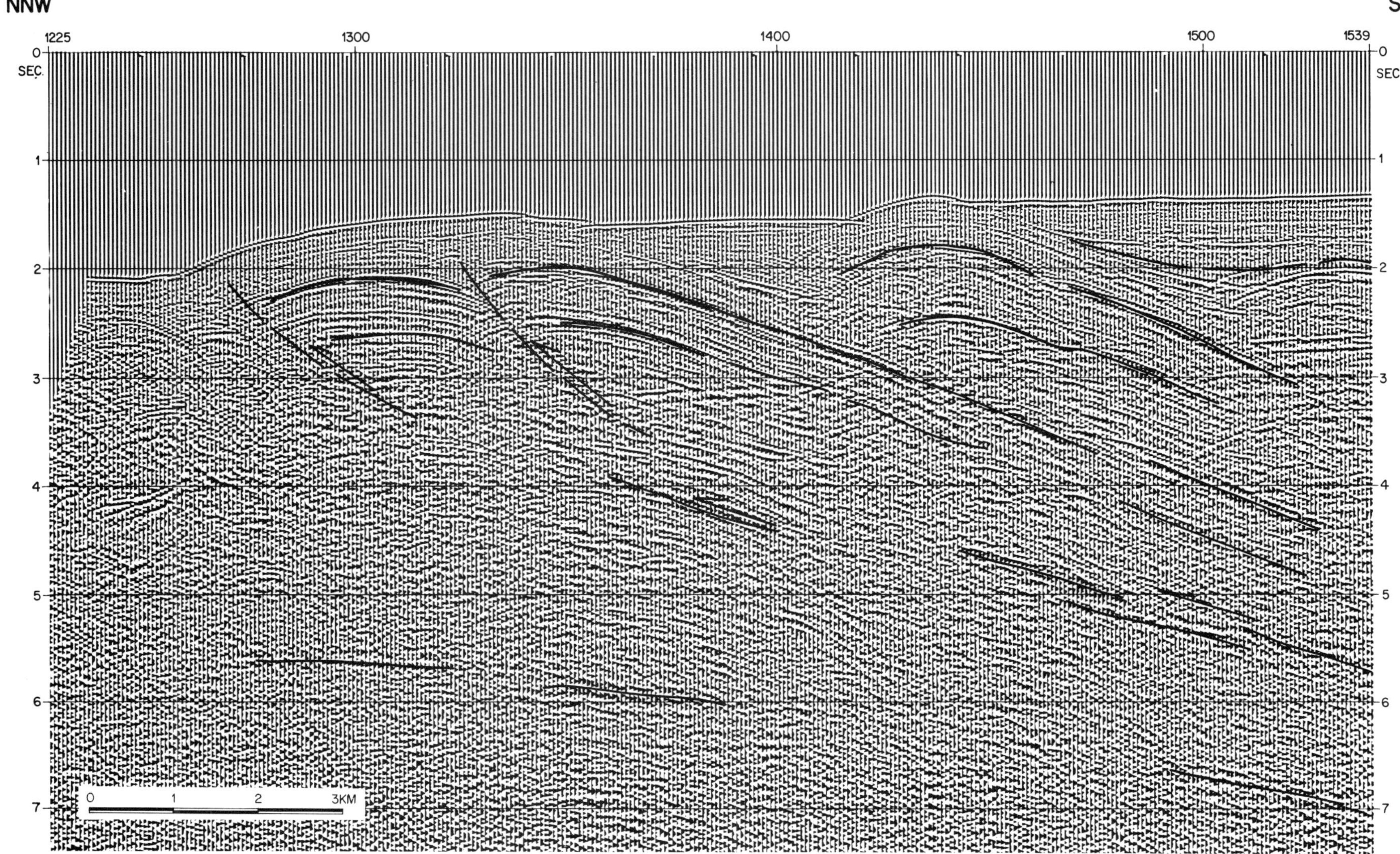

Fig. 2-32 (Lehner et al., 1983)—Seismic line from the Carribean offshore of Colombia showing an apparent forearc detached thrust-fold belt that may be related to a larger wrench system. Permission to publish by American Association of Petroleum Geologists.

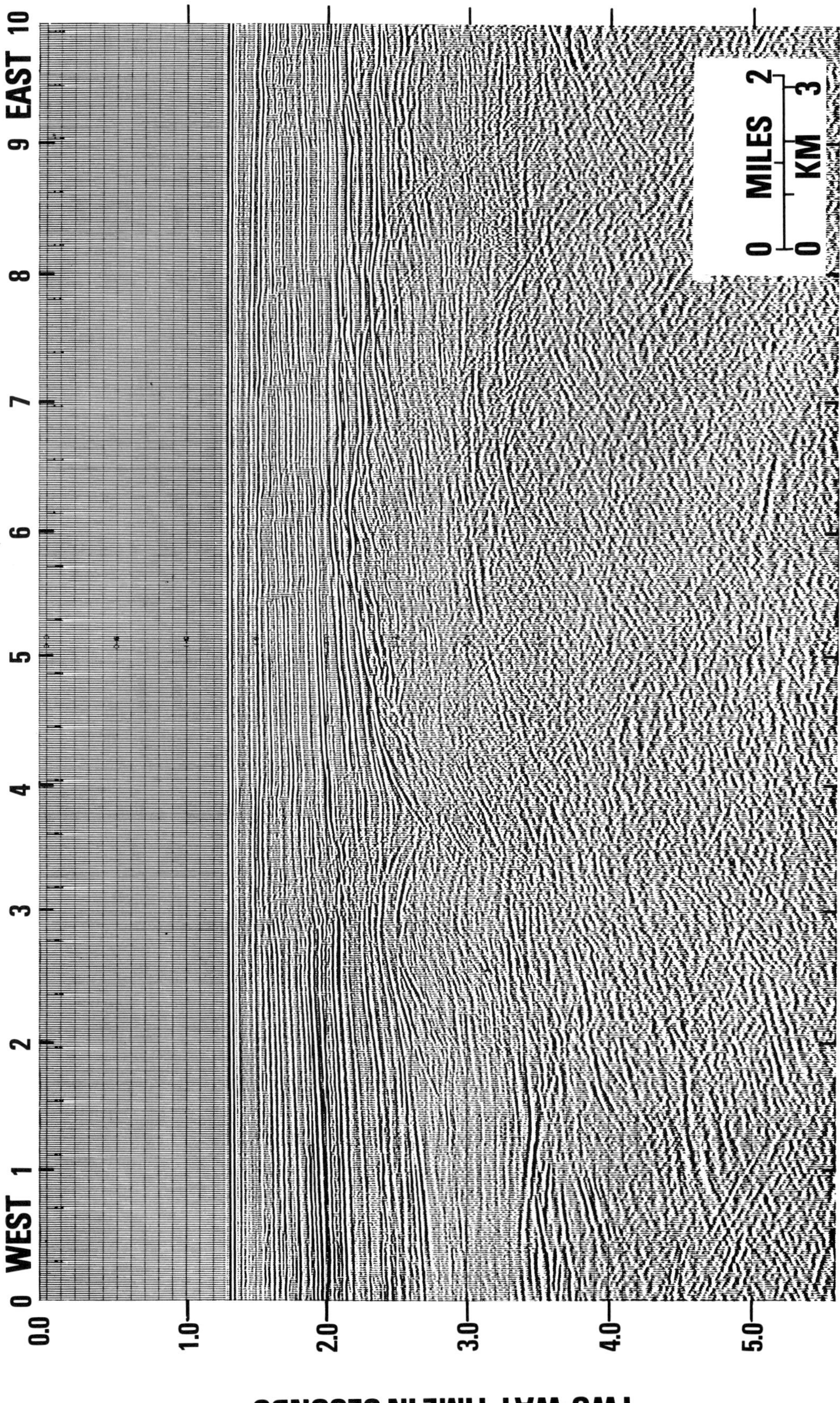
WEST
EAST
0 1 2 3 4 5 6 7 8 9 10
0.0
1.0
2.0
3.0
4.0
5.0
TWO-WAY TIME IN SECONDS
0 MILES 2
0 KM 3

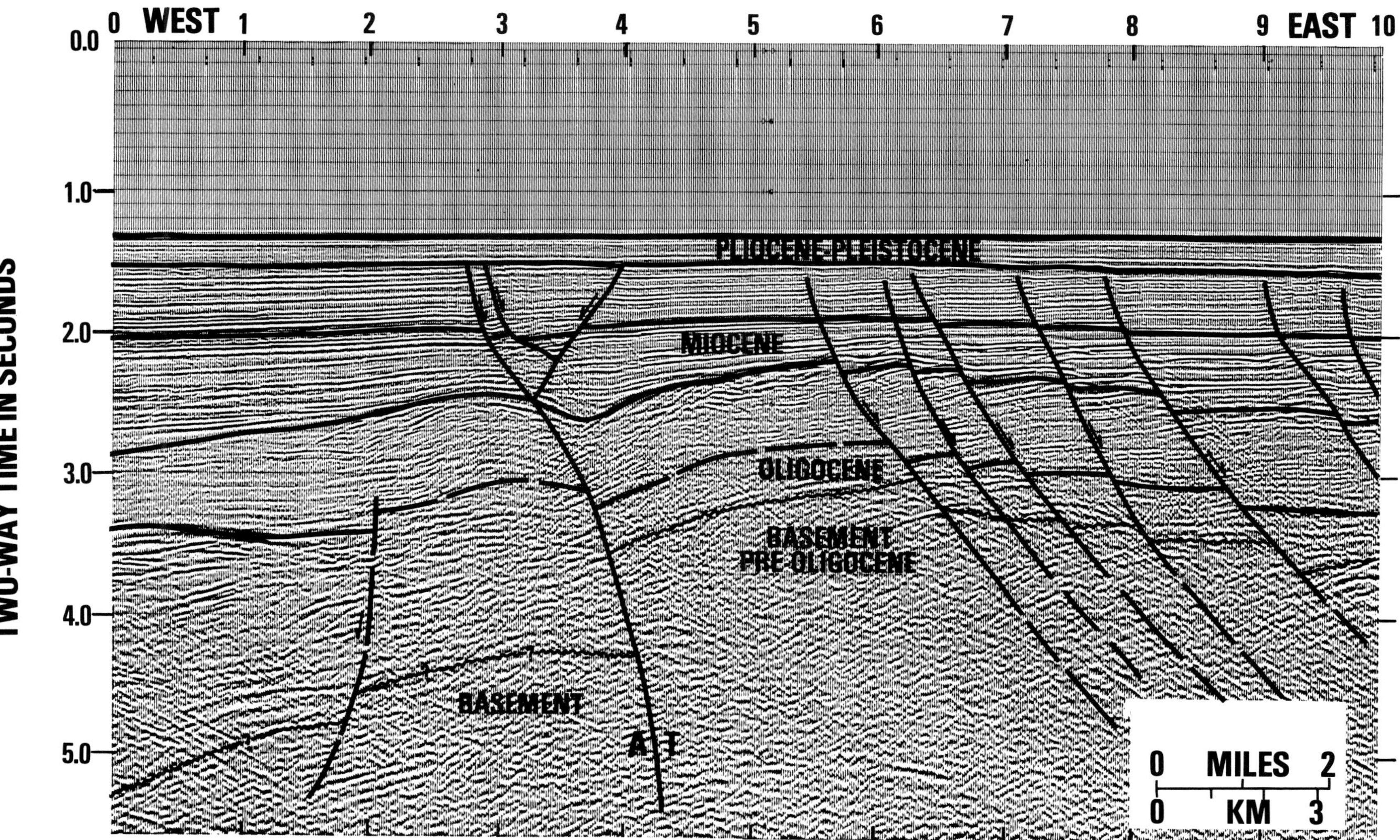

Fig. 2-33 (Harding, 1983)—Seismic line from Andaman Sea showing a divergent wrench fault. In contrast to figures 1-6 and 2-29 through 2-31, cycles (beds) are flat or dip toward the fault to define a sag, rather than dipping away to define an uplift or welt. Wrench fault is inclined into high-angle normal fault orientation and shows a change in vertical movement sense. En echelon normal faults present on east side of section. Permission to publish by American Association of Petroleum Geologists.

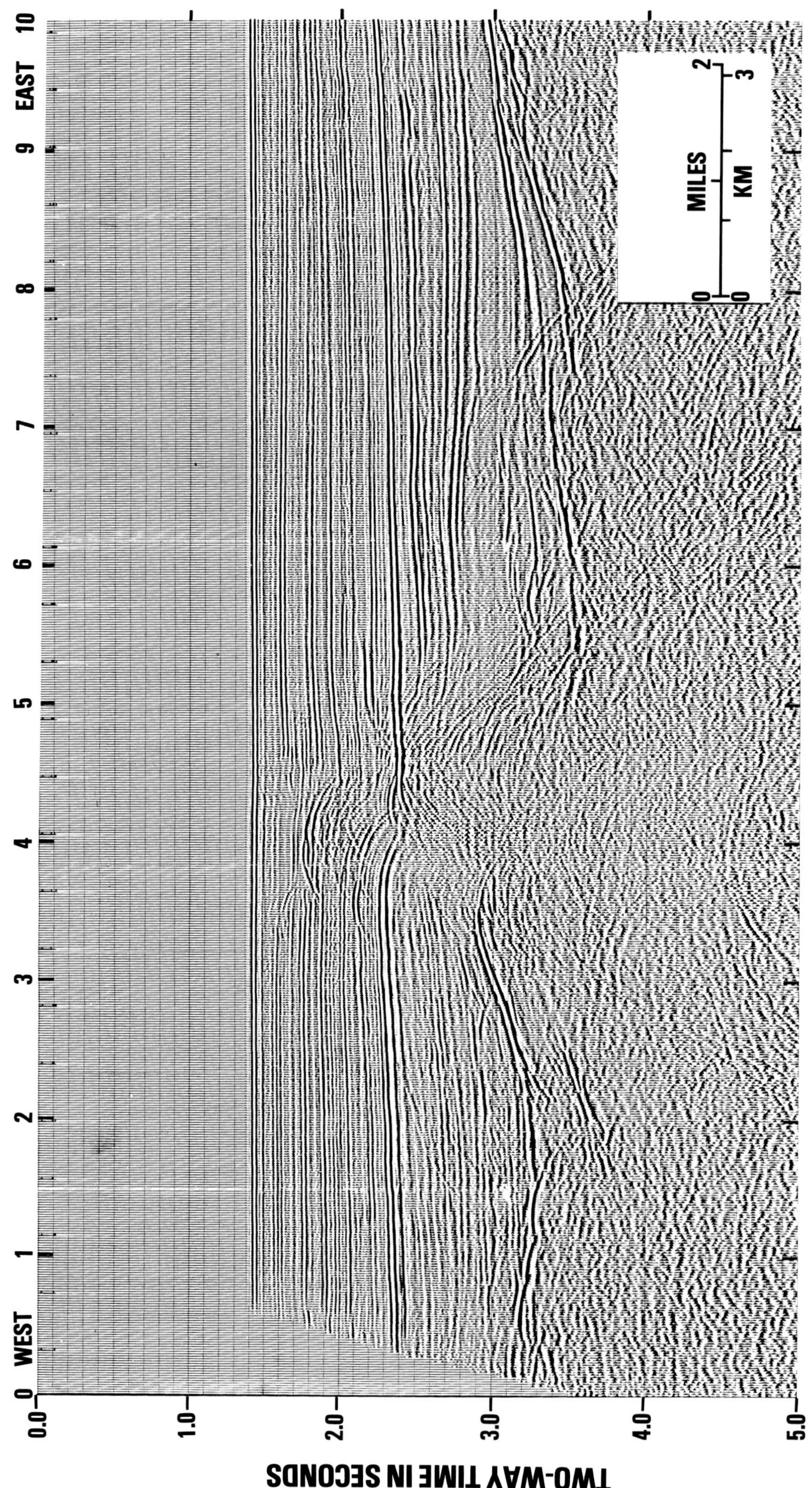

WEST
EAST
0
1
2
3
4
5
6
7
8
9
10
0.0
1.0
2.0
3.0
4.0
5.0
TWO-WAY TIME IN SECONDS
MILES
KM
0
2
0
3

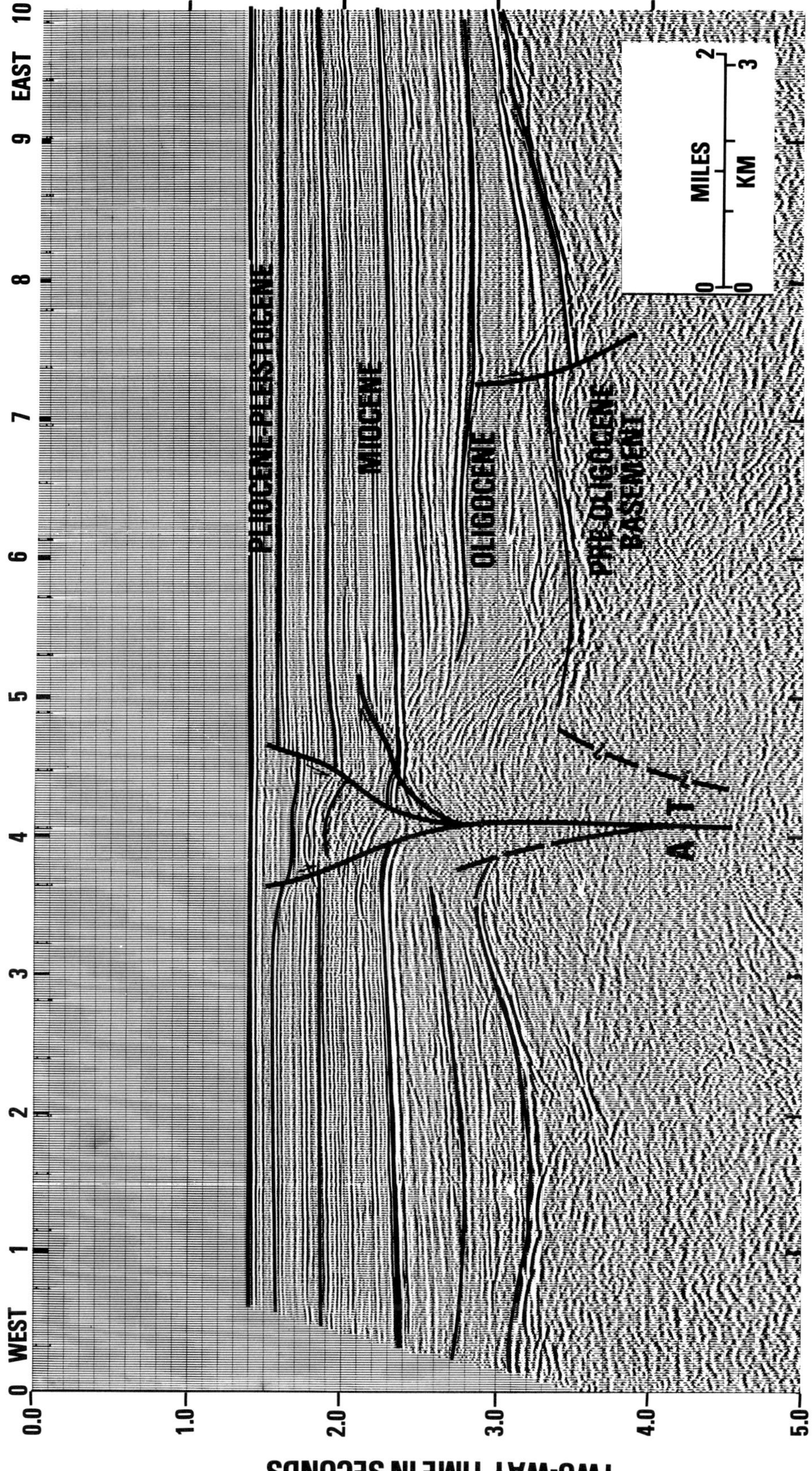

Fig. 2-34 (Harding, 1983)—Seismic line from Andaman Sea showing a divergent wrench fault with a "negative flower structure." A shallow sag occupies the area between bifurcated strands of the wrench fault. Note reverse separation below 2 sec just to right of reference point 4 and compare with figure 2-24. Permission to publish by American Association of Petroleum Geologists.

PRODUCING EXAMPLES

Petroleum plays related to the evolutionary stage of, and amount of displacement on a wrench fault, have been discussed in Chapter 1 (Fig. 1-5). In brief, early stage features can have their prospective en echelon anticlines little deformed and intact, whereas uplift and erosion along the more mature, larger displacement wrench zones can disrupt both fold and reservoir continuity. In the latter case less deformed folds and potential source, reservoir, and seal can be present basinward. Convergence and divergence in the system can also influence traps.

The Inglewood-Newport trend (Figs. 2-35 through 2-43) in the Los Angeles Basin, California, is considered a small displacement (<1.6 km; 1 mi) wrench system, probably with mostly parallel movement. At the Inglewood field (Fig. 2-36) the fold axis is at an acute angle to the main strike-slip fault (double line). This relationship seems to be unique to the wrench assemblage; the fold and the associated major fault are parallel in all other styles. The Inglewood anticline is roughly symmetrical, in contrast to most other styles in which folds (or drape or roll-over features) are markedly asymmetrical. The Inglewood-Newport trend represents a very young oil system with source rocks of late Miocene age, and maturation, migration, seal, and structure all of Pliocene and younger age.

Across the Los Angeles Basin from the Inglewood-Newport trend, the Whittier fault (Figs. 2-44 through 2-49) with approximately 4.8 km (3 mi) of convergent right-lateral slip is treated as an intermediate displacement wrench. Owing to the convergent component, the Whittier fault invariably appears as a high-angle reverse fault.

The San Andreas fault along the southwest side of the San Joaquin Valley is a large displacement wrench trend with some 298 km (185 mi) of right-lateral slip (Figs. 2-50 through 2-54). Harding (1976) found that the growth history of structures on the southwest side of the San Joaquin Basin showed southwest to northeast progression of deformation outward from the San Andreas fault through time. However, movement continued on the San Andreas as growth occurred on the northeastwardmost structures presenting the problem of transmitting stress from the San Andreas through an intervening dead zone to the area to the northeast. Possibly other deep-seated faults were (are) present and active beneath the northeast area.

Finally, all well developed en echelon folds are not always prospective. Folds in central Panama (Fig. 2-55) are formed on uplifted oceanic crust which is devoid of conventional reservoir, source, and seal rocks.

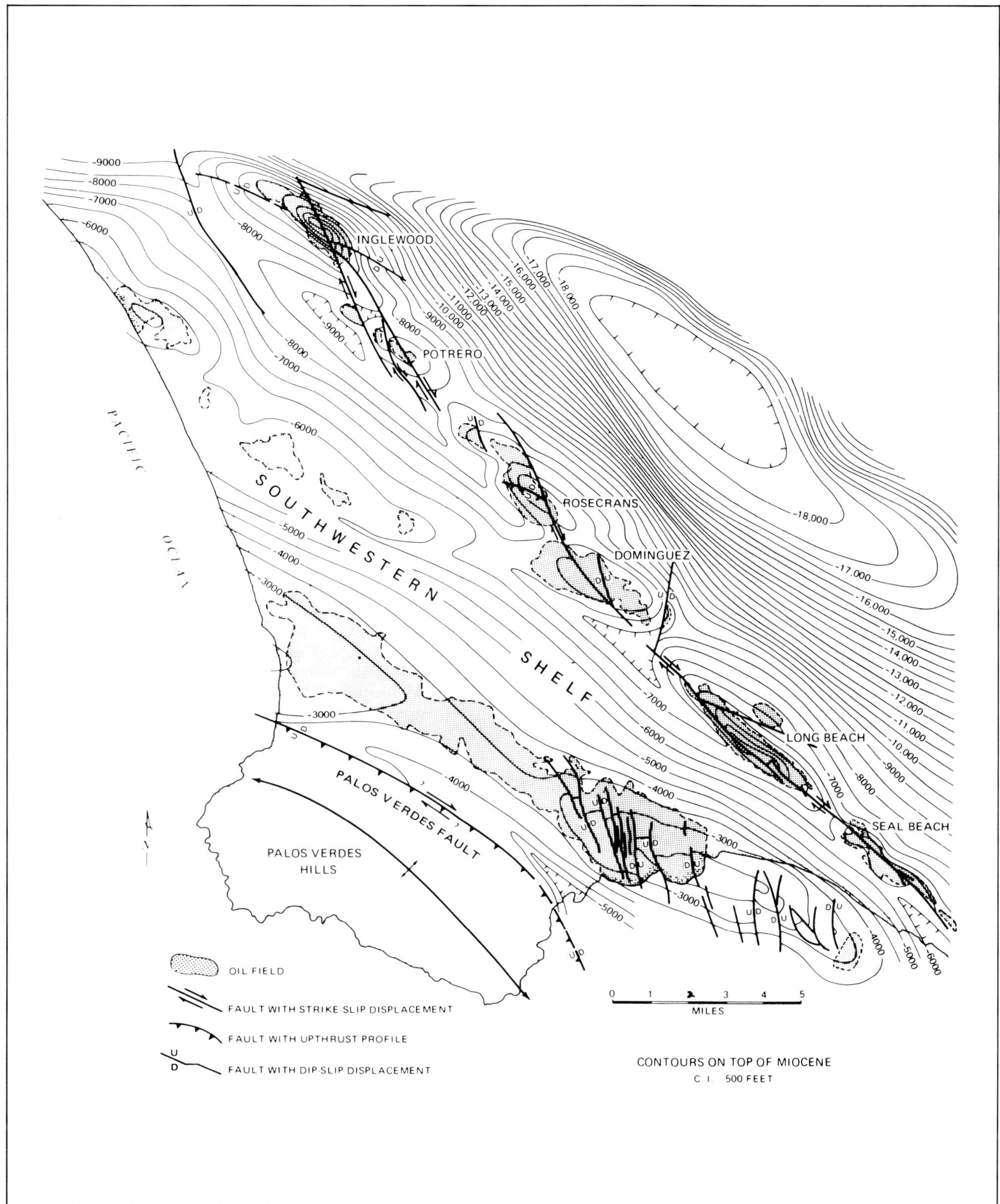

Fig. 2-35 (Harding, 1973)—Structure of northwest and central part of Inglewood-Newport trend (with oil fields named) and southwestern shelf of Los Angeles Basin. Permission to publish by American Association of Petroleum Geologists.

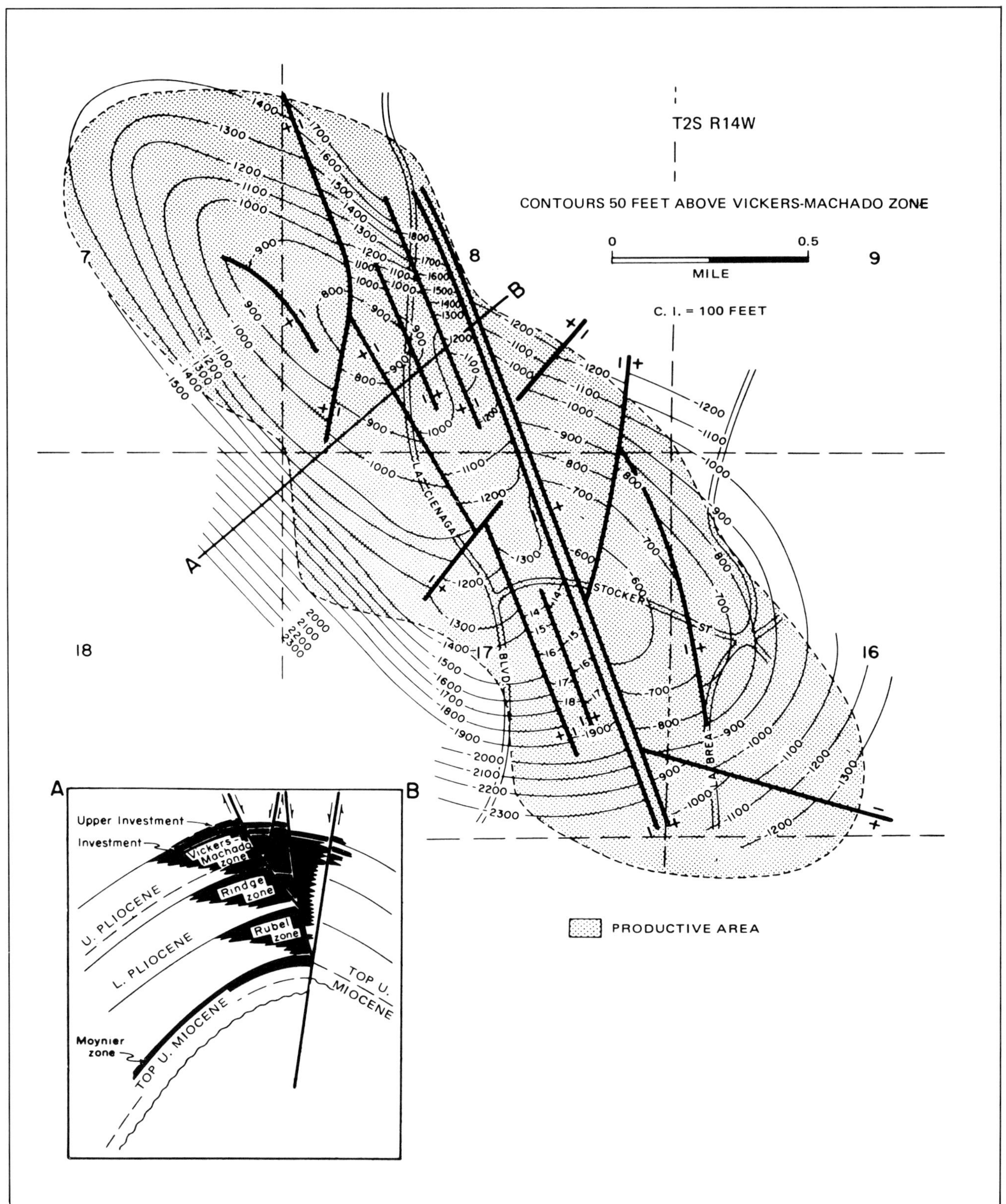

Fig. 2-36 (Harding, 1973)—Inglewood oil-field structure (after California Div. Oil & Gas, 1961). Strike-slip faulting is evidenced by approximate 1,500 ft (457 m) offset of fold axis by regional faults and by common occurrence of horizontal slickensides in well cores. Permission to publish by American Association of Petroleum Geologists.

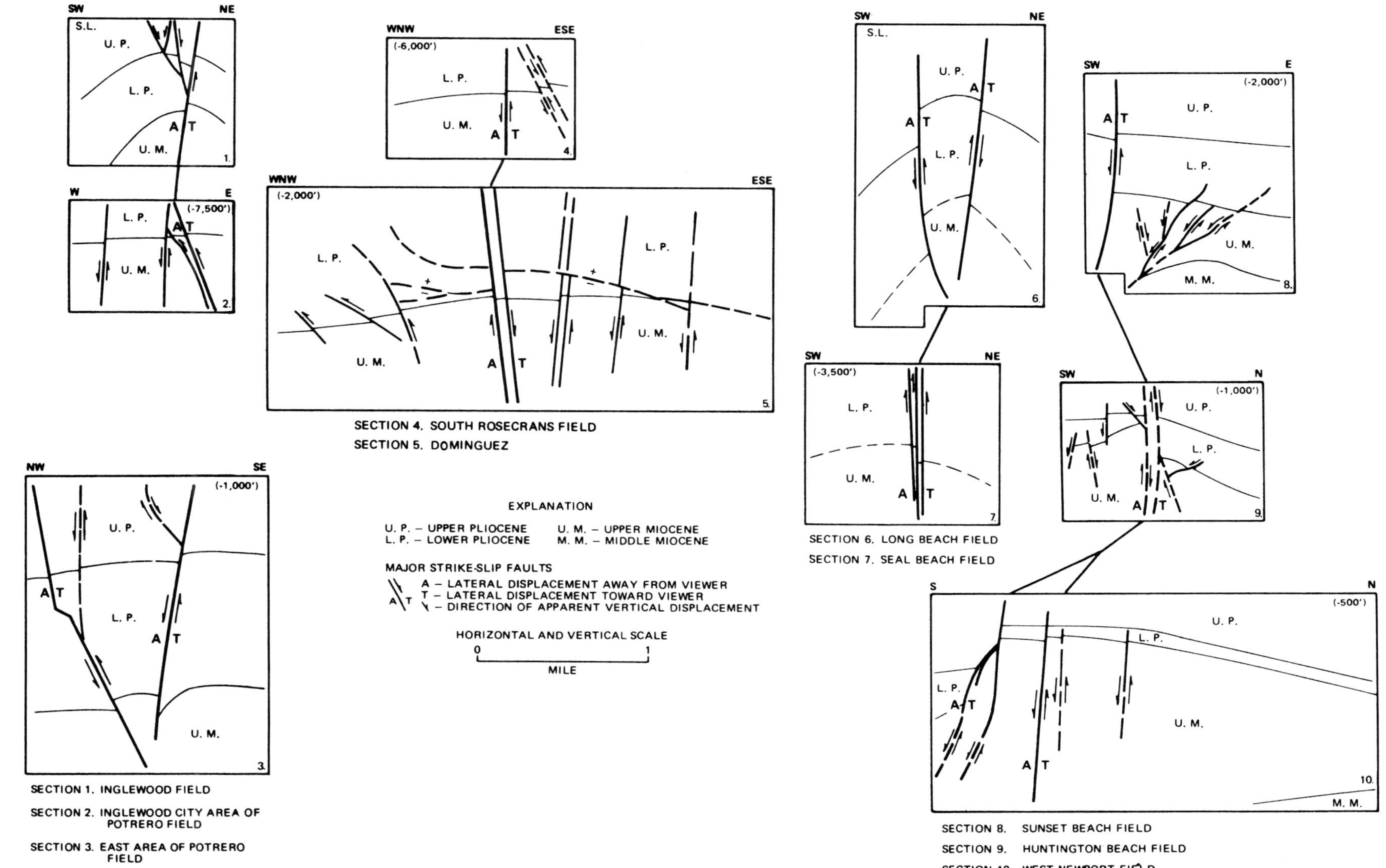

Fig. 2-37 (Harding, 1973)—Cross sections in geographic sequence from northwest to southeast across major oil fields and pools of Inglewood-Newport trend showing variable dip and vertical displacement. Permission to publish by American Association of Petroleum Geologists.

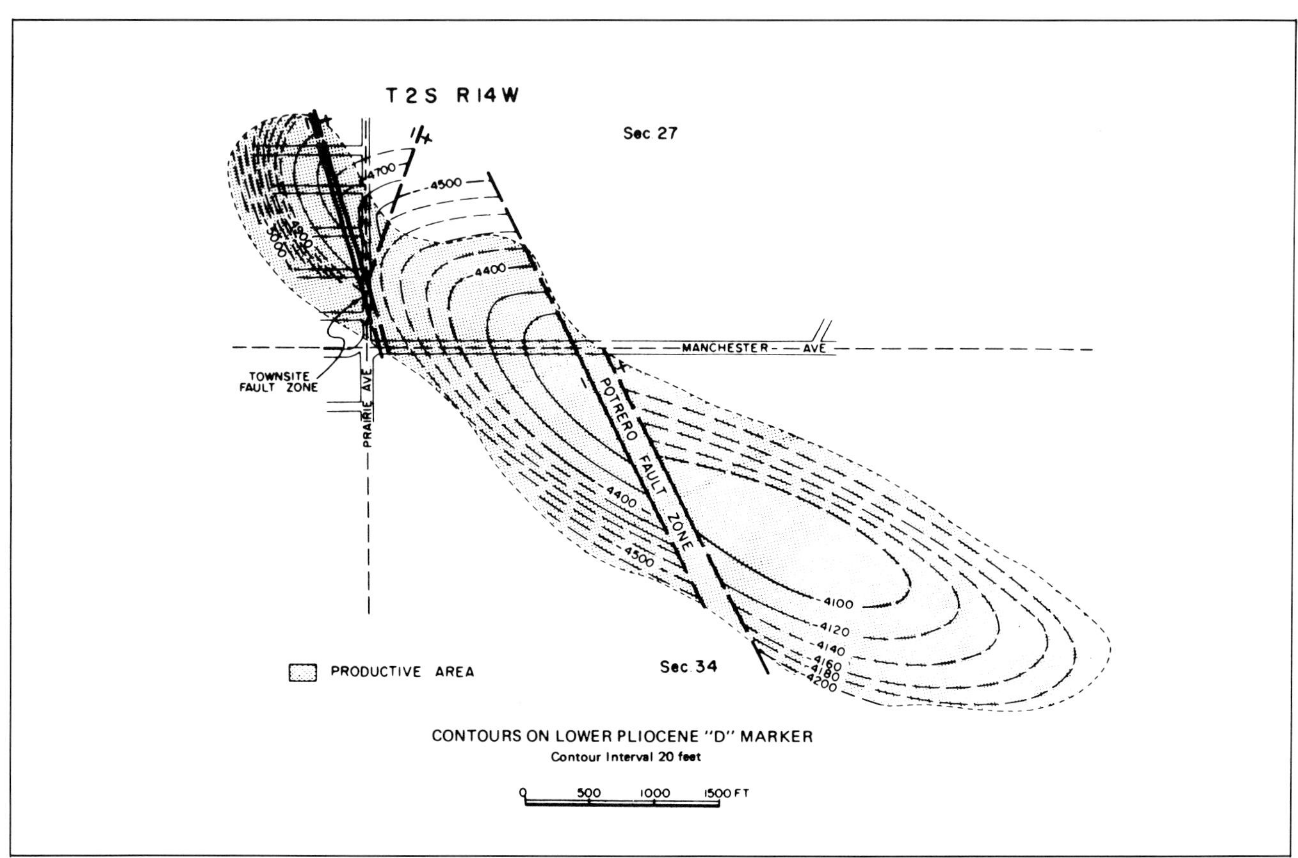

Fig. 2-38 (Harding, 1973)—Potrero oil-field structure (after California Div. Oil & Gas, 1961). En echelon strike-slip faults have offset fold axis 366 m (1,200 ft) laterally on Potrero fault and approximately 183 m (600 ft) on Townsite fault complex. Note acute-angle orientation of fold to fault. Permission to publish by American Association of Petroleum Geologists.

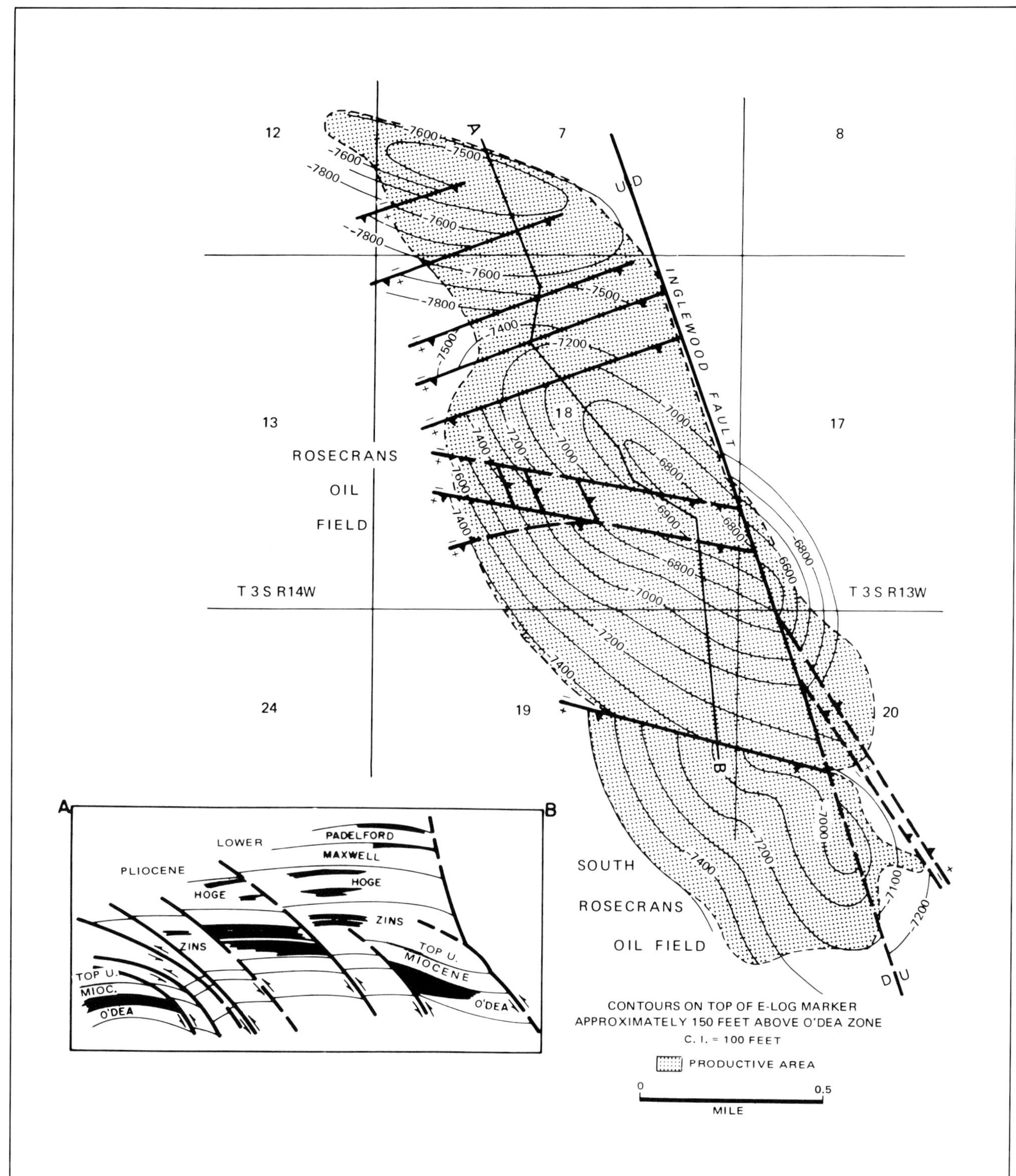

Fig. 2-39 (Harding, 1973)—Rosecrans field structure (after California Div. Oil & Gas, 1961) displays distinct pattern of en echelon fold culminations oriented obliquely and in right-handed sense to Inglewood fault. Small right-lateral displacement is evidenced by approximate 183 m (600 ft) offset of axis of middle anticline. Orientation of reverse faults across north end of structure related to compression directed NW-SE in gap of Inglewood fault trend. Permission to publish by American Association of Petroleum Geologists.

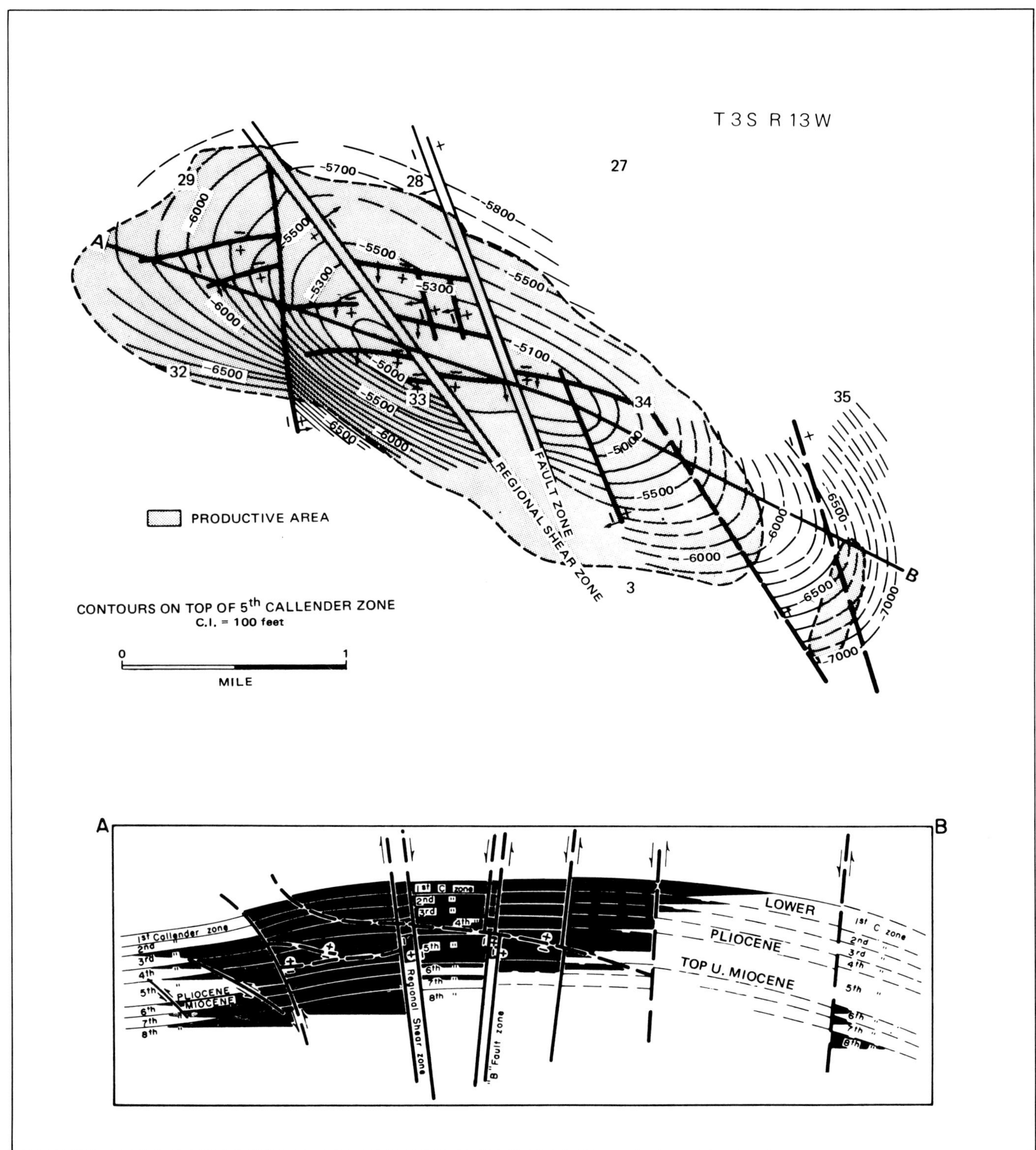

Fig. 2-40 (Harding, 1973)—Dominguez oil-field structure (after California Div. Oil and Gas, 1961). Normal faults trend north-northwest in response to east-west extensional components generated by right-lateral wrenching; reverse faults trend mostly east-west in response to compressive components. Absence of observable strike-slip offset of fold axis here may be attributed to position of structure between dying-out ends of en echelon strike-slip faults. Note angle between fold axis and major faults. Permission to publish by American Association of Petroleum Geologists.

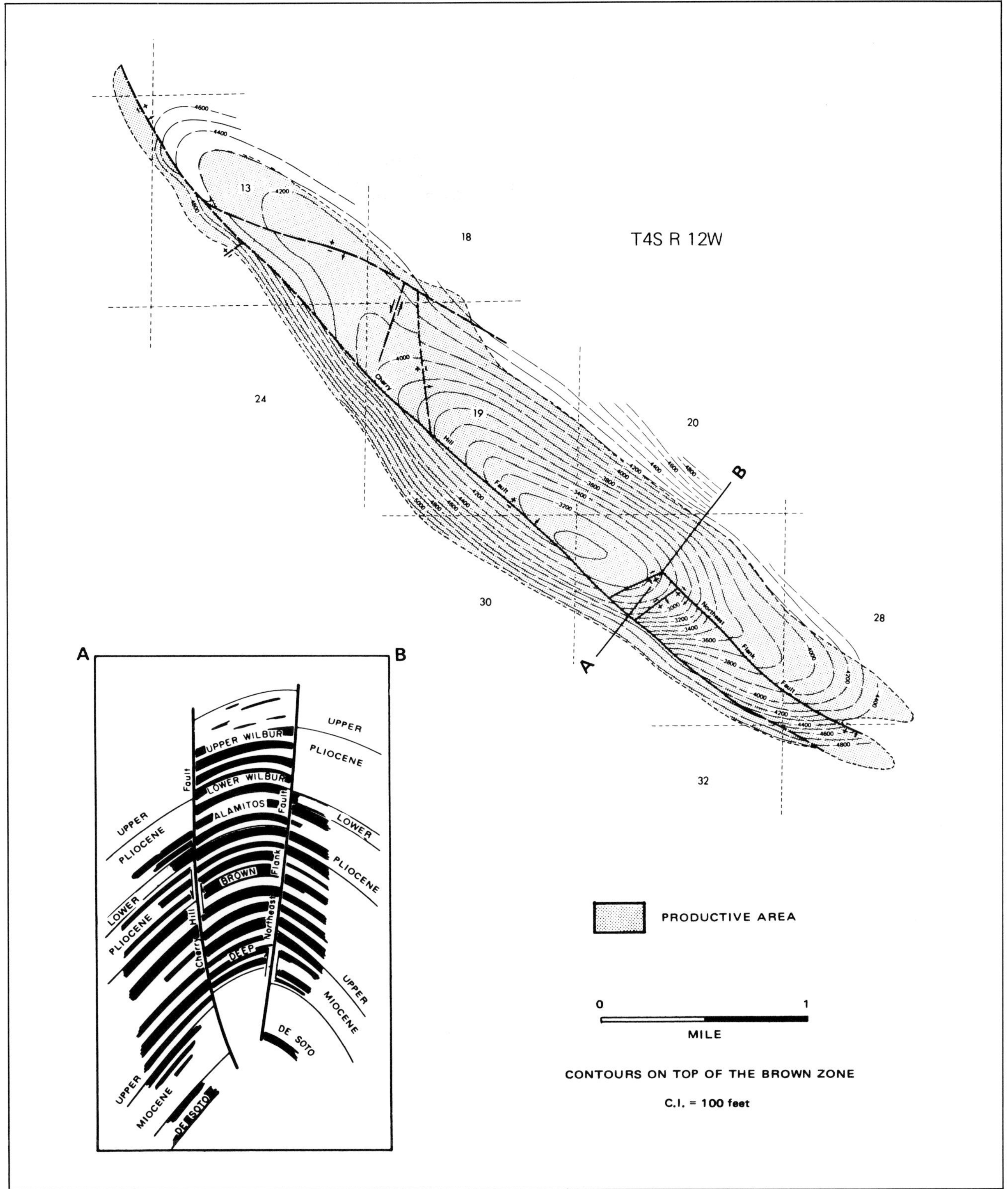

Fig. 2-41 (Harding, 1973)—Long Beach field structure demonstrates over 610 m (2,000 ft) of right-lateral offset of fold crest and plunge elements from corresponding bulges in field's southwest flank. En echelon folding is suggested by oblique orientation of axes of the two culminations to Cherry Hill and Northeast flank faults. Permission to publish by American Association of Petroleum Geologists.

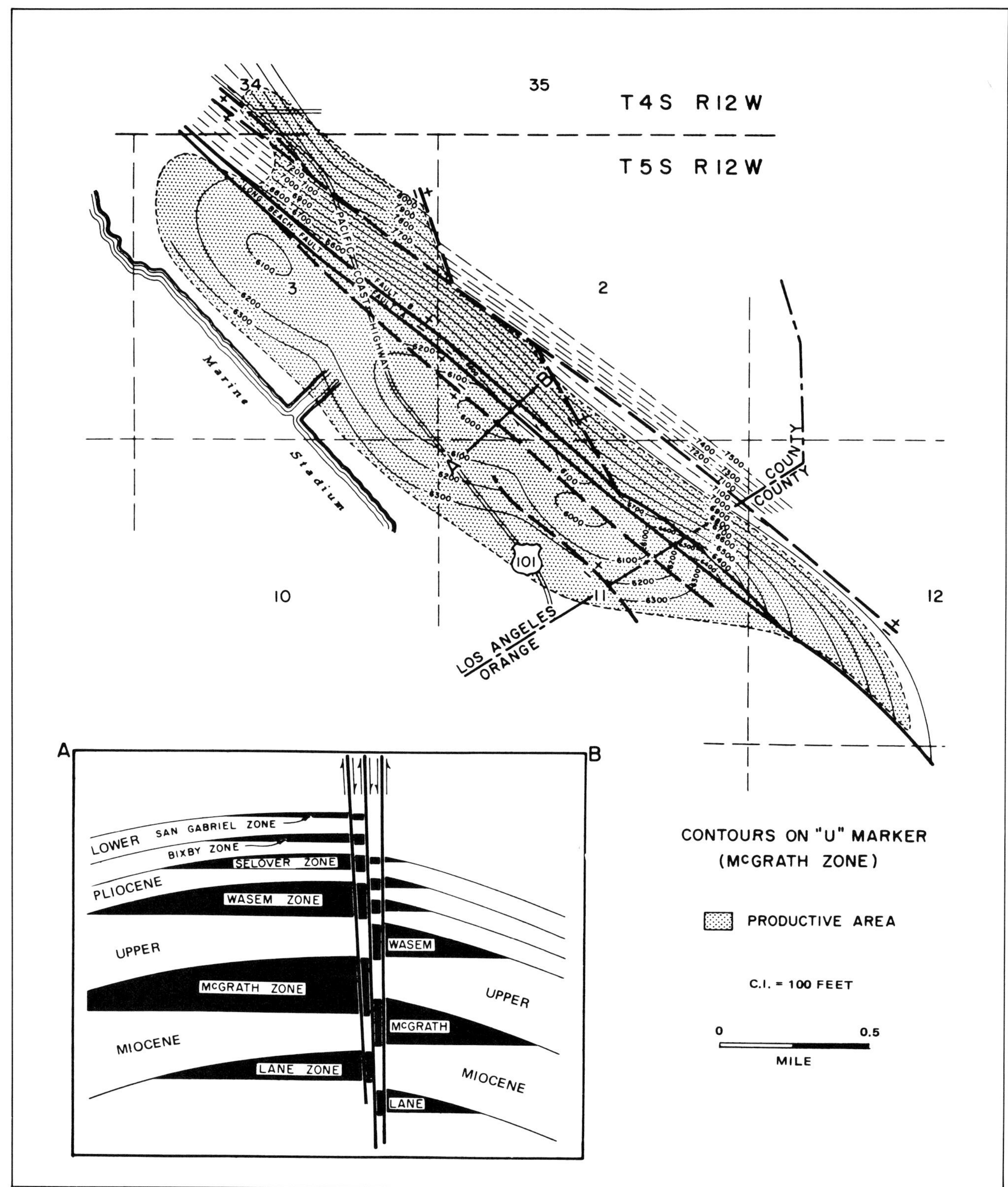

Fig. 2-42 (Harding, 1973)—Seal Beach oil-field structure (after California Div. Oil & Gas, 1961). Rough correlation of fold culminations with flanking bulges in northeast block indicates 762 m (2,500 ft) of right-lateral offset. Apparent dip-slip components on horst and graben slices are nonsystematic, i.e., they do not step down from fold axis in consistent direction. Permission to publish by American Association of Petroleum Geologists.

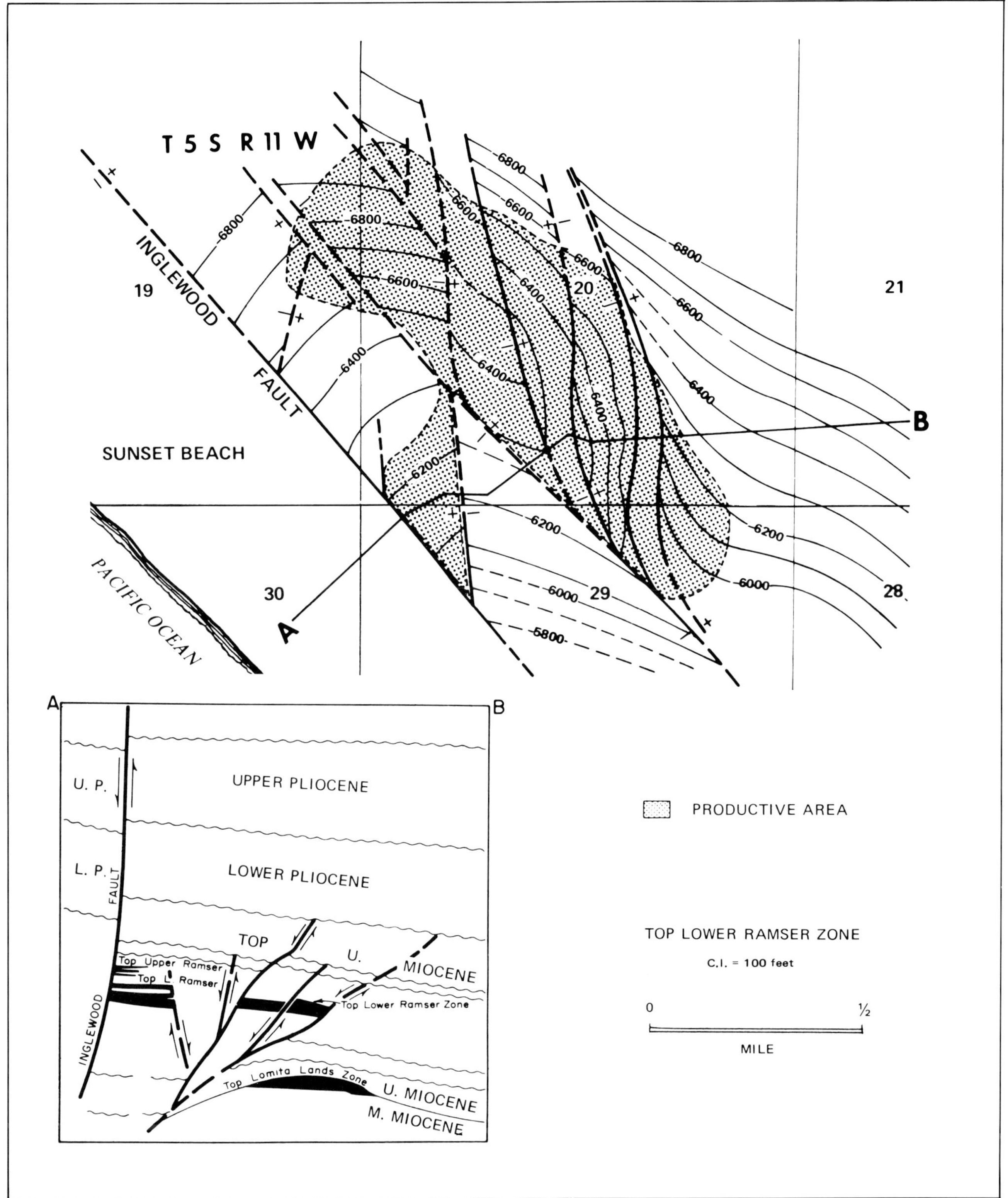

Fig. 2-43 (Harding, 1973)—Sunset Beach oil field structure. En echelon normal faults form traps in shallower producing zones, and in one area regional wrench fault itself provides part of closure. Permission to publish by American Association of Petroleum Geologists.

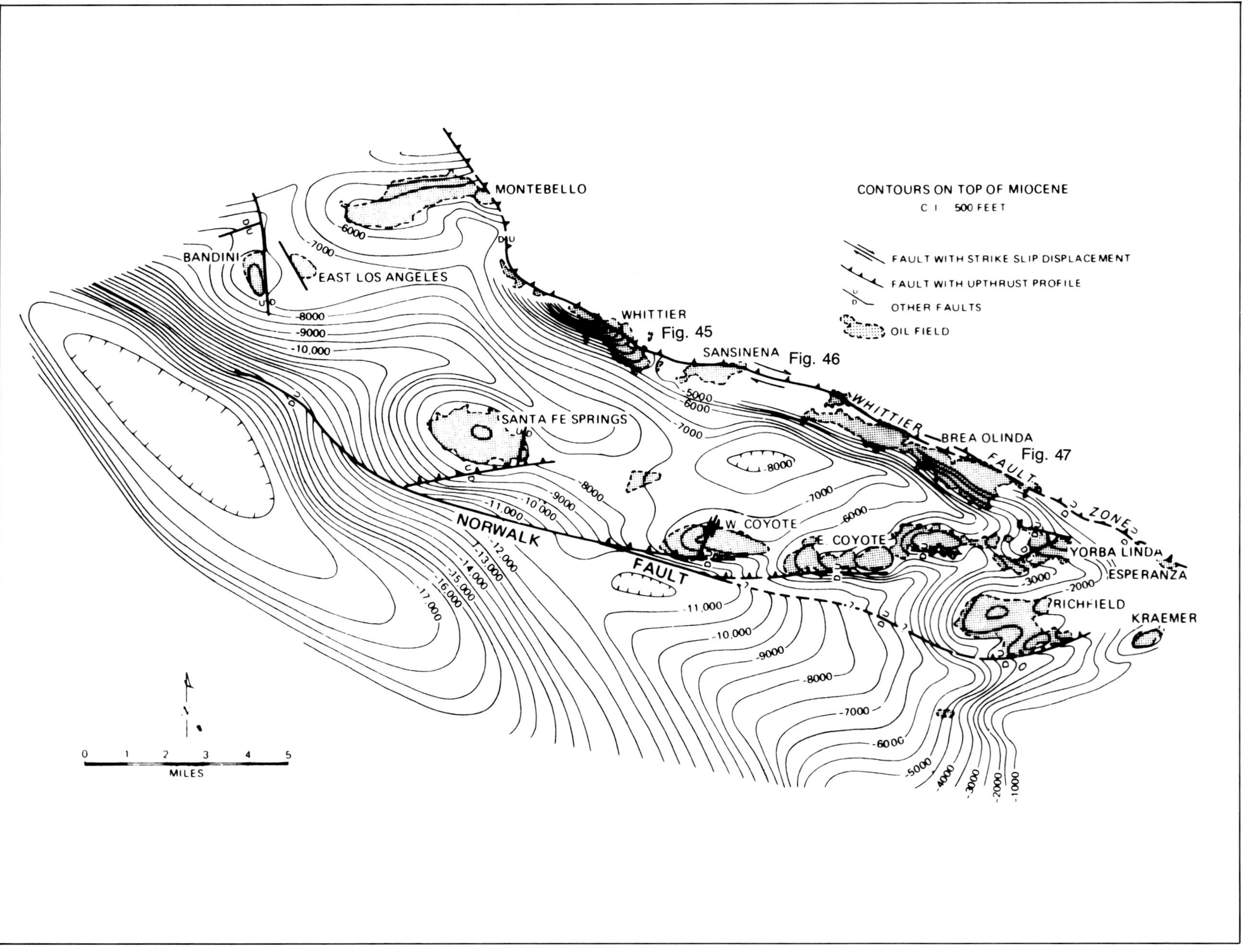

Fig. 2-44 (Harding, 1974a)—Structure of eastern shelf with Whittier fault, Los Angeles Basin (oil fields named). Permission to publish by American Association of Petroleum Geologists.

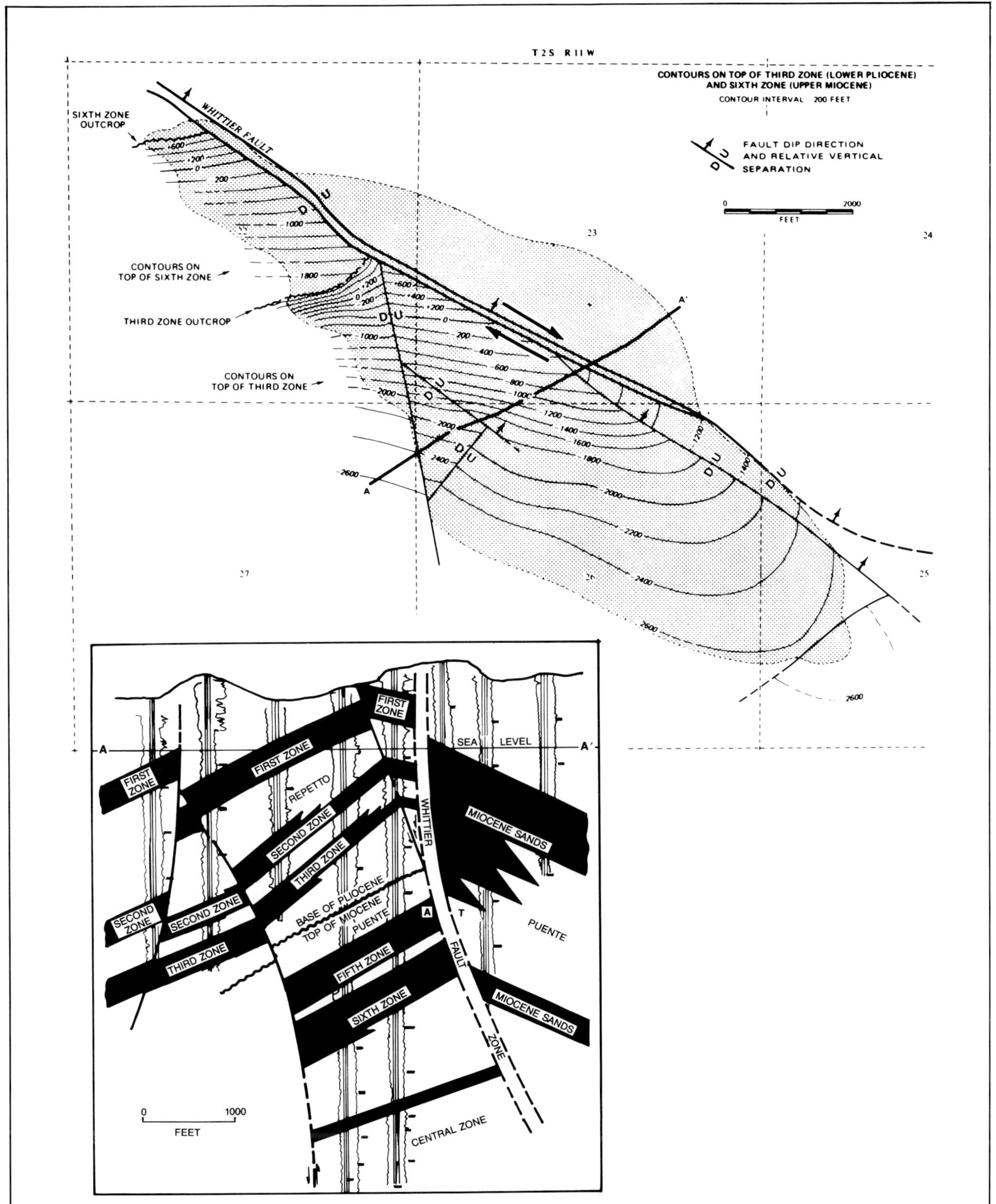

Fig. 2-45 (Harding, 1974a)—Structure of Whittier oil field. Northeast closure is provided by Whittier fault, up-plunge closure by tar seal in outcrops of producing sands. Production on northeast side of fault is marginal and comes from tight, older Miocene sands. Permission to publish by American Association of Petroleum Geologists.

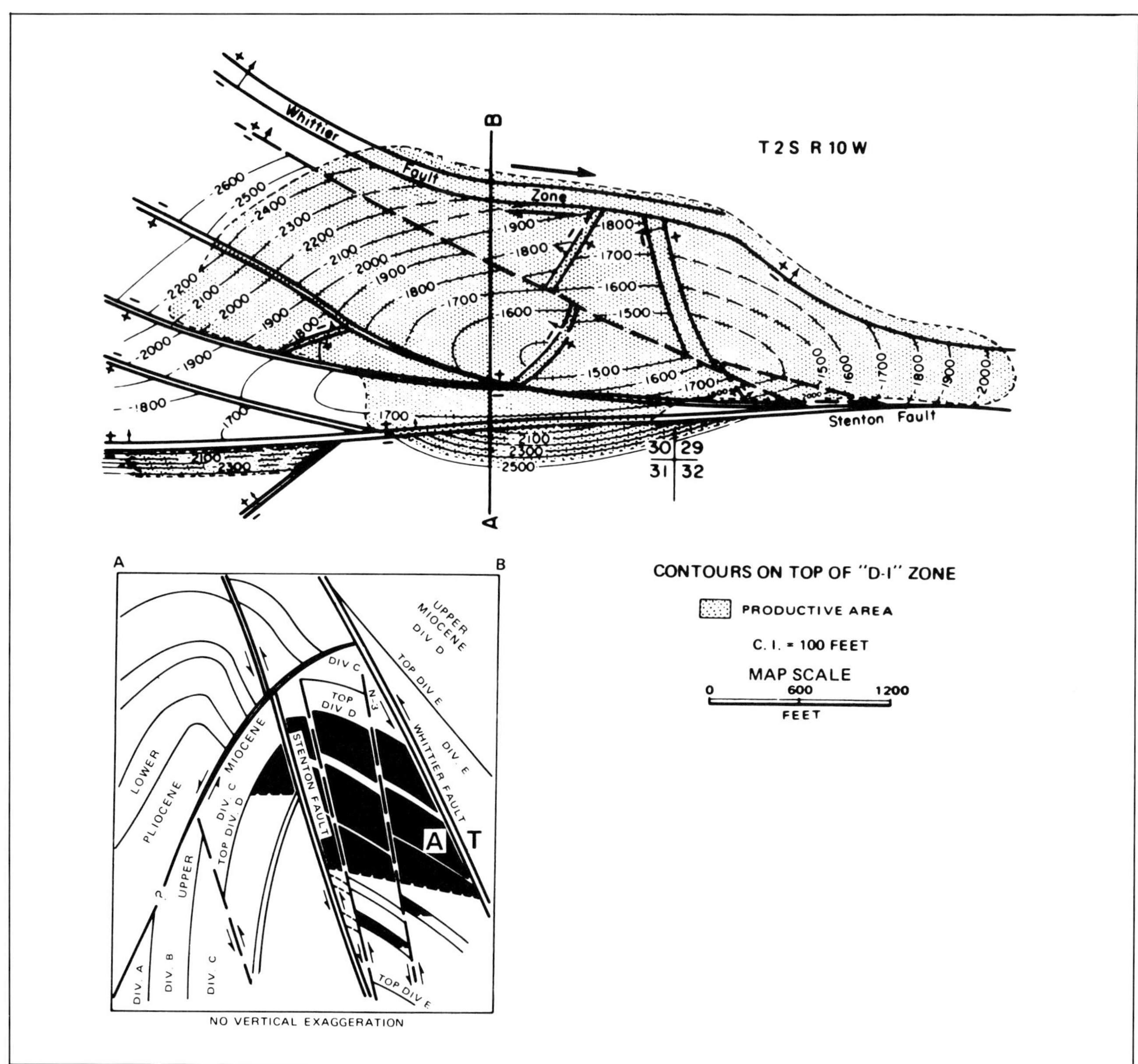

Fig. 2-46 (Harding, 1974a)—Sansinena oil-field structure. East-west trend of drag fold repeats orientation of similar en echelon folds exposed along high side of Whittier fault and is supportive evidence of right-lateral wrench deformation. Internal faulting and squeezed-fold profile document highly compressive aspect of deformation. Permission to publish by American Association of Petroleum Geologists.

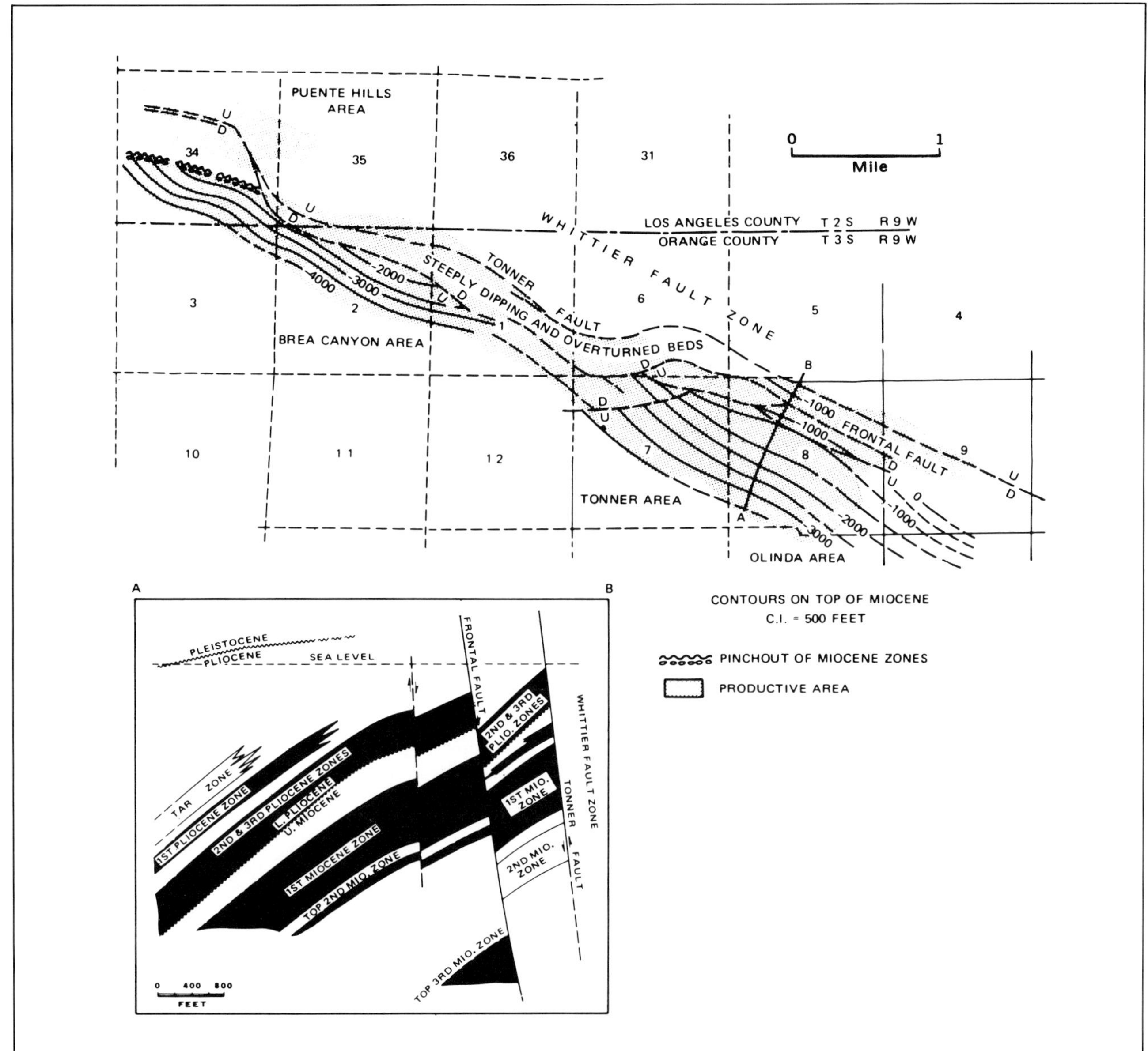

Fig. 2-47 (Harding, 1974a)—Brea-Olinda oil field. Lateral closure is provided by broad, structural bowing against wrench zone, secondary cross faults, and porosity terminations. Frontal fault is anomalous and has been interpreted variously as having reverse or normal apparent dip-slip separations. The fault bounds an extensive slice whose dominant movement may have been strike slip. Permission to publish by American Association of Petroleum Geologists.

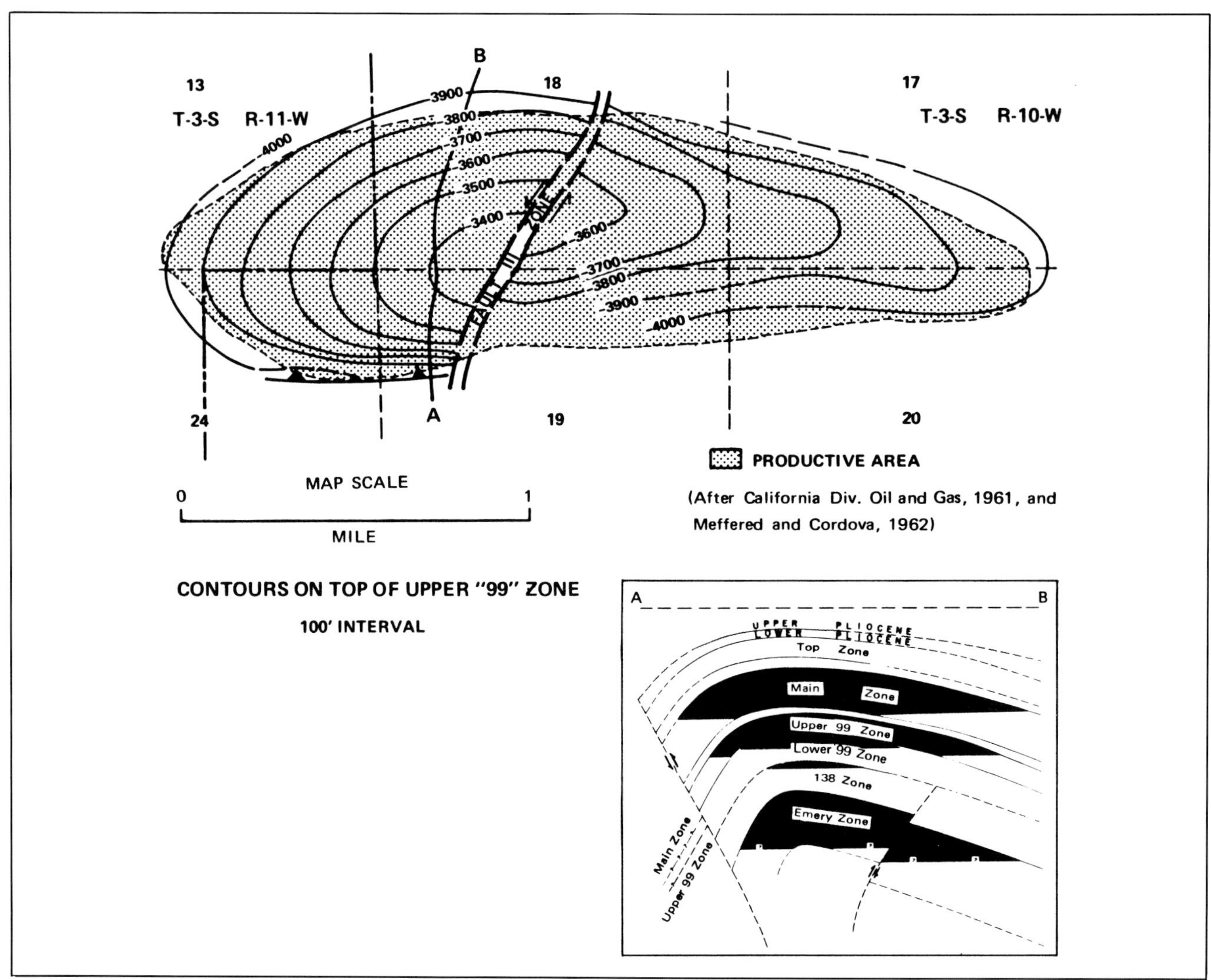

Fig. 2-48 (Harding, 1974b)—West Coyote oil field illustrating culmination closure typical of oil fields on eastern shelf of Los Angeles Basin. Main reverse fault is downthrown to south in concert with basin's subsidence. Trend of fold is consistent with right-lateral displacement in the area. Relatively simple structure in more basinward direction contrasts with tightly folded Sansinena structure (cf Fig. 2-46). Reprinted by permission.

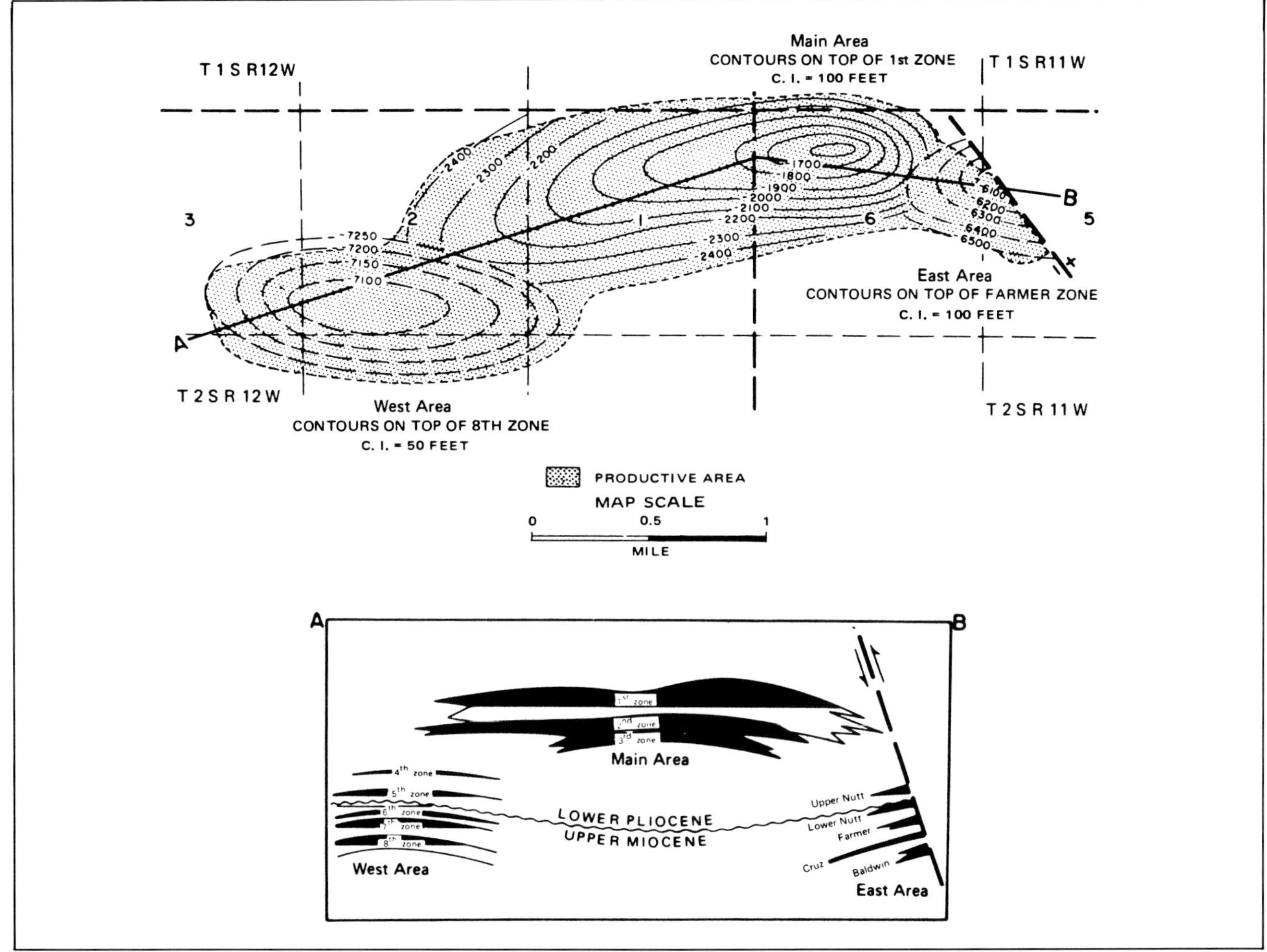

Fig. 2-49 (Harding, 1974b)—Structure of Montebello oil field. In the upper Miocene a large nose is closed up plunge by a fault with reverse separation that marks the northwestern end of the Whittier fault zone. Pronounced depositional irregularities cause a marked shift of fold crests at lower Pliocene horizons (Main Area of field). Reprinted by permission.

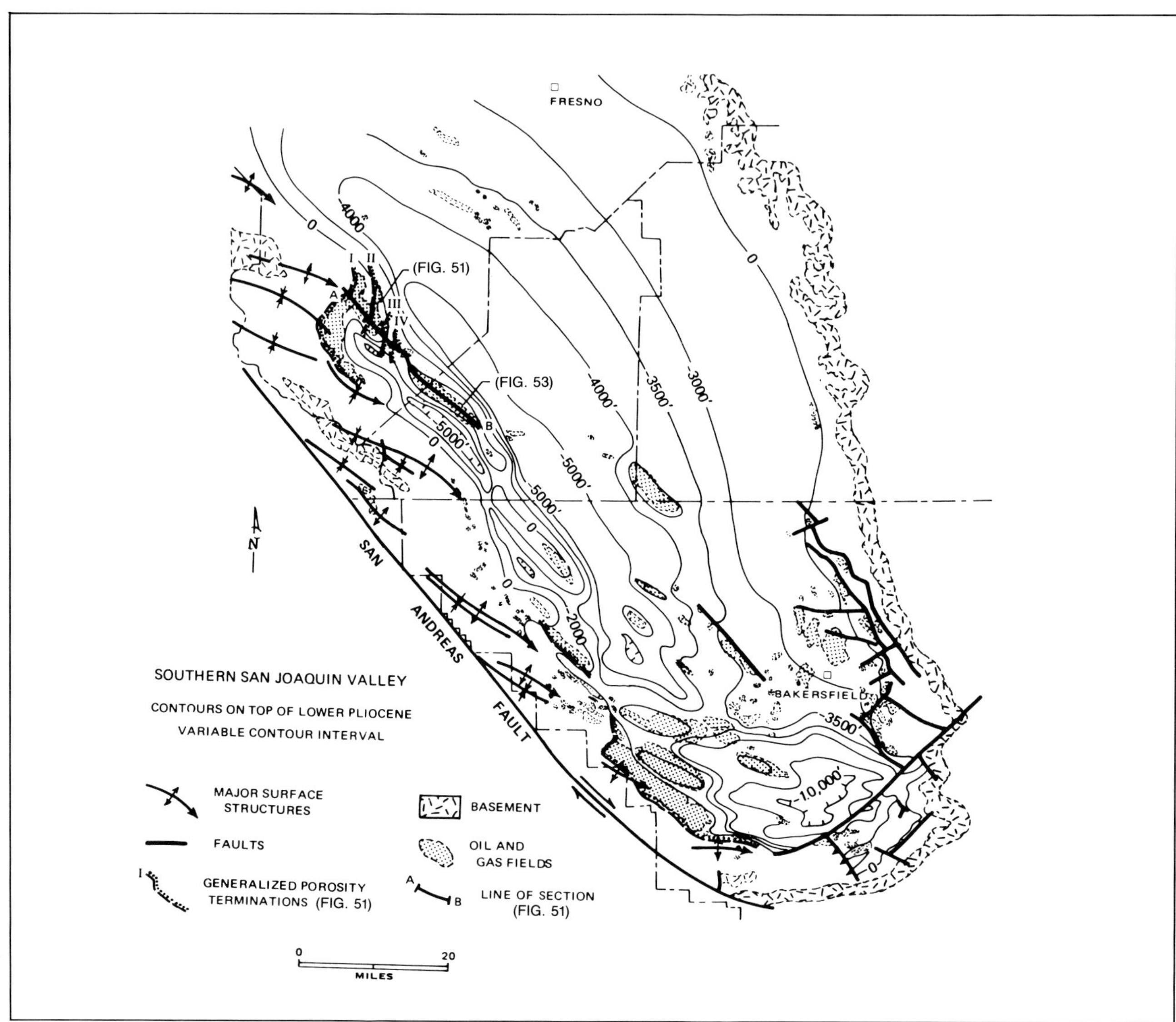

Fig. 2-50 (Harding, 1974a)—Major structures and oil fields, San Joaquin Valley, California (after Hoots et al., 1954). Figure numbers refer to subsequent illustrations in text, Roman numerals to porosity terminations on Coalinga nose shown in figure 2-51. Permission to publish by American Association of Petroleum Geologists.

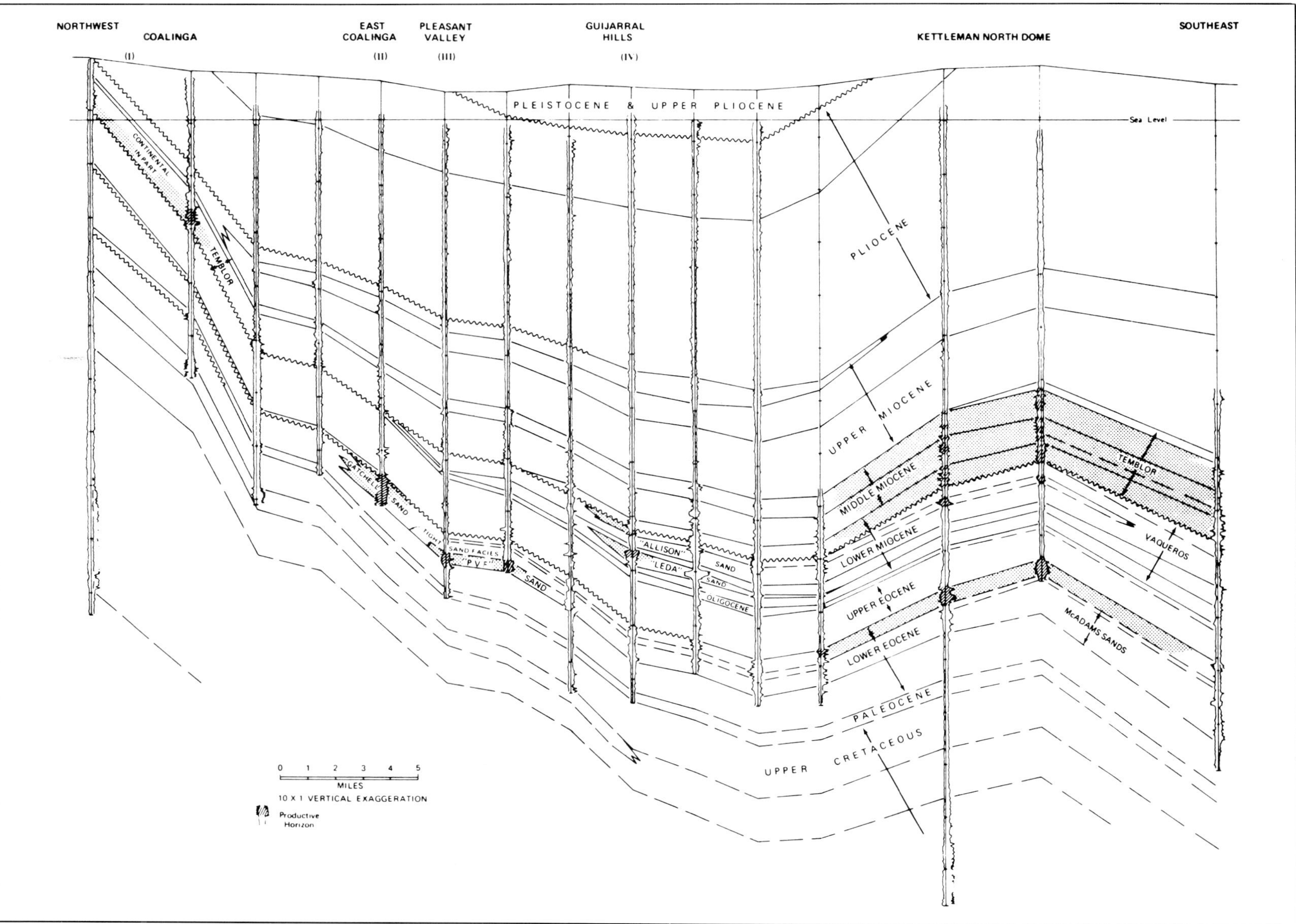

Fig. 2-51 (Harding, 1974a)—Cross section up plunge of Coalinga nose; see figure 2-50 for location of section and stratigraphic oil fields identified with Roman numerals. Effective stratigraphic closures included (I) surface tar seal, sand truncation, and permeability loss caused by transition to nonmarine facies; (II) sand shaleout (see also Fig. 2-52); (III) permeability barriers; and (IV) sand shaleout, sand wedgeout, and multiple permeability barriers in Eocene sands (latter not indicated). Productive limits are generalized; stratigraphic traps had yielded nearly 1.2 billion BO by end of 1971 (California Div. Oil & Gas, 1971). Permission to publish by American Association of Petroleum Geologists.

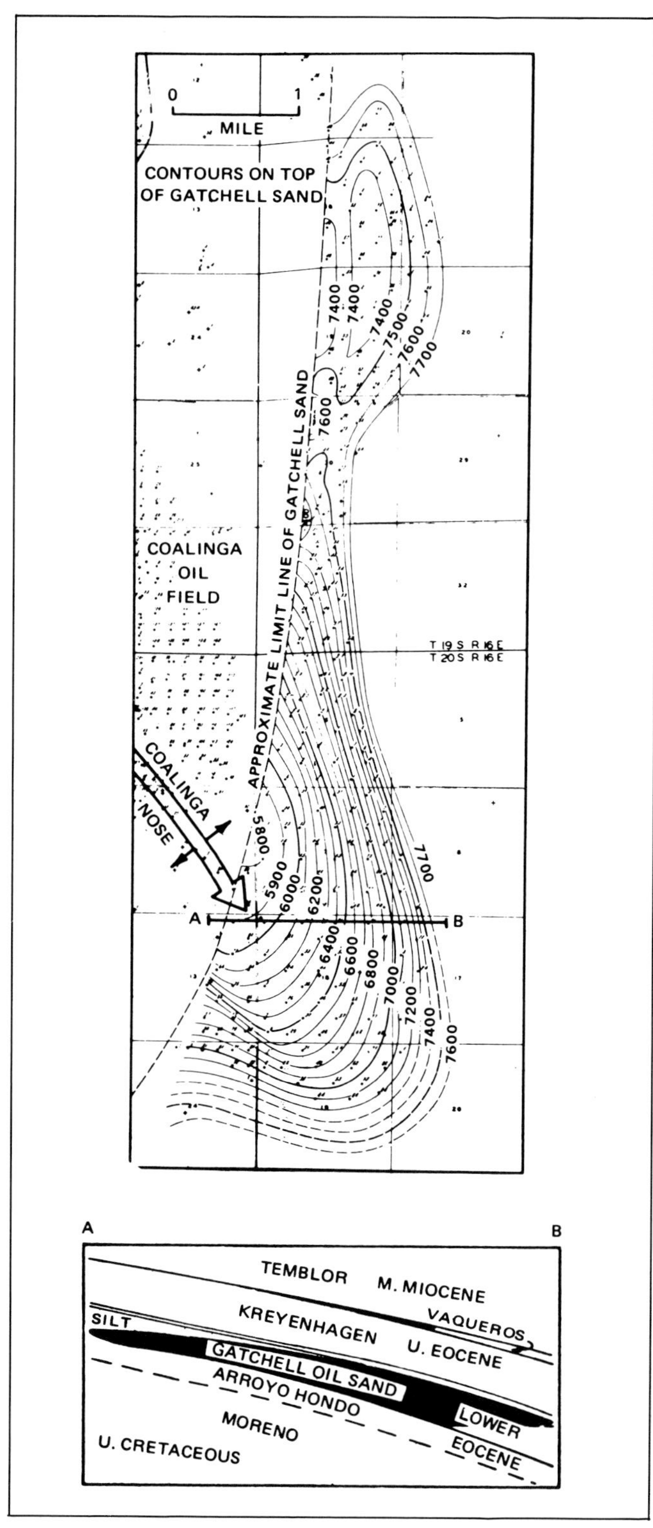

Fig. 2-52 (Harding, 1974a)—Structure of major East Coalinga extension oil field. Relative absence of internal structural complexities at folds well away from wrench fault is shown here. Permission to publish by American Association of Petroleum Geologists.

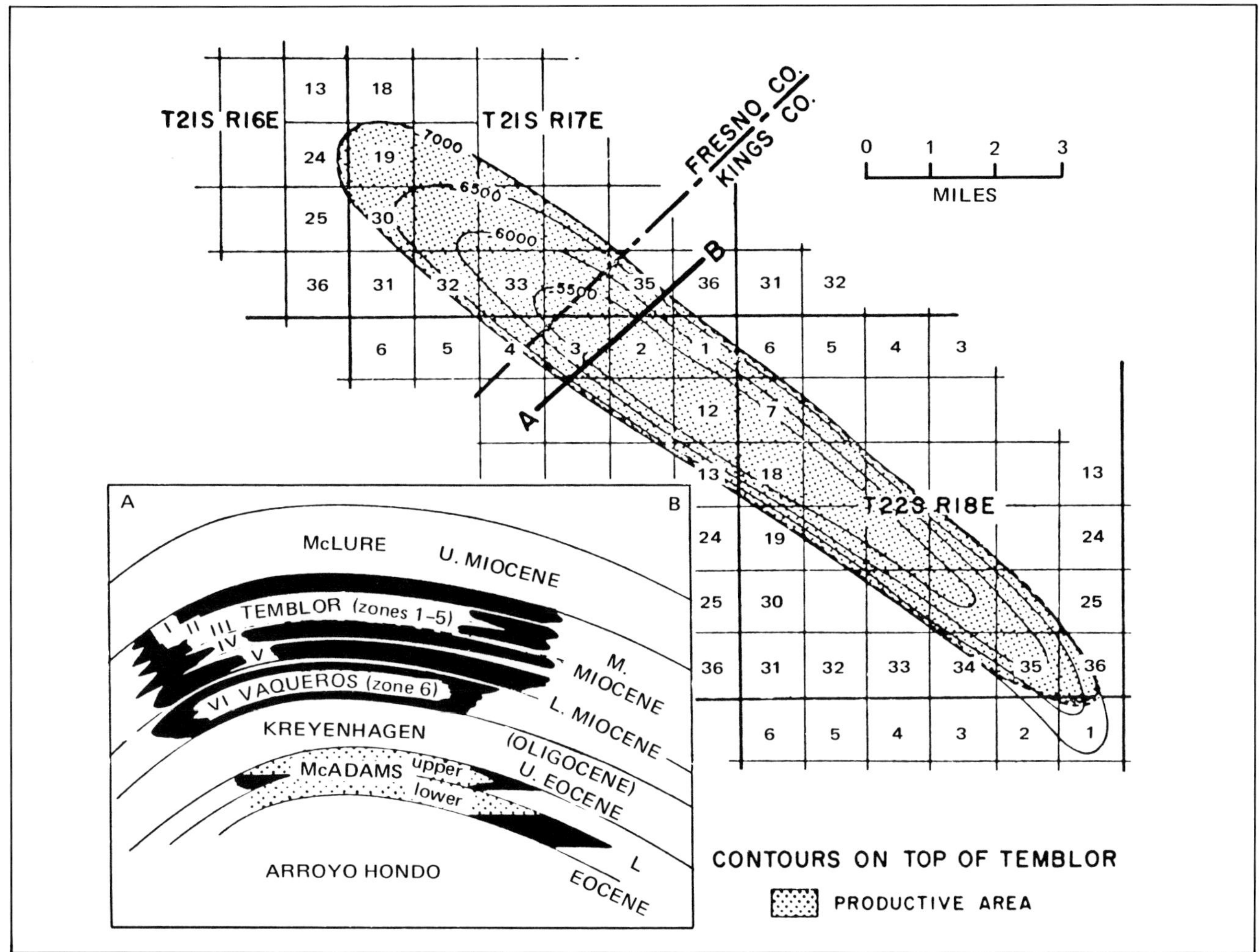

Fig. 2-53 (Harding, 1974b)—Structure of Kettleman North Dome oil field. Nearly all horizons productive up structure in stratigraphic traps are present here and produce. Cumulative production through 1971 totaled 450 million BO.

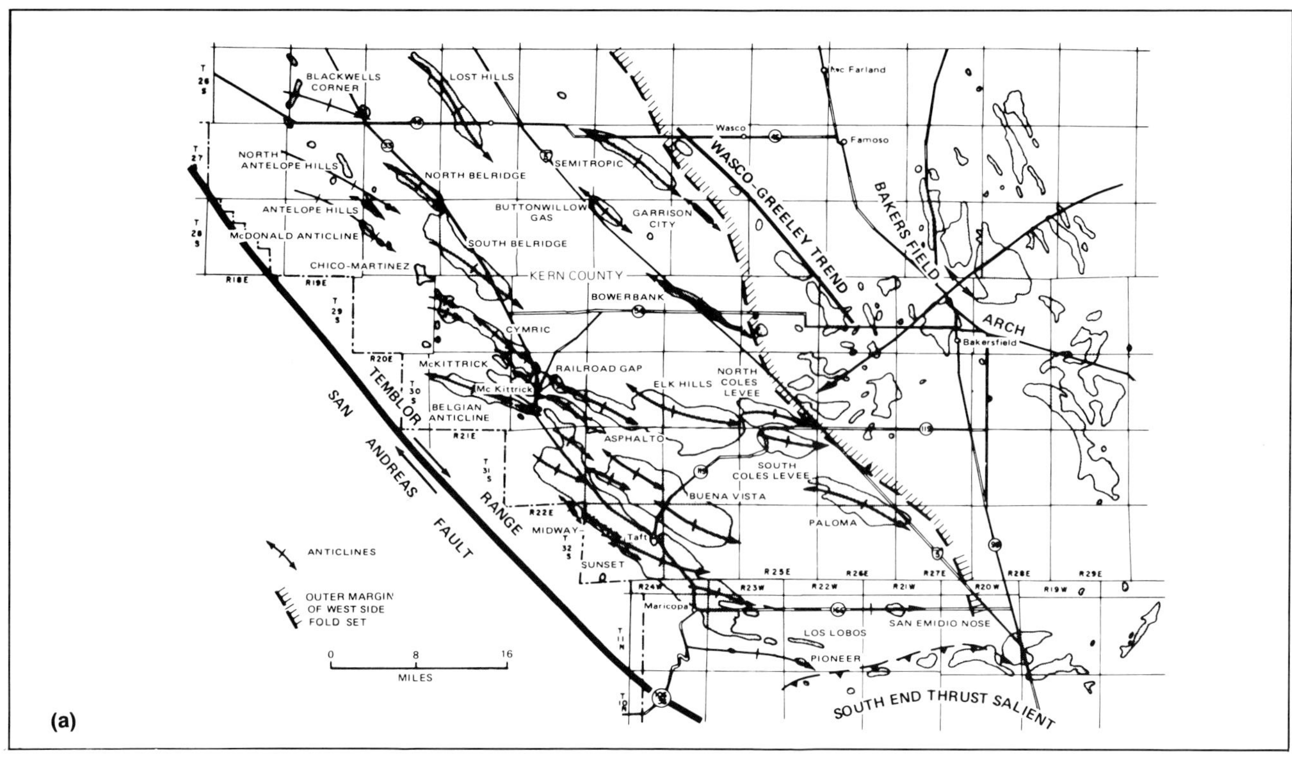

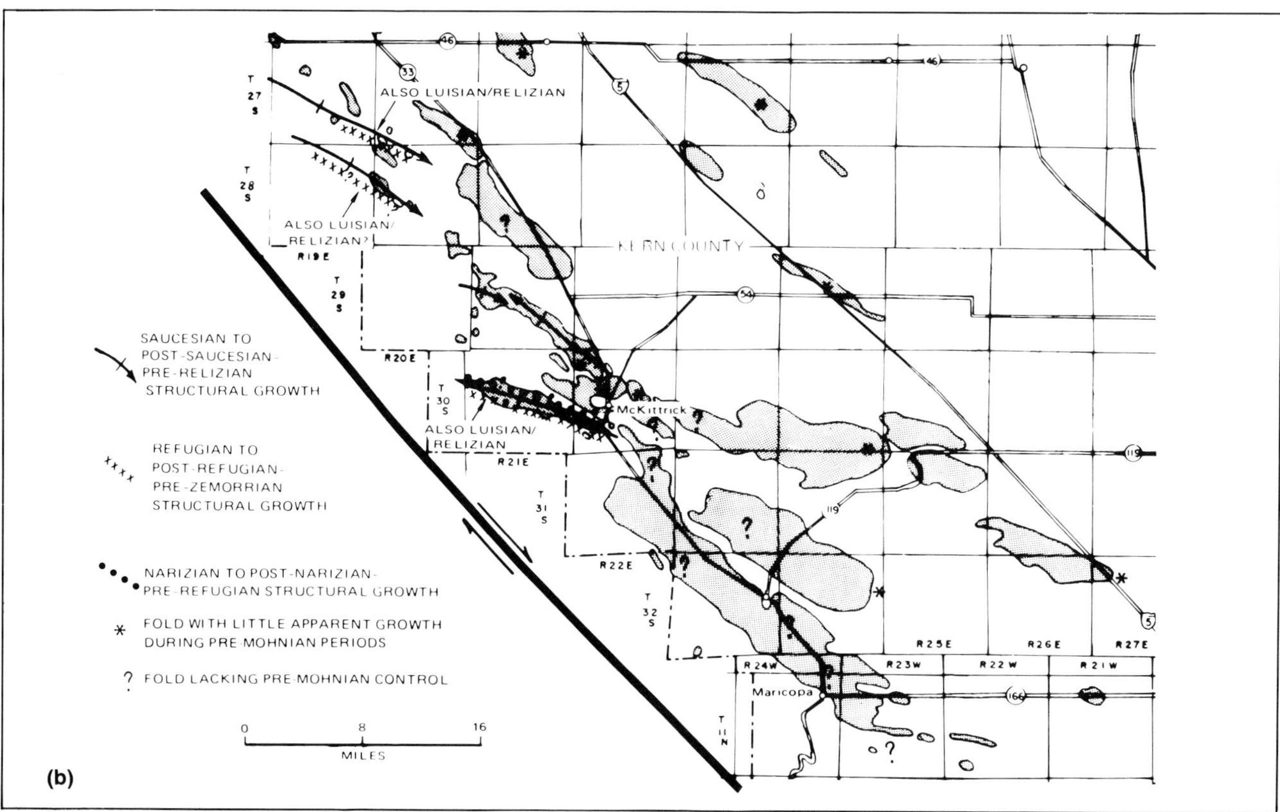

Fig. 2-54 (Harding, 1976)—(a). Index map of west side oil fields and main producing structures, San Joaquin Valley. Permission to publish by American Association of Petroleum Geologists.

(b). Pre-Mohnian folding. In this and subsequent maps discernible structural growth periods of main producing anticlines in southwest part of San Joaquin Valley. Structures questionably or imprecisely dated are denoted by axes symbols combined with small question marks. In cases of ambiguous time spans all possible episodes included by time span are indicated in this manner. Permission to publish by American Association of Petroleum Geologists.

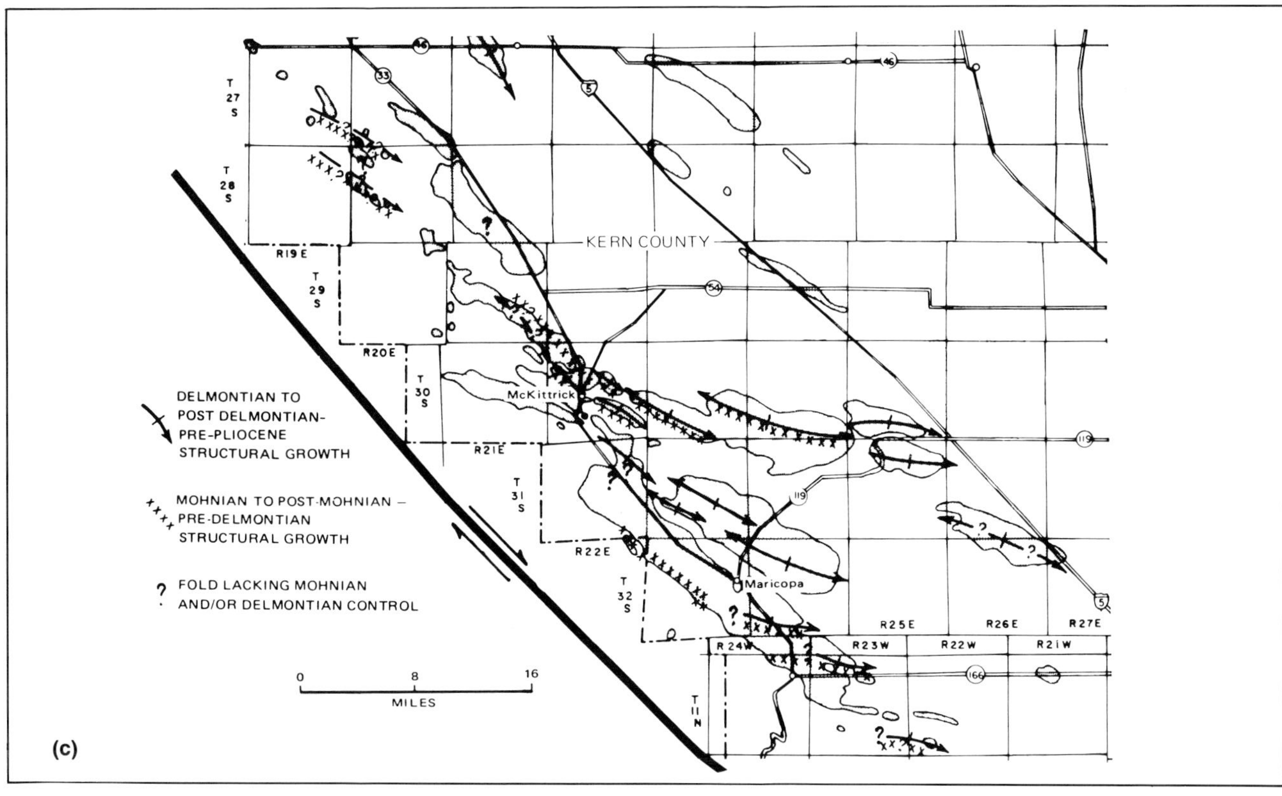

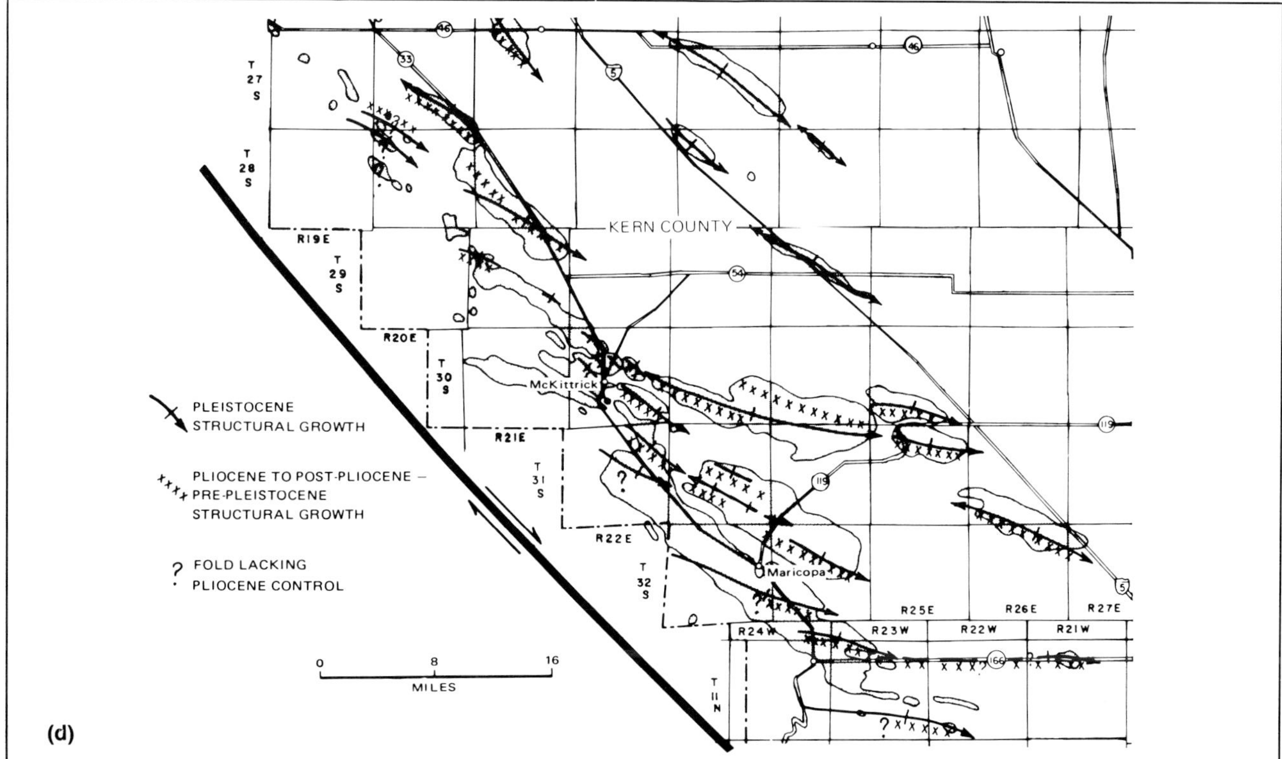

(c). Mohnian and Delmontian folding. Permission to publish by American Association of Petroleum Geologists.
(d). Pliocene and Pleistocene folding. Permission to publish by American Association of Petroleum Geologists.

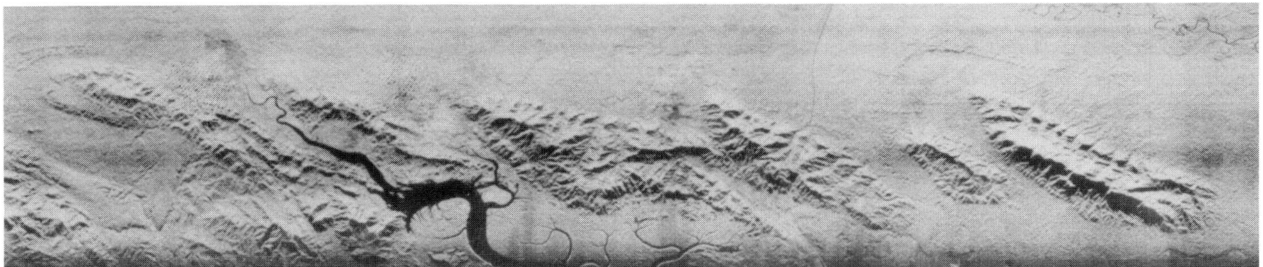

Fig. 2-55 (Wilcox et al., 1973)—Radar imagery showing en echelon folds caused by left-lateral couple, Darien Basin, Panama. Permission to publish by American Association of Petroleum Geologists.

PLATE TECTONIC HABITATS

Transform Boundaries

The primary plate tectonic habitat of the wrench assemblage is the transform boundary where the kinematics is that of wrenching (Fig 1-2). Examples include the Dead Sea, San Andreas, Queen Charlotte, and Alpine fault zones.

Transform faults are, of course, critical in showing directions of plate motion. This arises because plate motion can be analyzed according to Euler's theorem that states that the motion of any rigid body on a sphere can be described by rotation about a pole. Transform faults act as small circles about a pole of spreading and hence establish the azimuth of plate movement (Fig. 2-56).

Poles of spreading can be identified by the location of the intersection of great circles, each great circle being struck perpendicular to a small circle transform (Fig. 2-57). The pole is usually located within a circle of error. A pole of spreading can also be located by the spreading rate method where the pole is at the top of an arc of 90 drawn on a great circle to that transform that is coincident with the maximum rate of spreading. Some poles of spreading for large plates are shown in figure 2-58.

Since transform faults behave as small circles to a particular pole of spreading, they appear as lines of latitude on Mercator map projections that are cast on that pole. All other transforms appear as something other than lines of latitude inasmuch as they have different poles of spreading. Examples are given for the Central Atlantic (Fig. 2-59), South Pacific (Fig. 2-60), and "Fossil" Pacific (Fig. 2-61).

Changes in the direction of transform faults reflect changes in the direction of spreading. An example from the northern Atlantic ocean shows the change in azimuth of the Gibbs fracture zone (Fig. 2-62).

Convergent Boundaries

Longitudinal wrench faults and oblique wrench faults or foreland conjugate shears are known in convergent boundary settings (Fig. 1-2). A fault is termed longitudinal where it parallels the regional structural grain.

The Hoetanapan or Semangko fault of Sumatra is considered a longitudinal wrench (Fig. 2-63). It probably developed, and is developing, as northward movement of the Indian Ocean plate carried, and carries, northwestward that part of the continental lithosphere (from trench to wrench) that was and is in contact with the underlying and subducting oceanic lithosphere (Fig. 2-64).

Foreland conjugate shears (oblique wrench faults) develop as nonrotational (pure) shears (Fig. 2-65) from a horizontally directed maximum principal compressive stress in the forelands of detached thrust-fold belts (Figs. 2-65, 2-66). If one set is better developed than the other, as seems to be true for those in the WNW-ESE quadrant of the Rocky Mountain foreland, some external rotation is implied.

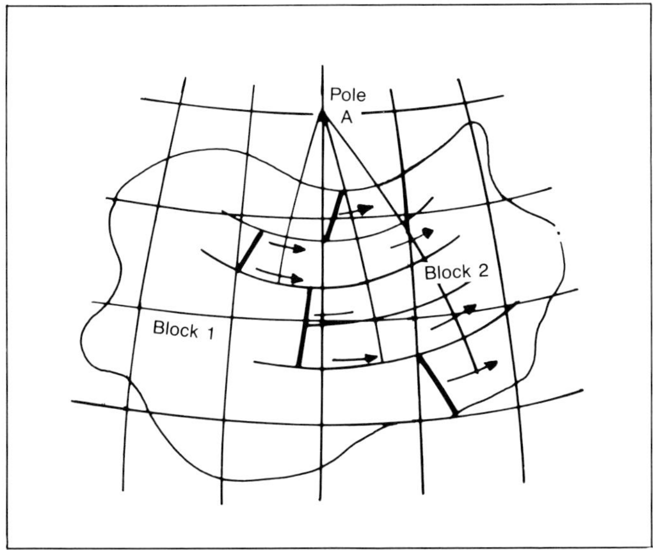

Fig. 2-56 (Morgan, 1968)—On a sphere, the motion of block 2 relative to block 1 must be a rotation about some pole. All faults on the boundary between 1 and 2 must be small circles concentric about the pole A. Copyright by the American Geophysical Union. Reprinted by permission.

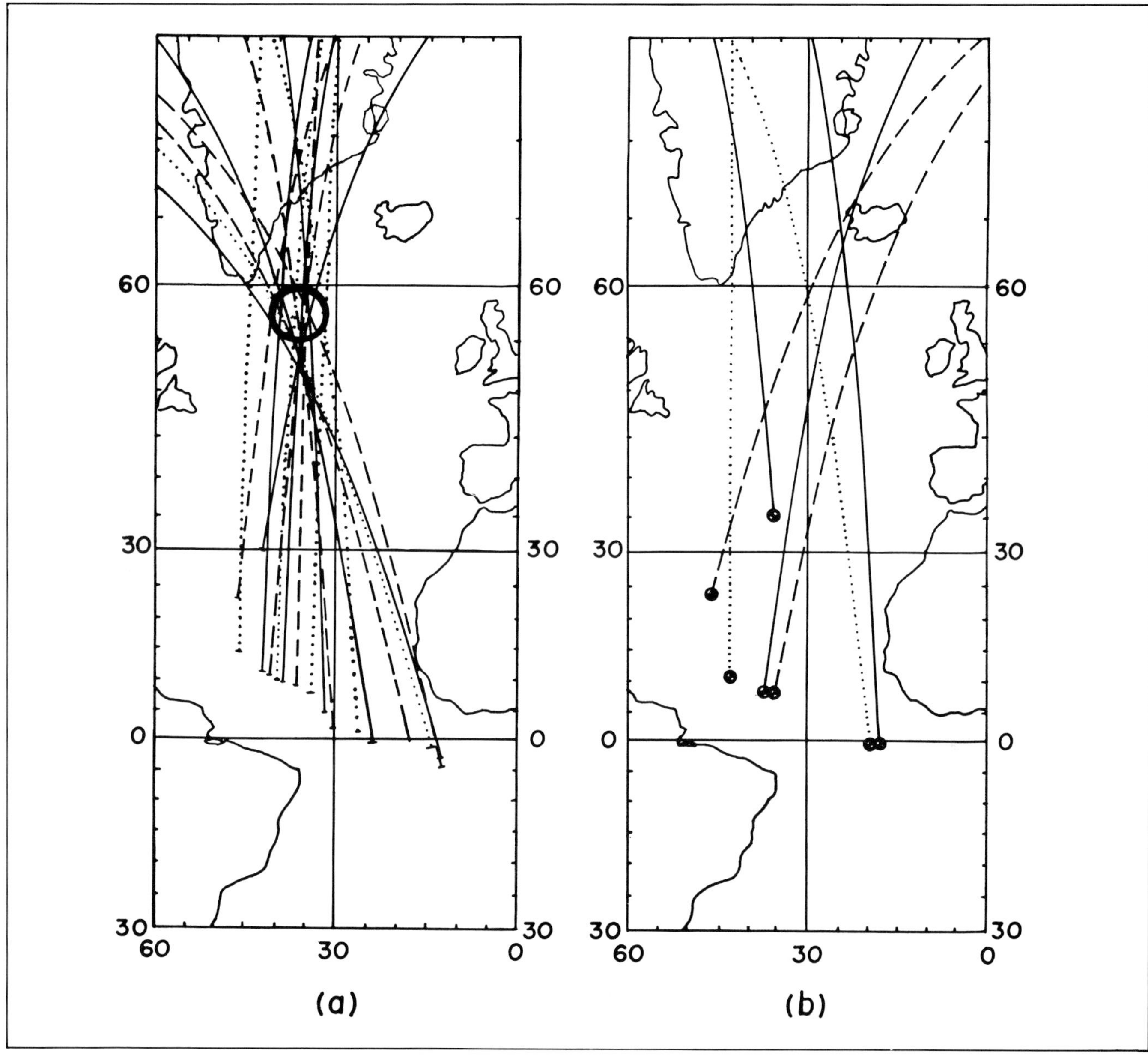

Fig. 2-57 (Morgan, 1968)—Great circles perpendicular to the strike of offsets of the mid-Atlantic ridge are shown in (a). With one exception, all of these lines pass within the circle center at 58°N, 36°W. Great circles perpendicular to the strike determined by earthquake mechanism solutions are shown in (b). Copyright by the American Geophysical Union. Reprinted by permission.

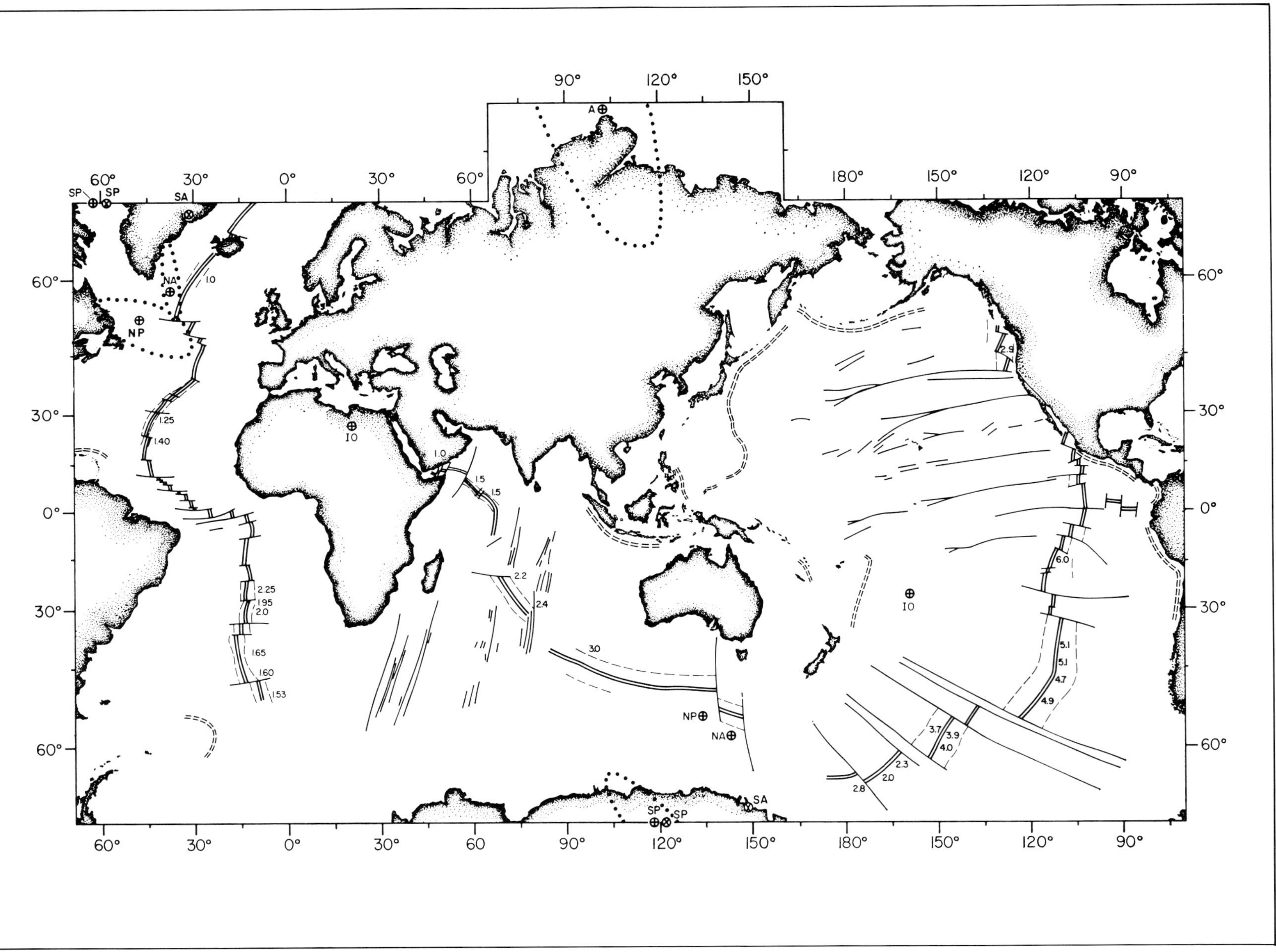

Fig. 2-58 (Le Pichon, 1968)—The axes of the actively spreading mid-ocean ridges are shown by a double line; the fracture zones or transform faults by a single line; anomaly 5 (≈ 10 m.y. old) by a single dashed line; the active trenches by a double dashed line. The spreading rates are given in centimeters per year. The locations of the centers of rotation obtained from spreading rates are shown by ×; those obtained from the azimuths of the fracture zones by +. NA stands for North Atlantic; SA for South Atlantic; NP for North Pacific; SP for South Pacific; IO for Indian Ocean; and A for Arctic. Copyright by the American Geophysical Union. Reprinted by permission.

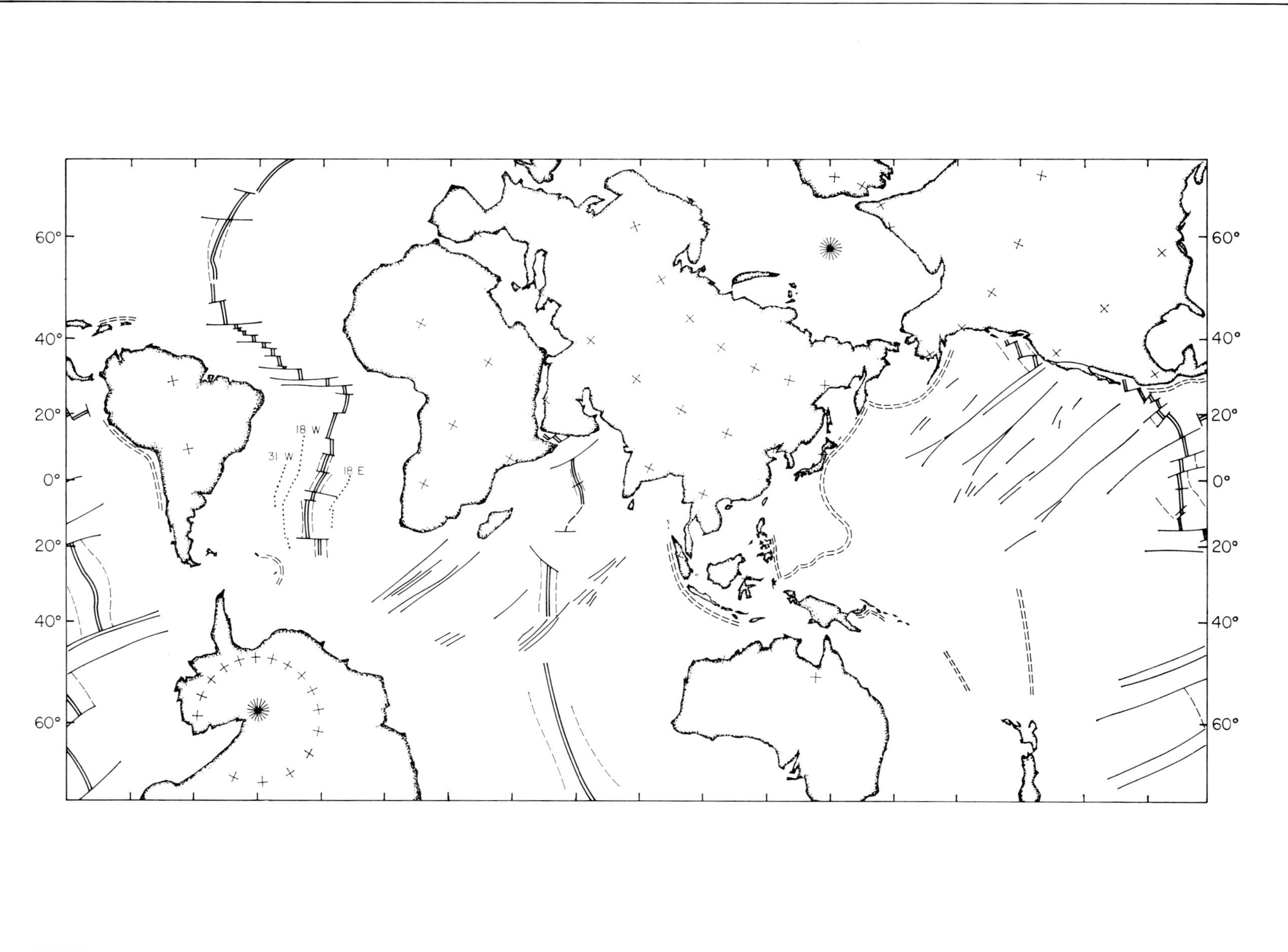

Fig. 2-59 (Le Pichon, 1968)—Mercator projection on Central Atlantic pole of spreading. Copyright by the American Geophysical Union. Reprinted by permission.

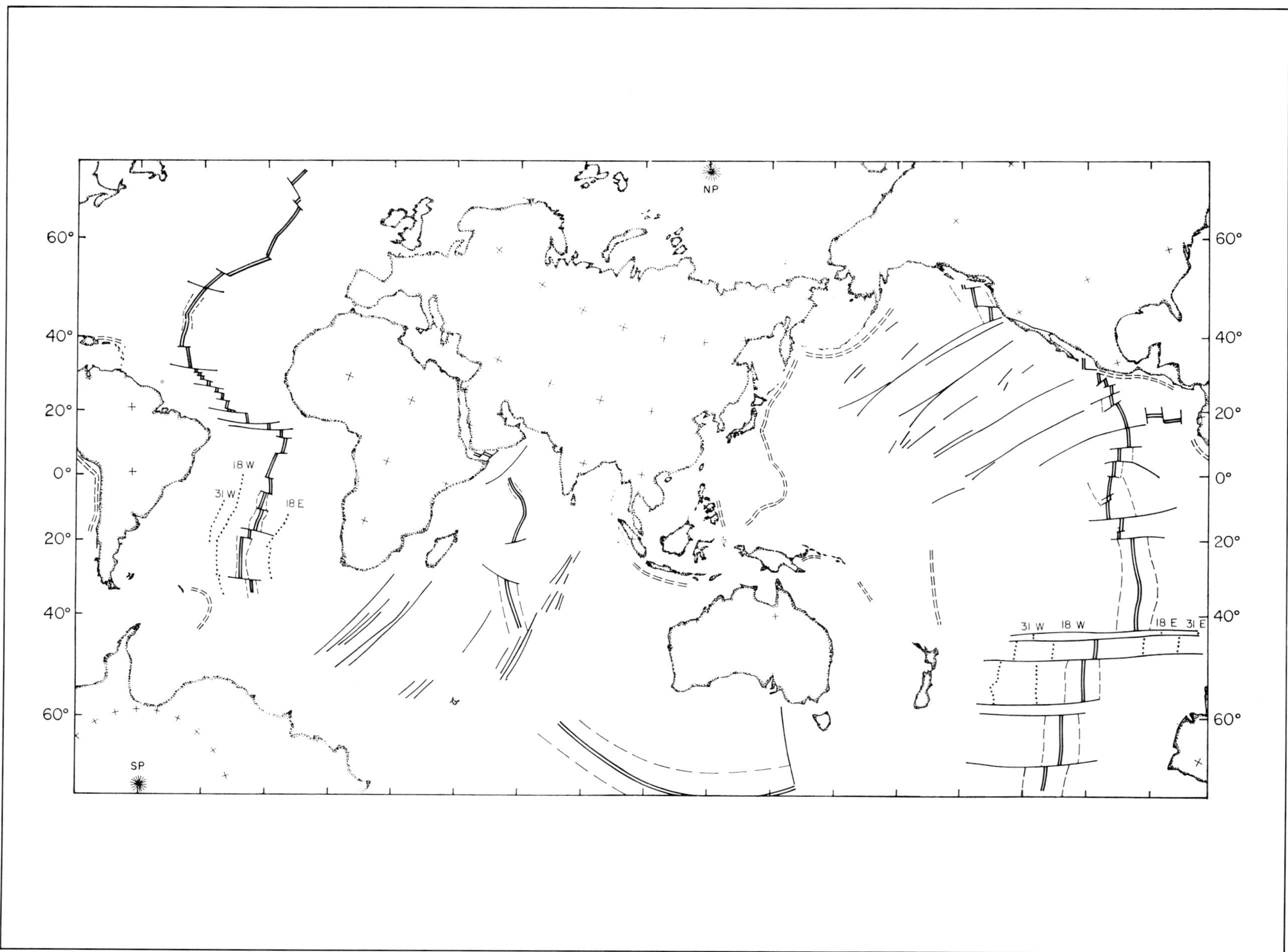

Fig. 2-60 (Le Pichon, 1968)—Mercator projection on South Pacific pole of spreading. Copyright by the American Geophysical Union. Reprinted by permission.

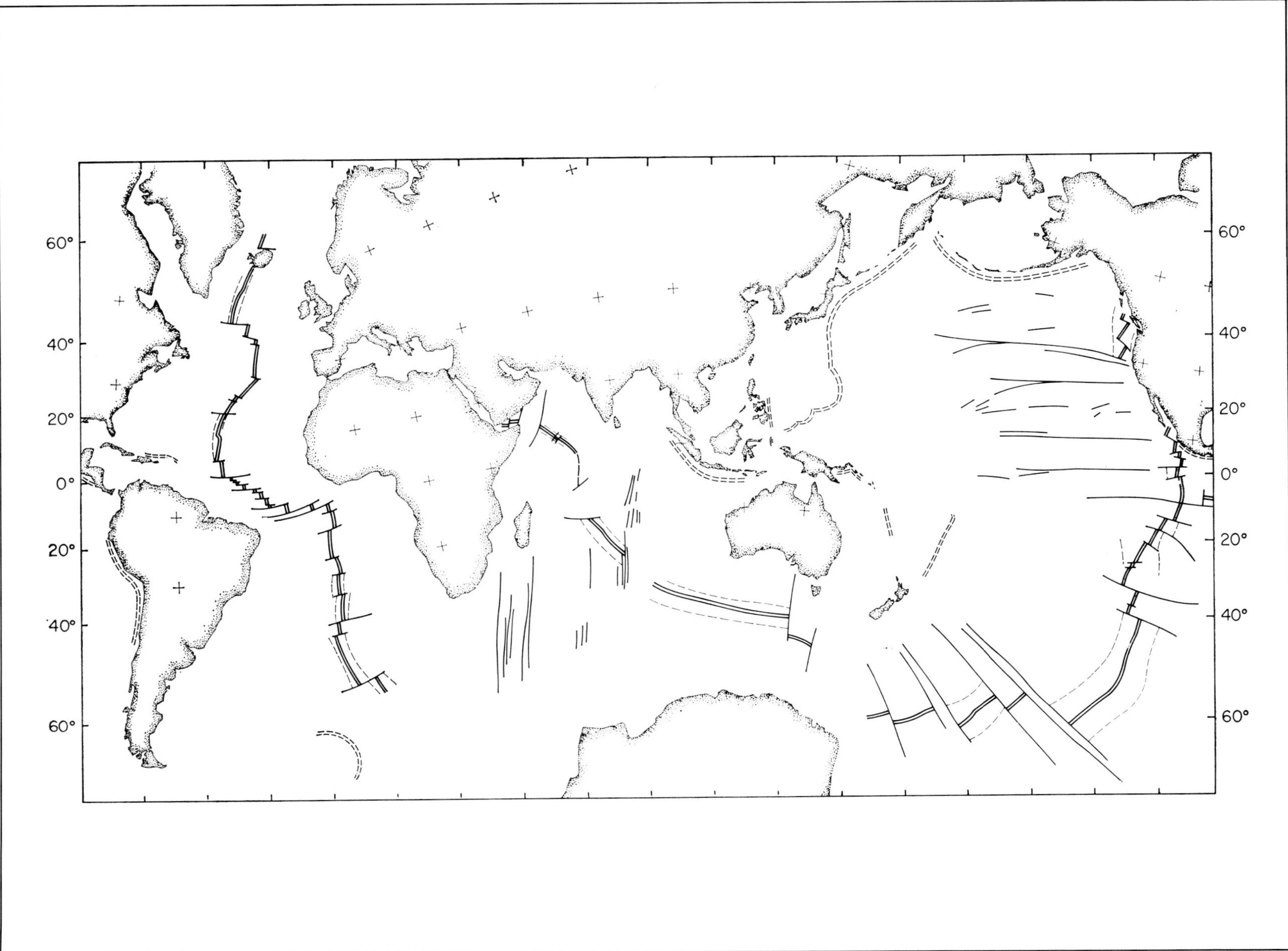

Fig. 2-61 (Le Pichon, 1968)—Mercator projection on "Fossil" Pacific pole of spreading. Copyright by the American Geophysical Union. Reprinted by permission.

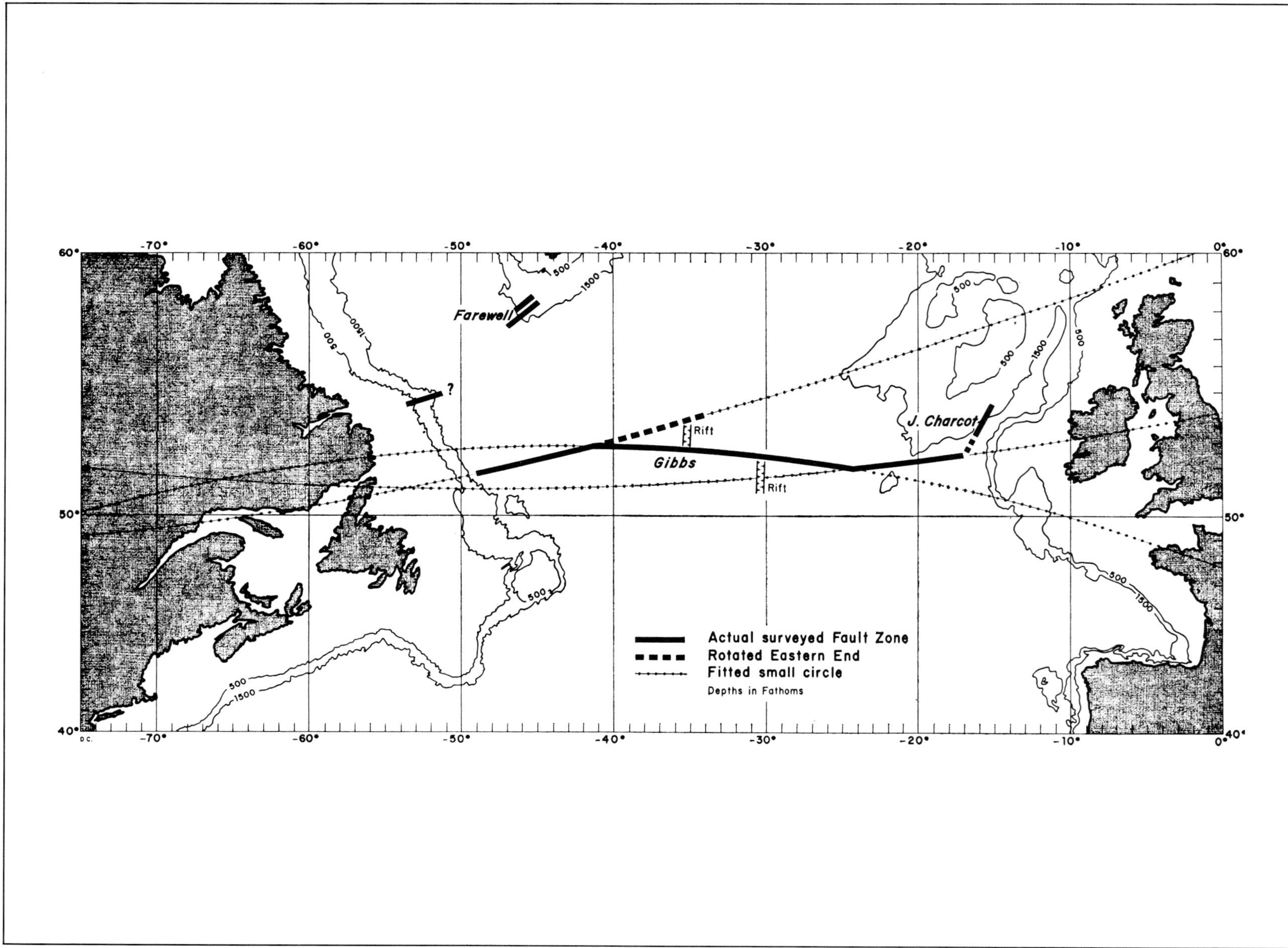

Fig. 2-62 (Le Pichon et al., 1971)—Major fracture zones in the northern Atlantic. The central part of the Gibbs fracture zone has been fitted to a small circle. Then the western end has been rotated to the eastern end by the rotation thus determined (62.8°N, 137.0°E, 11.54°). A small circle is then fitted to the two ends of the Gibbs fracture zone (79.9°N, 89.9°W). Copyright by the American Geophysical Union. Reprinted by permission.

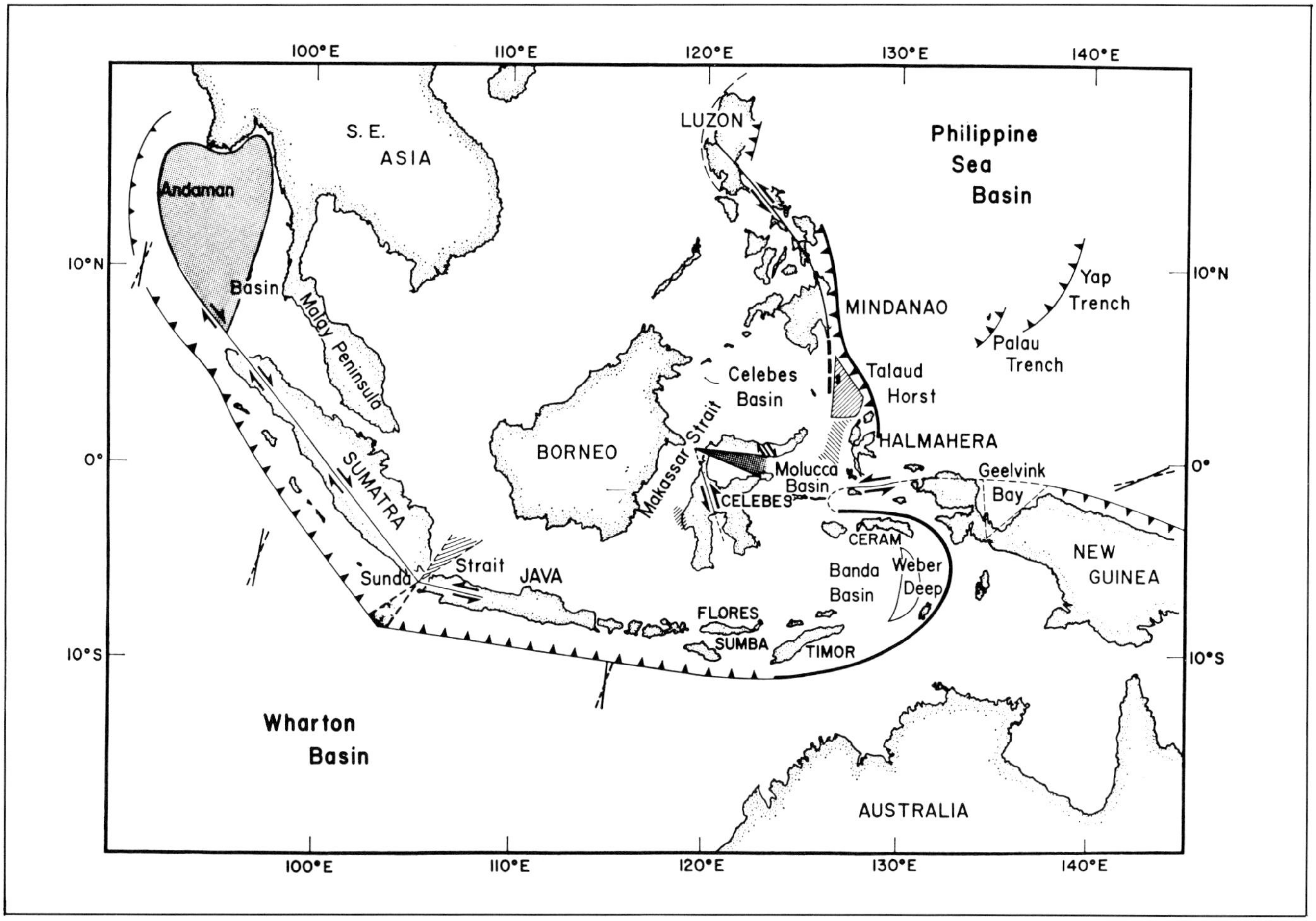

Fig. 2-63 (Fitch, 1972)—Sumatra fault as longitudinal wrench fault on convergent boundary. Shading and striping mark regions of known or suspected extension and shortening, respectively. Locations of transcurrent faults in Celebes and Java and the sense of motion on them are from Katili (1970). Fault on Java probably does not exist. Directions of relative motion between the major plates computed from rigid plate models are given by solid lines (Morgan, 1972) and dashed lines (Le Pichon, 1968). Copyright by the American Geophysical Union. Reprinted by permission.

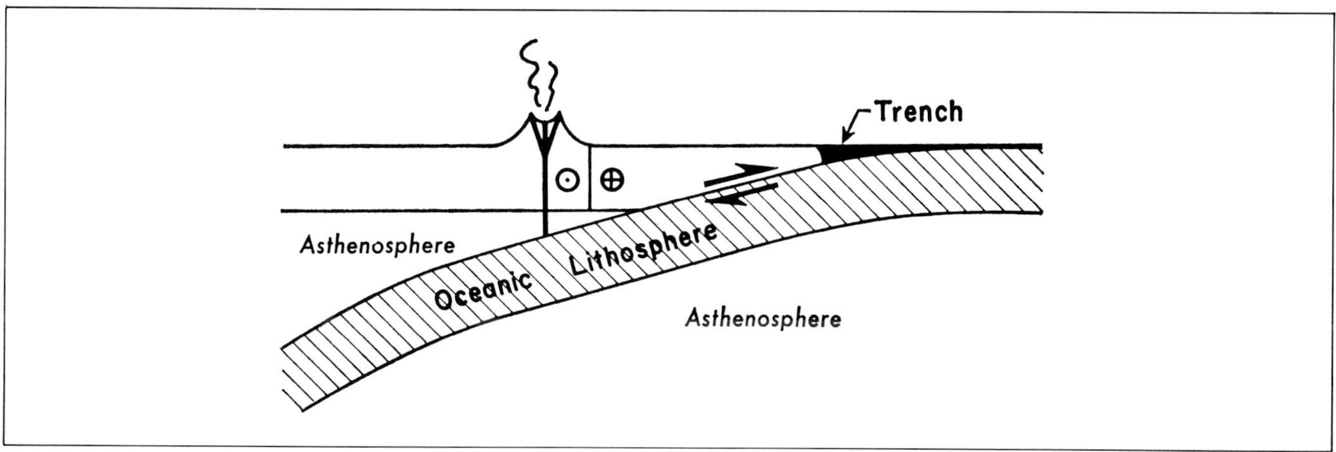

Fig. 2-64 (Fitch, 1972)—Mechanism for development of longitudinal wrench fault. Copyright by the American Geophysical Union. Reprinted by permission.

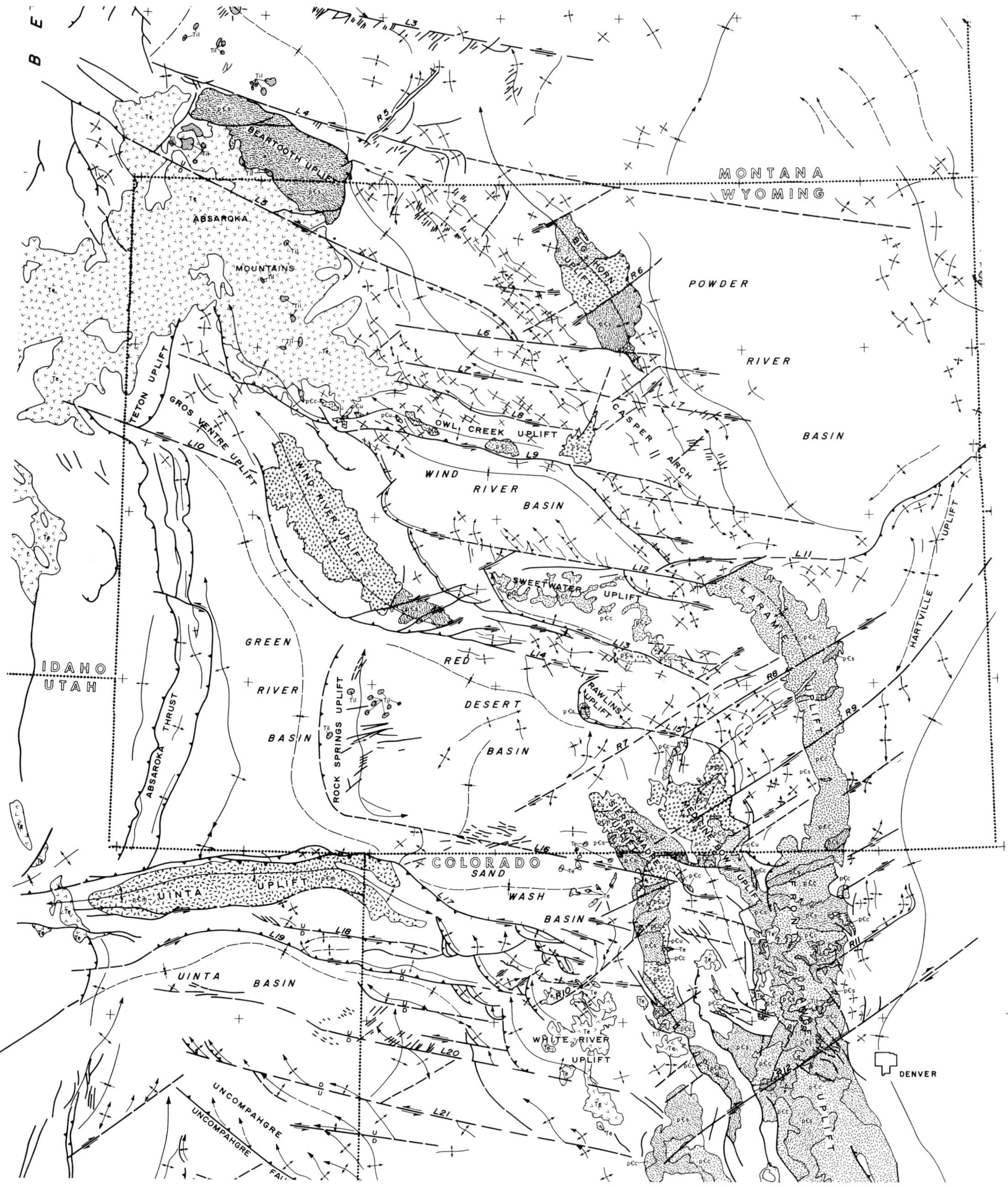
MONTANA
WYOMING
IDAHO
UTAH
COLORADO
BEARTOOTH UPLIFT
ABSAROKA
MOUNTAINS
BIG HORN UPLIFT
POWDER
RIVER
BASIN
TETON UPLIFT
GROS VENTRE UPLIFT
OWL CREEK UPLIFT
CASPER ARCH
WIND
RIVER
BASIN
WIND RIVER UPLIFT
SWEETWATER UPLIFT
LARAMIE UPLIFT
HARTVILLE UPLIFT
GREEN
RIVER
BASIN
ROCK SPRINGS UPLIFT
RED
DESERT
BASIN
RAWLINS UPLIFT
ABSAROKA THRUST
SIERRA MADRE UPLIFT
MEDICINE BOW UPLIFT
UINTA UPLIFT
SAND
WASH
BASIN
UINTA
BASIN
WHITE RIVER
UPLIFT
UNCOMPAHGRE
UNCOMPAHGRE FAULT
FRONT RANGE UPLIFT
DENVER

SOME PROBABLE WRENCH FAULT ZONES

LEFT WRENCH		RIGHT WRENCH
L1 CAT CREEK	L12 FISH CREEK	R1 PENDROY
L2 WILLOW CREEK	L13 SEMINOLE	R2 SCAPEGOAT-BANNATYNE
L3 LAKE BASIN	L14 WIND RIVER	R3 HINSDALE
L4 NYE-BOWLER	L15 RAWLINS-ELK MT.	R4 WELDON-BROCKTON
L5 GARDINER	L16 KING SOLOMON CREEK	R5 FROMBERG
L6 TENSLEEP	L17 UINTA	R6 TONGUE RIVER
L7 NEIBER-SUSSEX	L18 YAMPA	R7 MILLER HILL-FLAT TOP
L8 THERMOPOLIS	L19 BLUE MT.	R8 COMO RIDGE
L9 OWL CREEK	L20 DUCHESE-DOUGLAS CREEK	R9 NASHFORK-HARTVILLE
L10 GROS VENTRE	L21 GARMESA	R10 WILLIAMS PARK
L11 CASPER MT.	L22 APISHAPA	R11 PIERCE-BLACK HOLLOW
		R12 COLORADO MINERAL BELT

TECTONIC DIAGRAM FOR CENTRAL ROCKY MOUNTAIN AREA

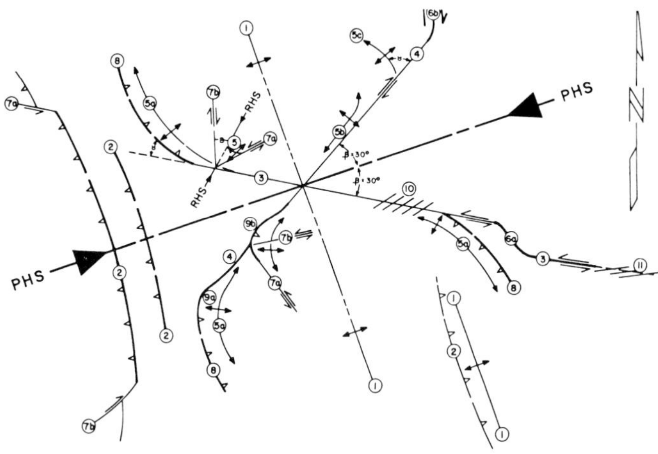

SYMBOLS

PHS Principal horizontal stress direction

RHS Reoriented horizontal stress direction
Angle between PHS or RHS and wrench fault traces
angle between wrench fault and associated parafolds

TECTONIC ELEMENTS: LEGEND FOR FIGURE

1. Primary fold
2. Primary thrusts
3. First order left wrench faults
4. First order right wrench faults
5. Parafolds
   a) Acute (<90°)
   b) "Compression ridge", or parallel ($< \delta \approx 0°$)
   c) Prependicular ($< \delta \approx 90° \pm$)
6. Secondary order faults (after Chinnery)
   a) Left wrench
   b) Right wrench
7. Second order faults (after Moody and Hill)
   a) Left wrench
   b) Right wrench
8. First order thrust
9. Trapdoor structures, formed by the joining of:
   a) Type I: First order wrench with first order thrust fault
   b) Type II: First order wrench fault with second order wrench fault having opposite sense of lateral slip
10. Pinnate tension fractures
11. Pinnate shear fractures

EXPLANATION

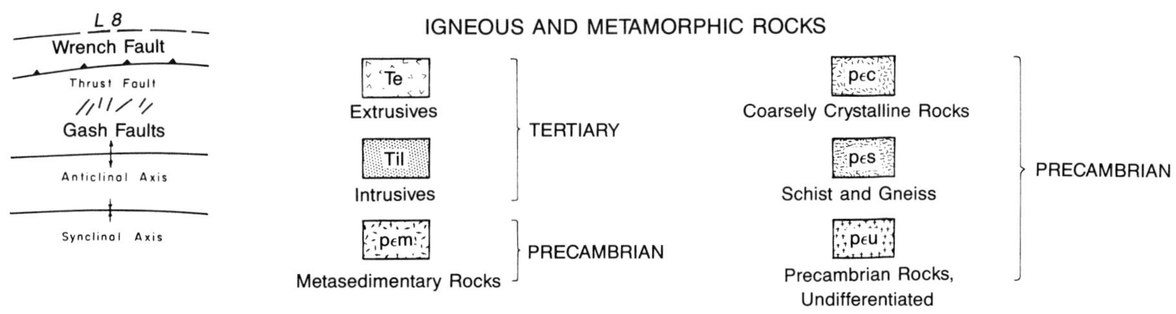

Fig. 2-65 (Stone, 1969)—Tectonic sketch map of the central Rocky Mountains foreland interpreting numerous structures as a result of wrenching. Though wrenching has undoubtedly occurred (e.g., see Fig. 2-66), most of the structures may be a result of shortening caused by compression where σ_3 is vertical instead of horizontal. Permission to publish by Rocky Mountain Association of Geologists.

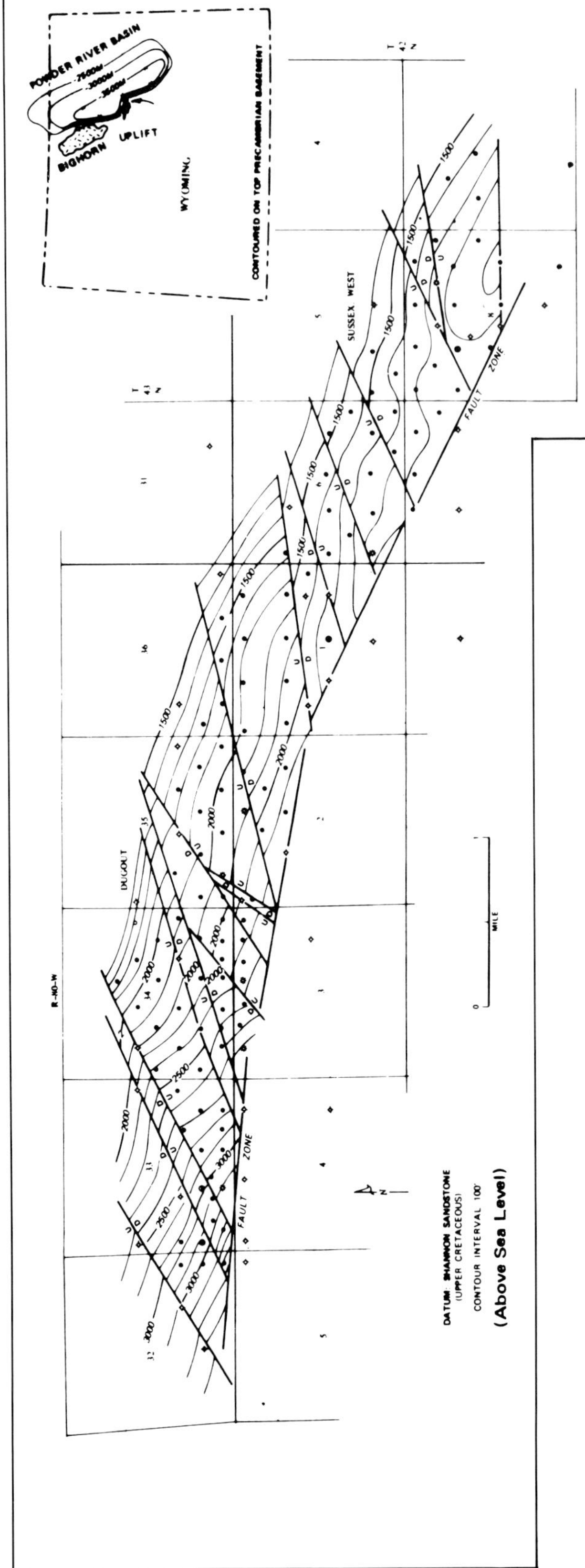

Fig. 2-66 (Harding, 1974)—West Sussex-Dugout trend at the north end of the Casper arch interpreted as foreland conjugate shear with NE-SW-trending en echelon normal faults (in tension fracture orientation) requiring left-lateral slip. The trend also coincides with a left-stepping deflection in the axis of the Powder River Basin and with a zone of suspected left-lateral slip along the southern margin of the Big Horn Basin. Permission to publish by American Association of Petroleum Geologists.

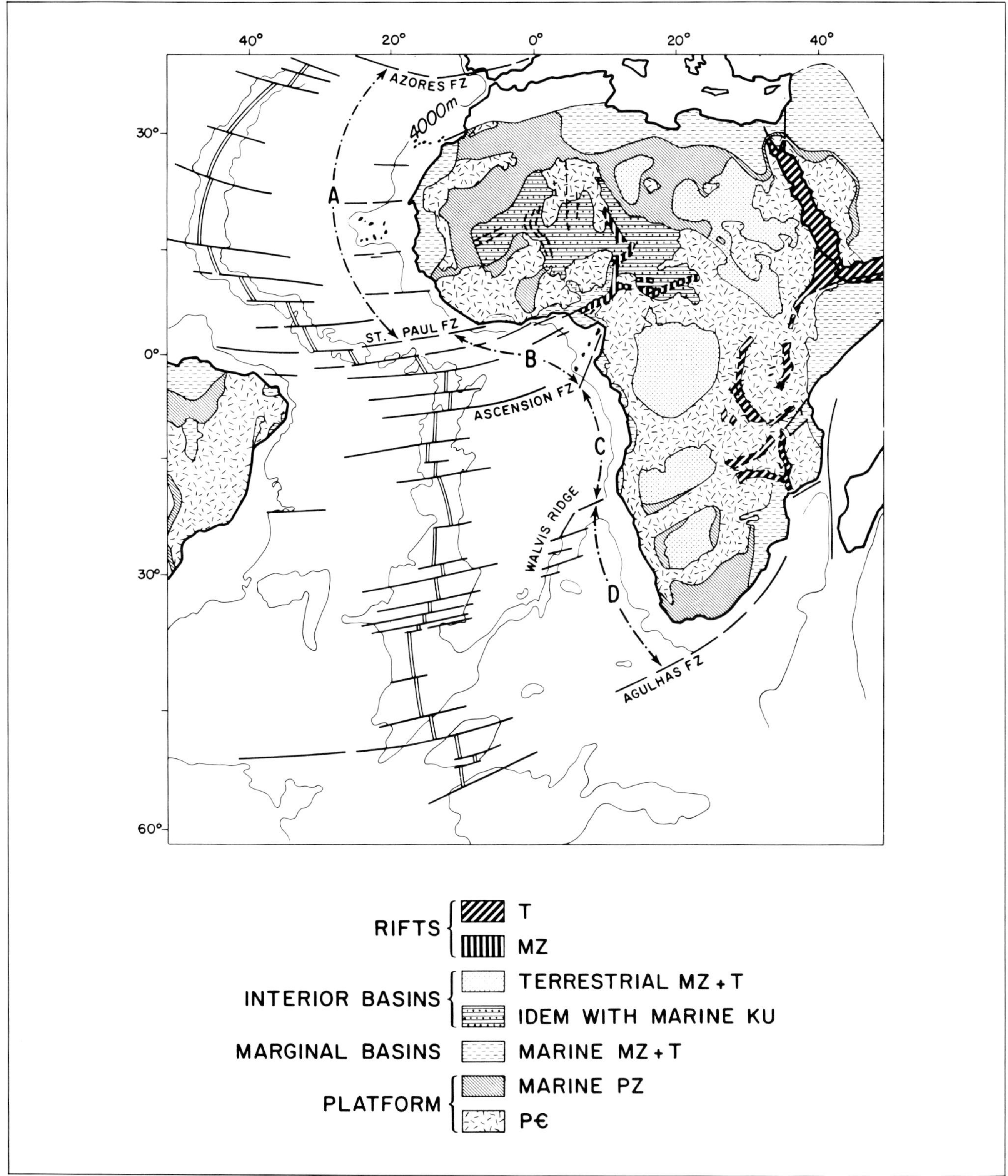

Fig. 2-67 (Lehner and DeRuiter, 1977)—Fracture zones and tectonic trends onshore allow subdivision of Atlantic margin of Africa into four segments which differ in basin history and structure: A, northwest Africa segment, between Azores-Gibraltar and St. Paul-Las Palmas fractures; B, equatorial segment in Gulf of Guinea, between St. Paul fracture zone and Cameroon slope; C, Cameroon-Gabon-Angola segment, between Cameroon slope and Walvis Ridge; D, Walvis Ridge-Cape segment, off Namibia and South Africa, between Walvis Ridge and Agulhas fractures. Permission to publish by American Association of Petroleum Geologists.

Divergent Boundaries

Oceanic spreading centers are usually laced by transform faults. On oceanic crust the transforms are relatively young (usually Cretaceous and younger), but many of them, or their projections, intersect the continental margin slopes and shelves. The affect of these intersections is to compartmentalize the continental margin into basins and subbasins each having slightly different structural and stratigraphic histories. A large scale example is provided by the continental margin of west Africa (Fig. 2-67) and a finer scale example from east Greenland (Fig. 2-68).

The fact that Jurassic age basins in east Greenland are partially outlined by present-day transform faults (Surlyk, 1978) constitutes irrefutable evidence that transform fault directions are determined at an early time. This timing is effectively explained by the block diagram of figure 2-69 that shows an incipient transform fault offsetting two different centers of rifting and initiated at the same time as the rifts. Subsequent break-up, separation, and sea-floor spreading would lead to transform faults on oceanic crust, but the transform direction was predetermined in the early rifting stage.

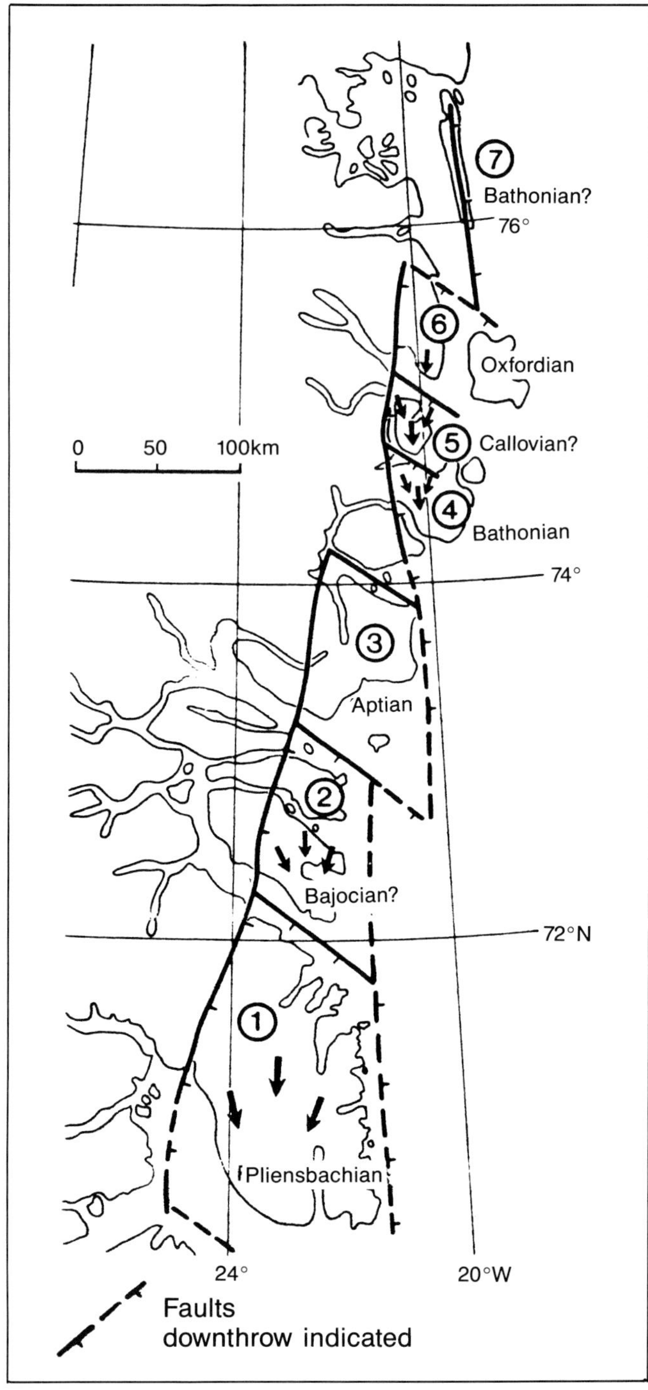

Fig. 2-68 (Surlyk, 1978)—Map of East Greenland with Jurassic fault blocks (numbered) and dominant sediment transport direction (arrows). The time of first post-Triassic transgression is indicated on each block. Jurassic cross faults appear to have localized much younger oceanic transform faults. Cross faults are: between blocks 2 and 3 projected into Jan Mayen fracture zone; between blocks 2 and 1 projected into un-named fracture zone at 70.5°N; hypothetically limiting block 1 projected into Spar fracture zone. Reprinted by permission from Nature. Copyright © Macmillan Journals Limited.

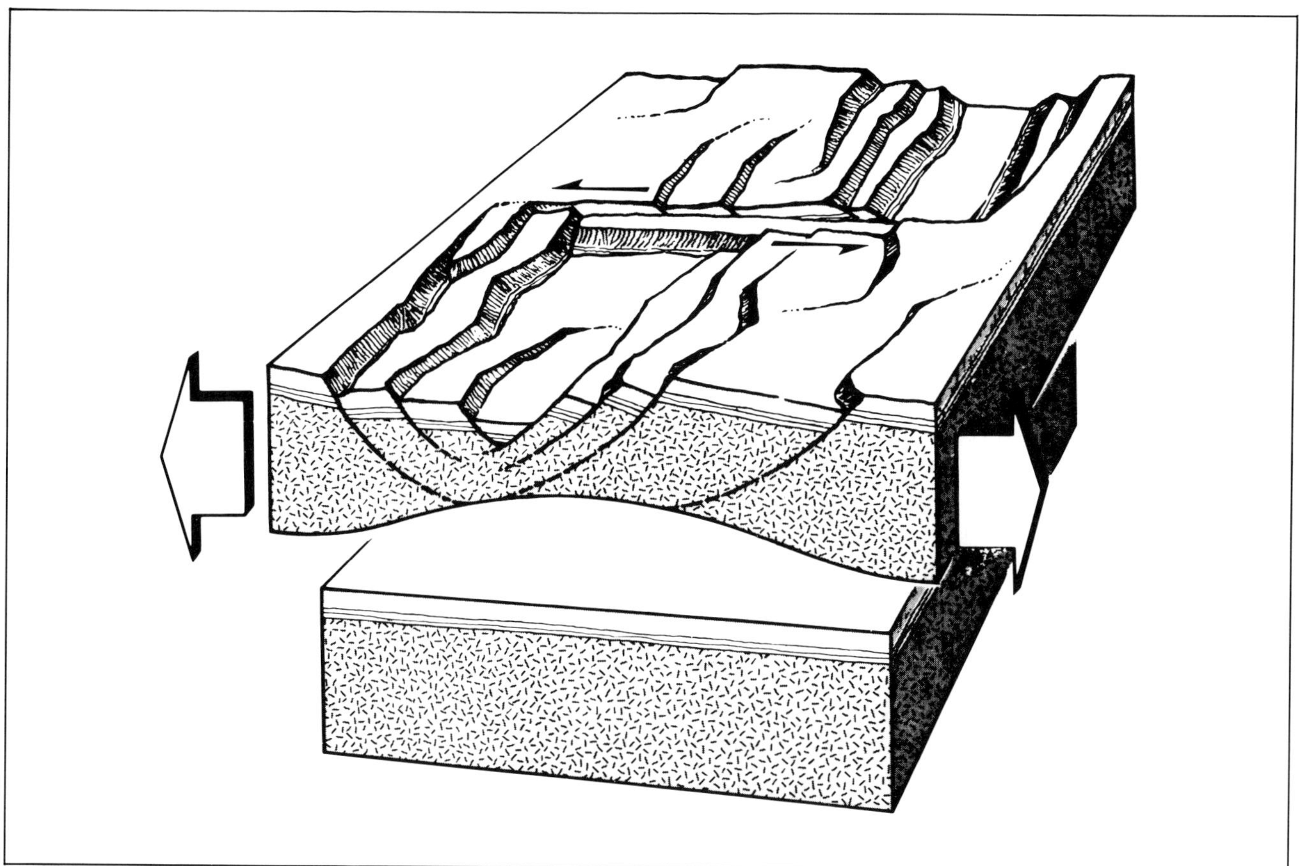

Fig. 2-69 (Bally and Snelson, 1980)—Block diagram which illustrates concept of "early transform faulting" accompanying listric normal faults that effect rifting and provides a mechanism for transverse trends in virtually all stages of rift history. Permission to publish by Canadian Society of Petroleum Geologists.

In the article that follows "Wrench vs. compressional structures with applications to southeast Asia", I wish to correct the statement that the gross mismatch of the Pematang Formation across the Pungut-Tandun fault at Tandun (Fig. 2-75) is solely an indication of wrenching. I now consider it more likely that the thick Pematang Formation was deposited on the northeast side of a normal fault trough (Robert Lambert, pers. comm., October, 1983) and subsequently inverted by a combination of compression and wrenching (see Chapter 10—Structural Inversion).

WRENCH VS. COMPRESSIONAL STRUCTURES WITH APPLICATION TO SOUTHEAST ASIA[3]

Abstract For differentiating between wrench and compressional structures in southeast Asia, criteria are established to distinguish three basic structural types. These are:

1– wrench structures (basement-involved);
2– compressional structures (basement-involved);
3– compressional structures (detached).

Structures in northeast Java are interpreted as compressional in origin, taking the Kawengan oil fields as an example. The Pungut and Tandun oil fields in the central Sumatra basin are interpreted as wrench structures. Regional compression in Java and regional wrench deformation in Sumatra appear to be related to the change in orientation of the subduction zone marking the boundary between the east Indian Ocean and Asian plates.

INTRODUCTION

Both wrench and compressional motion can occur: 1– within basement and its overlying sedimentary cover, or 2– solely within the detached sedimentary cover.

Wrench structures usually involve the basement; detached wrench features are relatively restricted,[4] occurring for example as tear faults in detached thrust-fold belts, and presumably are not common in southeast Asia. In contrast, compressional structures are commonly found in both basement and detached environments, which in southeast Asia are manifested in back-arc and in forearc-trench landward wall settings, respectively. Thus, in differentiating between wrench and compressional structures in southeast Asia, it is necessary to establish criteria to distinguish three basic structural types: 1– wrench structures (basement involved); 2– compressional structures (basement involved); 3– compressional structures (detached).

RECOGNITION CRITERIA

Differentiation between wrench and compressional structures presents inherent difficulties for all wrench motion involves some component of compression. Moreover, when wrench motion is convergent, the compressional dynamics, the maximum principal compressive stress has a horizontal orientation. Minimum principal compressive stress is horizontal in wrenching and vertical in compression. However, the minimum stress can interchange between horizontal and vertical orientation while the maximum stress remains horizontal, so that wrench and compressional motion are not mutually exclusive. Realizing that wrench and compressional deformation can be gradational, it is nevertheless usually possible to discern the differences between the two types (Table 2-2).

TABLE 2-2.
Recognition of Wrench vs. Compressional Structures (partly adapted from Harding and Lowell, 1979).

	BASEMENT INVOLVED		DETACHED
	WRENCH	COMPRESSIONAL	COMPRESSIONAL
Dominant Transport Expression	Strike slip (side-by-side, convergent, or divergent) of sub-regional to regional plates along more or less vertical boundaries.	High to low angle convergent dip slip of blocks and slabs	Subhorizontal to high-angle convergent dip slip of sedimentary cover in sheets and slabs.
Displacement Sense	Inconsistent	Consistent	Consistent
Trends	En echelon[5] array of: folds, synthetic, antithetic and P shears, normal faults.	Relay[6] and Dogleg[7] three directional.	Relay—one directional.
Fold Form (when present)	Symmetric ± and fold axis oblique to associated fault.	Asymmetric and fold axis parallel to associated fault.	Asymmetric and fold axis parallel to associated fault.

[3]SEAPEX Proceedings, V. v, 1980, p. 63–70.
[4]Detachment in the wrench style may be more common than previously suspected. See figure 2-32.

[5]En echelon—Consistently overlapped structures aligned parallel with one another but oblique to the zone of deformation in which they occur.
[6]Relay—Inconsistently overlapped elements aligned parallel with one another and with the zone of deformation in which they occur.
[7]Dog-leg—Zigzag, obtuse-angle trends.

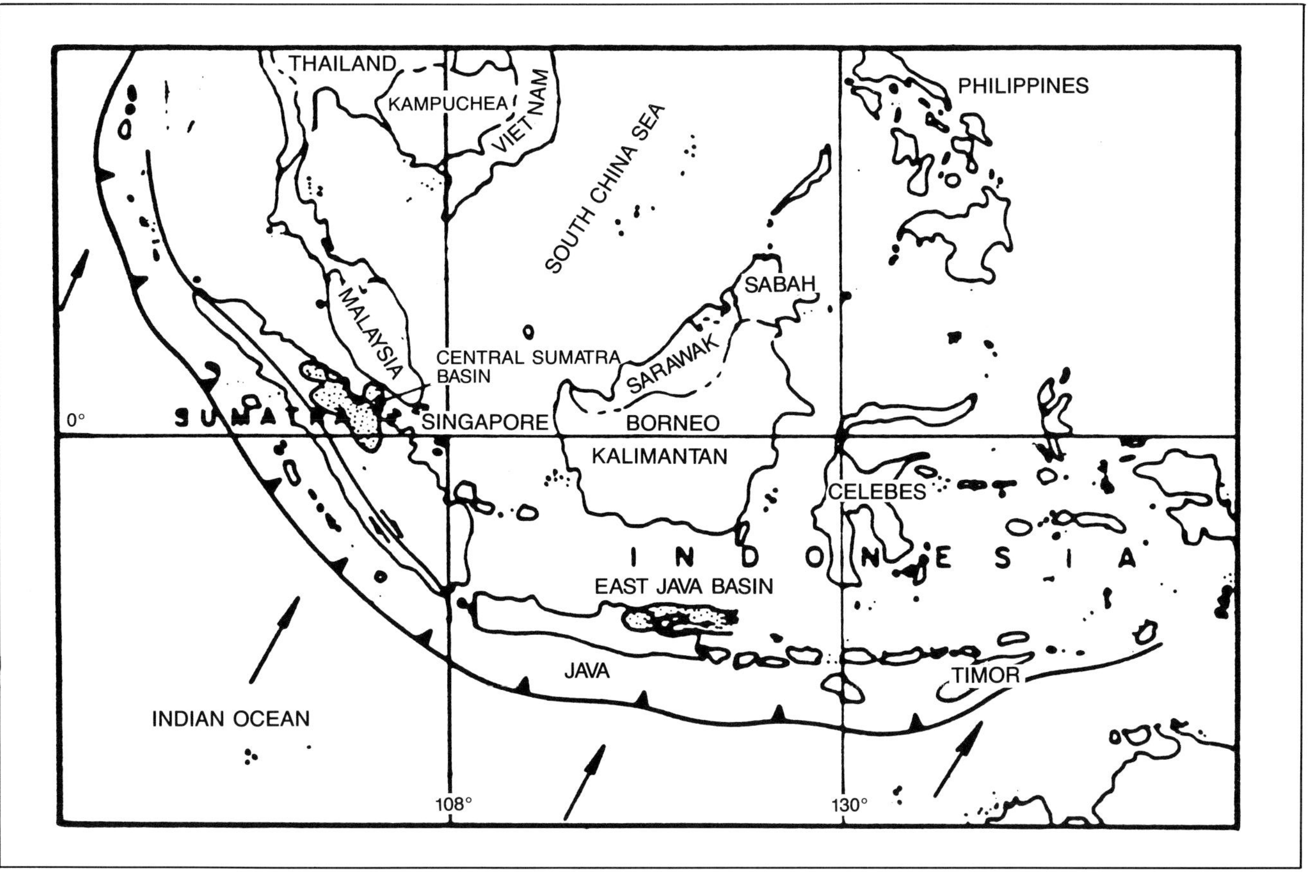

Fig. 2-70—Map of southeast Asia showing locations of central Sumatra and east Java basins, trace of subduction zone, and direction of motion of Indian Ocean plate relative to southeastern Eurasia.

Examples from Indonesia

Examples of possible compressional and wrench structures are taken from northeast Java and the central Sumatra basin, respectively, in Indonesia (Fig. 2-70).

On the basis of fold asymmetry and fold parallelism with associated faults, structures in northeast Java appear to be caused by compression. Anticlinal trends in northeast Java are shown in Figure 2-71, adapted from Soetantri et al. (1973). The structures are of Pliocene and Pleistocene age and involve Oligocene to Pleistocene sediments. Mapped oilbearing structures appear to be simple anticlines with gentle plunge and often show asymmetric limbs. They are often bounded by longitudinal faults which may become reverse faults and thrusts. A good example is the structural trend containing the Kawengan oil fields (Fig. 2-72). It is not known whether this deformation is basement-involved, but the structural modification by shale diapirism in this basin (Sutarso and Suyitno, 1976) suggests some degree of detachment.

A compressive interpretation of the northeast Java structures differs from the wrench origin suggested by Situmorang et al. (1976), but seems consistent with the results of later seismic surveys in the area (Kariyoso et al., 1977).

In contrast, the Pungut and Tandun oil fields, described by Mertosono (1975) from the central Sumatra basin, show several indications of wrenching. The locations of these structures are shown in figures 3-73 and 3-74. The structures are associated with a major NNW-SSE fault cutting Palaeogene to Pliocene sediments. Fault movement occurred during the early to middle Miocene and was reactivated during the Plio-Pleistocene.

The indications of wrenching include the inconsistent attitude of the main fault and inconsistent displacement across the fault shown by seismic data from the two fields. The fault is reverse, downthrown to the west, at the Tandun field (Fig. 2-75) and normal, downthrown to the east, at the Pungut field (Fig. 2-76). This pattern is consistent with right-lateral wrenching, where changes in the orientation of the fault trend (Fig. 2-74) cause zones of convergent wrenching (Tandun) and divergent wrenching (Pungut).

Another indication of wrenching is the gross mismatch of the Pematang Formation across the fault at Tandun (Fig. 2-75). The expected oblique relationship between the fold axes and the fault trend is not well developed but is suggested by the Tandun structure.

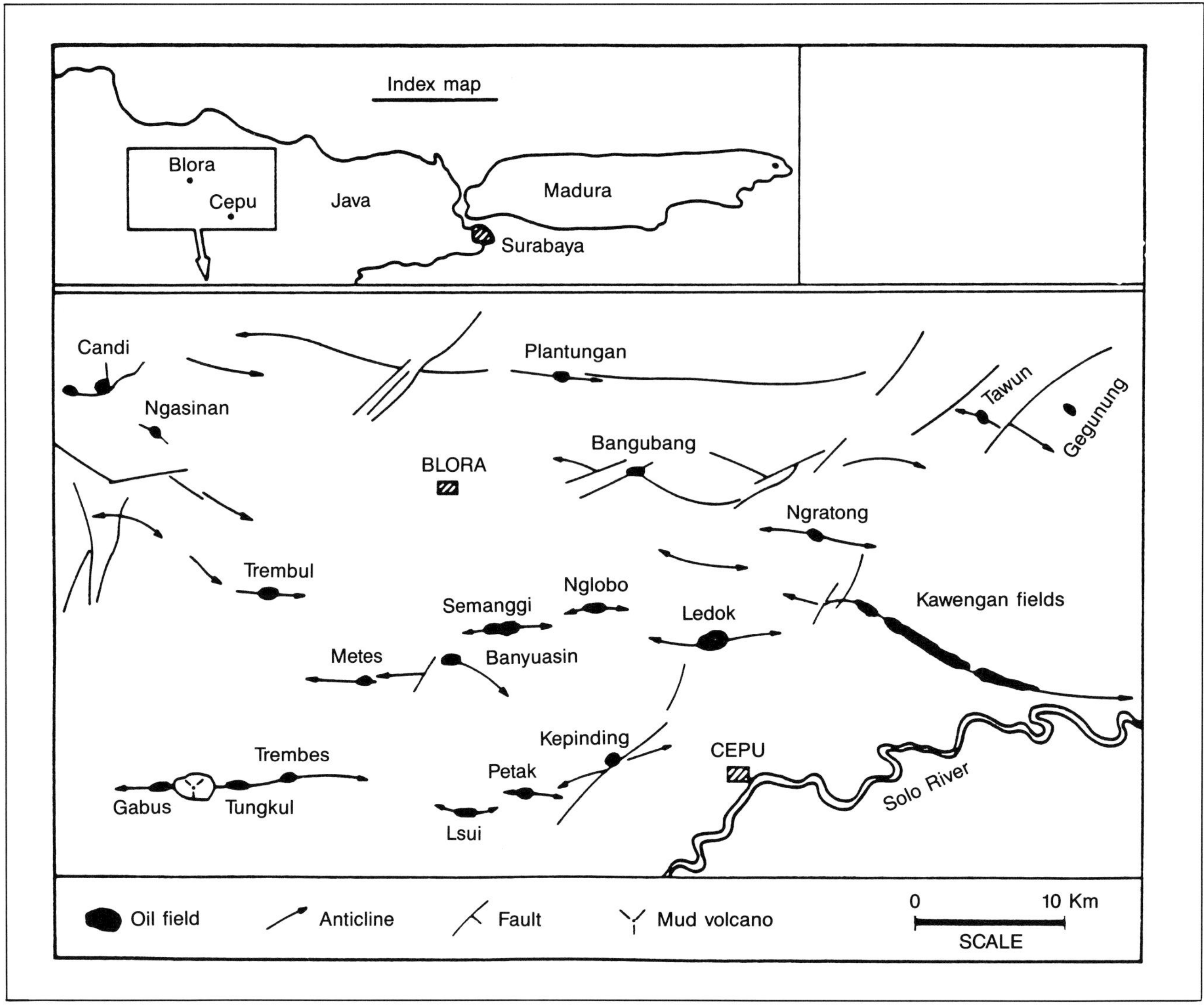

Fig. 2-71—Structural trends and oil fields, northeast Java (from Soetantri et al., 1973).

REGIONAL RELATIONSHIPS

This difference in structural style from northeast Java to central Sumatra appears to be related to the direct convergence and attendant compression between the east Indian Ocean plate and the Asian plate over the entire length of Java, in contrast to the component of wrenching, superimposed on the convergence, between the two plates in the area of central Sumatra (Fig. 2-70). According to this interpretation, the consequences of plate interaction are manifested as much as 500 km (312 mi) from the actual plate boundaries.

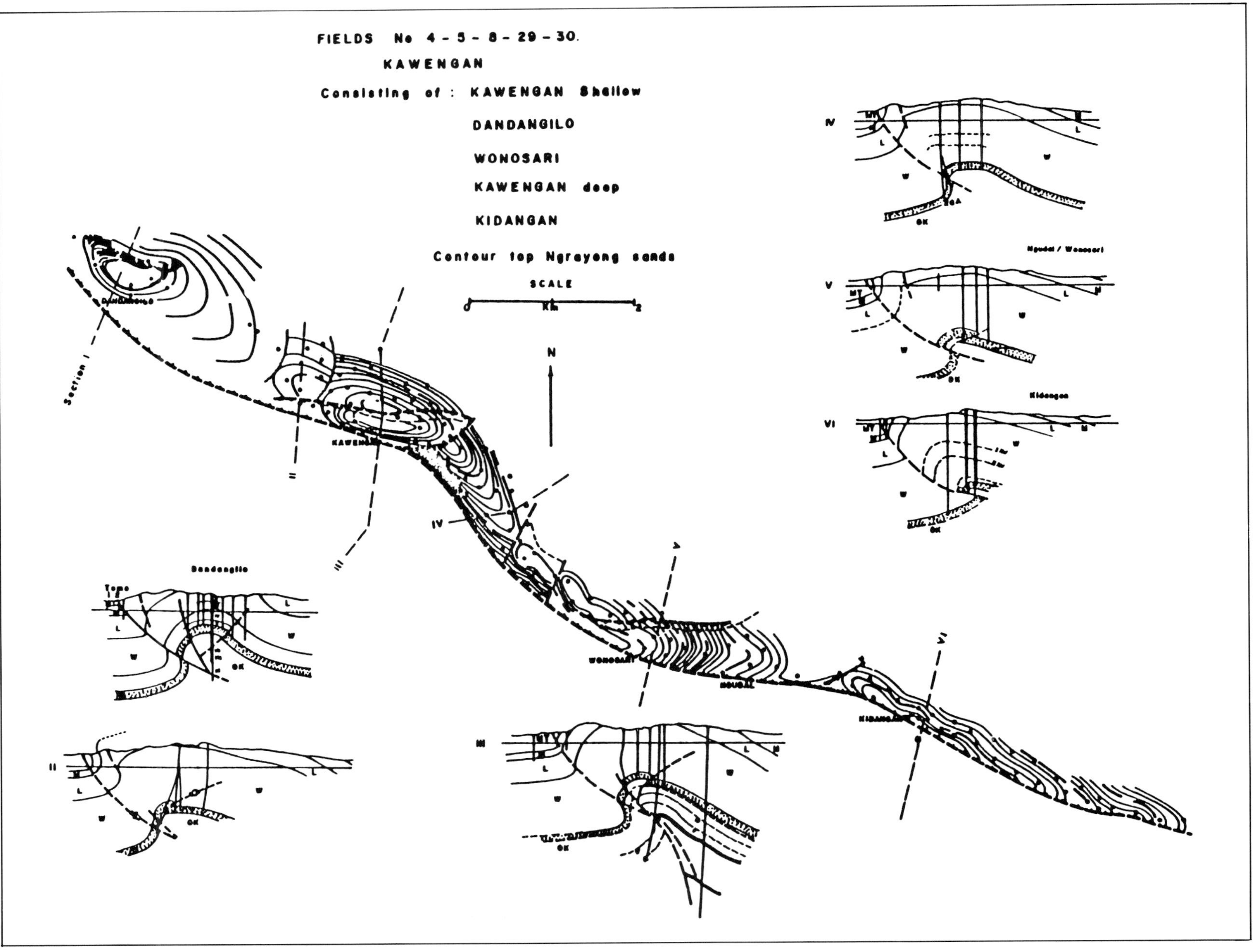

Fig. 2-72—Kawengan fields: structural contours on top Ngrayong sands and structural cross sections I to VI.

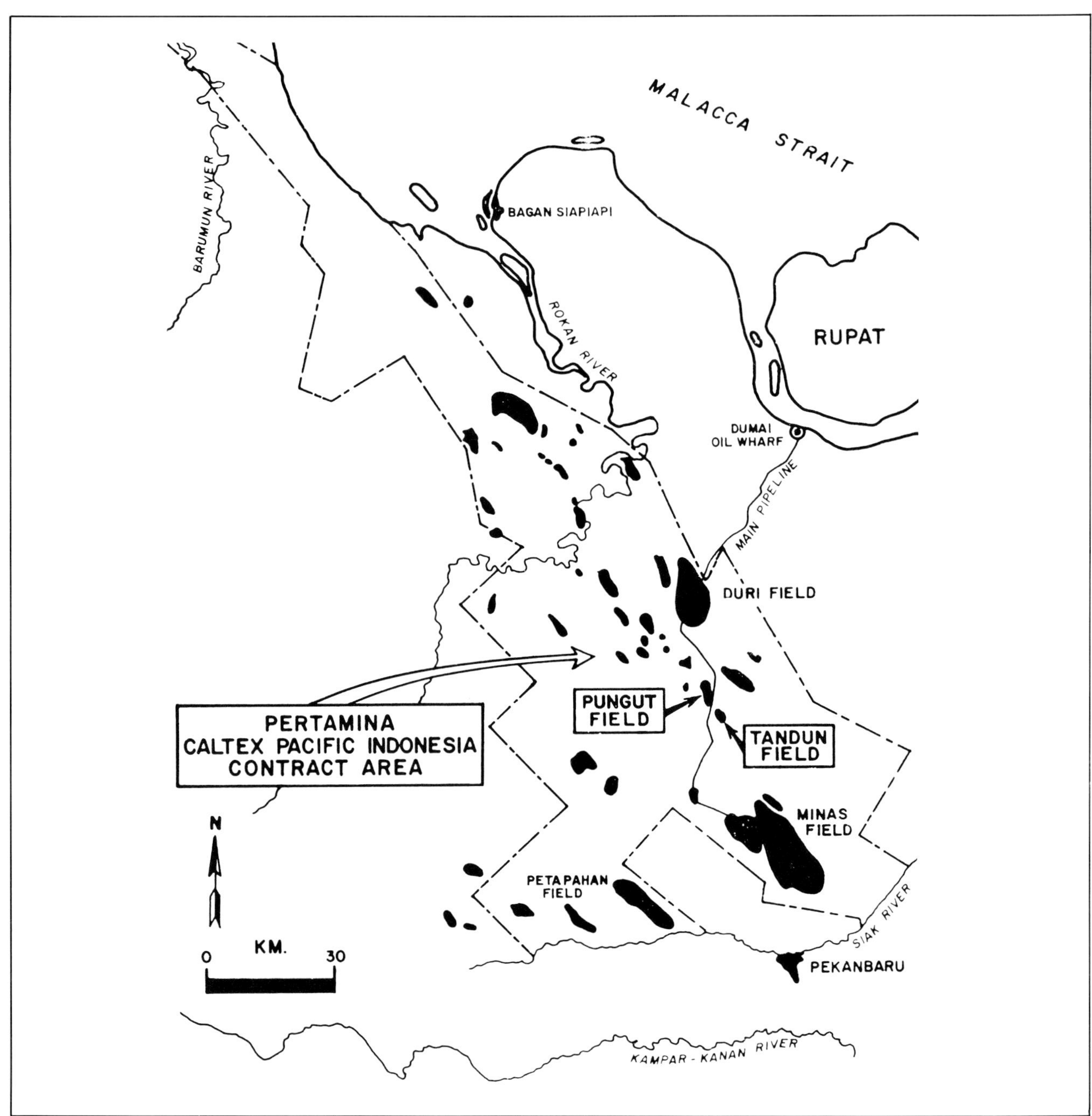

Fig. 2-73—Location map, Pungut and Tandun fields, central Sumatra.

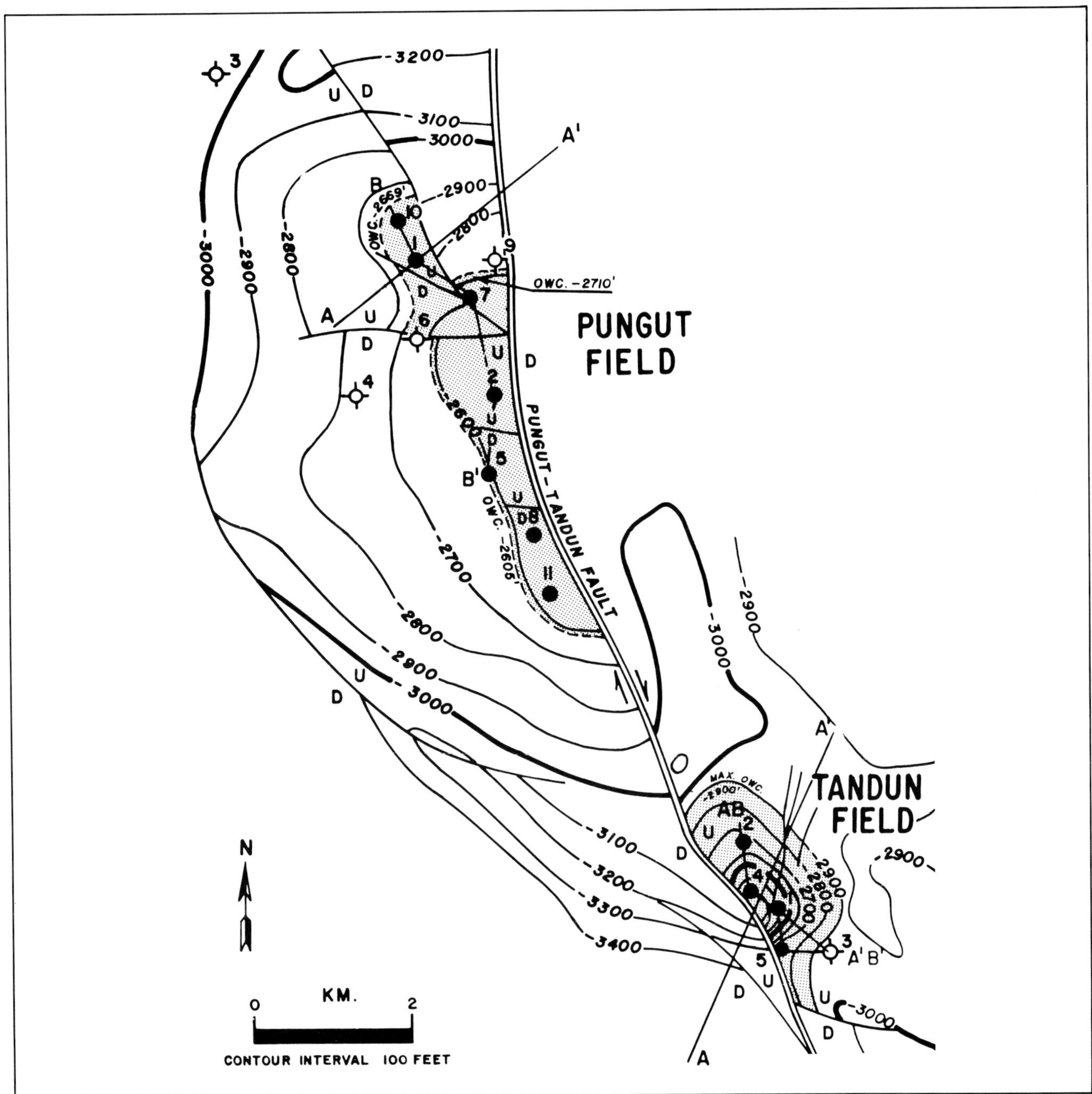

Fig. 2-74—Pungut and Tandun fields: structural contour map on top Sihapas Group.

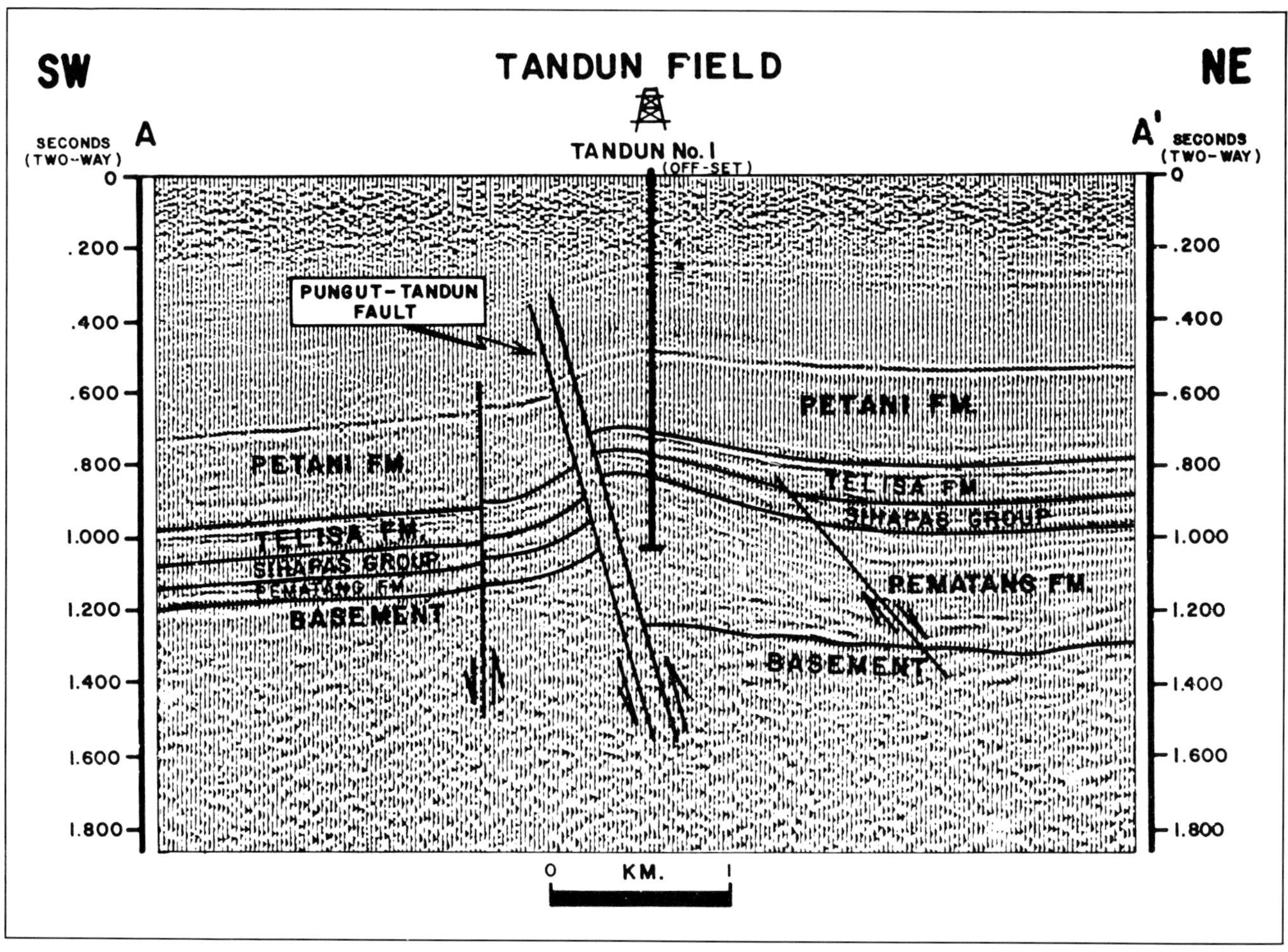

Fig. 2-75—SW-NE seismic section A-A' across Tandun field (location shown in Fig. 2-74).

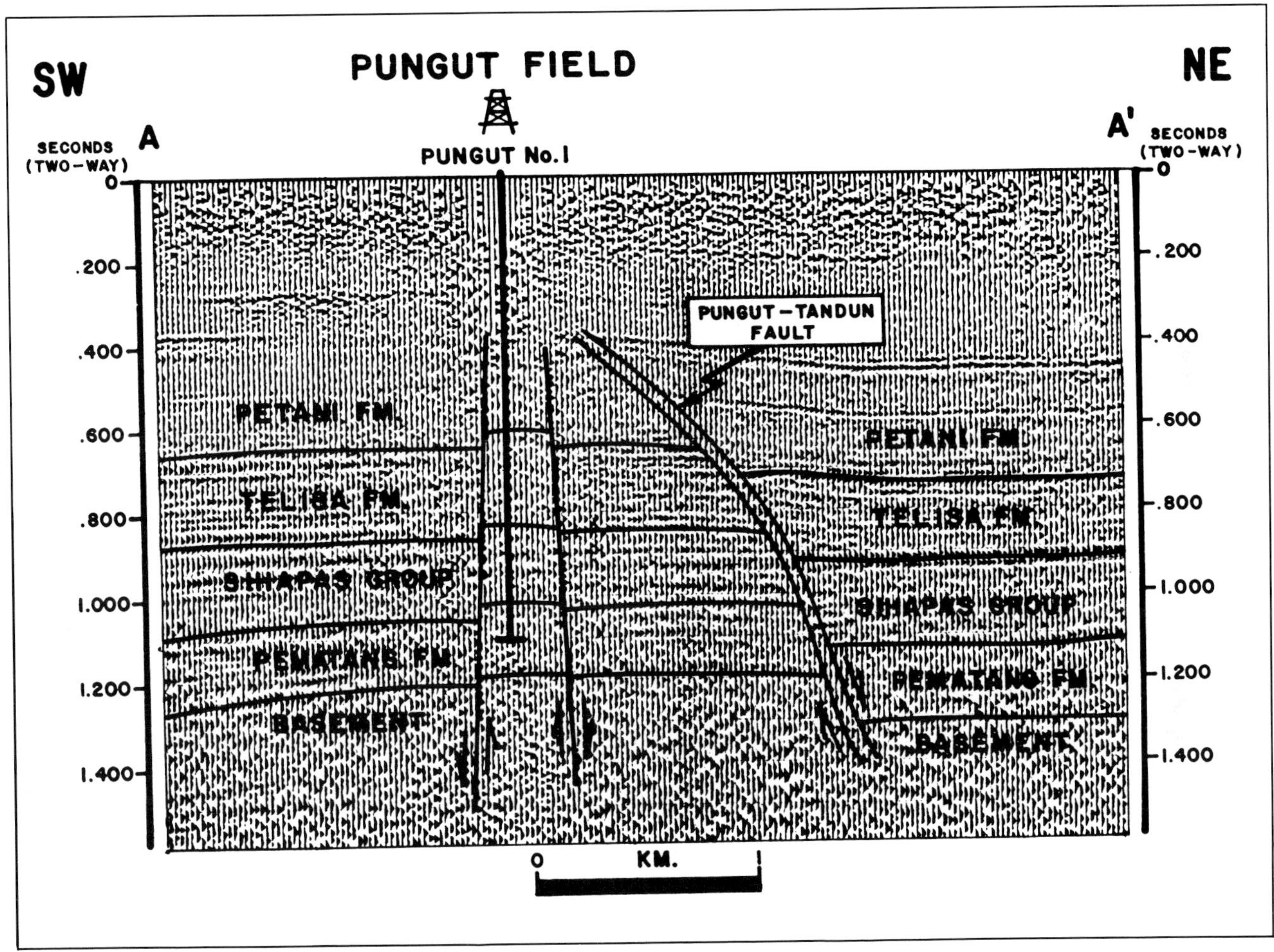

Fig. 2-76—SW-NE seismic Section A-A' across Pungut field (location shown in Fig. 2-74).

SUMMARY

Numerous criteria serve to distinguish the wrench assemblage from other structural styles:

1– all wrench assemblage structures, except upthrust-bounded welts that are parallel to the deep-seated wrench zone, occur in an en echelon arrangement and this pattern appears to be peculiar to the wrench style.

2– the acute angular relationship of wrench-related folds to major associated and causative faults also seems unique to wrenching.

3– wrench assemblage folds show greater symmetry than those of other styles.

4– it is likely that most faults of all other styles are listric, either normal or thrust, whereas most wrench-related faults are closer to vertical.

5– change of displacement (propellering or scissoring) along a dipping fault requires the unlikely mechanical condition that both normal and reverse movement have occurred on the same fault. Change in displacement on a vertical fault, however, is mechanically quite simple. I believe that a true change in displacement sense along the same fault strongly implies at least some wrench deformation.

Finally, numerous characteristics of the wrench assemblage are given in table 1-1 (Chapter 1).

REFERENCES

Allen, C. R., 1965, Transcurrent faults in continental areas, *in* A symposium on continental drift: *Royal Society of London Philosophical Transactions,* n. 1088, p. 82–89.

Alverson, D. C., Woloshin, D. P., Terman, M. J., and Woo, C. C., 1967, Atlas of Asia and eastern Europe to support detection of underground nuclear testing, V. II, *Tectonics:* Washington, D. C. United States Geological Survey.

Anderson, E. M., 1951, The dynamics of faulting and dyke formation, with applications to Britain: Edinburgh, Oliver and Boyd, 2nd ed., 206 p.

Atkinson, D. J., 1963, Tertiary rocks of Spitsbergen: *American Association Petroleum Geologists Bull.,* V. 47, p. 302–323.

Bally, A. W., Gordy, P. L., and Stewart, G. A., 1966, Structure, seismic data, and orogenic evolution of southern Canadian Rocky Mountains: *Bulletin Canadian Petroleum Geology,* V. 14, p. 337–381.

Bally, A. W., and Snelson, S., 1980, Facts and principles of world petroleum occurrence: realms of subisdence: *Canadian Society of Petroleum Geologists, Mem. 6,* p. 9–90.

Bartlett, W. L., Friedman, M., and Logan, J. M., 1981, Experimental folding and faulting of rocks under confining pressure, Part IX. Wrench faults in limestone layers: *Tectonophysics,* V. 79, p. 255–277.

Beck-Mannagetta, P., 1964, Geologische Ubersichtskarte der Republik Osterreich, 1:1,000,000: Wein, Geologische Bundensanstalt.

Beloussov, V. V., 1962, Basic problems in geotectonics: New York, McGraw-Hill Book Co., Inc., 816 p.

Bol, A. J., and Van Hoorn, B., 1978, Structural styles in western Sabah offshore: Geological Society Malaysia, 2nd Petroleum Seminar, Kuala Lumpur.

Brown, W. G., 1981, Washita Valley fault system—a new look at an old fault (abs.): *American Association Petroleum Geologists Bull.,* V. 65, p. 1496.

California Division of Oil and Gas, 1961, California oil and gas fields, maps, and data sheets, pt. 2, Los Angeles-Ventura basins and central coastal regions: California Division Oil and Gas, p. 495–913.

———, 1971, Oil and gas statistics: California Division Oil and Gas, California Oil Fields—Summ. Operations, V. 57, n. 2, p. 68–79.

Challinor, A., 1967, The structure of Broggerhalvoya, Vestspitsbergen: *Geological Magazine,* V. 104, p. 322–336.

Chapin, C. E., and Cather, S. M., 1981, Eocene tectonics and sedimentation in the Colorado Plateau-Rocky Mountain area, *in* Dickinson, W. R., and Payne, W. D., eds., Relations of tectonics to ore deposits in the southern Cordillera: *Arizona Geological Society Digest,* V. 14, p. 173–198.

Cloos, Ernst, 1955, Experimental analysis of fracture patterns: *Geological Society America Bulletin,* V. 66, p. 241–256.

Crouch, J. K., Bachman, S. B., and Shay, J. T., 1984, Post-Miocene compressional tectonics along the central California margin *in* Crouch, J. K., and Bachman, S. B., eds., Tectonics and sedimentation along the California margin: Society Economic Paleontologists and Mineralogists, Pacific section, V. 38, p. 37–54.

Deratmiroff, G. N., 1971, Late Cenozoic imbricate thrusting in Venezuelan Andes: *American Association Petroleum Geologists Bull.* V. 55, p. 1336–1355.

Dewey, J. F., and Bird, J. M., 1971, Origin and emplacement of the ophiolite suite: Appalachian ophiolites in Newfoundland: *Journal Geophysical Research,* V. 76, p. 3179–3206.

Emmons, R. C., 1969, Strike-slip fracture patterns in sand models: *Tectonophysics,* V. 7, p. 71–87.

Fitch, T. J., 1972, Plate convergence, transcurrent faults, and internal deformation adjacent to Southeast Asia and western Pacific: *Journal Geophysical Research,* V. 77, p. 4432–4460.

Flood, B., Nagy, J., and Winsnes, T. S., 1971, Geological map of Svalard 1:500,000, Sheet IG, Spitsbergen southern part: *Norsk Polarinstitutt Skrifter* Nr. 154A.

Gibbs, A. D., 1984, Structural evolution of extensional basin margins: Journal Geological Society, v. 141, Part 4, p. 609–620.

Harding, T. P., 1973, Newport-Inglewood trend, California—an example of wrenching style of deformation: *American Association Petroleum Geologists Bull.* V. 57, p. 97–116.

———, 1974a, Petroleum traps associated with wrench faults: *American Association Petroleum Geologists Bull.,* V. 58, p. 1290–1304.

———, 1974b, Acumulaciones importantes de hidrocarburos originades por deformaciones causadas por fallas laterales: Petrotecnia, Instituto Argentino del Petroleo, no 2–3, p. 17–22; n. 4, p. 12–19.

———, 1976, Tectonic significance and hydrocarbon trapping consequences of sequential folding synchronous with San Andreas faulting, San Joaquin Valley, California: *American Association Petroleum Geologists Bull.,* V. 60, p. 356–378.

———, 1983, Divergent wrench fault and negative flower structure, Andaman Sea, *in* Bally, A. W., ed., Seismic expression of structural styles: *American Association Petroleum Geologists Studies in Geology #15,* V. 3.

Harding, T. P., and Lowell, J. D., 1979, Structural styles, their plate tectonic habitats, and hydrocarbon traps in petroleum provinces: *American Association Petroleum Geologists Bull.,* V. 63, p. 1016–1058.

Harding, T. P., Vierbuchen, R. C., and Christie-Blick, N., 1984, Structural styles and plate tectonic settings of divergent (transtensional) wrench faults (*abs*): *American Association Petroleum Geologists Bull.* V. 68/4. p. 484.

Harland, W. B., 1965, The tectonic evolution of the Artic-North Atlantic region: *Royal Society of London Philosophical Transactions,* V. 258, ser. A, p. 59–75.

———, 1969, Contribution of Spitsbergen to understanding of tectonic evolution of North Atlantic region: *American Association Petroleum Geologists Mem. 12,* p. 817–851.

———, 1971, Tectonic transpression in Caledonian Spitsbergen: *Geological Magazine,* V. 108, p. 27–42.

Hoots, H. W., Bear, T. L., and Kleinpell, W. D., 1954, Geological summary of the San Joaquin Valley, California, *in* Jahns, R. H., ed., Geology of southern California: *California Division Mines Bull.* 170, chap. 2, pt. 8, p. 113–129.

Horsfield, W. T., and Maton, P. I., 1970, Transform faulting along the De Geer line: *Nature,* V. 226, p. 256–257.

Howell, B. F., Jr., 1954, Geology of the Little Tujunga area, Los Angeles County, *in* Jahns, R. H., ed., Geology of southern California: *California Division of Mines and Geology Bull.,* 170, map sheet no. 10.

Jennings, C. W., and Troxel, B. W., 1954, Geology guide through the Ventura basin and adjacent areas, *in* Jahns, R. H., ed., Geology of southern California: *California Division of Mines and Geology Bull.,* 170, *Geology,* Guide no. 2, p. 63.

Julivert, M., 1970, Cover and basement tectonics in the Cordillera Oriental of Colombia, South America, and a comparison with some other folded chains: *Geological Society America Bull.,* V. 81, p. 3623–3646.

Kariyoso, G., Effendi, R., and Sugijanto, 1977, Seismic survey in the northeast Java Basin: Indonesian Petroleum Association, Proc. 6th Annual Conv., V. II, p. 13–41.

Katili, J. A., 1970, Large transcurrent faults in Southeast Asia with special reference to Indonesia: *Geologische Rundschau,* V. 59, p. 581–600.

Lehner, P., and DeRuiter, P. A. C., 1977, Structural history of Atlantic margin of Africa: *American Association Petroleum Geologists Bull.,* V. 61, p. 961–981.

Lehner, P., Doust, H., Bakker, E., Allenbach, P., and Guneau, J., 1983, Active margins, part 5—South American trench, profiles P-1304, P-1307, and P-1017, *in* Bally, A. W., ed., Seismic expression of structural styles: *American Association Petroleum Geologists Studies in Geology #15,* V. 3.

LePichon, X., 1968, Sea-floor spreading and continental drift: *Journal Geophysical Research,* V. 73, p. 3661–3697.

———, Hyndman, R. D., and Pautot, G., 1971, Geophysical study of the opening of the Labrador Sea: *Journal Geophysical Research,* V. 76, p. 4724–4743.

Lovering, T. S., 1928, The fracturing of incompetent beds: *Journal Geology,* V. 36, p. 709–717.

Lowell, J. D., 1969, On the origins of upthrusts: *Geological Society America, Abs. with Programs for 1969,* Pt. 7 (Ann. Mtg.), p. 138.

———, and Genik, G. J., 1972, Sea-floor spreading and structural evolution of southern Red Sea: *American Association Petroleum Geologists Bull.,* V. 56, n. 2, p. 247–259.

Lowell, J. D., Genik, G. J., Tucker, P. M., and Nelson, T. H., 1975, Petroleum and plate tectonics of the southern Red Sea, *in* Fisher, A. G., and Judson, S., eds., Petroleum and global tectonics: Princeton University Press, p. 129–153.

Mertosono, S., 1975, Geology of the Pungut and Tandun oil fields: *Indonesian Petroleum Association, Proc.* 4th Annual Conv., V. I, p. 165–179.

Morgan, W. J., 1968, Rises, trenches, great faults, and crustal blocks: *Journal Geophysical Research,* V. 73, p. 1959–1982.

———, 1972, Plate motion and deep mantle convection:

Geological Society America, Mem. 132.

Nordenskiold, A. E., 1866, Utkast till Spitsbergens geologi, Stockholm 1866: Kgl. *Svenska Vetenskapsakad.* Handl., B. 6, nr. 7, 35 p.

Orvin, A. K., 1940, Outline of the geological history of Spitsbergen: *Skrifter om Svalbard og Ishavet,* no. 78.

Parker, J. R., 1967, The Jurassic and Cretaceous sequence in Spitsbergen: *Geological Magazine,* V. 104, p. 487–505.

Pitman, W. C., III, and Talwani, M., 1972, Sea-floor spreading in the North Atlantic: *Geological Society America Bull.,* V. 83, p. 619–646.

Rozycki, S. Z., 1959, Geology of the north-western part of Torell Land Vestspitsbergen: *Studia Geologica Polonica,* V. 2, p. 1–98.

Situmorang, B., Siswoyo, Thajib, E., and Paltrinieri, F., 1976, Wrench fault tectonics and aspects of hydrocarbon accumulation in Java: Indonesian Petroleum Association, Proc. 5th Annual Conv., v. II, p. 53–67.

Soetantri, B., Samuel, L., and Nayoan, G. A. S., 1973, The geology of the oilfields in northeast Java: Indonesian Petroleum Association, Proc. 2nd Annual Conv., p. 149–175.

Stocklin, J., 1968, Structural history and tectonics of Iran: A review: *American Association Petroleum Geologists Bull.,* V. 52, p. 1229–1258.

Stone, D. S., 1969, Wrench faulting and Rocky Mountain tectonics: *Mountain Geologist,* V. 6, n. 2, p. 67–79.

Surlyk, F., 1978, Jurassic basin evolution of East Greenland: *Nature,* V. 274, n. 5667, p. 130–133.

Sutarso, B., and Suyitno, P., 1976, The diapiric structures and relation to the occurrence of hydrocarbon in northeast Java Basin: Indonesian Association Geology, 5th Annual Meeting, Jogjakarta.

Tanner, J. H. III, 1967, Wrench fault movements along Washita Valley fault, Arbuckle Mountain area, Oklahoma: *American Association Petroleum Geologists Bull.,* V. 51, p. 126–134.

Temple, P. G., and Zimmerman, J., 1969, Tectonic significance of alpine ophiolites in Greece and Turkey: *Geological Society America, Abs.* with Programs for 1969, Pt. 7 (Ann. Mtg.), p. 221–222.

United States Geological Survey and American Association of Petroleum Geologists, 1962, Tectonic map of the United States: Washington, D. C., U. S. Geological Survey.

Vogt, P. R., Ostenso, N. A., and Johnson, G. L., 1970, Magnetic and bathymetric data bearing on sea-floor spreading north of Iceland: *Journal Geophyscial Research,* V. 75, p. 903–920.

White, D. A., Roeder, D. H., Nelson, T. H., and Crowell, J. C., 1970, Subduction: *Geological Society America Bull.,* V. 81, p. 3431–3432.

Wilcox, R. E., Harding, T. P., and Seely, D. R., 1973, Basic wrench tectonics: *American Association Petroleum Geologists Bull.,* V. 57, p. 74–96.

3 COMPRESSIVE BLOCKS AND BASEMENT THRUSTS

INTRODUCTION

Except for inversion structures, thrusts and reverse faults involving the basement are almost always associated with convergent plate boundaries (Plate 1-1). Common habitats of basement thrusts are the forearc setting—deep in the landward walls of oceanic trenches and the backarc setting—the internal sides of detached thrust-fold belts. Neither setting is prospective for oil and gas, the former mainly because of high pressure metamorphism, the latter because of high temperature metamorphism.

Compressive blocks are characteristically, if not exclusively, found in the forelands of detached thrust-fold belts. "Foreland Deformation" and "Plate Tectonics and Foreland Basement Deformation," reprinted in the following pages, provide an introduction, background, plate tectonic view, and up-date of the compressive block style.

FORELAND DEFORMATION[1]

FORELAND STRUCTURES

Basement-involved

Compression that telescopes and shortens the pre-existing sedimentary wedge is generally accepted as the dominant deformation in thrust-fold belts (Fig. 3-1). Shortening is also believed by most to have occurred in the basement of the foreland, but the cause is much more controversial. Shortening in the foreland has been attributed to compression in the basement, but also to vertical uplift of basement. Even where compression may have occurred in the deep crust there is controversy as to whether the horizontal component was dominant over the vertical component at the level of the contact between basement and sedimentary rocks and in the rocks above.

This controversy is very difficult to resolve because most of the data gathered in foreland studies (such as surface attitudes of faults, block asymmetry, trends, etc.) permit more than one interpretation. In my opinion, three observations of basement structures are pertinent to the problem of dominant horizontal vs. vertical component. First, on many mountain front faults that part of the fault surface between the lead edge (marked by surface or subcrop traces) and a well or wells drilled through basement one to several miles behind the lead edge usually dips much less than 45° (Gries, 1981), so that horizontal component is dominant. Second, the dip of the fault mountainward from the well(s) is usually unknown, however high quality reflection seismic lines across the Wind River Range (Fig. 3-2) and the Casper arch (Gries, 1983; Sprague, 1983) demonstrate low-angle basement thrusts and thrust overlap of 16 km (10 mi) or more. Third, the presence of overturned section beneath the basement thrust wedge for significant distances behind the lip of the wedge (Gries, 1981) is most easily accomplished by low-angle thrusting, as exemplified in detached thrust-fold belts.

In addition, the observation by Gries (1983a, figs. 11, 16, and 17) that opposing low-angle reverse and thrust faults occur on both flanks of foreland uplifts, such as the Owl Creek Range, Granite, and Uinta mountains,

[1] Article by James D. Lowell. From Rocky Mountain Foreland Basins and Uplifts, 1983, p. 1–8. Reprinted with permission from the Rocky Mountain Association of Geologists.

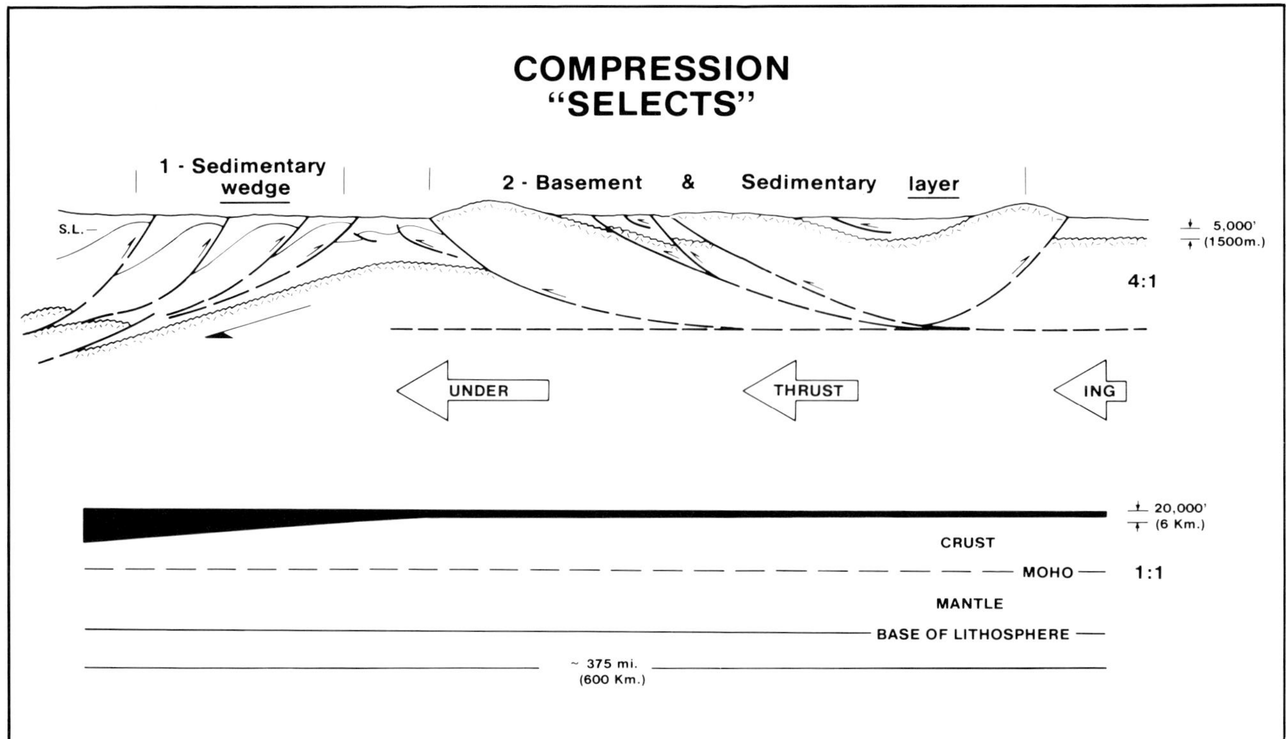

Fig. 3-1—Detached thrust-fold belt and foreland structures as a result of compression. Compression concentrated in sedimentary wedge of thrust-fold belt, except for basement thrusts at far left; compression of both basement and sedimentary cover of foreland to account for both basement-involved and detached structures there. Shortening in foreland is envisioned as a response to underthrusting from east to west as compression "works" eastward and finally dissipates to the east or right side of section (Lowell, 1977); as foreland structures finally terminate eastward, compression may be manifested mainly as vertical uplift, as in the Big Horn Mountains (Stearns and Stearns, 1978). In contrast to vertical uplift a key point in my interpretation is that faults are shown to flatten, rather than steepen (upthrust), with depth; nevertheless a substantial amount of uplift can be accommodated on both thrusts and moderate to steep reverse faults by this mechanism. A deep detachment or glide horizon is shown diagrammatically in the basement. It is interesting to compare the deep detachment with a shallower one in the Franklin Mountains of the Northwest Territories of Canada (Cook, 1983); both should facilitate shortening.

constitutes compelling evidence for shortening of the basement by compression with horizontal component dominant over vertical component.

Detached

Detached structures have been recognized in localized settings in the foreland for many years (see, for example, Beckwith, 1941 and Murphy and Roberts, 1954). Because they are dwarfed by the larger relief basement structures (Fig. 3-1), they have never been treated systematically or put in a proper framework until very recently (Petersen, 1983; Lowell, 1983). Low-angle and bedding-plane thrusts and associated folds are widespread in the sedimentary section above basement. Moreover, they are distributed in a variety of structural settings ranging from gentle to steep structural flanks (Pierce, 1966; Petersen, 1983) and structural crests (Murphy et al., 1956; Petersen, 1983).

Detachment is not simply a mechanism that locally accomodates bed-length problems of drape folds above vertically uplifted basement blocks (Stearns, 1971), but is a pervasive structural style in Rocky Mountain forelend basins. Some of the variations of this style are illustrated by examples from the Big Horn Basin. At Pitchfork anticline detachment features form three times more structural relief in the sedimentary section than on the basement (Petersen, 1983); clearly, a large underlying basement structure was not a pre-requisite for detachment. Reflection seismic on the west side of the basin shows that Heart Mountain anticline is detached from deep basement by a major fault and a subsidiary fault opposing one another (Fig. 3-3). This detachment structure, formed by opposing thrusts, is an analog to the Owl Creek, Granite, and Uinta basement features previously mentioned. Finally a well exposed, sharp fold in the Triassic Chugwater Formation at Red Butte on the northeast side of Rattlesnake Mountain (Pierce, 1966) appears to be detached above a planar basement surface that from surface observations dips consistently about 14° to the northeast (Fig. 3-4). Detached structures are here interpreted to reflect shortening caused by compression in the sedimentary cover concomitant with compression in the basement and presumably the entire lithosphere.

FORELAND ORIGINS

If all foreland structures were caused by compression operating in the crust and the cover above, then virtually all detached thrust-fold belts should have adjacent foreland deformation. However, relatively few of the world's detached thrust-fold belts have well developed structures in their forelands.

Thrust-fold-belt—foreland pairs with which I have some personal familiarity are listed in table 3-1. Undoubtedly more exist, but shallow cover, deep burial, and lack of exploration have prevented their recognition. Thrust-fold belts and forelands are paired both geographically and temporally. But because many thrust-fold belts do not have deformed forelands, the inference can be made that their causative forces are not entirely the same or at least are not equally operative. It is particularly curious, for example, that the world's best developed foreland, the Rocky Mountains, does not continue northward into Canada as does the adjacent Utah-Idaho-Wyoming-Montana-Alberta thrust-fold belt. Tangential compression responsible for the thrust-fold belt was not the sole cause of foreland deformation (Lowell, 1974).[2]

Forelands that may have formed purely by unidirectional compression and accompanying flat-plate subduction (Jordan et al., 1983), as is possible for the Andean foreland in Argentina, average only about one-fourth the width (Vasquez and Gorrono, 1980) of the Rocky Mountain foreland. Though compression must be the dominant force of Rocky Mountain foreland deformation, it is probable that the large areal dimensions, the immense structural relief, and the diverse trends of the province required a greater variety of deformational forces and sources of compression than has affected other forelands.

In 1974, I concluded that foreland basement deformation was a response to a complexly varying stress system that included tangential crustal compression (later viewed as "underthrusting", Fig. 3-1), wrenching, and vertical uplift, partly in response to an overridden lithospheric slab. The system was envisioned as operating with compression maintaining a more or less constant east-northeast-west-southwest orientation and strike slip also occurring as least principal maximum compressive stress changed from vertical to horizontal.

Work over the past decade reveals that deformation was even more complex. As theorized so ably by Gries (1983a), the direction of compression changed through time so that a late phase of Laramide orogeny (early Eocene) was marked by north-south directed compression that created east-west-trending structures, which in turn cut across earlier north-south and northwest-southeast Laramide features. Gries has shown the importance of careful consideration of timing of Laramide structural events.

TABLE 3-1—Thrust fold belts and adjacent forelands.

PAIRED THRUST-FOLD BELTS/FORELANDS		AGE
Wyoming-Utah-Idaho	Wyoming Province	"Laramide"
Marathon	Permian Basin	Penn-Perm
Andes	Pampean-S. Mendoza	Tertiary
Apennines	Adriatic	Tertiary
Taurus	Syrian Arc	Late Cretaceous

In further advancing the case for northward movement of the Colorado Plateau, Chapin and Cather (1981) have demonstrated that strike slip was not simply a passive side-effect of compression. The large scale en echelon arrangement of basins and uplifts along the east side of the Plateau in Colorado and New Mexico (Chapin and Cather, 1981) is impressive evidence for right-lateral slip.

The concept of an eastward-dipping, subducted, lithospheric slab overridden by the Rocky Mountain foreland has also had modifications. The original idea of an overridden, thinned continental lithospheric plate exerting buoyant forces differentially against the superjacent plate containing the foreland (Lowell, 1974) has been modified by:

1– Dickinson and Snyder (1978) who postulated the formation of foreland structures by a shallow descending lithospheric slab which slid along, maintained contact with, and scraped against the overriding upper plate; and
2– Livaccari et al. (1981) who thought that the overridden lithosphere was "a large oceanic plateau of an anomalously thick and buoyant oceanic crust"

Whatever the character of the effected overridden slab, its termination to the north in central Montana would have deep-seated left-lateral movement (Lowell, 1974, p. 277), which may have imparted rotational (simple) shear to the foreland to the south (Sales, 1968), thus, in turn, explaining the dominant asymmetry of west-northwest-trending left-lateral fault zones over northeast-trending right-lateral zones (Stone, 1969).

Jordan et al. (1983) have inferred that foreland basement deformation and associated magmatic gaps in the Andes of northern Argentina (lat 28°–33°S) and in the North America Cordilleran are both related to flat-plate (nearly horizontal) subduction. The close similarities in structures of the Pampaenas Ranges of Argentina and the Rocky Mountain foreland make this an attractive hypothesis. However, it should be noted that comparable foreland structures both north (Sierra Santa Barbara, Sierra

[2]To substantiate this contention the foreland of the Alpine thrust-fold belt in southwest Bavaria, Germany, shows normal faults in basement rather than compressive features (Fig. 6-23).

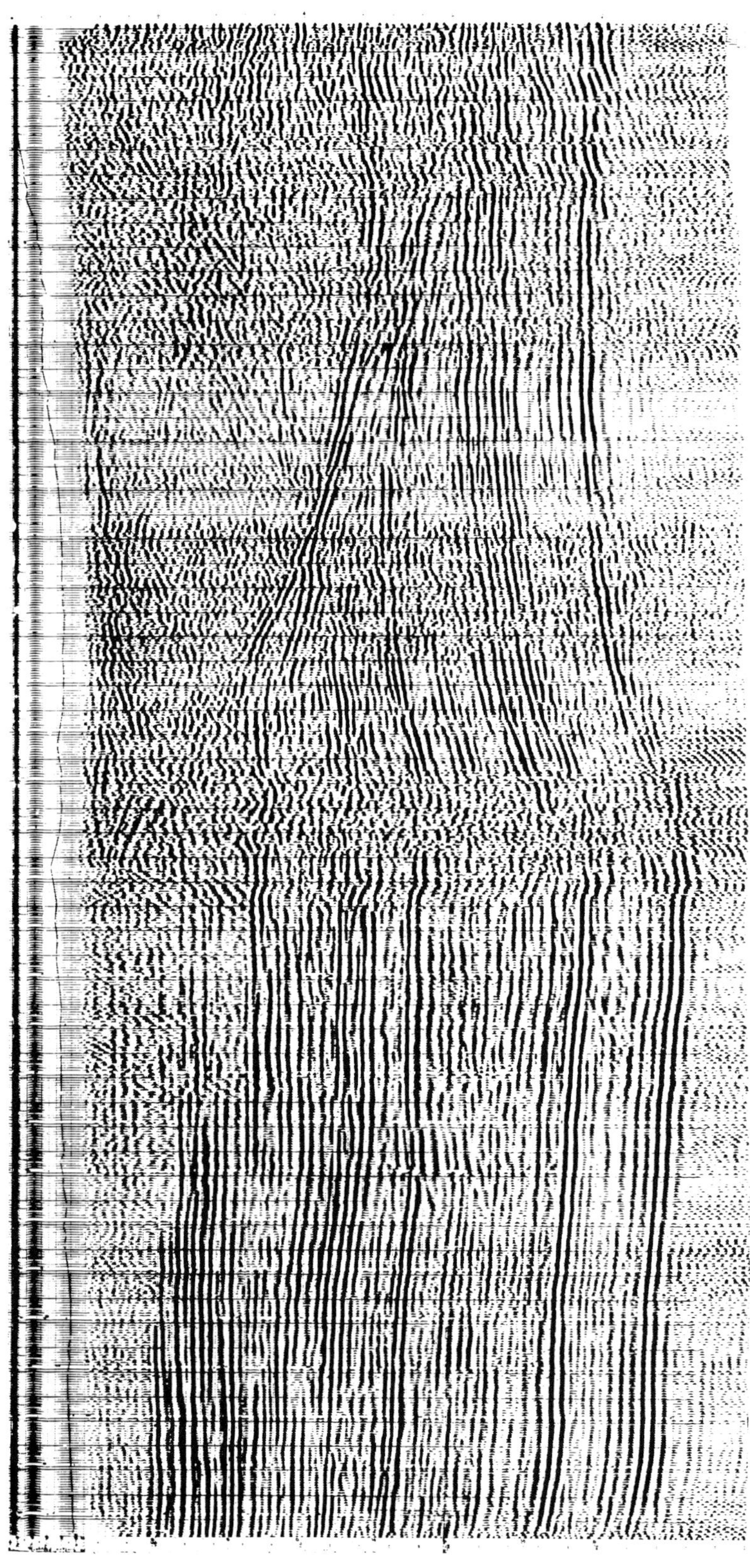

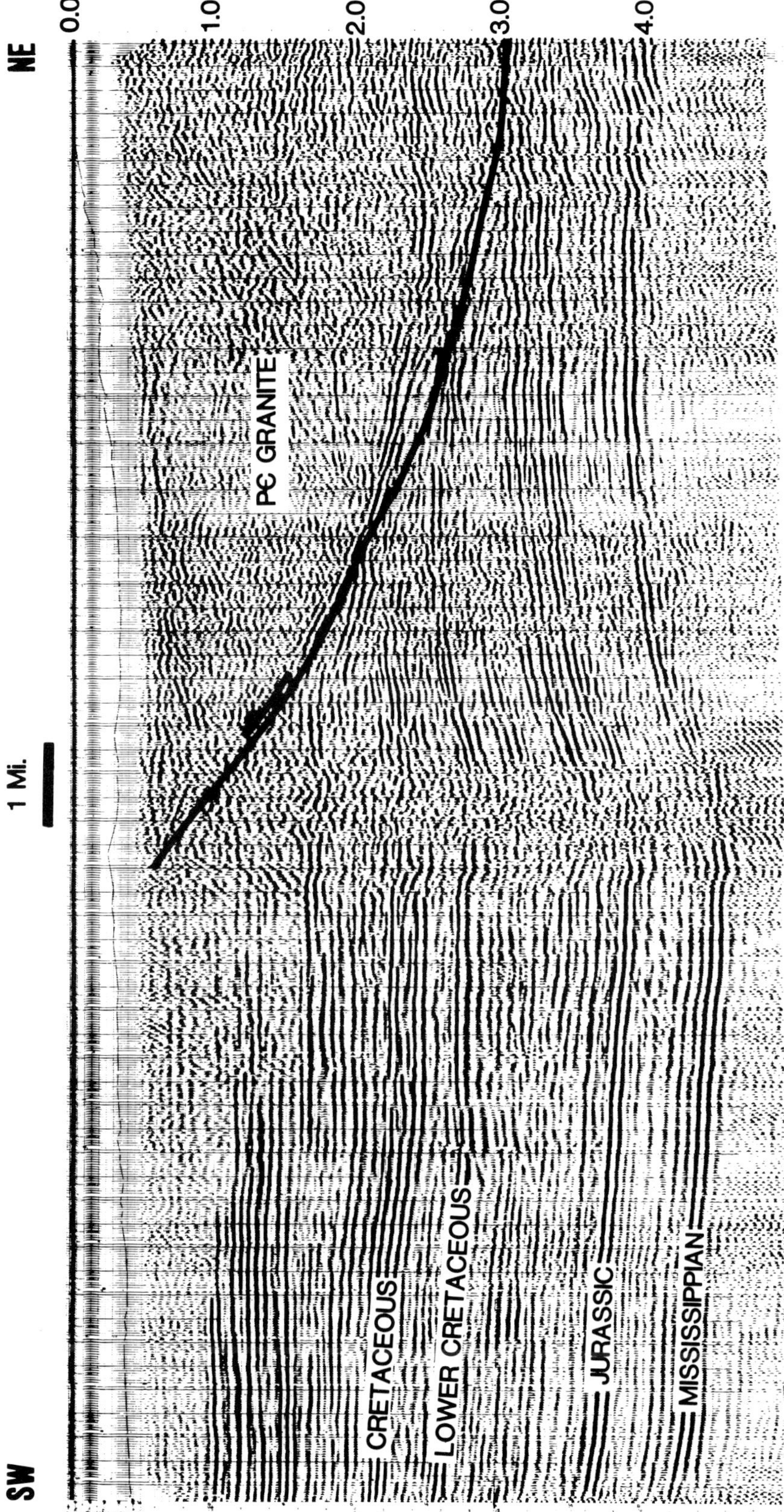

Fig. 3-2—Reflection seismic section (unannotated and annotated) from northeastern Green River Basin to southwest part of Wind River Range, Wyoming, showing 16 km (10 mi) of thrust overlap of Precambrian crystalline rocks over Phanerozoic sedimentary rocks (courtesy of Conoco). Note velocity pull-up beneath granite.

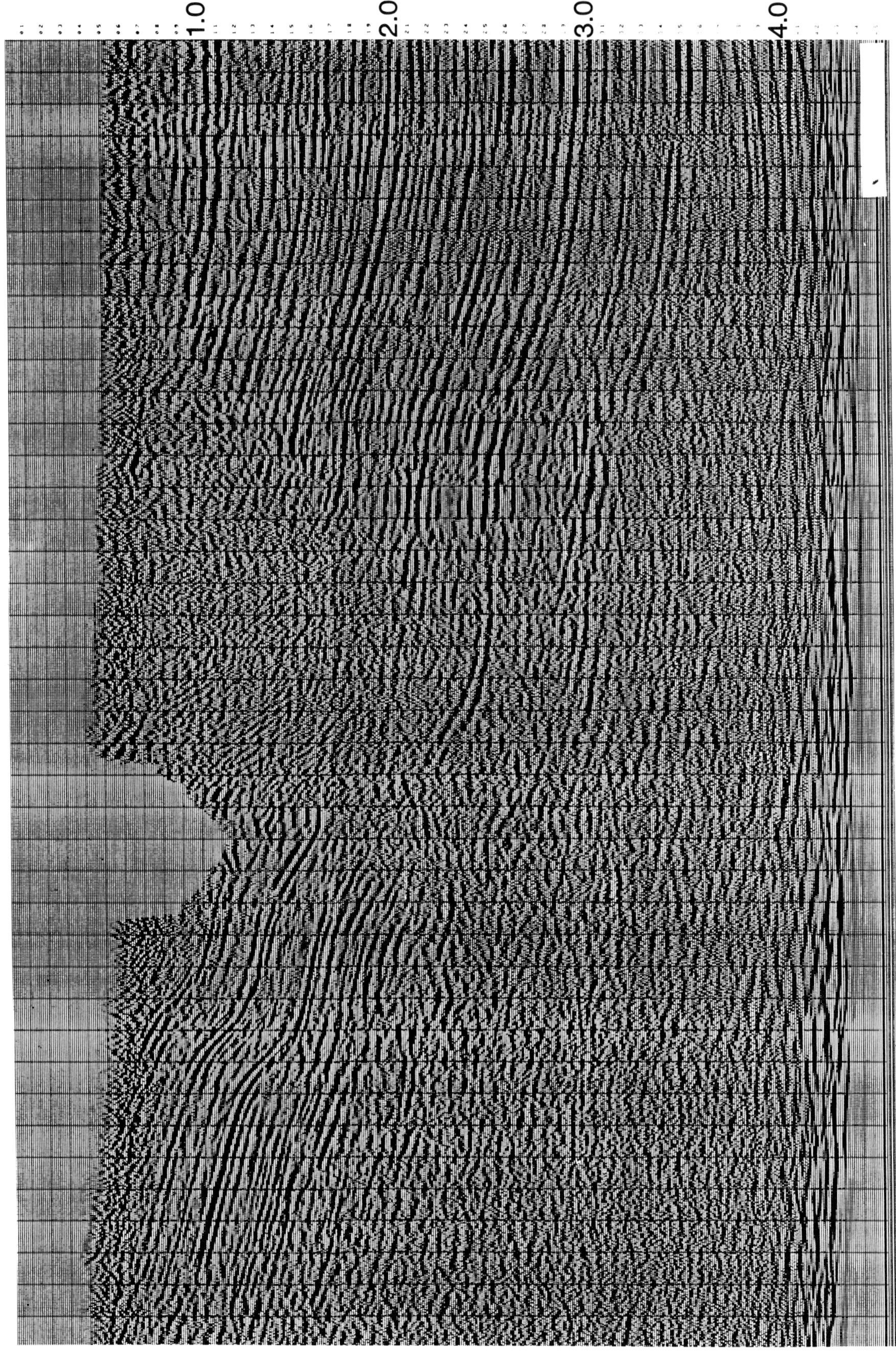
1.0
2.0
3.0
4.0

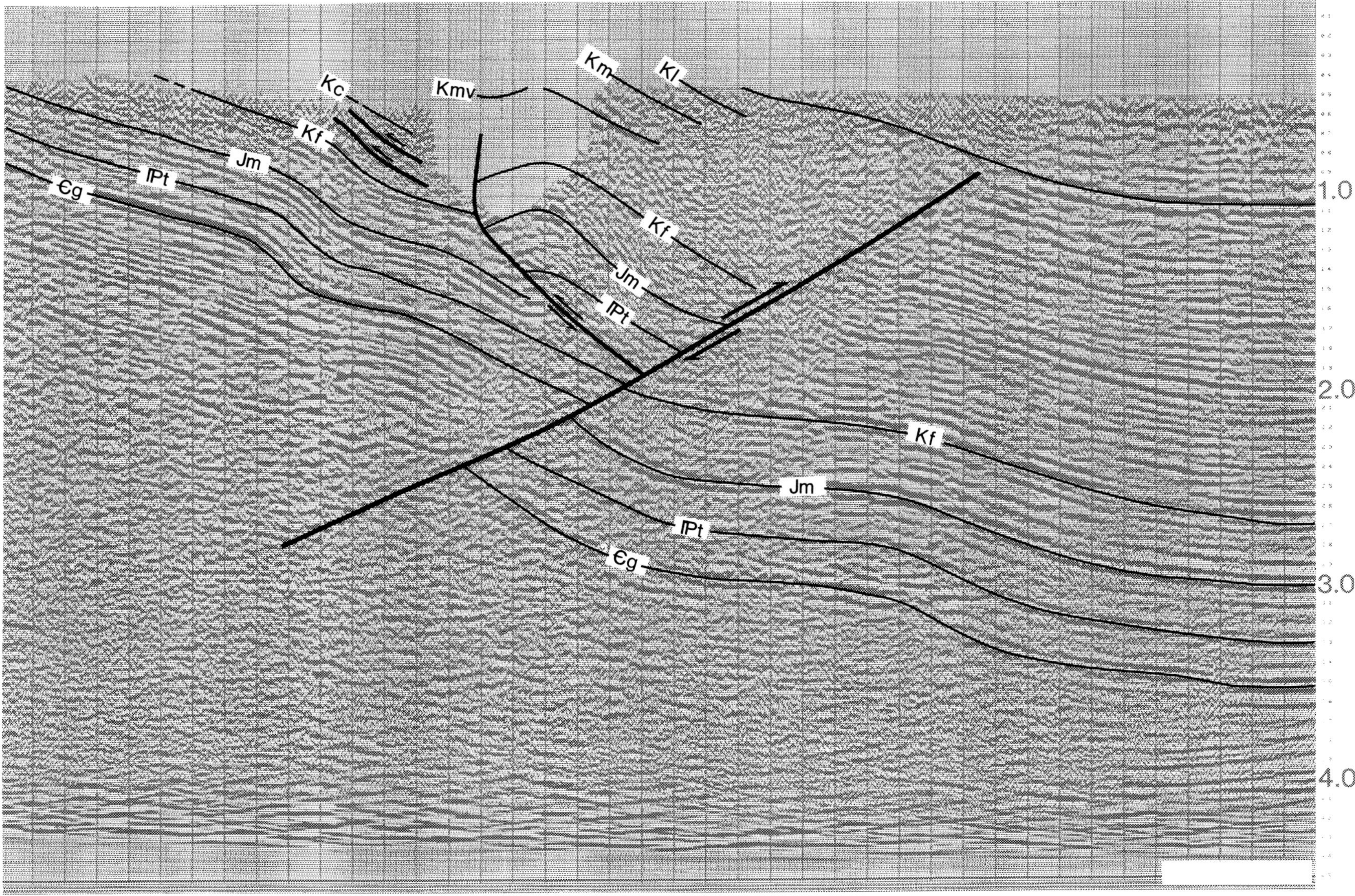

Fig. 3-3—Reflection seismic section (migrated-unannotated and annotated) showing Heart Mountain detachment structure on west side of Big Horn Basin, Wyoming. Heart Mountain anticline is disharmonic to both the lower and higher basement levels, just beneath Cg. Down-structure viewing of the surface geologic map (Pierce, 1966) supports a detachment interpretation. Orientation of opposing faults indicates horizontally directed maximum principal compressive stress. Ꞓg = Cambrian Gallatin Formation; Pt = Pennsylvanian Tensleep Sandstone; Jm = Jurassic Morrison Formation; Kf, Kc, Kmv, Km, and Kl = Cretaceous Frontier, Cody, Mesaverde, Meeteetse, and Lance formations.

Fig. 3-4—View northward of Rattlesnake Mountain anticline and fold in the Triassic Chugwater Formation at Red Butte (arrow) detached above northeast-dipping planar basement surface.

del Centinela, and Sierra del Maiz Gordo from lat 24°–24°30′S) and south (Sierra de le Cara Cura and Sierra do Reyes from lat 36°30′–37°S) of the Pampaenas Ranges are not associated with flat-plate subduction. This indicates that the relationship between attitude of a subducted plate and development of foreland structures is not fully understood at the Andean type locality, let alone the earlier Rocky Mountain counterpart.

In summary, the Rocky Mountain foreland appears unique in having undergone far more diverse deformation than other forelands.

CONCLUSION

The amount of thrust overlap that conceals prospective sedimentary rocks beneath the edges of large foreland structures is of great economic importance. The distance of suspected overlap influences both shooting programs and subsequent interpretation of seismic surveys, drilling practices and procedures, and, of course, structural interpretations. Models of foreland deformation that accommodate great shortening, also afford the greatest thrust overlap and, hence, the most prospective section.

If more wells are drilled in settings such as, and comparable to, those discussed along the Front Range (Bieber, 1983; Jacob, 1983), Wet Mountains (Jacob, 1983), Sangre de Cristo Mountains (Lindsey et al., 1983), Nacimiento uplift (Woodward, 1983), Uncompahgres (Frahme and Vaughn, 1983), Wind River Range (Berg, 1983; Steidtmann et al., 1983), Washakie Range (Winterfeld and Conard, 1983), and in southwest Montana (Schmidt and Garihan, 1983; Garihan et al., 1983), then surely more papers like Sprague (1983) will record future major petroleum discoveries in mountain front settings. In this same vein, there should be more exploration beneath the bounding thrusts or reverse faults of major fields, such as Elk Basin (Stone, 1983d).

From a global perspective the sheer length of detached thrustfold belts of at least 80,000 linear kilometers (50,000 mi) for combined Caledonian (early-mid-Paleozoic), Variscan (late Paleozoic), Laramide, and Tertiary orogenic belts gives rise to conservatively 6,500,000 sq km (2,500,000 sq mi) or .64 billion ha (1.6 billion acres) of potential forelands to explore. We have seen that much of this will not be as structured as the Rocky Mountain foreland, but clearly much remains to be investigated.

PLATE TECTONICS AND FORELAND BASEMENT DEFORMATION[3]

Abstract The extent and complexity of Laramide foreland basement deformation in the Wyoming province adjacent to the Idaho-Wyoming thrust-fold belt have led to numerous tectonic hypotheses, including tangential crustal compression (manifest as thrusting), vertical uplift, and strike slip; actually, all three types of deformation have occurred. With the advent of plate tectonics, tangential compression related to subduction has been mentioned most frequently as the ultimate cause of foreland deformation. Tangential compression is presumably operative in all subduction-related orogenic belts, but because most orogenic belts do not have foreland basement deformation comparable with that of the Wyoming province, compression *alone* does not seem sufficient to account for foreland basement structuring. Apparently a more specific condition is necessary. It is here proposed that *in addition* to tangential compression and strike slip, a lithospheric slab subducted well beneath the foreland is a fundamental requisite for deformation there, providing the buoyancy necessary for uplift of basement blocks. Recent work based on K_2O/S_iO_2 ratios of Cenozoic andesites suggests that paleosubduction did occur far east within the Wyoming province (Lipman et al., 1971).

INTRODUCTION

The most famous and extensive area of foreland basement deformation known along any orogenic belt in the world is the Wyoming province (Prucha et al., 1965), where Laramide basement block movement has occurred more than 400 km (250 mi) cratonward from the Idaho-Wyoming thrust-fold belt of partially the same age (Fig. 3-5). The structural style of the Wyoming province has been given many different interpretations, though all concede that uplift has been an important part of the deformation. Blackstone (1963) summarized published orientations of fault planes bounding basement blocks by saying they vary from normal through vertical and upthrust to low-angle thrust. (In fact, fault attitudes are highly variable. At Rattlesnake Mountain west of Cody, Wyoming, a low-angle thrust passes immediately into a high-angle normal fault at depth [Pierce, 1966]; whereas along the southwest flank of the Wind River Range, a low-angle thrust persists with miles of overlap of Precambrian crystalline basement [Berg, 1962]). In view of the differences in interpretation of structural style, it is not surprising that at least as many hypotheses of types of deformation have been proposed. Berg (1962) reviewed hypotheses of foreland basement deformation by block uplift (the bounding fault is a simple upthrust), thrust uplift, and fold-thrust uplift, preferring the last. The ultimate mode of uplift has generally been considered the result of one of three stress systems:

1– tangential crustal compression with important vertical adjustments or wedge uplifts (Chamberlin, 1945; Thom, 1955; Grose, 1972) or with important lateral coupling (Sales, 1968; Thomas, 1971),
2– vertical uplift (Bengtson, 1956; Osterwald, 1961; Eardley, 1963; Harms, 1964); and
3– wrenching or strike slip (Stone, 1969).

Almost certainly, no single hypothesis is sufficient to satisfy the structural complexity of the Wyoming province, and elements of all three seem to be needed. Tangential compression seems required for the Wind River Range but not for Rattlesnake Mountain; conversely, vertical uplift probably works for Rattlesnake Mountain but does not explain the thrust overlap of the Wind River Mountains. Moreover, strike slip seems necessary, for example, for the en echelon structures along the north and south sides of the Wind River Basin (Keefer, 1970).

Although much attention has been given in the past few years to the relations between plate tectonics and the evolution of orogenic belts, little attempt has been made to relate plate motion and the stresses that cause basement deformation in the foreland immediately adjacent to an orogenic belt. The purpose of this brief paper is to try to set foreland basement deformation in a plate-tectonics perspective.

FORELAND BASEMENT DEFORMATION AND SUBDUCTION

Equally as remarkable as the basement structuring of the Wyoming foreland province adjacent to the Idaho-Wyoming thrust-fold belt, is that along the same orogenic belt, at the latitude of the U.S.–Canadian border, and northward there is no foreland basement deformation whatever. Indeed, observable foreland basement deformation is rare along most of the world's orogenic belts. In the context of plate tectonics, many orogenic belts have been attributed to crustal compression produced by the interaction of continental and oceanic lithosphere. The paucity of foreland basement deformation adjacent to orogenic belts worldwide and, specifically, its absence northward adjacent to the same orogenic belt where it is best developed, mean that tangential compression was probably not the sole cause of foreland basement structuring. Though compression is undoubtedly important (Fig. 3-6), apparently a more specific condition must be satisfied, and I propose it is subduction directly beneath the foreland (Fig. 3-7).

On the basis of K_2O/S_iO_2 ratios of Cenozoic andesites, Lipman et al. (1971) have postulated that subduction occurred beneath the western United States in early and middle Cenozoic time along two subparallel imbricate zones (Fig. 3-7). Although the use of potassium content versus depth curves for a given silica content as a basis for determining the depth to proposed paleosubduction

[3]From *Geology*, June, 1974, by James D. Lowell, p. 275–278.

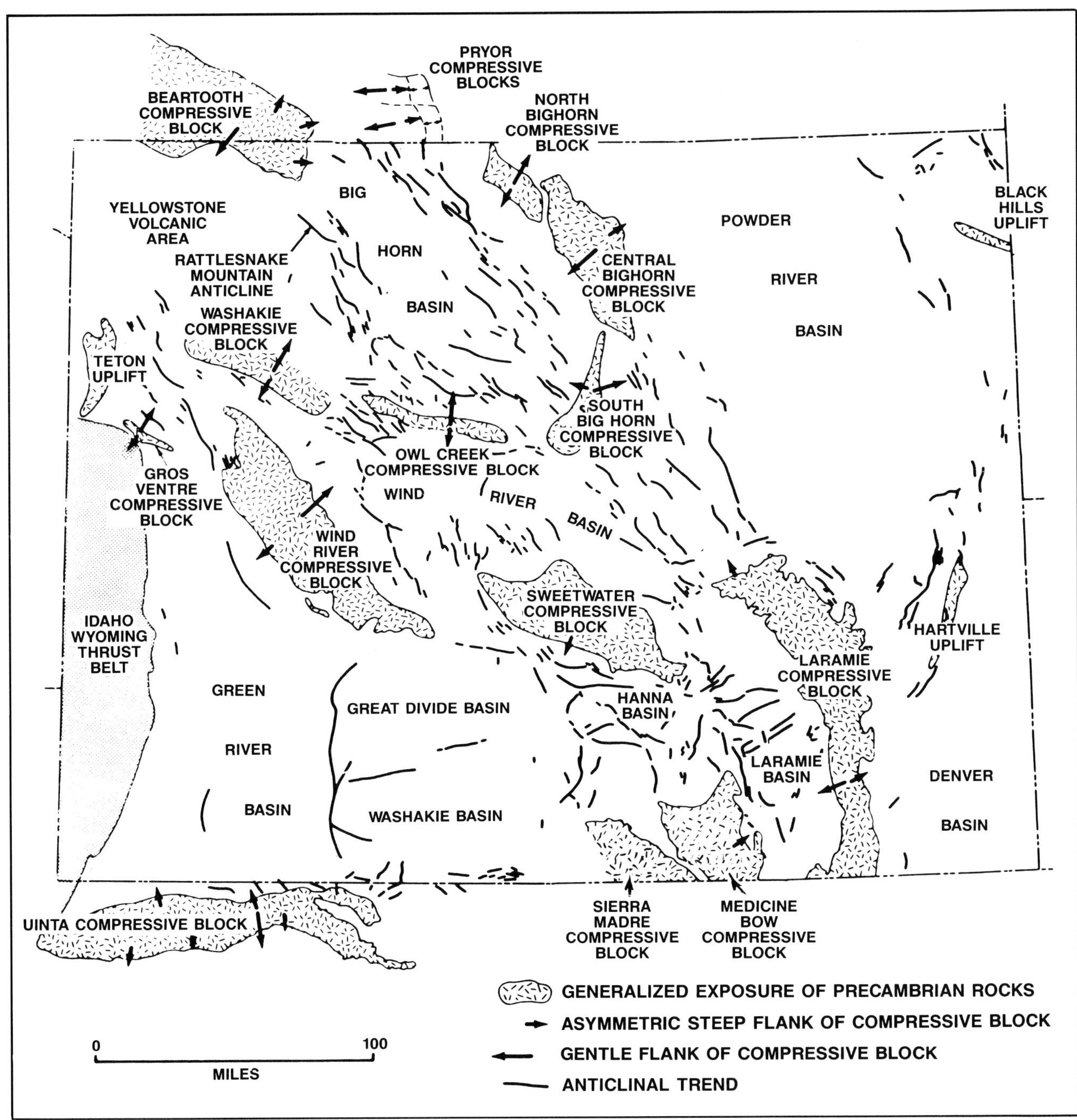

Fig. 3-5—Map of portion of Wyoming province showing position of foreland basement structures (major mountain ranges and anticlines) with respect to Idaho-Wyoming thrust belt. The most salient cross-sectional characteristic of compressive blocks is asymmetry which is so ubiquitous as to seem almost inherent. Asymmetry, manifested by a short, steep reverse-faulted or thrusted flank and a long, gentle, tilted flank, obtains in the Wyoming province at all scales, from large mountain ranges with 12,195 m (40,000 ft) of structural relief on the Precambrian basement to anticlines with 122 m (400 ft) of relief.

zones have recently been questioned (Nielson and Stoiber, 1973), the possible structural implications of this technique applied to the western United States seem too great to ignore, and the assumption is made in this paper that Lipman et al. (1971) are correct.

According to Lipman et al. (1971), a western subduction zone emerged at the continental margin, while an eastern zone was entirely beneath and decoupled from the continental plate; both zones dipped about 20° eastward. The eastern zone essentially coincides with the

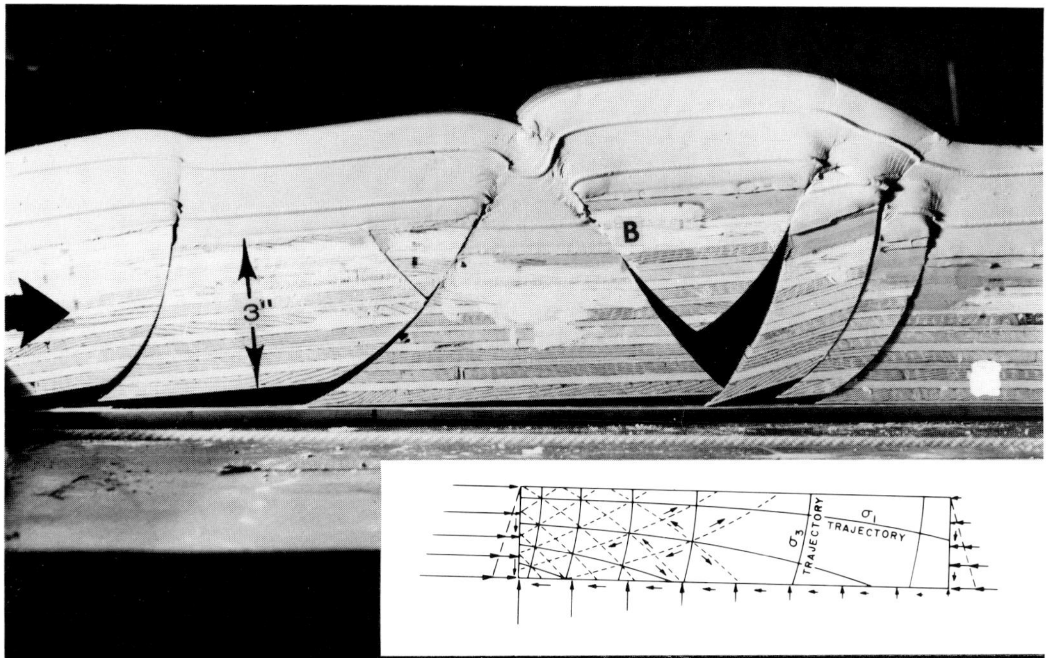

Fig. 3-6—Reverse movement on pre-existing listric faults attained by tangential compression of rigid base as indicated by arrow. In context of this paper, model is likened to horizontal continental lithospheric plate of figures 3-7b and 3-8a. Compression commences at one end of plate or rigid base at an early stage of subduction and is transmitted toward interior. Deformation begins in foreland closest to adjacent orogenic belt and progresses cratonward as subducted slab descends and moves eastward. Wind River Range, having greatest amount of thrust overlap (such as that simulated at "B" in model) of any of Wyoming province foreland basement blocks, is also closest to Idaho-Wyoming thrust belt, which presumably was locus of compression. For lithospheric plate, pre-existing curved fault surfaces would not necessarily be a prerequisite to thrusting, because stresses will produce curving thrusts. Inset shows for a homogeneous elastic medium the trajectories of principal stresses (solid lines) and potential reverse fault surfaces (broken lines) compatible with indicated boundary stresses of lateral compression (Hubbert, 1951). Model requires pre-existing surfaces, because no internal deformation could occur in clay if block below was not segmented. Response to compression is mainly upward movement of blocks, and compressional structures are restricted mostly to zone above edge of block. Asymmetry of structures is natural consequence of thrust-fault configuration. (If any pre-existing normal faults were later compressed, asymmetry should result.) Model somewhat reminiscent of wedge-uplift hypothesis of Thom (1955), in which vertical uplift is main response to deep-seated compression. Note that compression accumulated through entire rigid base of model prevents formation of large tensional features in overlying clay, which roughly corresponds to sedimentary rocks above crystalline basement.

foreland upthrusting of New Mexico, Colorado, Wyoming, and Montana, and it tapers in Montana to possible absence in Alberta. The frequent association of magmatic and volcanic activity with modern subduction zones is a priori evidence of buoyancy forces related to the latter. Moreover, the large Bouguer gravity minima (Woollard and Joesting, 1964) roughly coincident with the proposed eastern paleosubduction zone may even suggest the possible lingering existence of a relatively low-density and, therefore, buoyant plate at depth. Finally, the eastern subducted plate may have been sialic in part (see Fig. 3-8 caption). The presence of a detached lithospheric slab at depth may have exerted an appreciable buoyant force directed more or less vertically against the superjacent continental lithospheric plate. Tectonic implications are considerable:

1– Buoyancy acting on the continental plate, which includes pre-existing steep to vertical basement faults and factures, should be manifested as uplifted, differentially rotated and tilted, upthrust-bounded basement blocks (Lowell, 1969).

2– If a gently dipping lithospheric slab beneath the foreland is a requisite for basement block movement and is at the same time a requirement that is achieved only rarely in the geologic record (for example, subduction occurs far landward around the present Pacific Ocean, but zones with gentle dips—15° to 20°—are not common), then foreland basement deformation would not be a common event. Basement deformation also need not be present continuously in the foreland along the same thrust

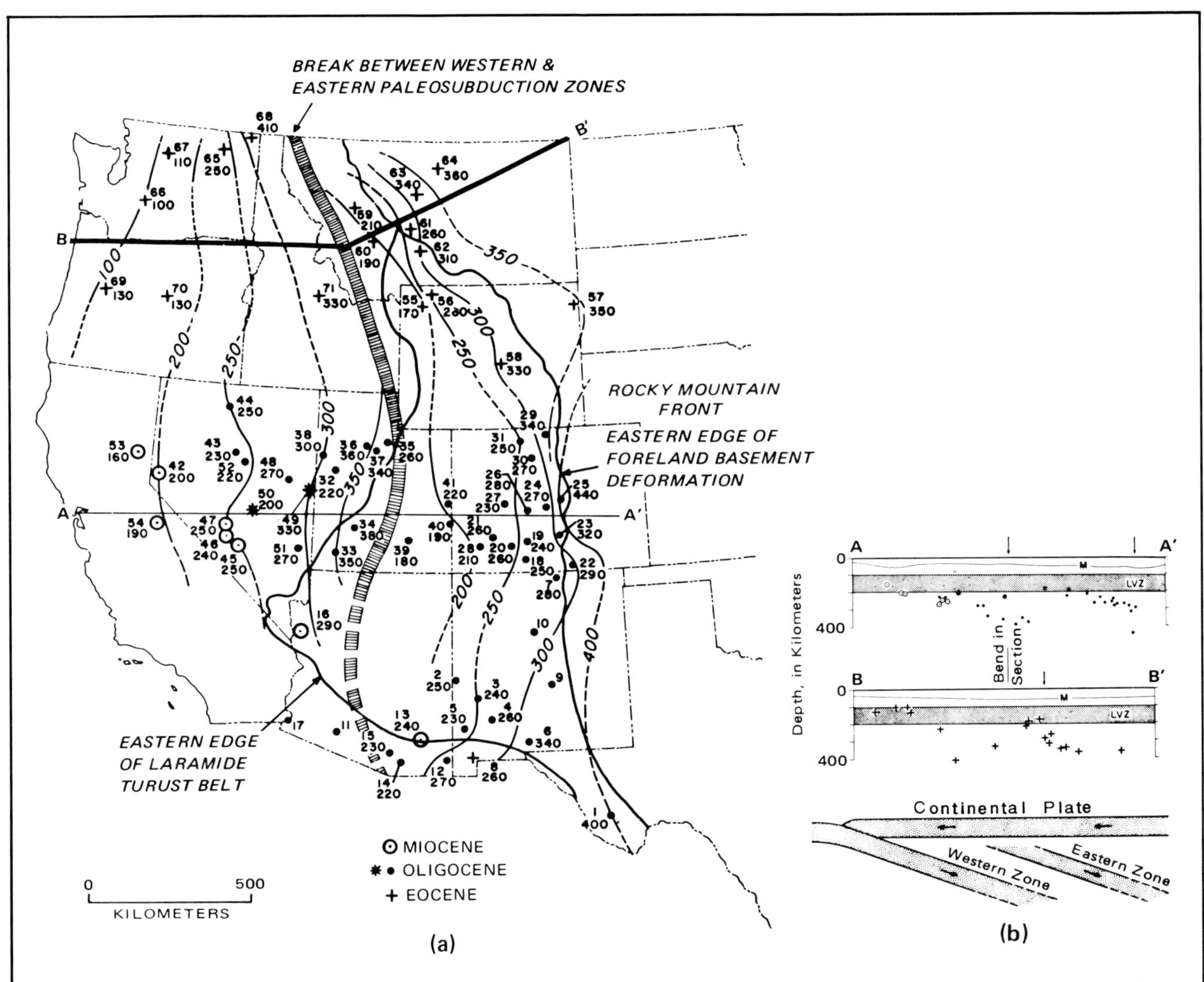

Fig. 3-7 (Lipman et al., 1971)—a. Contoured depths (in kilometers) to inferred middle Cenozoic subduction zones of western United States. Upper number at each point identifies locality (see original reference); lower number gives depth to subduction zone, obtained from K_2O-depth plots for modern arc volcanics. Hachured line indicates discontinuity in contours. Most of samples are from Eocene and Oligocene rocks that are somewhat younger than thrust-belt-foreland deformation of Laramide age that terminated in Eocene time. A lag could be expected, however, between emplacement of an underthrust lithospheric slab and time that slab furnished volcanic material at or near surface by partial melting. Note that although Black Hills (loc. 57) is shown to be east of limit of foreland basement deformation (Rocky Mountain Front), it may be a foreland structure. b. Sections showing inferred early and middle Cenozoic subduction zones. Horizontal scale same as vertical scale; surface topography not shown. All Oligocene (solid dot) and Miocene (circle with dot) points within 100 km (62 mi) north or south of section A-A are projected into this section. All points (+) for Eocene igneous rocks in northwestern United States are projected into section B-B'. M = Mohorovicic discontinuity, LVZ = approximate location of a low-velocity zone, and arrows are structural province boundaries from (a). Bottom section is schematic interpretation of lithospheric plate geometry suggested by sections A-A' and B-B'. Presumably, eastern zone is older of two. Permission to publish by P. W. Lipman.

belt; for example, the lithospheric slab underlying Wyoming may terminate northward by a deep-seated left-lateral arc-arc-type transform fault, surficially expressed by the various central Montana lineaments such as the Lake Basin fault zone. Hence, deformation would not have occurred to the north; presumably a comparable (right-lateral) termination also existed on the south.

3– The age of deformation in the Wyoming province foreland proper should have progressed from west to east as the lithospheric slab descended and moved eastward. This suggested age progression should be checked and would constitute one test of the present hypothesis.

CONCLUSIONS

Foreland basement deformation is envisioned as the response to a complexly varying stress system that includes tangential crustal compression, wrenching or strike slip, and vertical uplift. Perhaps most important, however, is the possible requirement well beneath the foreland of a subducted lithospheric slab that provided much of the impetus for uplift. Horizontally directed maximum principal compressive stress, generated at the earliest stages of subduction or before any significant lithospheric underthrusting (Figs. 3-6, 3-8) and subsequently, equally or more important vertical maximum principal compressive stress produced by the buoyancy of an underthrust lithospheric slab (Figs. 3-7, 3-8), would have both been expressed mainly as dip-slip movement on reverse faults and upthrusts. In the former case, as least principal compressive stress varied from vertical to horizontal, lesser amounts of strike slip could have also occurred, especially along pre-existing structural elements oriented at approximately 30° to the direction of any horizontal maximum principal compressive stress (compare Stone, 1969, pl. 1). Limited amounts of strike slip in a generally east-west direction might be provided by segmentation of the underthrust lithospheric slab by transform faults that would in turn impart lateral movement to the overlying lithosphere somewhat as proposed by Hoppin and Jennings (1971). Such a complex stress system with so many attendant possibilities for structural variety would not only explain the geology but also the proliferation of hypotheses and disagreement between hypotheses of foreland basement deformation.

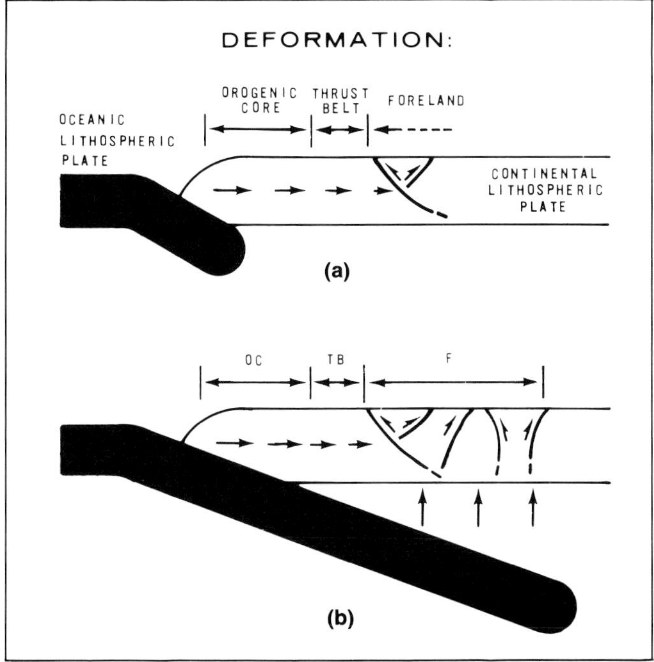

Fig. 3-8—Conceptual diagram of Laramide foreland basement deformation. a. Initial underthrusting of oceanic plate causes tangential compression and beginning of foreland deformation in continental plate. Horizontal arrows represent direction and amount of compression; solid lines with half arrows indicate thrusting with direction of displacement. *(Character of eastern underthrust plate [oceanic versus continental] was never distinguished by Lipman et al. [1971]. It is here termed "oceanic" by convention; it may have had a significant continental aspect. Eastern subducted slab might have included greatly thinned continental lithosphere [compare Lowell and Genik, 1972] that originated during a much earlier period of rifting [Stewart, 1972]. Greater sialic component would have increased buoyancy potential of the underthrust plate and also helped to explain intracontinental position of Idaho-Wyoming orogenic belt [Burchfiel and Davis, 1972].) b. With greater length of underthrusting, buoyancy (vertical arrows) of underthrust slab causes uplift and deformation farther continentward in foreland. Actual deformation in orogenic core and adjacent thrust belt (which preceded foreland deformation) is not shown. Subsequent to stage b, continental plate overrides oceanic plate and another subduction zone is formed at lead edge of continental plate; buoyancy from subducted detached oceanic slab may be effective long after it has been decoupled, providing explanation for late Cenozoic uplift that also affected region.*

FORUM

PLATE TECTONICS AND FORELAND BASEMENT DEFORMATION: COMMENT[4]

Lee A. Woodward, Department of Geology, University of New Mexico, Albuquerque, New Mexico.

Lowell (Geology, 1974, V. 2, n. 6) has pointed out the interplay and contemporaneity of horizontal compression, vertical movement, and strike-slip movement during Laramide basement deformation of the Rocky Mountain foreland. This conclusion seems to me to be well substantiated, and the evidence need not be reviewed here; the reader is referred to the summary of evidence presented by Lowell (1974, p. 275–276).

[4]From *Geology*, December, 1974, Vol. 2, n. 12. Used with permission.

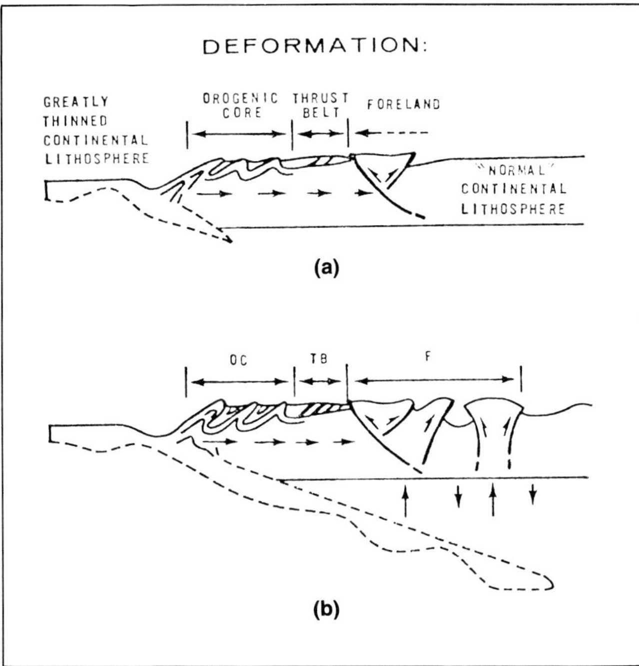

Fig. 3-9—Conceptual diagram of Laramide foreland basement deformation. a. Initial underthrusting of thinned continental lithospheric plate causes tangential compression and beginning of foreland deformation in overlying plate. Horizontal arrows represent direction and amount of compression[1]; solid lines with half arrows indicate thrusting with direction of displacement. b. With greater length of underthrusting, buoyancy (vertical arrows) of underthrust slab causes differential uplift and subsidence and deformation farther continentward in the foreland. Subsequent to stage b, normal continental plate overrides greatly thinned continental plate, and another subduction zone, this time involving a descending oceanic plate, is formed to left (west). Igneous activity related to subduction is not shown.

Developing a model that accounts for dominantly vertical movements with subsidiary crustal shortening and strike-slip movements in a primarily compressional stress field is the crucial problem in discussing the mechanics of Laramide foreland deformation. Lowell (1974) has suggested that an inferred subducted lithospheric slab beneath the crust of the western United States (Lipman et al., 1971, 1972) rose buoyantly to provide the component of vertical movement while the crust was under tangential compression.

Although a strong vertical component of movement is required by the geometry of structures in the Rocky Mountain foreland, I think there are several objections to the mechanism of a buoyant, low density, lithospheric slab beneath the basement:

1– A detailed study by Love (1972) of igneous rocks in the Absaroka-Gallatin volcanic province of Wyoming indicates that there are serious problems in that area with the subduction model proposed by Lipman et al. (1971, 1972); an unusual manipulation of the hypothetical subduction zone would be necessary to explain the volcanic subprovinces and the temporal, chemical, and isotopic relations observed by Love (1972). Gilluly (1971, p. 2391) has pointed out that with the expected dip of the subduction zone, it would be likely that any such zone under the foreland would generate kimberlite magma rather than the more silicic rocks noted by Lipman et al. (1972). Thus, the existence of a subducted lithospheric slab beneath the foreland is open to serious doubt.

2– Basin subsidence (with reference to sea level) is responsible for a least a few thousand feet of the structural relief formed during Laramide time in the Bighorn Basin (Mackin, 1947, p. 116) and perhaps as much as 5,183 m (17,000 ft) in the Wind River Basin (Keefer, 1965, p. 1889). Active subsidence of basins seems incompatible with a buoyant subducted lithospheric slab beneath the foreland.

3– Round to elliptical basins and elongate uplifts that trend north, northwest, and east do not form the pattern that one would expect from buoyancy beneath the entire foreland. Rather, uniform epeirogenic rise of the region seems more likely. To call upon the compressional element of the stress field to determine the outlines and positions of the uplifts and basins ignores the basic problem of dominant vertical movements.

4– Episodic vertical movements that began as early as Paleozoic time characterize some of the present-day uplifts and basins of the foreland (Read and Wood, 1947). Rejuvenation of the uplifts at various intervals during Phanerozoic time implies a mechanism that predates the inferred Cenozoic subducted slabs. There is a possibility that pre-Cenozoic subducted slabs may have caused the pre-Laramide vertical movements, but the coincidence of repeated displacement of uplifts seems more than fortuitous. Calling upon different mechanisms at different times to explain the rejuvenation of uplifts seems equally unlikely. Therefore, I think that buoyant subducted lithospheric slabs fail to account satisfactorily for Laramide deformation as well as older deformations in the foreland.

I suggest that we consider several other possible mechanisms to account for the vertical component of movement observed in Laramide foreland deformation, inasmuch as the model proposed by Lowell (1974) has several deficiencies. Other mechanisms that I think have merit and can be tested as new geophysical and geochemical data become available include lateral transfer of sial by

[1]These arrows should now be replaced by an arrow directed from right to left denoting underthrusting within the upper plate of "normal" continental lithosphere (cf. Fig. 3-1).

metamorphic flowage (Gilluly, 1973) and phase changes in the lower crust and (or) upper mantle (Kennedy, 1959).

PLATE TECTONICS AND FORELAND BASEMENT DEFORMATION: REPLY[5]

Woodward's (1974) comments are well taken, to the point, and of the type that will help resolve more of the questions about foreland basement deformation. They have already forced me to think further on the subject. I will answer his numbered points individually.

1– Questioning the very existence of a subducted lithospheric slab is, of course, the most critical point. If there is no subducted slab beneath the foreland, then the model collapses. Woodward's doubt is not the first one raised, for in addition to the reservations I expressed (Lowell, 1974, p. 276), Gilluly (1971, p. 2391) found the proposed easterly subduction zone impossible to accept: However, if Lipman et al. (1971) are correct that a deep-seated subduction zone did exist, I believe that it would have had significance for foreland structuring.

2– Regarding the incompatibility of a buoyant subducted lithospheric slab with active basin subsidence, I would envision the slab as exerting buoyant forces *differentially* against the superjacent foreland, a distinction that should have been made originally. If the subducted slab included thinned continental lithosphere, a possibility suggested in the caption for figure 3-8 (Lowell, 1974) but not actually illustrated in that figure, then perhaps differences in sialic thickness within the slab could have led to differential uplift and subsidence in the lithosphere above, as shown here in figure 3-9. Thicker parts of subducted sial might cause uplift in the overlying plate, whereas thinner parts of sial, or simatic layers lacking sial, might induce subsidence. As a precedent for subduction of sialic crust, the term subduction as originally introduced in the Alps referred to the descent of crustal material; further, the need to dispose of sialic basement crust in several orogenic belts, possibly by a subduction mechanism, has been reviewed recently by Elliott (1973).

3– In my view, the trends of foreland structures in relation to differential buoyancy beneath the foreland do not present much of a problem. Not only would I call upon the compressional element of the stress field to determine foreland structural trends (for both dip-slip and strike-slip features), but I would reiterate (Lowell, 1974, p. 276) the possible importance of pre-existing faults and fractures in this regard.

4– The episodic movements of foreland uplifts and basins since Paleozoic time are indeed a troublesome point. But I believe that this is a problem of all models, not just mine. The ancestral Rocky Mountains notwithstanding, I can only say that no foreland deformational event was as profound as the Laramide orogeny, and perhaps, therefore, it did have a special origin.

I agree that additional geophysical and geochemical data are needed to test any hypothesis of foreland basement deformation. I suspect, too, that we have much to learn about the behavior of subduction zones and their structural significance.

[5]From *Geology*, 1974.

HORIZONTAL VS. VERTICAL ORIGINS

The horizontal versus vertical argument for foreland block origins has been fanned by the many different observations of fault attitudes in the Rocky Mountain foreland (Fig. 3-10) and by the economic implications of sedimentary rocks prospective for oil and gas being present beneath fold-thrust uplifts and lacking beneath block uplifts (Fig. 3-11). I have reasoned that well control, some seismic, overturned beds, wedge uplifts, and detachment structures favor a compressive origin (Lowell, 1983). To these could be added the bed-length problem created by vertical uplift (Fig. 3-12).

Points that have been used against compression and for vertical uplift include the brittle behavior of basement such that it cannot fold, multiple trends, and block corners.

Brittle behavior of basement at the relatively low temperatures and confining pressures that characterized Rocky Mountain Laramide deformation assumes an essentially homogeneous, isotropic crystalline rock, which should only undergo planar rotation and should not fold. In fact, basement in the Rocky Mountain foreland is often heterogeneous and anisotropic, consisting of metamorphic rocks with different, layered lithologies. Together with sheeting and fractures these basement rocks can demonstrably bend, e.g., at Five Springs Creek on the west side of the north Big Horn Mountains block and at Clark's Fork of the Yellowstone River on the east flank of the Beartooth Mountains, and can locally participate in flexural-slip folding. Even where basement is more isotropic, as in granitic batholiths, pervasive fractures have facilitated bending, bowing, or warping. In my opinion the concept of a homogeneous, isotropic basement is an illusion.

Multiple trends of Rocky mountain foreland structures have been cited as evidence against compression. Presumably unidirectional compressive forces could not resolve the conglomeration of northwest-southeast, north-south, east-west, northeast-southwest, and west-northwest—east-southeast trends present in the foreland (Fig. 1-8). Evidence has ac-

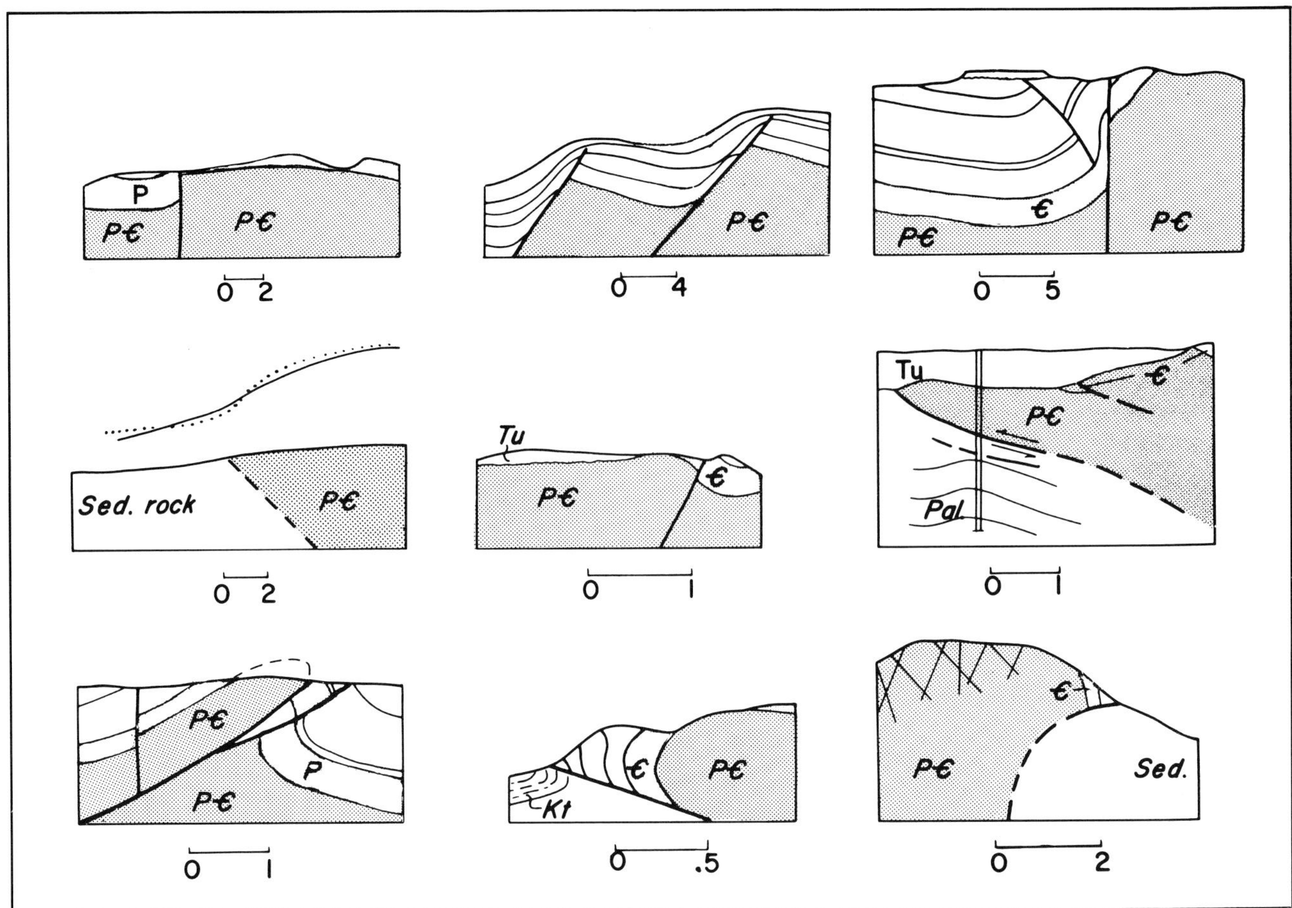

Fig. 3-10 (Blackstone, 1963)—Interpretations of fault-plane dip for Rocky Mountain foreland Precambrian basement blocks. Permission to publish by American Association of Petroleum Geologists.

cumulated (Chapin and Cather, 1981; Gries, 1983), however, that not all foreland structures are of the same age and more than one episode and direction of Laramide compression have occurred (Figs. 3-13, 3-14). It is possible that the direction of compression rotated through time from southeast-northwest to east-west to northeast-southwest to north-south, thus accounting for the many foreland trends. In this interpretation vertical uplift does not as adequately explain multiple trends as does compression, for the trends are simply considered as random in the verticalist view, whereas rotating compression gives some order to them.

Block corners have been cited in support of vertical uplift. Block corners or trapdoor structures are created at the intersections of faults having multiple trends (Fig. 3-15). Inasmuch as the two intersecting faults seem to be of the same reverse dip-slip type, they necessarily limit the amount of horizontal displacement (Fig. 3-16) and are probably a fair indicator of dominantly vertical movement (Fig. 3-17). My interpretation is that areas with block corners may have a dominant vertical component. One large area with trapdoor features in northwest Wyoming and adjacent southern Montana, including the north Big Horn block (Fig. 3-18), Pryor Mountains (Fig. 1-8), and Buffalo Fork area (Fig. 3-19), may be characterized by dominantly vertical movement because it lies opposite a recess in the Idaho-Wyoming detached thrust-fold belt and presumably was less affected by compression.

Finally it should be noted that thrusts caused by vertical uplift can be modeled very easily (Figs. 3-20 through 3-22), but so can thrusts caused by compressive blocks (Fig. 3-6); thus, experimental data is not at all definitive of origin. Faults of upthrust trajectory can form from purely dip-slip (Figs. 3-20 through 3-22) and it can be difficult to differentiate them from half-flower structures caused by wrenching.

FORELAND DETACHMENT STRUCTURES

In an area having perhaps more relief on the top of the Precambrian surface than any other structural province in the world, attention understandably has been focused on the basement-cored structures of the Rocky Mountain fore-

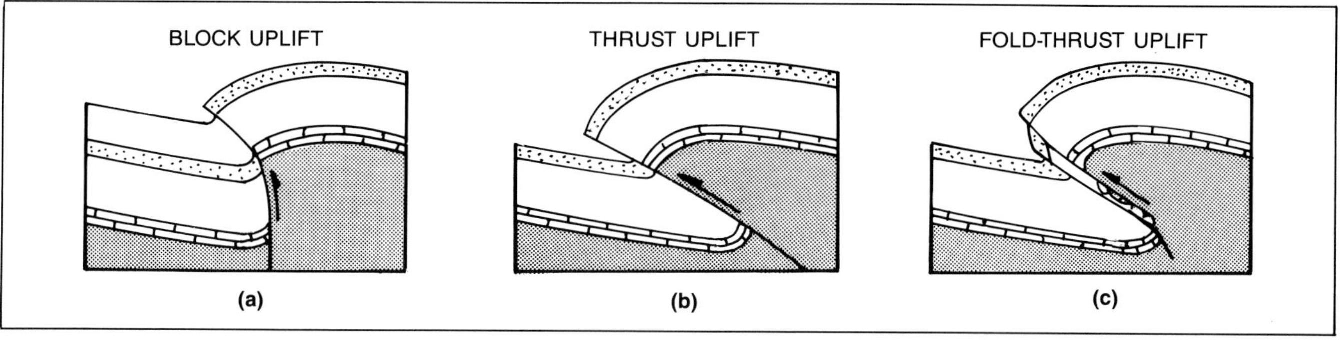

Fig. 3-11 (Berg, 1962)—Block (vertical) uplift vs. thrust and fold-thrust (compressive) uplift. Permission to publish by American Association of Petroleum Geologists.

land. Deep-seated compression interpreted as responsible for the shortened basement features has also created detached structures in the overlying sedimentary cover (Fig. 3-1). These latter structures have never been the object of systematic study, yet they provide important additional evidence for shortening of the Rocky Mountain foreland, and other forelands, and are also prospective for oil and gas.

Detached structures can be attributed to at least two types and stages of deformation. First, compression operating early in the development but prior to the differentiation of the foreland into blocks can create small scale fold and thrust structures. Probable examples of early compressional structures include those formed on what are now the gentle tilted flanks of the Owl Creek Range in Wyoming and the Cara Cura Range in Argentina (Fig. 3-23). Axes of folds and trends of faults in this early compressional phase are at distinct angles to the trends of the blocks on which they now occur. Second, after differentiation of the foreland into blocks, the flexural slip mode of folding in competent sedimentary layers dictated that space problems in both anticlines and synclines be accomodated by the creation of decollement surfaces and associated detachment structures. Examples have been documented from virtually every Rocky Mountain foreland basin. Specifically cited are the North Park Basin in Colorado (Fig. 3-24), Elk Mountain area (Fig. 3-25), Hanna and Laramie basins (Blackstone, 1983), northwestern Wind River Basin (Fig. 3-26), and Big Horn Basin (Figs. 3-27, 3-28) in Wyoming, Crazy Mountains Basin in Montana (Fig. 3-29), and the Cara Cura mountains

Fig. 3-12—Bed-length accomodation mechanisms for forced (drape) folds. In competent beds above a vertical fault, thinning is rarely observed, so that flowage, fracturing, and extended hinges are minor. Detachment maintains a constant thickness of competent beds through the hinge of the monocline, but is implausible in that each foreland basement structure would require local detachment and is impossible in that compensating extensional or pull-apart zones would be required and these are simply not recognized. Compression alleviates the bed length problem because the basement and overlying cover move in concert.

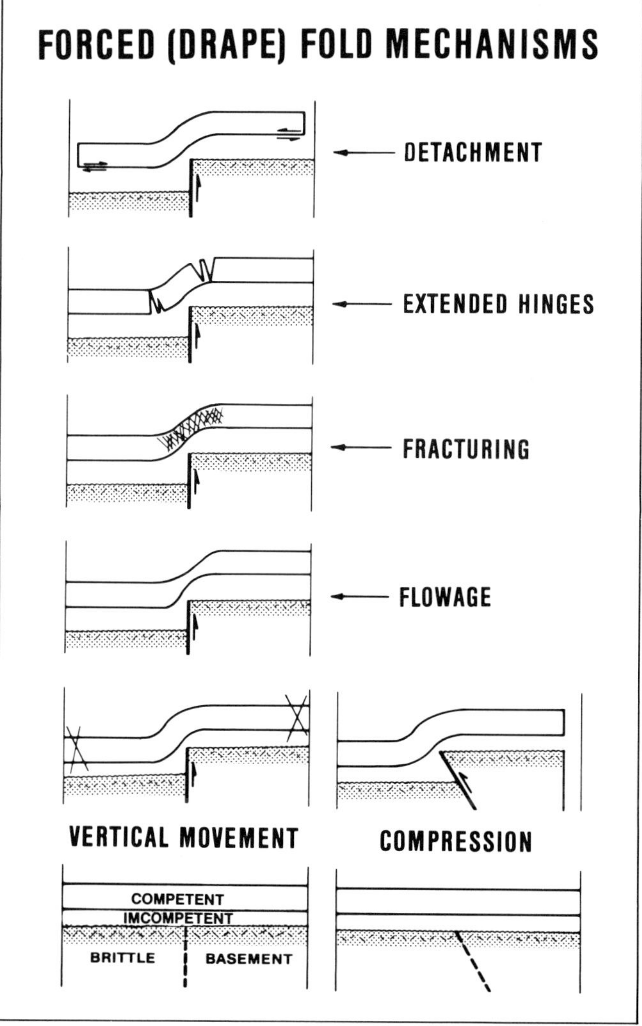

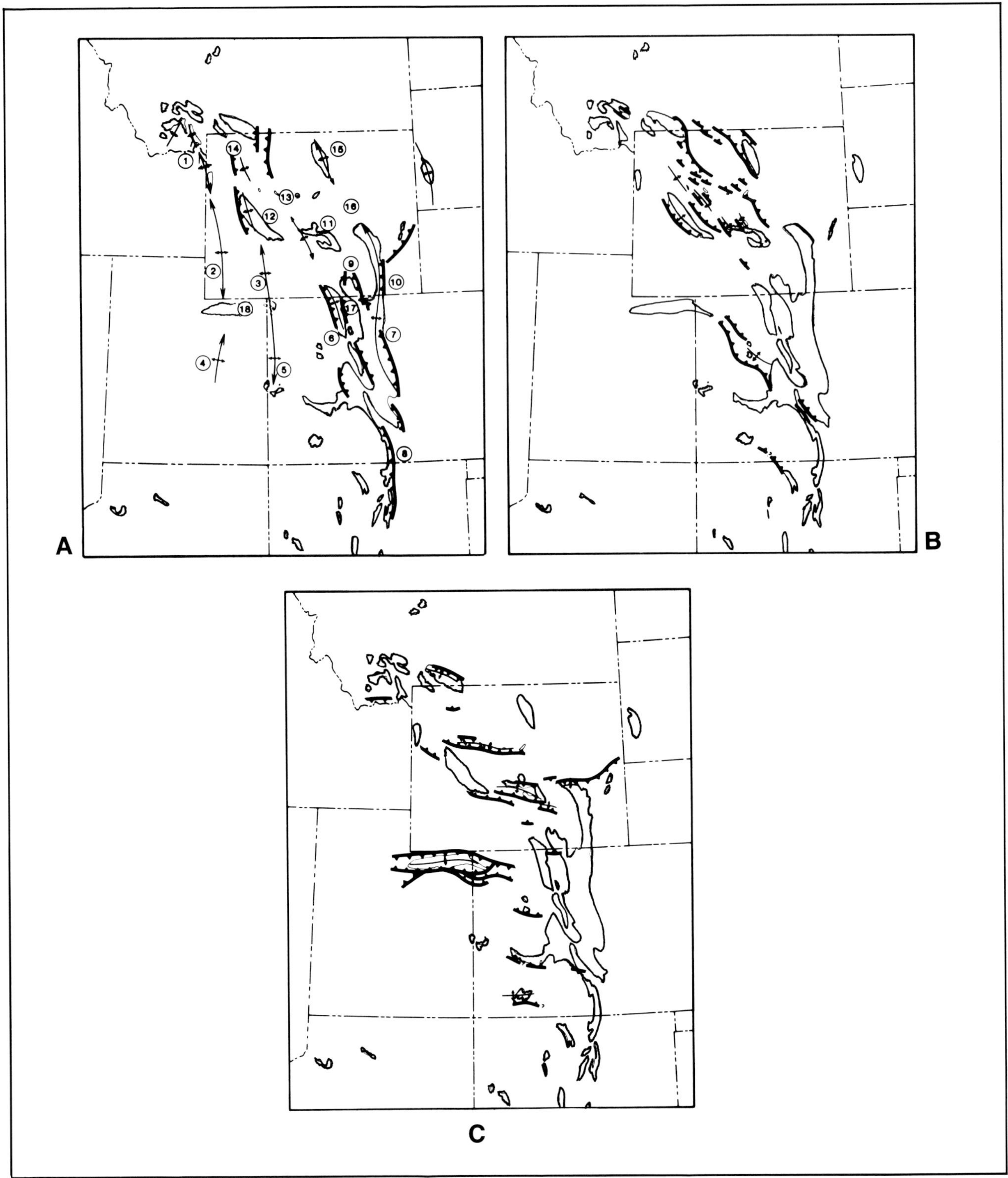

Fig. 3-13 (Gries, 1983b)—A. North-south-trending arches and uplifts are early Laramide features and are progressively younger from the west (Campanian) to the east (Paleocene). The numbers correspond with uplifts on the graph in figure 3-14. B. Northwest-trending features are most dominant in the northern foreland and are associated with Paleocene movement. C. East-west trending structures are the youngest Laramide features (Eocene) and are associated with the strongest period of compression in the foreland. Permission to publish by Rocky Mountain Association of Geologists.

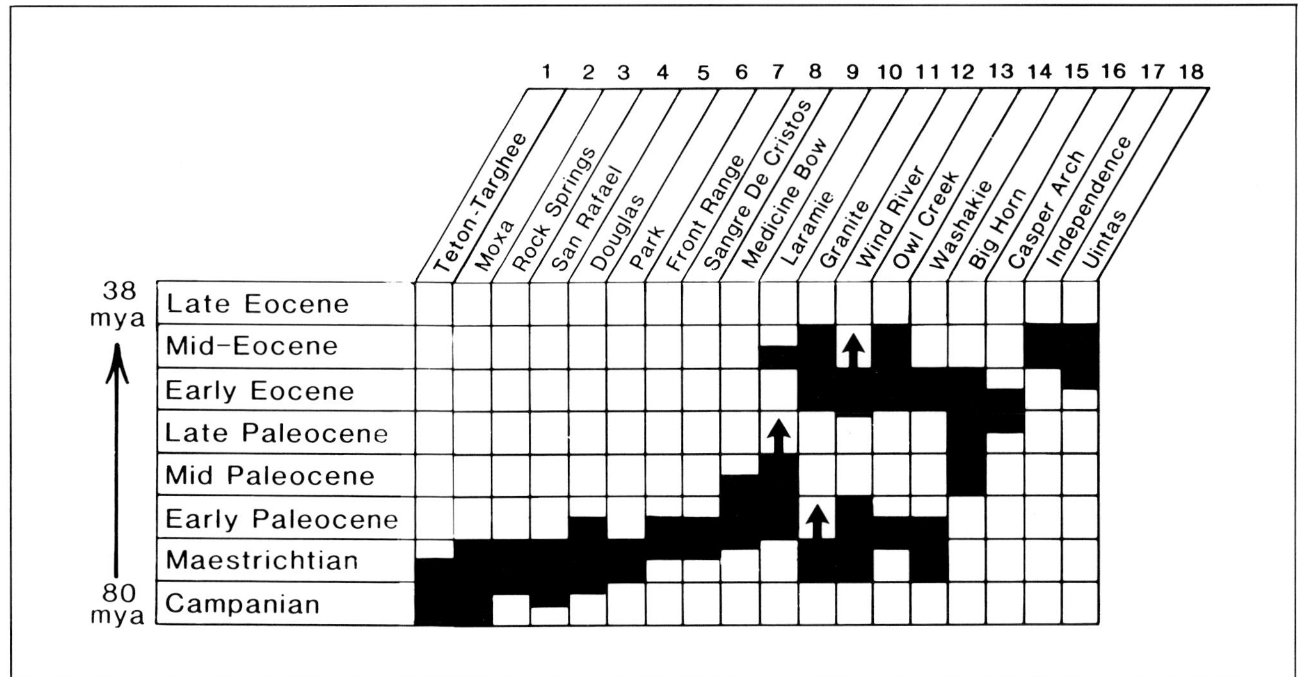

Fig. 3-14 (Gries, 1983b)—Timing of uplift on Laramide ranges shows an early Laramide pulse of deformation (Campanian through middle Paleocene) and a late Laramide pulse (early and middle Eocene) separated by a late Paleocene lull. Permission to publish by Rocky Mountain Association of Geologists.

in Argentina. I interpret detachment as a regional, not a local, phenomenon. Prominent detachment horizons in the Rocky Mountain foreland are shales of Cambrian, Triassic (Chugwater), and Cretaceous (Mowry and Cody and equivalent) age.

Oil and gas have been produced from detached folds. A negative aspect is that otherwise prospective beds beneath a completely detached structure do not have closure unless they are affected by deeper faults. More optimistically, closures related to detachment are themselves prospective. Both hanging walls (Fig. 3-3) and footwalls (Fig. 3-26A) of detached thrusts have been productive. Figure 3-26A shows a hanging wall structure, Dry Creek anticline, thrust over a footwall structure, Rolfe Lake, that has produced over 1MMBO.

ANCILLARY STRUCTURES OF COMPRESSIVE BLOCKS

Additional features which are characteristic of compressive block structural style and which affect structural interpretation in this terrain include normal faults, surficial gravity slides, tear faults, and shear zones with vertical fold axes. These features develop during more advanced stages of, and are incidental to the more fundamental faulting; were it not for the stresses that originally generated the compressive block, they would not exist. Late collapse of compressive blocks or slabs is perhaps the ultimate ancillary

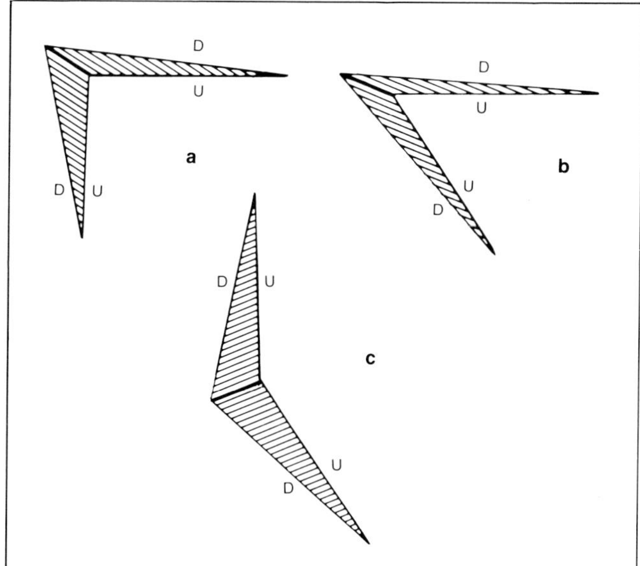

Fig. 3-15—Faults intersect to form trapdoor structures of a.- right angle, b.- acute angle, and c.- obtuse angle types. Maximum fault displacement is at the point of intersection, dying out toward the ends. Some trapdoors, particularly the obtuse angle types, may appear to have one curving fault that is difficult to differentiate from two intersecting faults (Fig. 1-10). Areal closure is greatest on obtuse angle trapdoors and progressively decreases through right-angle to acute angle trapdoors.

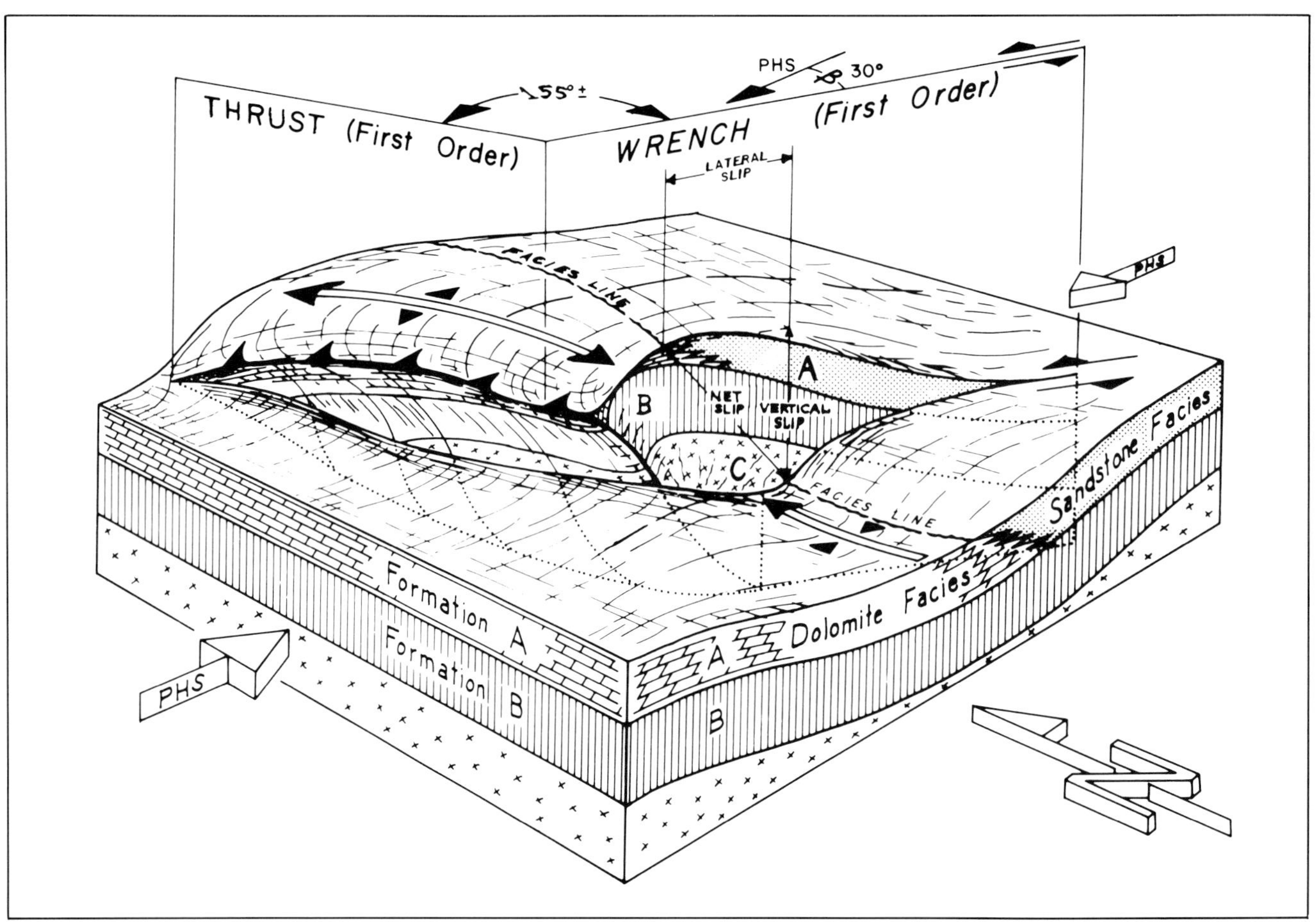

Fig. 3-16 (Stone, 1969)—Conceptual "wrench-thrust" model of trapdoor structure. Though this may occur on a limited basis, the two intersecting faults are more often of the same type (cf. Fig. 3-17). Permission to publish by D. S. Stone.

structure in which the secondary feature exceeds the original (if only in surface expression).

Stretching and arching behind (i.e., upliftward of) the main bounding fault of a compressive block in the crestal region of the high block creates downward propagating normal faults ranging in displacement from a few to hundreds of feet (Figs. 3-30, 3-31). The latter can significantly offset reservoirs on producing anticlines and can be recognized on and affect the structural interpretation of seismic sections (Fig. 3-32). Alternatively, normal faults can be the result of collapse of the frontal wedge of crystalline rocks caused by the wedge not being able to support itself above the less dense sedimentary rocks below (note Fig. 3-33).

Compressive block faulting can create significant relief (up to several thousand feet) which gives rise to unstable slopes, particularly when incompetent materials are involved. Gravity slides have occurred on the south flank of the Owl Creek Mountains where massive middle Paleozoic carbonates have been undercut by erosion of soft underlying Cambrian shales and transported basinward (Figs. 3-30, 3-31, 3-33). Examples of possibly structurally similar gravity slides are known in the subsurface in the Cayanosa and Rojo Caballos areas of west Texas (Guinan, 1971).

Vertical uplift and basinward movement of compressive blocks proceed at different rates leading to tear faults and lateral shear, which in turn lead to vertical fold axes, along some block edges. This structural response has been studied along the northeast front of the Beartooth Mountains, Montana (Figs. 3-34, 3-35). Tear faults can be anticipated in the subsurface on the short asymmetric flank of any compressive block of large structural relief; they should trend approximately perpendicular to the main, block-bounding fault.

Collapse of some foreland basement uplifts has occurred. According to Sales (1983) uplifts usually collapse "back down the root-zone of the thrust that raised the uplift because isostatic stresses were concentrated there... Uplifts have 'antiroots' and strong positive gravity anomalies with slabs held up by strength rather than buoyancy, making them susceptible to collapse" (Fig. 3-36). Sales (1983) has

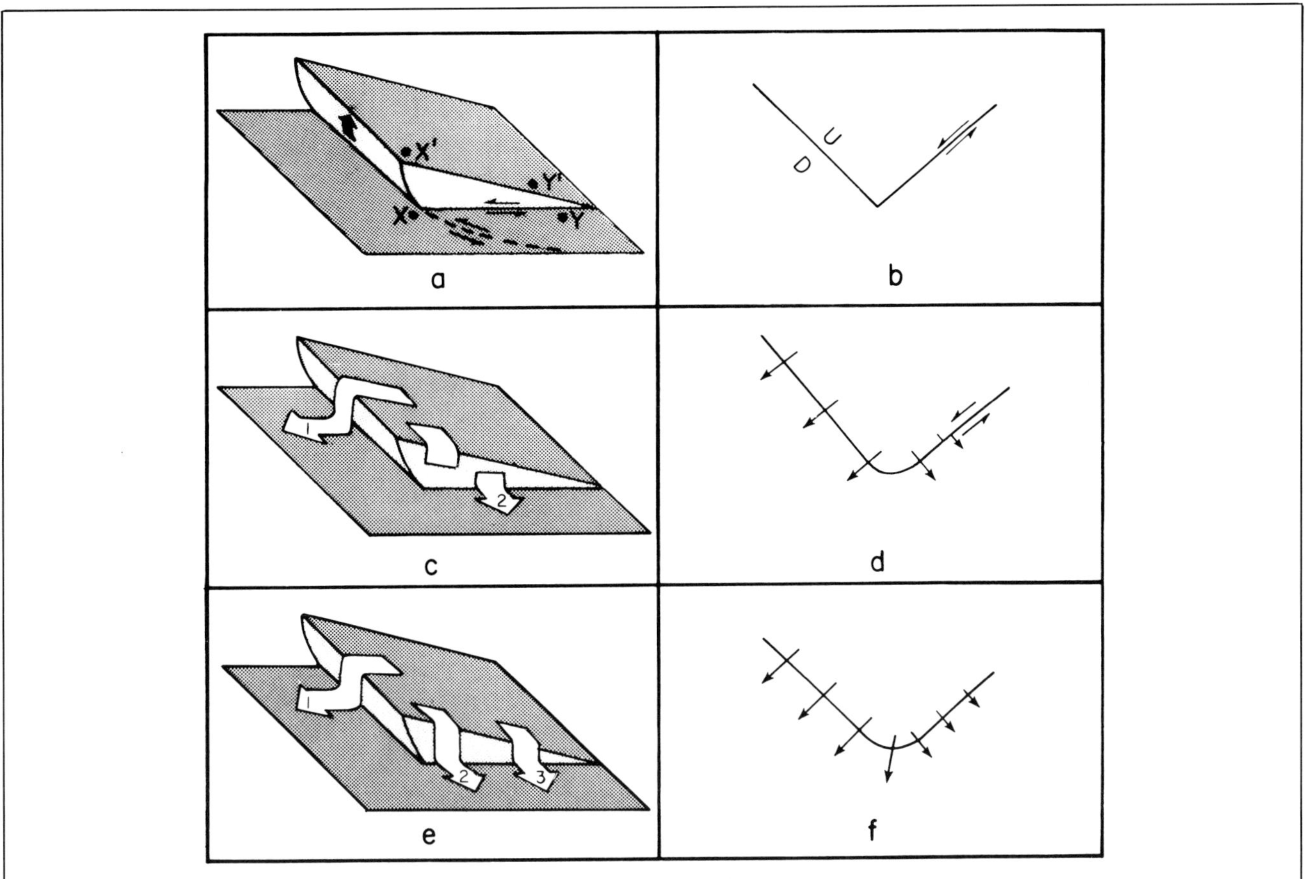

Fig. 3-17 (Stearns, 1978)—Possible trapdoor structures. e and f seem to be the most common and signify dominantly vertical movement. From Geological Society of America Memoir 151. Used with permission.

suggested that the distance between the earlier thrust fault and the later normal fault can be used to estimate the amount of overhang of the thrust wedge over the subjacent sediments.

Cobb and Poth (1980) also used the idea of later (Tertiary Basin and Range) collapse of an earlier (Laramide) wrench and thrust-uplifted feature to interpret the structure of the Santiago-northern Del Carmen mountains north of Big Bend National Park, west Texas (Fig. 3-37).

FOLDS

Both compressive block and thrust-fold belt styles are particularly well suited to illustrate types of folds common in petroleum provinces. The subject is introduced here and additional aspects of folding are covered in Chapter 6. A most utilitarian fold classification for the petroleum geologist is that of Donath and Parker (1964) who differentiated flexural-slip, flexural-flow, passive-slip, and passive-flow folds.

Flexural Slip

Flexural-slip folding can be simulated by bending any layered sequence such as playing cards, a sheaf of paper, or competent, well indurated, stratified rocks (Fig. 3-38, 3-39). Each successively higher layer overrides its lower neighbor by slip along bedding surfaces with the effect that there is no actual change (only bending) of original bed length and no change in volume (or area with reference to cross sections). Since bed thickness remains constant, sides stay parallel, and curvature in successive layers is more or less concentric, flexural-slip folds are also referred to as "parallel" or "concentric" folds (Fig. 3-40). It should be noted, however, that parallel and concentric folds are not necessarily the same. Though both maintain constant thickness, only concentric folds have constant arcs. Thus, all concentric folds are parallel, but not all parallel folds are concentric. For example, some parallel folds have sharp angular hinges. In any case, the term flexural-slip fold is preferred as it is more descriptive.

A basic assumption in drawing balanced structural cross

Fig. 3-18—West side of north Big Horn compressive block. A. View eastward of right-angle corner. Note wrap around of sedimentary rocks without offset. B. View southward of oblique corner.

sections (Dahlstrom, 1969) is that folding has been by flexural slip. This is because the definition of a balanced cross section, that hanging-wall and foot-wall cutoffs of the same stratigraphic unit always match, requires no change in volume during folding. Implications of no volume change are consistent bed lengths and associated consistency of shortening, which lead to the premise that fault displacements should also be consistent. These relationships are further explained in the reprint on "Balanced Cross Sections" at the end of Chapter 6.

Flexural-slip behavior (Fig. 3-41) associated with foreland compressive block deformation often leads to detachment structures which have previously been discussed.

Flexural Flow

Flow that takes place *within* layers is flexural-flow folding (Fig. 3-42). Incompetent rocks, such as shales, evaporites, and some limestones, are invariably involved and, indeed, are requisites of flexural flow. The more ductile the

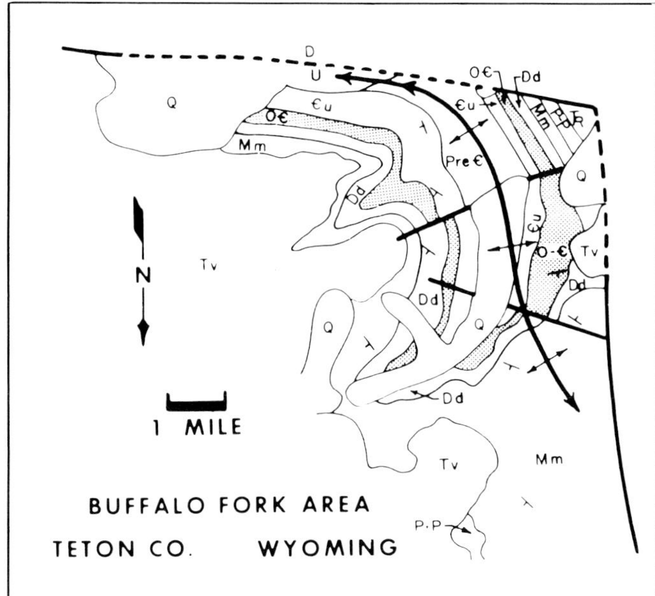

rock, the greater the tendency for flow. Flexural-flow folds have also been called "similar" folds. Cross sections cannot be drawn as balanced in flexural-flow folds because of volume changes caused by the folding (Fig. 3-43).

Flexural-flow and flexural-slip folds are often found in the same exposures, their occurrence dictated by rock competence and anisotropy and bed thickness and spacing (Figs. 3-44, 3-45).

Passive Folding

Passive folding involves slip or flow *across* layer boundaries, e.g., sedimentary bedding (Fig. 3-46). As Donath

Fig. 3-19 (Bengtson, 1956)—Trapdoor uplift in northwestern Wyoming. Two reverse faults intersect at approximate right angles, paralleled by a causally related, arcuate anticline asymmetric towards the faults. Note transverse faults. Permission to publish by Wyoming Geological Association.

Fig. 3-20—Clay model showing upthrust developed from vertical differential rotation of blocks along a 70° dipping saw-cut which simulates a preexisting normal fault. The wooden blocks were originally horizontal with the clay cake across them. Block 1 was rotated 15°, and block 2 was rotated 5° during the uplift. The greater the differential uplift (rotation) between blocks, the greater the compression in the clay and the better the development of the upthrust. Repeated experiments have shown that for lower dips of the saw-cut between the blocks, the creation of upthrusts requires greater differential block rotation. If the differential rotation on the two blocks is about the same, the clay is under mild extension and normal faults develop.

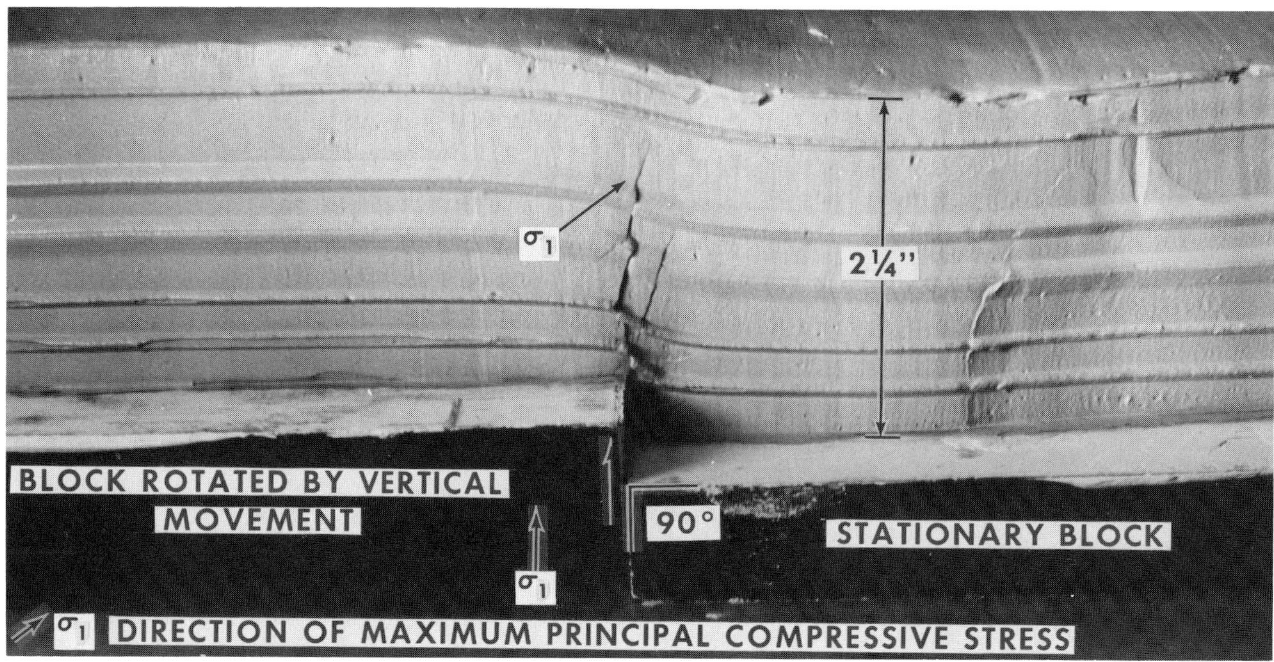

Fig. 3-21—Clay model of an asymmetric block showing the steep upthrust flank on the right created by purely vertical rotation of a rigid wooden board along a pre-existing vertical surface beneath the clay.

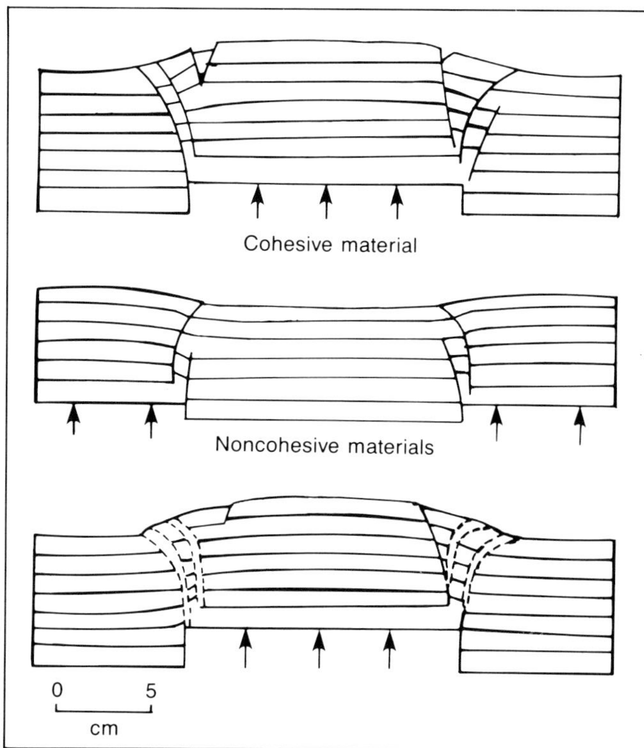

Fig. 3-22 (Sanford, 1959)—Line drawing of a model, consisting of granular materials with a small mixture of clay, showing upthrusts created by vertical displacement along a pre-existing vertical step. Note that upthrusts always flare outward toward the relatively down-dropped block. From Geological Society of America Bulletin, V. 70, p. 19–52. Used with permission.

and Parker (1964, p.45) have stated, "the layering does not control the configuration of the deformed mass—it merely reflects the deformation". Passive folding is most likely to occur by an increase in ductility, which can be brought about by higher temperature or pressure. Thus passive folds are most common in metamorphic rocks, such as slates, marbles, gneisses, and schists.

SEISMIC EXPRESSION

Seismic examples are divided into those that cover prospective anticlinial features and those that show the configurations of the larger block or mountain fronts.

Anticlines

A seismic profile across the Rangely oil field in northwest Colorado (Fig. 1-7) shows many of the characteristics of foreland anticlines. The fault or fault zone on the steep, asymmetric flank of the structure is difficult to accurately position, both laterally and vertically. Lateral positioning is complicated by diffractions having similar attitudes as those anticipated for dip of beds into the fault zone. Since compressive blocks are characterized by faults that offset the basement, but often die out upward in the sedimentary section, it is also difficult to determine the vertical extent of faults on seismic records (Fig. 3-47).

Mountain Fronts

Geologic models and interpretations of mountain front geometries are shown in figures 3-48 through 3-51. A con-

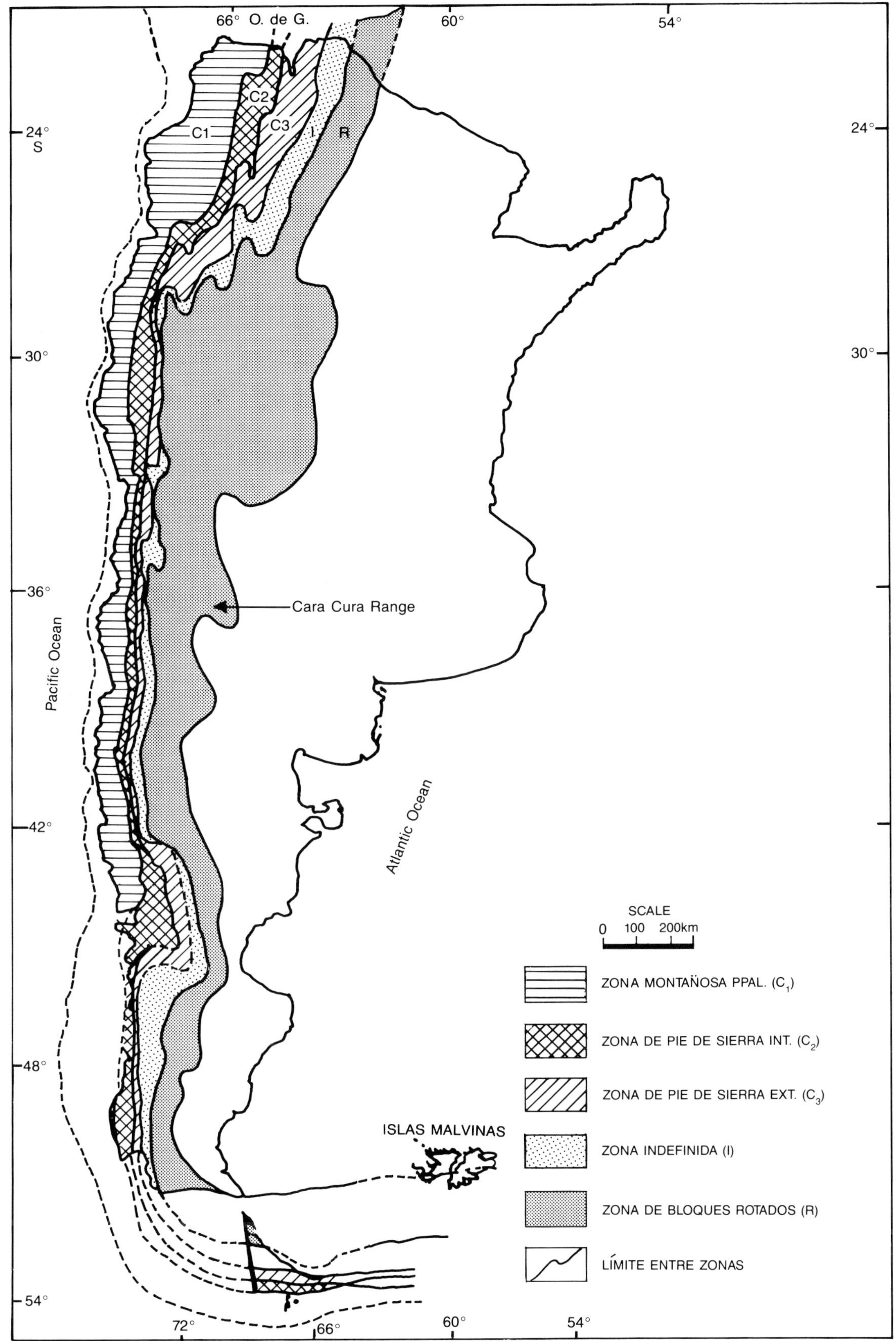

Fig. 3-23 (Vasquez and Gorrono, 1980)—Zonation of Argentinian Andes showing zone of compressive (rotated) blocks (R) in foreland. Permission to publish from Asociación Geológica Argentina.

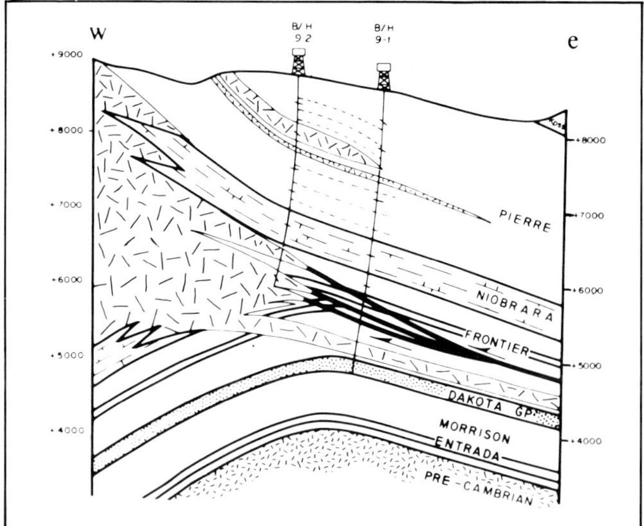

Fig. 3-24 (Wellborn, 1977)—Detachment faults in Mowry Formation (below Frontier) at Granny's Nipple anticline, west side of North Park Basin, Colorado. In this and succeeding examples (Figs. 3-25 through 3-29) good well control or other data virtually preclude that the faults could go into basement as they would have to take impossible trajectories; in addition there is no hint of basement uplift where the fault realistically could go to basement. Permission to publish by Rocky Mountain Association of Geologists.

ceptual view of the evolution of a fold-thrust uplift is illustrated in figure 3-48, while figure 3-49 shows greater detail of the fold-thrust model; figure 3-49 would correspond roughly to stage II of figure 3-48. Based primarily on well control figures 3-50 and 3-51 show relatively large and small amounts of thrust overlap, respectively.

Reflection raypaths are less distorted where compressive blocks have large thrust overlaps, measured in miles, rather than small overlaps, measured in hundreds or thousands of feet. Consequently, the structures of figures 3-48 (stage IV) and 3-50 would likely facilitate good records comparable to figure 3-2. In contrast, the interpretation of structures with steep and vertical dips (Figs. 3-48—stages II and III, 3-49, 3-51) is almost always complicated by poor quality records (Fig. 3-52).

Good seismic records in the footwalls of compressive block frontal thrusts having significant overlap have also been obtained along the Uncompahgre uplift in east-central Utah (Figs. 3-53, 3-54) and along the northwest end of the Casper arch fronting the northeast side of the Wind River Basin in central Wyoming (Fig. 3-55).

In summary, exploration in most foreland block subthrust plays has been confounded by velocity pull-up problems that make it difficult to demonstrate four-way closure. It should also be noted that fault and fracture zones within the Precambrian basement wedge are frequently better reflectors than the main thrust plane or surface itself (Ray et al., 1983). Skeen and Ray (1983) have constructed seismic models for a variety of thrust conditions and the reader is referred to their work for more detail on seismic expression of foreland mountain front faults.

PRODUCING EXAMPLES

Many fields of the compressive block style show trapdoor configurations (Fig. 3-15). Keystone field of the Permian Basin foreland of the Marathon thrust-fold belt in west Texas (Fig. 1-10) and Grass Creek field in the Big Horn Basin (Fig. 3-56) are prime examples.

Asymmetry of structure also plays an important part in trapping petroleum in the Rocky Mountain foreland. The vast majority of structural fields there have steep faulted flanks out-of-the-basin with gentle tilted flanks that dip into

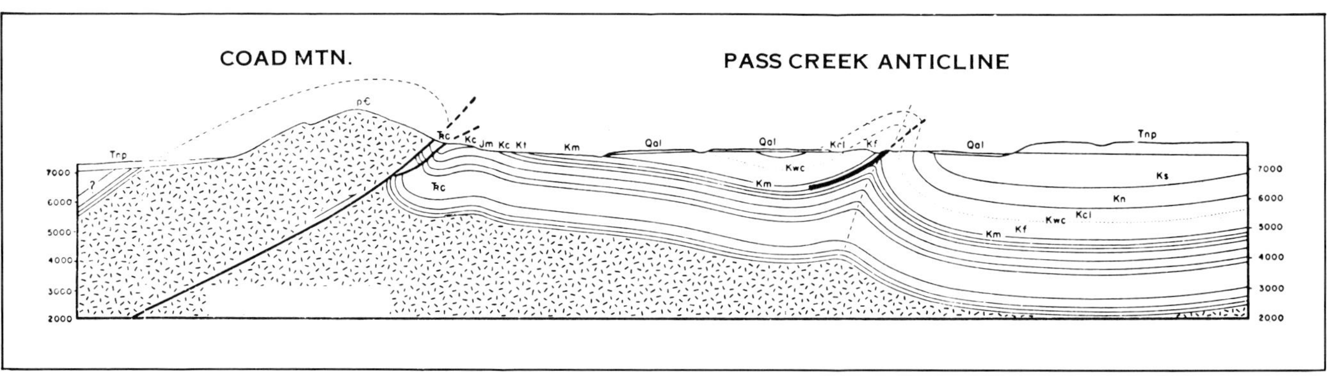

Fig. 3-25 (Beckwith, 1941)—Detachment of Pass Creek anticline, Elk Mountain area at north end of Laramie Basin, southeast Wyoming. Juxtaposition of Mowry (Km) on Frontier (Kf) can be seen in field. Soleing of fault in Mowry could also be observed (Beckwith, 1941) until recently when the critical exposure was covered with mud and cuttings from a nearby well. From Geological Society of America Bulletin, V. 52. Used with permission.

Illustration 3-26 begins on next page ⟶

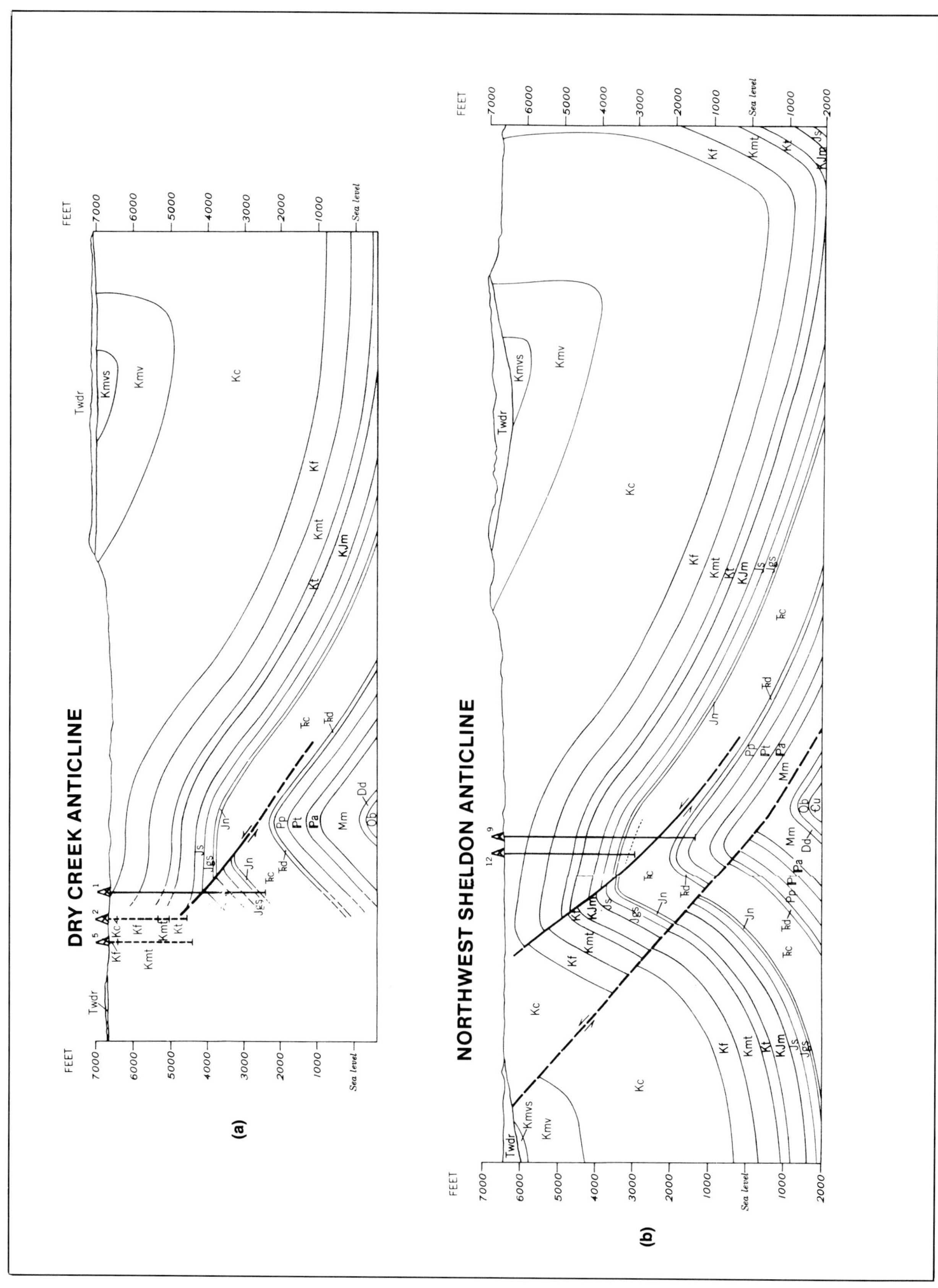
DRY CREEK ANTICLINE
(a)
FEET
7000
6000
5000
4000
3000
2000
1000
Sea level
NORTHWEST SHELDON ANTICLINE
(b)
FEET
7000
6000
5000
4000
3000
2000
1000
Sea level
1000
2000

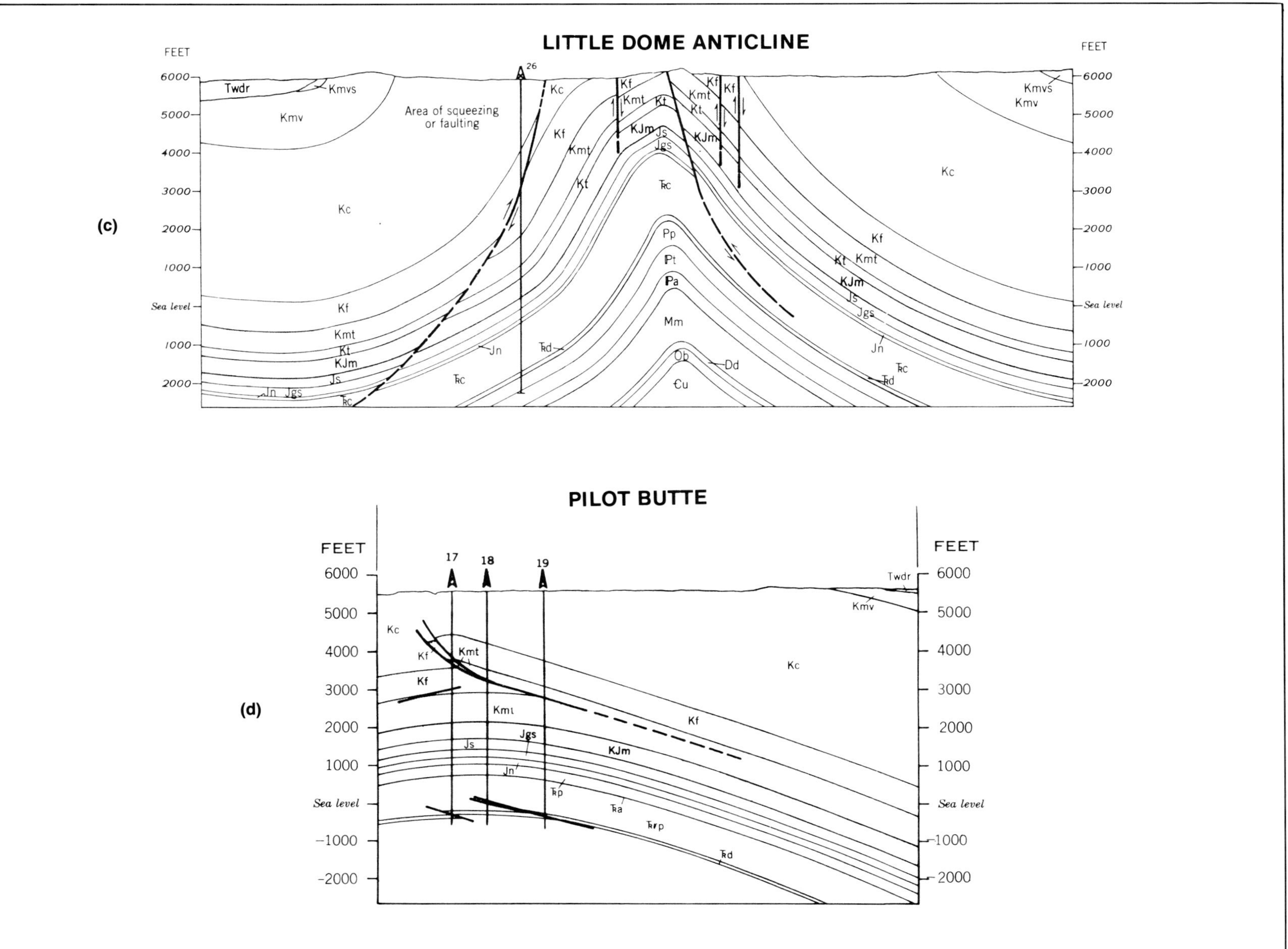

Fig. 3-26 (Murphy et al., 1956 A through C)—Detachment structures in northwest end of Wind River Basin, central Wyoming. Sole horizon is Triassic Chugwater Formation at A– Dry Creek anticline, B– Northwest Sheldon anticline, and C– Little Dome anticline. At D– (Murphy and Roberts, 1954) Pilot Butte three wells demonstrate that the Frontier (Kf) is repeated by a fault that soles in Mowry shales (Kmt). Courtesy United States Department of the Interior.

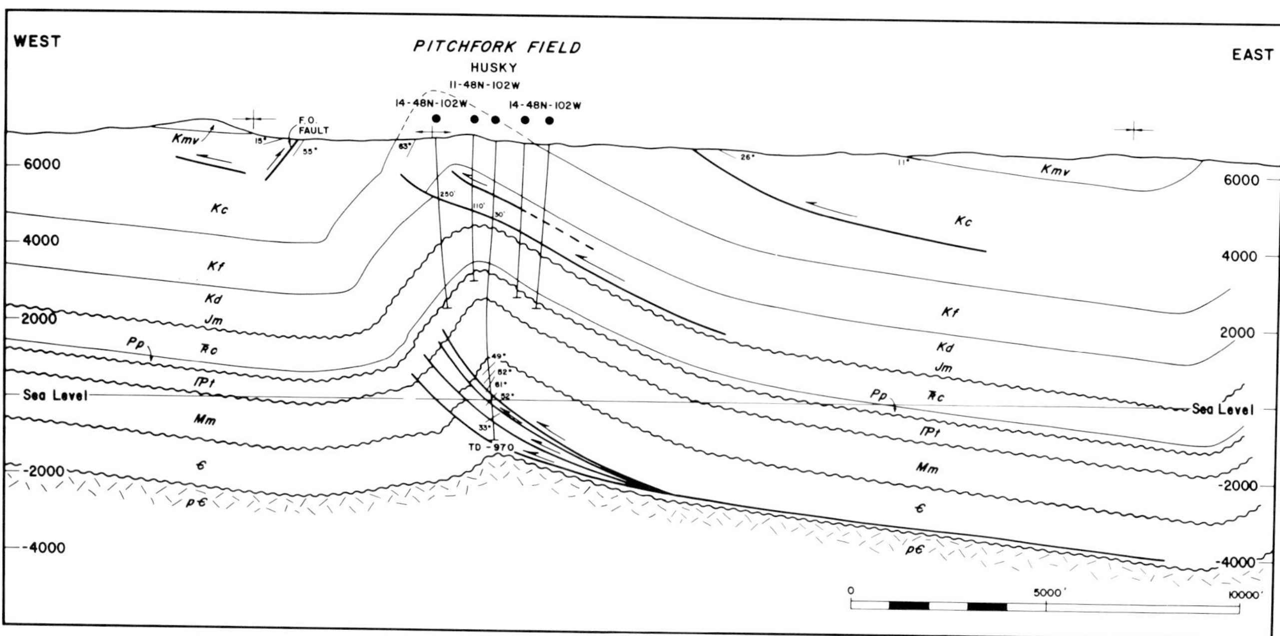

Fig. 3-27 (Petersen, 1983)—Detachment faults at two levels (Jurassic and Cambrian) at Pitchfork field, west side Big Horn Basin, northwest Wyoming. Note much greater structural relief, due to thrust stacking, in sedimentary section than on basement. Permission to publish by Rocky Mountain Association of Geologists.

Fig. 3-28—View southward of Cedar Mountain (south plunge of Rattlesnake Mountain anticline) showing repetition of Ordovician Big Horn Dolomite by detachment fault at center of picture; fault soles in Cambrian shales toward lower right corner of photograph. Carinate fold in Mississippian Madison Limestone at top of mountain effects a fault-fold interchange. (See also Fig. 3-44.)

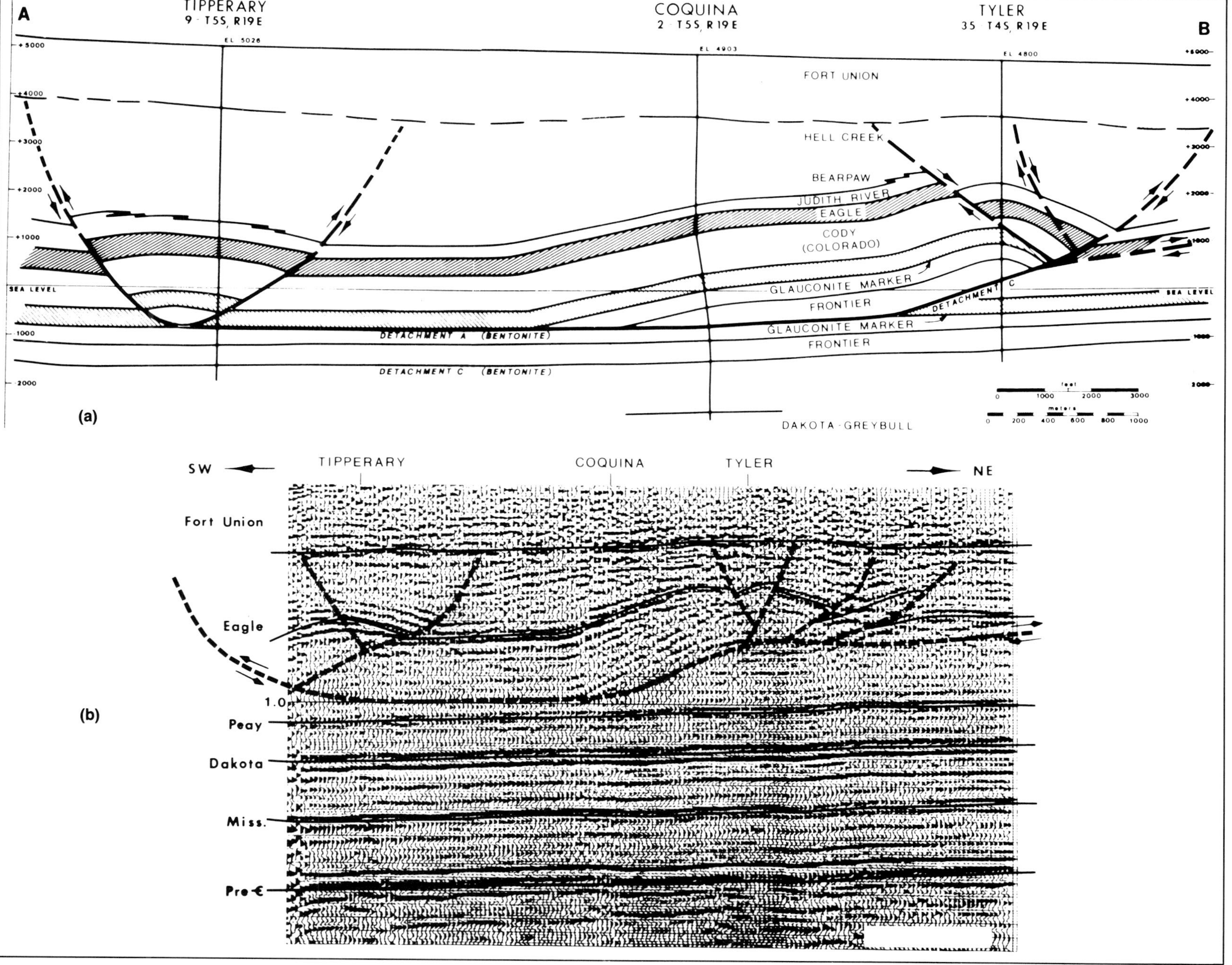

Fig. 3-29 (Stone, 1983a)—Detachment faults and folds in Crazy Mountains Basin (Reed Pt. syncline), south-central Montana. Cross section (a) and seismic section (b) are in approximately same location. Note good well and seismic control on continuity of beds beneath detachment. Permission to publish by D. S. Stone.

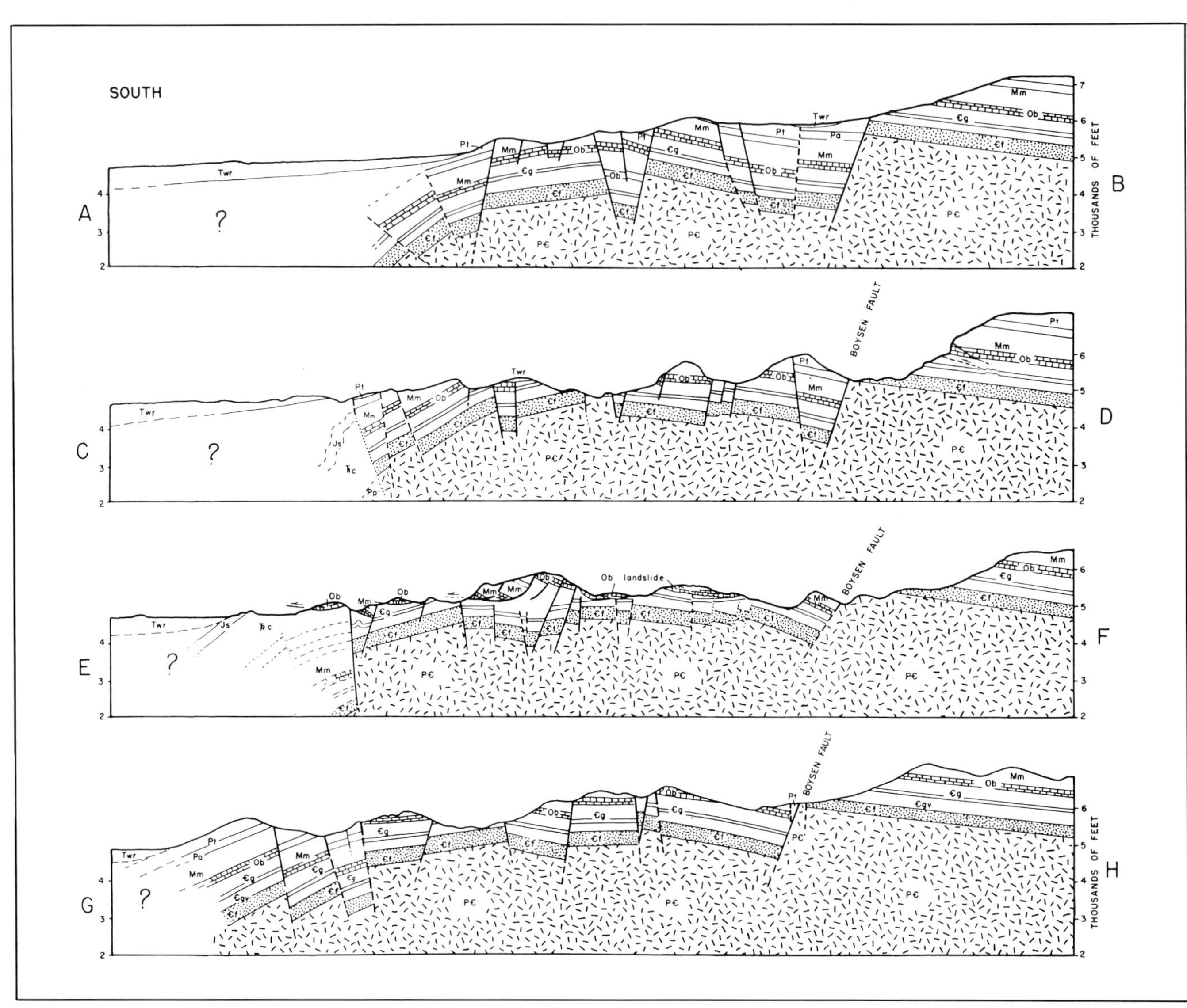

Fig. 3-30 (Wise, 1963)—Normal faults in Owl Creek Range, central Wyoming. Reverse fault in all cases is to the south and below the line of section. Permission to publish by American Association of Petroleum Geologists.

Fig. 3-31—Owl Creek Range view eastward above Boysen Dam of normal faults and gravity slump affecting Paleozoic carbonates. Ob is Ordovician Big Horn Dolomite.

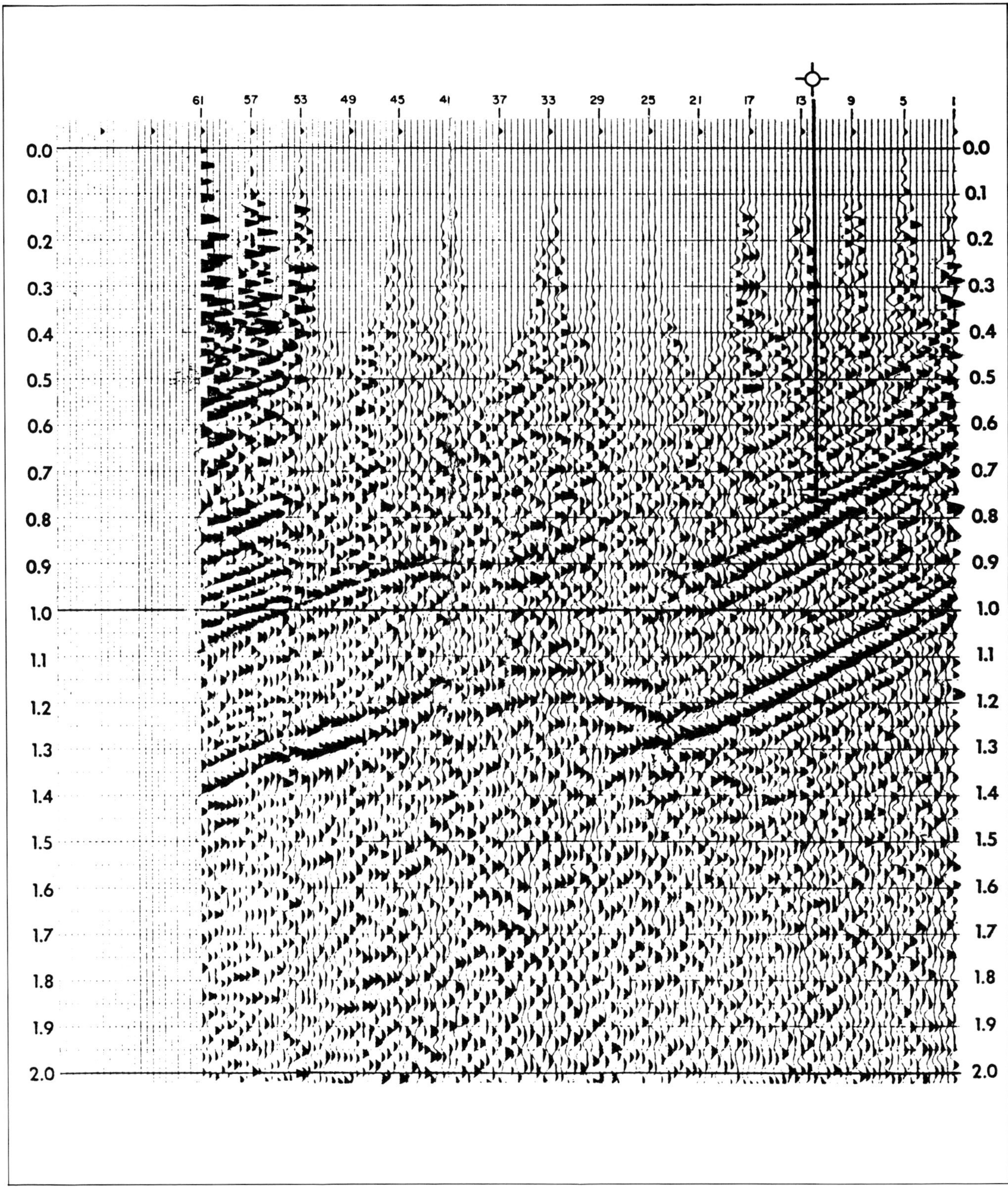

Fig. 3-32—Seismic section showing discontinuous nature of upper event on high block between 0.8 and 0.9 sec. vs. continuity of lower event between 1.1 and 1.2 sec. interpreted as caused by downward propagating keystone normal faults that do not affect lower event.

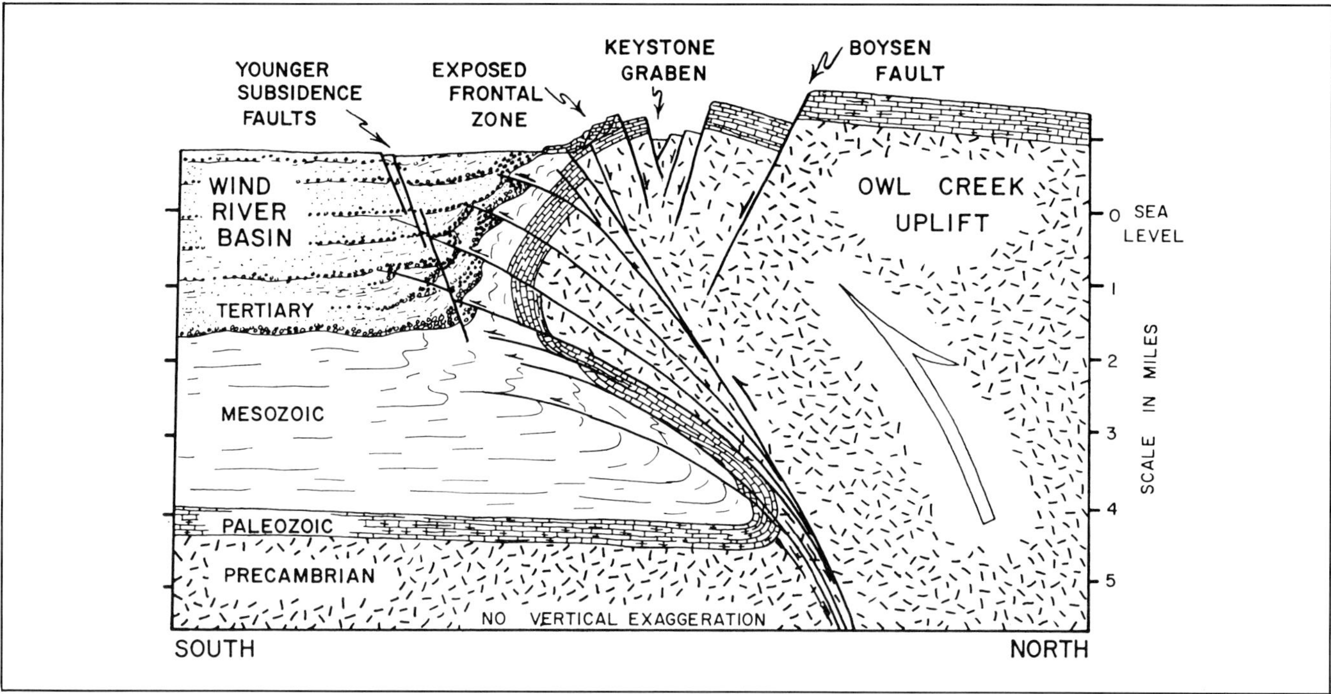

Fig. 3-33 (Wise, 1963)—Diagrammatic view of Owl Creek uplift with keystone faults and gravity slumps. Permission to publish by American Association of Petroleum Geologists.

the basin, thus facilitating excellent out-of-basin drainage (Figs. 3-56 through 3-58). Curiously the asymmetry of structural fields in west Texas is not nearly as systematic.

The most significant compressive block sub-thrust discovery made thus far in the Rocky Mountain foreland is the Tepee Flats field in central Wyoming, where a separate, asymmetric structure can be mapped beneath the Precambrian crystalline wedge (Figs. 3-55, 3-59, 3-60).

In addition to structural accumulations on compressive blocks, stratigraphic and combination traps, which are dependent on pinchouts, porosity, permeability, and hydrodynamics, can also occur.

REFERENCES

Beckwith, R. H., 1941, Structure of the Elk Mountain district, Carbon County, Wyoming: *Geological Society America Bull.*, V. 52, p. 1445–1486.

Bengtson, C. A., 1956, Structural geology of the Buffalo Fork area, northwestern Wyoming and its relation to the regional tectonic setting: *Wyoming Geological Association Guidebook* 11th Ann. Field Conf., p. 158–168.

Berg, R. R., 1962, Mountain flank thrusting in Rocky Mountain foreland, Wyoming and Colorado: *American Association Petroleum Geologists Bull.*, V. 46, p. 2019–2032.

———, 1983, Geometry of the Wind River thrust, *in* Lowell, J. D., ed., Rocky Mountain foreland basins and uplifts: Rocky Mountain Association Geologists, p. 257–262.

Bieber, D. W., 1983, Gravimetric evidence for thrusting and hydrocarbon potential of the east flank of the Front Range, Colorado, *in* Lowell, J. D., ed., Rocky Mountain foreland basins and uplifts: Rocky Mountain Association Geologists, p. 245–255.

Blackstone, D. L., 1963, Development of geologic structure in central Rocky Mountains, *in* Backbone of the Americas: *American Association Petroleum Geologists Mem. 2,* p. 160–179.

———, 1983, Laramide compressional tectonics, southeastern Wyoming: University of Wyoming *Contributions to Geology,* V. 22, n. 1, p. 1–38.

Brown, W.G., 1983, Sequential development of the fold-thrust model of foreland deformation, *in* Lowell, J. D., ed., Rocky Mountain foreland basins and uplifts: Rocky Mountain Association Geologists, p. 57–64.

Burchfiel, B. C., and Davis, G. A., 1972, Structural framework and evolution of the southern part of the Cordilleran orogen, western United States: *American Journal Science,* V. 272, p. 97–118.

Chamberlin, R. T., 1945, Basement control in Rocky Mountain deformation: *American Journal Science,* V. 243-A, p. 98–116.

Chapin, C. E., and Cather, S. M., 1981, Eocene tectonics and sedimentation in the Colorado Plateau-Rocky Moun-

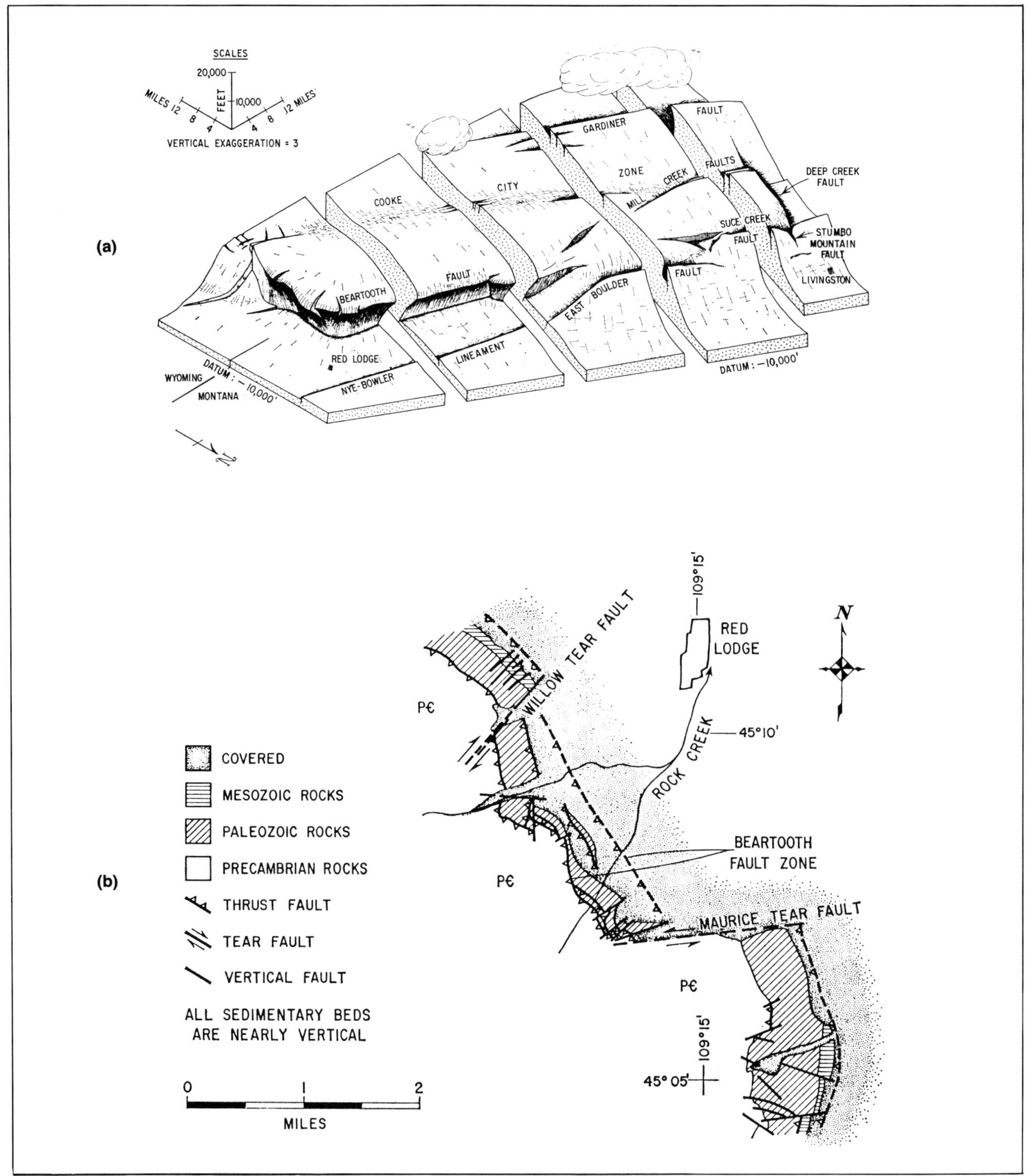

Fig. 3-34 (Foose et al., 1961)—Tear faults on northeast corner of Beartooth block, south-central Montana. a. Three dimensional block diagram interpretation; b. Map view. Permission to publish by American Association of Petroleum Geologists.

Fig. 3-35—View westward of Line Creek fold at mouth of Line Creek on east side of Beartooth block showing near vertical axes of folds in section from Ordovician Big Horn Dolomite through Mississippian Madison Limestone. Vertical fold axes are related to lateral movement and some have been referred to as "spindle" structures.

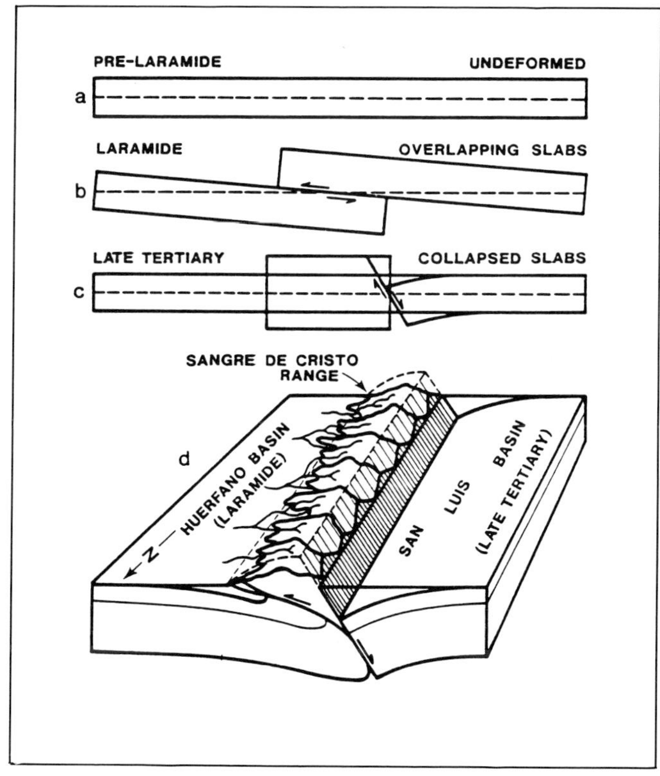

Fig. 3-36 (Sales, 1983)—"Perspective diagram of the interpreted relation between the Laramide thrust and Neogene normal fault that have blocked out the central Sangre de Cristo uplift" (south-central Colorado) as an illustration of uplift collapse. Note that "the outer thrust lip 'hung up' and is left standing higher than the once higher core of the uplift." Permission to publish by Rocky Mountain Association of Geologists.

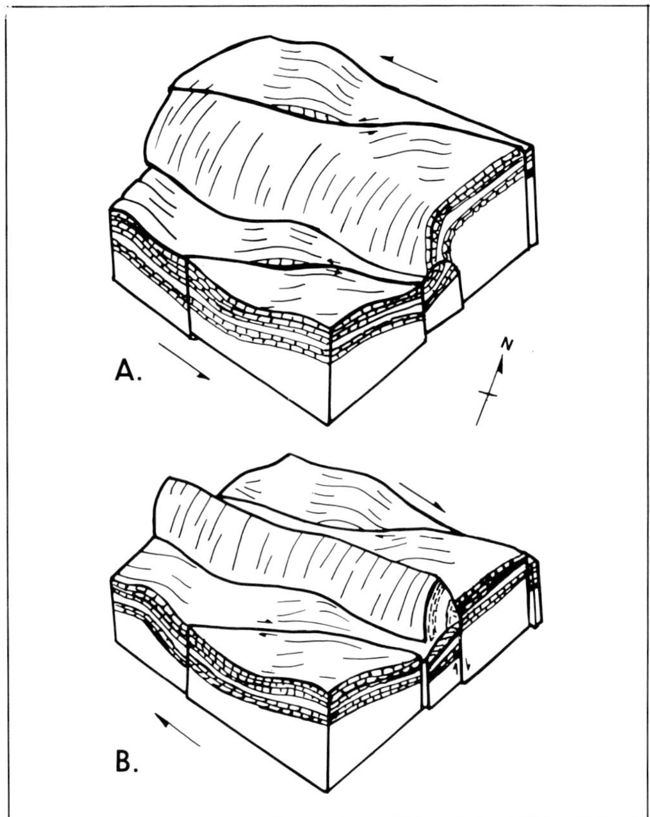

Fig. 3-37 (Cobb and Poth, 1980)—Block A– monocline created by Laramide deformation. Block B– the steep portions of Laramide fault planes were used during Basin and Range deformation to downdrop the rocks north of the monocline leaving the narrow, northwest-trending range standing in relief. Permission to publish by New Mexico Geological Society, Inc.

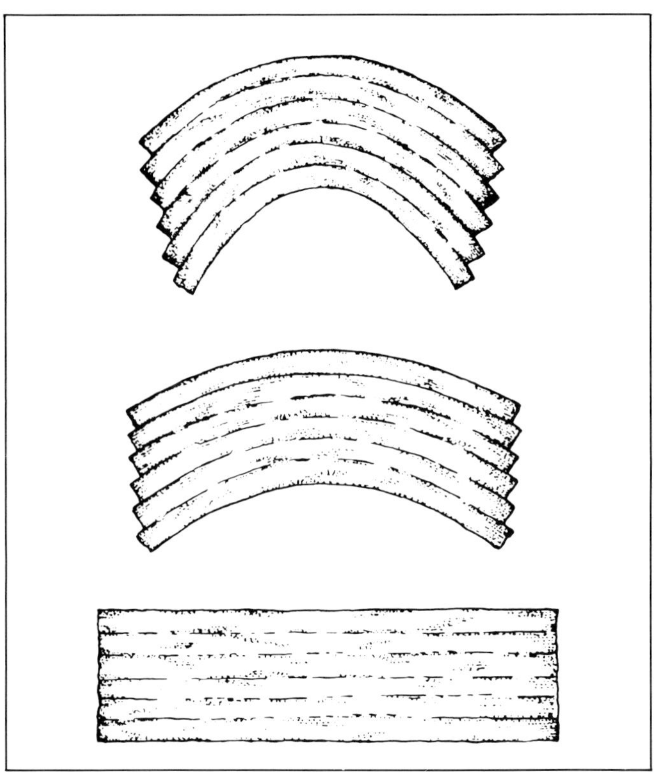

Fig. 3-38 (Donath and Parker, 1964)—Flexural-slip folding. Note that integrity of bedding is not only maintained but that bedding helps dictate the form of the fold. From Geological Society of America Bulletin, V. 75. Used with permission.

Fig. 3-39—Flexural-slip fold in Madison Limestone on southwest flank of Rattlesnake Mountain anticline, northwest Wyoming.

Fig. 3-40 (Dennis, 1967)—Flexural-slip (parallel or concentric) folds (after Van Hise, 1896a). Permission to publish by American Association of Petroleum Geologists.

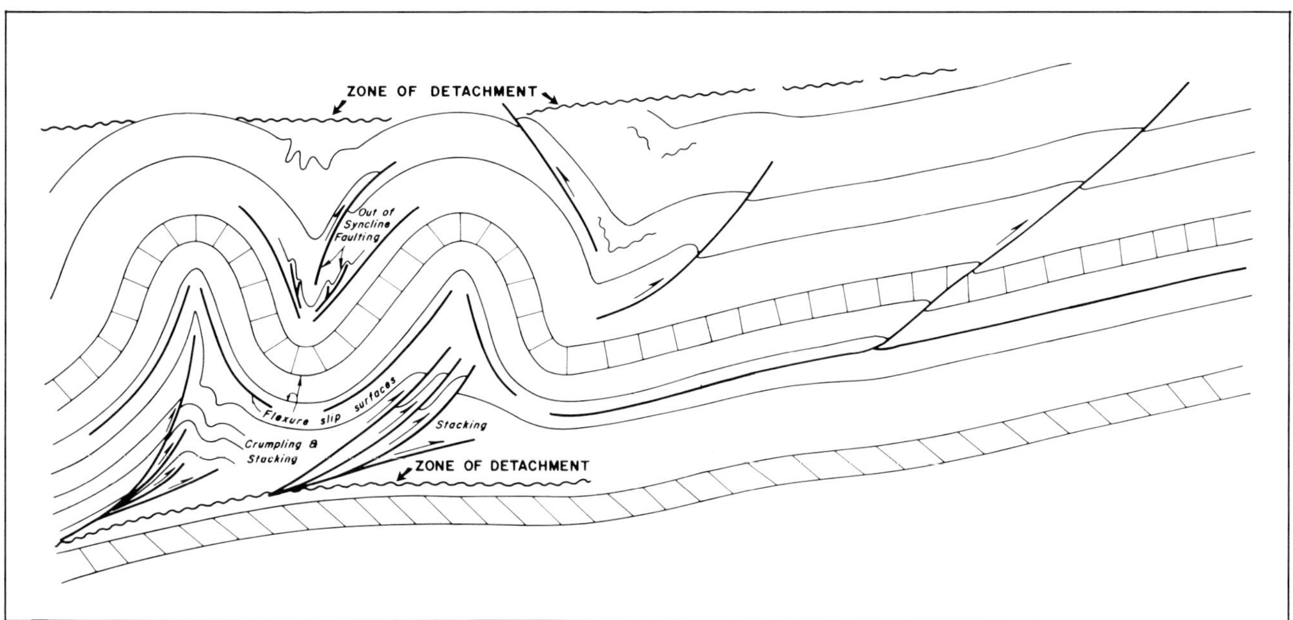

Fig. 3-41 (Petersen, 1983)—Behavior diagram for flexural slip folding. Permission to publish by Rocky Mountain Association of Geologists.

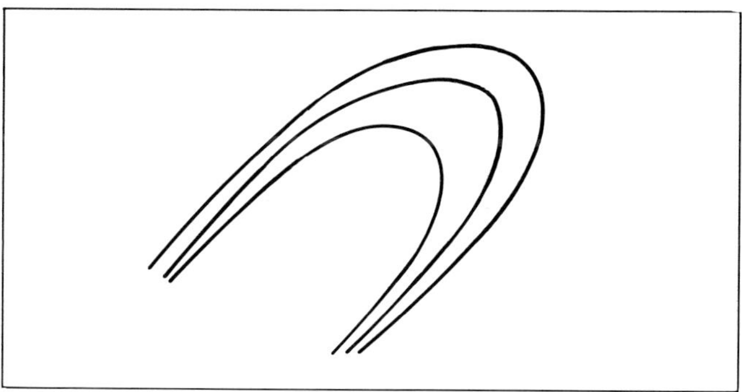

Fig. 3-42—One type of flexural-flow fold that involves thinning of the limbs and thickening in the crest. Flowage occurs within, not across, beds.

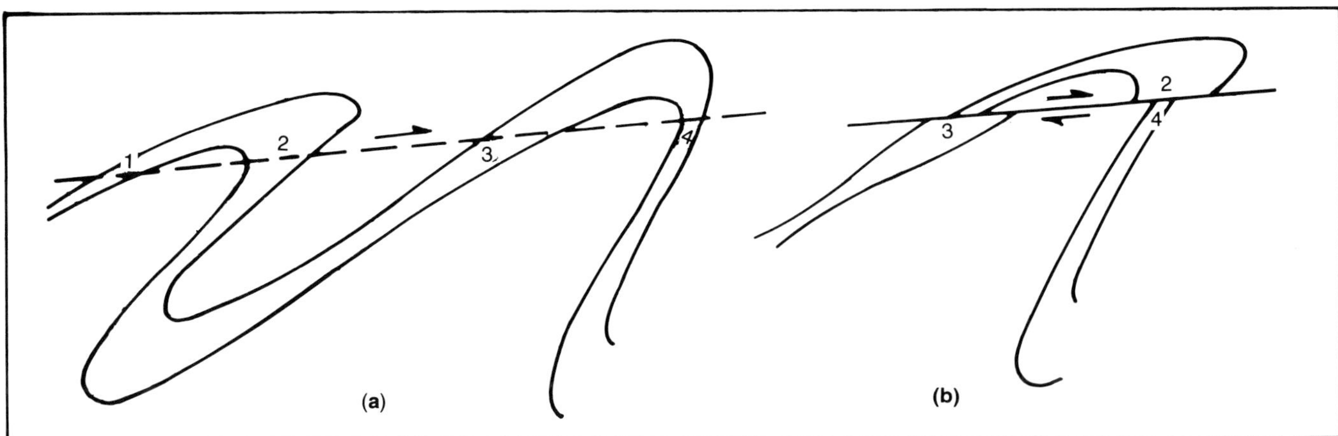

Fig. 3-43—Flexural-flow fold (a) where, when faulted, (b) hanging-wall cutoff does not match foot-wall cutoff because of volume change in unit during folding.

Fig. 3-44—View southeast of Cedar Mountain (south plunge of Rattlesnake Mountain anticline northwest Wyoming) showing monoclinal flexural-slip fold and associated out-of-syncline detachment fault (cf. Fig. 3-41) in competent Big Horn Dolomite and carbonate rocks above, and extreme flexural flow of shales between Gallatin Formation and Big Horn. Note splay of Precambrian block-bounding fault with attendant rotation and repetition of strata.

Fig. 3-45 (Prucha et al., 1965)—View north of Cretaceous and Precambrian at Buckhorn Mountain on east side of Front Range, north of Boulder, Colorado. Nearly vertical beds just to right of center in lower picture are massive Dakota sandstone (zone 3 of upper photo) turned up against fault to right. Massive sandstone is overlain by shale (2) succeeded by thinly interbedded sandstone and shale (1). In upper photo unit 1 has undergone complex flexural-slip folding owing to interlayered character and close bed spacing; unit 2 has been deformed by flexural flow; and unit 3 has behaved as a tilted slab, unable to fold because of massiveness and lack of bedding (i.e., slippage) planes. Permission to publish by American Association of Petroleum Geologists.

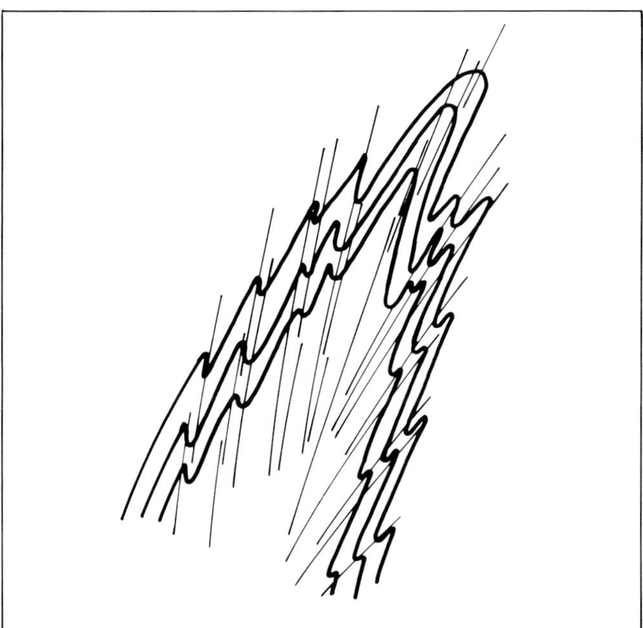

Fig. 3-46—Passive fold. Passive-slip and passive-flow folds are geometrically the same differing only if movement is microscopically continuous (flow) or discontinuous (slip). Cleavage (which is pervasive rather than restricted to a particular lithology) or foliation may effect slippage.

tain area, *in* Dickinson, W. R., and Payne, W. D., eds., Relations of tectonics to ore deposits in the southern Cordillera: *Arizona Geological Society Digest,* V. 14, p. 173–198.

Cobb, R. C., and Poth S., 1980, Superposed deformation in the Santiago and northern Del Carmen mountains, Trans-Pecos Texas: *New Mexico Geological Society Guidebook,* 31st Field Conf., Trans-Pecos region, p. 71–75.

Cook, D. G., 1983, The Northern Franklin Mountains, Northwest Territories, Canada—a scale model of the Wyoming province, *in* Lowell, J. D., ed., Rocky Mountain foreland basins and uplifts: Rocky Mountain Association Geologists, p. 315–338.

Dahlstrom, C. D. A., 1969, Balanced cross-sections: *Canadian Journal Earth Sciences,* V. 6, n. 4, p. 743–757.

Dennis, J. G., 1967, International tectonic dictionary: *American Association Petroleum Geologists Mem. 7,* 196 p.

Dickinson, W. R., and Snyder, W. S., 1978, Plate tectonics of the Laramide orogeny, *in* Matthews, Vincent III, Laramide folding associated with basement block faulting in the western United States: *Geological Society America Mem. 151,* p. 355–366.

Donath, F. A., and Parker, R. B., 1964, Folds and folding: *Geological Society America Bull.,* V. 75, p. 45–62.

Eardley, A. J., 1963, Relation of uplifts to thrusts in Rocky Mountains, *in* Backbone of the Americas: *American Association Petroleum Geologists Mem. 2,* p. 209–219.

Elliot, D., 1973, Plate tectonics and the problem of too much granitic basement: *Geology,* V. 1, p. 111.

Foose, R. M., Wise, D. U., and Garbarini, G. S., 1961, Structural geology of the Beartooth Mountains, Montana and Wyoming: *Geological Society Amercia Bull.,* V. 72, p. 1143–1172.

Frahme, C. W., and Vaughn, E. B., 1983, Paleozoic geology and seismic stratigraphy of the northern Uncompahgre front, Grand County, Utah; *in* Lowell, J. D., ed., Rocky Mountain foreland basins and uplifts: Rocky Mountain Association Geologists, p. 201–211.

Garihan, J. M., Schmidt, C. J., Young, S. W., and Williams, M. A., 1983, Geology and recurrent movement history of the Bismark-Spanish Peaks-Gardiner fault system, southwest Montana, *in* Lowell, J. D., ed., Rocky Mountain foreland basins and uplifts: Rocky Mountain Association Geologists, p. 295–314.

Gilluly, J. 1971, Plate tectonics and magmatic evolution: *Geological Society America Bull.,* V. 82, p. 2383–2396.

———, 1973, Steady plate motion and episodic orogeny and magmatism: *Geological Society America Bull.,* V. 84, p. 499–514.

Gries, Robbie, 1981, Oil and gas prospecting beneath the Precambrian of foreland thrust plates in the Rocky Mountains: *Mountain Geologist,* V. 18, n. l, p. 1–18.

———, 1983a, Oil and gas prospecting beneath the Precambrian of foreland thrust plates in the Rocky Mountains: *American Association Petroleum Geologists Bull.,* V. 67, p. 1–28.

———, 1983b, North-south compression of Rocky Mountain foreland structures; *in* Lowell, J. D., ed., Rocky Mountain foreland basins and uplifts: Rocky Mountain Association Geologists, p. 9–32.

Grose, L. T., 1972, Regional geology, tectonics, *in* Geologic Atlas of the Rocky Mountain Region, United States of America: Rocky Mountain Association Geologists, p. 35–44.

Guinan, M.A., 1971, Slide-block geology, Cayanosa and adjacent areas, Pecos and Reeves counties, Texas (abs.): *American Association Petroleum Geologists Bull.,* V. 55, p. 340.

Harms, J. C., 1964, Structural history of the southern Front Range: *Mountain Geologist,* V. 1, n. 3, p. 93–101.

Hoppin, R. A., and Jennings, T. V., 1971, Cenozoic tectonic elements Big Horn Mountain region, Wyoming-Montana: *Wyoming Geological Association Guidebook* 23d Ann. Field Conf., p. 39–47.

Hubbert, M. K., 1951, Mechanical basis for certain familiar geologic structures: *Geological Society America Bull.,* V. 72, p. 355–372.

Jacob, A. F., 1983, Mountain front thrust, southeastern Front Range and northeastern Wet Mountains, Colo-

Illustration 3-47 begins on next page ⟶

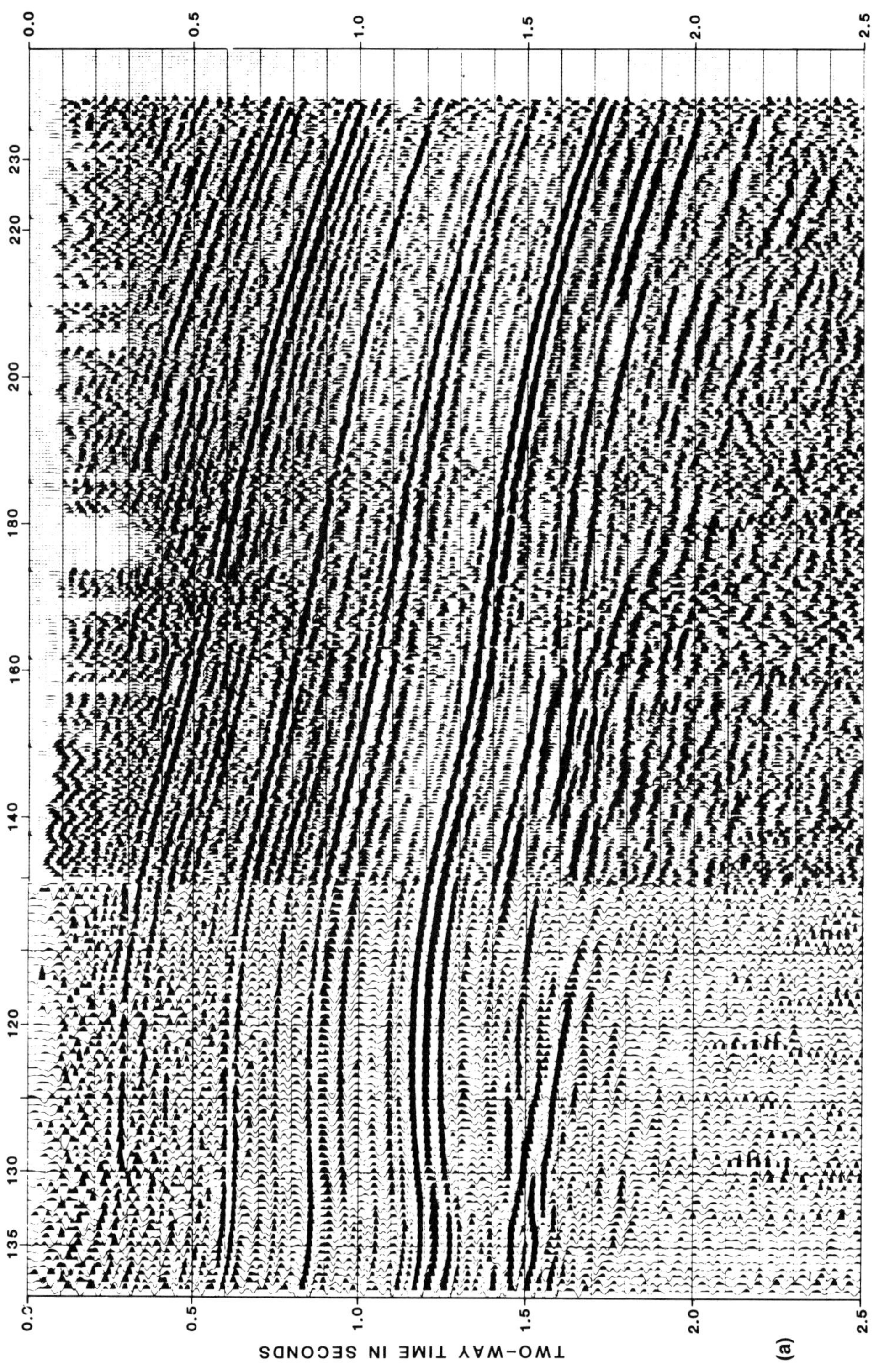
0.0
0.5
1.0
1.5
2.0
2.5
230
220
200
180
160
140
120
130
135
TWO-WAY TIME IN SECONDS

(a)

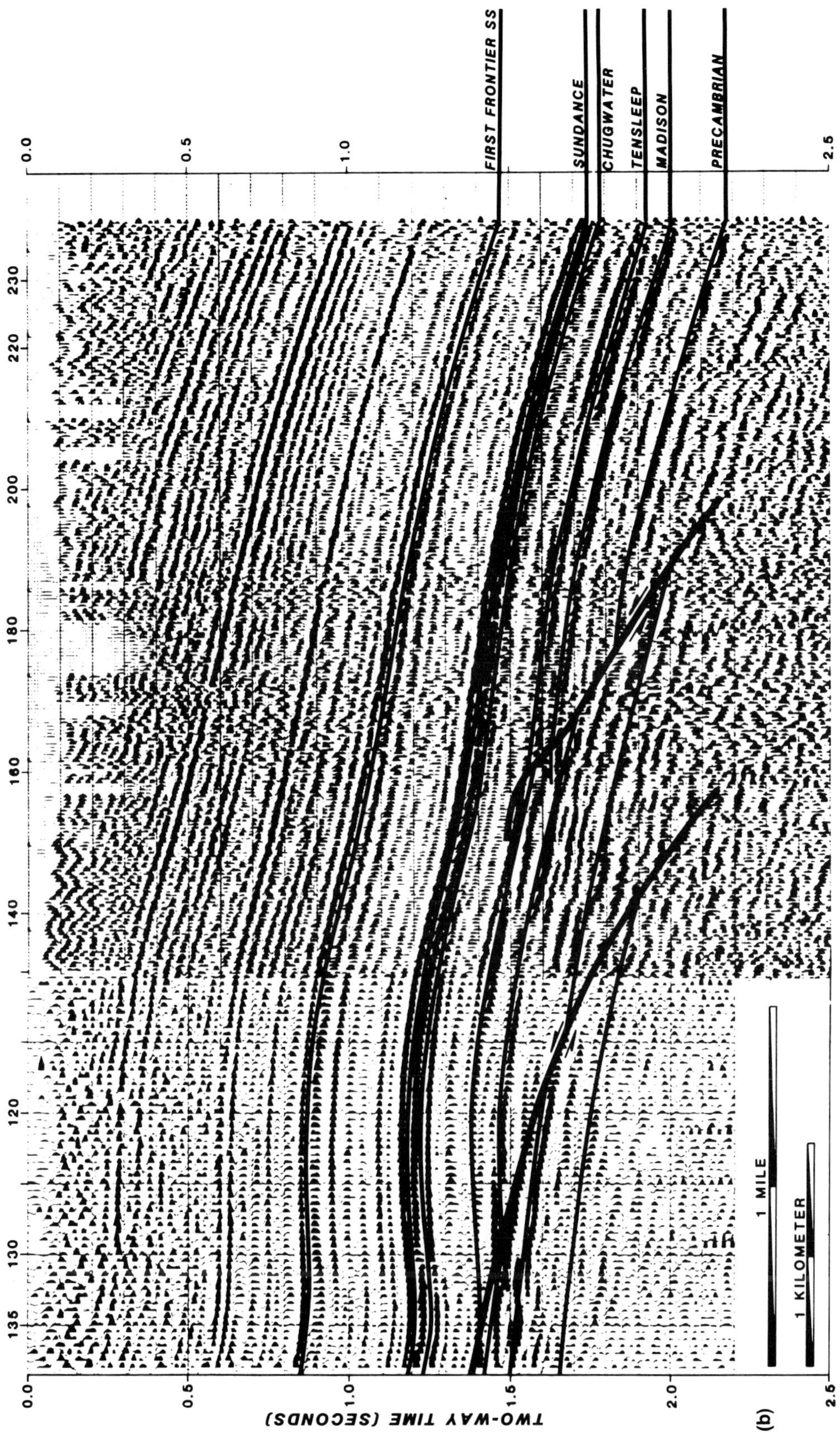

Fig. 3-47 (Stone, 1983b)—Seismic line (a–uninterpreted, b–interpreted) from North Fork area, western Powder River Basin, Wyoming, showing oil-productive North Fork (left) and Kaycee (right) compressive-block anticlines with basement interpreted to be offset by thrust fault dipping about 45° and flattening upward within an evaporite of the Goose Egg Formation (between Chugwater and Tensleep) such that upper beds are not affected by fault. The upward flattening is reminiscent of detachment structure discussed previously in this chapter. Both basement involved structures have residual gravity expression. c (next page)–geologic interpretation of seismic section. Permission to publish by D. S. Stone.

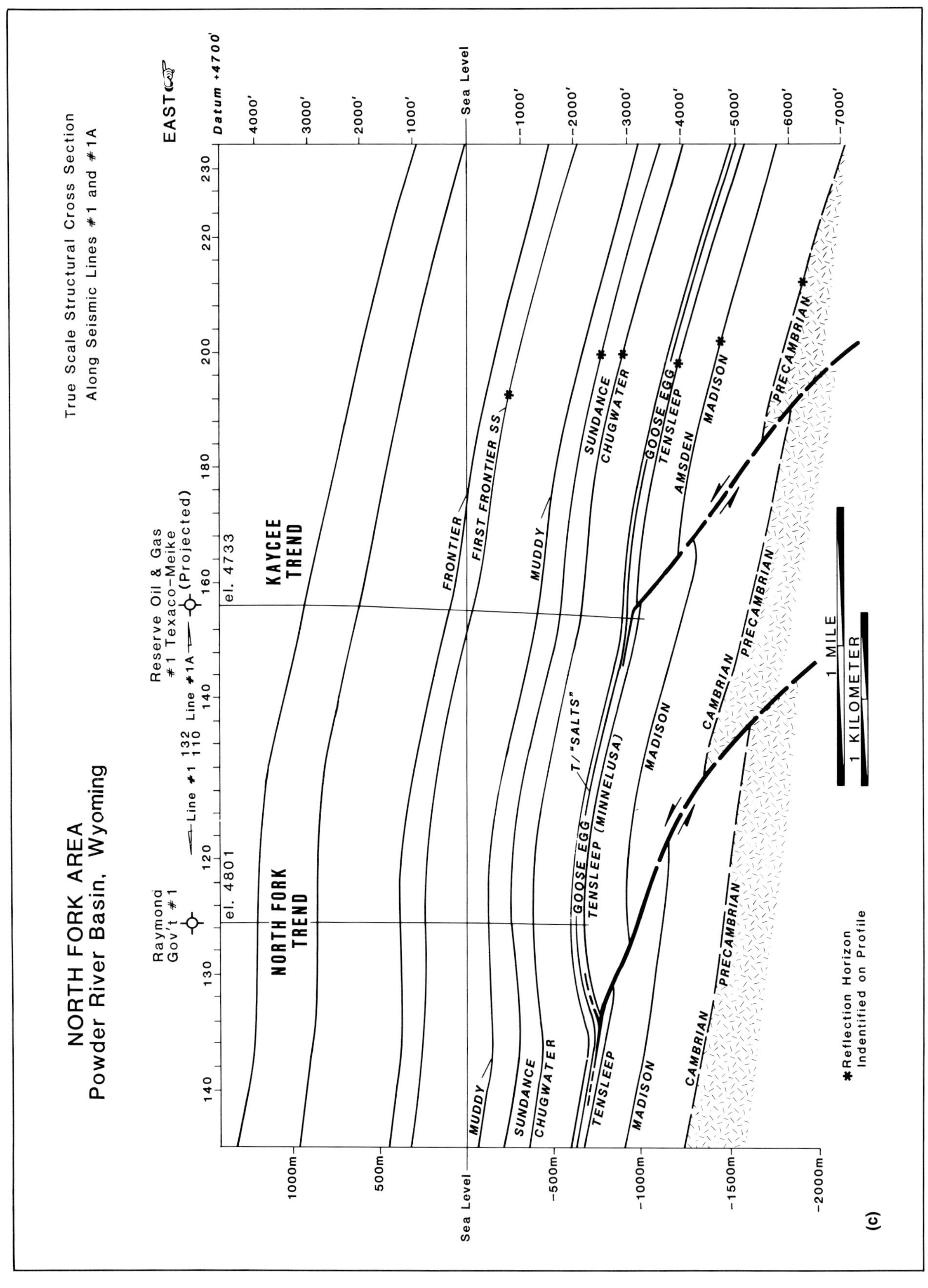

NORTH FORK AREA
Powder River Basin, Wyoming
True Scale Structural Cross Section
Along Seismic Lines #1 and #1A
Raymond Gov't #1
Reserve Oil & Gas #1 Texaco-Meike (Projected)
Line #1
Line #1A
el. 4801
el. 4733
NORTH FORK TREND
KAYCEE TREND
EAST
Datum +4700'
4000'
3000'
2000'
1000'
Sea Level
1000'
2000'
3000'
4000'
5000'
6000'
7000'
1000m
500m
Sea Level
-500m
-1000m
-1500m
-2000m
140
130
120
110
132
140
160
180
200
220
230
FRONTIER
FIRST FRONTIER SS
MUDDY
SUNDANCE
CHUGWATER
GOOSE EGG
TENSLEEP
TENSLEEP (MINNELUSA)
T/"SALTS"
AMSDEN
MADISON
CAMBRIAN
PRECAMBRIAN
1 MILE
1 KILOMETER
*Reflection Horizon Indentified on Profile
(c)

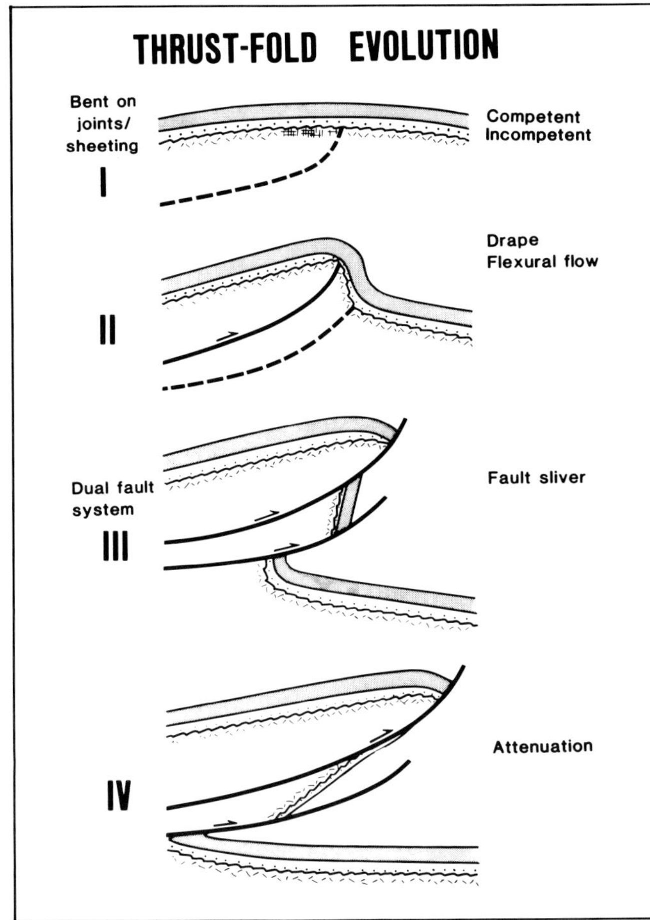

Fig. 3-48—Evolution of fold-thrust uplift. The most controversial of this and other similar models may be the initial attitude and subsequent extreme rotation and elongation of the middle limb or sliver between the dual faults.

rado, *in* Lowell, J. D., ed., Rocky Mountain foreland basins and uplifts: Rocky Mountain Association Geologists, p. 229–244.

Jordan, T. E., Isacks, B. L., Allmendinger, R. W., Brewer, J. A., Ramos, V. A., and Ando, C. J., 1983, Andean tectonics related to geometry of subducted Nazca plate: *Geological Society America Bull.,* V. 94, p. 341–361.

Keefer, W. R., 1965, Geologic history of Wind River Basin, central Wyoming: *American Association Petroleum Geologists Bull.,* V. 49, p. 1878–1892.

_______, 1970, Structural geology of the Wind River Basin, Wyoming: *United States Geological Survey Prof. Paper 495-D,* 35 p.

Kennedy, G. C., 1959, The origin of continents, mountain ranges, and ocean basins: *American Scientist,* V. 47, p. 491–504.

Lindsey, D. A., Johnson, B. R., and Andriessen, P. A. M., 1983, Laramide and Neogene structure of the northern Sangre de Cristo Range, south-central Colorado, *in* Lowell, J. D., ed., Rocky Mountain foreland basins and uplifts: Rocky Mountain Association Geologists, p. 219–228.

Lipman, P. W., Prostka, H. J., and Christiansen, R. L., 1971, Evolving subduction zones in the western United States: *Science,* V. 174, p. 821–825.

_______, 1972, Cenozoic volcanism and plate-tectonic evolution of the western United States: 1. Early and middle Cenozoic: *Royal Society London Philosophical Trans.,* V. 271, p. 217–248.

Livaccari, R. F., Burke, K., and Sengor, A. M. C., 1981, Was the Laramide orogeny related to subduction of an oceanic plateau: *Nature,* V. 289, n. 5795, p. 276–278.

Love, L. L. A., 1972, The dacites of the Washakie Needles, Bunsen Peak, and the Birch Hills, Wyoming and their relationship to the Absaroka-Gallatin volcanic province [M.S. thesis]: Univ. of New Mexico, Albuquerque, 86 p.

Lowell, J. D., 1969, On the origins of upthrusts: *Geological Society of America, Abs.* with Programs for 1969, Pt. 7 (Ann. Mtg.), p. 138.

_______, 1974, Plate tectonics and foreland basement deformation: *Geology,* V. 2, p. 275–278.

_______, 1977, Underthrusting origin for thrust-fold belts with applications to the Idaho-Wyoming belt: *Wyoming Geological Association 29th Annual Field Conference Guidebook,* p. 449–455.

_______, 1983, Foreland detached deformation (abs.): *American Association Petroleum Geologists Bull.,* V. 67, p. 1344.

Lowell, J. D., and Genik, G. J., 1972, Sea-floor spreading and structural evolution of southern Red Sea: *American Association Petroleum Geologists Bull.,* V. 56, n. 2, p. 247–259.

Mackin, J. H., 1947, Altitude and local relief of the Big Horn area during the Cenozoic: *Wyoming Geological Association Guidebook,* Field Conference, Big Horn Basin, Wyoming, p. 103–120.

Murphy, J. F., and Roberts, R. W., 1954, Geology of the Steamboat Butte-Pilot Butte area, Fremont County, Wyoming: *United States Geological Survey Oil and Gas Inv. Map, OM 151.*

Murphy, J. F., Privrasky, N. C., and Moerlein, G. A., 1956, Geology of the Sheldon-Little Dome area, Fremont County, Wyoming: *United States Geological Survey Oil and Gas Inv. Map OM 181.*

Nielson, D. R., and Stoiber, R. E., 1973, Relationship of potassium content in andesitic lavas and depth to the seismic zone: *Journal Geophysical Research,* V. 78, p. 6887–6892.

Osterwald, F. W., 1961, Critical review of some tectonic problems in Cordilleran foreland: *American Association Petroleum Geologists Bull.,* V, 45, n. 2, p. 219–237.

Peterson, F. A., 1983, Foreland detachment structures, *in* Lowell, J. D., ed., Rocky Mountain foreland basins and

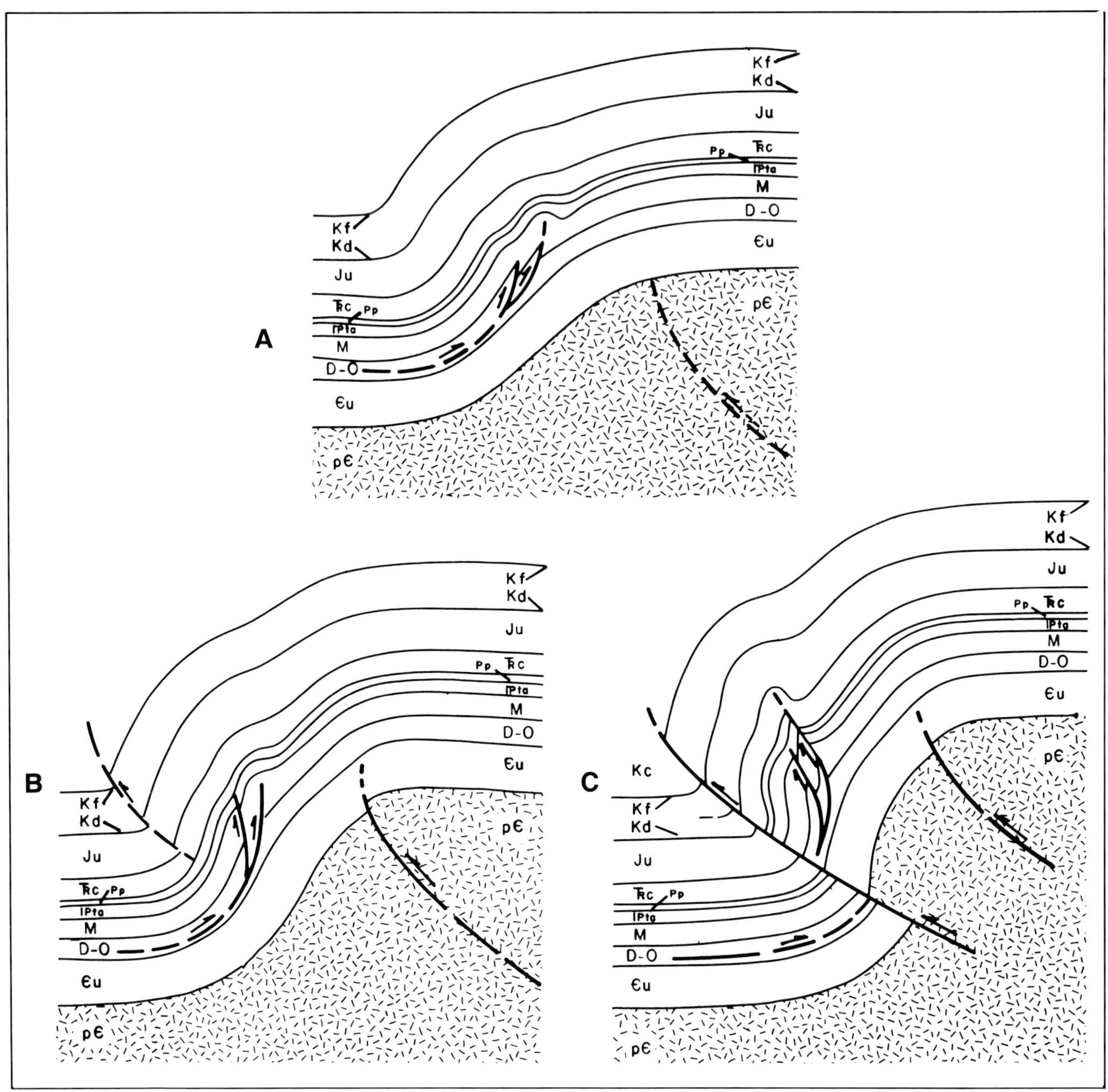

Fig. 3-49 (Brown, 1983)—Details of early evolution of fold-thrust feature including development of detached, faulted fold on hanging wall of fold thrust at C. Note "vertical" appearance of uplift even though sequence is part of a larger fold-thrust evolution. Permission to publish by Rocky Mountain Association of Geologists.

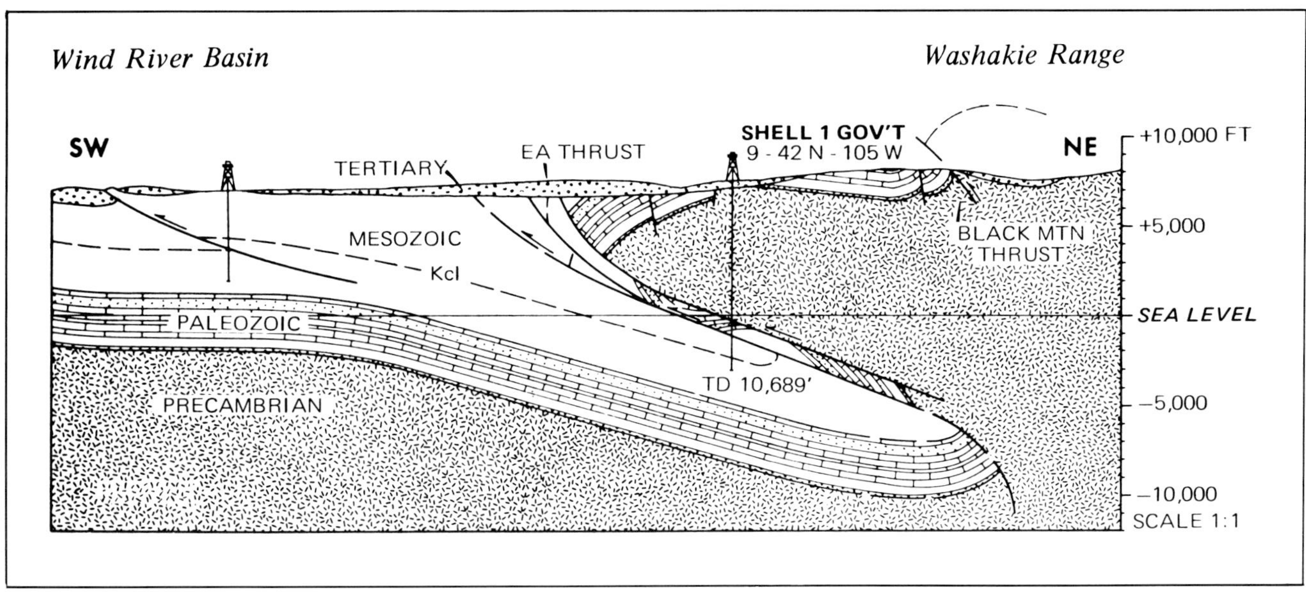

Fig. 3-50 (Berg, 1962)—Fold-thrust structure on southwest side of Washakie Range, northwest Wyoming. Permission to publish by American Association of Petroleum Geologists.

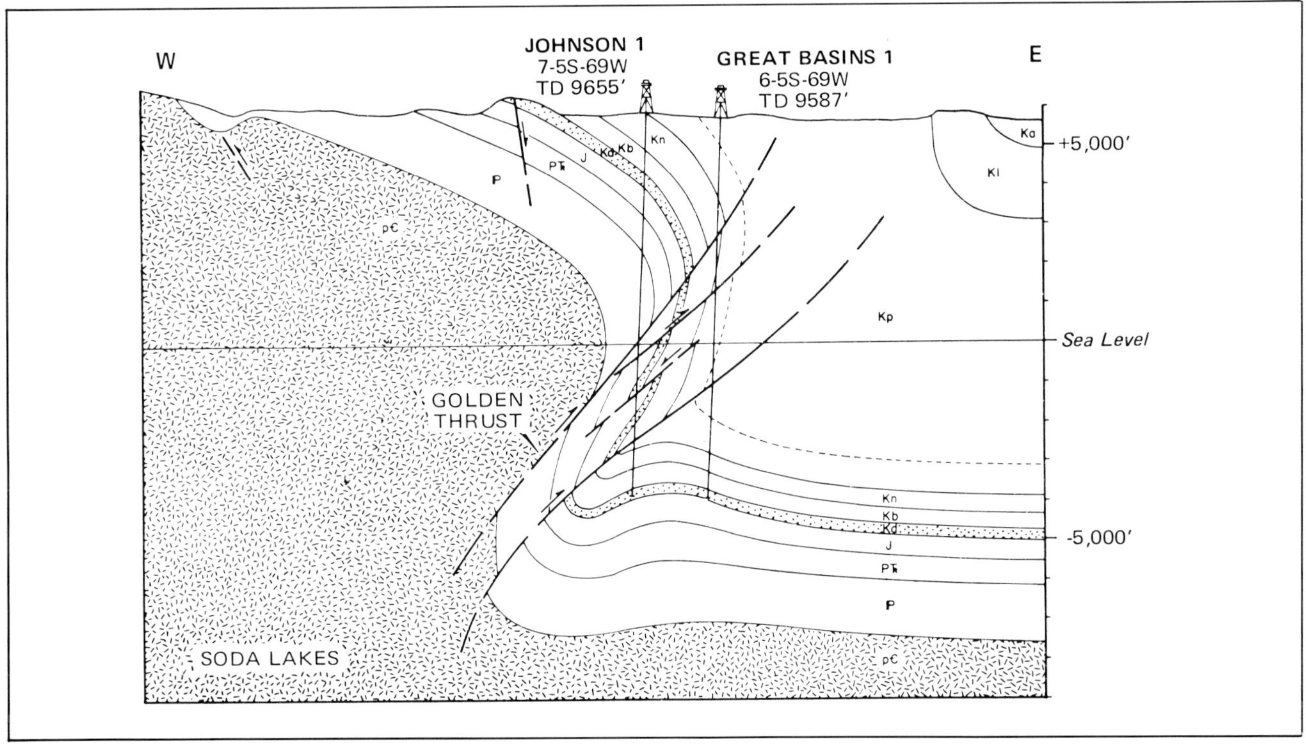

Fig. 3-51 (Berg, 1962)—Early to mid-stage fold-thrust uplift, Front Range, central Colorado. Permission to publish by American Association of Petroleum Geologists.

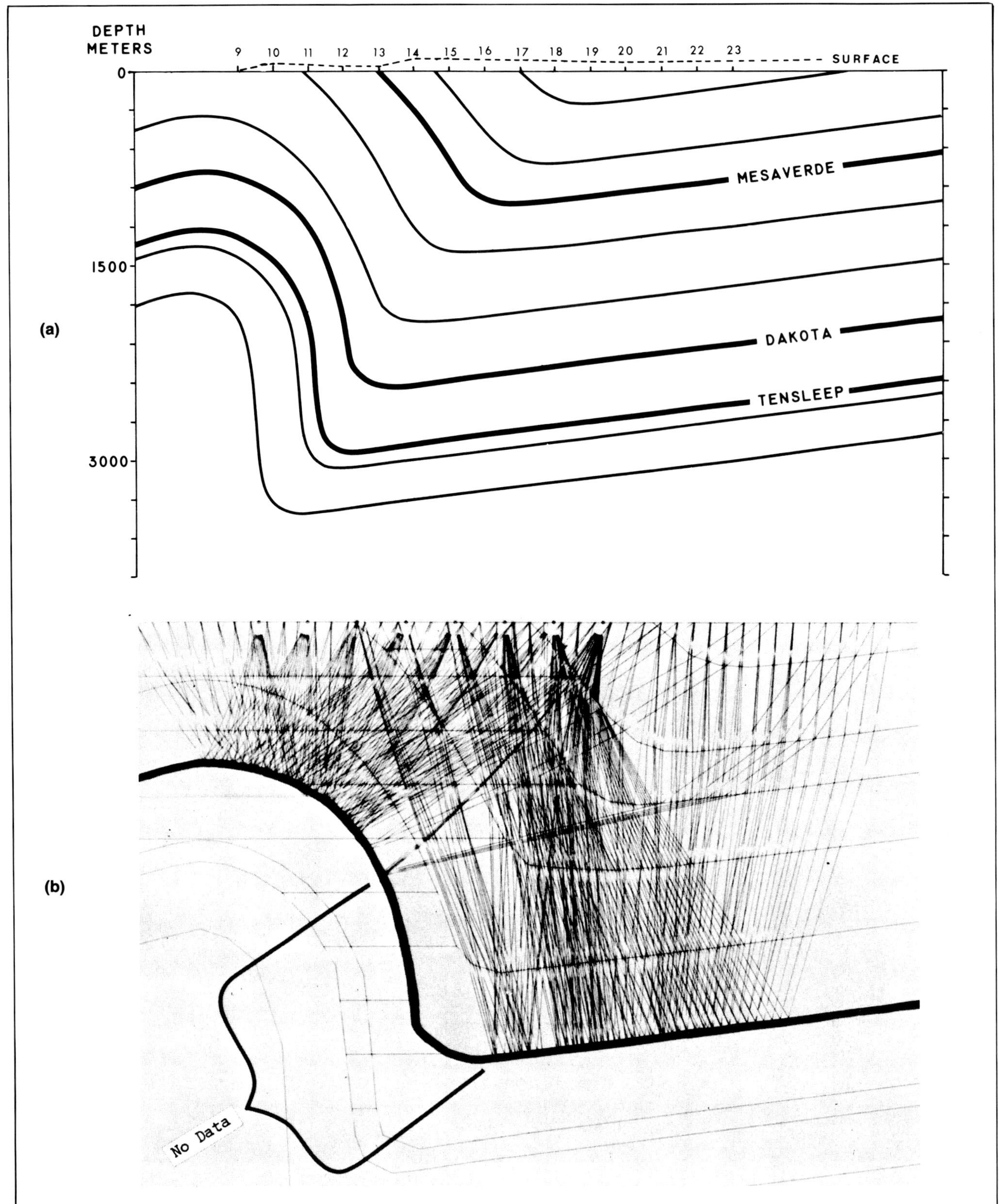

Fig. 3-52 (Sacrison, 1978)—a. Geologic model of fold above compressive block in northern Big Horn Basin, Wyoming. Model assumes no faulting at horizons shown. b. Ray path diagram for geologic model. Heavy line is Dakota. c. Synthetic seismic section for geologic model. d. Actual seismic section. From Geological Society of America Memoir. Used with permission.

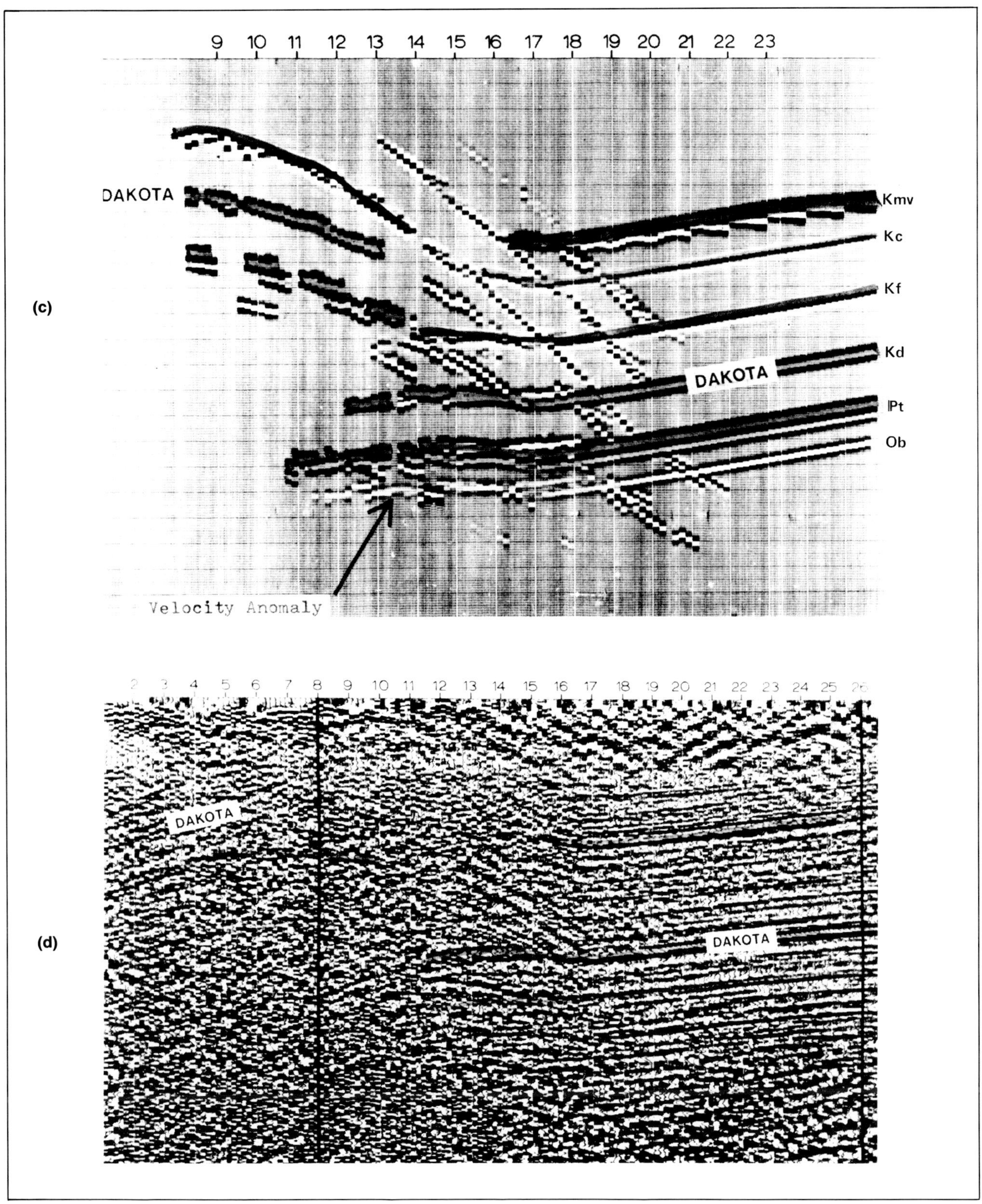
(c)
9 10 11 12 13 14 15 16 17 18 19 20 21 22 23
DAKOTA
Kmv
Kc
Kf
Kd
DAKOTA
IPt
Ob
Velocity Anomaly
(d)
2 3 4 5 6 7 8 9 10 11 12 13 14 15 16 17 18 19 20 21 22 23 24 25 26
DAKOTA
DAKOTA

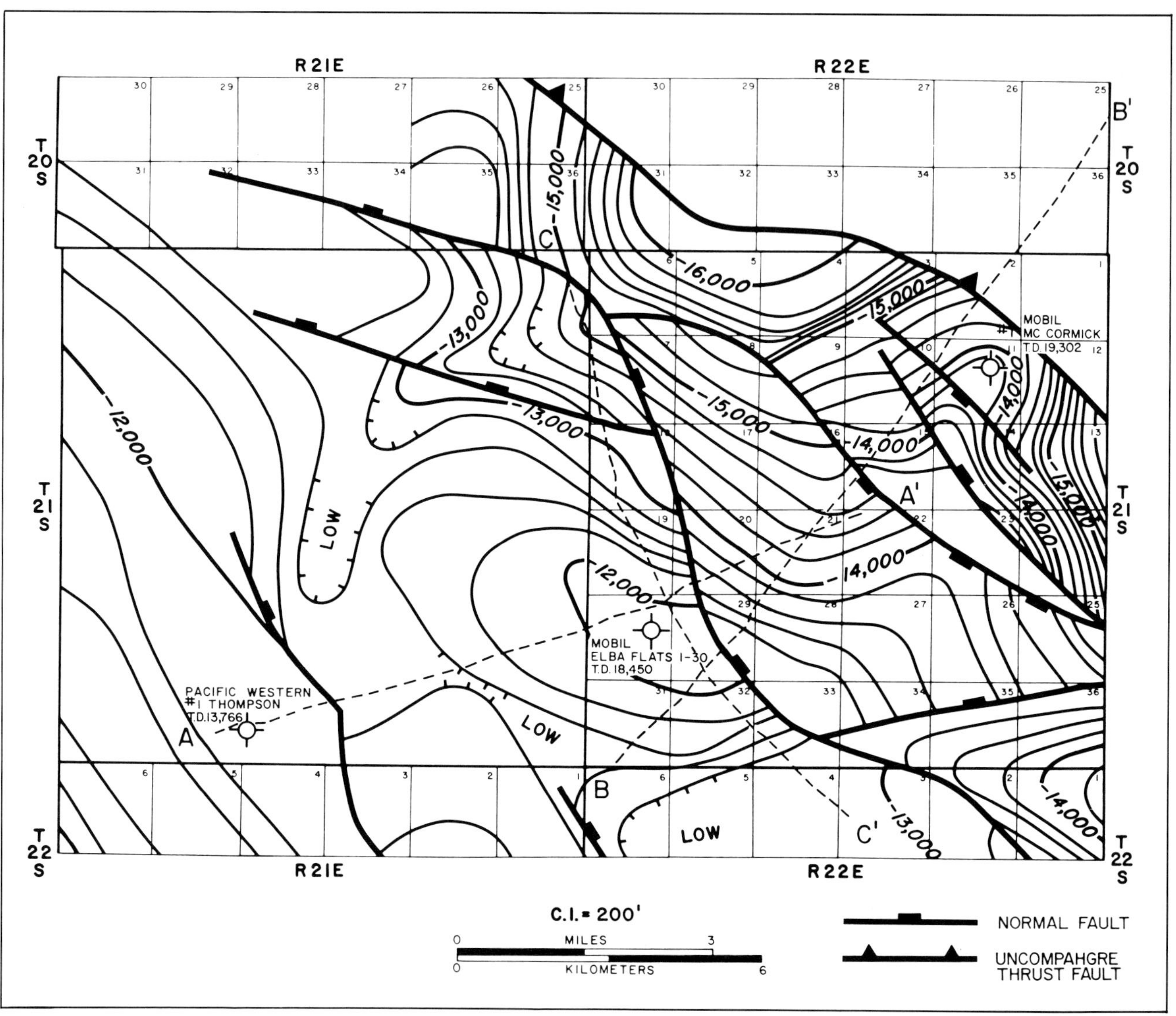

Fig. 3-53 (Frahme and Vaughn, 1983)—Structure contour map on top of Mississippian, northeast side of Paradox Basin, east-central Utah showing line of seismic section B-B', figure 3-54. Permission to publish by Rocky Mountain Association of Geologists.

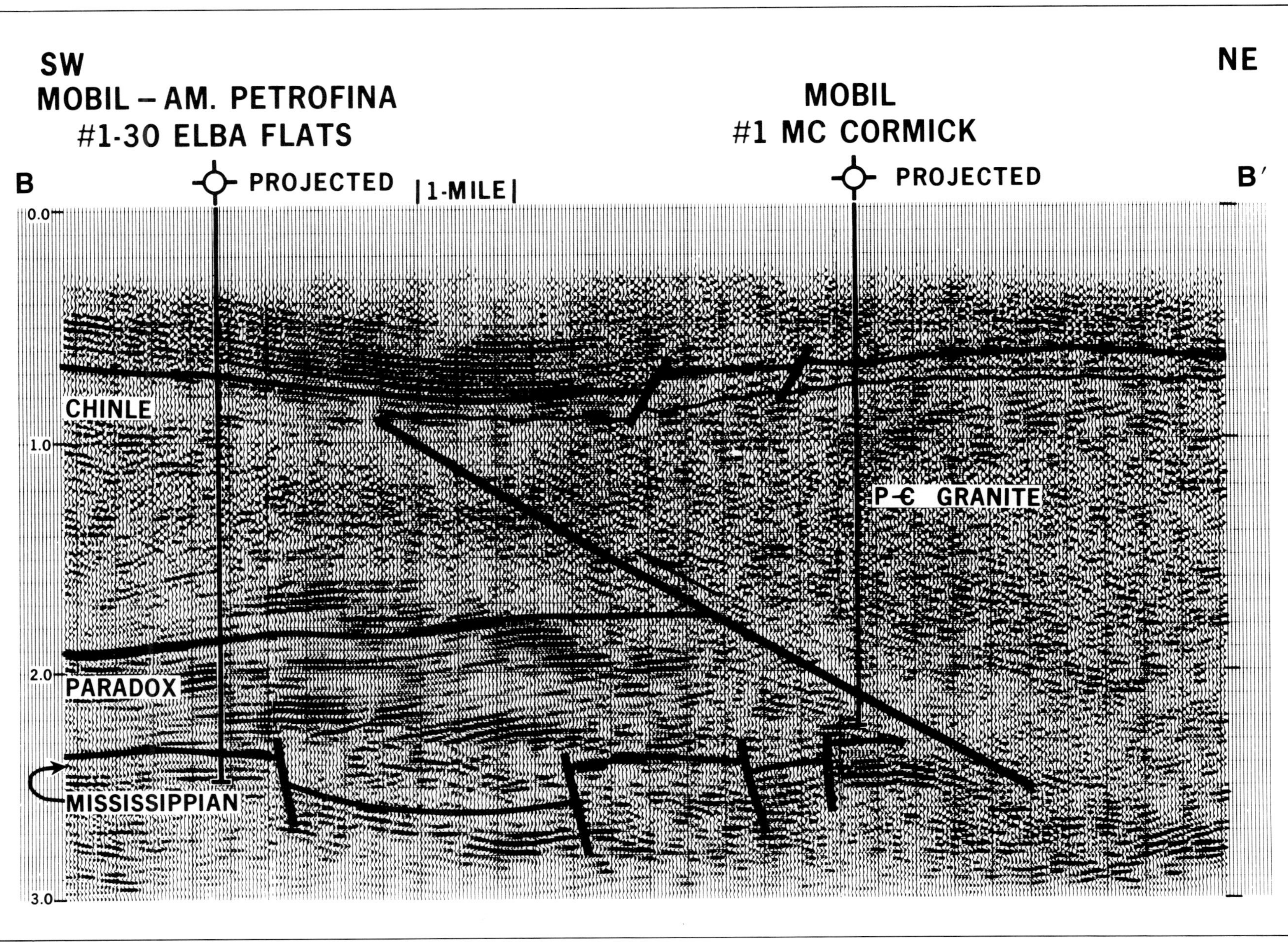

Fig. 3-54 (Frahme and Vaughn, 1983)—Seismic line B-B′ showing at least six miles (10 km) of thrust overlap on the Uncompahgre fault zone and the structure in the rocks of the footwall. Location of line shown in figure 3-53. Permission to publish by Rocky Mountain Association of Geologists.

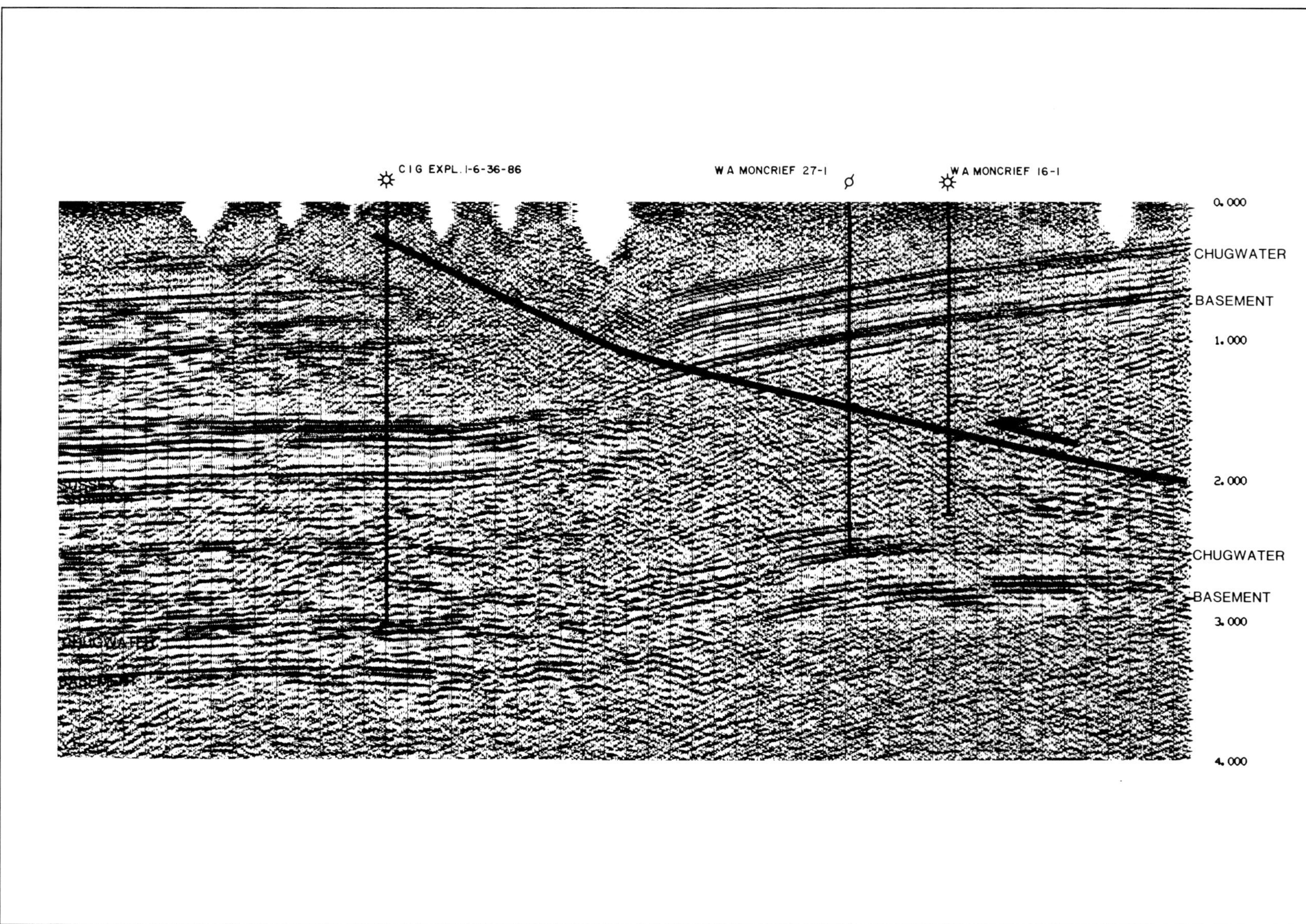

Fig. 3-55 (Sprague, 1983)—Seismic section across Casper arch frontal thrust, central Wyoming, showing minimum 10 km (6 mi) of overlap and the structure in rocks below Precambrian basement wedge. Note velocity pull-up beneath basement. Section approximately parallel to structural cross section A-A' located on figure 3-59. Permission to publish by Rocky Mountain Association of Geologists.

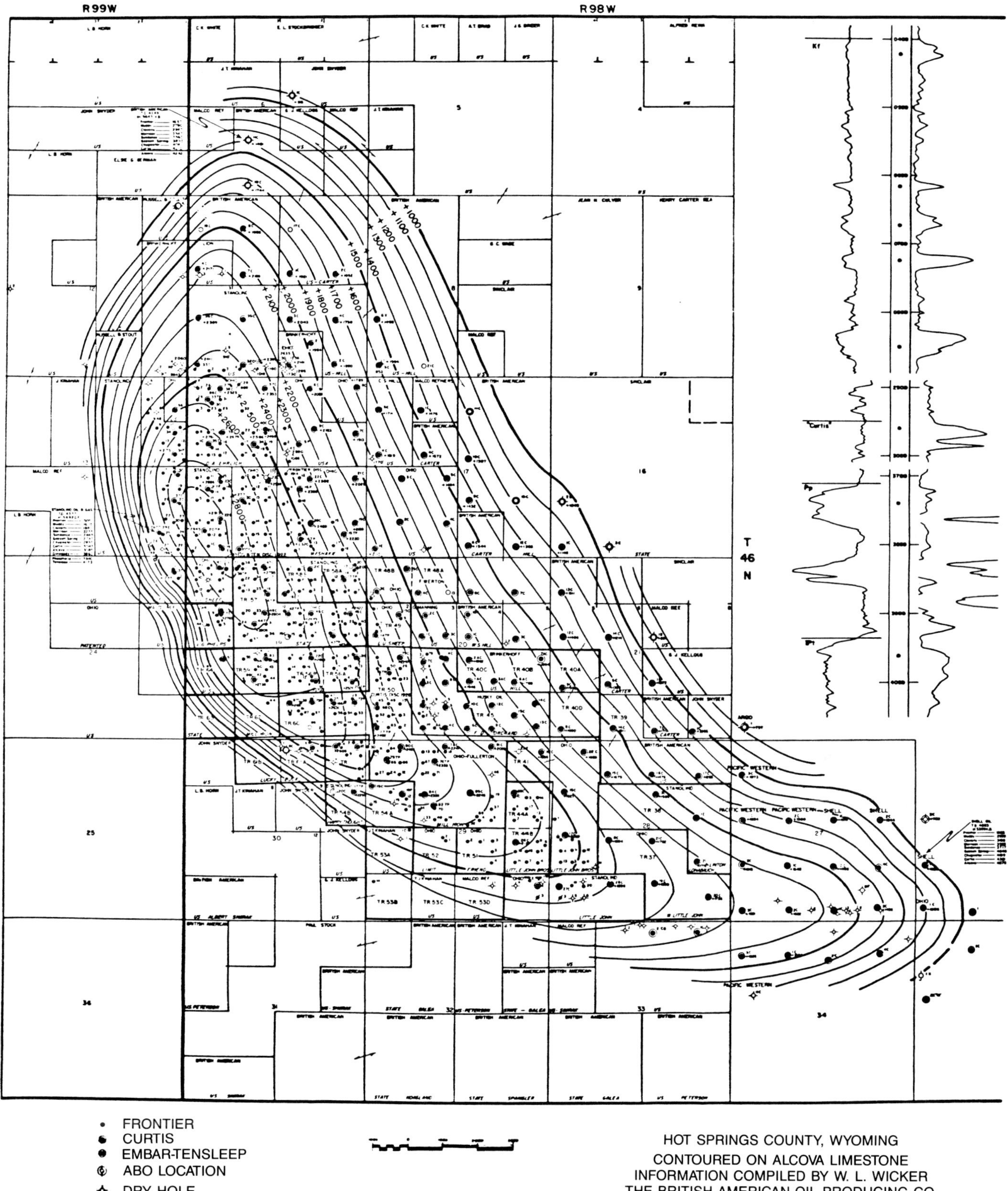

Fig. 3-56 (Wyoming Geological Association, 1957)—Grass Creek field, Big Horn Basin. Note WNW and NNW trends (dog-leg) shown by close spacing of contours on steep (SW) side of structure which is fault controlled at depth. Permission to publish by Wyoming Geological Association.

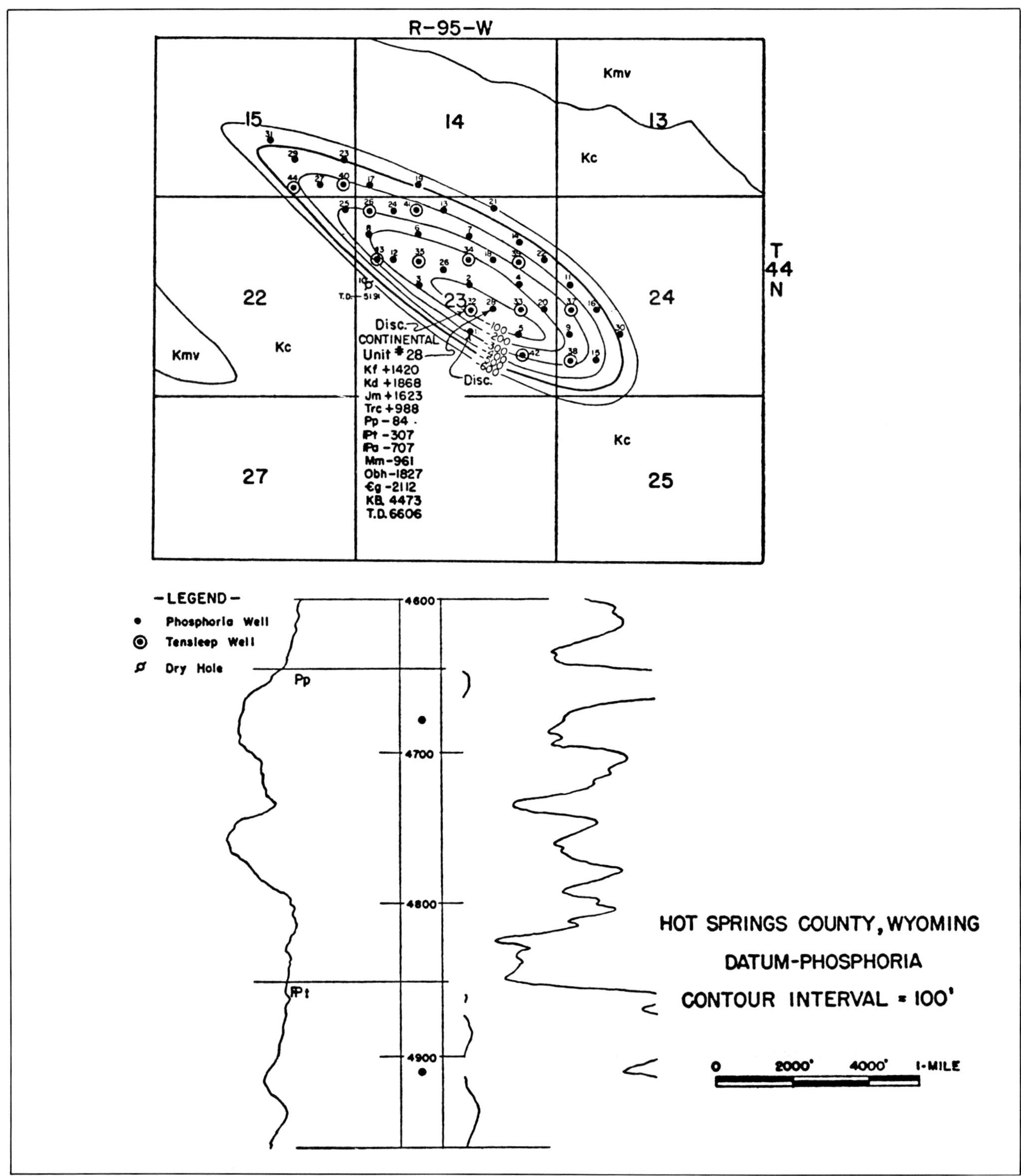

Fig. 3-57 (Wyoming Geological Association, 1957)—Gebo is asymmetric to the SW so that gentle flank dips NE into the basin facilitating excellent out-of-the-basin drainage. Note that production is established much farther down the gentle flank than the steep flank because of hydrodynamic flow from the SW. Gebo, like most Rocky Mountain foreland basement structures, is considerably fractured so that reservoir porosity and permeability are increased. Permission to publish by Wyoming Geological Association.

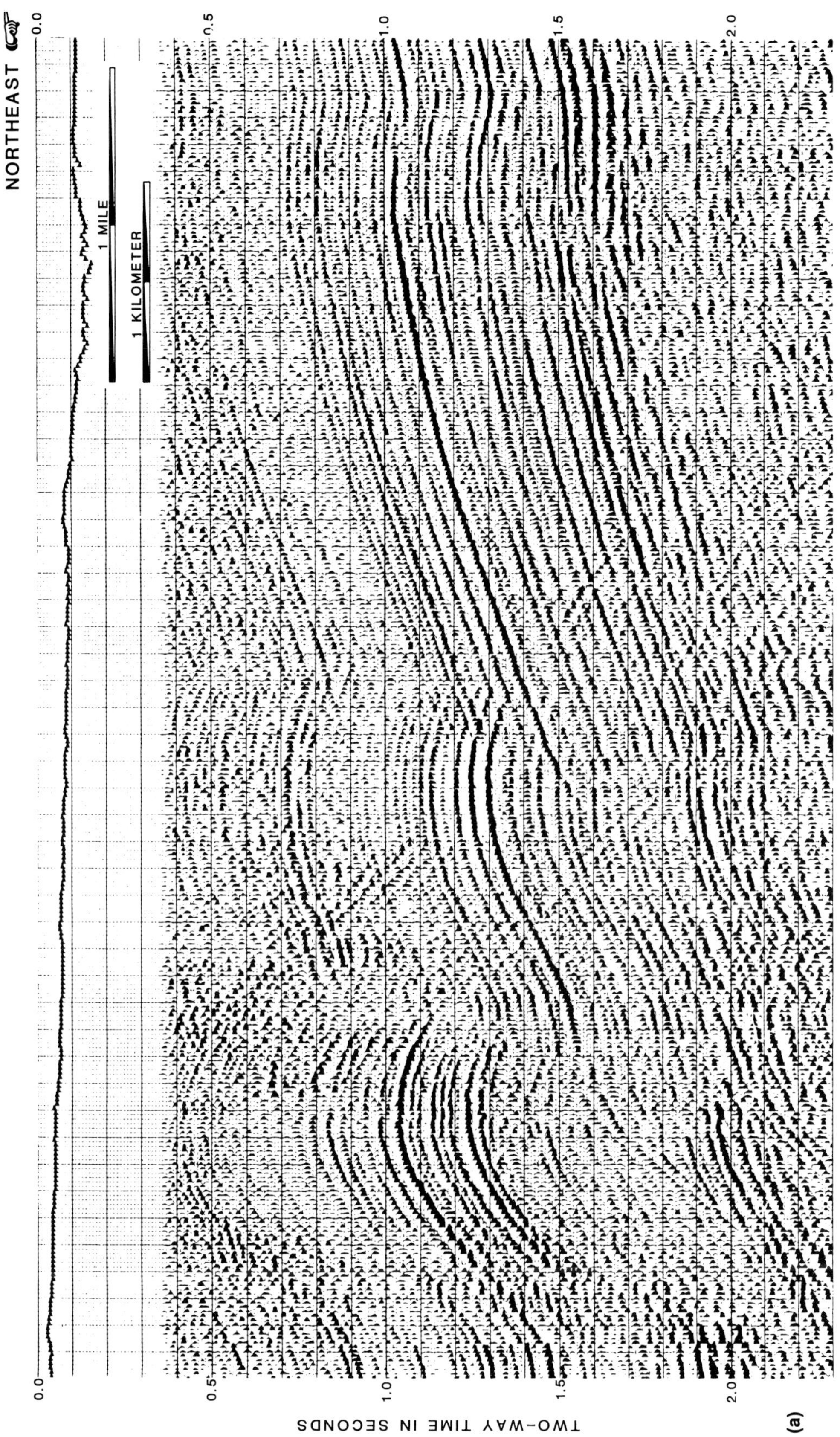
NORTHEAST
1 MILE
1 KILOMETER
0.0
0.5
1.0
1.5
2.0
TWO-WAY TIME IN SECONDS

(a)

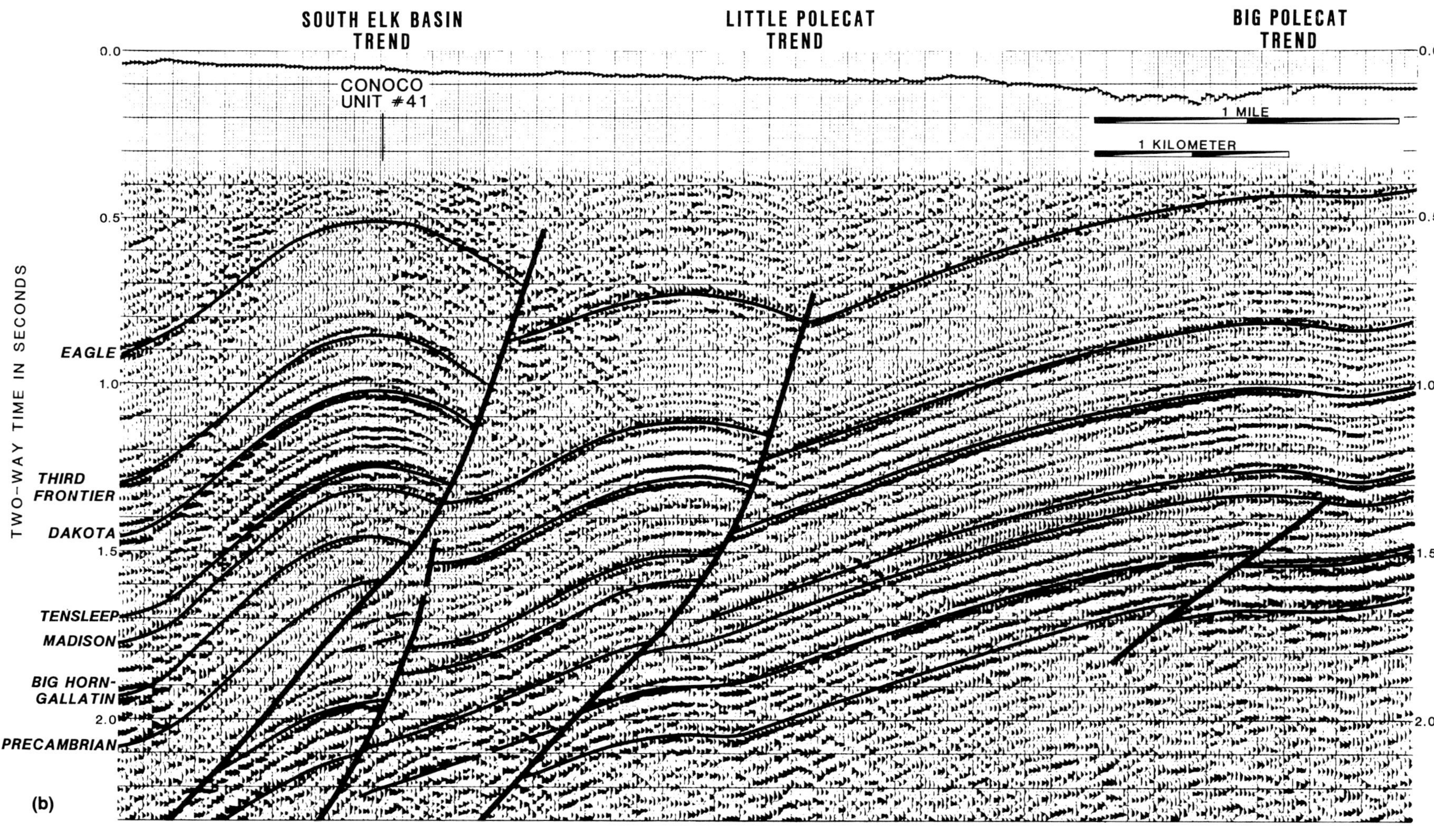

Fig. 3-58 (Stone, 1983c)—Seismic line (a and b) and structure section (c) across South Elk Basin field, Little Polecat trend, and Big Polecat trend, northeast side Big Horn Basin, Wyoming, showing out-of-basin fault and gentle basinward-dipping flanks that should facilitate effective out-of-basin drainage. All features have oil production. Permission to publish by D. S. Stone.

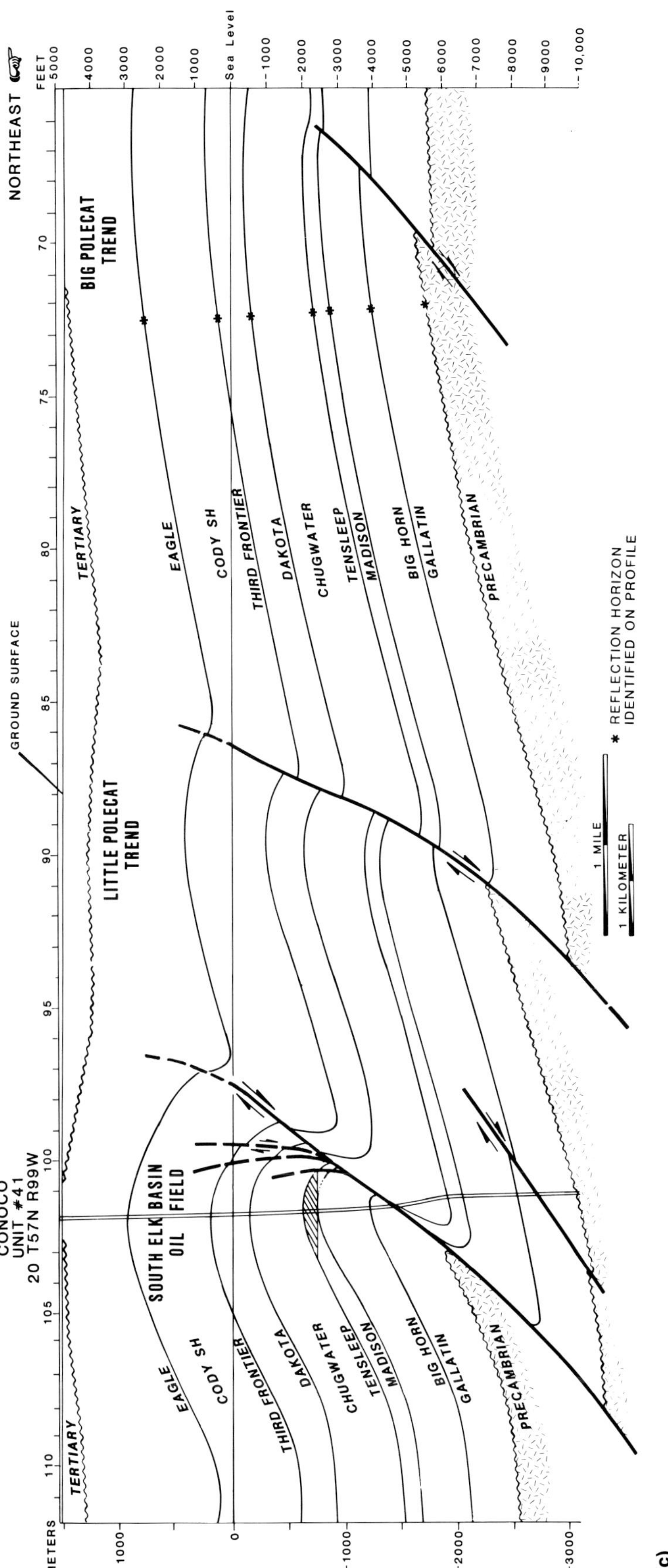

NORTHEAST
FEET
5000
4000
3000
2000
1000
Sea Level
-1000
-2000
-3000
-4000
-5000
-6000
-7000
-8000
-9000
-10,000
BIG POLECAT TREND
LITTLE POLECAT TREND
GROUND SURFACE
CONOCO UNIT #41
20 T57N R99W
SOUTH ELK BASIN OIL FIELD
TERTIARY
EAGLE
CODY SH
THIRD FRONTIER
DAKOTA
CHUGWATER
TENSLEEP
MADISON
BIG HORN
GALLATIN
PRECAMBRIAN
METERS
1000
0
-1000
-2000
-3000
110
105
100
95
90
85
80
75
70
1 MILE
1 KILOMETER
* REFLECTION HORIZON IDENTIFIED ON PROFILE

(c)

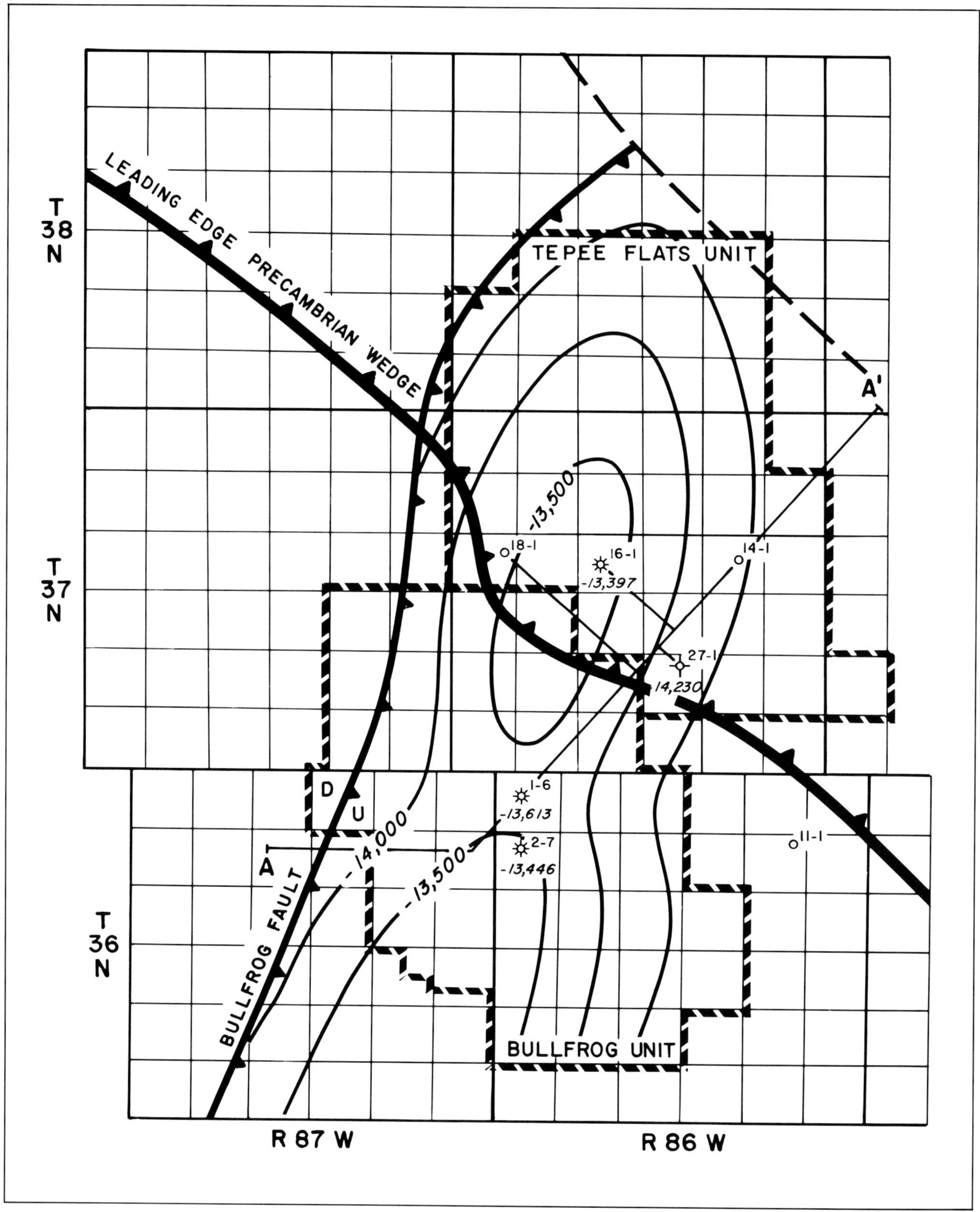

Fig. 3-59 (Sprague, 1983)—Structure contour map of Tepee Flats-Bullfrog field, central Wyoming, on Lower Cretaceous Muddy Formation. Permission to publish by Rocky Mountain Association of Geologists.

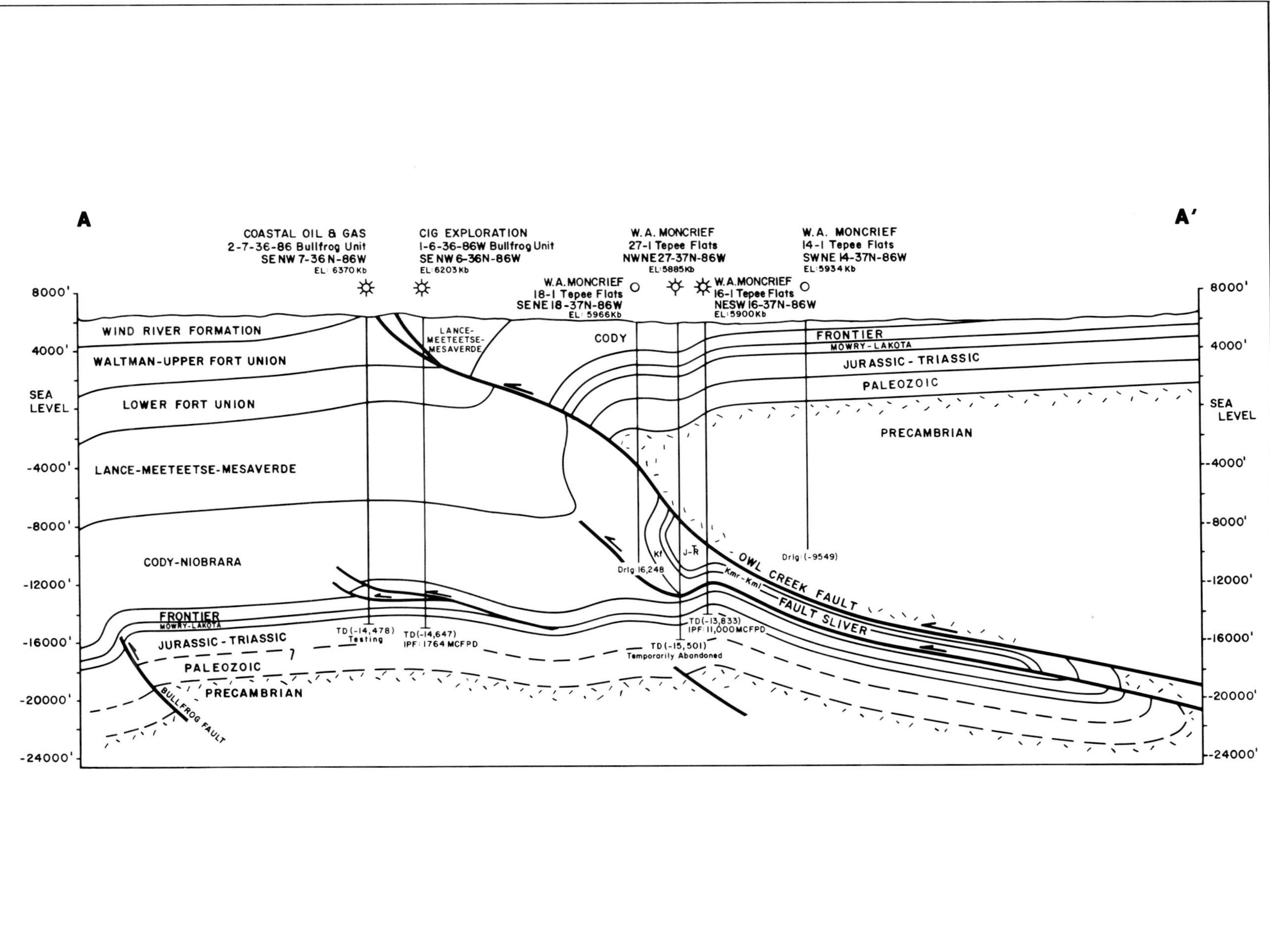

Fig. 3-60 (Sprague, 1983)—Structural cross section A-A' through Bullfrog and Tepee Flats units. True scale. See figure 3-59 for location. Permission to publish by Rocky Mountain Association of Geologists.

uplifts: Rocky Mountain Association Geologists, p. 65–77.

Pierce, W. G., 1966, Geologic map of the Cody quadrangle, Park County, Wyoming: *United States Survey Geological Quadrangle Map GQ-542.*

Prucha, J. J., Graham, J. A., and Nickelsen R. P., Basement controlled deformation in Wyoming province of Rocky Mountain foreland: *American Association Petroleum Geologists Bull.,* V. 49, n. 7, p. 966–992.

Ray, R. R., Gries, Robbie, and Babcock, J. W., 1983, Acoustic velocities, synthetic seismograms and lithologies of thrusted Precambrian rocks, Rocky Mountain foreland, *in* Lowell, J. D., ed., Rocky Mountain foreland basins and uplifts: Rocky Mountain Association Geologists, p. 125–135.

Read, C. B., and Wood, G. H., 1947, Distribution and correlation of Pennsylvanian rocks in late Paleozoic sedimentary basins of northern New Mexico: *Journal Geology,* V. 55, p. 220–236.

Sacrison, W. R., 1978, Seismic interpretation of basement block faults and associated deformation, *in* Matthews, Vincent, III, ed., Laramide folding associated with basement block faulting in the western United States: *Geological Society America Mem.* 151, p. 39–49.

Sales, J. K.,1968, Cordilleran foreland deformation: *American Association Petroleum Geologists Bull.,* V. 52, p. 2016–2044.

———, 1983, Collapse of Rocky Mountain basement uplifts, *in* Lowell, J. D., ed., Rocky Mountain foreland basins and uplifts: Rocky Mountain Association Geologists, p. 79–97.

Sanford, A. R., 1959, Analytical and experimental study of simple geologic structures: *Geological Society America Bull.,* V. 70, p. 19–52.

Schmidt, C.J., and Garihan, J. M., 1983, Laramide tectonic development of the Rocky Mountain foreland of south western Montana, *in* Lowell, J. D., ed., Rocky Mountain foreland basins and uplifts: Rocky Mountain Association Geologists, p. 271–294.

Skeen, R. C., and Ray, R. R., 1983, Seismic models and interpretation of the Casper arch thrust: Application to Rocky Mountain foreland structure, *in* Lowell, J. D., ed., Rocky Mountain foreland basins and uplifts: Rocky Mountain Association Geologists, p. 99–124.

Sprague, E. L., 1983, Geology of the Tepee Flats-Bullfrog fields, Natrona County, Wyoming, *in* Lowell, J. D., ed., Rocky Mountain foreland basins and uplifts: Rocky Mountain Association Geologists, p. 339–343.

Stearns, D. W., 1971, Mechanisms of drape folding in the Wyoming province: *Wyoming Geological Association 23d Annual Field Conference Guidebook,* p. 125–143.

———, 1978, Faulting and forced folding in the Rocky Mountains foreland, *in* Matthews, Vincent III, ed., Laramide folding associated with basement block faulting in the western United States: *Geological Society America Mem. 151,* p. 1–37.

———, and Stearns, M. T., 1978, Geometric analysis of multiple drape folds along the northwest Big Horn Mountains front, Wyoming, *in* Matthews, Vincent III, ed., Laramide folding associated with basement block faulting in the western United States: *Geological Society America Mem. 151,* p. 139–156.

Steidtmann, J. R., McGee, L. C., and Middleton, L. T., 1983, Laramide sedimentation, folding, and faulting in the southern Wind River Range, Wyoming, *in* Lowell, J. D., ed., Rocky Mountain foreland basins and uplifts: Rocky Mountain Association Geologists, p. 161–167.

Stewart, J. H., 1972, Initial deposits in the Cordilleran geosyncline: Evidence of a Late Precambrian (>850 m.y.) continental separation: *Geological Society America Bull.,* V. 83, p. 1345–1360.

Stone, D. S., 1969, Wrench faulting and Rocky Mountain tectonics: *Mountain Geologist,* V. 6, n. 2, p. 67–79.

———, 1983a, Detachment thrust faulting in the Reed Point syncline south-central Montana: *Mountain Geologist,* V. 20, n. 4, p. 107–112.

———, 1983b, Seismic profile: North Fork area, Powder River Basin, Wyoming, *in* Bally, A. W., ed., Seismic expression of structural styles: *American Association Petroleum Geologists Studies in Geology Series #15,* V. 3.

———, 1983c, Seismic profile: South Elk Basin, *in* Bally, A. W., ed., Seismic expression of structural styles: *American Association Petroleum Geologists Studies in Geology Series #15,* V. 3.

———, 1983d, The Greybull Sandstone pool (Lower Cretaceous) on the Elk Basin thrust-fold complex, Wyoming and Montana, *in* Lowell, J. D., ed., Rocky Mountain foreland basins and uplifts: Rocky Mountain Association Geologists, p. 345–356.

Thom, W. T., Jr., 1955, Wedge uplifts and their tectonic significance, *in* Poldervaart, A., ed., Crust of the Earth—A symposium: *Geological Society America Spec. Paper 62,* p. 369–376.

Thomas, G. E., 1971, Continental plate tectonics: *Wyoming Geological Association Guidebook 23d Ann. Field Conf.,* p. 103–123.

Vasquez, J., and Gorrono, R., 1980, Limite de la faja plegada en la rep. Argentina: *Associacion Geologica Argentina Revista,* XXXV(4), p. 582–585.

Wellborn, R. E., 1977, Structural style in relation to oil and gas exploration in North Park-Middle Park Basin, Colorado, *in* Veal, H. K., ed., Exploration frontiers of the central and southern Rockies: Rocky Mountain Association Geologists, p. 41–60.

Winterfeld, G. F., and Conard, J. B., 1983, Laramide tectonics and deposition, Washakie Range and northwestern Wind River Basin, Wyoming, *in* Lowell, J. D., ed., Rocky Mountain foreland basins and uplifts: Rocky Mountain Association Geologists, p. 137–148.

Wise, D. U., 1963, Keystone faulting and gravity sliding driven by basement uplift of Owl Creek Mountains, Wyoming: *American Association Petroleum Geologists Bull.*, V. 47, n. 4, p. 586–598.

Woodward, L. A., 1974, Plate tectonics and foreland basement deformation: Comment: *Geology,* V. 2, n. 12, p. 570.

Woollard, G. P., and Joesting, H. R., 1964, Bouguer gravity anomaly map of the United States: American Geophysical Union and United States Geological Survey, scale 1:2,500,000.

Wyoming Geological Association, 1957, Oil and gas field volume.

4 EXTENSIONAL FAULT BLOCKS

OCCURRENCE

The plate tectonic habitats of extensional fault blocks are summarized in Chapter 1. Extensional blocks may be the most common of all the structural styles in line with the following rationale. The oceanic spreading system encircles the globe like the seam of a tennis ball for a linear distance of approximately 50,000 km (31,200 mi) (Fig. 4-1). Normal faults are the dominant style at depth for much of this system. Considering further that the now passive continental margins on either side of this spreading system were normal faulted during continental rifting and separation, something less than, but approaching 100,000 km (62,400 mi) of continental shelves and slopes have a dominantly extensional block style. Behind-the-arc spreading in the western Pacific Ocean has also imparted an extensional block style to the western side of many island arcs and their opposing continental margins farther west. Finally, intra-plate rift systems including the East African, Russian Baikal, and U.S. Basin-Range have an extensional block style. Most of these aforementioned features are presently evolving or have recently evolved. In addition fossil rift systems, such as the Suez of Egypt, Sirte of Libya, Rhine of Germany, Viking and Central of the North Sea, Reconcavo of Brazil, and Grand Banks of Canada, have an extensional block style. In my opinion, extensional blocks are the most commonly occurring of all styles.

MECHANICS

If extensional blocks are the most frequent of all structural styles, a model (Fig. 4-2) is desirable to explain them, particularly with respect to the evolution of rifts and continental margins. Much of the following discussion is based on work done in the southern Red Sea (Lowell and Genik, 1972; Lowell et al., 1975).

Initiation of Rifting

Rifts can be initiated as arches or sags. Precambrian crystalline rocks are exposed extensively around the perimeter of the Red Sea and define an arch of large regional dimensions, 2,000 km (1,248 mi) by 500 km (312 mi); younger sedimentary strata dip gently off the flanks (Fig. 4-3). I believe this regional arching of the continental lithosphere reflects the first step in the structural evolution of the Red Sea. On the basis of established structural behavior in the bending of a competent layer (Fig. 4-4), arching should be accompanied by the propagation of normal faults from the top of the lithosphere downward (Fig. 4-2). Erosion and non-marine or continental sedimentation were other concurrent events.

In contrast, rifts initiated as sags should be accompanied by the propagation of listric normal faults from the base of the brittle lithosphere upward (Fig. 4-5). Concurrent events would be less erosion, greater subsidence, and, normally, earlier marine sedimentation than would be expected from arching.

Arching may be caused by a thermally activated volume increase attendant on convective upwelling in the asthenosphere, whereas sagging may be a result of immediate divergent convective flow in the asthenosphere, which effectively bypasses the arching stage. The two processes may not be mutually exclusive. The configuration of the Mohorovicic discontinuity (Moho) beneath the Rhine graben is a bullseye reflecting uplift in the south and an elongate ridge possibly reflecting a sag in the north (Fig. 4-6).

Rifting

In sagging, the presumed divergent convective flow just below the base of the brittle lithosphere leads to rifting. In

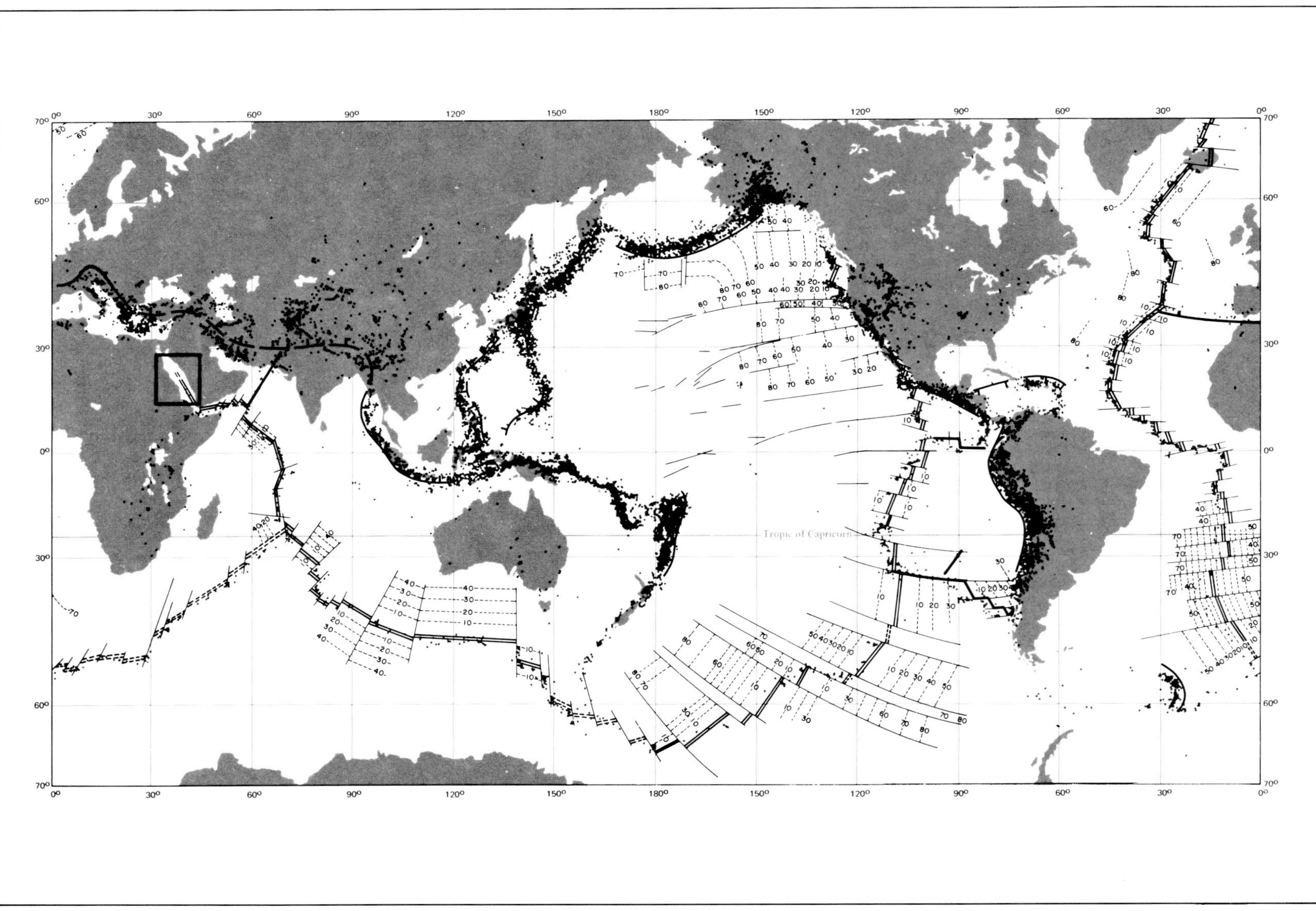

Fig. 4-1 (Lowell and Genik, 1972)—Global rift system. Midocean ridges (solid double line, dashed where uncertain) are defined by earthquake epicenters (dots) and isochrons (thin dashed lines) which were derived from magnetic patterns and give age of sea floor in millions of years; ridges are offset by transform faults (thin solid lines). Subduction zones are indicated by heavy hachured line. (From Sea-Floor Spreading, by J. R. Heirtzler.

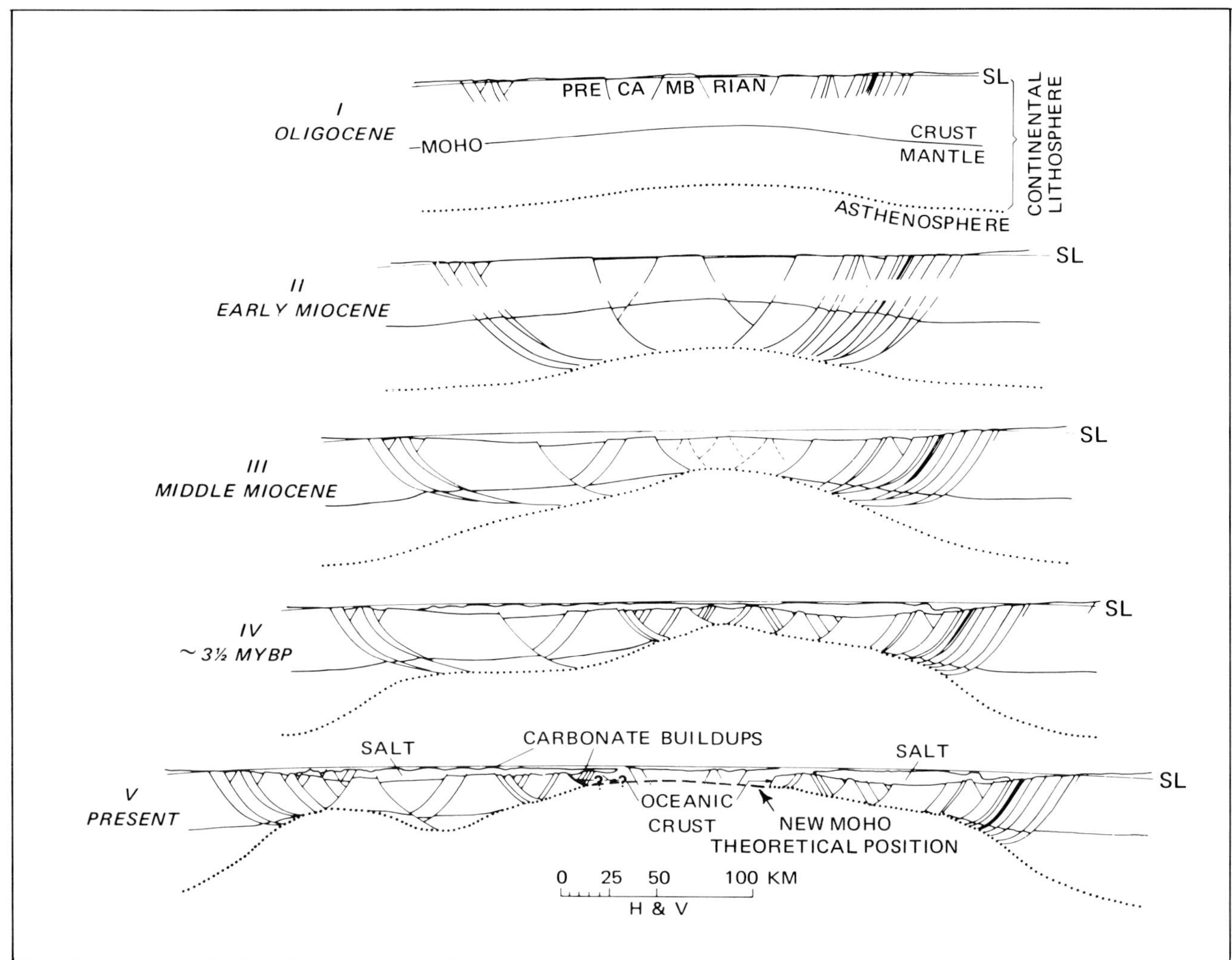

Fig. 4-2 (Lowell et al., 1975)—Thinning and breaching of lithosphere are believed to occur by combination of listric normal faulting, extensional (plastic) deformation internal to individual fault blocks possibly facilitated by deep-seated heating, and thermal and chemical reactions at base of lithosphere. Evolution could abort at any stage, or between stages. Note that significant vertical movement and associated dip-slip are natural consequences of arching, breakup, and separation of a lithospheric plate and occur at all stages of the model. (Section V is simplified reduction of cross section on figure 4-7B.) Permission to publish by American Association Petroleum Geologists.

the initial arching model that divergent flow replaces the earlier convective upwelling and also starts the rifting process.

In the southern Red Sea (Fig. 4-2) after the arching of the continental lithosphere, lateral extension is interpreted to have thinned it. Thinning could have been accomplished in large part by shear on listric normal faults. Continued rapid lithospheric thinning and rifting by normal faulting caused subsidence on horst and graben and tilted blocks, which led to an extensive marine incursion and thick sedimentary sequence deposited in Miocene time. The presence of continental clastics above the marine Miocene probably resulted from a second period of arching with an attendant flood of terrigenous material. Finally, the upper part of the continental crust was sufficiently thinned and breached to allow insertion of oceanic basalts. The latter have magnetic anomalies that can be dated to 3.5 my (late Pliocene); and a sea-floor spreading half-rate of 1 cm/year has been estimated at 16°N by Vine (1966). It is noteworthy that the continental lithosphere was extended considerably, beginning in the Oligocene or earlier and continuing through the Miocene and much of Pliocene, before it was sundered to permit emplacement of material, either of melted lithosphere or sublithosphere origin, and the rate of separation by sea-floor spreading could actually be measured. Large amounts of extension can be achieved by rotation of beds into listric normal faults (Fig. 4-2). Still greater extension, over 200%, can be accommodated by rotation of the faults

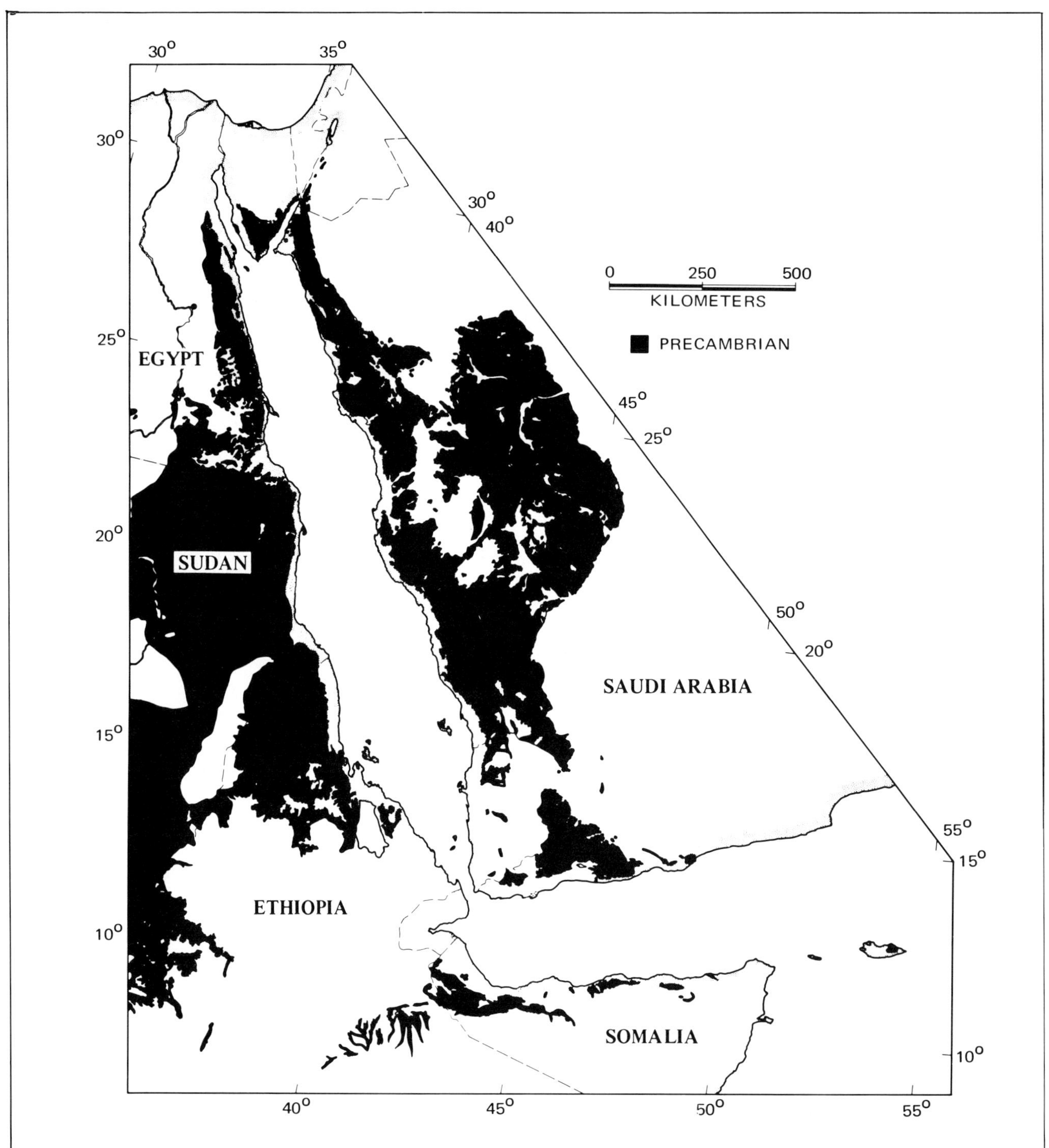

Fig. 4-3 (Lowell and Genik, 1972)—Outcrop distribution of Precambrian rocks defining regional Red Sea arch. Permission to publish by American Association Petroleum Geologists.

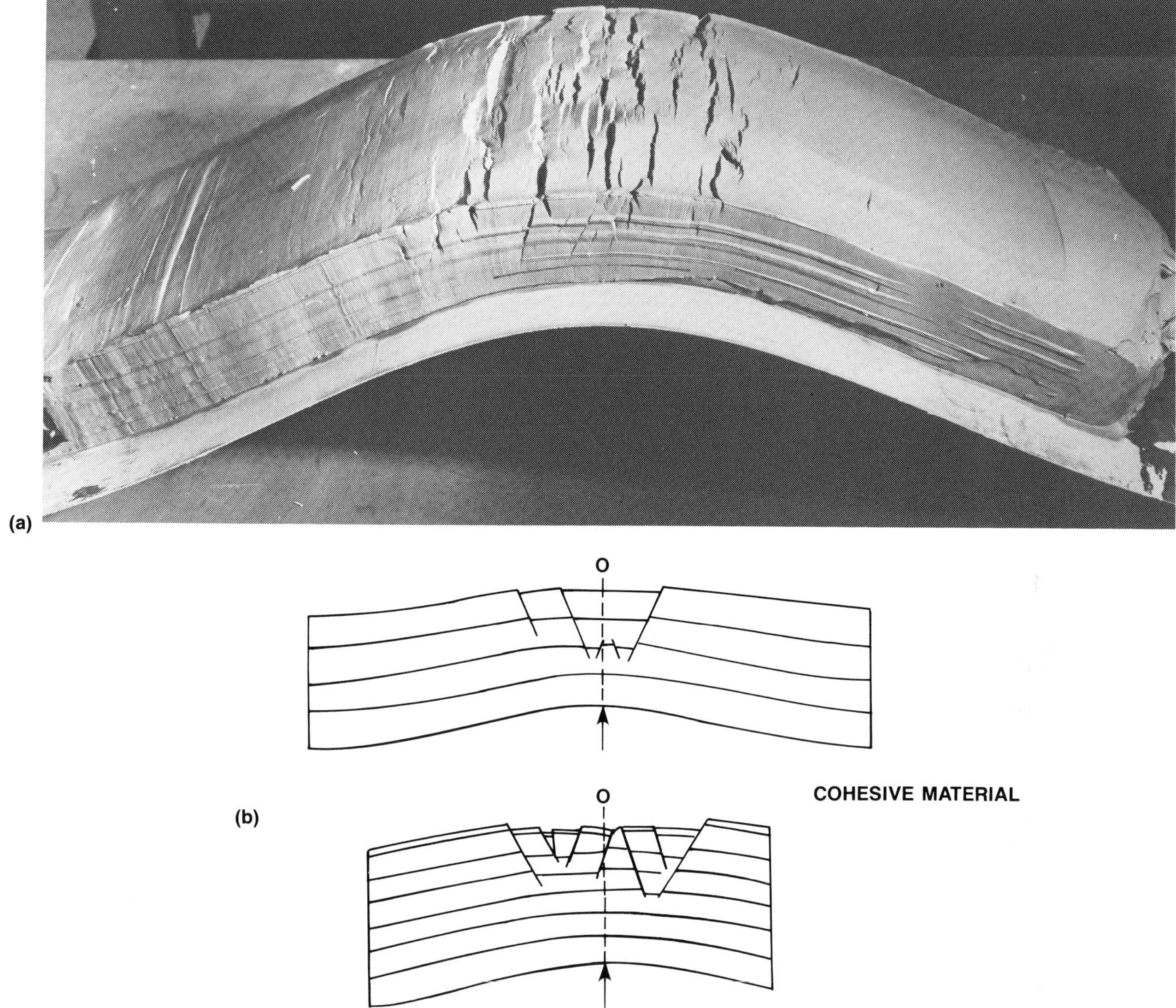

Fig. 4-4—A. arched clay model, B. (Sanford, 1959)—Normal faults (conjugate shears) by arching form progressively outward from crest of arch, have maximum displacement at surface, and die out at depth. (This response is also followed in stress analyses of assumed elastic behavior). Note that a perfect arch is never achieved because of collapse of the central part by normal faulting. From Geological Society of America Bulletin, V. 70. Used with permission.

themselves, in addition to the rotation of beds (Davis, 1983).

Rifting in the Red Sea is centered along the axial trough. This is portrayed on the cross sections in the southern Red Sea (Fig. 4-7 A, B) by positive Bouguer gravity, a magnetic anomaly pattern, the presence of tens of kilometers of oceanic crust between lithospheric plates capped by continental crust, and an interpreted rise of the Moho and the base of the lithosphere. A second (western) rift is indicated by another inferred elevation of the Moho and the base of the lithosphere, and is shown west of the axial trough on both cross sections.

The western rift is the northern end of the Danakil depression (Fig. 4-7A). This feature, an asymmetric graben, is deepest and most severely faulted on the west side adjacent to the Ethiopian plateau and has the Danakil horst on the east (Holwerda and Hutchinson, 1968). The Danakil depression is a Pliocene rift that is still evolving (Bannert et al., 1970). It contains at least 1000 m (3,280 ft) of Quaternary section that is not present in the surrounding area (Holwerda and Hutchinson, 1968). The depression contains active volcanoes (Mohr, 1967a; Girdler, 1969; Tazieff, 1968, 1969, 1970) and igneous intrusives (Holwerda and Hutchinson, 1968). Evidence of recent faulting is the presence of faulted alluvial fans and faulted recent volcanics (Tazieff and Varet, 1969; Tazieff, 1970); the area is seismically active (Fairhead and Girdler, 1970). More-

EXTENSION

LEAST PRINCIPAL

COMPRESSIVE STRESS

8.2 CM.

(a)

MAXIMUM PRINCIPAL COMPRESSIVE STRESS

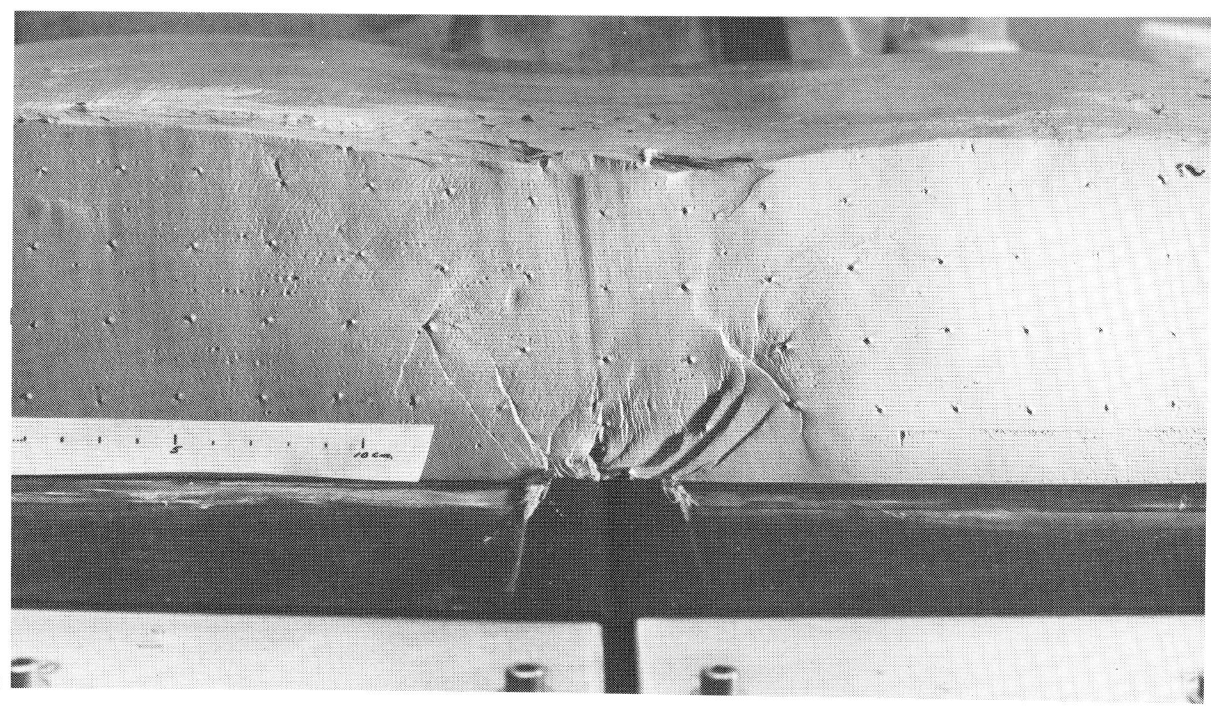

(b)

Fig. 4-5—Two experiments (A, B) showing pull apart of clay cake by extension at base. Listric normal faults propagate upward from base and top is sagged. Faulted homoclines and monoclines can thus be explained by this mechanism. Note in each case maximum principal compressive stress (σ_1) is vertical and least principal compressive stress (σ_3) is horizontal.

over, the Danakil depression was covered by seawater as recently as some tens of thousands of years ago (Tazieff, 1970, Lalou et al., 1970). The Danakil depression easily could be flooded by marine waters, considering it is only 40 km (25 mi) from the Red Sea and has a large area below sea level with a maximum depth of 117 m (384 ft). The present subaerial condition of the Danakil depression is believed to be transient, and I agree with Tazieff (1968, 1969, 1970), who considered the depression a temporarily emergent part of the Red Sea.

A western rift also is present farther north in the southern Red Sea (Fig. 4-7B). This feature, which is partly expressed by the north Massawa channel, is spectacularly demonstrated on reflection seismic (Fig. 4-8) by large normal faults that have displaced the sea bottom by as much as 400 m (1,312 ft) and basement by more than 1000 m (3,288 ft).

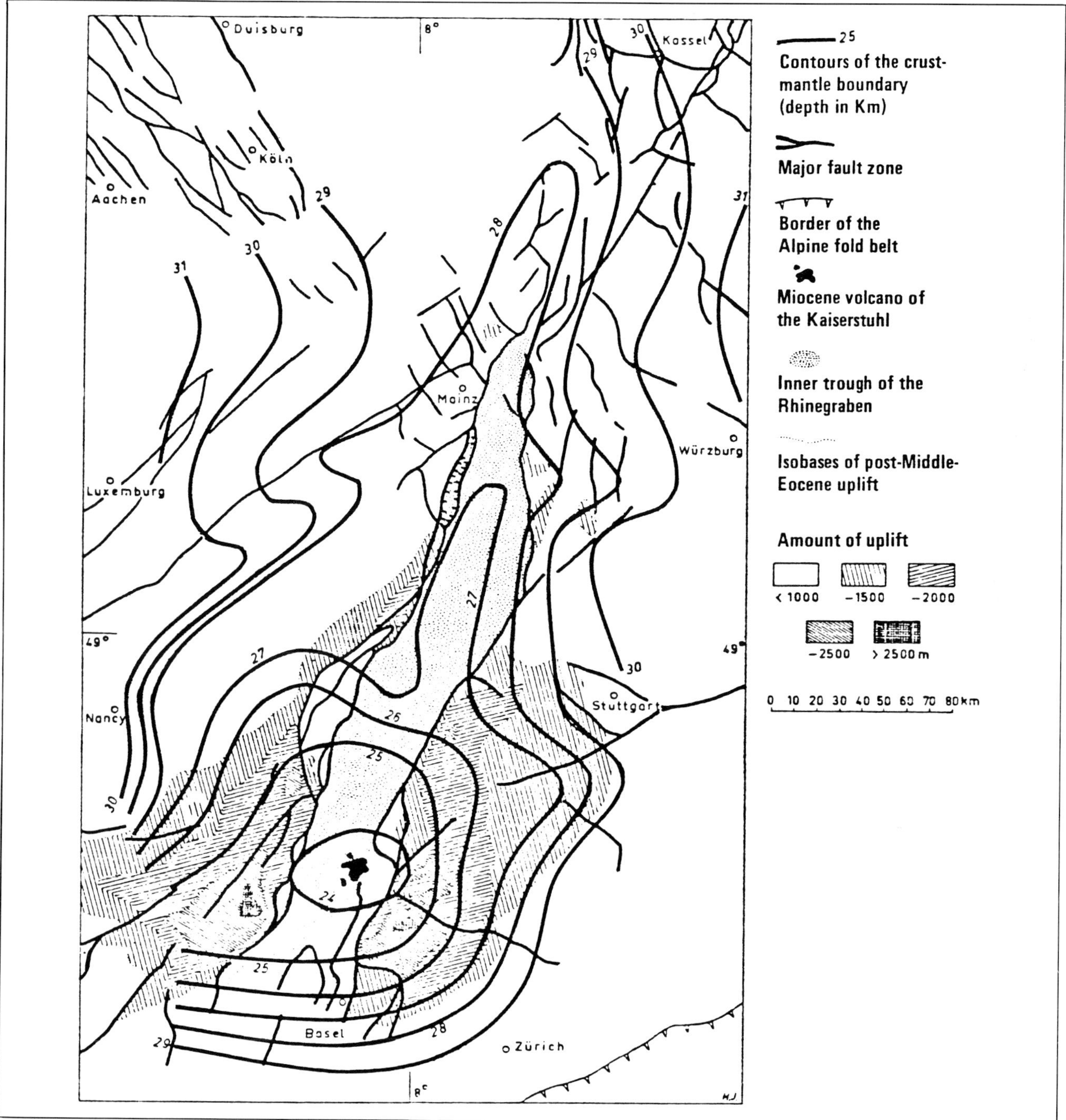

Fig. 4-6 (Illies, 1975)—Rhine graben, contours on the Moho. Permission to publish by Geologische Rundschau.

The magnetic field is almost flat across the rift, presumably because the continental lithosphere has not been breached to allow the insertion of oceanic crust.

About midway between the line of the two cross sections (Fig. 4-7) is the north-northwest-trending Gulf of Zula. It is bounded by the Buri Peninsula on the east and on the west by the Red Sea coastal plain, which is downfaulted from the Ethiopian Plateau lying still farther west. A large down-to-the west normal fault bounds the northern projection of the Buri Peninsula on its west side (Frazier, 1970), making the Gulf of Zula a large graben. Landward, just beyond the south end of the Gulf of Zula, two active volcanoes are known, and earthquakes have occurred in the vicinity. The graben structure of the Gulf of Zula likely continues northward as the western rift, and southward as the Danakil depression, probably with a branch west along the

(a)

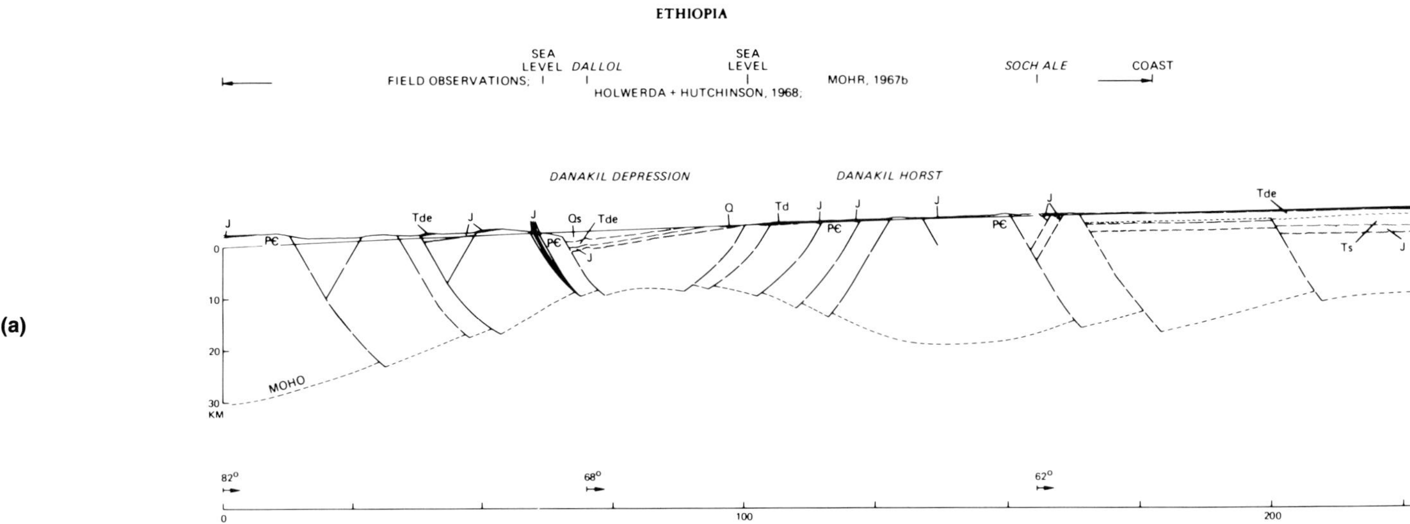

(b)

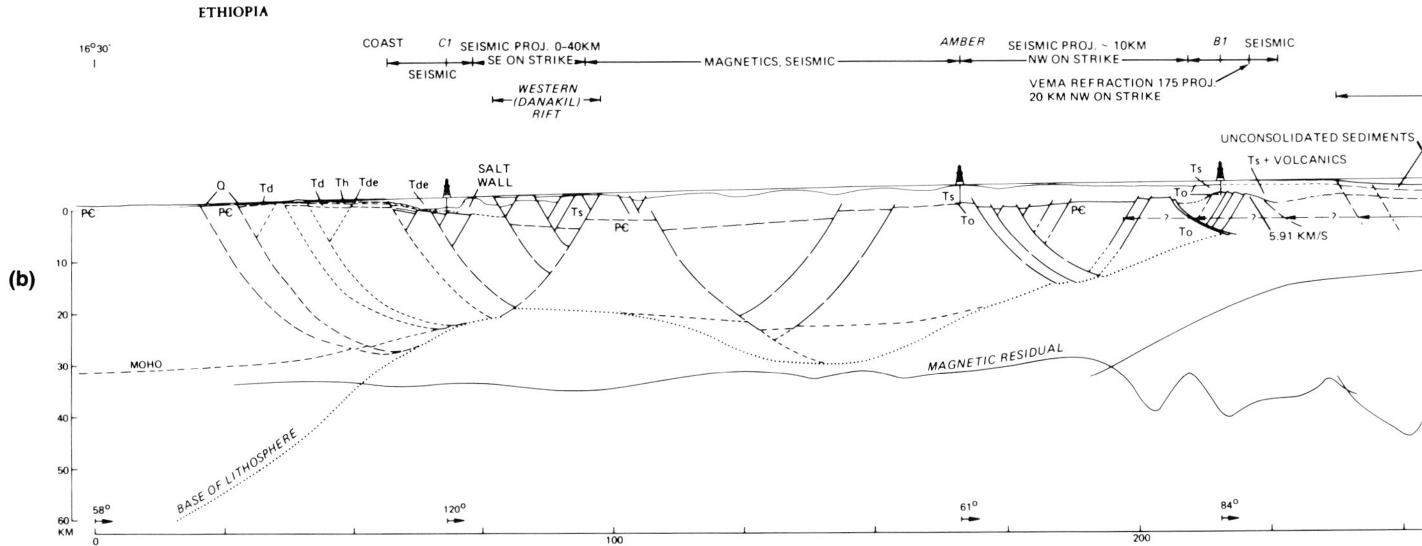

Fig. 4-7 (Lowell and Genik, 1972)—Structure sections, index map, and legend. A. Structure section from Ethiopian Plateau across Danakil depression, Danakil horst, and southern Red Sea at Jabal Zubayr to western Yemen. B. Structure section from northern part of Ethiopian Plateau across southern Red Sea through vicinity of Dahlak and Farasan islands to southern corner of Saudi Arabia. (Bathymetry from bathymetric map; topography from USAF Jet Navigation charts; gravity and, in part, magnetics from Allan, 1970.) C. Stratigraphic legend and index map. Permission to publish by American Association of Petroleum Geologists.

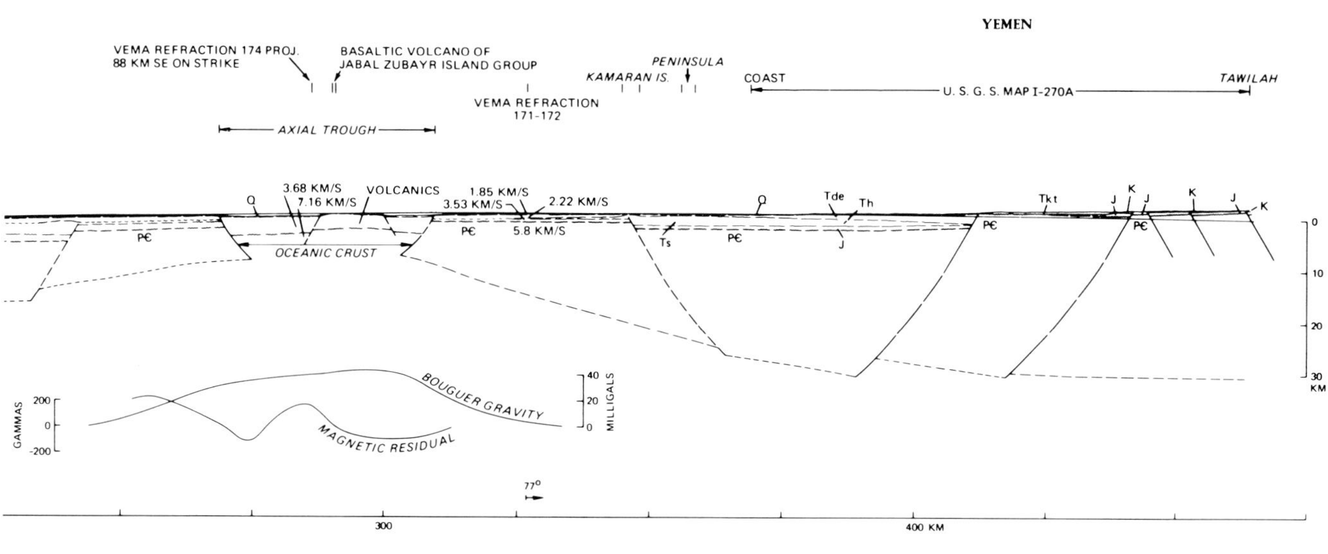
YEMEN
VEMA REFRACTION 174 PROJ.
88 KM SE ON STRIKE
BASALTIC VOLCANO OF
JABAL ZUBAYR ISLAND GROUP
PENINSULA
KAMARAN IS.
COAST
TAWILAH
U. S. G. S. MAP I-270A
VEMA REFRACTION
171-172
AXIAL TROUGH
3.68 KM/S
7.16 KM/S
VOLCANICS
1.85 KM/S
3.53 KM/S
2.22 KM/S
5.8 KM/S
OCEANIC CRUST
BOUGUER GRAVITY
MAGNETIC RESIDUAL
GAMMAS
MILLIGALS
77°
300
400 KM

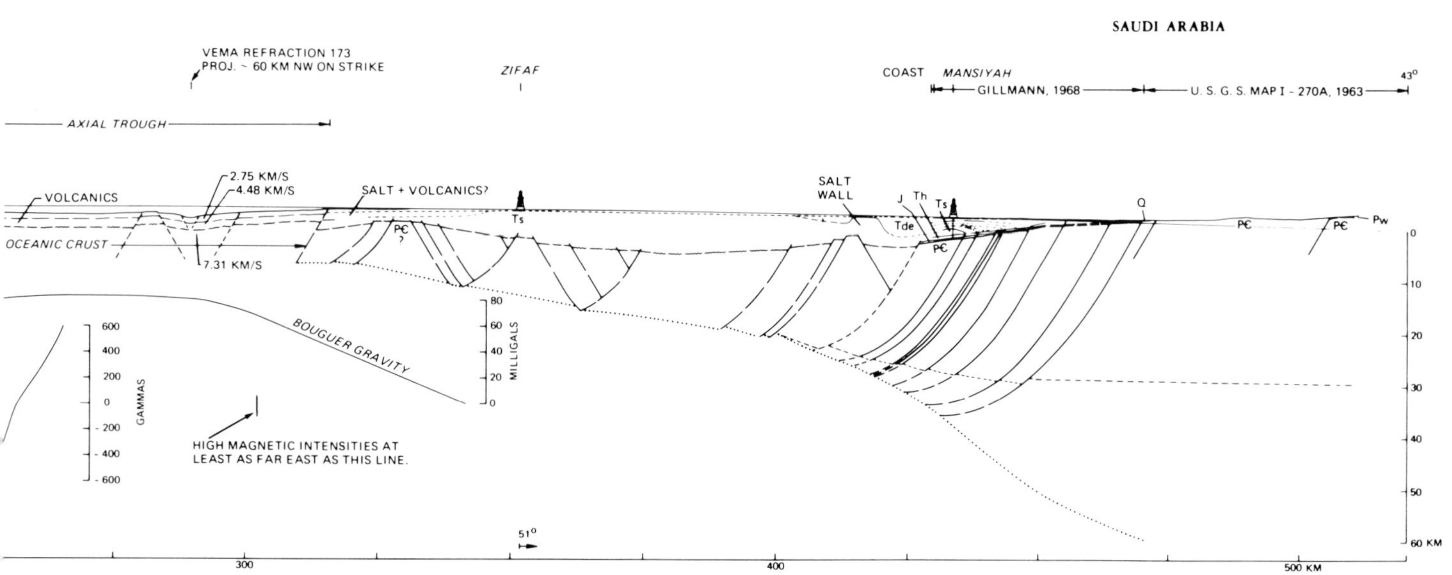
SAUDI ARABIA
VEMA REFRACTION 173
PROJ. ~ 60 KM NW ON STRIKE
ZIFAF
COAST
MANSIYAH
GILLMANN, 1968
U. S. G. S. MAP I - 270A, 1963
43°
AXIAL TROUGH
VOLCANICS
2.75 KM/S
4.48 KM/S
SALT + VOLCANICS?
SALT
WALL
OCEANIC CRUST
7.31 KM/S
BOUGUER GRAVITY
MILLIGALS
GAMMAS
HIGH MAGNETIC INTENSITIES AT
LEAST AS FAR EAST AS THIS LINE.
51°
300
400
500 KM
60 KM

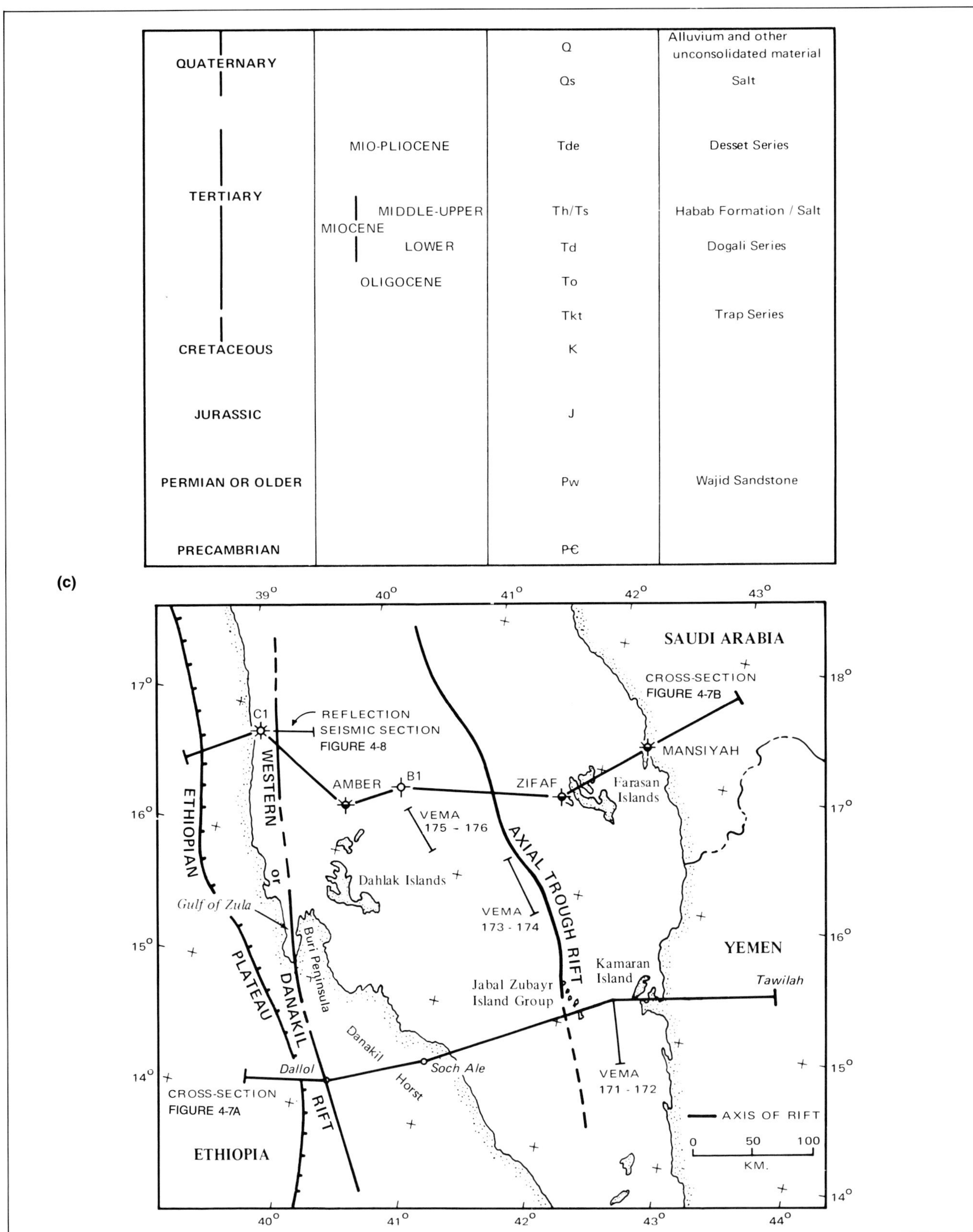

QUATERNARY		Q	Alluvium and other unconsolidated material
		Qs	Salt
TERTIARY	MIO-PLIOCENE	Tde	Desset Series
	MIOCENE MIDDLE-UPPER	Th/Ts	Habab Formation / Salt
	MIOCENE LOWER	Td	Dogali Series
	OLIGOCENE	To	
		Tkt	Trap Series
CRETACEOUS		K	
JURASSIC		J	
PERMIAN OR OLDER		Pw	Wajid Sandstone
PRECAMBRIAN		P€	

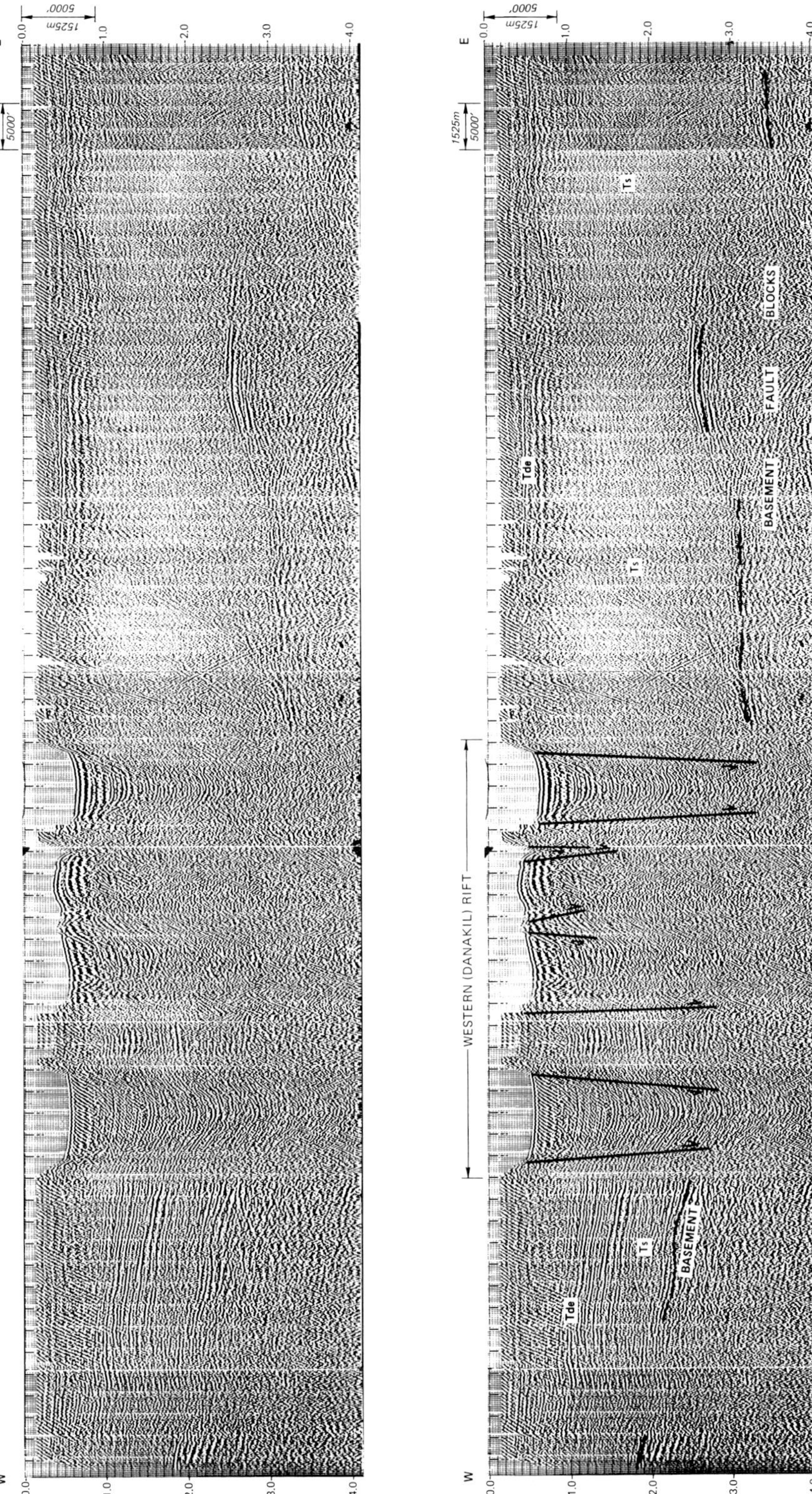

Fig. 4-8 (Lowell and Genik, 1972)—Reflection seismic section in western part of southern Red Sea showing western (Danakil) rift. (See figure 4-7C for line of section.) Permission to publish by American Association of Petroleum Geologists.

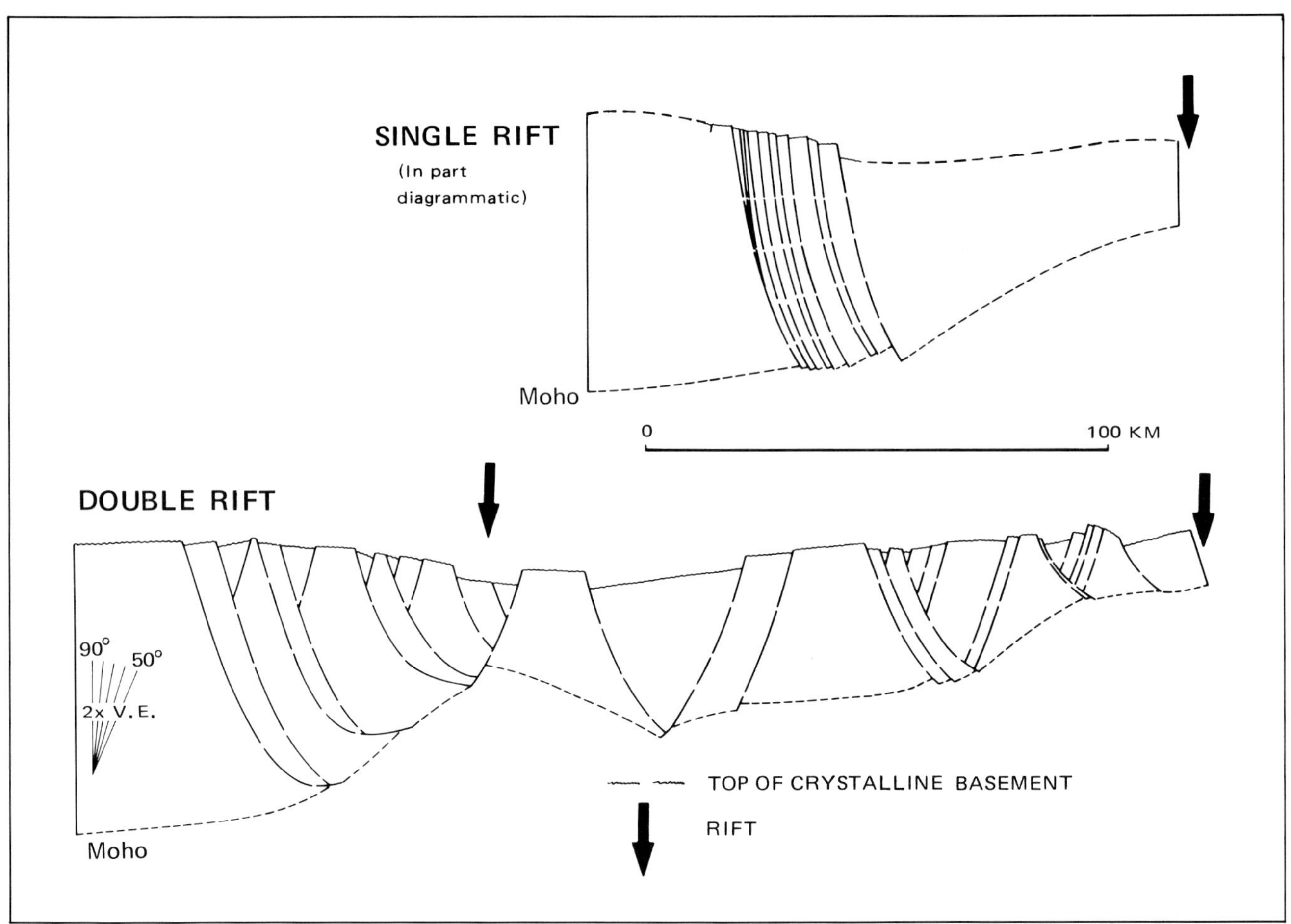

Fig. 4-9 (Lowell et al., 1975)—Contrast of double and single rifting. Thinning of crust occurs over a wider area by double rifting than by single rifting and apparently results in higher heat flow from sublithospheric sources for the former. The double-rifted section is the west side of the structure section of figure 4-7B.) From Petroleum and Global Tectonics, copyright © Princeton University Press. Used with permission.

Ethiopian plateau, where Bannert et al. (1970) showed a discrete graben system.

The Danakil depression and the western rift of the southern Red Sea thus seem to be one continuous rift complex. This feature and the axial rift of the Red Sea constitute two concurrently developing rifts that run parallel for 400 km (250 mi) and possibly much more. The north-northwestward extent of the western rift is unknown, but bathymetry suggests that it easily could continue another 250 km (156 mi) to beyond 19° N. Tazieff (1968) proposed that the northwest trends and horst-and-graben structures of the northern Afar are related to similar structures along the coast of Sudan reported by Carella and Scarpa (1962) and Sestini (1965), and also to structures on the west side of the Gulf of Suez (Said, 1962).

Depending on which rift, the axial or western, is ultimately dominant as a locus of sea-floor spreading, the intervening large fragment of continental crust (including the Danakil horst at the south end) will move either with the Nubian or Arabian plates, respectively, and will be detached to some degree from the main plate. As the two rifts converge slightly on the north (Fig. 4-7), some counterclockwise rotation of the continental fragment has occurred; counterclockwise rotation of the Danakil horst has been proposed by Laughton (1966) and amplified by Roberts (1969), Burek (1970), and Mohr (1970).

A double ridge system has been postulated for the Reykjanes ridge southwest of Iceland (Godby et al., 1968). Another instance of double rifting in the geologic record may have occurred in the North Atlantic, where the Rockall Plateau represents an elongate piece of continental crust separated from the British continental shelf by a relatively narrow band of oceanic crust or greatly thinned continental crust (Bullard et al., 1965; Roberts et al., 1970).

The effects of double rifting have had a significant impact on petroleum occurrence in the southern Red Sea. Double rifting causes extension of continental crust over a broader area than does single rifting, which effects a nar-

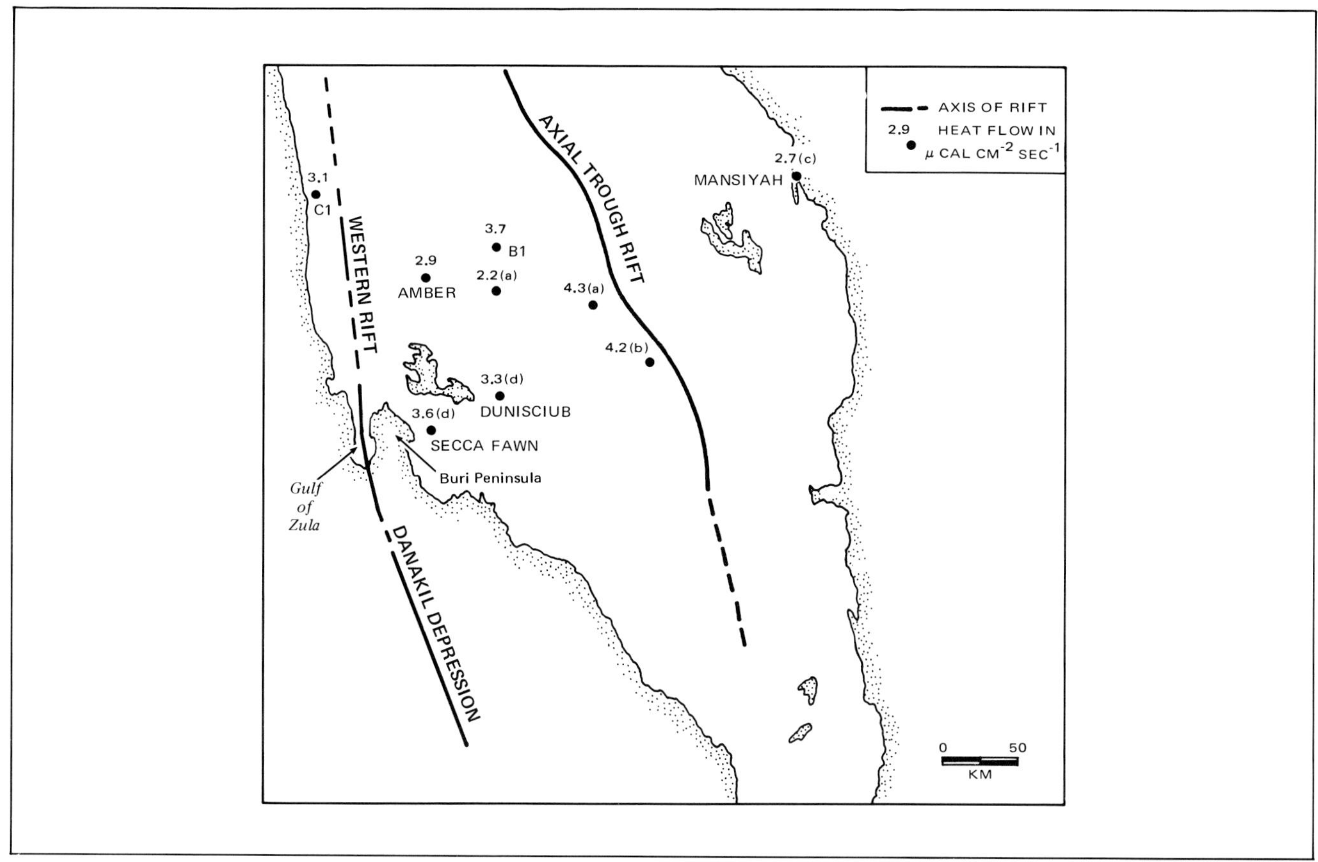

Fig. 4-10 (Lowell et al., 1975)—Heat flow in southern Red Sea. Values are (a) from Langseth and Taylor, 1967, (b) from Sclater, 1966, (c) from Girdler, 1970, and (d) calculated from Frazier, 1970. From Petroleum and Global Tectonics, copyright © Princeton University Press. Used with permission.

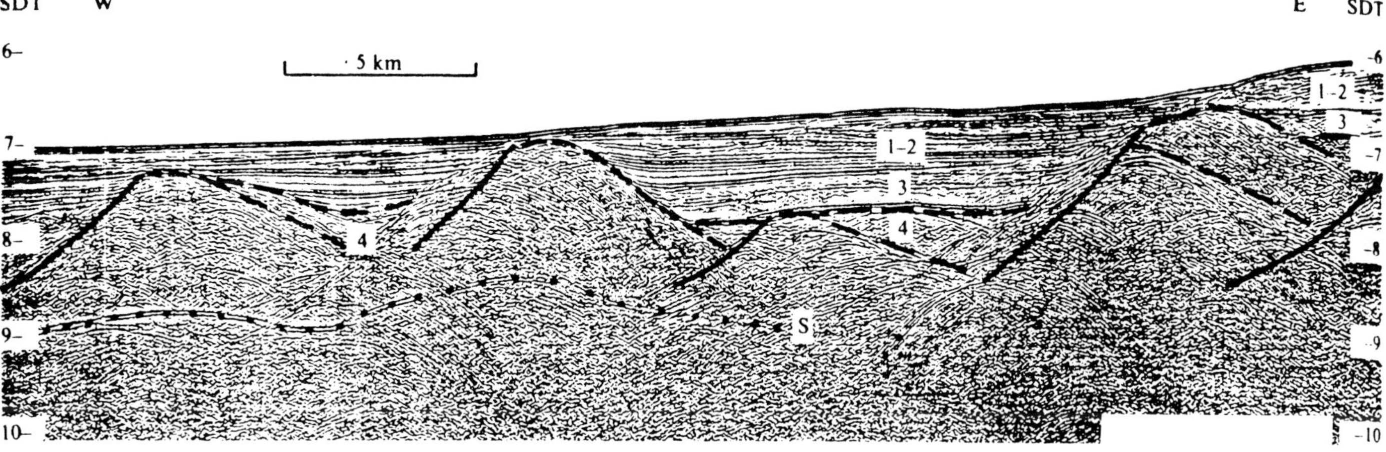

Fig. 4-11 (de Charpal et al., 1978)—Seismic profile from western approaches to the English Channel showing blocks bounded by listric normal faults that may sole at undulating horizon "S" at or near the level of the Moho. Vertical scale is 2-way time in seconds. Reprinted by permission from Nature, V. 275. Copyright © Macmillan Journals Limited.

rower tapered crustal wedge (Fig. 4-9). Assuming a deep heat source, double rifting with its thinner continental crust would lead to higher temperatures in the sedimentary section. Heat flow in the southern Red Sea apparently increases toward each rift from a cooler intervening area (Fig. 4-10). Moreover, heat flow is very high, often greater than 4.0 microcalories, and geothermal gradients in the southern Red Sea range from 28° to more than 50° C/km (2.5°–3.7°F/100 ft). In the thermal maturation process, oil subject to this kind of heat might soon be converted to gas. A gas blowout beneath the salt in the C1 well partially confirms that the area may be gas prone, at least at depths of 3000 m (9,840 ft) and greater.

Salt Occurrence

A thick accumulation of salt is a natural consequence of continental breakup and sea-floor spreading at low latitudes. Enormous thicknesses of salt have accumulated in the southern Red Sea. From reflection seismic information, the deepest part of the basin west of the axial rift (Figs. 4-7B, 4-8) is estimated to have 5000 m (16,400 ft) of bedded evaporites. The Amber well in this basin penetrated 3340 m (10,955 ft) of evaporite, almost entirely halite. This is not a true stratigraphic thickness because of salt flowage, but it further documents a considerable amount of salt. The age of the salt is Miocene.

The Miocene marine trough that formed during the earlier stages of rifting was elongate and periodically restricted, e.g., silled at times by lava flows at its narrow ends. Further, drainage controlled by the earlier arching stage was directed away from the trough, so that relatively little clastic deposition occurred there. In the hot, arid, low latitudes of the Miocene Red Sea trough, continuous evaporation of normal seawater, which was periodically replenished, could account for great thicknesses of salt.

Evaporites are also present in the Danakil depression (Fig. 4-7A), where at least 1000 m (3,280 ft) of virtually pure, bedded halite of marine origin and Quaternary age has been drilled in the deep asymmetric west side of the basin (Holwerda and Hutchinson, 1968). The evaporites must have precipitated during periodic incursions of the Red Sea into the downfaulted Danakil depression. The Danakil evaporite is remarkable in that it is a modern example of salt accumulating in a narrow, restricted, rifted trough that is forming as a consequence of continental breakup. As mentioned, this environment is a very dynamic one, replete with active faulting, igneous intrusion, volcanism, and hot springs.

We have seen the deleterious effect high heat flow has had on deep (sub-salt) reservoirs in extensional blocks of the southern Red Sea. However, oil seeps have been reported from the southern Red Sea, presumably emanating from sediments at relatively shallow depths above the Miocene salt. In this case high temperatures no doubt hastened

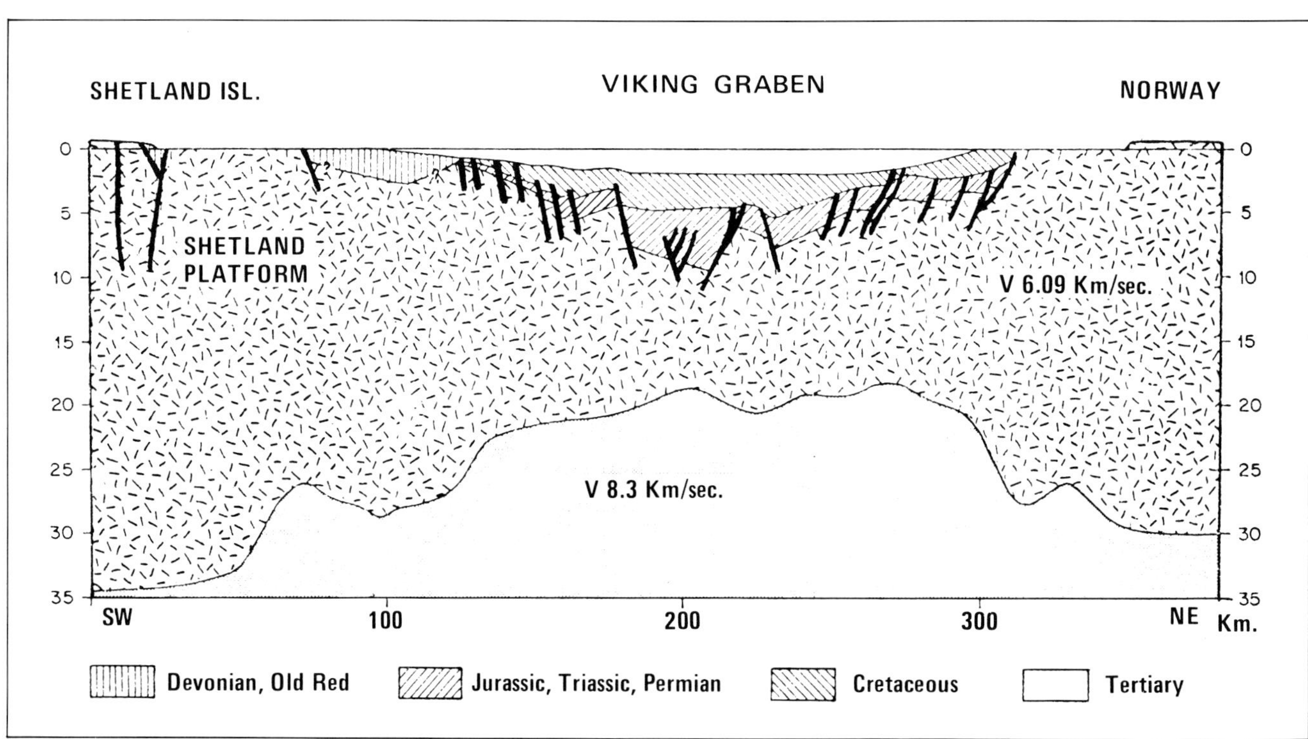

Fig. 4-12 (Solli, 1976)—Refraction and reflection profile across Viking graben showing thinned, but continuous, crust by rise on Moho (boundary between 6.09 and 8.3 km/sec velocity layers). Permission to publish by University of Bergen.

the maturation process of organic material to oil. Structural reversals are common above the salt, but, unfortunately, potential source and seal rocks are not very plentiful.

Applications

The rift or extensional block model for the southern Red Sea (Fig. 4-2) can be applied to other areas. Listric normal faults must effect lithospheric thinning on all pull-apart continental margins and their development was probably much like that proposed for the Red Sea. Faults are shown to sole at the base of the brittle lithosphere in the proposed model. In some instances that level could be the Moho. It is of great interest that in the area of the English Channel de Charpal et al.(1978) have interpreted listric normal faults to flatten above or at the Moho along a sub-horizontal seismic reflector (Fig. 4-11). Listric normal faults also effect rotation of fault blocks, whereas planar normal faults create only horizontal horst and graben. The difference is critical in petroleum exploration because the largest accumulations typically have been found in the highest parts of tilted blocks. Since most normal fault blocks are tilted, we can assume that listric normal faulting has been the dominant process.

Sea-floor spreading could continue in the Red Sea to form a full fledged ocean. Conversely, the evolutionary process could have stopped at any time, such that stage II of figure 4-2 would approximate the non-marine Triassic graben system of the eastern United States, or stage III would be comparable to many interior rift systems which have continental crust continuous beneath them (Fig. 4-12). Comparisons are practically limitless.

Occurrence of salt in the southern Red Sea is analogous to several salt basins that formed during continental breakup along the west coast of Africa (Belmonte et al., 1965; Ayme, 1965; Pautot et al., 1970).

Finally, carbonates in the southern Red Sea (Fig. 4-2) have accumulated to several thousand feet in thickness as subsidence accompanied horizontal movement of the newly evolving continental margin away from a spreading center at low latitudes favorable for carbonate deposition. In this regard the carbonates of the Red Sea may be of similar origin to those of the Bahamas in the western Atlantic which are much thicker owing to their longer period of slow subsidence and buildup as the Atlantic Ocean evolved.

PROFILE VIEW

From the foregoing section on mechanics it has been shown that the major normal faults almost certainly are listric and sole at depth. This configuration has been summarized by Proffett (1977) who has shown normal faults to range from vertical through 60°–70° to almost flat (Fig. 4-13). At the most common levels of investigation in petroleum exploration, the 60°–70° dip range is probably the most common. At these levels normal faults can appear as horst and graben (Fig. 4-14) or, more commonly, as tilted or rotated blocks that sometimes are expressed as half graben (Fig. 4-15). Cylindrical or convex upward faults are relatively rare (Fig. 4-14).

As are compressive blocks, extensional blocks invariably are markedly asymmetric with a gentle, tilted, relatively unfaulted flank and a steep faulted flank. The steep flank fault is always a normal fault, however, and no thrust or reverse-fault overlap and attendant repetition are known. Moreover, extensional blocks do not display steep, vertical, or overturned beds on their steep flanks.

PATTERNS OF RIFTING

Rift systems and their associated extensional blocks can have relay, dog-leg, and transverse fault patterns and be

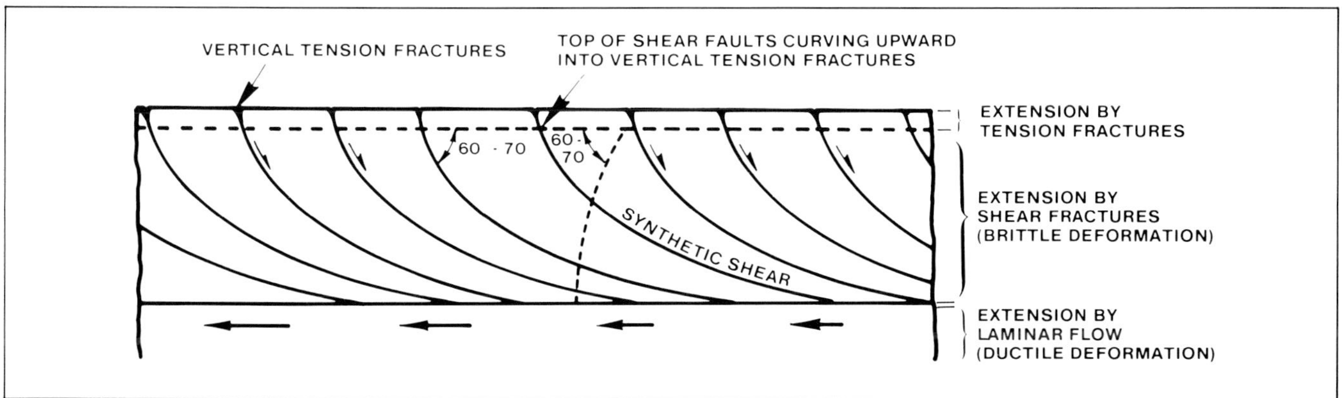

Fig. 4-13 (Proffett, 1977)—Most failure on normal faults is by shear. Vertical tension fractures are restricted to surficial levels of low confining pressure and are infrequently observed because of erosion; however, some excellent examples are known from Iceland. Faults of brittle domain are shown to sole at depth into laminar flow. From Geological Society of America Bulletin, V. 88. Used with permission.

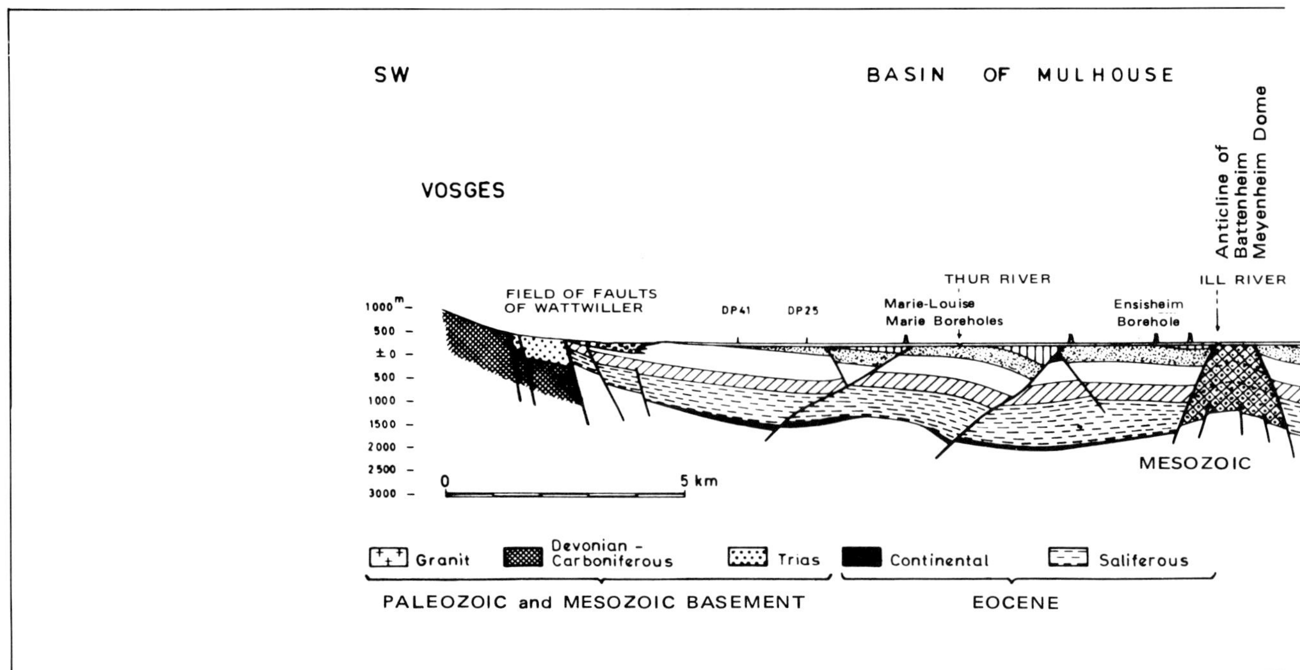

Fig. 4-14 (Sittler, 1969)—Rhine graben showing horst and graben configuration and a cylindrical fault below the point marked as DP75. Permission to publish by Elsevier Science Publishers B.V. Amsterdam.

developed in aulacogen and extensional prong habitats in addition to more conventional graben and pull-apart continental margin settings. Aulacogens are long-lived deeply subsiding troughs that extend from continental margins far into adjacent platforms. Extensional prongs are aborted extensional basins created by the projection of pull-apart margin structures beyond a break that eventually becomes a transform plate boundary.

Plan Views

Rift or graben systems are almost always shown in cross section to be bounded by normal faults ranging from high to very low angle (Figs. 4-14, 4-15 and 4-16), in partial accord with conventional shear failure criteria (Fig. 2-1A) which predicts a relay pattern in map view. A relay pattern consists of randomly overlapped elements (faults, folds) aligned parallel with one another and with the zone of deformation in which they occur. Relay patterns consistently develop from arching (Fig. 4-17), while a slightly modified relay pattern can be modeled by pull apart (Fig. 4-18). In both instances the relay direction is perpendicular to σ_3 (least principal compressive stress, or, effectively, extension).

The relay system of faults, however, does not describe the full pattern of rifting, which, in turn, means that conventional shear failure criteria (Fig. 2-1A) do not adequately treat normal faulting. In addition to the relay pattern, a dog-leg pattern, (with faults trending approximately 30° on either side of the relay direction), is present and resembles a pattern of non-rotational shear (Fig. 2-5). However, dog-leg faults are normal, rather than vertical, in cross section and result from σ_1 oriented vertically, rather than horizontally (Fig. 2-1B). Dog-leg patterns from many of the world's great rift systems are illustrated in figures 4-19 through 4-27, which are compiled from various maps and arranged progressively from larger to smaller scale. Besides relay and dog-leg patterns, transverse faults are also present. When related to incipient transform motion, as discussed in Chapter 2 (Fig. 2-69), transverse faults should be vertical in profile. In the absence of transform motion transverse faults should be normal in profile.

I believe that all rift systems generally have relay, dog-leg, and transverse fault directions (Fig. 4-28). This is suggested, for example, from the Suez area where fault directional plots show well developed NNE (Akaba), NNW (Suez), and NW (unnamed) trends (Fig. 4-29). I interpret the NNW trend as relay and perpendicular to regional extension, the NNE and NW trends as doglegs, and a transverse trend as only weakly developed.

Explanations

Several possibilities have been put forth to explain the rifting pattern of figure 4-28:

1– DeSitter (1956, p. 471, 472), noting the discrepancy in surface and cross-sectional fault patterns in both the Rhine and East African rift systems, sug-

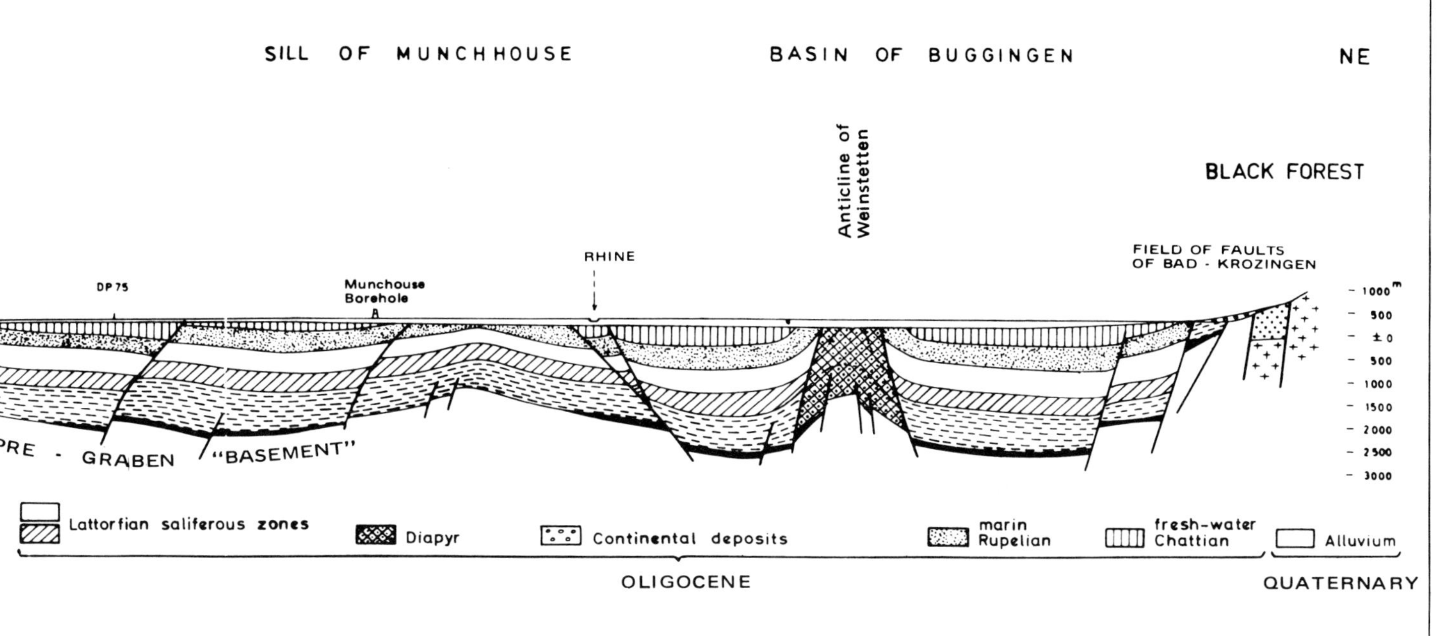

gested that the surface fault pattern was due to a compressional stress condition in an unspecified lower tier of the earth's crust, whereas the actual fault movement (normal sense displacement) resulted from dilation in an upper tier. It is extremely difficult to conceive of different tiers of the earth's crust reacting so disparately to an applied stress. This is particularly so in the context of plate tectonics where we are accustomed to thinking of the crust as simply the upper part of the lithosphere, and earthquake first-motion studies indicate that the entire rigid portion of the lithosphere behaves as a single unit.

2– Donath (1962) provided a very logical and highly probable explanation for an extensive and complex dog-leg fault pattern in basalt flows in south-central Oregon (Fig. 4-30). As illustrated in figure 4-31, Donath interpreted the faults as having

"originally developed as conjugate strike-slip shears [due to north-south maximum principal stress and east-west minimum principal stress] to form a rhombic pattern of [vertical] fractures and incipient fractures. Subsequent to the development of this fracture system, block faulting occurred [through a redistribution of surface forces acting on individual fault blocks] in which the dominant movement on the shears was dip slip. This block faulting resulted in the depression, elevation, and tilting of individual blocks..."

Thus, dip slip occurred on vertical fractures that were earlier created as non-rotational conjugate strike-slip shears, but along which there was virtually no actual lateral movement.

Donath (1962, p. 13, 14) stated that

"it is not possible to determine how much time elapsed between the development of the fracture system and the stage of block faulting. Block faulting may not have occurred for

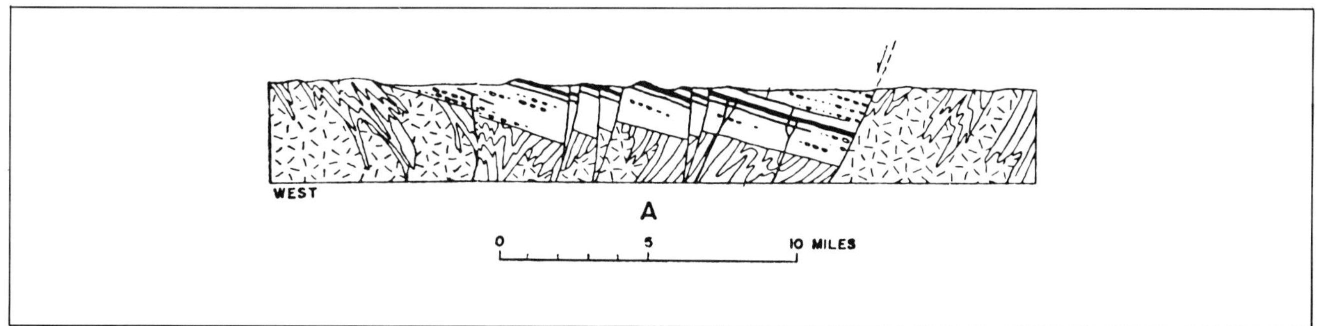

Fig. 4-15 (Longwell, 1933)—Connecticut Valley half-graben system. Permission to publish by Bureau de Recherches Geologiques et Minieres for 16th International Geological Congress.

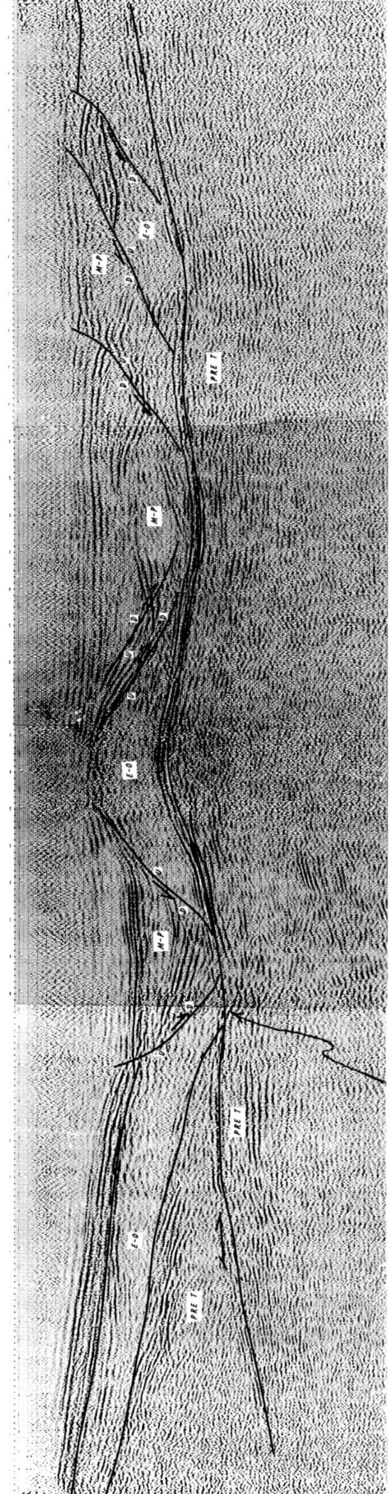

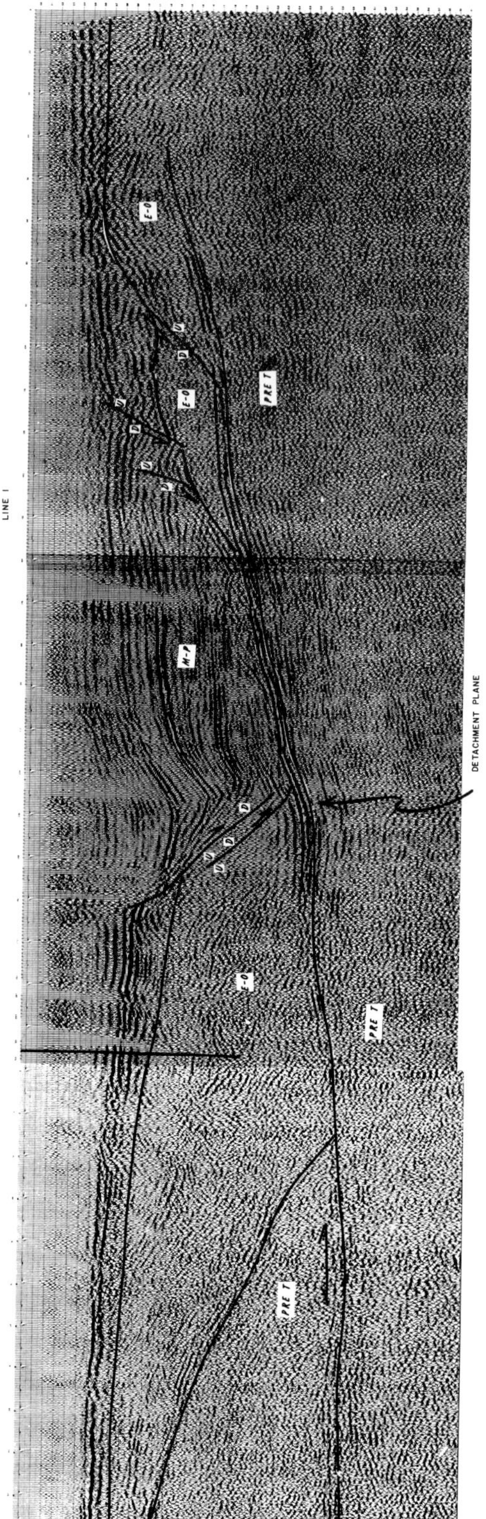

Fig. 4-16 (McDonald, 1976)—Seismic lines from Sevier Desert basin, Millard County, west-central Utah, showing detachment plane thought to represent earlier thrust with subsequent extension of Tertiary strata during Basin-Range deformation. Vertical line to left of center on Line 1 is Gulf Oil Co. No. 1 Gronning well, T.D. 2457 m (8061 ft). PRE T - Pre-Tertiary, E-O - Eocene and Oligocene, M-P - Miocene and Pliocene. Permission to publish by Rocky Mountain Association of Geologists.

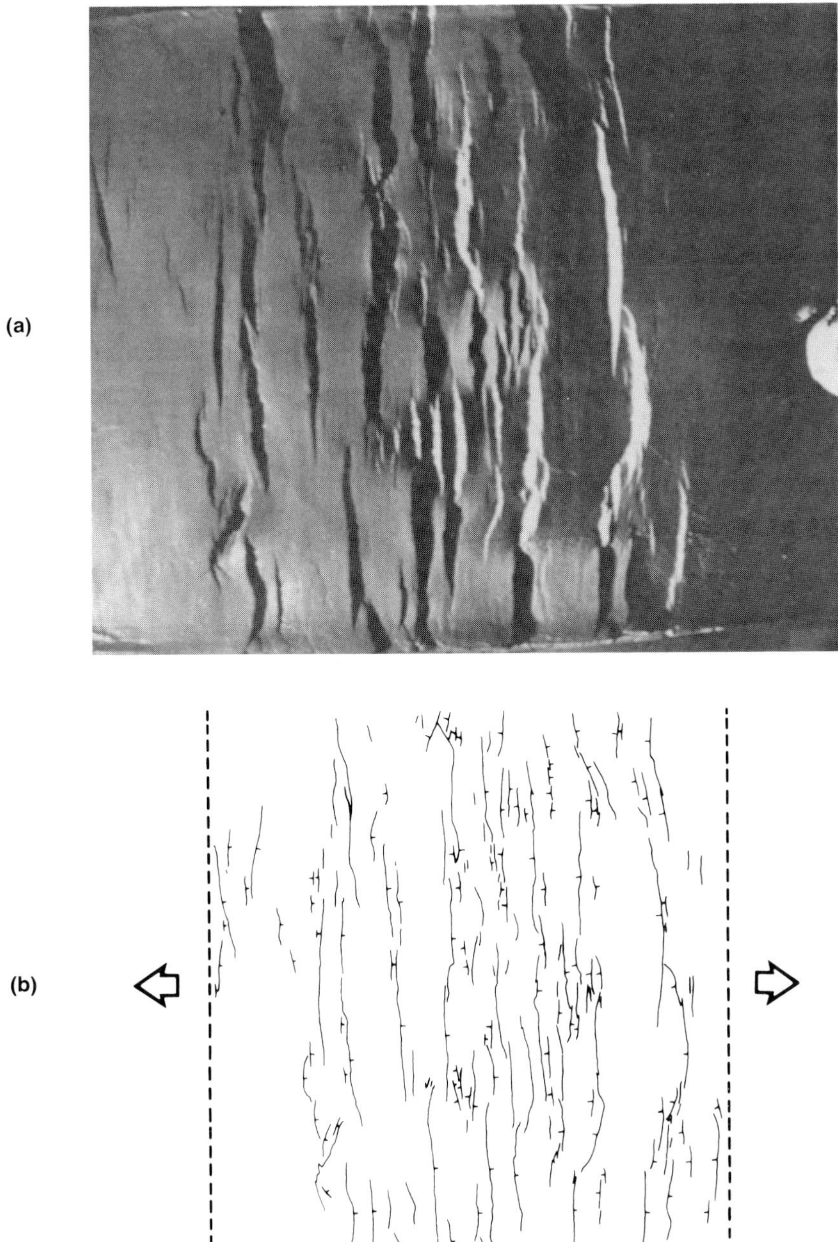

Fig. 4-17—Clay model (A) with line drawing (B) showing relay pattern created by arching. Effective direction of extension shown by arrows. Movement on all faults was purely dip slip with barbs on downthrown side. Note that random overlap (relaying) provides a mechanism for both distributing and terminating displacement.

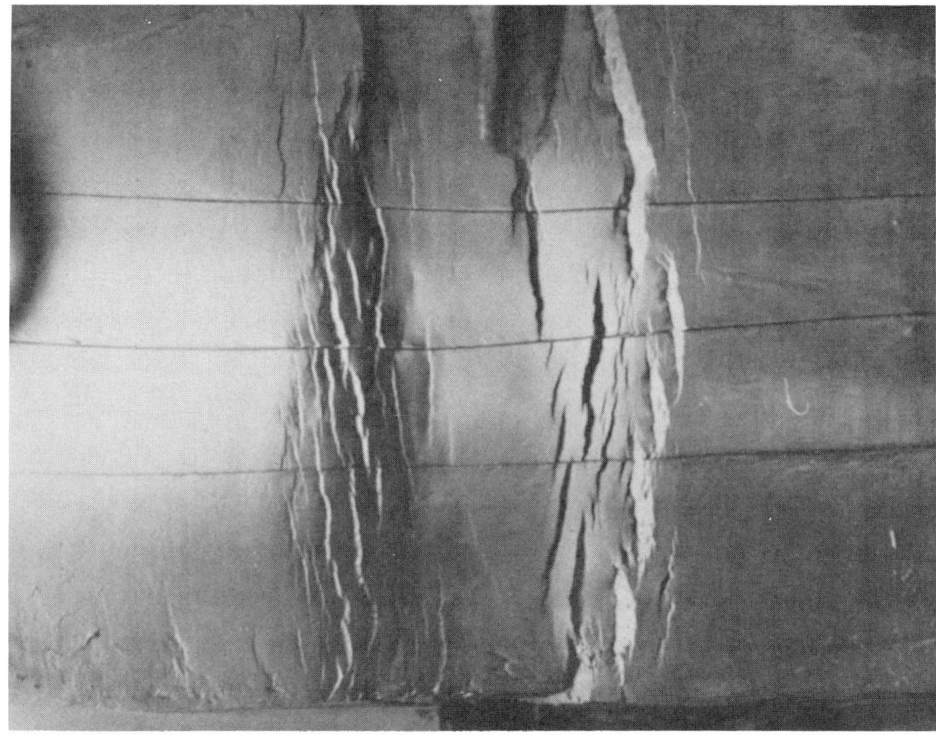

Fig. 4-18—Clay model showing modified relay pattern from pull apart. Numerous faults are skewed or not quite parallel to trend of zone of deformation. Extension is "east-west". Illumination is from the left.

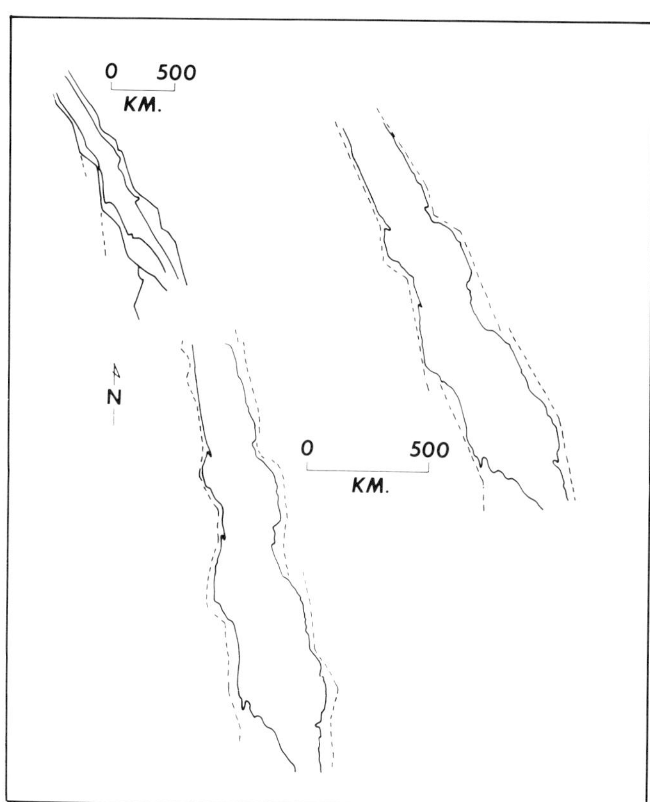

Fig. 4-19—Red Sea bounding fault trends.

some time after the formation of the strike-slip shears, or it may have followed closely in a nearly continuous sequence of deformation."

Donath's interpretation does not solve the present problem for, although it explains the dog-leg pattern, in cross section the faults are vertical rather than in normal fault orientation.

3– In his description of the Ottawa-Bonnechere graben in Ontario and southwestern Quebec, Canada, Kay (1942) had a wholly satisfactory explanation, at least at this locality, for the dog-leg versus normal-fault dilemma. He interpreted the dog-leg pattern (Fig. 4-32) as a result of movement on pre-existing joints expressed in Precambrian rocks, the joints, incidentally, having greater frequency in proximity to the faults. The faulting was thought to be of early Tertiary age, and the joints to be the result of pre-Paleozoic movements. Thus there was a 500-my interval between the development of the joints and the faults whose direction the former was believed to influence. Though exposures are few, Kay continually refered to the faults as normal faults; the attitude of the joints was not given. Presumably block rotation on joints led to high-angle normal faults.

Kay's interpretation of the faulting in the Ottawa-Bonnechere graben undoubtedly applies in other

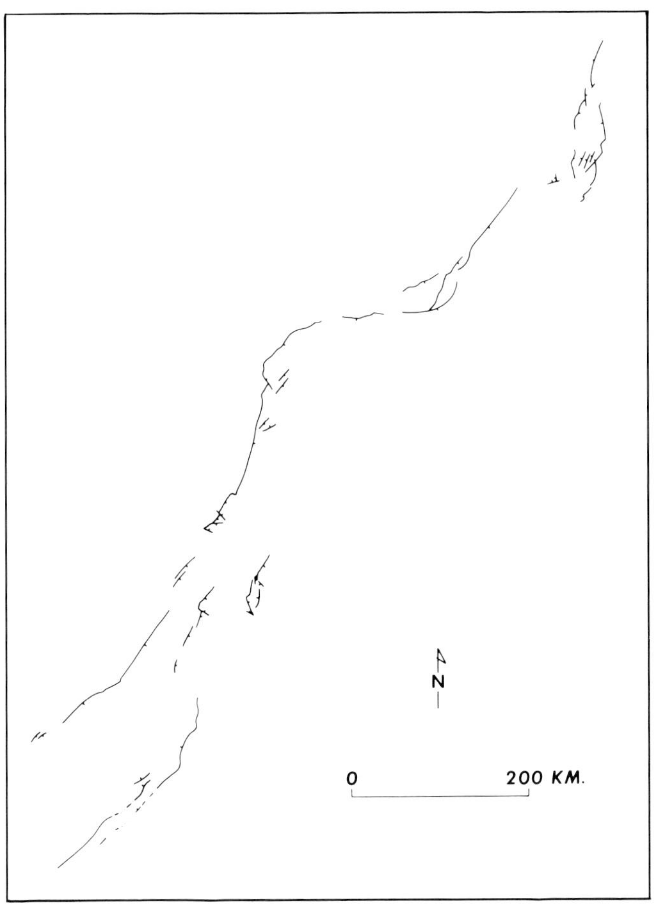

Fig. 4-20—Triassic graben system, eastern United States.

Fig. 4-21—Rhine graben fault system, Germany.

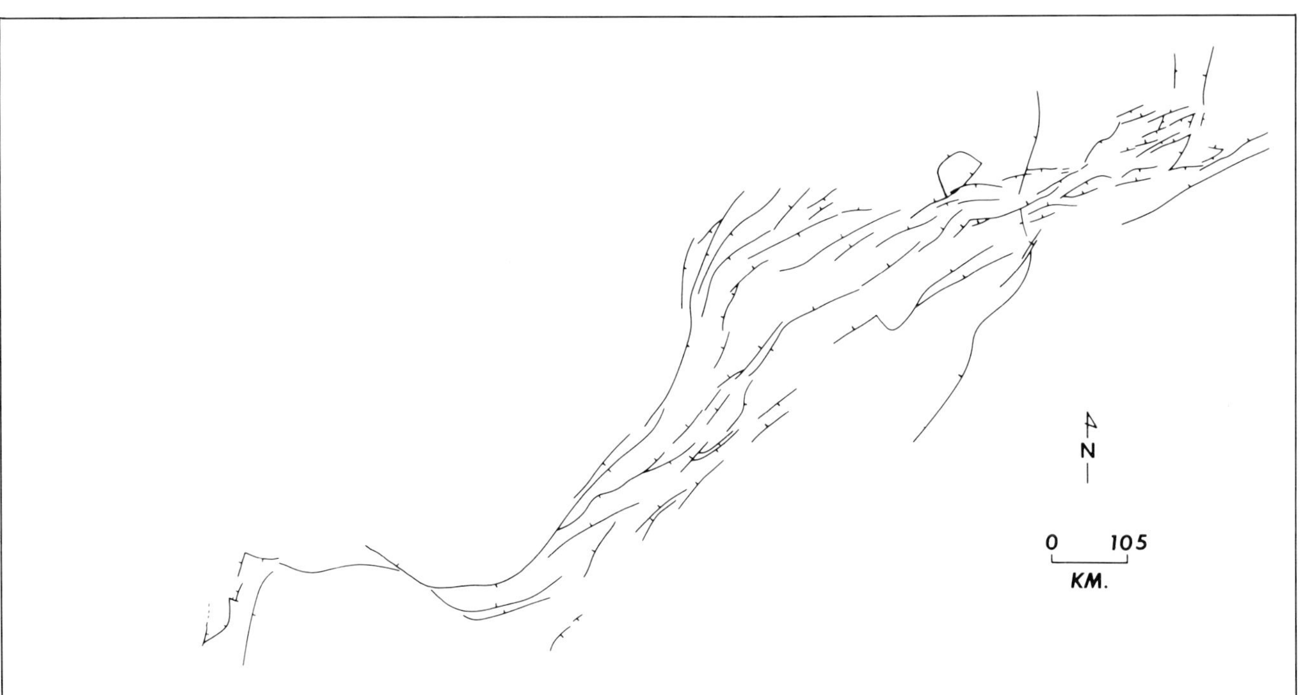

Fig. 4-22—Baikal rift system, U.S.S.R.

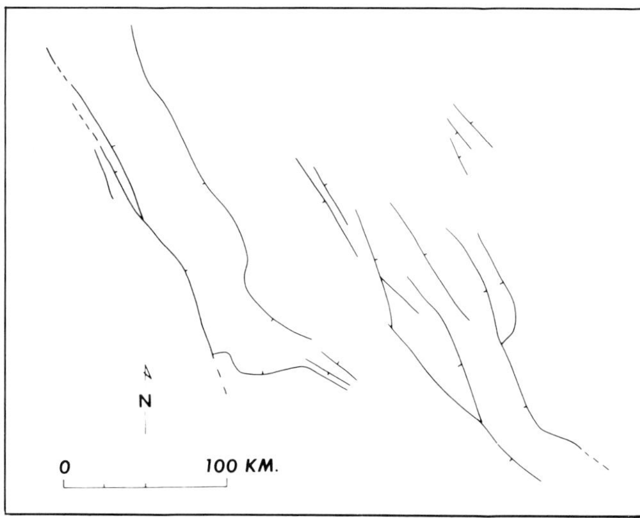

Fig. 4-23—Sirte Basin fault system, Libya.

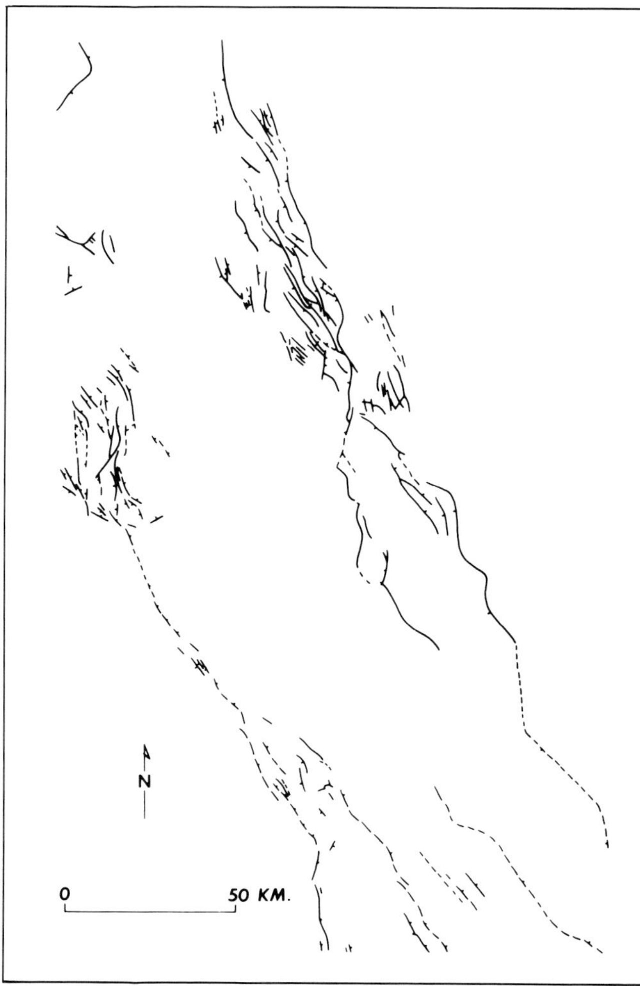

Fig. 4-24—Suez graben fault system, Egypt.

graben systems. To be a universal explanation, however, each area of graben development must have the proper coincidental pre-existing joint pattern. This, of course, is far from proven and, thus, should at least lead us to investigate other possible interpretations.

4— Although not proposed originally as an explanation for the general pattern of rifting (Fig. 4-28), Dewey and Burke's (1974) concept of coalescing triple junctions does explain doglegs on the large scale of continental margins (Fig. 4-33). For this mechanism to preserve a systematic dog-leg pattern on strike, one arm of the triple junction would always have to be either missing, very poorly developed, or not recognized as such. Though this could explain the dog-leg bends on some continental margins and aulacogens, it seems rather contrived and can hardly be applied to the worldwide and smaller scale occurrence of doglegs.

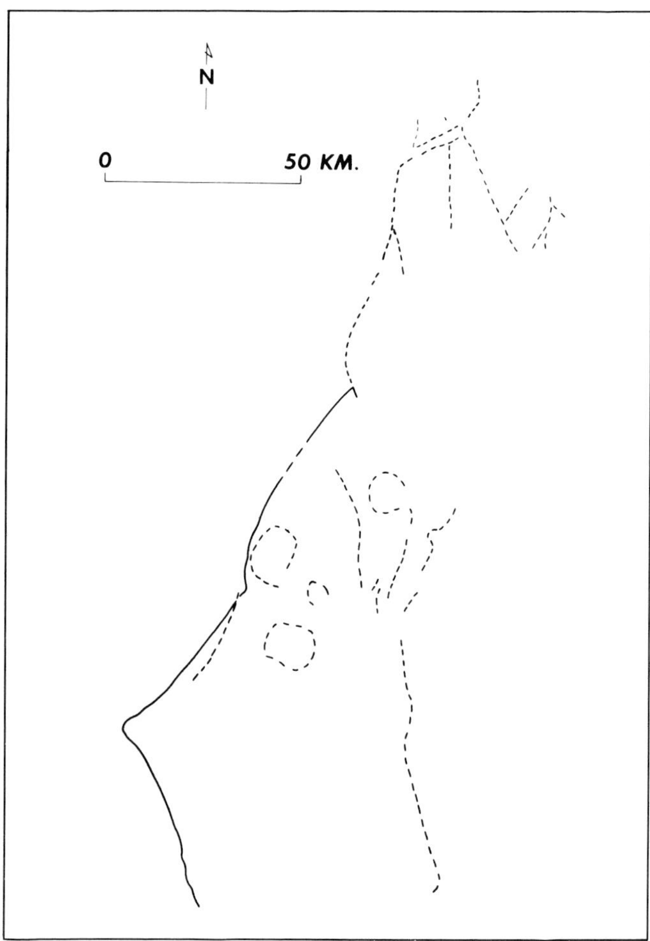

Fig. 4-25—Oslo graben fault system, Norway.

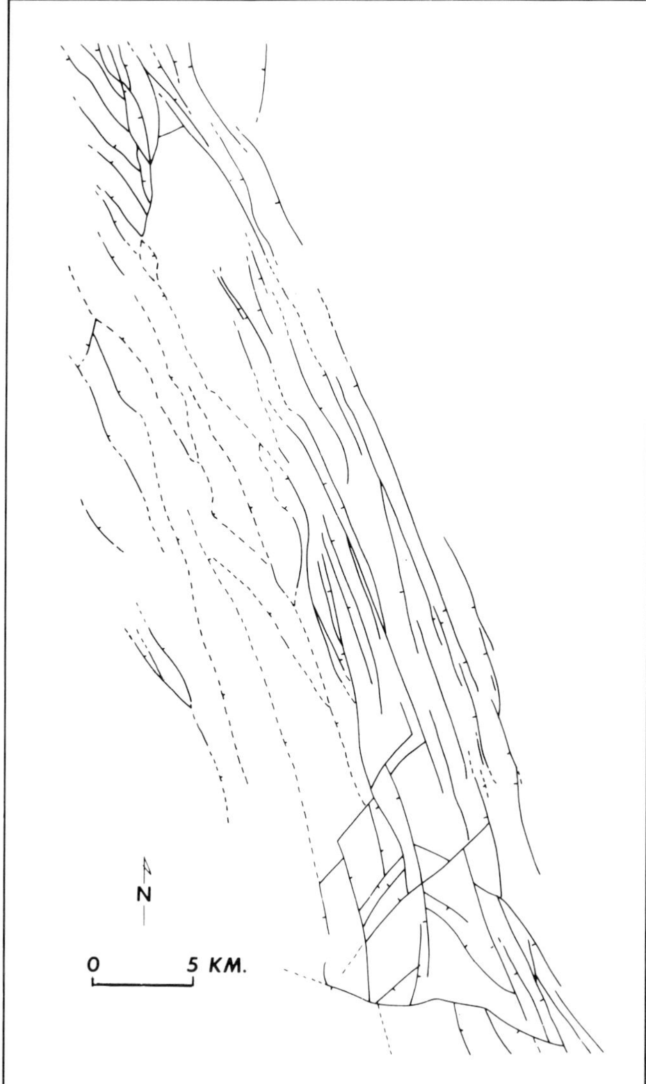

Fig. 4-26—Salt Flat graben fault system, west Texas.

Simultaneous Deformation

A single phase of deformation that explains the dog-leg, relay, transverse, and normal-fault aspects of rifting is attractive, since it obviates the need for pre-existing structure in each area. Geologic evidence reveals that many graben systems and pull-apart continental margins have had a history of arching and lateral extension. Further, a salt dome in West Germany reveals an enlightening plan and profile pattern as a result of arching and lateral extension (Figs. 4-34, 4-35). Experimentally, normal faults exhibiting both dog-leg and relay patterns can be achieved with precisely these movements.

Figure 4-36 shows a range of normal fault relay and dog-leg patterns that can be produced in clay model ex-

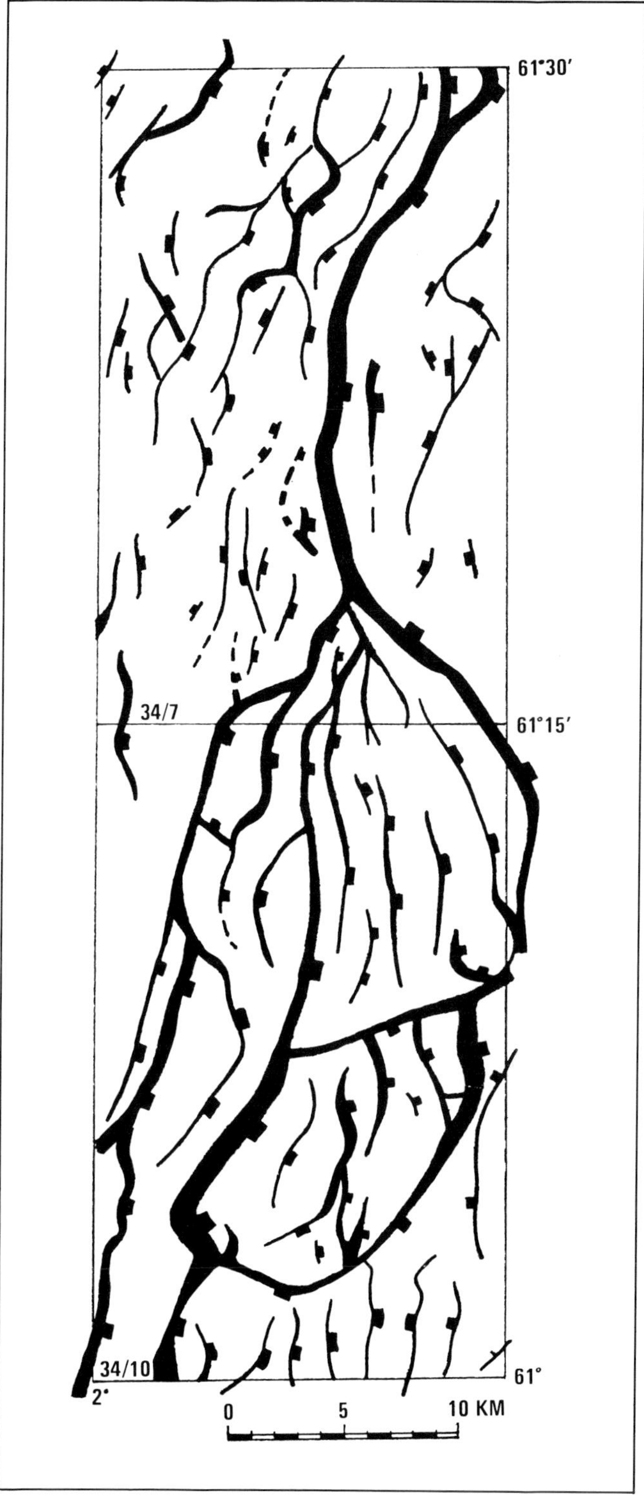

Fig. 4-27 (Halstead, 1975)—Fault pattern in the Norwegian North Sea based on 1 km (.6 mi) seismic grid. Permission to publish by Norwegian Petroleum Society.

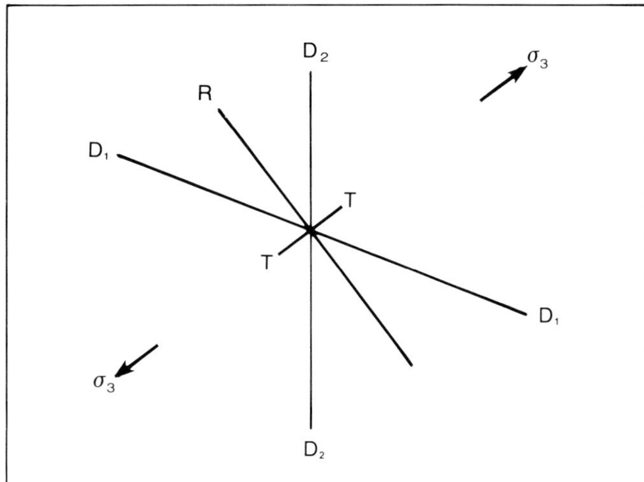

Fig. 4-28—Ideal normal fault directions from SW-NE regional extension (σ_3). R = relay perpendicular to extension, D_1 and D_2 = dogleg, and T = transverse. Faults can intersect to form a variety of acute, right, and obtuse angle trapdoor structures comparable to compressive blocks (cf. Fig. 3-15).

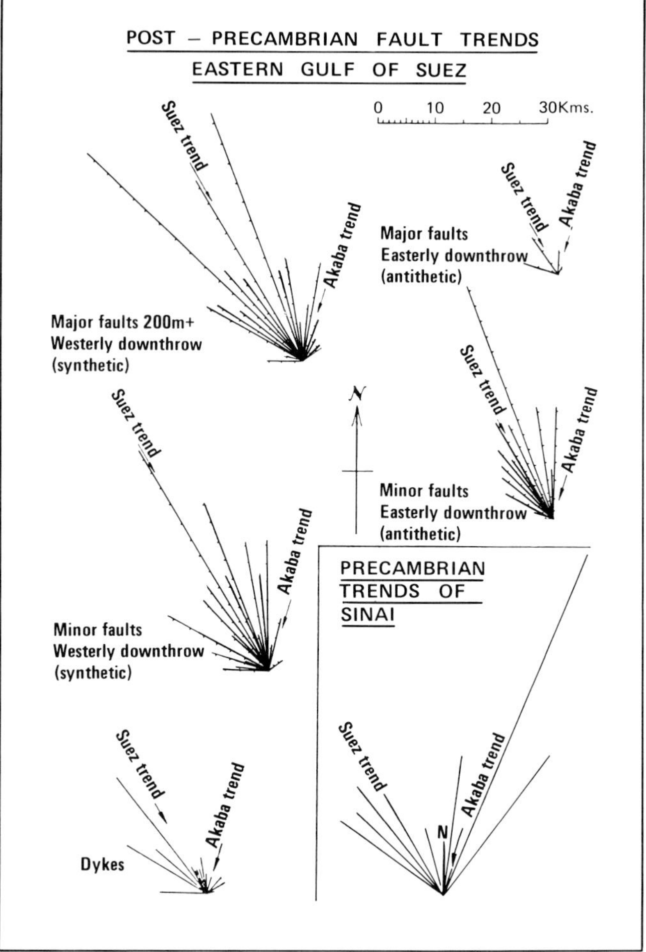

Fig. 4-29 (Robson, 1971)—Rose diagrams of fault trends in post Precambrian sediments of east Suez rift. Dyke trends also shown. Inset—Rose diagram of trends in the Precambrian of Sinai Peninsula. Permission to publish by Blackwell Scientific Publications Ltd.

periments when maximum principal compressive stress is varied between vertical (arching) and horizontal (lateral extension). The clay model is built on a flexible wire mesh which is in turn positioned above a centerjack; arching is achieved by uplift of the centerjack, and lateral extension is achieved by lengthening of the wire mesh in one direction with corresponding shortening at right angles. Fault patterns vary from pure relay, where the wire mesh remains laterally fixed but is arched under the clay (Fig. 4-36a), progressively to a combination of relay and dogleg, as lateral extension accompanies arching (Fig. 4-36b–d).

Whether the incipient lateral shortening (occurring as a consequence of lateral extension of the wire mesh) required in figure 4-36 is compatible with rifting mechanics in nature remains to be determined. A more appealing, and probably more realistic, experiment would be arching simultaneous with extension that does not require related shortening. Under these conditions maximum principal compressive stress would remain vertical. Such experiments were conducted by Withjack and Schiener (1982), who achieved relay, dog-leg, and transverse trends in single experiments (Fig. 4-37). The interplay of arching and lateral extension, then, is attractive as a mechanism of graben development, since it accomplishes the deformation in essentially one step. Instances are known (Erte-Ale volcano, northern Afar triangle, Ethiopia) where a combination of regional extension and arching has created relay and dog-leg trends in pristine, previously undeformed rocks, proving definitively that the mechanism works. Pre-existing grain no doubt modifies and even controls some patterns of rifting, but clearly it is not a pre-requisite for those patterns to develop. It even can be argued that the consistent pattern of rifting reflects common deformation rather than pre-existing fracture control, and, thus, that rifting, like wrenching, has its own map-view signature.

Simultaneous arching and extension is not a universal exploration for the patterns of rifting, however, simply because it is doubtful that all rift systems have been both arched and extended at the same time. Interestingly, the single operation of extension can produce both a dog-leg pattern (Ramberg et al., 1978; Fig. 4-38) and a modified relay pattern (Fig. 4-18).

In contrast, arching invarably seems to lead to a relay pattern (Figs. 4-4A, 4-17, 4-36a). Perhaps rifts with a well developed relay pattern had an initial history of arching, whereas rifts with a more dominant dog-leg pattern originated as extensional sag features.

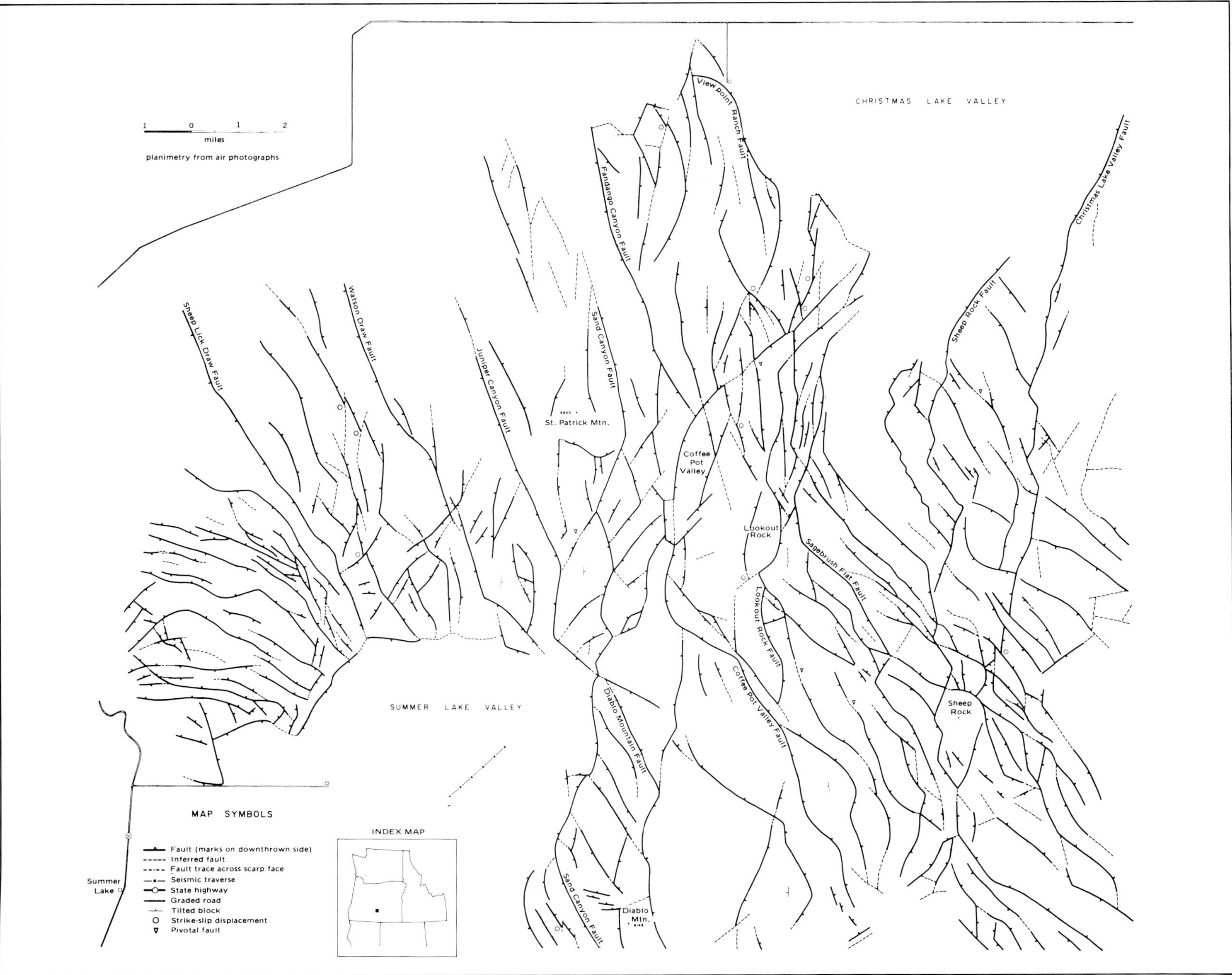

Fig. 4-30 (Donath, 1962)—Structural geologic map showing dog-leg pattern in south-central Oregon. From Geological Society of America Bulletin, V. 73. Used with permission.

LEAST
PRINCIPAL
COMPRESSIVE
STRESS
MAXIMUM PRINCIPAL
COMPRESSIVE STRESS
REDISTRIBUTION
OF SURFACE
FORCES

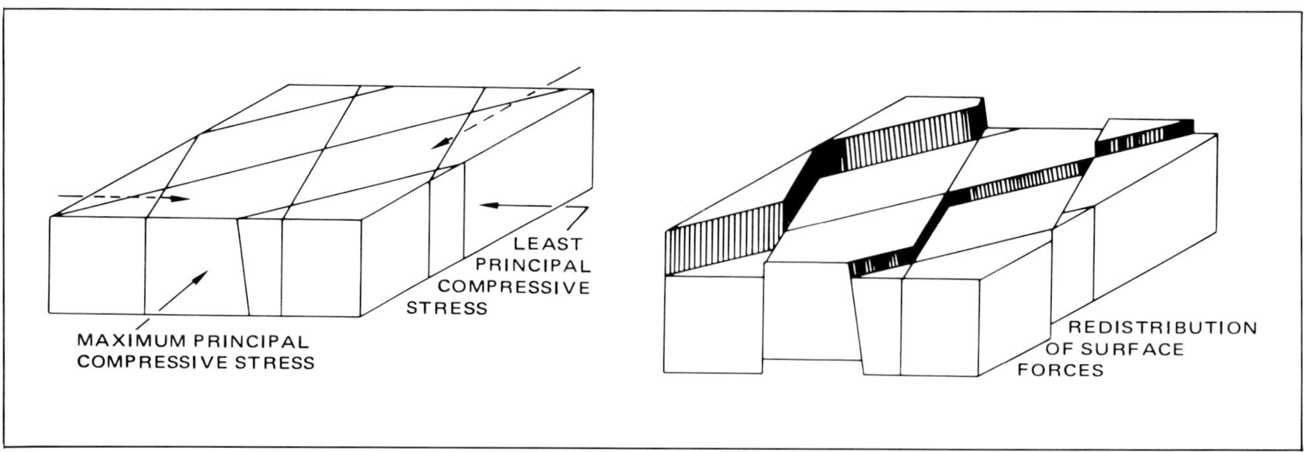

Fig. 4-31 (Donath, 1962)—Development of vertical dip-slip faults from conjugate strike-slip shears. From Geological Society of America Bulletin, V. 73. Used with permission.

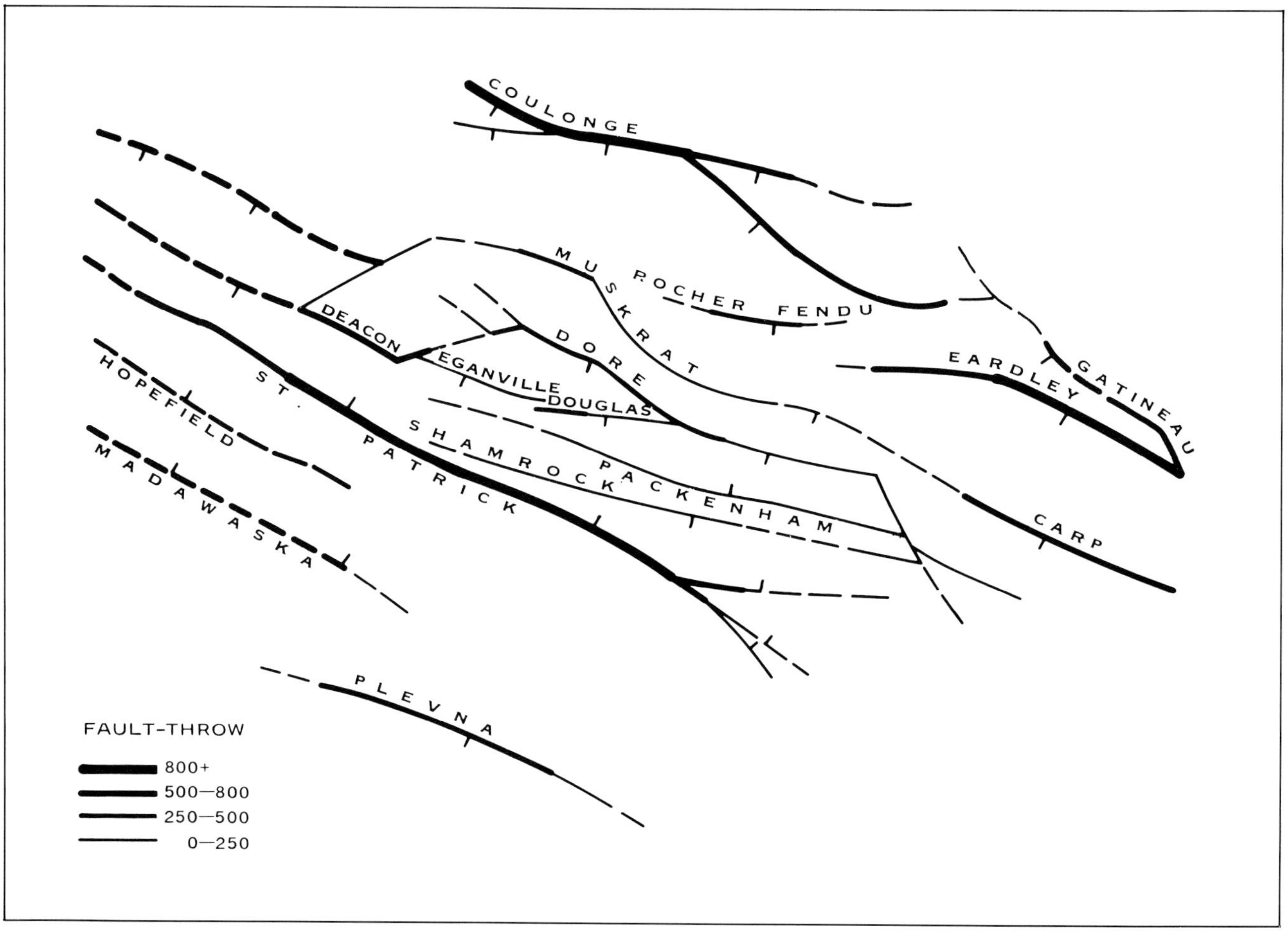

COULONGE
MUSKRAT
ROCHER FENDU
DEACON
DORE
EGANVILLE
DOUGLAS
EARDLEY
GATINEAU
HOPEFIELD
ST. PATRICK
SHAMROCK
PACKENHAM
CARP
MADAWASKA
PLEVNA
FAULT-THROW
800+
500—800
250—500
0—250

Fig. 4-32 (Kay, 1942)—Map of Ottawa-Bonnechere graben with dog-leg fault pattern interpreted as having been caused by reactivation of a pre-existing joint system. From Geological Society of America Bulletin, V. 53. Used with permission.

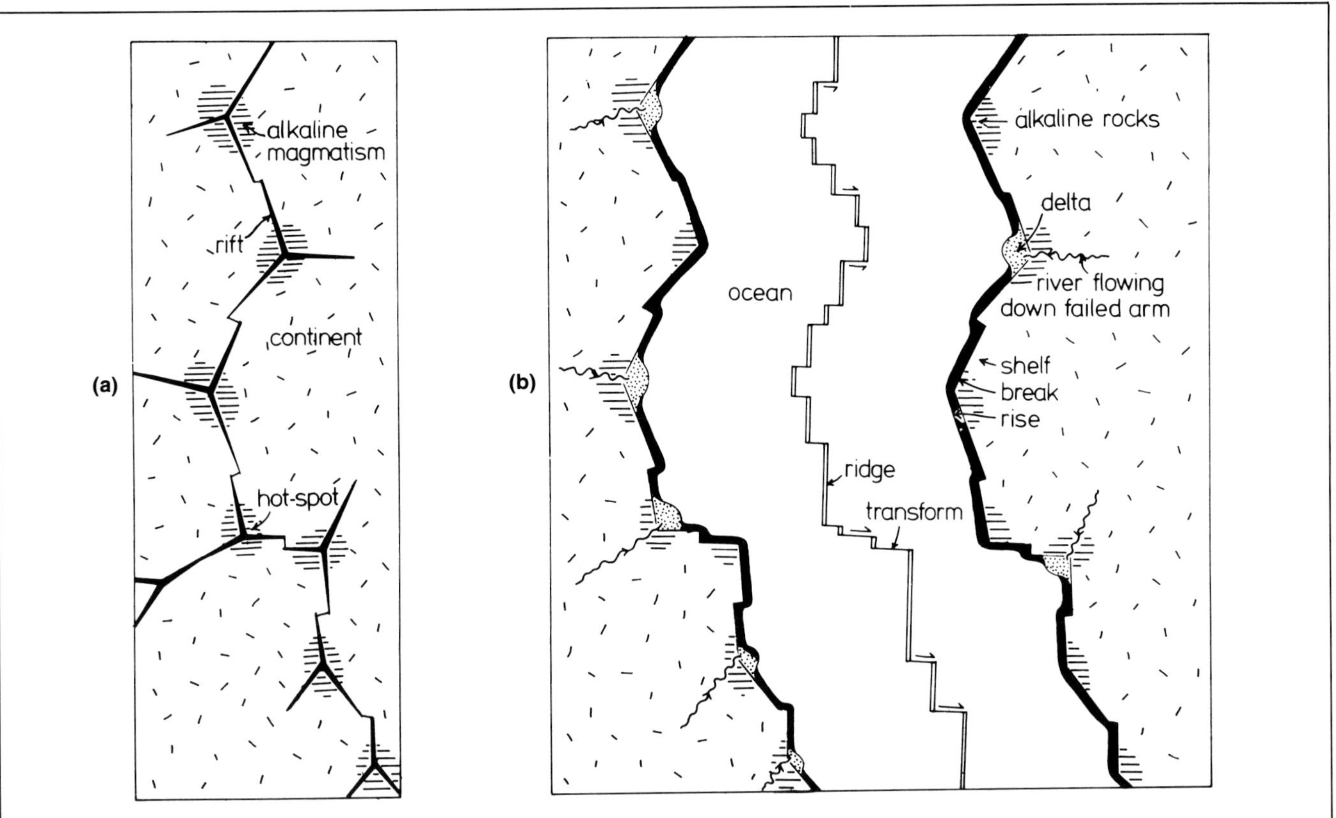

Fig. 4-33 (Dewey and Burke, 1974)—A. Continental crust ruptures along rifts that meet at rrr (rift-rift-rift) junctions over hot spots characterized by alkaline magmatism. Major irregularities in continental margins mark sites of former rrr triple junctions; minor irregularities mark misfit connections between propagating rifts. B. Ocean in advanced Atlantic stage. Rivers flow down failed arms to feed deltas at re-entrants on continental margins. Shapes of continental margins retain pattern of continental rupture. Reprinted by permission.

Aulacogens

Aulacogens are long-lived deeply subsided transverse troughs that extend from continental margins far into adjacent platforms. Hoffman (1972) postulated that the abandoned arms of ridge-ridge-ridge (RRR) triple junctions could explain the origins of aulacogens (Fig. 4-39). An aulacogen could also form as the abandoned arm of a ridge-ridge-fracture (RRF) triple junction. In any case the two active arms of the triple junction may become a continental margin geosyncline (Fig. 4-40). Since the two arms form a dog-leg, aulacogens are normally located where the continental margin or geosyncline makes a re-entrant angle into the platform. Inasmuch as the continental margin and the aulacogen originate at the same time, the oldest sediments of the aulacogen, especially near its mouth, are much like those of the contineta1 margin. Through time, however, as the continental margin evolves to a geosyncline or unconfined depositional site, the aulacogen remains relatively narrow (100–200 km, 62–124 mi, across), and sedimentation accordingly becomes different. Sediment thickness in aulacogens is on the order of several kilometers, diminishing somewhat platformward where the aulacogen itself eventually dies out. Aulacogens are thus elongate fault troughs in a cratonic setting with substantial sedimentary fill and can be highly prospective basins for oil and gas.

Aulacogens were first described from the Russian platform by Shatzky (1955) and further discussed by Salop and Scheinmann (1969), who gave the Donetz depression as a typical example. Hoffman (1972) described the Athapuscow aulacogen in 2-billion-year-old rocks of the northwest Canadian shield in the region of Great Slave Lake and offered as other examples of aulacogens the Anakarko Basin in Oklahoma, the Afar trough in Ethiopia, and the Benue trough in Nigeria (Fig. 4-41). The Richardson trough straddling the northernmost part of the boundary between the Yukon and Northwest Territories in Canada appears to me to be another possible example of an aulacogen.

During the orogenic stage of the adjacent geosyncline Hoffman described the Athapuscow aulacogen as having been mildly compressed prior to a final stage involving transcurrent faulting. He showed the Athapuscow aulaco-

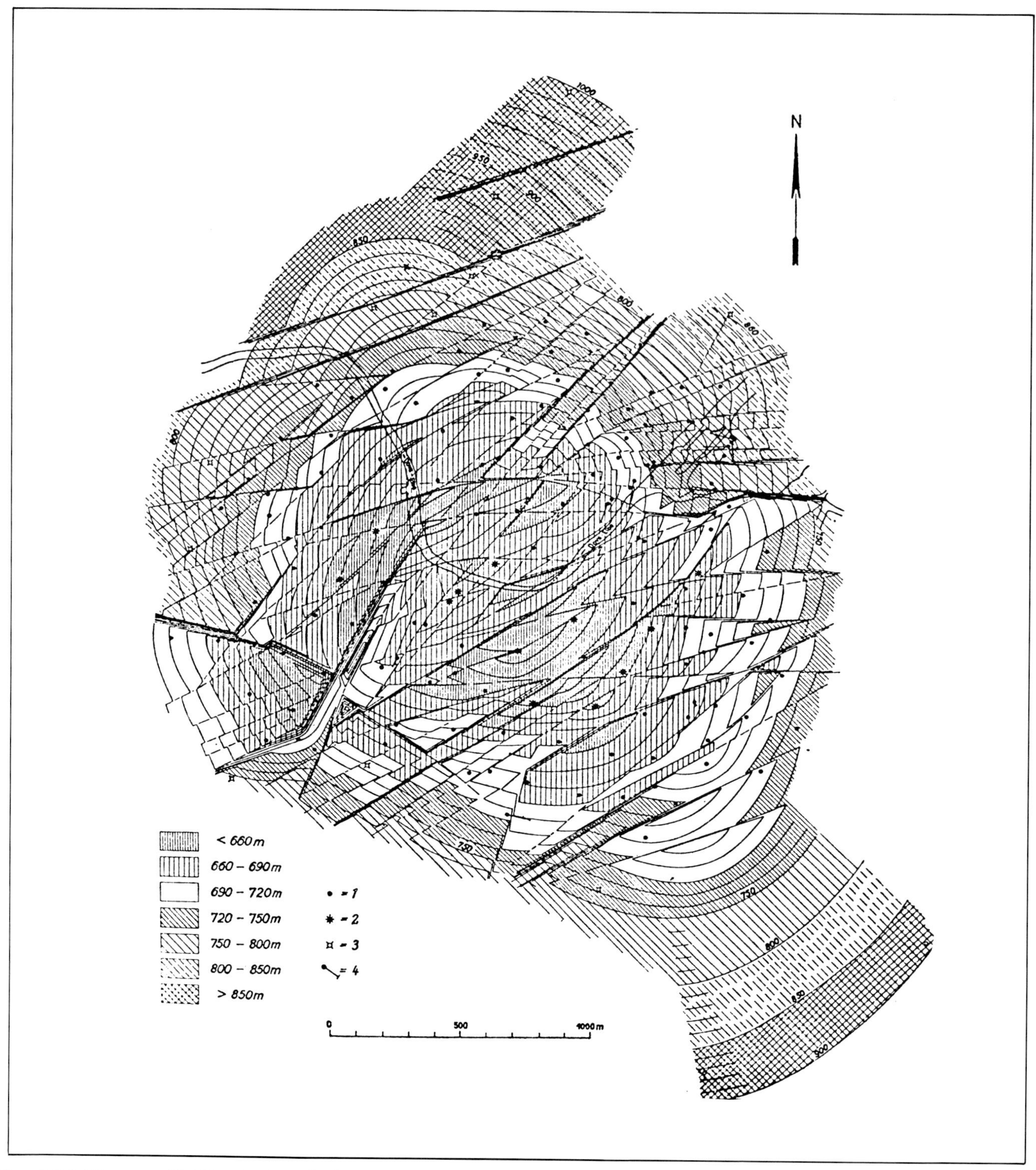

Fig. 4-34 (Gussow, 1968)—Reitbrook dome, near Hamburg, Germany. A nearly circular dome should show radial fracture pattern in response to uplift from salt below. However, radial trend is overwhelmed by preferential NE-SW trend suggesting strong NW-SE component of extension. Symbols denote (1) oil well, (2) gas well, (3) dry hole, and (4) directionally drilled well. Permission to publish by American Association of Petroleum Geologists.

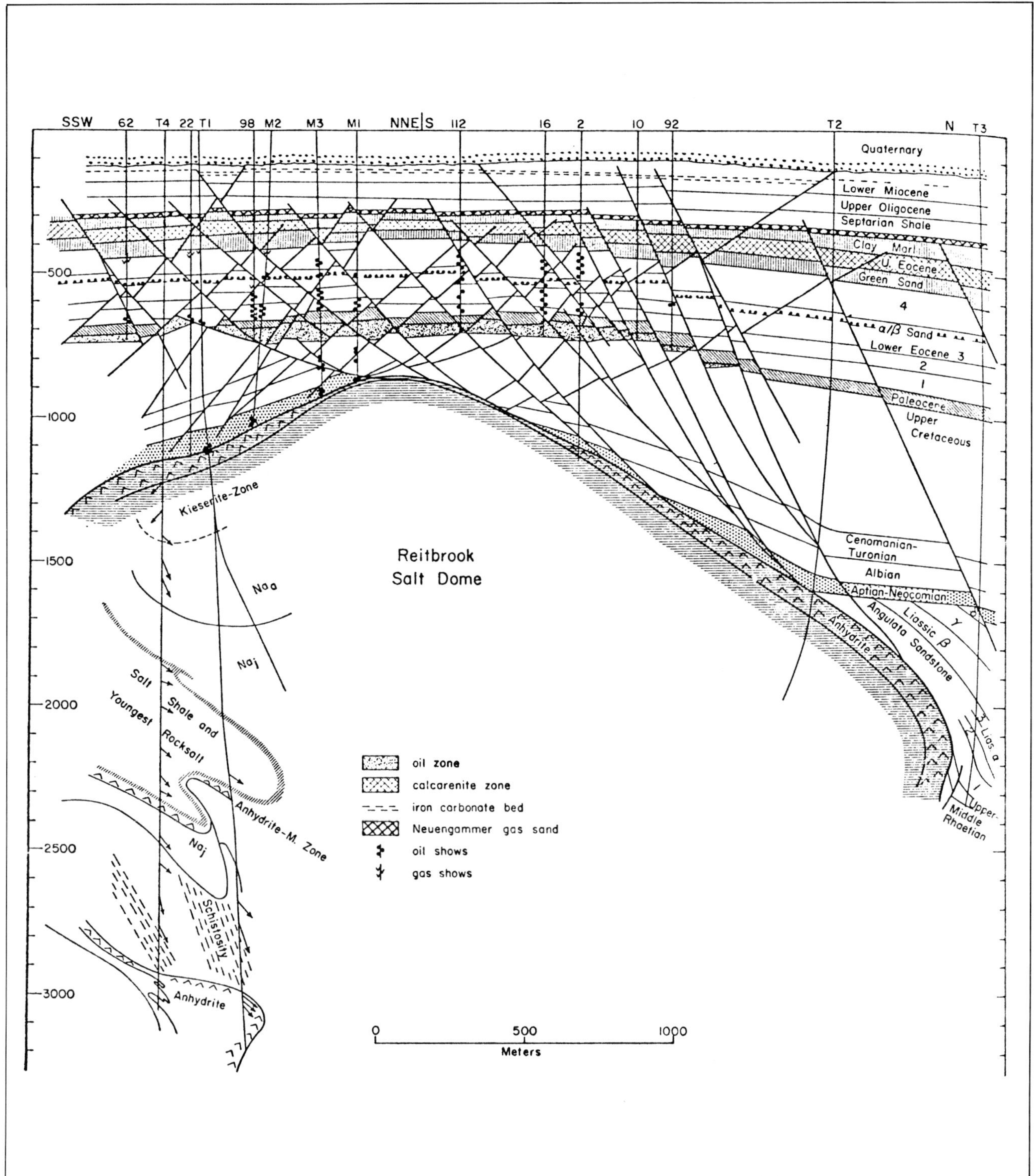

Fig. 4-35 (Gussow, 1968)—Cross section of Reitbrook oil field, showing fault pattern in strata overlying salt dome. Note that all faults are normal and, although the north-dipping set cross cuts the south-dipping set, may have been roughly concurrent. Permission to publish by American Association of Petroleum Geologists.

(a)

(b)

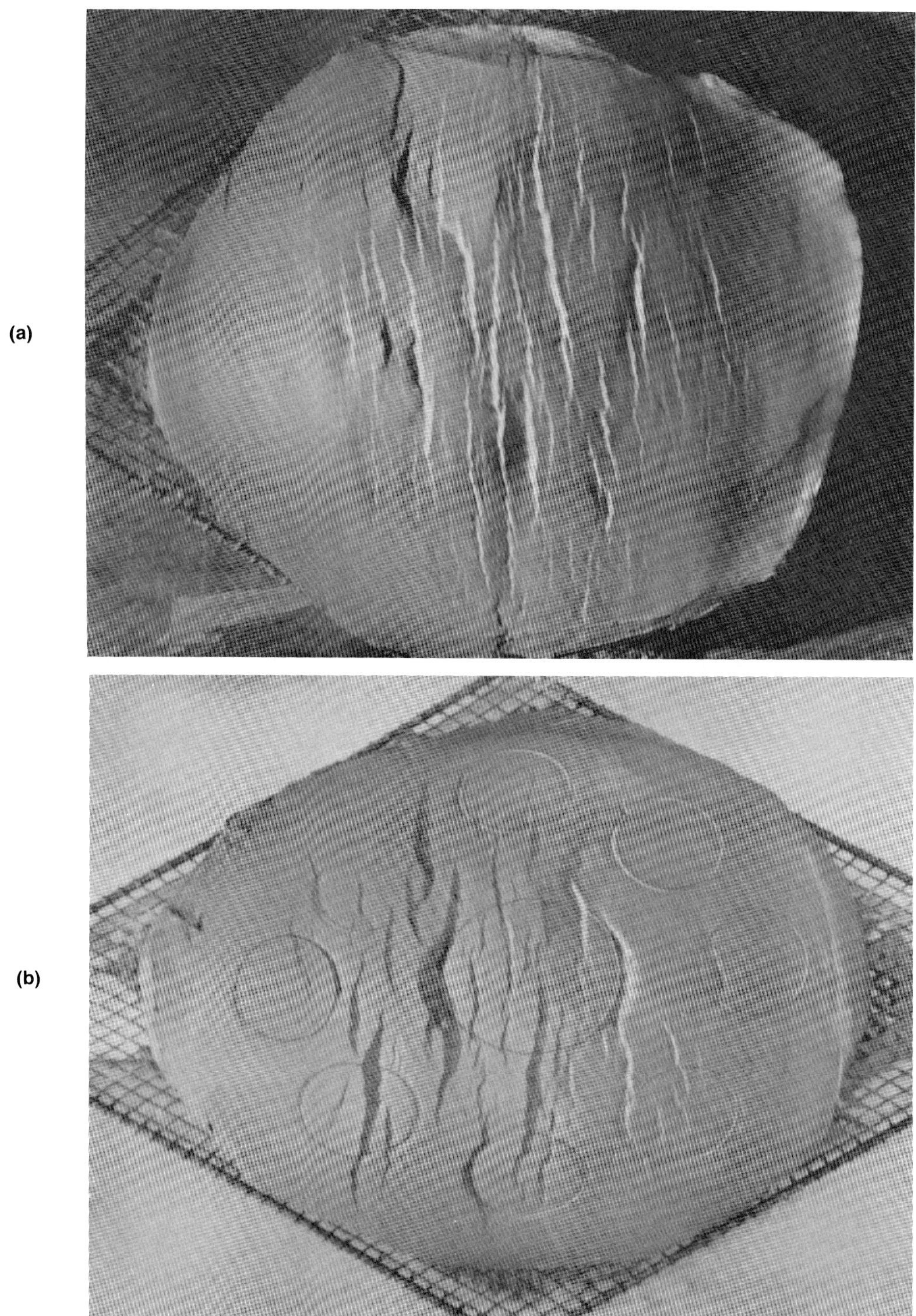

Fig. 4-36—Clay models showing relay pattern from arching (a) with progressively better developed dog-leg patterns from combination of arching and extension (b,c,d). Dominant slip on the faults is in the dip direction. Note that both normal and vertical faults are created by this mechanism but that even the vertical faults (incipient strike-slip conjugate shears) would be rotated by arching to lower dips, i.e., high-angle normal faults. Illumination is from the left.

(c)

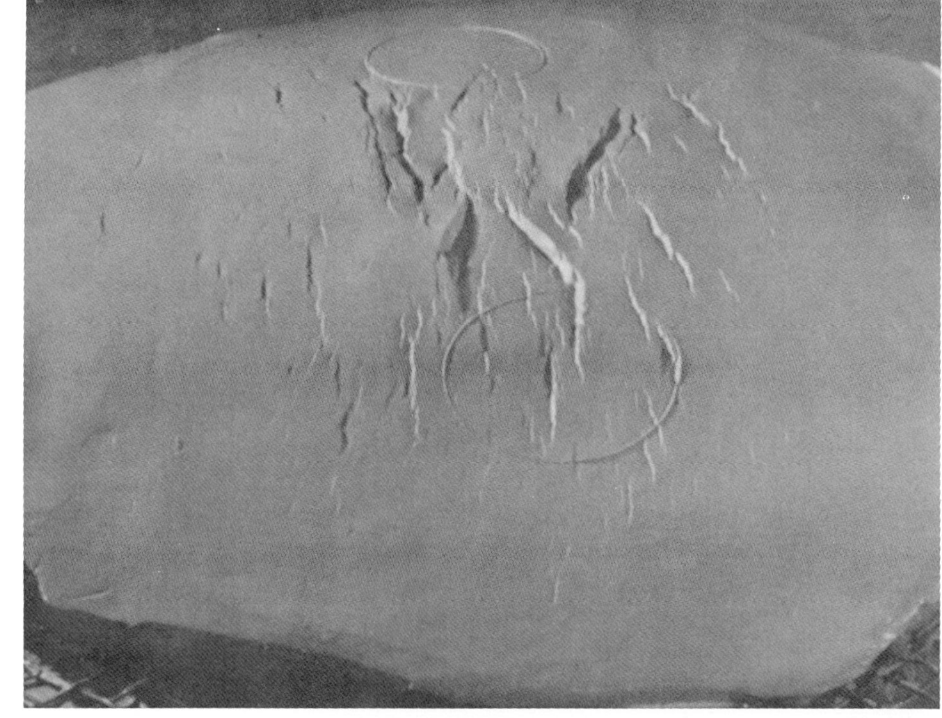

(d)

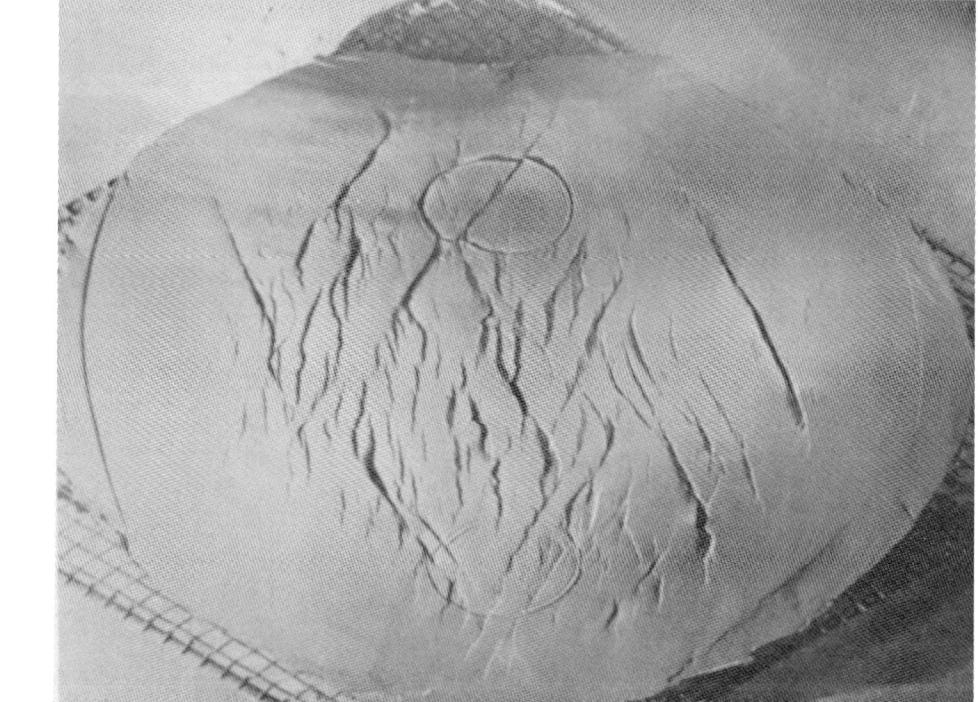

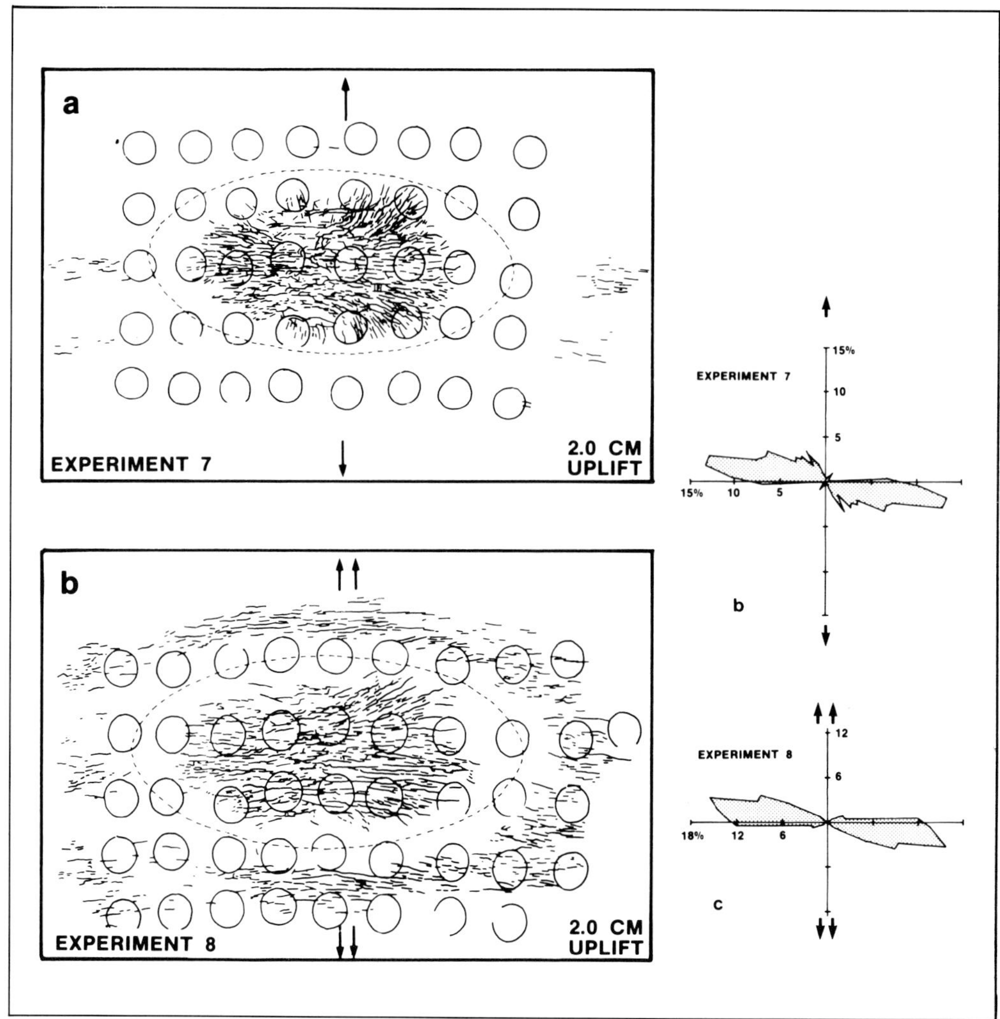

Fig. 4-37 (Withjack and Schiener, 1982)—Line drawings and rose diagrams of fault patterns produced by simultaneous arching and extension (as indicated by arrows). Experiment 7 extension rate slightly more than half arch rate. Experiment 8 extension 1.5 × arch. All faults are normal. Transverse direction is not developed, but was in some other experiments. Permission to publish by American Association of Petroleum Geologists.

gen as being upthrust-bounded, the upthrusts once having been normal faults that became curved during further subsidence and mild compression (Fig. 4-40). In all the previously cited examples (except for the Afar trough, which is still in a relatively early evolutionary stage), however, strike-slip faulting is well known. It may be that during plate convergence, subduction and associated compression occur along the continental edge, but that pre-existing transverse zones of weakness (aulacogens with thinner crust than their surroundings) are subject to differential mega-tearing or strike slip with associated upthrusts.

Extensional Prongs

Extensional prongs have some of the character of aulacogens from which they are not always distinguishable. Extensional prongs are aborted features created by the projection of pull-apart margin structuring beyond a break that eventually becomes a transform plate boundary. The Gulf of Suez can be considered a type example. It experienced marked basining and sedimentation in Miocene time, as did the Red Sea. Pull-apart aborted in the Suez region, however, and continued Red Sea opening was instead transformed farther south. The Suez rift is thus an on-trend offshoot of the Red Sea beyond a transform plate boundary through the Gulf of Aqaba and the Dead Sea.

The northeast coast of Canada demonstrates the existence of extensional prongs in a most remarkable way (Fig. 4-42). In the drift of the then continuous European and Greenland plate away from the Canadian portion of the North American plate, the southwest margin of the Grand Banks, Davis Strait, and Nares Strait can be readily iden-

Fig. 4-38 (Ramberg et al., 1978)—Dog-leg fault pattern in plaster of Paris under extension. From Geological Society of America Bulletin, V. 89. Used with permission.

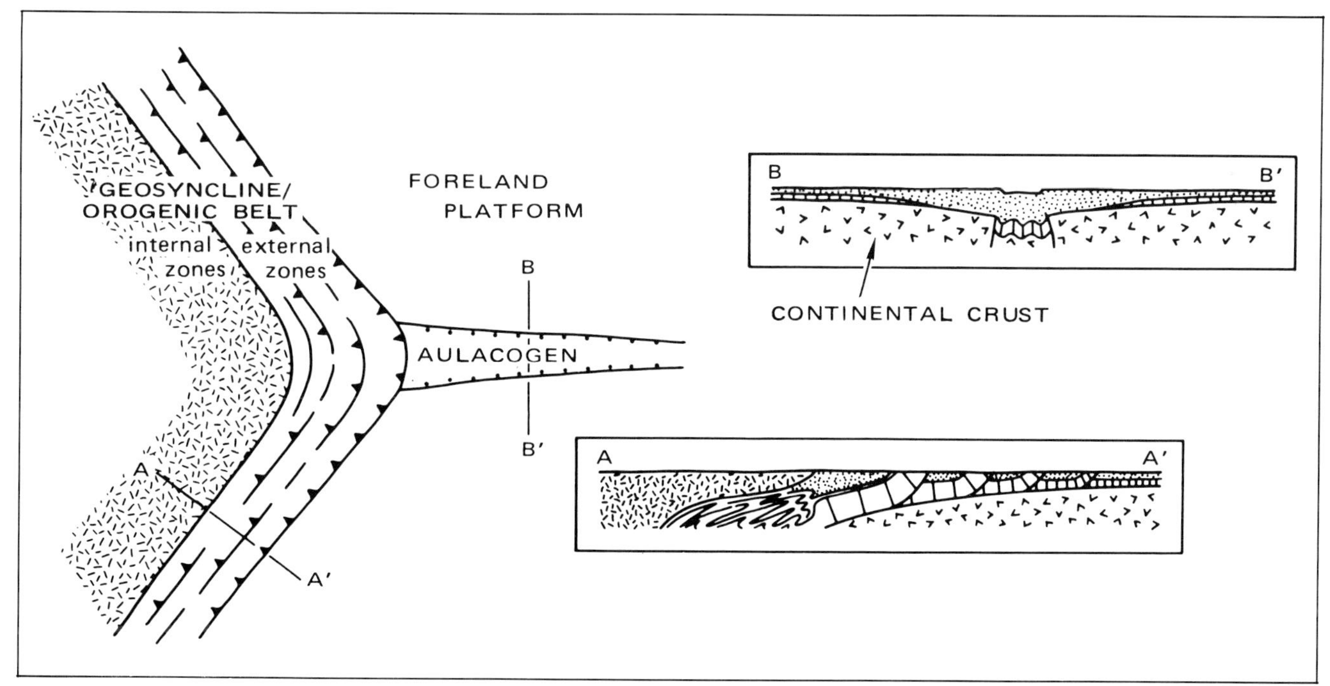

Fig. 4-39 (Hoffman, 1972)—Map and cross-sectional view (B-B') of an aulacogen. Permission to publish by The Royal Society of London.

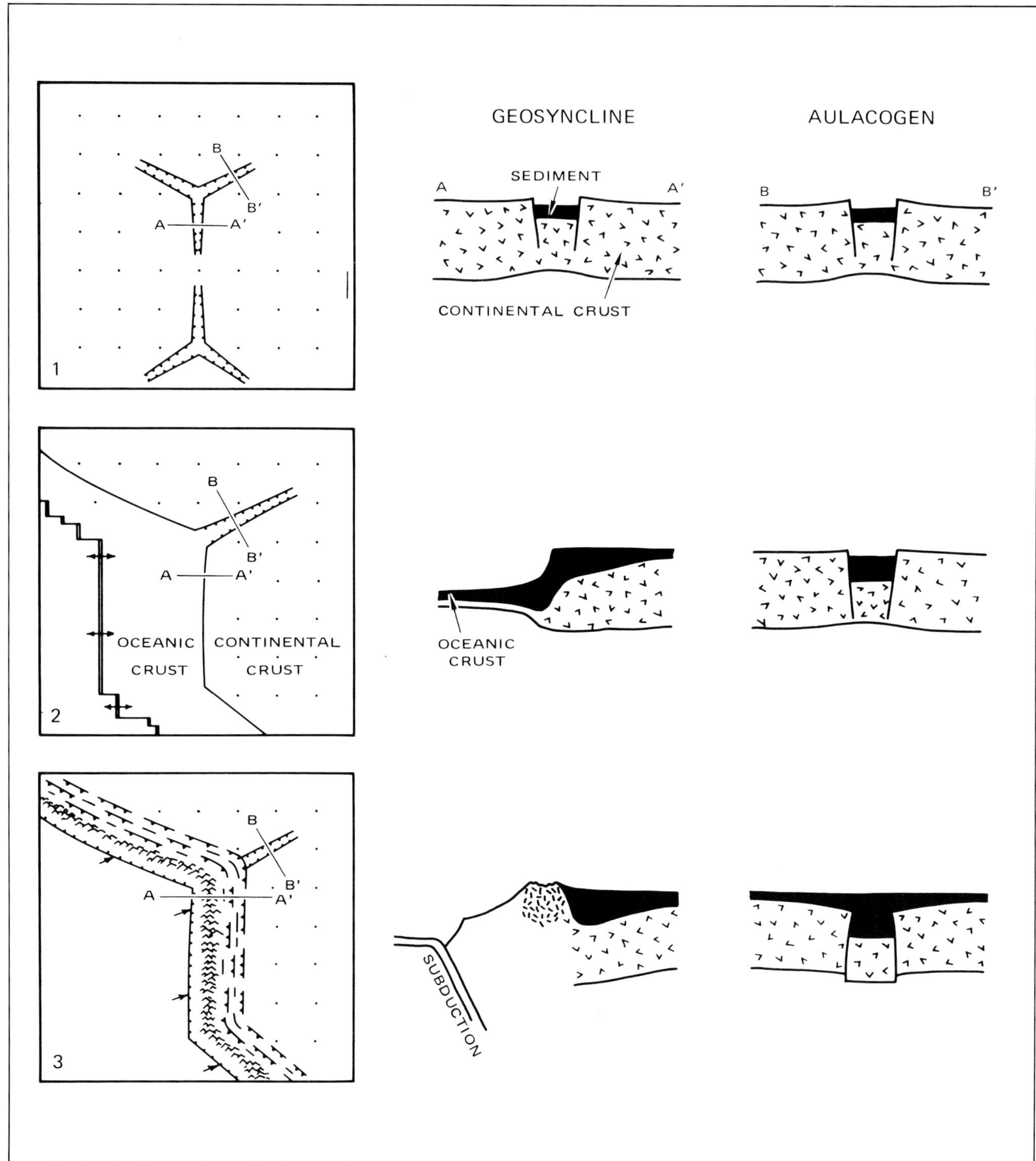

Fig. 4-40 (Hoffman, 1972)—Maps and cross sections diagrammatically showing evolution of an aulacogen and an associated geosyncline. Permission to publish by The Royal Society of London.

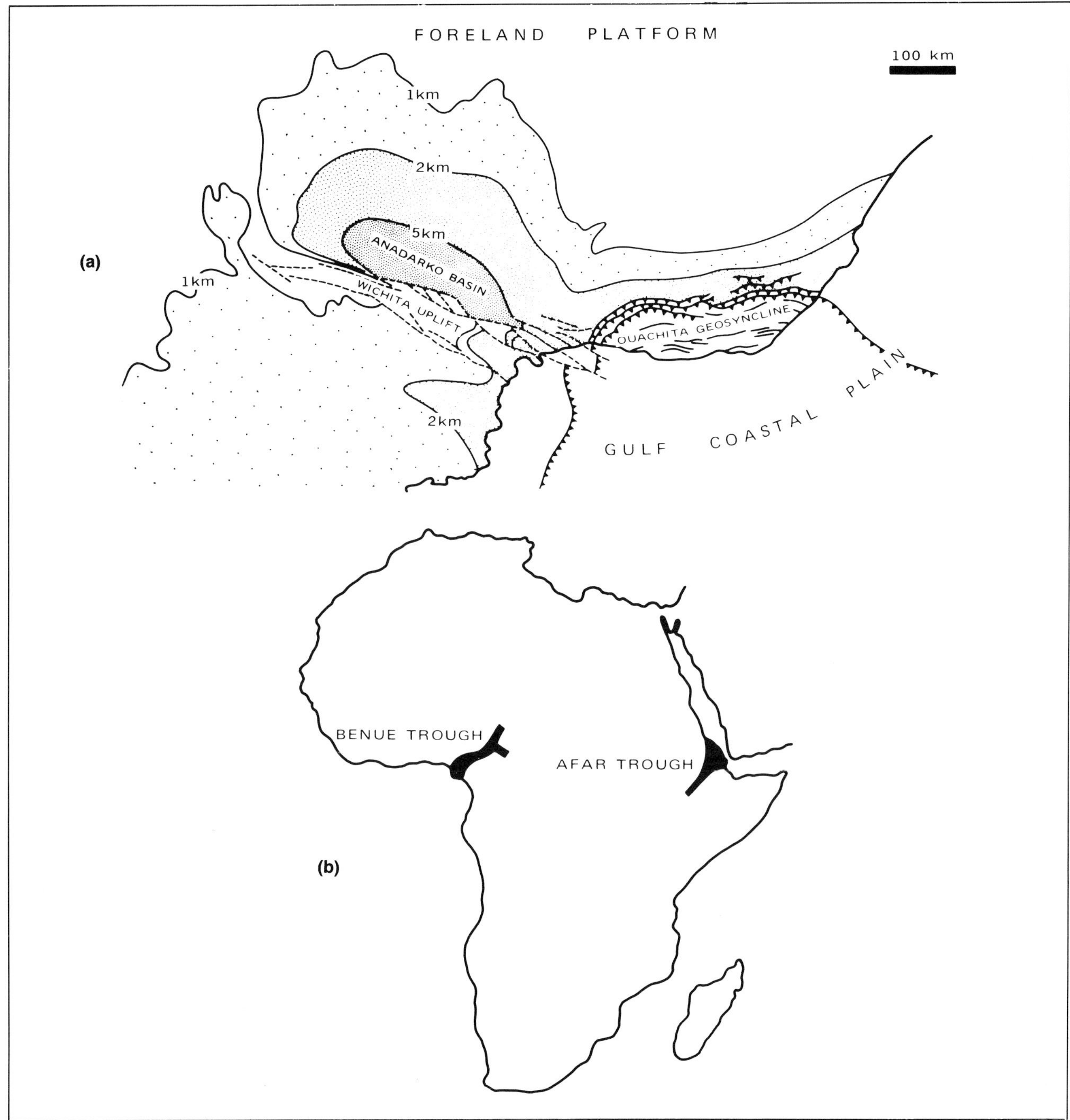

Fig. 4-41 (Hoffman, 1972)—a. Anadarko Basin aulacogen. b. Benue and Afar aulacogens. Benue trough is aborted arm of RRF triple junction. Afar trough can be viewed as a northward continuation of the East Africa rift system. Permission to publish by The Royal Society of London.

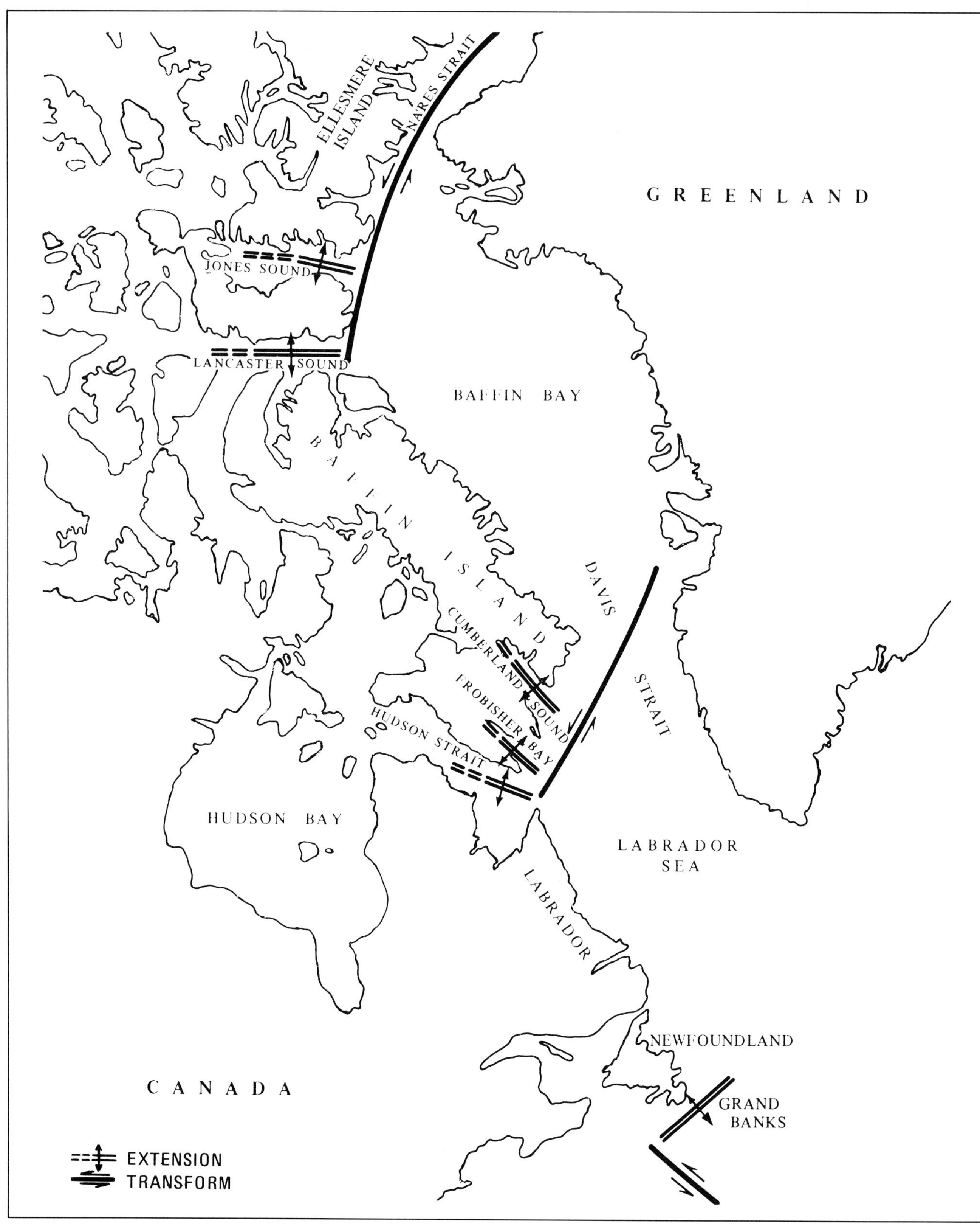

Fig. 4-42—Extensional prong fault troughs, northeast coast of Canada.

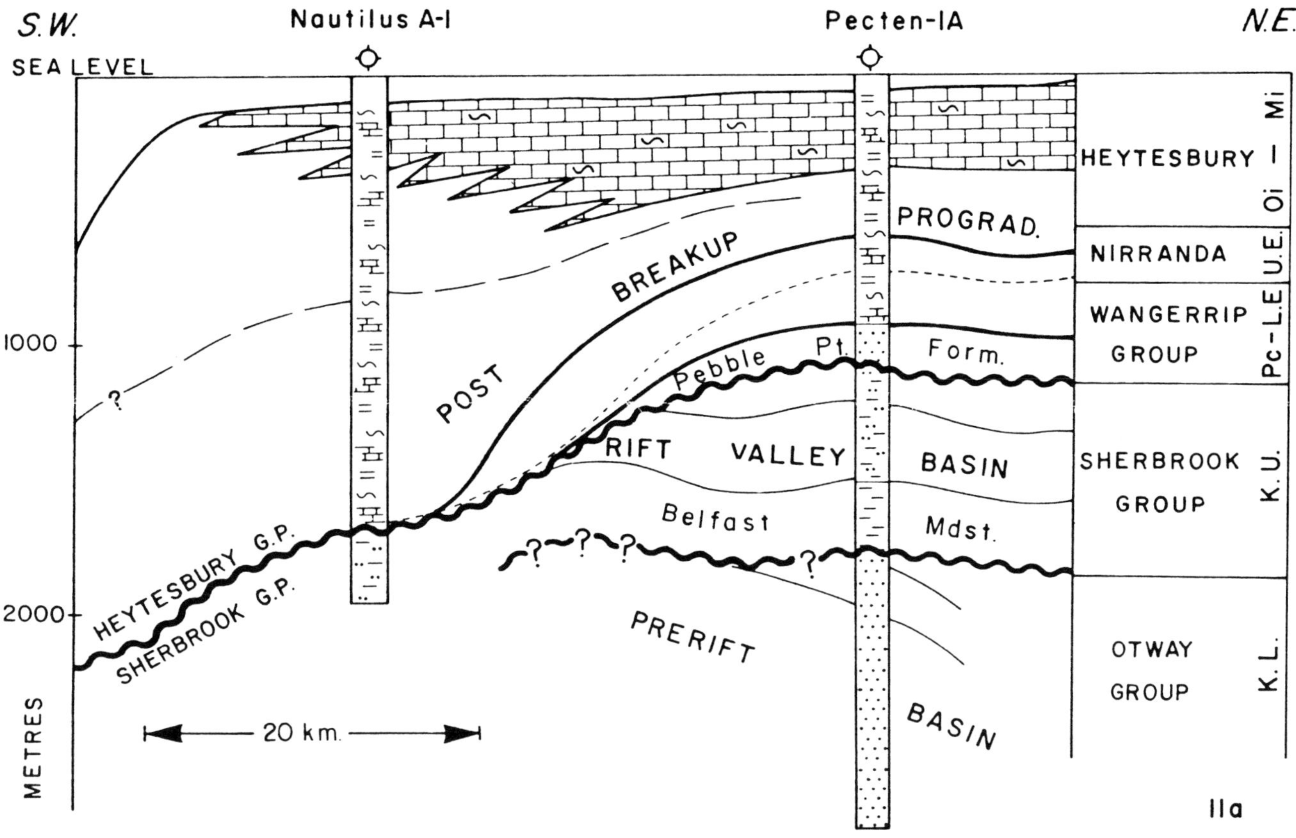

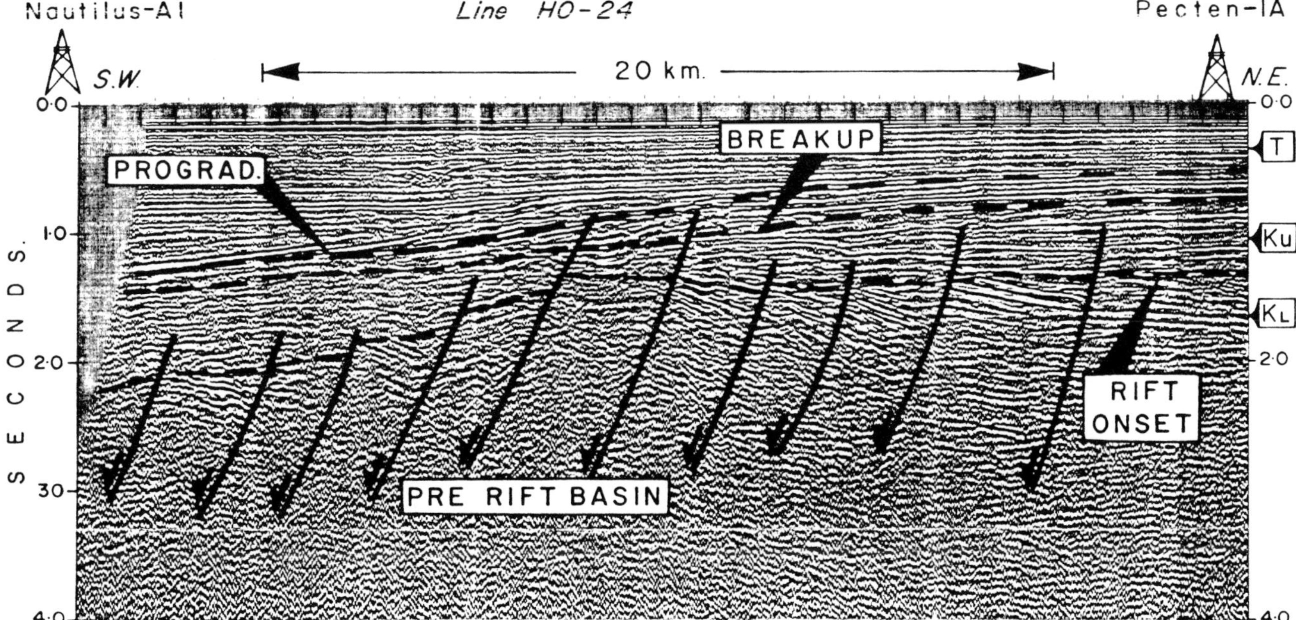

Fig. 4-43 (Falvey, 1974)—Seismic and stratigraphic sections from Otway Basin off southwestern Victoria, Australia showing prerift, rift, and postrift-progradtional structural-stratigraphic sequences. Marine conditions occurred early in rifting. Permission to publish from Australian Petroleum Exploration Association Limited.

tified as sites of transform fault movement. South of each of these transform features is a segment of pull-apart continental margin. At the time of each respective opening, however, the tendency for extension did not terminate at the transform margin but projected considerably beyond the transform boundary such that extensional structuring took place: beyond the southwest Grand Banks across the entire inner Grand Banks and into the Celtic Sea on the British continental shelf; beyond Davis Strait into Hudson Strait, Frobisher Bay, and Cumberland Sound; and beyond Nares Strait into Lancaster and Jones sounds.

Relay, dog-leg, and transverse faults could be expected in both aulacogens and extensional prongs. Both aulacogens and extensional prongs are in favorable cratonic settings for oil and gas accumulation.

SEISMIC EXPRESSION—STRUCTURAL AND STRATIGRAPHIC

Extensional blocks in rifted and subsequent contintental margin settings have a reasonably predictable structural and stratigraphic evolution. Certain events are common to these settings on a global scale, probably more so than for other structural styles.

Prerift

Prerift sedimentary sections are recorded in virtually every rift and pull-apart continental margin situation (Figs. 4-43 through 4-46). The prerift section is totally unrelated to the subsequent rift phase and can be of any lithology. For example, in the Sirte Basin of Libya the Cambro-Ordovician Gargaf quartzite directly underlies a Cretaceous rift sequence; clastics of Cambrian through Cretaceous age (including the Nubian Sandstone) and Eocene limestones are present beneath the Miocene rift fill in the Suez graben of Egypt.

The prerift section is faulted during rifting and theoretically has a better chance of preservation when the rift basin was initiated as a sag rather than as an arch, for the section is more susceptible to erosion from high fault blocks in the latter origin. The top of the prerift section is usually marked by an angular unconformity that marks the onset of rifting and is visible on seismic sections (Figs. 4-43, 4-44). Other prerift and rift sections seem indistinguishable on seismic (Fig. 4-45).

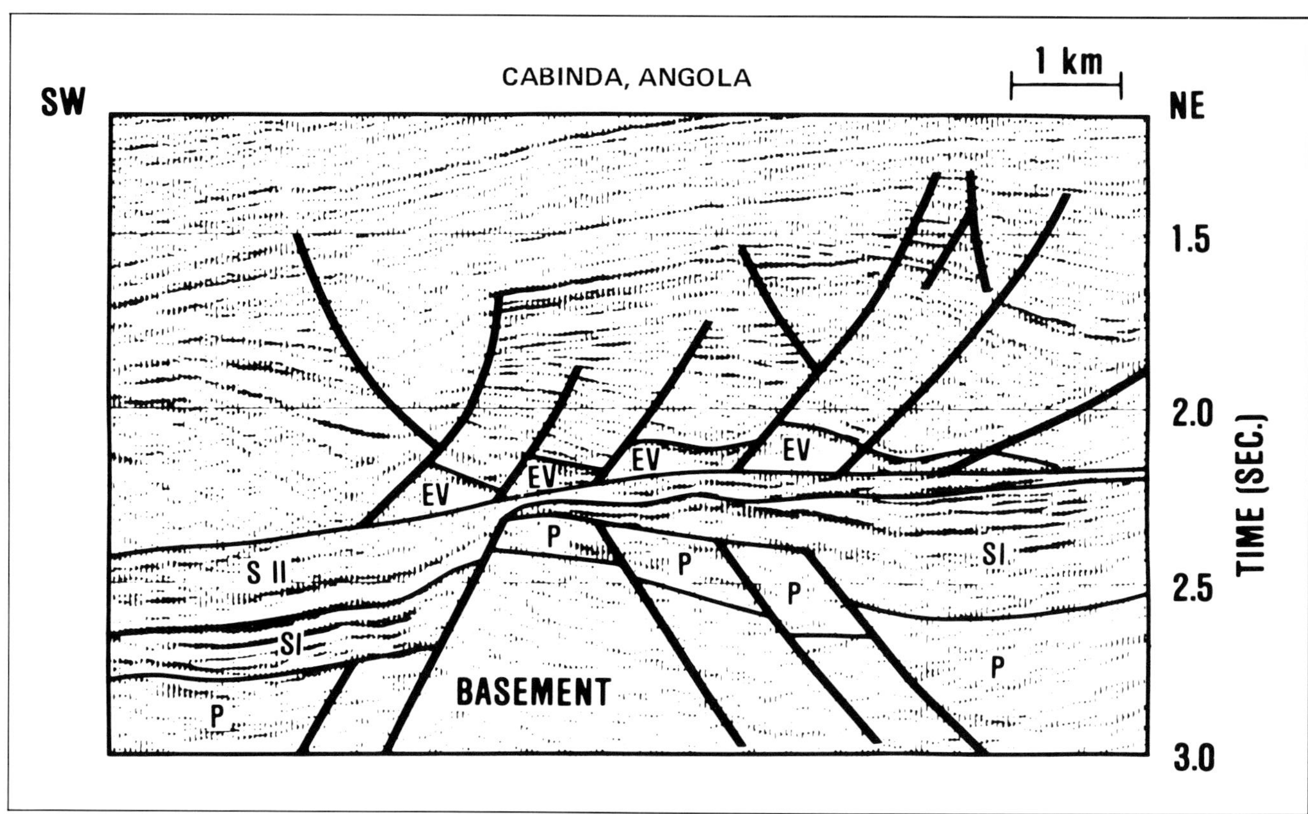

Fig. 4-44 (Brice et al., 1982)—Seismic sections from Cabinda, Angola, West Africa showing prerift (P), rift (S1 and SII), evaporites (EV), and postrift-progradational structural-stratigraphic sequences. Marine in upper part of SII. Permission to publish by American Association of Petroleum Geologists.

Rift

The most common sediments that accumulate during rifting are proximal coarse clastics, including conglomerates and red beds shed from rising fault blocks, and distal finer-grained clastics, including lacustrine lithologies, all in a continental setting. Several rifted basins have source and reservoir rocks entirely of continental origin with no marine rocks present. Examples include rifted basins in China, Brazil (Reconcavo-Netto, 1984), and Sudan (Schull, 1984).

The presence of evaporites in a rifted setting has already been discussed in this chapter in connection with the Red Sea. The prevalence of marine conditions in the rift phase depends on arch vs. sag origin, rate of rifting and associated basin depth, and proximity to salt water. As with the prerift section, rift fill is more likely to be preserved in low, down-faulted areas.

According to Barker (1981) rifts are also characterized by a predictable occurrence of organic material.

> "Rifts... in the early stages of continental breakup receive high percentages of terrestrially-derived organic matter (gas or waxy crude prone) initially, but aquatic organic matter (normal crude prone) becomes quantitatively more significant as the rift widens and marine conditions develop."

Normal faults tend to die out upward in the rift sequence (Figs. 4-43, 4-44, 4-46).

Postrift

The postrift section is marked by gradual subsidence expressed by generally parallel unbroken cycles on reflection seismic (Figs. 4-43 through 4-46). The lower part of the section is characterized by gently dipping reflections that represent the final establishment of a marine transgression. The later postrift sequence is often marked by cycles of sigmoidal shape, which reflect progradation of sediments in a seaward direction.

Possible traps for petroleum within the fundamentally unstructured postrift section include drape inherited from the underlying fault blocks related to rifting, differential compaction features, evaporite-associated structures both concordant gentle reversals and discordant diapirs, and depositional features such as turbididte buildups and reefs. Finally an entirely separate style of detached normal faults or extensional glide-plane faults can occur in a postrift sequence particularly where it is a deltaic regressive stratigraphic sequence. Where rifting has aborted, Harding (1984) has referred to the postrift section as an interior sag basin (Fig. 4-47).

PRODUCING EXAMPLES

Historically, the bulk of petroleum has been discovered on the upthrown sides of extensional blocks. More specifically most production has come from the highest part of a tilted block and almost all blocks are tilted as demanded by rotation on listric normal faults. Flat-topped, horizontal blocks are rare. Trapdoor structures are important habitats for oil and gas in the extensional block style. In syn-depositional deformation the higher parts of normal-fault blocks are often ideally situated to localize reefs, facies changes, and unconformities, which, in turn, can localize hydrocarbon accumulations.

The closure effected by faulting is the dominant closure in trapping petroleum in extensional blocks. Fold closures result primarily from differential uplift and subsidence within the confines of the extensional block and from drape of beds over the edge of the block. Fold axes characteristically parallel block edges and their associated and causative normal faults. This is in marked contrast to the wrench assemblage style where fold axes intersect associated faults at an acute angle.

Far lesser amounts of petroleum have been discovered on the downthrown sides of normal blocks. Examples include entrapment in sand lenses at Piraworth, Roggendorf, and Matzen fields in the Vienna Basin and in closures caused by reverse drag and antithetic faulting at Muehlberg also in the Vienna Basin. More recently the discovery in fractured volcanics at Trap Spring, Railroad Valley, Nevada, was made on the downthrown side of a normal block. West Heather field in the North Sea is another recent low-side discovery. More exploration is clearly needed in this setting if major reserves are to be added to the extensional block style; the upthrown sides have been rather thoroughly explored.

The most prolific fields have been found in rift systems that received extensional structuring, but stopped at varying stages short of continental breakup, e.g., Suez graben, Viking graben, Sirte Basin, Reconcavo Basin, and Grand Banks. Some of these are illustrated in the following pages.

The Suez graben is a classic area for oil trapped in prerift and rift-associated reservoirs in the higher parts of tilted extensional blocks beneath an evaporite seal (Figs. 4-48 through 4-52). A very great problem in exploration, however, has been the masking effect of the salt on seismic sections to the extent that sub-salt (South Gharib Formation, figures 4-50 through 4-52) mapping was virtually impossible. Of particular interest recently has been the discovery of oil on the downthrown side of a major normal fault in the West Abu Rudeis trend (Fig. 4-49), later named the October field. The GS173-1 well (Fig. 4-53) tested 4900 BOPD from a 100-ft-thick (30 m) oil zone in the basal Miocene Nukhul sandstones and congomerates. This unit was presumably derived from erosion of an adjacent fault block elevated during the Miocene rifting phase. The discovery has enormous implications for exploration in extensional block terrane, because potential fanglomerate reservoirs depos-

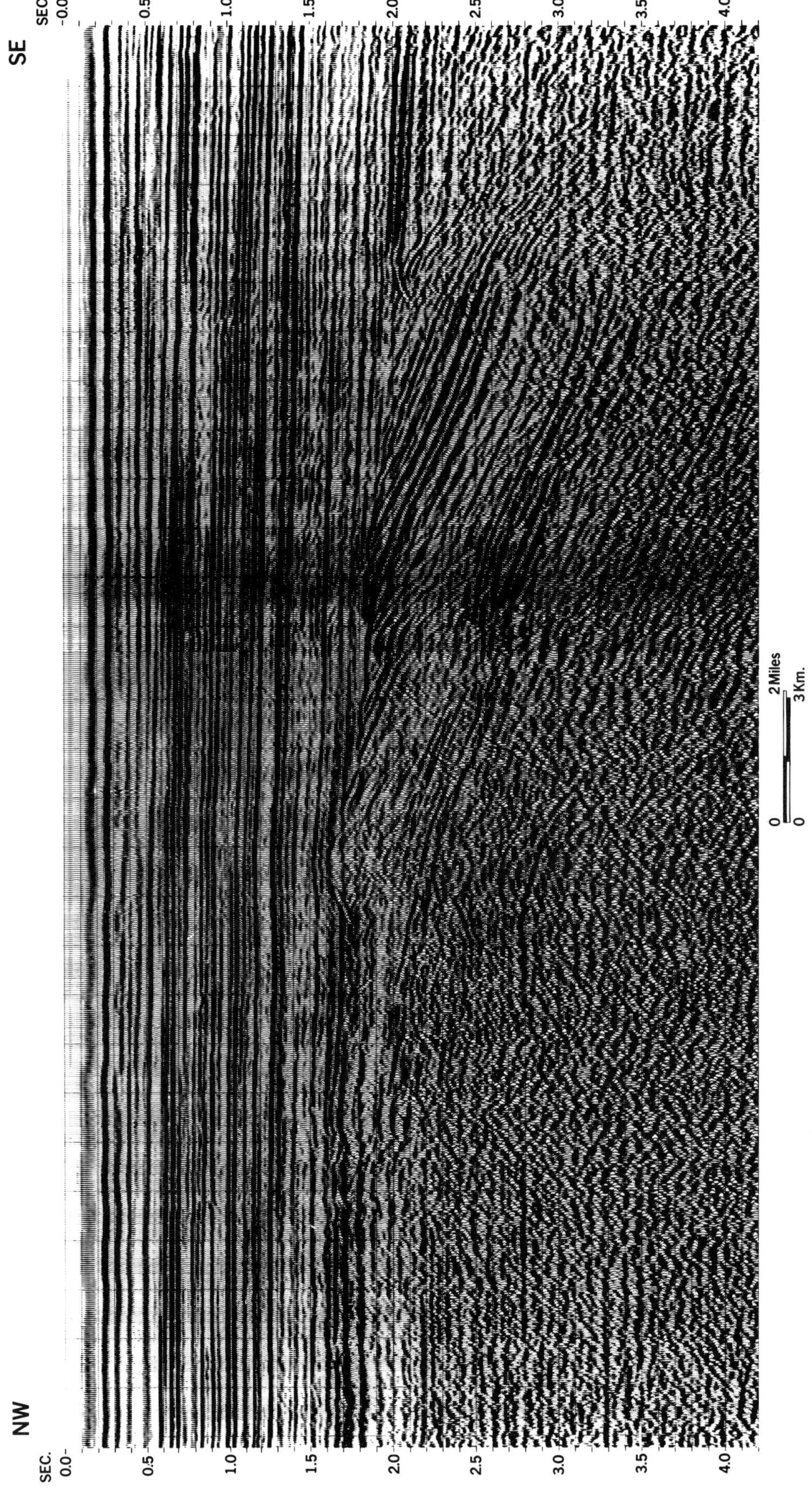
SE
NW
SEC.
0.0
0.5
1.0
1.5
2.0
2.5
3.0
3.5
4.0
0
2 Miles
0
3 Km.

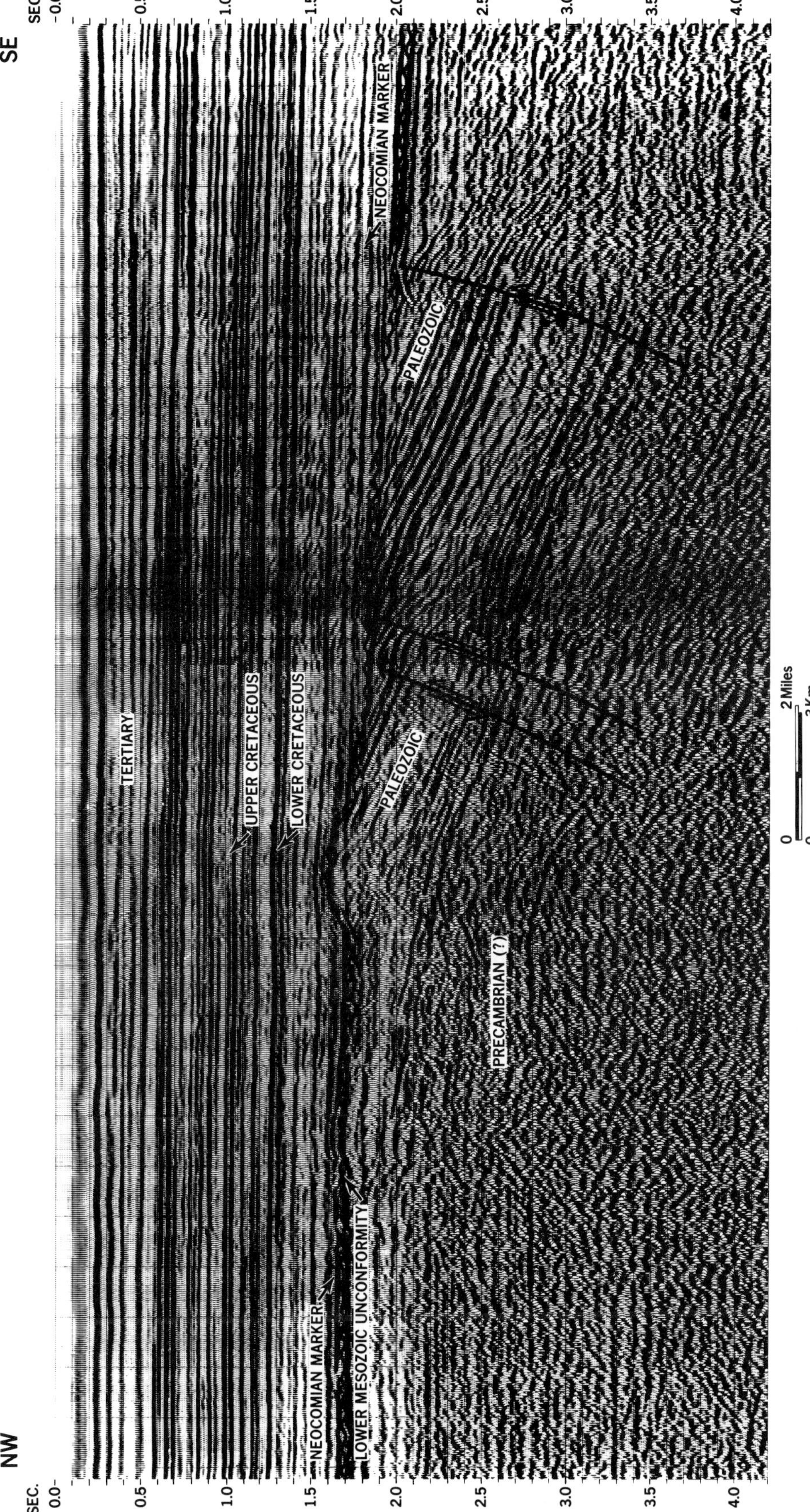

Fig. 4-45 (Crutcher, 1983)—Seismic section from southeast Georgia embayment, U.S. Atlantic offshore showing tilted, half-graben, Paleozoic and probably Triassic prerift and rift sections below early Mesozoic angular unconformity that marks end of active rifting. No intergraben unconformity separating the prerift Paleozoic from the rift Triassic is distinguishable. Permission to publish by American Association of Petroleum Geologists.

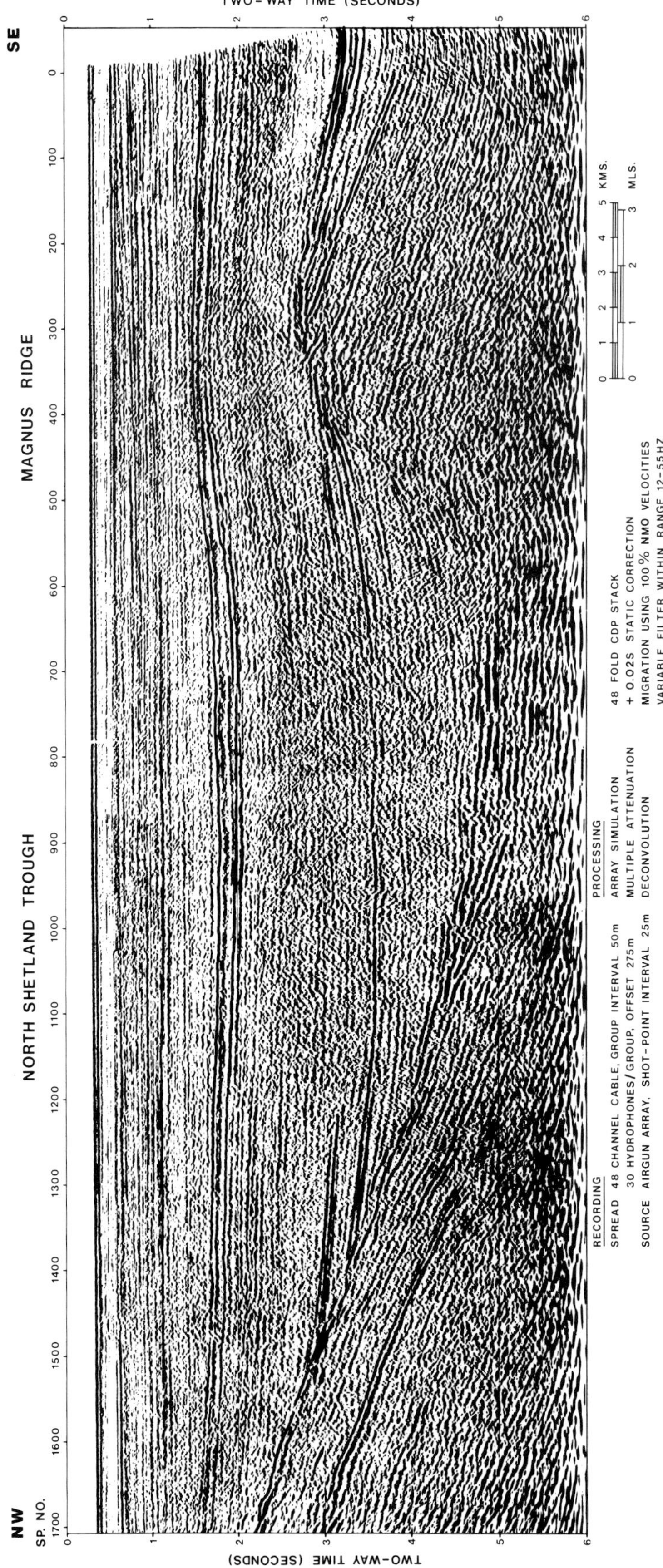
TWO-WAY TIME (SECONDS)
SE
NW
SP. NO.
MAGNUS RIDGE
NORTH SHETLAND TROUGH
KMS.
MLS.
RECORDING
SPREAD 48 CHANNEL CABLE, GROUP INTERVAL 50m
30 HYDROPHONES/GROUP, OFFSET 275m
SOURCE AIRGUN ARRAY, SHOT-POINT INTERVAL 25m
PROCESSING
ARRAY SIMULATION
MULTIPLE ATTENUATION
DECONVOLUTION
48 FOLD CDP STACK
+ 0.02S STATIC CORRECTION
MIGRATION USING 100% NMO VELOCITIES
VARIABLE FILTER WITHIN RANGE 12-55HZ
TWO-WAY TIME (SECONDS)

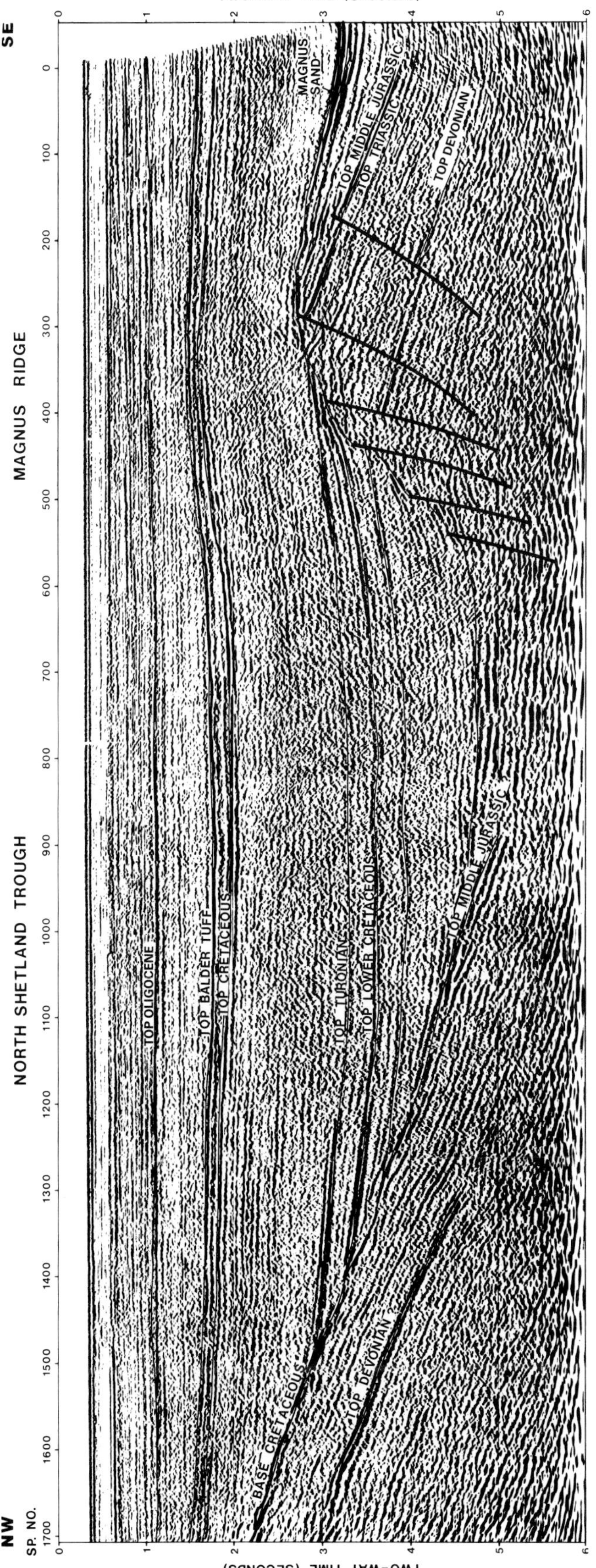

Fig. 4-46 (Evans and Parkinson, 1983)—Seismic line through extensional North Shetland trough and Magnus ridge in northern North Sea. Prerift section appears to extend all the way from top Devonian through Jurassic without a seismic break; however, extension began much earlier in the remainder of the North Sea. Major rift growth took place in Early Cretaceous time. Permission to publish by American Association of Petroleum Geologists.

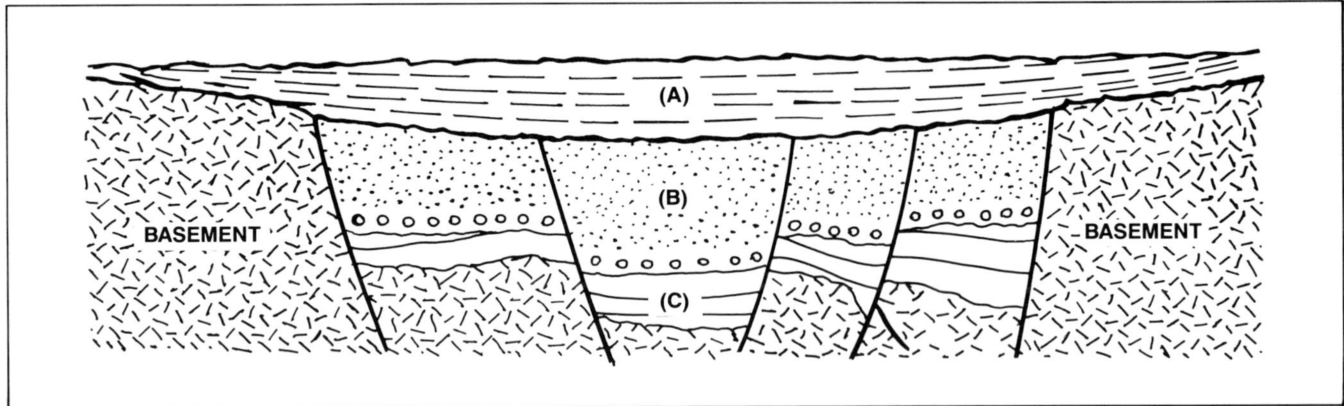

Fig. 4-47 (Harding, 1984)—Interior sag basin (A) caused by regional subsidence after rifting (cf. Fig. 4-12). Permission to publish by American Association of Petroleum Geologists.

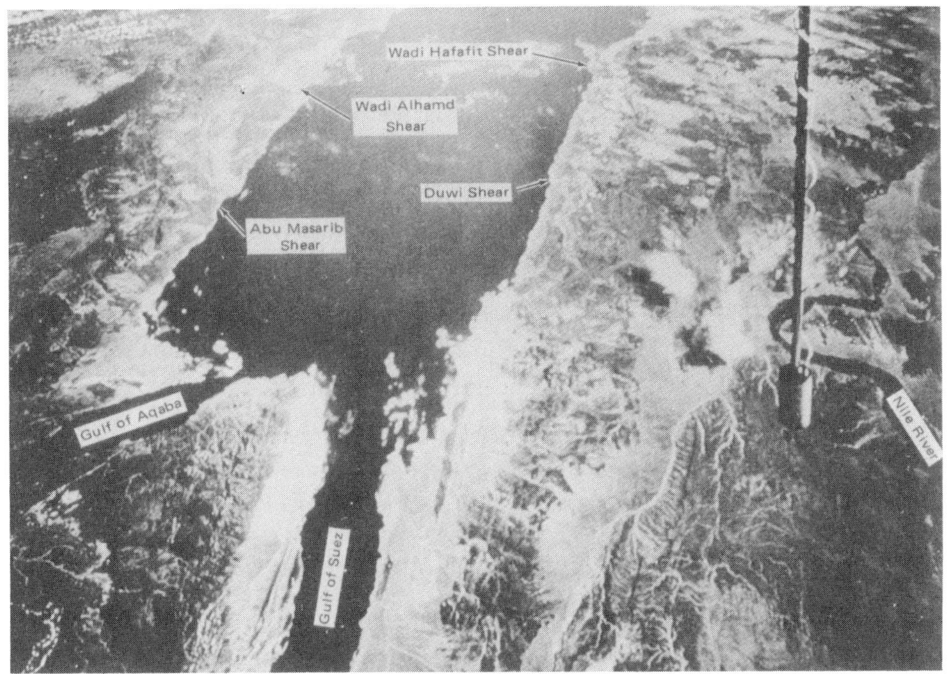

Fig. 4-48 (Lowell et al., 1975)—Gemini XII photograph (looking south). The Suez graben, though on trend with the Red Sea and the site of considerable Miocene rifting and evaporite accumulation (Heybroek, 1965), may be an aborted rift, for it is not now on the plate boundary associated with the Red Sea; that boundary turns up the Gulf of Aqaba into the Dead Sea fault zone instead of following from the northern Red Sea into the Gulf of Suez. The Suez graben is in a very favorable tectonic setting for oil occurrence. It is a Miocene cratonic basin in which subsidence has led to large normal fault blocks, some with reservoir rocks present under an evaporite seal, that contain major oil fields. High temperature has been neither a geologic nor a drilling factor in Suez graben petroleum development. This is attributed to the cessation of extension (which ultimately leads to active sea-floor spreading) when the lithosphere was still of almost normal thickness and provided an insulator to the deleterious effects of heat flow on preservations of oil. Photo courtesy NASA.

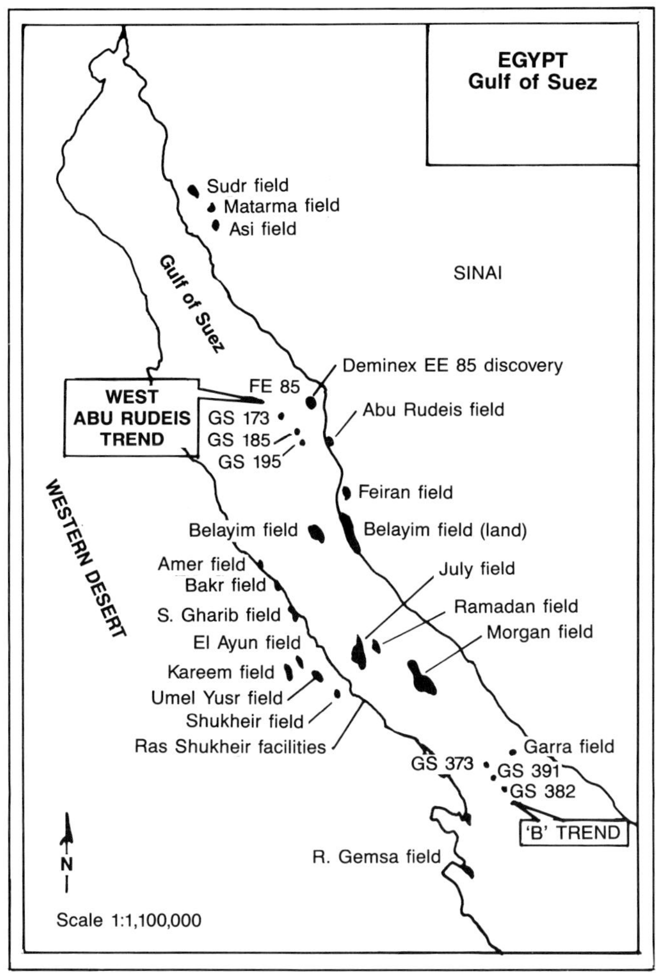

Fig. 4-49 (Abdine, 1981)—Index map for Gulf of Suez. Reprinted from "Egypt's Petroleum Geology: Good Grounds for Optimism" by A. S. Abdine, WORLD OIL, October, 1981.

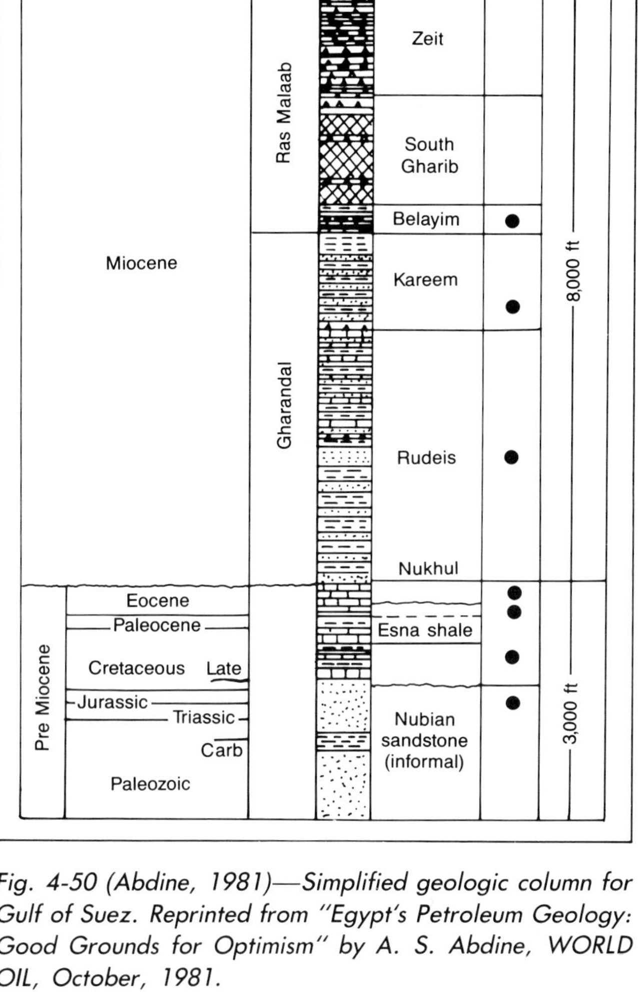

Fig. 4-50 (Abdine, 1981)—Simplified geologic column for Gulf of Suez. Reprinted from "Egypt's Petroleum Geology: Good Grounds for Optimism" by A. S. Abdine, WORLD OIL, October, 1981.

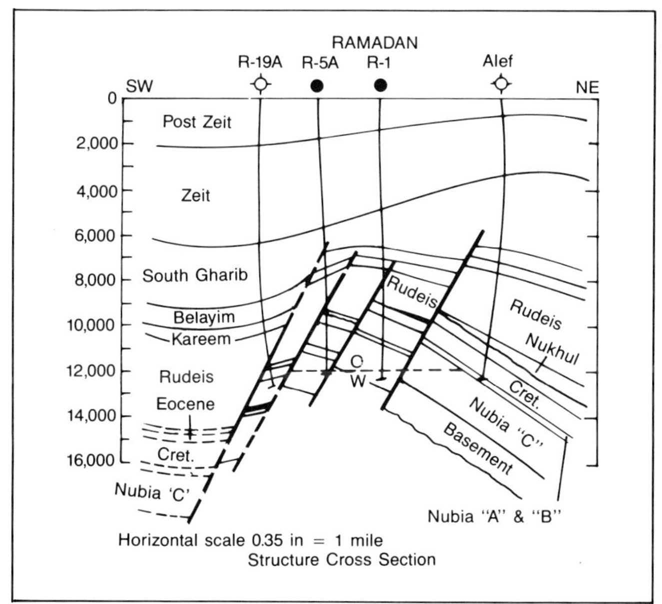

Fig. 4-51 (Abdine, 1981)—Ramadan field. Note common oil-water contact and offset of high on salt (South Gharib) from structures below. Note SW dip in "R-19A" fault block. Reprinted from "Egypt's Geology: Good Grounds for Optimism" by A. S. Abdine, WORLD OIL, October, 1981.

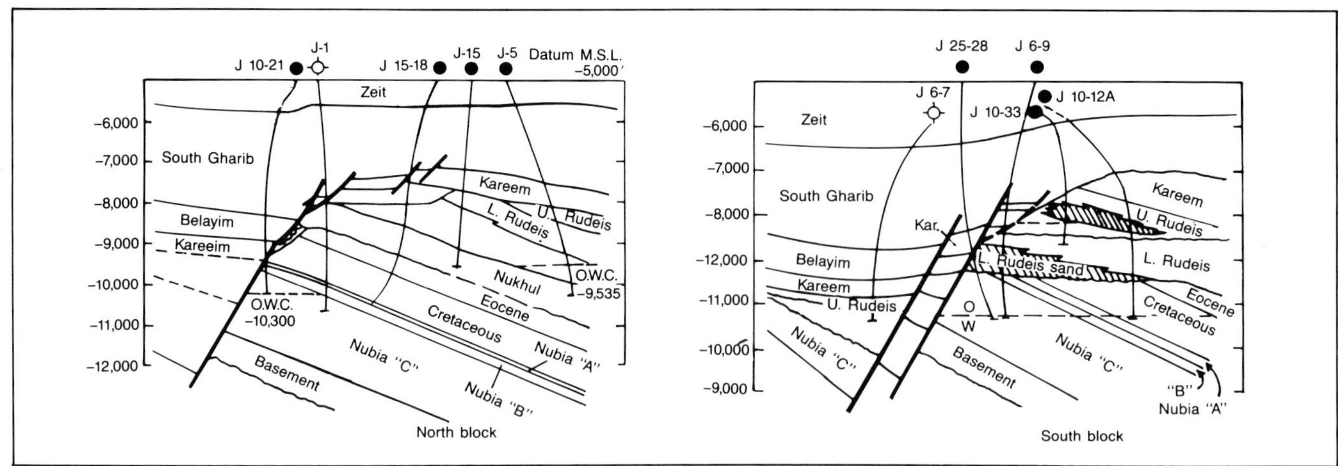

Fig. 4-52 (Abdine, 1981)—July field. High on top of salt does not persist at depth. Note position of J-1 dry hole between Nubian and Rudeis production on either side in north part of field. Also note pay established in Upper Rudeis sand (south block) down dip on gentle flank of extensional block. Reprinted from "Egypt's Petroleum Geology: Good Grounds for Optimism" by A. S. Abdine, WORLD OIL, October, 1981.

ited on the low sides of fault blocks should be common in this style.

Extensional block trapdoors are well demonstrated from the Reconcavo Basin of Brasil (Fig. 4-54). Part of the Dom Joao field is an acute angle trapdoor (Fig. 4-55) while a right angle trapdoor contributes to trapping at the Agua Grande field (Fig. 1-13). The Mata-Catu fault extends beyond the Agua Grande field area however, possibly representing an incipient transform feature transverse to the trend of the basin (cf. Fig. 2-69).

Heather field in the Viking graben of the northern North Sea (Fig. 4-56) is an obtuse angle trapdoor structure (Fig. 4-57). The Heather structure (Fig. 4-58) is one of several selected in the northern North Sea, including Beatrice (Fig. 4-59), Piper (Figs. 4-60, 4-61), Brent (Fig. 4-62), and Statfjord (Fig. 4-63), to show production from a rift-associated reservoir in the highest part of a tilted extensional block. In addition each of these fields is overlain by the same prolific source rock, the Upper Jurassic Kimmeridge Shale.

Brent (Fig. 4-62) and Statfjord (Fig. 4-63) are two of the largest offshore fields in the world and have an optimum combination of trap (extensional block), reservoir (porous Jurassic sands that were deposited during rifting), source (Kimmeridge Shale), seal (Kimmeridge Shale and fine-grained rocks above Cimmerian unconformity), temperature history (in part related to rifting), and burial history (thick Cretaceous and Tertiary).

As in the Suez graben, a very significant low-side field has been found in the northern North Sea. Brae (Figs. 4-

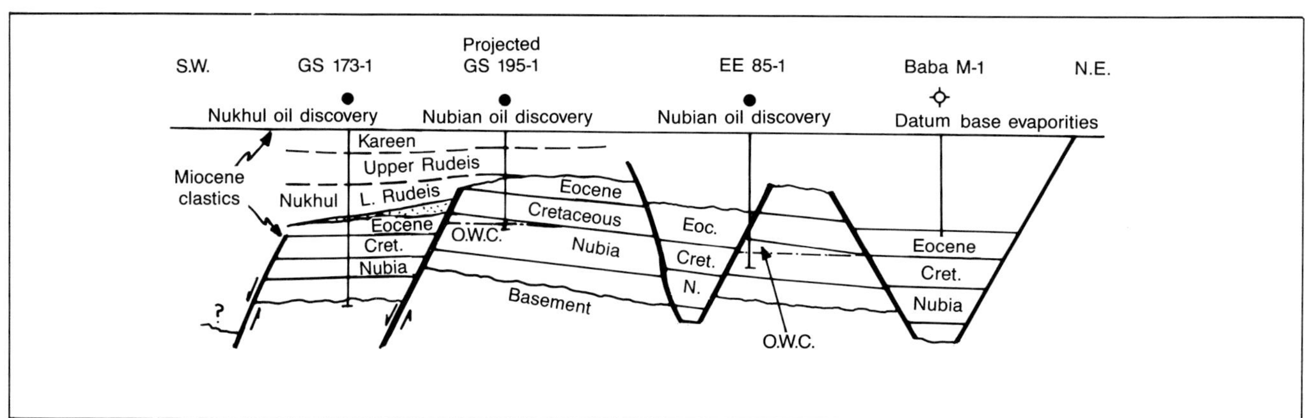

Fig. 4-53 (Abdine, 1981)—Cross section in West Abu Rudeis area showing Nukhul oil discovery from coarse clastics on downthrown side of normal fault block. Reprinted from "Egypt's Petroleum Geology: Good Grounds for Optimism" by A. S. Abdine, WORLD OIL, October, 1981.

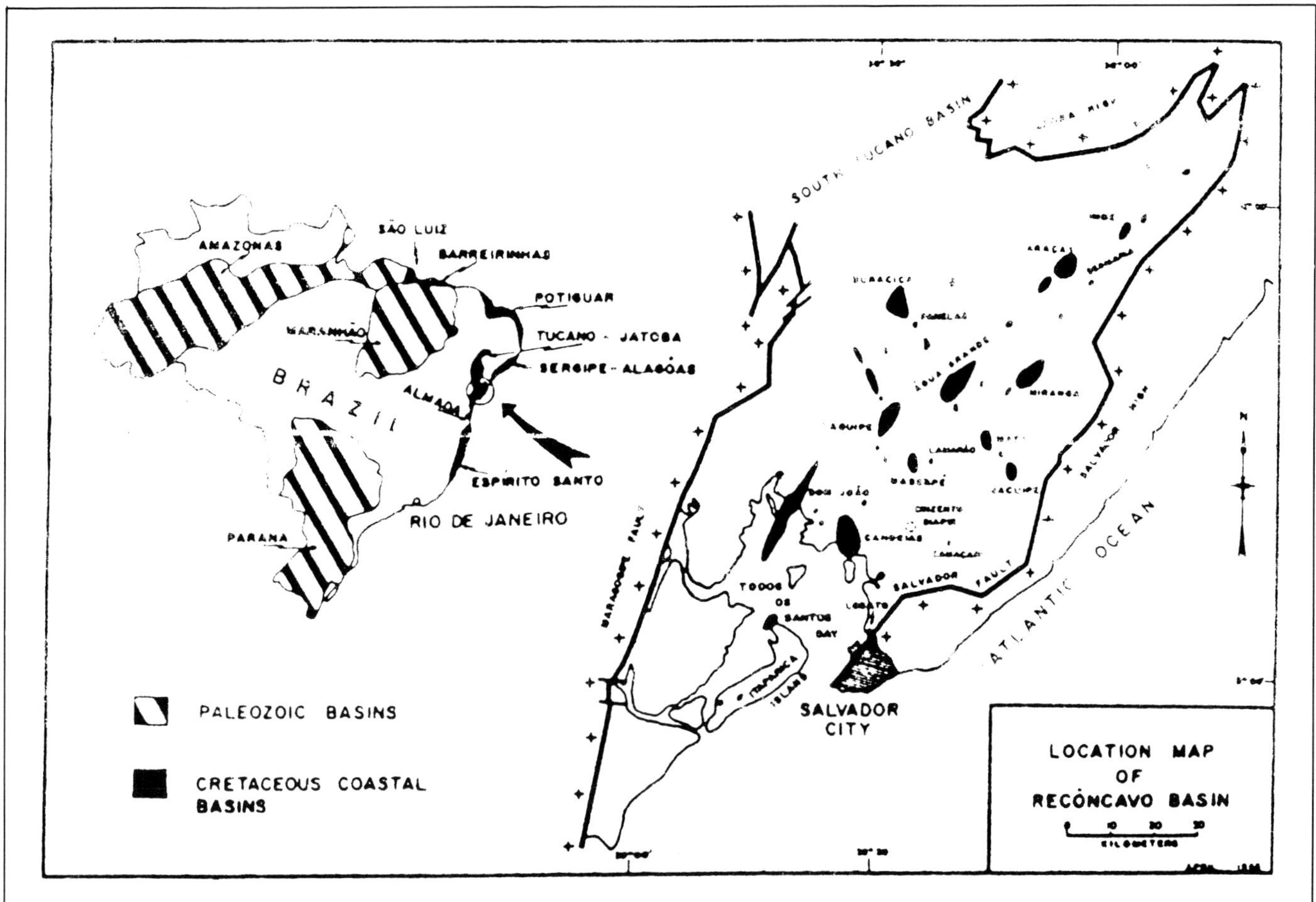

Fig. 4-54 (Ghignone and de Andrade, 1970)—Location map, Recôncavo Basin, Brazil. Oil and gas fields are shown. Note dog-leg trends of major basin-bounding faults. Permission to publish by American Association of Petroleum Geologists.

64, 4-65) is located on the downthrown side of a major normal fault. The Upper Jurassic reservoir at Brae is interpreted as a fan-delta complex (Fig. 4-66) derived from erosion of pre-Jurassic rocks, probably Devonian, of the Fladen Ground Spur on the west (high) side of the Viking graben.

REFERENCES

Abdine, A. S., 1981, Egypt's petroleum geology: Good grounds for optimism: *World Oil* (December), p. 99–106.

Allan, T. D., 1970, Magnetic and gravity fields over the Red Sea: *Royal Society London Philosophical Transactions* ser. A., V. 267, n. 1181, p. 153–180.

Aymé, J. M., 1965, The Senegal salt basin, *in* Salt basins around Africa: London Institute Petroleum, p. 83–90.

Bannert, D., Brinckmann, J., Käding, K. Ch., Knetsch, G., Kuersten, M., and Mayrhoffer, H., 1970, Zur Geologie der Danakil-Senke (nordliches Afar-Gebiet, NE-Athiopien): *Geologische Rundschau,* V. 59, n. 2, p. 409–443.

Barker, Colin, 1981, Plate tectonics, organic matter and basin evaluation for petroleum potential, American Association Petroleum Geologists Distinguished lecture series abstract: *Rocky Mountain Association Geologists Newsletter,* February, p. 3.

Belmonte, Y., Hirtz, P., and Wenger, R., 1965, The salt basins of the Gabon and Congo (Brazzaville); a tentative paleogeographic interpretation, *in* Salt basins around Africa: London Institute Petroleum, p. 55–74.

Blair, D. G., 1975, Structural styles in North Sea oil and gas fields, *in* Woodland, A. W., ed., Petroleum and the continental shelf north west Europe: V. 1 Geology, Halstead Press, p. 327–338.

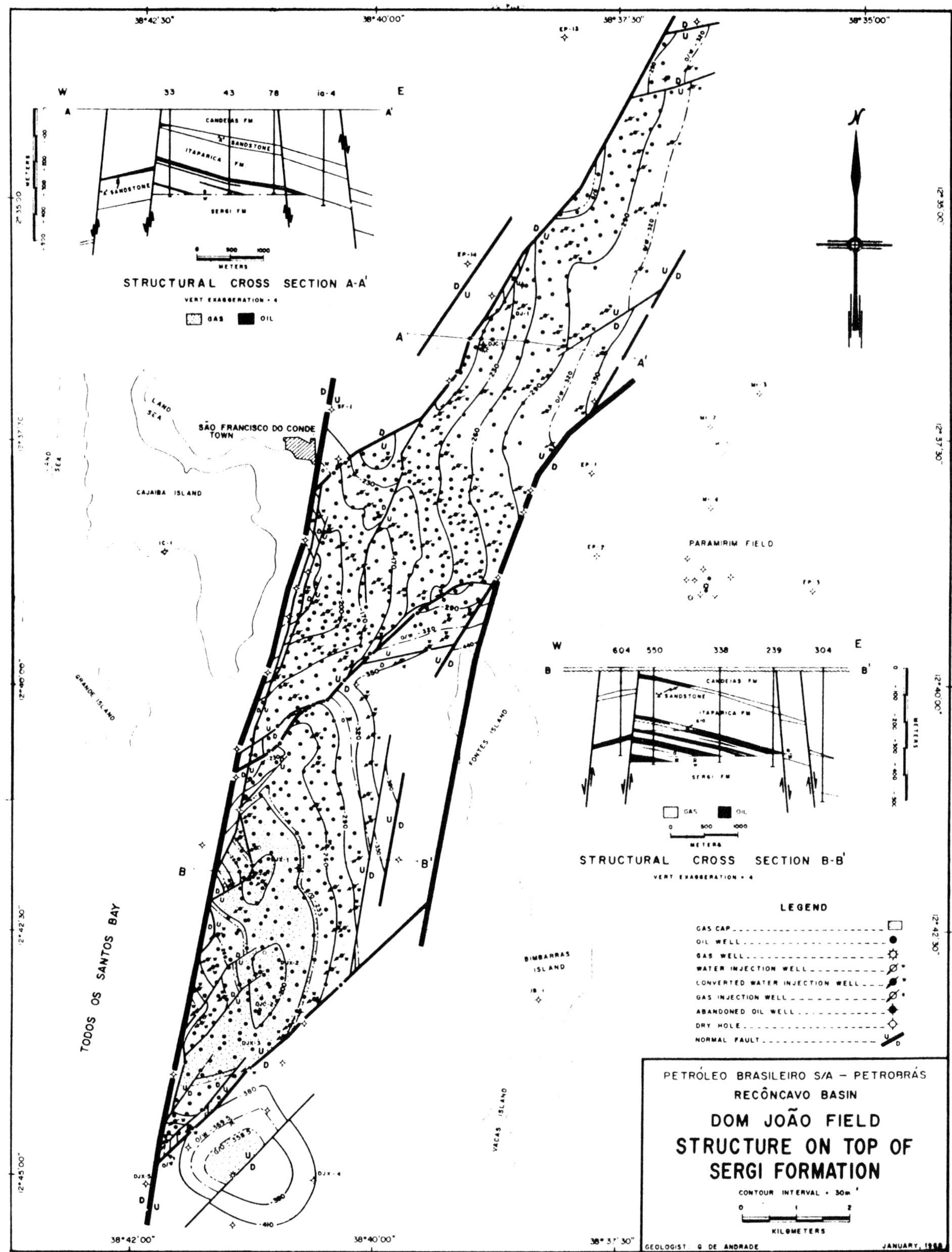

Fig. 4-55 (Ghignone and de Andrade, 1970)—Dom João field, Recôncavo Basin. Acute angle trapdoor structure at south end of field. Permission to publish by American Association of Petroleum Geologists.

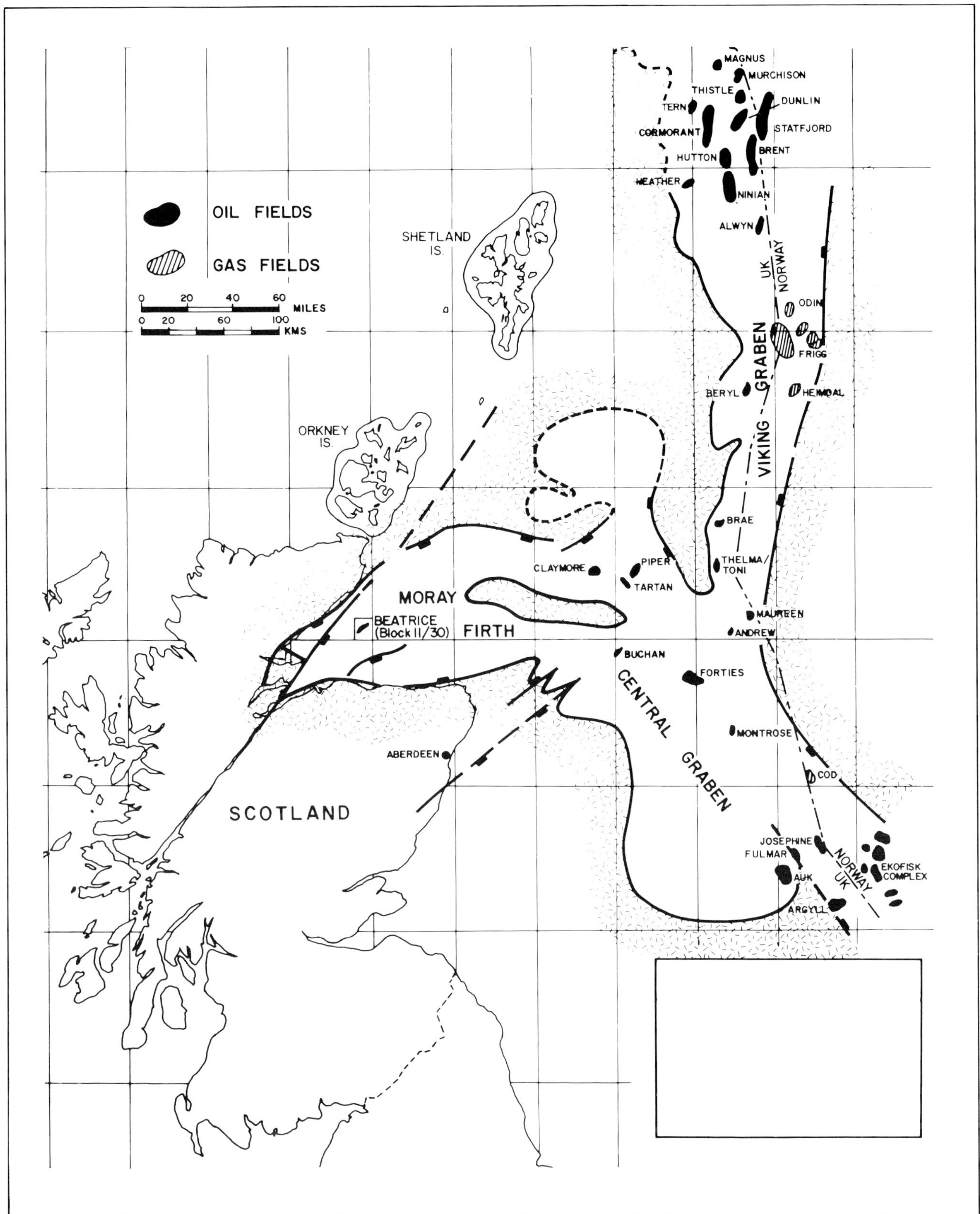

Fig. 4-56 (Linsley et al., 1980)—Index map for northern North Sea and figures 4-57 through 4-66 showing principal Mesozoic basins and major fields. Permission to publish by American Association of Petroleum Geologists.

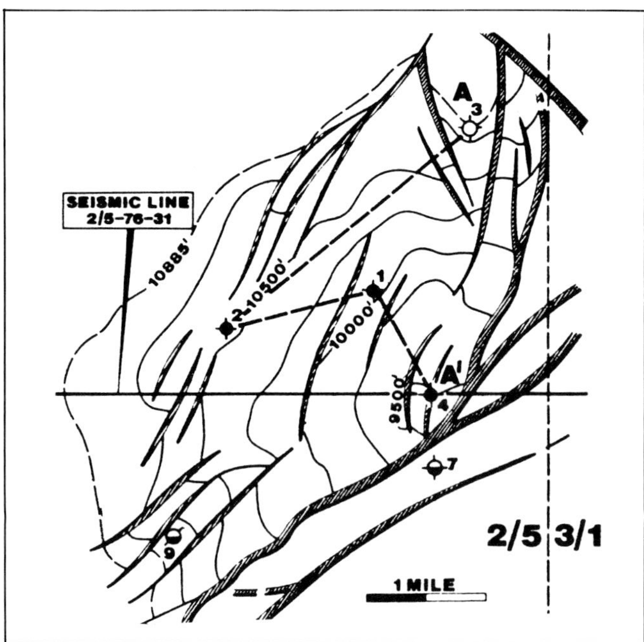

Fig. 4-57 (Gray and Barnes, 1981)—Heather field contoured on top of Middle Jurassic Brent Sandstone. Obtuse angle trapdoor defined by NE-trending fault on south and NNE-trending fault on east. Faults have effected a 427 m (1400 ft) oil column and 6000 acres of closure. Permission to publish by The Institute of Petroleum, London.

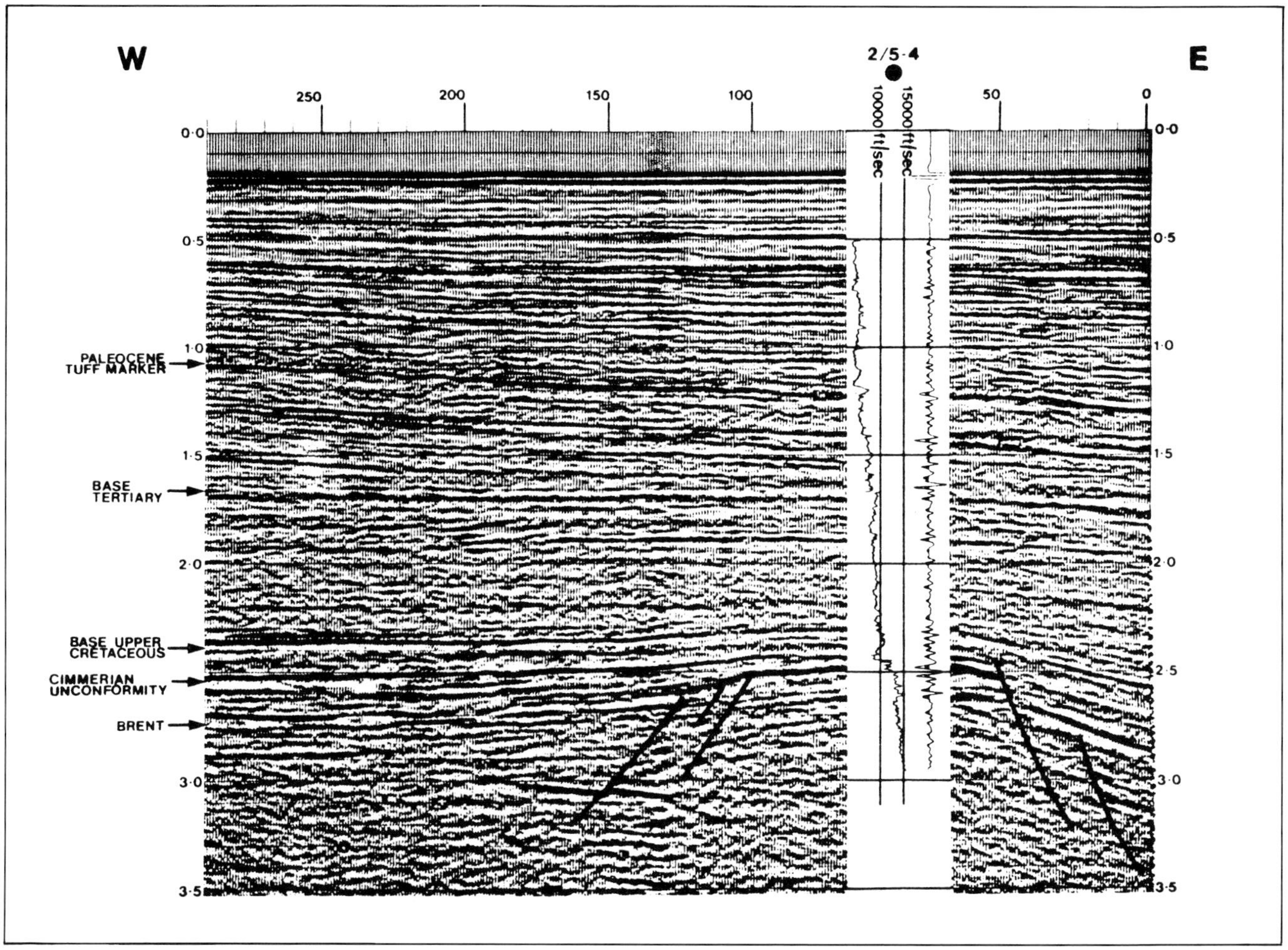

Fig. 4-58 (Gray and Barnes, 1981)—Seismic line 2/5-76-31 across Heather structure. See figure 4-57 for location. Permission to publish by The Institute of Petroleum, London.

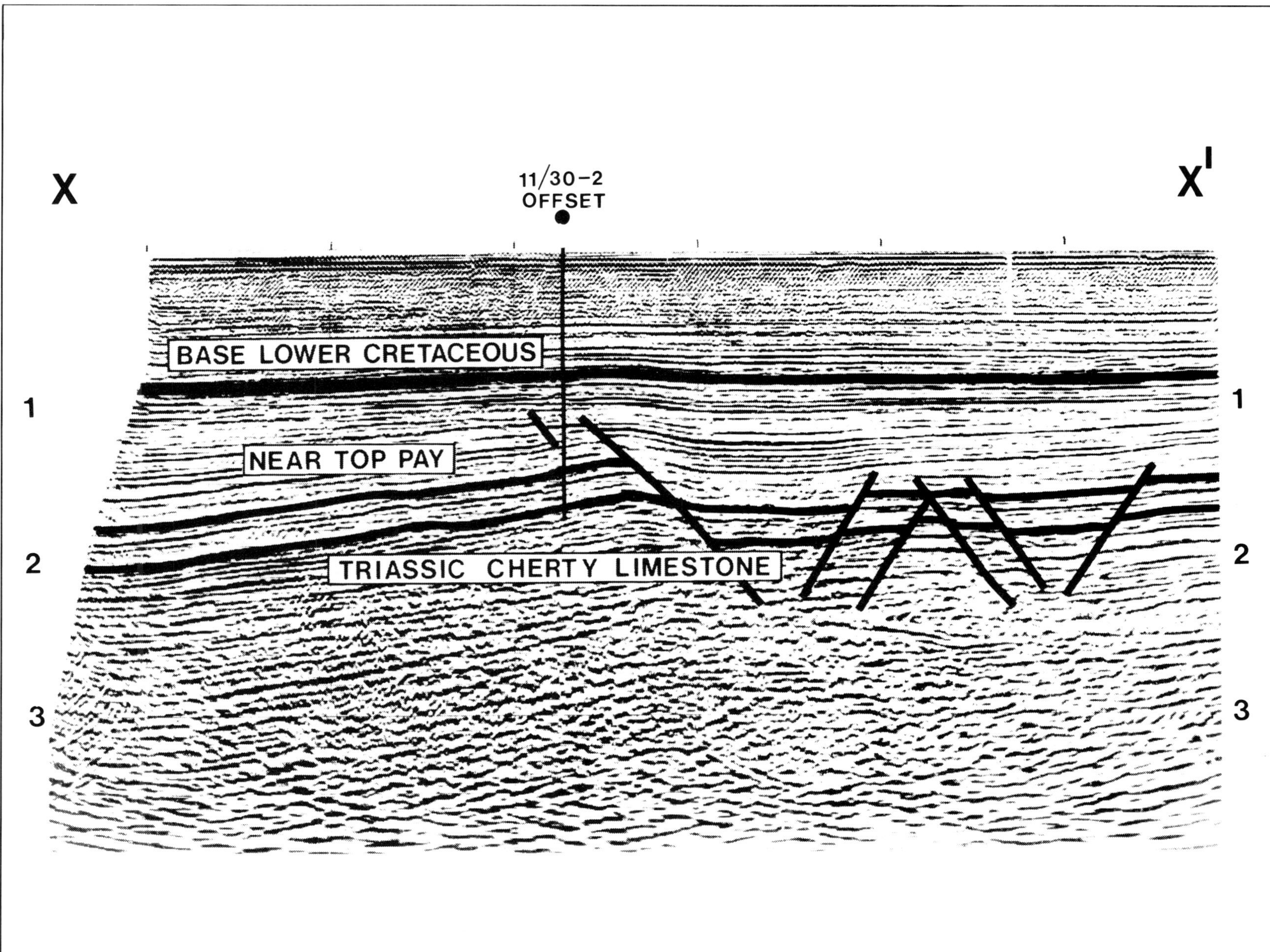

Fig. 4-59 (Linsley et al., 1980)—Seismic section through Beatrice field in Moray Forth showing accumulation from highest part of tilted block. Permission to publish by American Association of Petroleum Geologists.

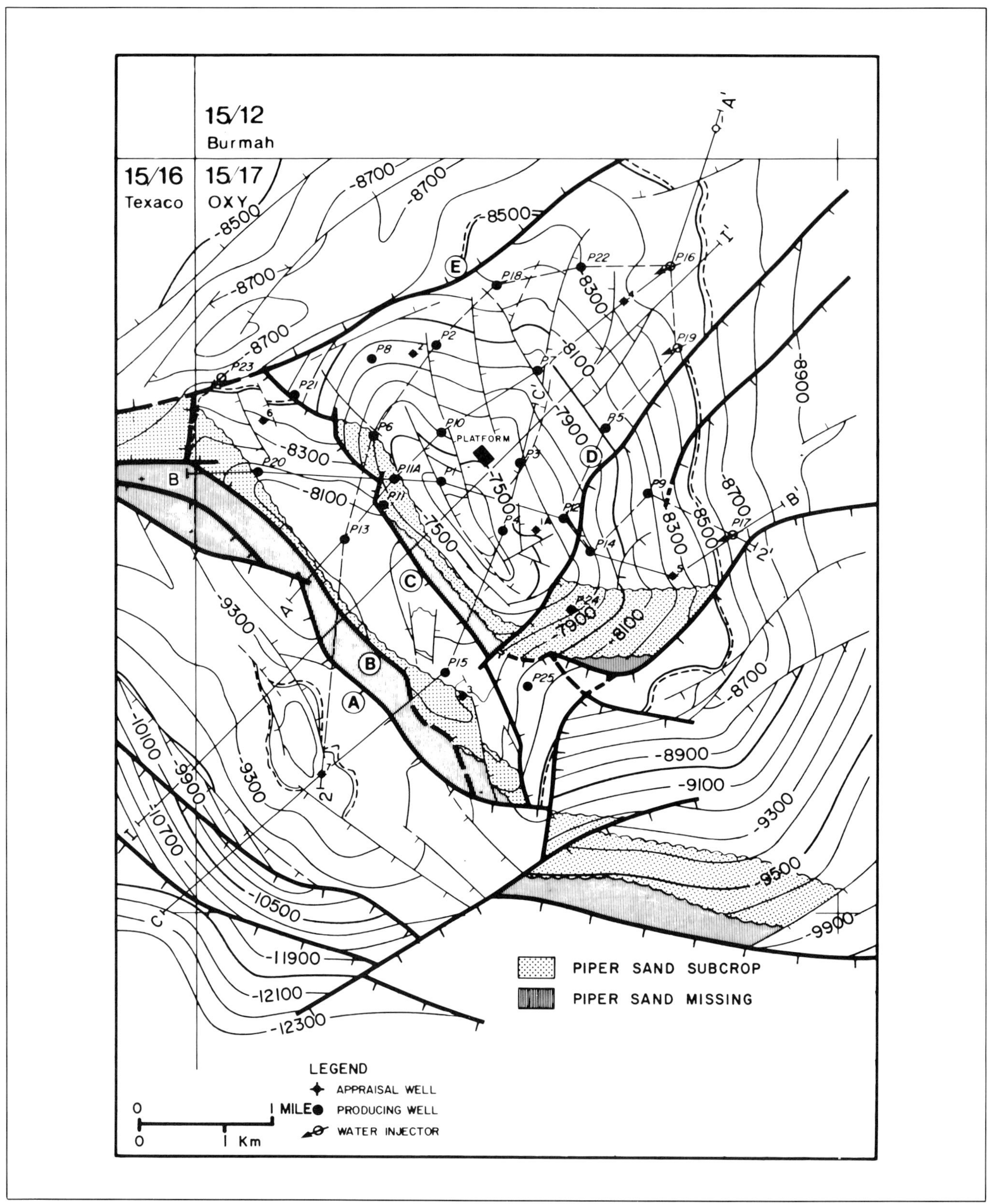

Fig. 4-60 (Maher, 1980)—Structure map top of Piper Sandstone showing complex internal faulted structure and associated erosion in Piper field. Permission to publish by American Association of Petroleum Geologists.

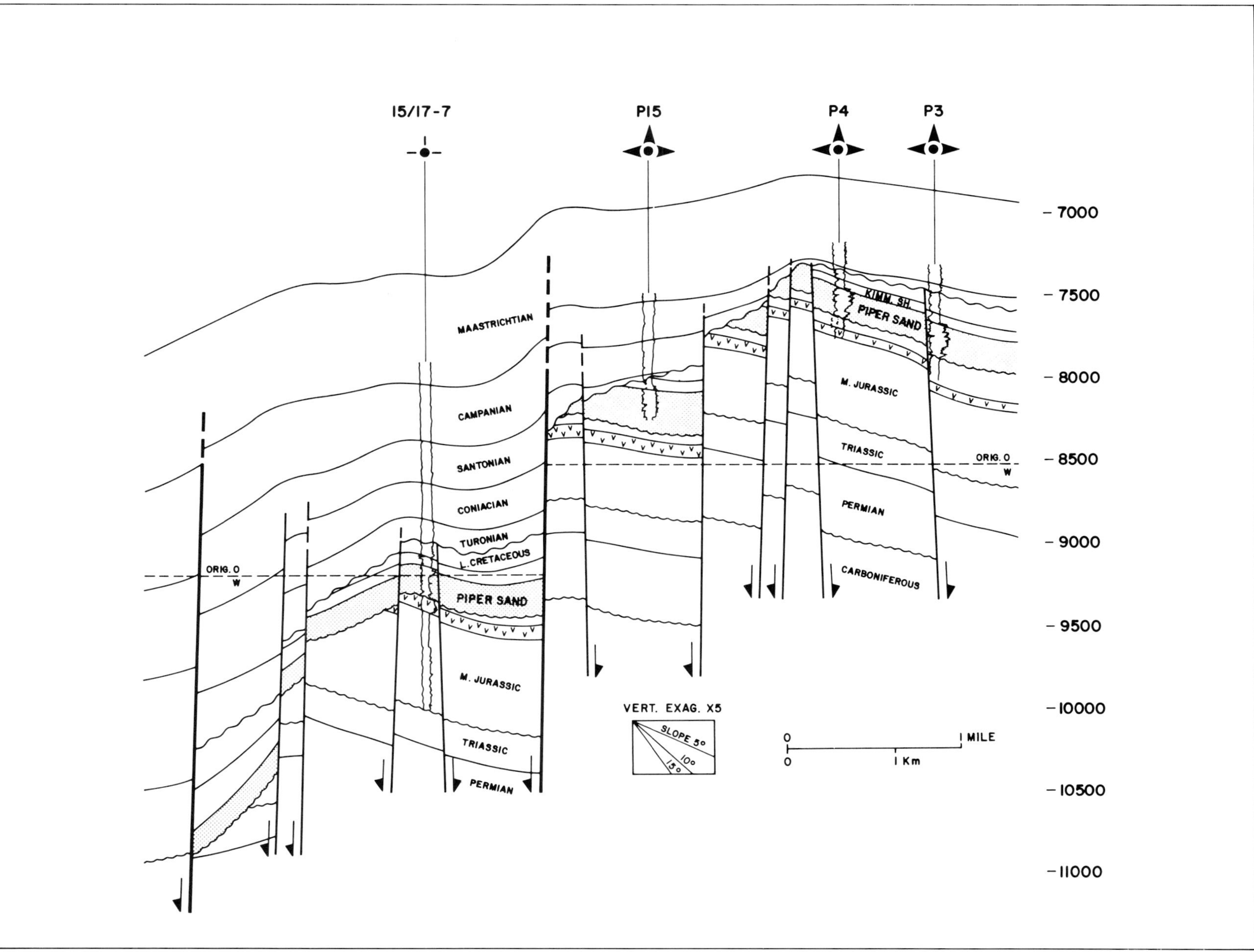

Fig. 4-61 (Maher, 1980)—Structural cross section C-C' through Piper field. Note complex history of unconformities, in part associated with rifting. See figure 4-60 for location. Permission to publish by American Association of Petroleum Geologists.

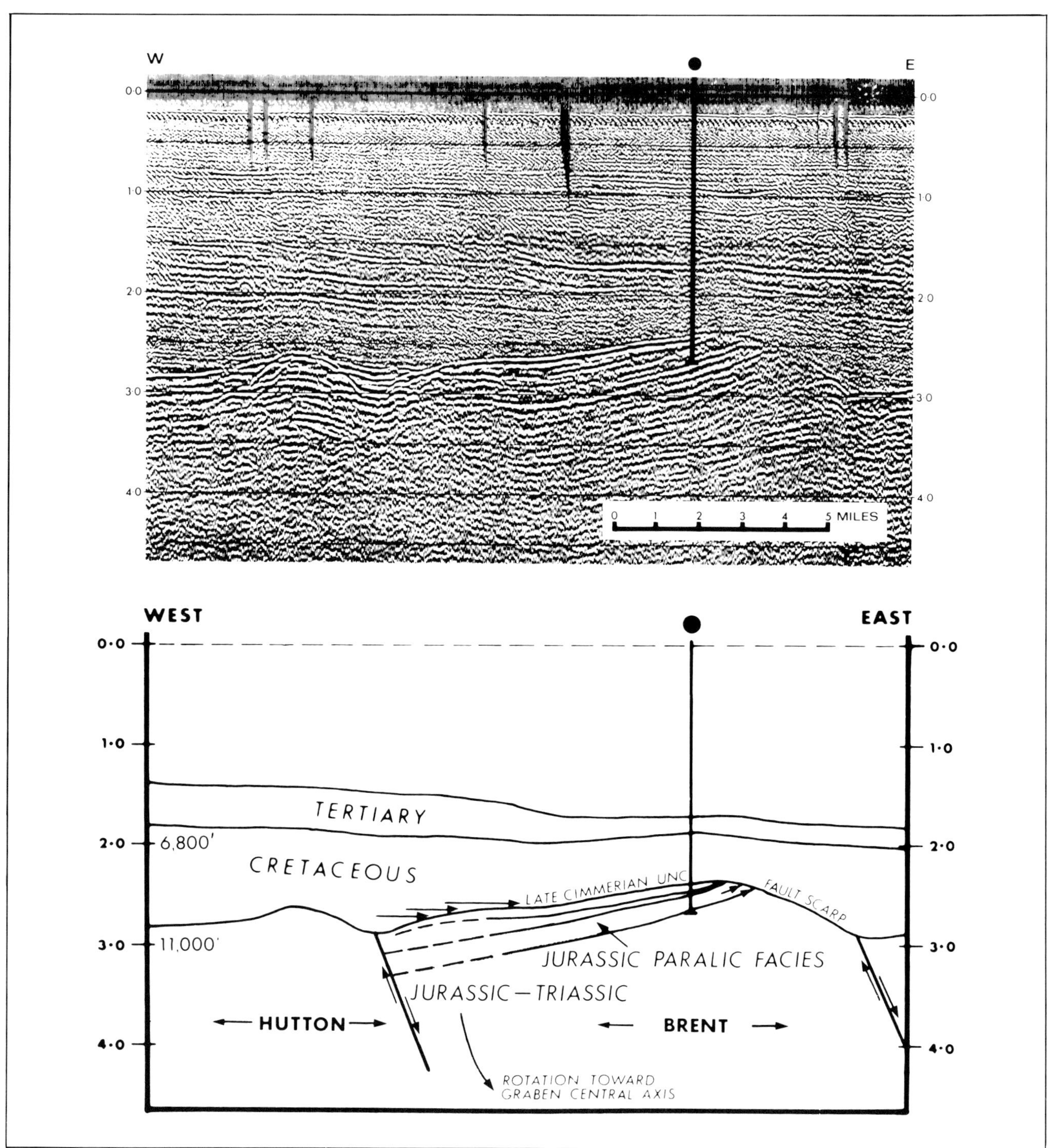

Fig. 4-62 (Blair, 1975)—West-east seismic section with geologic interpretation through the Brent oil field. Note that fault scarp has receded more than a mile (1.6 km) westward with attendent loss of highest part of tilted block. Permission to publish by Elsevier Applied Science Publishers Ltd.

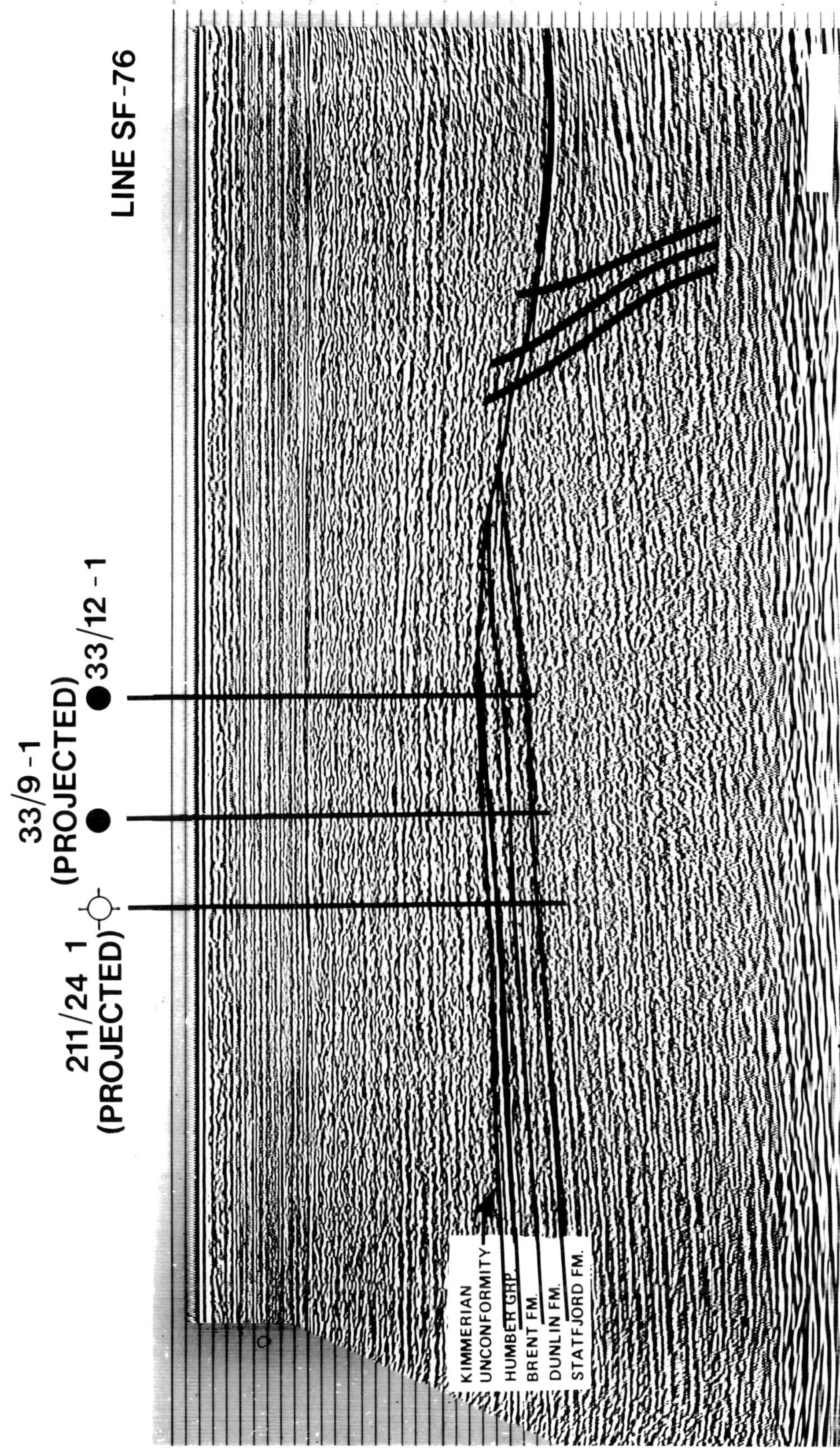

Fig. 4-63 (Kirk, 1980)—Dip-oriented sesmic line SF 76 through discovery well 33/12-1 of Statfjord field. As in Brent, note recession of fault scarp. Permission to publish by American Association of Petroleum Geologists.

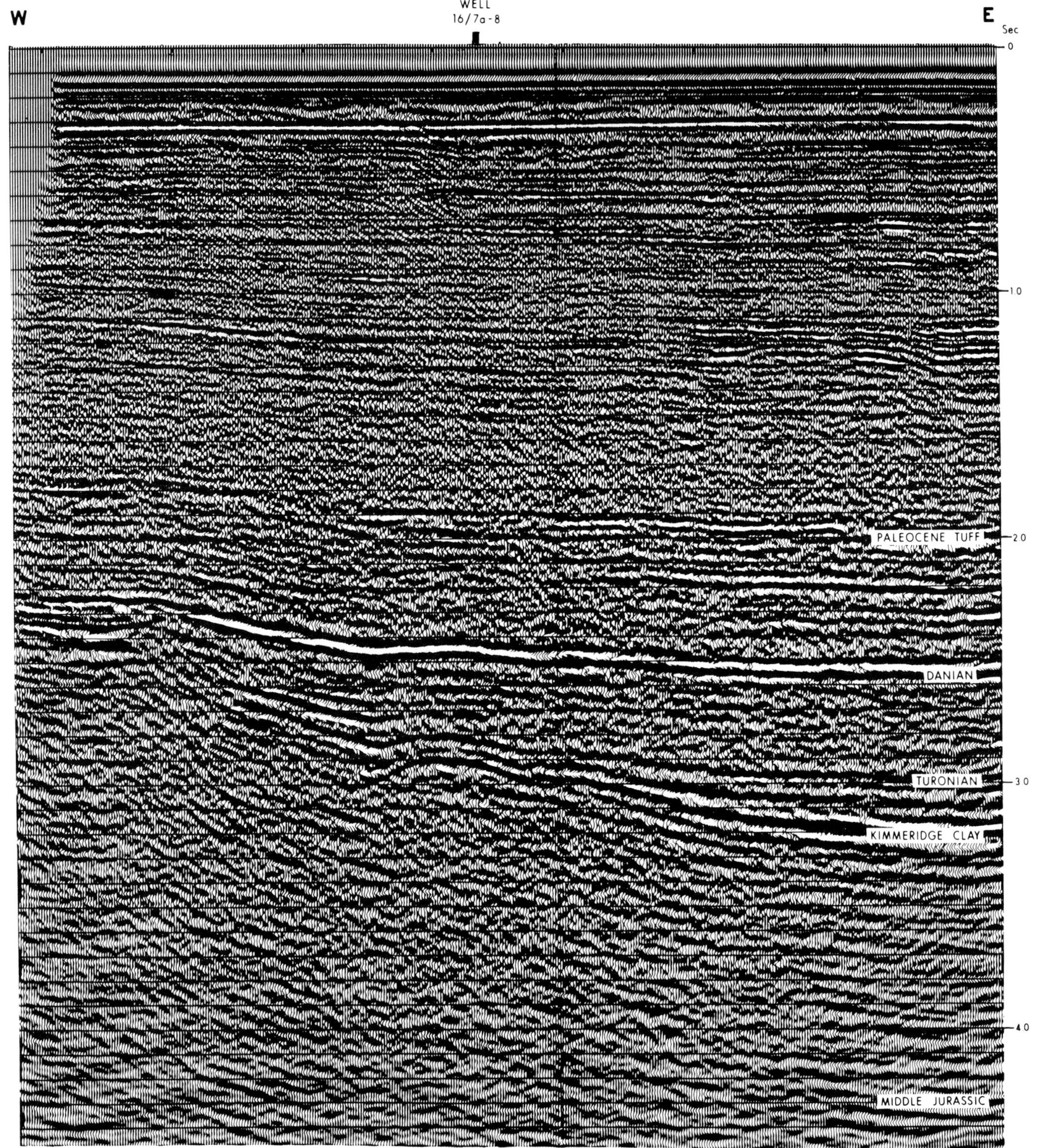

Fig. 4-64 (Harms et al., 1981)—Seismic section across south part of Brae field. Permission to publish by The Institute of Petroleum, London.

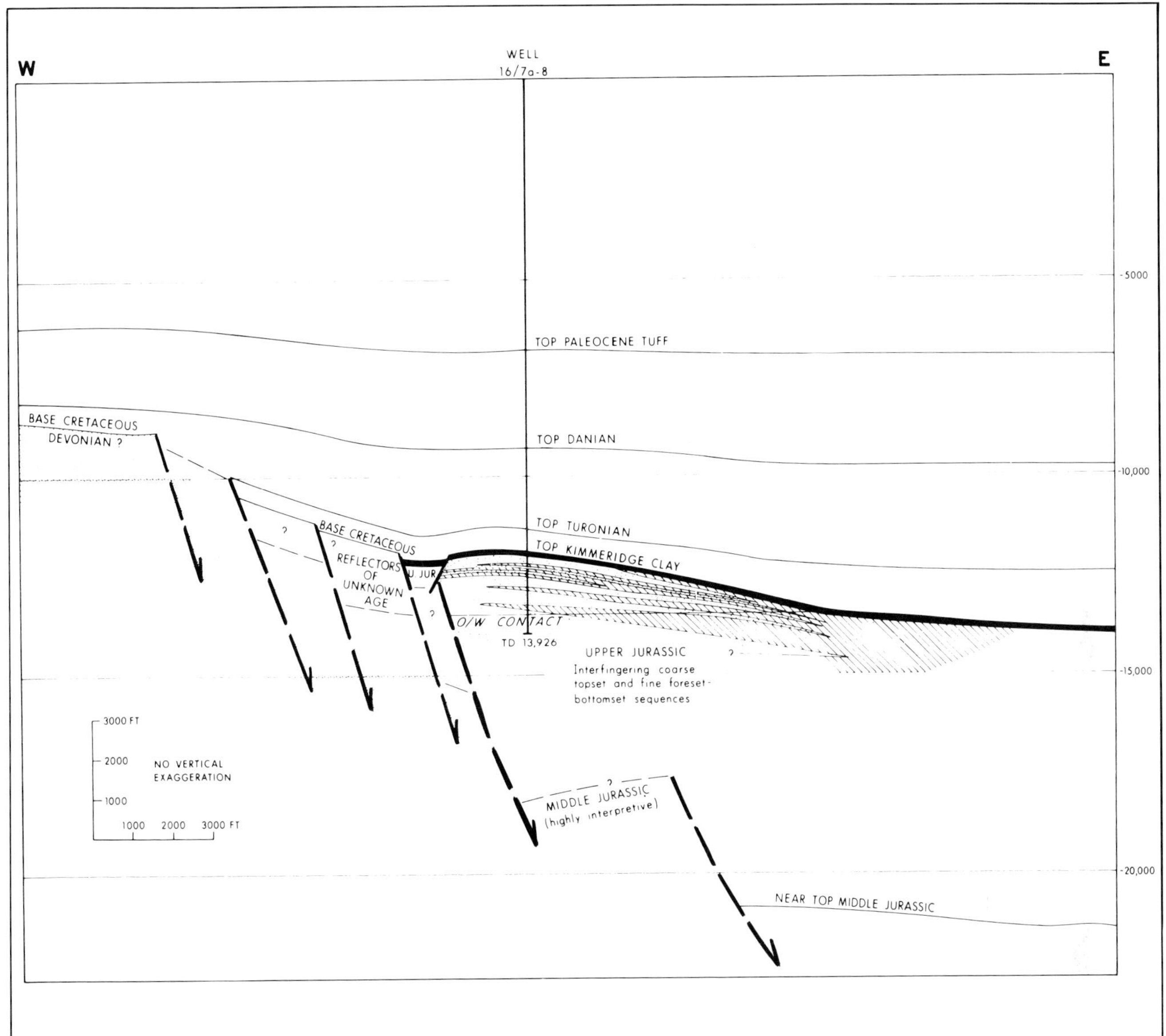

Fig. 4-65 (Harms et al., 1981)—Geologic interpretation of seismic section-figure 4-64. Note that oil accumulation is dependent on both fold and fault closure. Permission to publish by The Institute of Petroleum, London.

Brice, S. E., Cochran, M. D., Pardo, G., and Edwards, A. D., 1982, Tectonics and sedimentations of the south Atlantic rift sequence: Cabinda, Angola, *in* Watkins, J. S., and Drake, C. L., eds., Studies in continental margin geology: *American Association Petroleum Geologists Mem. 34,* p. 5–18.

Bullard, E. C., Everett, J. E., and Smith, A. G., 1965, The fit of the continents around the Atlantic, pt. 4, *in* Symposium on continental drift: *Royal Society London Philosophical Transactions,* ser. A, V. 258, n. 1088, p. 41–51.

Burek, P. J., 1970, Paleomagnetic evidence for an anti-clockwise rotation of the Danakil Alps, Ethiopia (abs): *Eos American Geophysical Union Transactions,* V. 51, n. 4, p. 271.

Carella, R., and Scarpa, N., 1962, Geological results of exploration in Sudan by Agip Mineraria: Beirut, 4th Arab Petroleum Congress, p. 23.

Crutcher, T. D., 1983, Southeast Georgia embayment, *in* Bally, A. W., ed., Seismic expression of structural styles: *American Association Petroleum Geologists Studies in Geology Ser. #15,* V. 2.

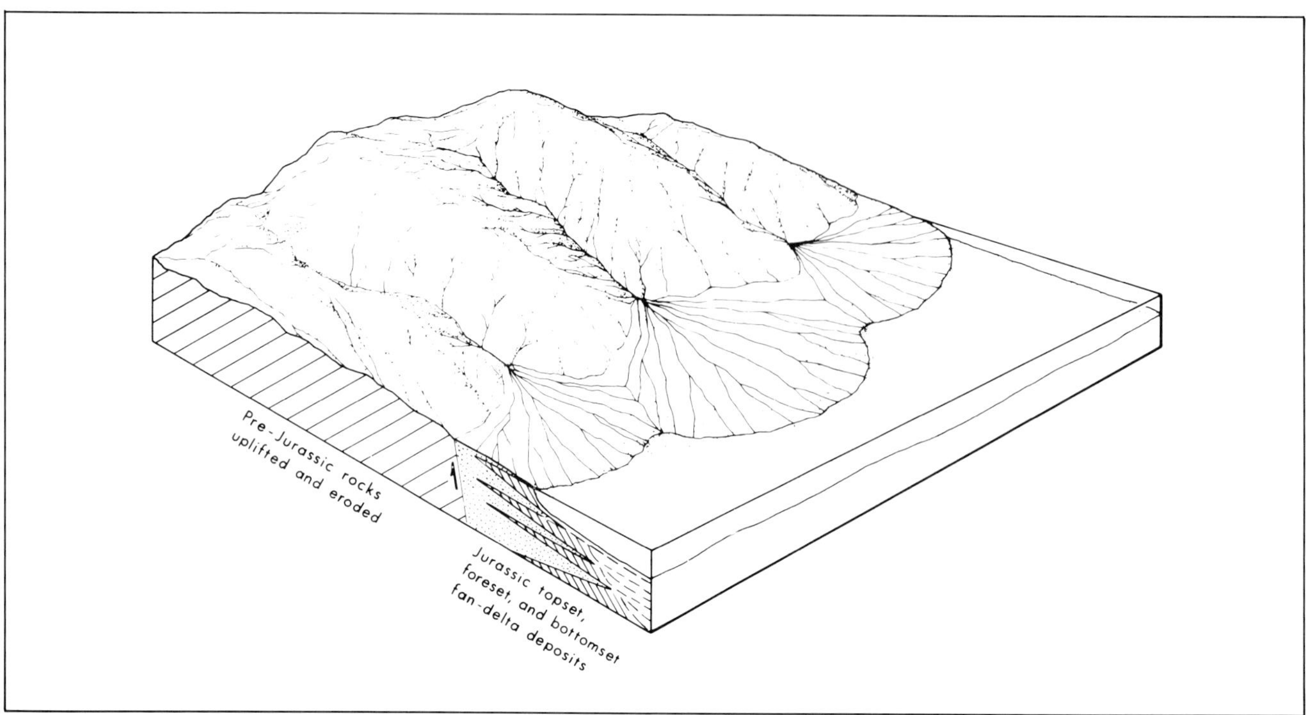

Fig. 4-66 (Harms et al., 1981)—Interpretation of Upper Jurassic reservoir at Brae as a fan-delta complex on low side of major, Viking graben bounding fault. Permission to publish by The Institute of Petroleum, London.

Davis, G. H., 1983, Shear zone model for the origin of metamorphic core complexes: *Geology,* V. 11, p. 342–347.

de Charpal, O., Guennoc, P., Montadert, L., and Roberts, D. G., 1978, Rifting, crustal attenuation and subsidence in the Bay of Biscay: *Nature* V. 275, p. 706–711.

DeSitter, L. U., 1956, Structural geology: London, New York, Toronto, McGraw-Hill, 1st ed., 552 p.

Dewey, J. F., and Burke, K., 1974, Hot spots and continental break-up: Implications for collisional orogeny: *Geology,* V. 2, p. 57–60.

Donath, F. A., 1962, Analysis of basin-range structure, south-central Oregon: *Geological Society America Bull.,* V. 73, p. 1–16.

Evans, A. C., and Parkinson, D. N., 1983, A half-graben and tilted fault block structure in the northern North Sea, *in* Bally, A. W., ed., Seismic expression of structural styles: *American Association Petroleum Geologists Studies in Geology #15,* V. 2.

Fairhead, J. D., and Girdler, R. W., 1970, The seismicity of the Red Sea, Gulf of Aden and Afar triangle: *Royal Society London Philosophical Transactions,* ser. A., V. 267, n. 1181, p. 49–71.

Falvey, D. A., 1974, The development of continental margins in plate tectonic theory: *Australian Petroleum Exploration Association Jour.,* p. 95–106.

Frazier, S. B., 1970, Adjacent structures of Ethiopia; that portion of the Red Sea coast including Dahlak Kebir Island and the Gulf of Zula: *Royal Society London Philosophical Transactions,* ser. A, V. 267, no. 1181, p. 131–141.

Ghignone, J. I., and de Andrade, G., 1970, General geology and major oil fields of Recôncavo Basin, Brazil: *American Association Petroleum Geologists Mem. 14,* p. 337–358.

Gillmann, M., 1968, Primary results of a geological and geophysical reconnaissance of the Jizan coastal plain in Saudi Arabia: 2nd Regional Technical Symposium Proc., Society Petroleum Engineers AIME, Saudi Arabia Sec., p. 189–212.

Girdler, R. W., 1969, The Red Sea; a geophysical background, *in* Degens, E. T., and Ross, D. A., eds., Hot brines and recent heavy metal deposits in the Red Sea: New York, Springer-Verlag, p. 38–58.

———, 1970, A review of Red Sea heat flow: *Royal Society London Philosophical Transactions,* ser. A, V. 267, n. 1181, p. 191–203.

Godby, E. A., Hood, P. J., and Bower, M. E., 1968, Aeromagnetic profiles across the Reykjanes ridge southwest of Iceland: *Journal Geophysical Research,* V. 73, n. 24, p. 7637–7649.

Gray, W. D. T., and Barnes, G., 1981, The Heather oil

field, *in* Illings, L. V., and Hobson, G. D., eds., Petroleum geology of the continental shelf of north west Europe, Institute of Petroleum, London, p. 335–341.

Gussow, W. C., 1968, Salt diapirism: Importance of temperature, and energy source of emplacement: *American Association Petroleum Geologists Mem. 8,* p. 16–52.

Halstead, P. H., 1975, Northern North Sea faulting: Norwegian Petroleum Society, Jurassic northern North Sea symposium, JNNSS110.

Harding, T. P., 1984, Graben hydrocarbon occurrences and structural style: *American Association Petroleum Geologists Bull.,* V. 68/3, p. 333–362.

Harms, J. C., Tackenberg, P., Pickles, E., and Pollock, R. E., 1981, The Brae oilfield area *in* Illing, L. V., and Hobson, G. D., eds., Petroleum geology of the continental shelf of north west Europe: Institute of Petroleum, London, p. 352–357.

Heirtzler, J. R., 1968, Sea-floor spreading: *Scientific American,* V. 219, n. 6, p. 60–70.

Heybroek, F., 1965, The Red Sea Miocene evaporite basin, *in* Salt basins around Africa: The Institute of Petroleum, London, p. 17–40.

Hoffman, P., 1972, Evolution of an early Proterozoic continental margin: *Royal Society London Philosophical Transactions,* ser. A., V. 273, pp. 547–581.

Holwerda, J. G., and Hutchinson, R. W., 1968, Potash-bearing evaporites in the Danakil area, Ethiopia: *Economic Geology,* V. 63, n. 2, p. 124–150.

Illies, J. H., 1975, Intraplate tectonics in stable Europe as related to plate tectonics in the Alpine system: *Geologische Rundschau,* V. 64, p. 677–699.

Kay, G. M., 1942, Ottawa-Bonnechere graben and Lake Ontario homocline: *Geological Society America Bull.,* v. 53, p. 585–646.

Kirk, R. H., 1980, Statfjord field—a North Sea giant, *in* Halbouty, M. T., ed., Giant oil and gas fields of the decade 1968–1978: *American Association Petroleum Geologists Mem. 30,* p. 117–129.

Lalou, C., Nguyen, H. V., Faure, H., and Moreira, L., 1970, Datation par la methode uranium-thorium des hauts niveaux de coraux de la depression de l'Afar (Ethiopie): *Revue Geographie Physique et Geologie Dynamique,* v. 12, n. 1, p. 3–8.

Langseth, M. G., and Taylor, P. T., 1967, Recent heat flow measurements in the Indian Ocean: *Journal Geophysical Research,* V. 71, p. 5321–5355.

Laughton, A. S., 1966, The Gulf of Aden, in relation to the Red Sea and the Afar depression of Ethiopia: *Canadian Geological Survey Paper 66-14,* p. 78–97.

Linsley, P. N., Potter, H. C., McNab, G., and Racher, D., 1980, The Beatrice field, inner Moray Firth, U.K., North Sea, *in* Halbouty, M. T., ed., Giant oil and gas fields of the decade 1968–1978: *American Association Petroleum Geologists, Mem. 30,* p. 117–129.

Longwell, C. R., 1933, Eastern New York and western New England: Guidebook 1: International Geological Congress, Excursion A-1, 16th Session.

Lowell, J. D., and Genik, G. J., 1972, Sea-floor spreading and structural evolution of southern Red Sea; *American Association Petroleum Geologists Bull.,* v. 56, n. 2, p. 247–259.

Lowell, J. D., Genik, G. J., Nelson, T. H., and Tucker, P. M., 1975, Petroleum and plate tectonics of the southern Red Sea, *in* Fischer, A. G., and Judson, S., eds., Petroleum and global tectonics: Princeton University Press, p. 129–153.

Maher, C. E., 1980, Piper oil field, *in* Halbouty, M. T. ed., Giant oil and gas fields of the decade 1968–1978: *American Association Petroleum Geologists Mem. 30,* p. 131–172.

McDonald, R. E., 1976, Tertiary tectonics and sedimentary rocks along the transition: Basin and Range province to plateau and thrust belt province, Utah, *in* Hill, J. G., ed., Geology of the Cordilleran hingeline: Rocky Mountain Association Geologists, p. 281–317.

Mohr, P. A., 1967a, Major volcano-tectonic lineament in the Ethiopian rift system: *Nature,* V. 213, n. 5077, p. 664–665.

———, 1967b, The Ethiopian rift system: Addis Ababa, Haile Selassie I Univ. *Geophysics Observatory Bull.,* n. 11, 65 p.

———, 1970, The Afar triple junction and sea-floor spreading: *Journal Geophysical Research,* v. 75, p. 7340–7352.

Netto, A.S.T., 1984, Reexploration in Recôncavo Basin, Brazil (abs): *American Association Petroleum Geologists Bull.,* V. 68/4, p. 512.

Pautot, G., Auzende, J., and LePichon, X., 1970, Continuous deep sea salt layer along North Atlantic margins related to early phase of rifting: *Nature,* V. 227, n. 5256, p. 351–354.

Proffett, J. M., 1977, Cenozoic geology of the Yerington district, Nevada, and implications for the nature and origin of basin and range faulting: *Geological Society America Bull.,* V. 88, p. 247–266.

Ramberg, I. B., Cook, F. A., and Smithson, S. B., 1978, Structure of the Rio Grande rift in southern New Mexico and west Texas based on gravity interpretation: *Geological Society America Bull.,* V. 89, p. 107–123.

Roberts, D. G., 1969, Structural evolution of the rift zones in the Middle East: *Nature,* V. 223, n. 5201, p. 55–57.

———, Bishop, D. G., Laughton, A. S., Ziolkowski, A. M., Scruton, R. A., and Matthews, D. H., 1970, New sedimentary basins on Rockall Plateau: *Nature,* V. 225, n. 5228, p. 170–172.

Robson, D. A., 1971, The structure of the Gulf of Suez (clysmic) rift, with special reference to the eastern side: *Journal Geological Society,* V. 127, p. 247–276.

Said, R., 1962, The geology of Egypt: Elsevier Publishing Company, Amsterdam-New York, 377 p.

Salop, L. I., and Scheinmann, Y. M., 1969, Tectonic history and structures of platforms and shields: *Tectonophysics,* V. 7, n. 5-6, p. 565–597.

Sanford, A. R., 1959, Analytical and experimental study of simple geologic structures: *Geological Society America Bull.,* V. 70, p. 19–52.

Schull, T. J. 1984, Oil exploration in nonmarine rift basins of interior Sudan (abs.): *American Association Petroleum Geologists Bull.,* V. 68/4, p. 526.

Sclater, J. G., 1966, Heat flow in the northwest Indian Ocean and the Red Sea, *in* A discussion concerning the floor of the northwest Indian Ocean: *Royal Society London Philosophical Transactions,* ser. A, V. 259, n. 1099, p. 271–278.

Sestini, J., 1965, Cenozoic stratigraphy and depositional history, Red Sea coast, Sudan: *American Association Petroleum Geologists Bull.,* V. 49, n. 9, p. 1453–1472.

Shatzky, N. S., 1955, On the origin of the Pachelma trough: Moskov. Obshch Lyubiteley Prirody Byull., Otdel. Geol., V. 5, p. 5–26.

Sittler, C., 1969, The sedimentary trough of the Rhine graben: *Tectonophysics,* V. 8, n. 4–6, p. 543–560.

Solli, M., 1976, In seismisk skorpeudersokelse Norges-Shetland: University of Bergen Thesis.

Tazieff, H., 1968, Relations tectoniques entre l'Afar et la mer Rouge: *Societe Geologique France Bull.,* ser. 7, V. 10, n. 4, p. 468–477.

———, 1969, Tectonique de l'Afar septentrional (Ethiopie): *Academe Sciences Comptes Rendus,* ser. D, V. 268, n. 16, p. 2030–2033.

———, 1970, The Afar Triangle: *Scientific American,* V. 22, n. 2, p. 32–40.

———, and Varet, J., 1969, Signification tectonique et magmatique de l'Afar septentrional (Ethiopie): *Revue Geographie Physique et Geologie Dynamique,* V. 11, fasc. 4, p. 429–450.

Vine, F. J., 1966, Spreading of the ocean floor-new evidence: *Science,* V. 154, n. 3755, p. 1405–1415.

Withjack, M. O., and Schiener, C., 1982, Fault patterns associated with domes—an experimental and analytical study: *American Association Petroleum Geologists Bull.,* V. 66, p. 302–316.

5 WARPS—ARCHES, DOMES, AND SAGS

Though basement warps may be the simplest of all structural styles in form, they are not the simplest to understand as noted in Chapter 1. Part of their obscurity lies in the fact that warps are usually subsurface features and lack the surface exposures that afford structural analysis of other styles. In the same vein, the lack of faulting associated with warps makes it difficult to understand them; in contrast, the faults that effect the wrench, compressive block, and extensional block styles are essential to the understanding of those styles. Some hypotheses of origin of basement warps have been mentioned in Chapter 1 and some are summarized in figure 5-1. The relationships of lateral and convective subcrustal flow to uplift and subsidence are self evident. Phase changes, from basalt to eclogite or olivine to spinel, for example, involve a conversion from less to more dense matter with subsidence above the denser region as that region tends to sink or stand lower than its surroundings; uplift takes place with the reverse reaction.

In a plate tectonic context early arching that aborts prior to rifting (Fig. 4-2) could be a mechanism for broad basement arches. Aborted triple junctions, having created a crustal thin and consequently a point of weakness, could, upon subsequent sediment loading, focus circular subsidence or sag (Fig. 5-2). Sags can also develop in the late stages of rift basins (Fig. 4-47).

Arches and domes are characterized by a relatively thin sedimentary section (owing to their typically intra-plate setting), low dips, and low relief. Uplift and subsidence are expressed mainly by gentle warping; faulting is distinctly subordinate although it can be important locally (Fig. 5-3). Arches and domes can be syn-depositional or post-depositional, or some combination of both (Fig. 5-4). In syn-depositional structures, onlap, convergence, truncation, and unconformities are all rather predictable responses of the sedimentary sequence; isopach mapping reveals much about growth history. Sedimentary units remain almost constant in thickness between flanks and crest of post-depositional structures. Both syn-depositional and post-depositional features are amenable to differential compaction.

Arches and domes, being structurally mild features are usually not very spectacular, but they have trapped great quantities of petroleum principally because of their frequently large size (Figs. 5-5, 5-6). Where they have relatively little to sometimes no closure (noses), stratigraphic variations can play an important part in pooling hydrocarbons. Onlap, convergence, truncation, and unconformities have already been mentioned and are partially illustrated by figures 5-5 through 5-7. Buried hills and permeability reduction because of clay-filled solution valleys adjacent to porous carbonate reservoirs have also been cited (Walters, 1958).

REFERENCES

Balducchi, A., and Pommier, G., 1970, Cambrian oil field of Hassi Messaoud, Algeria: *American Association Petroleum Geologists Mem. 14,* p. 477–488.

Delclaud, D., and Martel, A., 1959, Resultats des travaux de reconnaissance sur le champ de gaz de Hassi er Rmel: *Revue de l'Institut Francais du Petrole,* V. 14, p. 457–465.

Dimian, M. V., 1983, Hudson Bay Basin, *in* Bally, A. W., ed., Seismic expression of structural styles: *American Association Petroleum Geologists Studies in Geology #15,* V. 2.

Eardley, A. J., 1957, Structural geology of North America: Harper & Brothers, New York, p. 624.

Fowler, J. H., and Kuenzi, D. W., 1978, Keweenawan

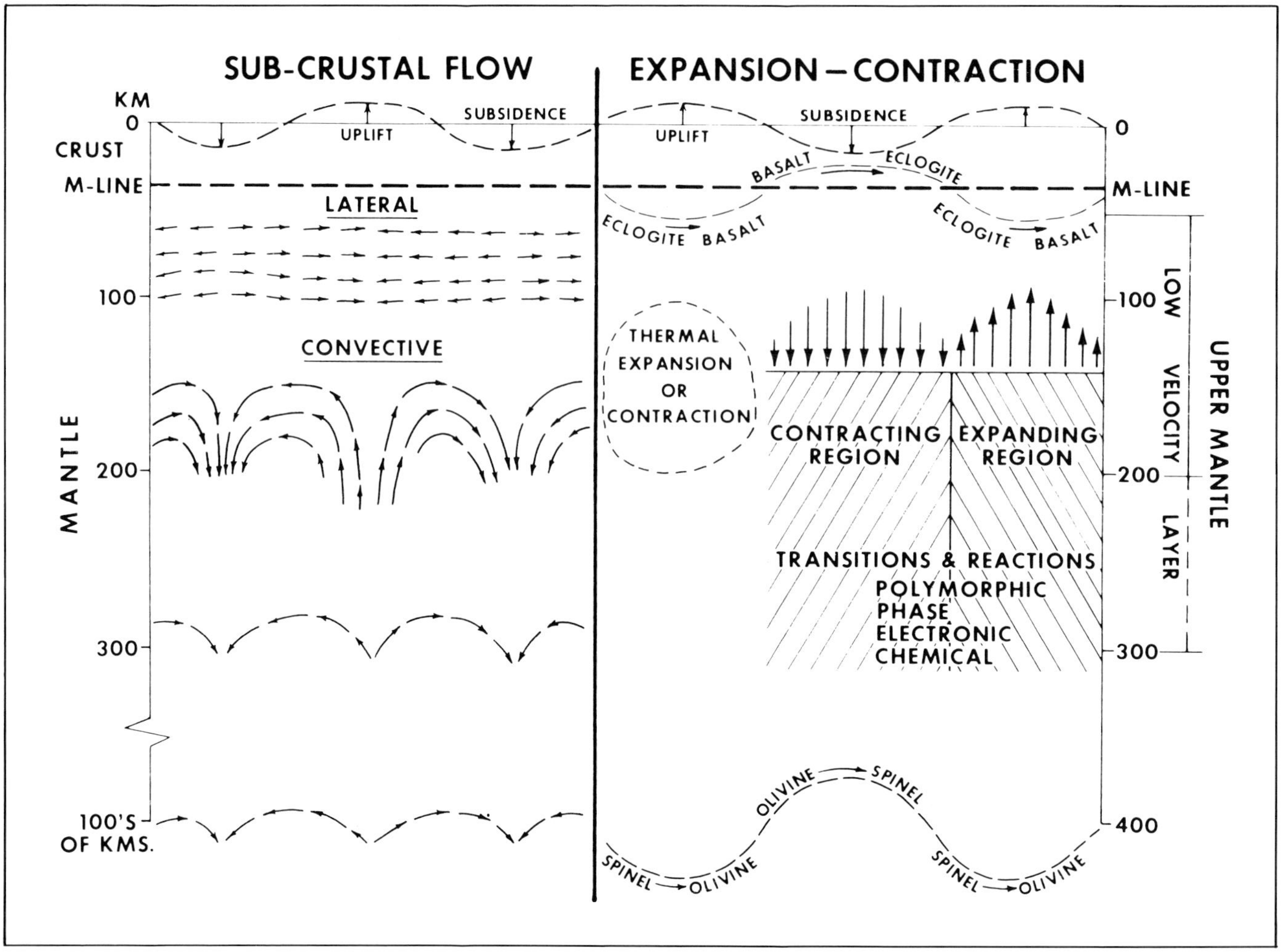

SUB-CRUSTAL FLOW
EXPANSION – CONTRACTION
KM
0
CRUST
M-LINE
UPLIFT
SUBSIDENCE
LATERAL
100
CONVECTIVE
200
300
100'S OF KMS.
MANTLE
BASALT
ECLOGITE
THERMAL EXPANSION OR CONTRACTION
CONTRACTING REGION
EXPANDING REGION
TRANSITIONS & REACTIONS
POLYMORPHIC
PHASE
ELECTRONIC
CHEMICAL
OLIVINE
SPINEL
400
LOW VELOCITY LAYER
UPPER MANTLE

Fig. 5-1—Sub-crustal flow and expansion-contraction as causes of crustal uplift and subsidence.

PALEOZOIC & MESOZOIC MICHIGAN BASIN SEQUENCE
APPROXIMATE LIMITS MID-MICHIGAN GRAVITY HIGH
SPARKS ET. AL. #1-8 MICHIGAN BASIN DEEP BOREHOLE
10 Km

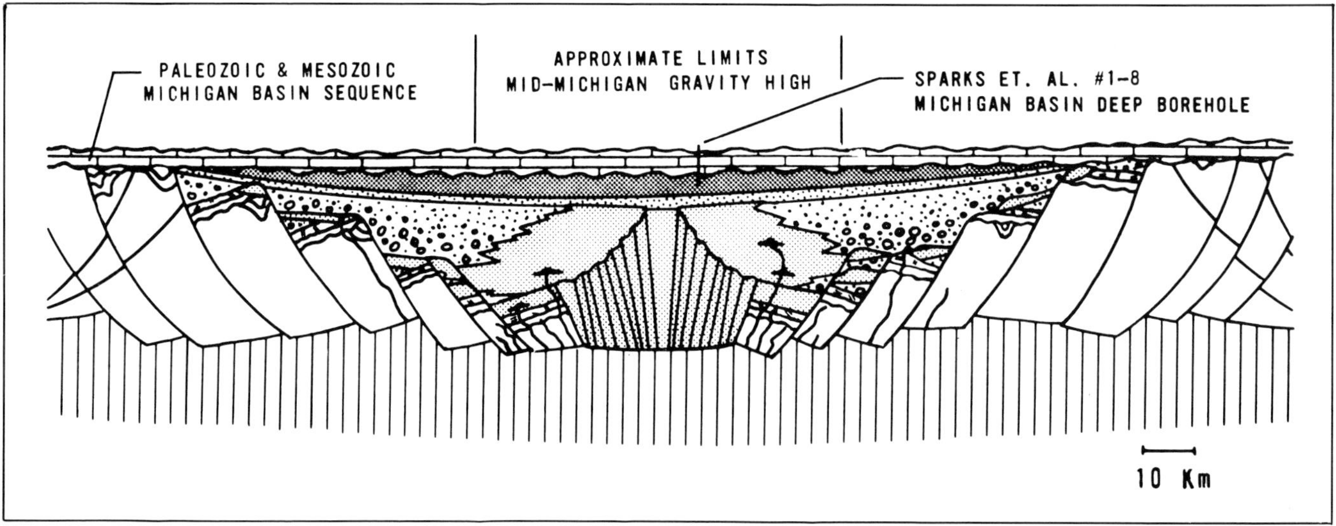

Fig. 5-2 (Fowler and Kuenzi, 1978)—Sag expressed mainly in Paleozoic rocks of Michigan Basin possibly created above an aborted triple point. From Journal of Geophysical Research, V. 83. Copyright by the American Geophysical Union. Reprinted by permission.

Illustration 5-3 begins on next page ⟶

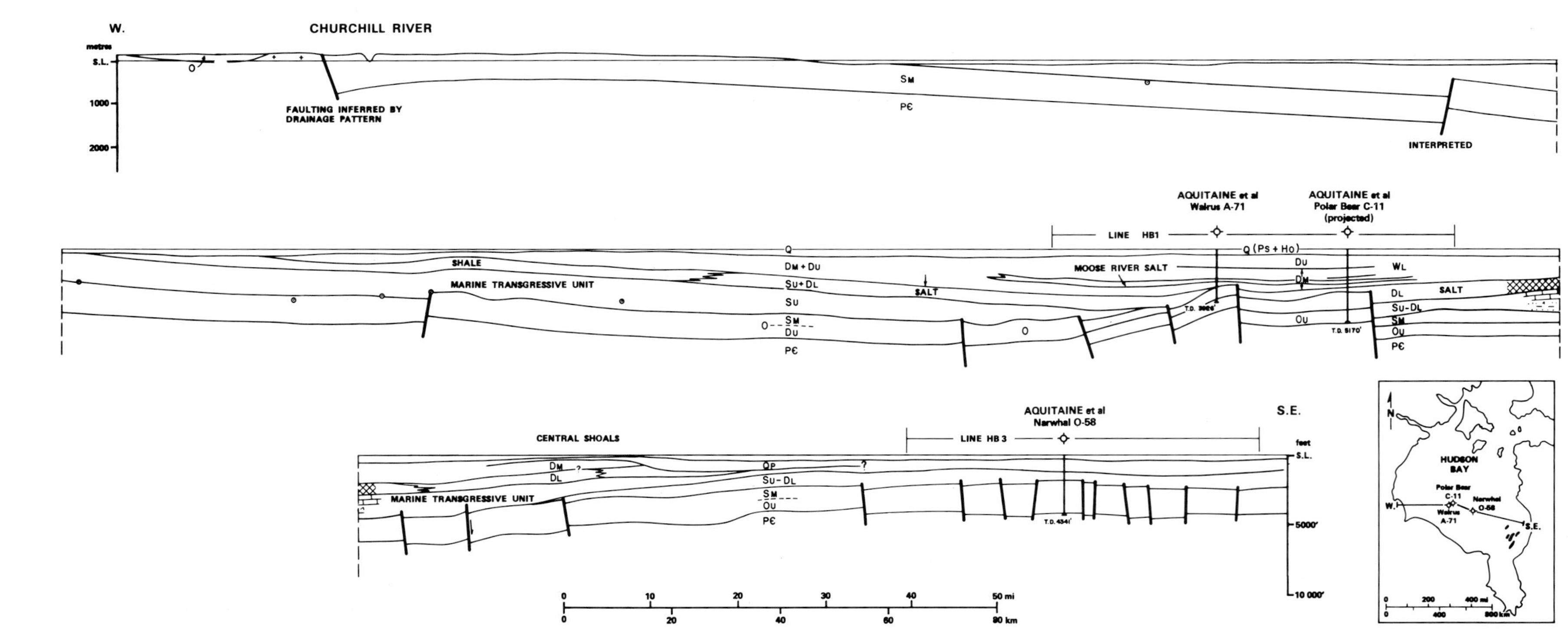

W.
CHURCHILL RIVER
metres
S.L.
1000
2000
FAULTING INFERRED BY DRAINAGE PATTERN
SM
PЄ
INTERPRETED
AQUITAINE et al Walrus A-71
AQUITAINE et al Polar Bear C-11 (projected)
LINE HB1
Q
Q (Ps + Ho)
SHALE
DM + DU
MOOSE RIVER SALT
DU
WL
MARINE TRANSGRESSIVE UNIT
SU + DL
SALT
DM
DL
SALT
SU
SU-DL
O
SM
DU
OU
T.D. 5170'
OU
PЄ
CENTRAL SHOALS
AQUITAINE et al Narwhal O-58
LINE HB 3
S.E.
feet
S.L.
DM
DL
QP
SU-DL
SM
OU
PЄ
T.D. 4341'
5000'
10 000'
0
10
20
30
40
50 mi
0
20
40
60
80 km
N
HUDSON BAY
Polar Bear C-11
Narwhal O-58
Walrus A-71
W.
S.E.
0
200
400 mi
0
400
800 km

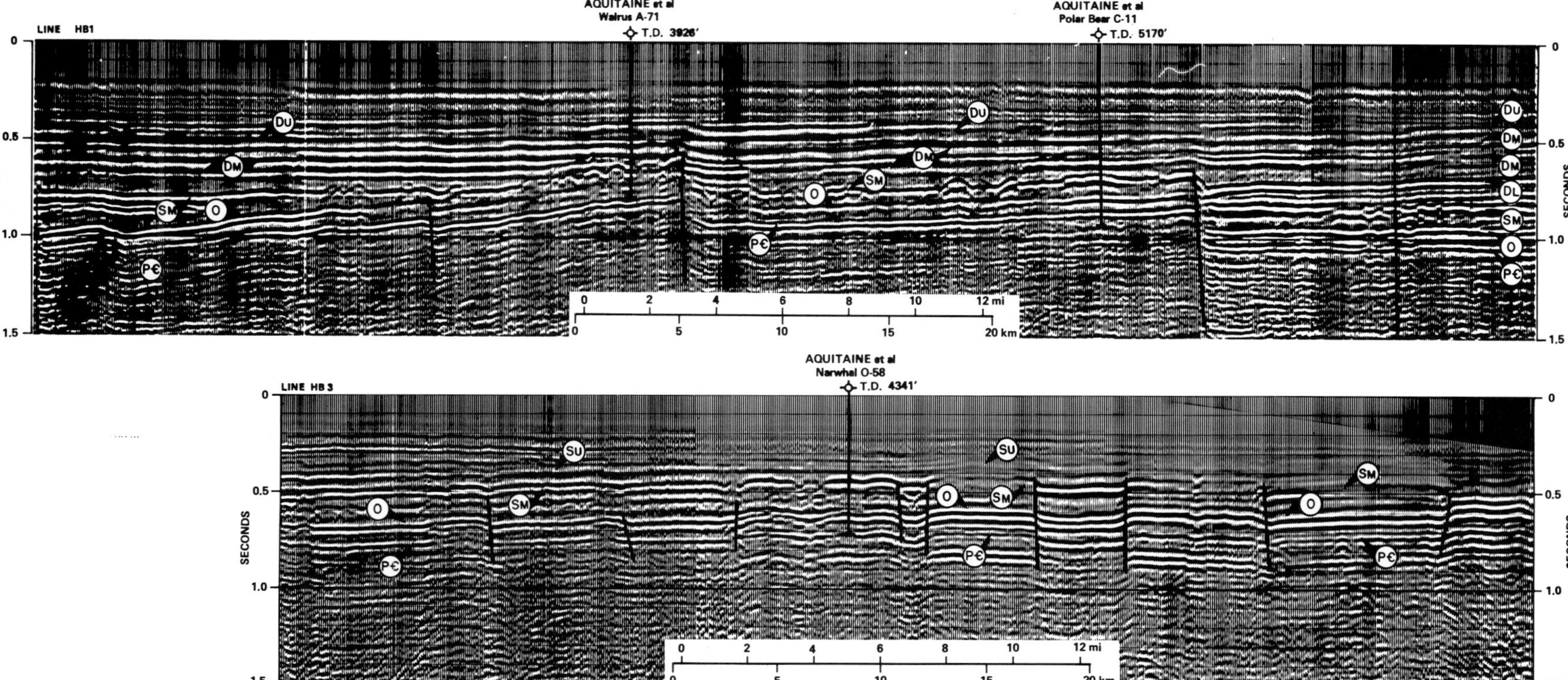

Fig. 5-3 (Dimian et al., 1983)—Seismic line and geological interpretation from cratonic Hudson Bay Basin. Faults have offset Precambrian basement and lower beds by a few hundred feet at most, but upper beds are only warped. Permission to publish by American Association of Petroleum Geologists.

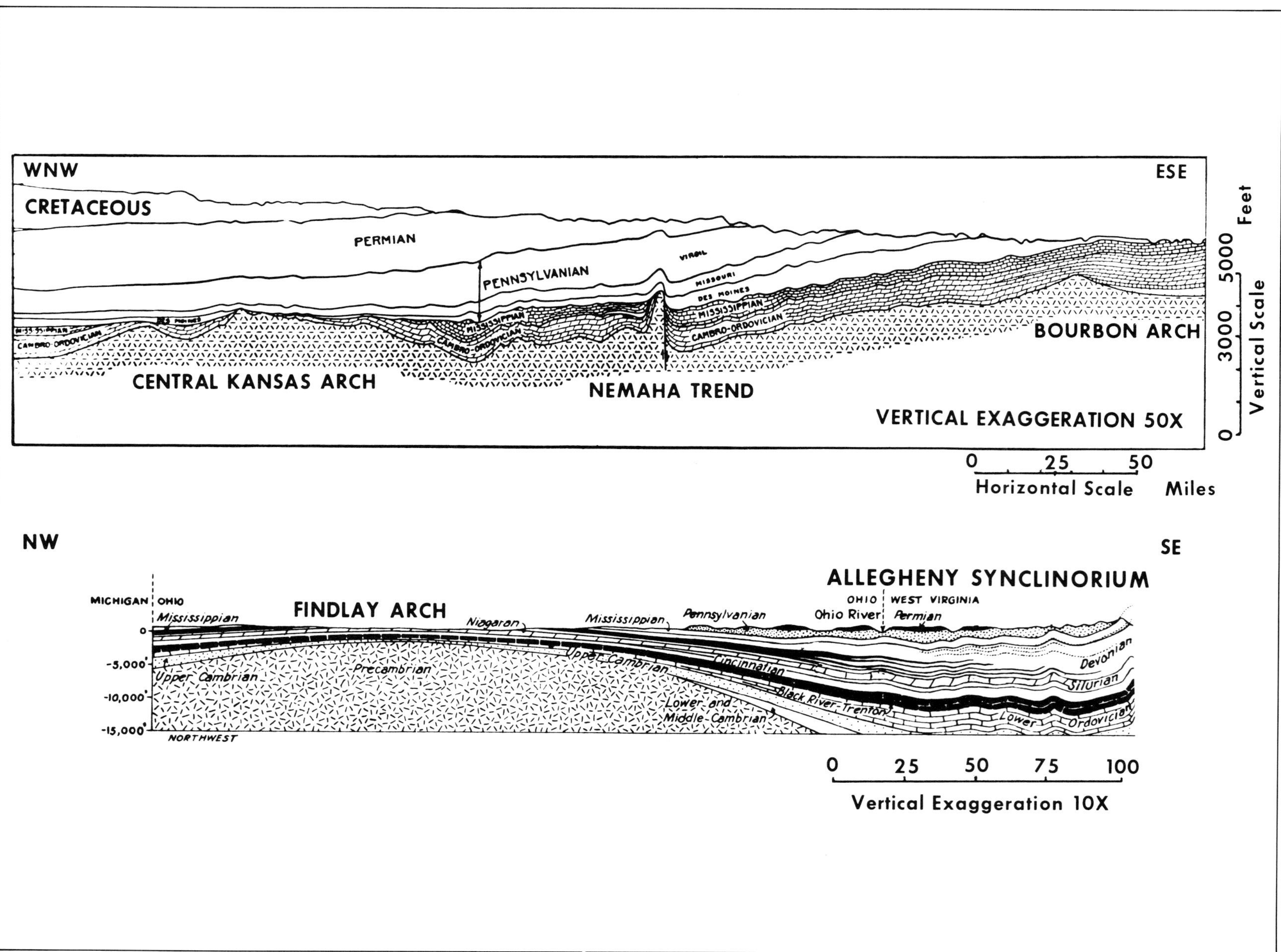

Fig. 5-4—The Findlay arch (Eardley, 1957), as a syn-depositional feature on the basis of onlap of Lower and Middle Cambrian and convergence of Upper Cambrian and Ordovician. Central Kansas arch (King, 1951) may have been, at least in part, post-depositional inasmuch as Cambro-Ordovician does not thin approaching the arch; however Pennsylvanian unconformity cuts through Cambro-Ordovician so relationship is not clear. Note that large vertical exaggeration is required to show the barest of details in such low relief structures.

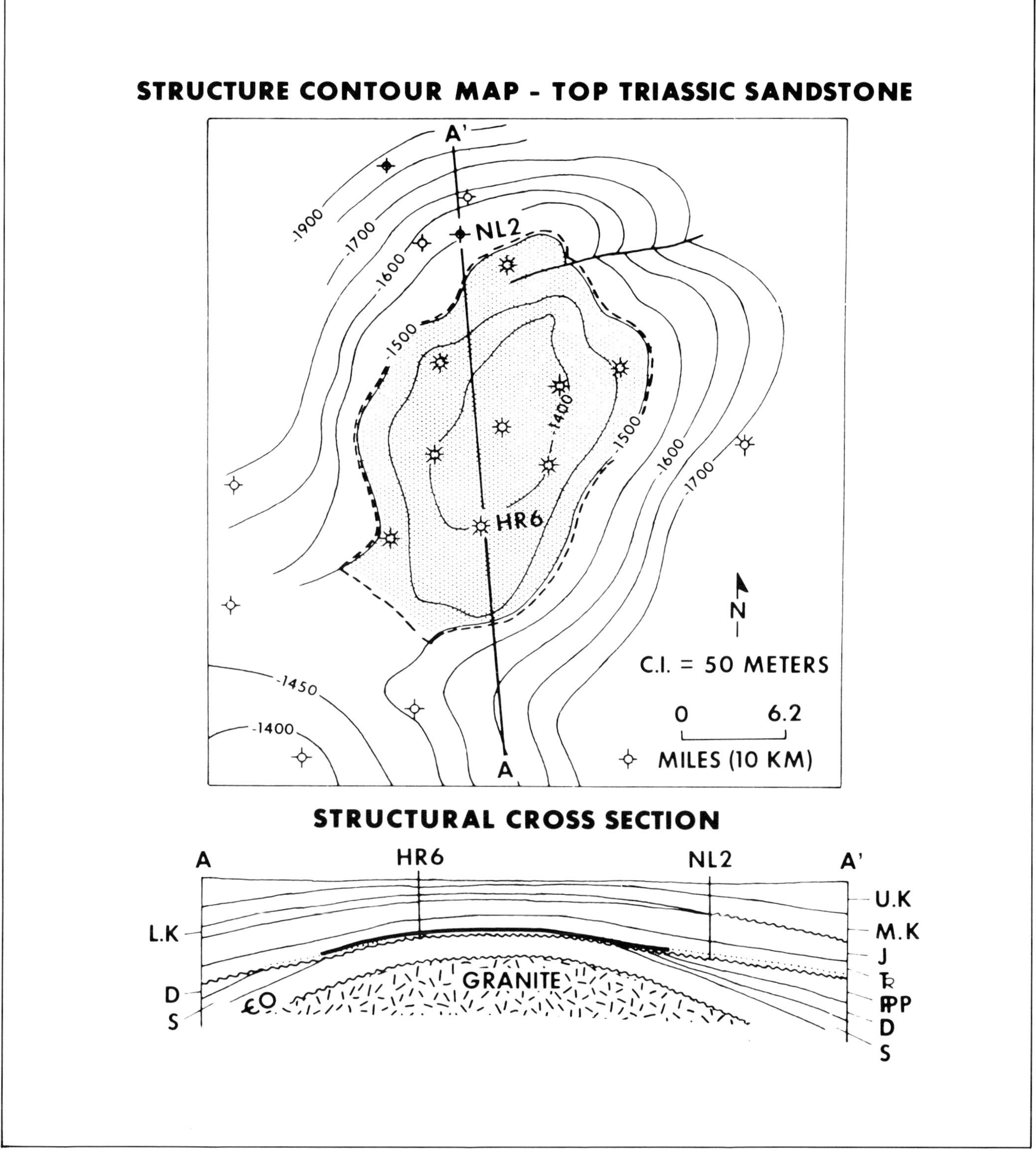

Fig. 5-5 (Delclaud and Martel, 1959)—Structure contour map (on top of Triassic) and cross section, Hassi Rmel field, Algeria. Relatively simple arch producing from a Triassic sandstone above truncated Paleozoic rocks and one of the world's largest gas fields, both in area (more than 700,000 acres) and EUR (70 TCF, 200 MMBO). Note lack of major faults. Permission to publish by l'Institut Francais du Petrole.

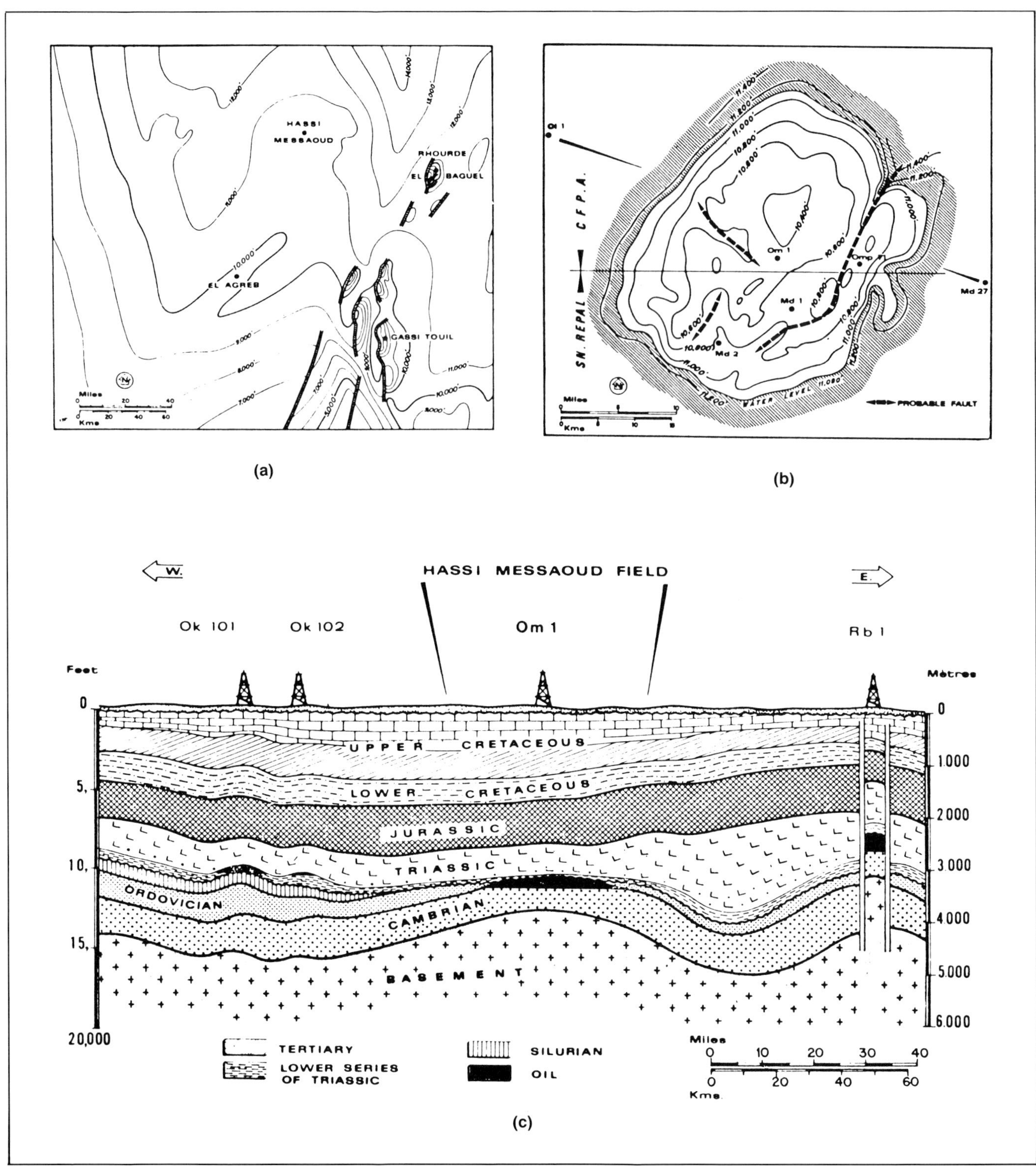

Fig. 5-6 (Balducchi and Pommier, 1970)—A. Regional structural contour map—datum top of Paleozoic—contour elevations below sea level — CI = 305 m (1000 ft) B. Structural contour map Hassi Messaoud Field—datum top of R2 Cambrian horizon-elevations below sea level — CI = 61 m (200 ft) C. Regional cross-section including Hassi Messaoud and Rhourde el Baguel (RB 1) fields, Algeria. Hassi Messaoud is a major field combining excellent trap (domal uplift), reservoir (Cambrian-Ordovician sandstone), source (Silurian shales), seal (Triassic lower series and evaporites), and migration from adjacent troughs. No major faults, but small faults (only a few shown) abound making reservoir tight in their vicinity owing to cementation. Vertical exaggeration approximately 16×. Permission to publish by American Association of Petroleum Geologists.

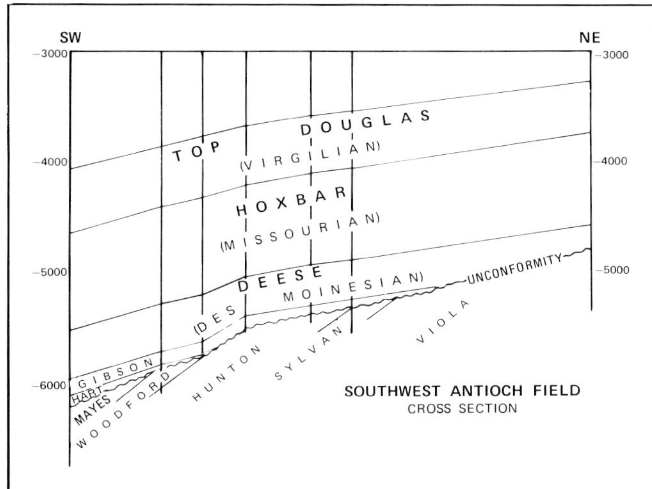

Fig. 5-7 (Selk, 1951)—The Southwest Antioch field in Garvin County, Oklahoma has oil pooled in up-dip limits of the Deese sand reservoir onlapping an unconformity on the flank of the Oklahoma City arch. Permission to publish by American Association of Petroleum Geologists.

turbidites in Michigan (deep borehole red beds): A foundered basin sequence developed during evolution of a protoceanic rift system: *Journal Geophysical Research,* V. 83, n. B12, p. 5833–5843.

King, P. B., 1951, The tectonics of middle North America: Princeton University Press, Princeton, New Jersey, 203 p.

Selk, E. L., 1951, Type of oil and gas traps in southern Oklahoma: *American Association Petroleum Geologists Bull.,* V. 35, p. 582–606.

Walters, R. F., 1958, Differential entrapment of oil and gas in Arbuckle Dolomite of central Kansas: *American Association Petroleum Geologists Bull.,* V. 42, p. 2133–2173.

6 THRUST-FOLD BELT ASSEMBLAGE

INTRODUCTION

Thrust-fold belts can be created by convergent wrench and by compressional-subduction processes. Criteria for differentiation between convergent wrench and compressional-subduction related belts have been given in Chapter 2. A relay vs. en echelon map pattern, listric thrust vs. upthrust profile, and wider vs. narrower zone of deformed sedimentary cover are among the most important characteristics to distinguish compressional-subduction from convergent wrench belts.

Strike slip can also take place in compressional-subduction orogenic belts, both in the form of longitudinal faults contemporaneous with subduction (Fitch, 1972) and as late lateral adjustments of colliding plates, as appears to be manifested along the Gailtal line which serves as a suture between the European plate and the southern Alps plate in the Austrian Alps. The potential for strike slip, subduction, and collision to occur in the same orogenic belt, coupled with the probability that subduction zones can flip and shift in time and space (Roeder, 1973), provide ample rationale in the framework of plate tectonics for the complexity of orogenesis.

Compressional-subduction belts, the subject of this chapter, are divided into forearc, backarc, and collisional types.

FOREARC BELTS

Forearc thrust-fold belts have only been recognized in the last 15 years. They represent deformation that has taken place in the landward walls of oceanic trenches in front of volcanic-magmatic arcs. Forearc belts are directly related to subducting oceanic lithosphere and, thus, are presently concentrated at the edges of the Pacific Ocean (Fig. 4-1), where they extend for a linear distance of many thousands of miles. Though prospectiveness for petroleum is very low in forearc thrust-fold belts, an understanding of their mechanics of deformation is instructive for the origins of collisional and backarc belts.

Similar models for oceanward thrusting in trench landward walls were proposed independently by Seely et al. (1974) and Beck and Lehner (1974). Both models were based on then new CDP deep penetration reflection seismic lines (Figs. 6-1 through 6-3), which represented an enormous improvement on earlier sparker records that masked the internal structure of the trench landward wall (Fig. 6-4). Both models are basically those of underthrusting[1].

The underthrusting process is probably better comprehended and conceptualized in the landward wall of an oceanic trench than in any other tectonic setting. Here an oceanic lithospheric plate, downgoing or subducting beneath a continental plate, can be viewed as the more active, non-yielding element in the deformation (Fig. 6-5). Drag and compression caused by the downgoing plate create compressional thrusts and folds in the trench landward

[1]Since the identical suite of structures (thrust, upright and overturned folds, and tear faults) with identical vergence is produced by either overthrusting or underthrusting (as first recognized by Van Hise in 1896), underthrusting properly must be restricted to genetic processes postulating "active agents" as was done by Hobbs (1914) and Lawson (1932). As Misch (1960) pointed out, "it is only on the basis of this distinction between an active unit, defined as stress-transmitting and non-yielding, and a passive unit, defined as stress-asorbing and yielding, that concepts such as 'underthrusting' and 'absolute direction of movement' are meaningful." Thus, although Lovering was among the early geologists to recognize the importance of underthrusting in 1932, his proposed field criteria for distinction between underthrusting and overthrusting on the basis of dynamically active and inactive blocks are not valid.

Misch (1960) postulated westward underthrusting of the basement to explain relatively eastward directed thrusts in northeastern Nevada, but it would still be difficult to conceive of basement being "strong enough to transmit stress without significant internal yielding," had not subsequent plate tectonic theory afforded a conveyor-belt type of movement for basement, and thereby removed the necessity of pushing it in some undefined way.

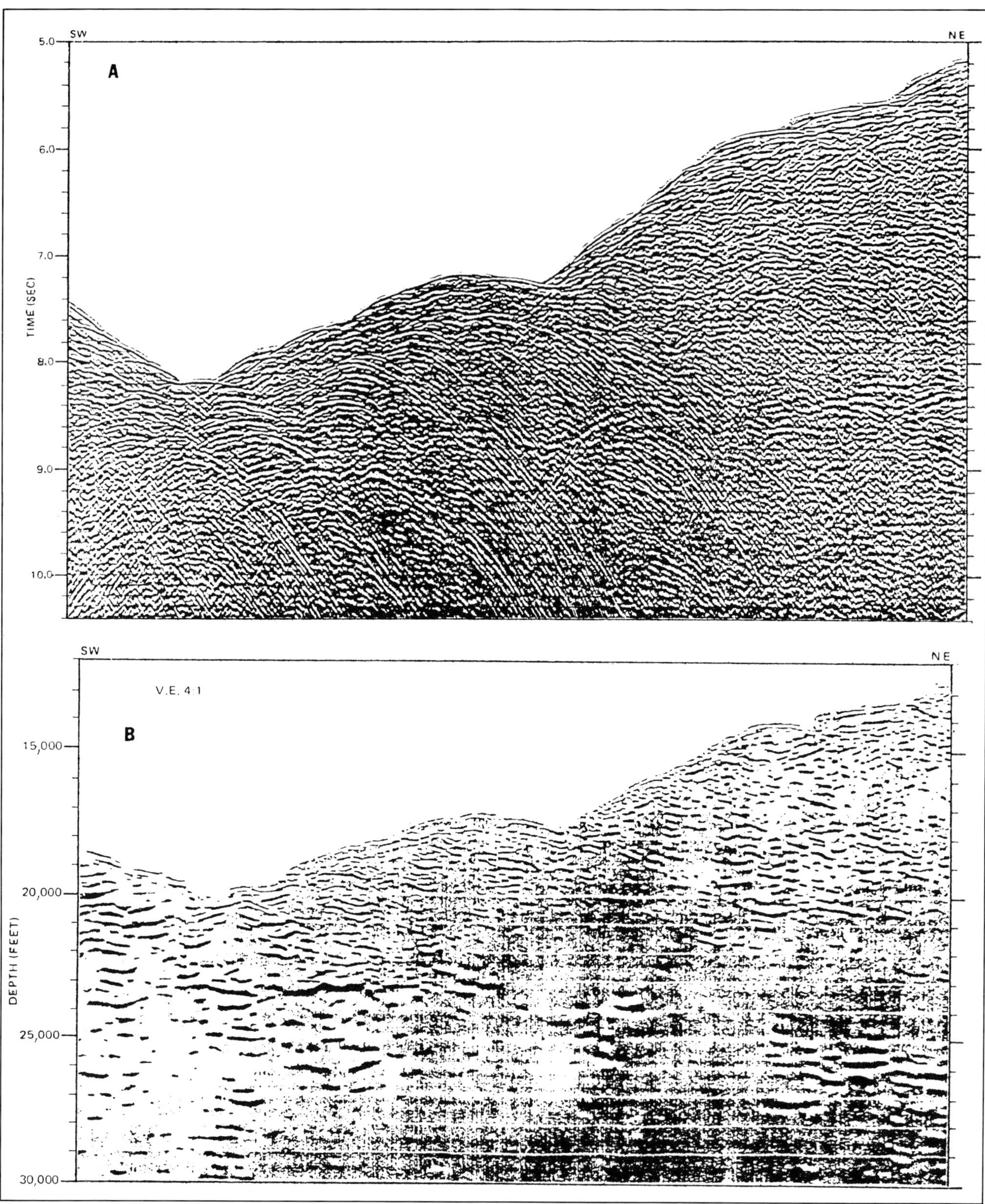

Fig. 6-1 (Seely et al., 1974)—Reflection seismic line from mid-America trench and slope. A. time display. B. clipped amplitude, migrated depth display showing detail of lower part of trench inner slope. From The Geology of Continental Margins, copyright © Springer-Verlag Publishers. Used with permission.

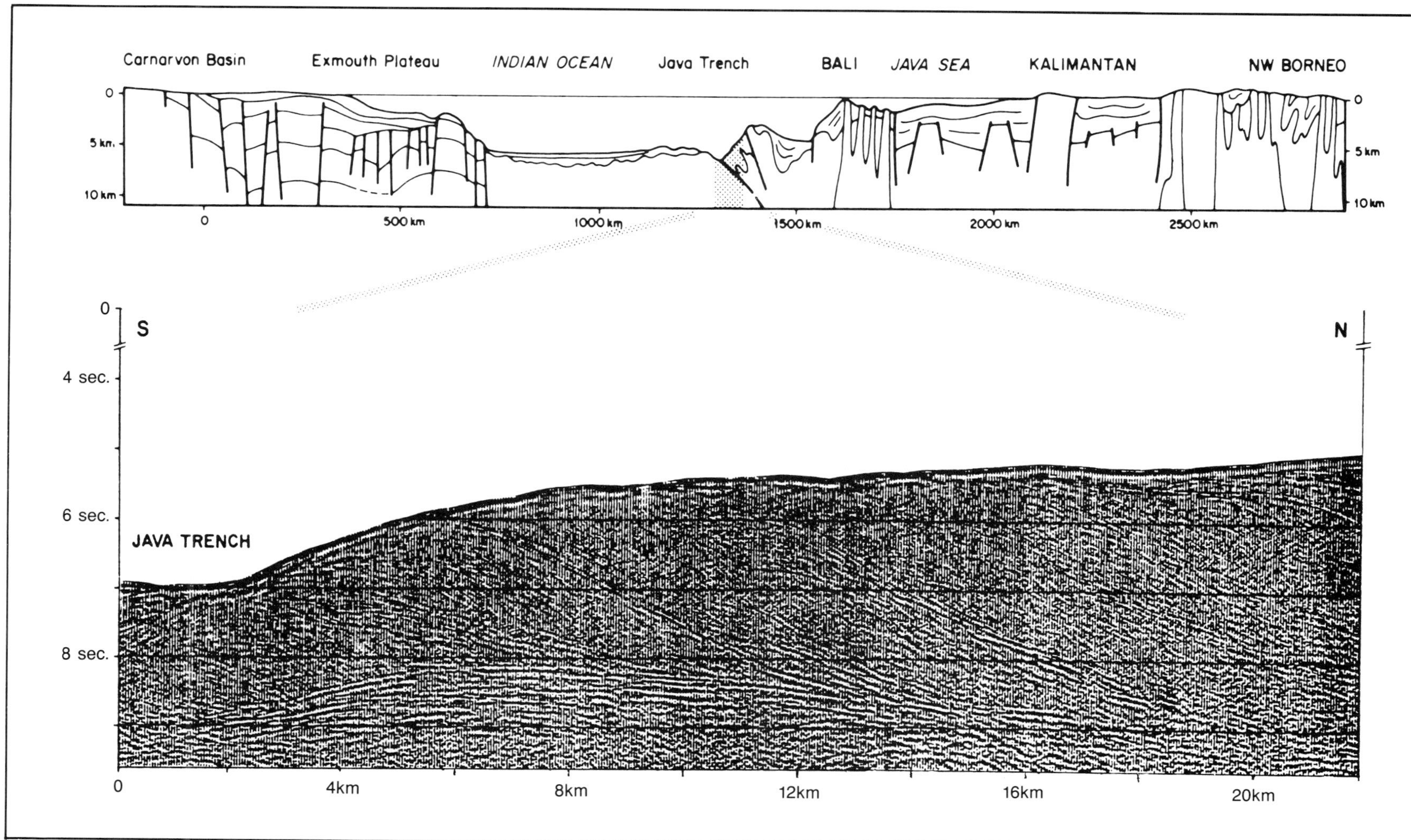

Fig. 6-2 (Beck and Lehner, 1974)—Structure of the Java trench inner slope from reflection seismic data. Permission to publish by American Association of Petroleum Geologists.

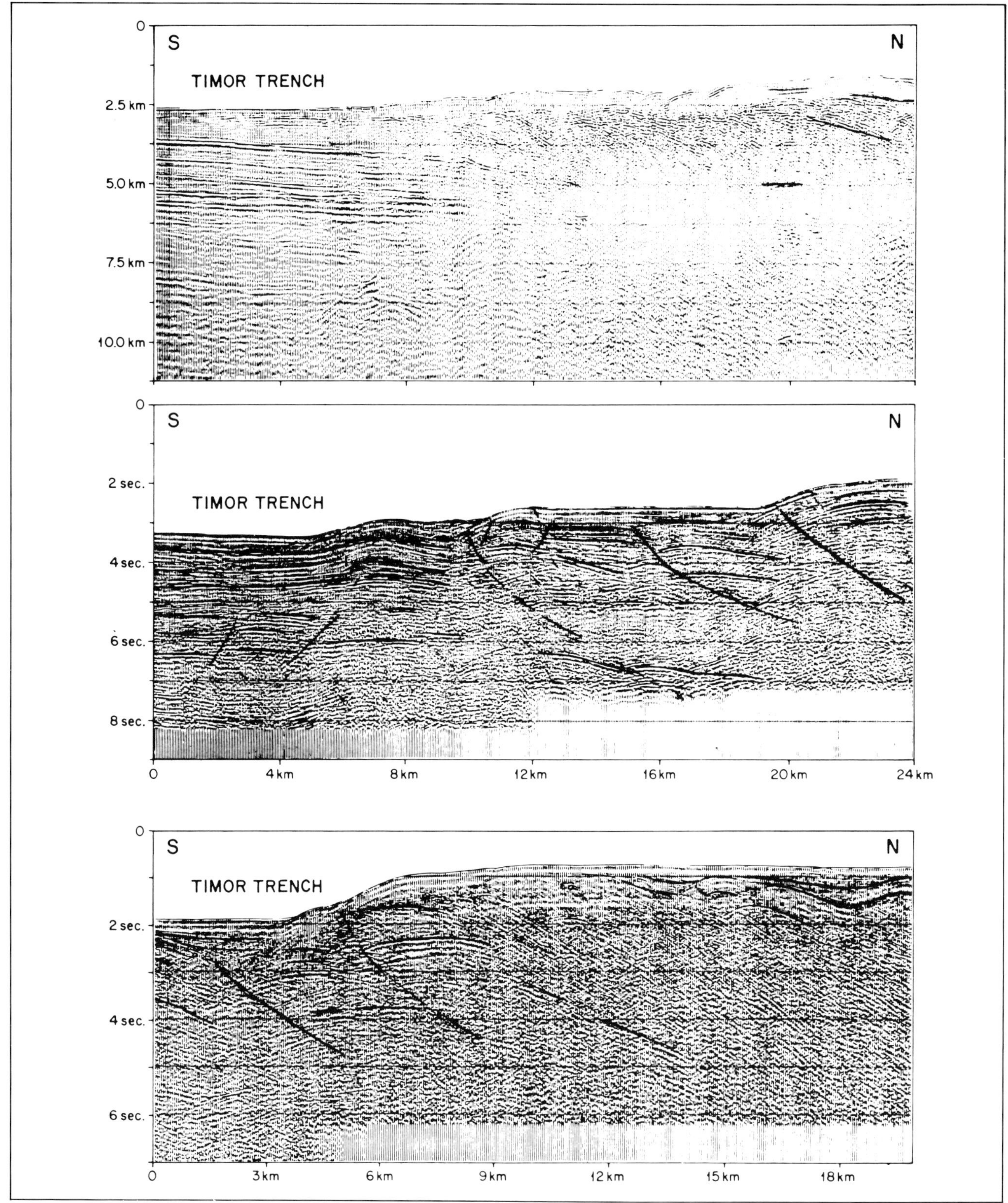

Fig. 6-3 (Beck and Lehner, 1974)—Three profiles across Timor trench south of Timor. North flank of trench is topographic expression of thrust front. Profiles show evidence of imbrications and compressional folding. Monoclinal nature of undisturbed sedimentary rocks on Australian (south) side of trench is visible on top profile. Strong reflector near 5 km (3 mi) could derive from Permian limestones. Permission to publish by American Association of Petroleum Geologists.

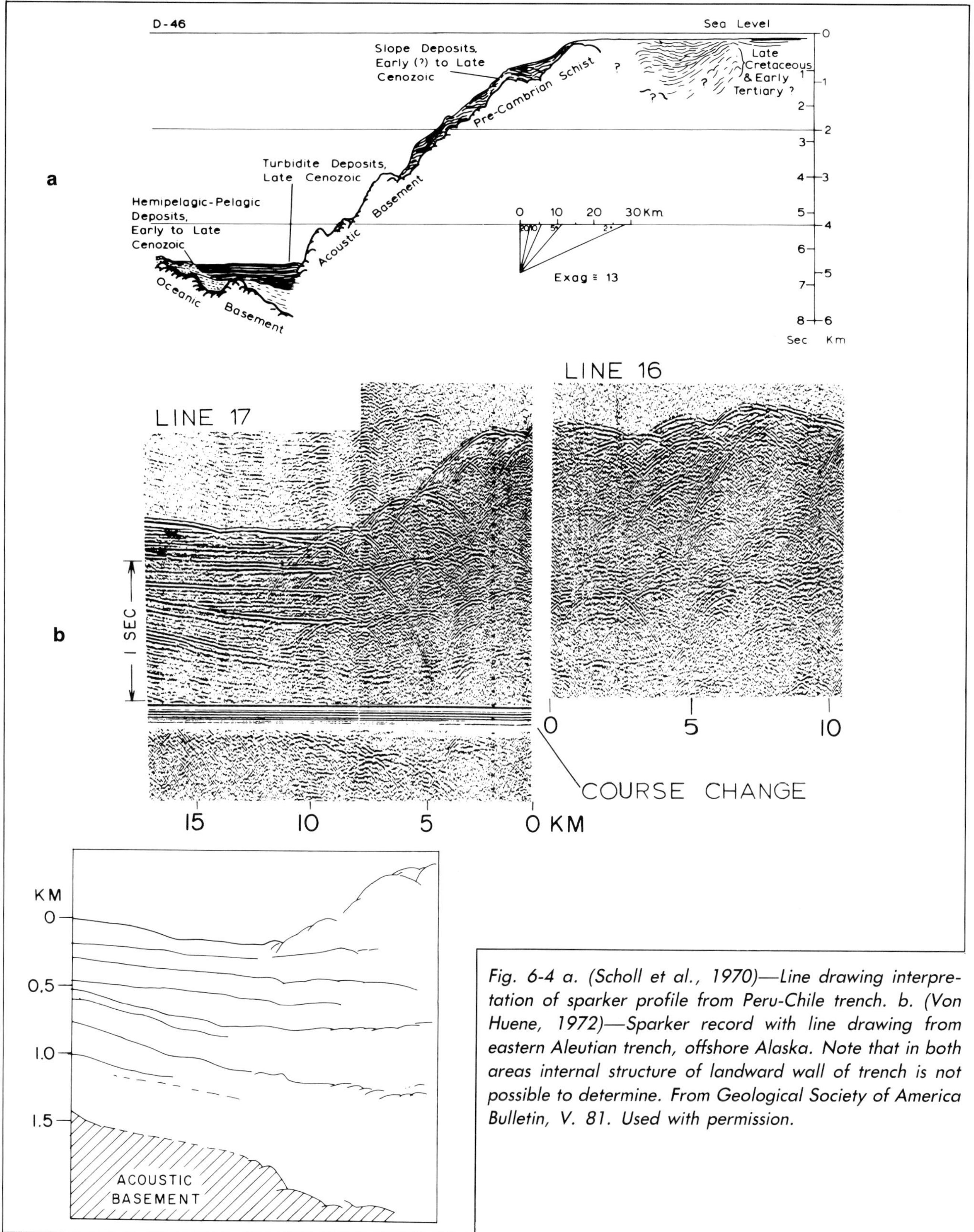

Fig. 6-4 a. (Scholl et al., 1970)—Line drawing interpretation of sparker profile from Peru-Chile trench. b. (Von Huene, 1972)—Sparker record with line drawing from eastern Aleutian trench, offshore Alaska. Note that in both areas internal structure of landward wall of trench is not possible to determine. From Geological Society of America Bulletin, V. 81. Used with permission.

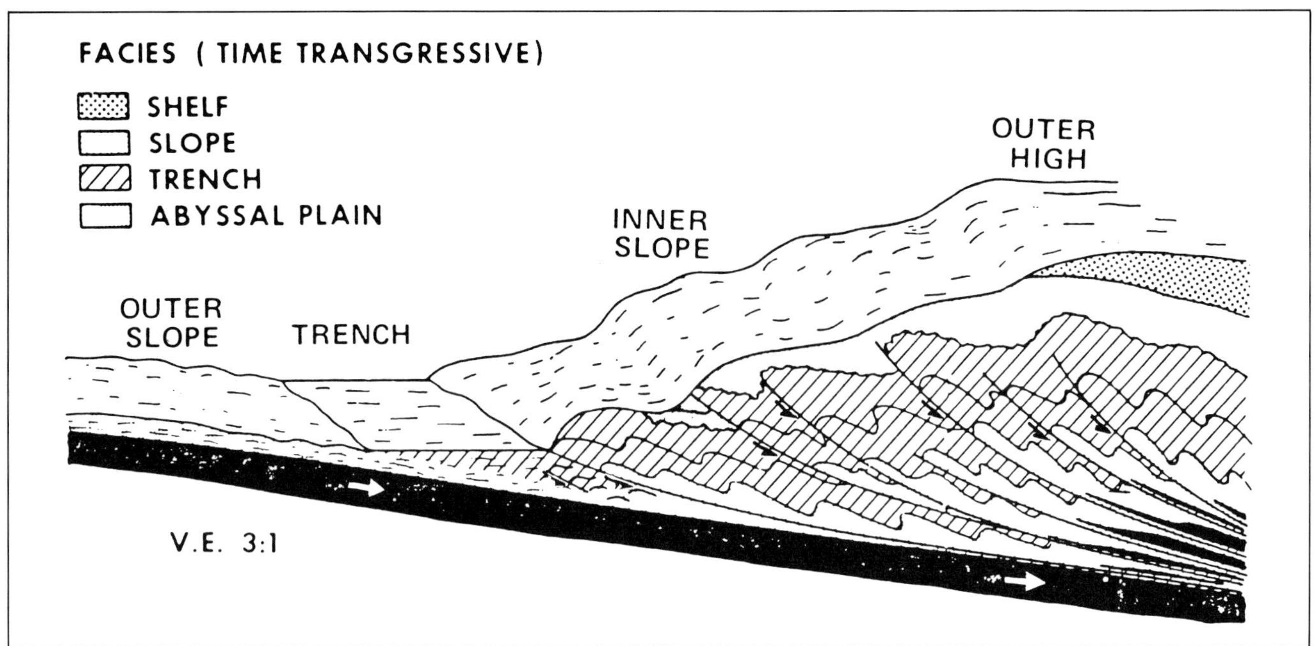

Fig. 6-5 (Seely et al., 1974)—Landward wall of oceanic trench with thrust-fold architecture caused by underthrusting of oceanic lithosperic plate. Note uplifted position of trench-deposited turbidites beneath the outer high. From The Geology of Continental Margins, copyright © Springer-Verlag Publishers. Used with permission.

wall. Besides accounting for the observed structures, salient points of the underthrust model are:

1– the sequence of thrusting is from oldest landward to youngest seaward, or decreasing in age toward the underthrust plate as the trench axis migrates seaward through time.

2– the wedging action provided by the insertion of successive thrust sheets from below uplifts and rotates earlier formed thrusts and folds and their contained rocks; axial surfaces of the folds may define fan structures (Fig. 6-6).

3– no deformation occurs to the subhorizontal strata of the trench until they encounter the seaward migrating zone of uncoupling at the intersection of the trench with its landward wall. Figure 6-7 shows remarkably clearly that trench sediments are undeformed until they are incorporated in and form the frontal folds of the landward wall. Thus, in not requiring that they be pushed from the rear, underthrusting obviates the material strength and stress transmission problems of long distance transport of thrust sheets.

Forearc thrust-fold belts are synthetic in that thrust direction and subduction have the same sense of motion.

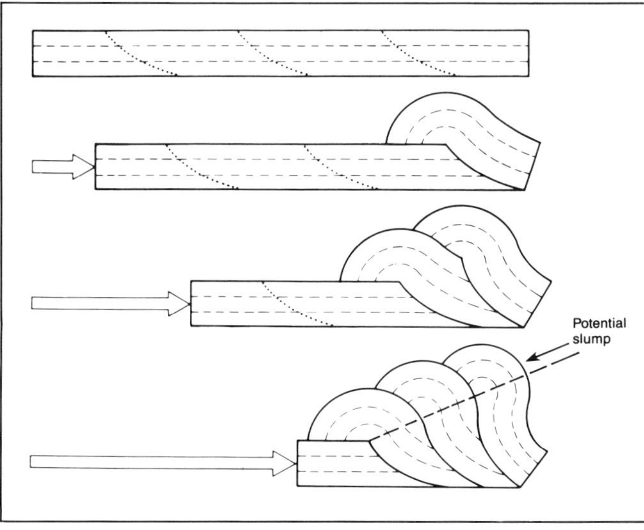

Fig. 6-6 (Seely et al., 1974)—Effect of wedging on thrust-fault dips and the development of fan structure in trench inner slope. The amount of rotation caused by wedging is related to the spacing between thrusts, the dips and configurations of the thrust surfaces, and the amounts of movement on those surfaces (up to a maximum, beyond which movement is not a factor; fault displacements shown equal that maximum). From The Geology of Continental Margins, copyright © Springer-Verlag Publishers. Used with permission.

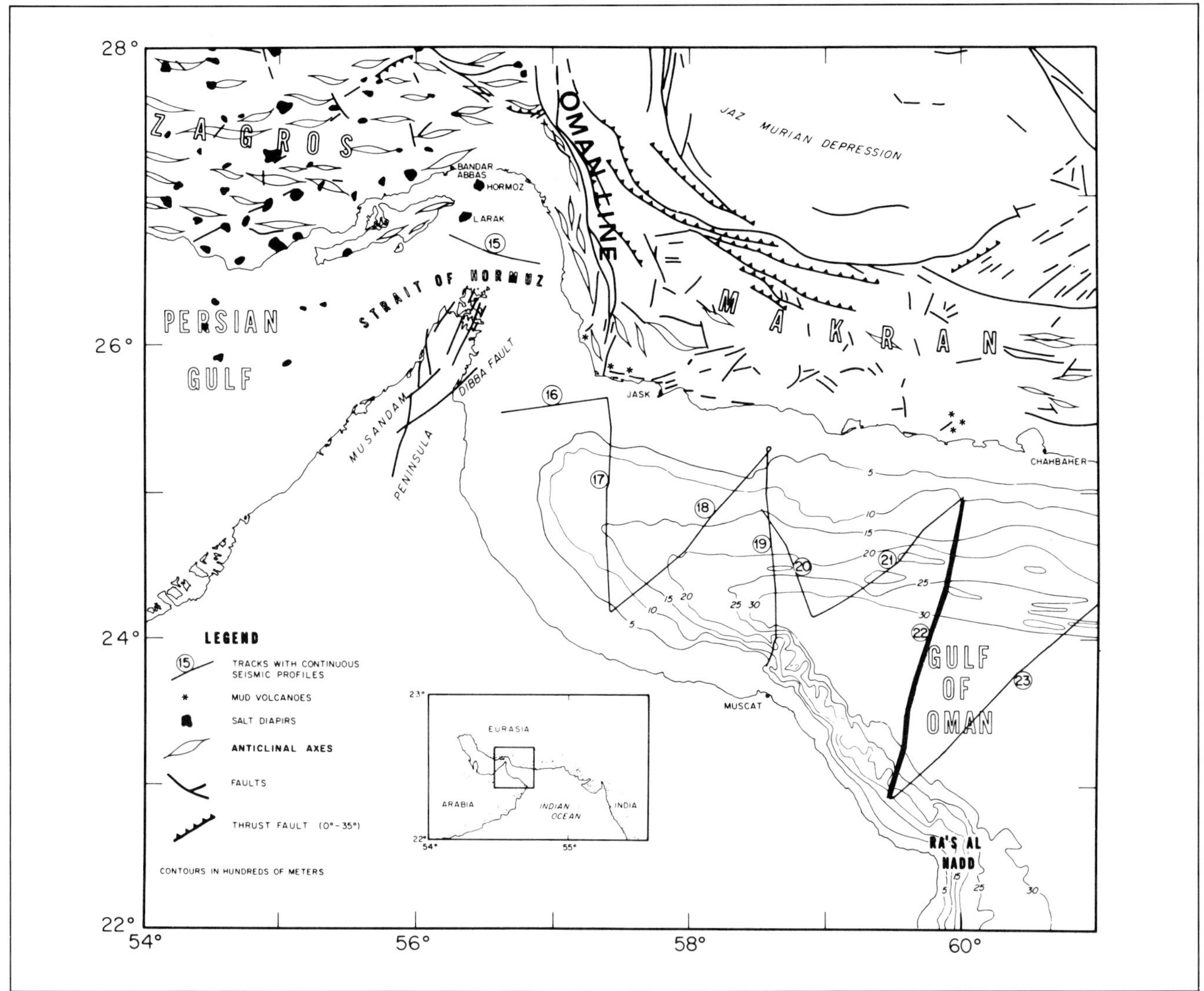

Fig. 6-7 (White and Ross, 1979)—A. Map of the Makran coast of Iran and adjacent Gulf of Oman. B. Seismic profile 22 (see map). Flat-lying deposits of the Gulf of Oman abyssal plain are initially deformed at the frontal fold and then incorporated into the accretionary prism of the Makran continental margin by a series of imbricate thrust faults. Vertical exaggeration approximately 13 times at sea level. From Journal of Geophysical Research. Copyright by the American Geophysical Union. Reprinted by permission.

BACKARC BELTS

Backarc thrust-fold belts occur behind or toward the plate interior from volcanic-magmatic arcs. They have been termed 'Cordilleran' belts by Dewey and Bird (1970). They have also been described as 'antithetic' by Roeder (1973) because directions are divergent between thrusting in the backarc and subduction of oceanic lithosphere; there is no direct compressional linkage from subduction to orogeny. The author believes that the process of underthrusting *within the continental lithosphere* explains backarc thrust-fold belts, but it is first necessary to consider other hypotheses of origin, including gravity.

The process of thrusting areally extensive thin sheets of sedimentary rocks for long distances has always been an enigma of structural geology. Perhaps the earliest hypothesis, overthrusting by compression, failed to account for effective transmission of stress through relatively weak materials. Later hypotheses of gravity movement invoked body forces to overcome the problems of stress transmission, but encountered other equally serious difficulties.

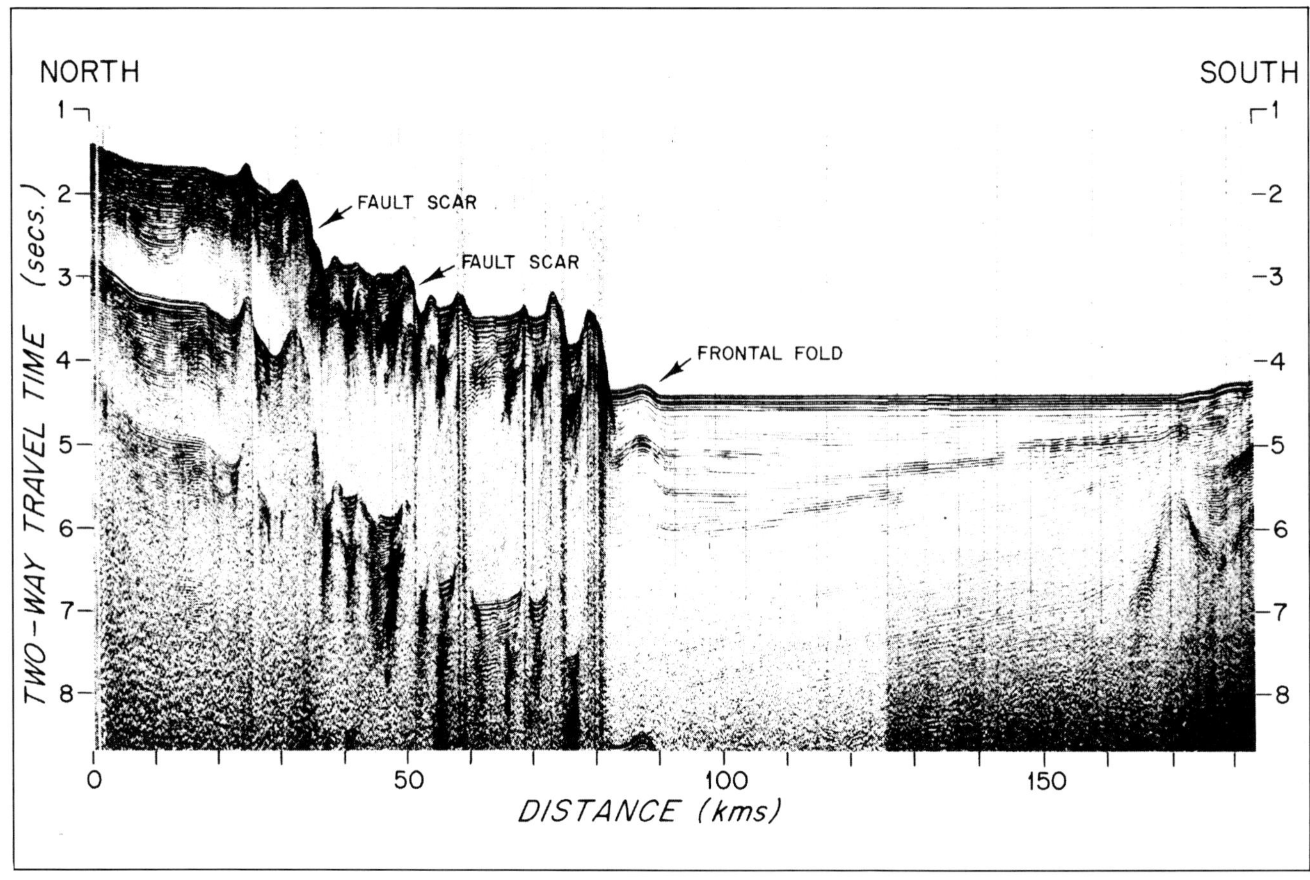

Fig. 6-7b (See caption for 6-7a, p. 267)

Gravity Sliding

Simple downslope gravity sliding was initially proposed for detached thrust-fold belts so that thrust sheets would be moved by body forces obviating the need for pushing them from the rear for long distances. However, serious objections can be raised to simple gravity slide models:

1– In the internal[2] portions of orogenic belts areas of tectonic denudation and associated normal faults required by gravity sliding are not observed; indeed synorogenic structures in that area are compressive.

2– Internal structures have a degree of penetrative deformation and, often, basement involvement, e.g., the Penninic nappes of the Alps, which indicate deformation under substantial load and crustal shortening.

3– Basement almost always slopes downward (toward the geosyncline and eventual orogenic core) throughout depositional and deformational history so that the gradient is actually opposed to that required for gravity sliding.

4– A globally well documented sequence of thrusting in thrust-fold belts is usually from older on the internal side to younger toward the external or cratonward side. Simple gravity models reverse this sequence, e.g., where proposed for the Montana Disturbed Belt thrusts are said to become younger internally (to the west) instead of older (Mudge, 1970, p. 384; Fig. 6-8).

5– The laterally extensive nature of thrust-fold belts following plate boundaries for thousands of kilometers is contrary to the more random or local aprons that would be expected from simple gravity sliding of sedimentary cover from uplifted regions. On the other hand the association of late Cenozoic thrust-fold belts with active subduction zones and older belts with presumed paleo-subduction zones is virtual proof of crustal convergence.

[2]"External" and "internal" are used here simply to indicate polarity of a compressional orogenic belt or thrust belt. "External" refers to the foreland or cratonward side, and "internal" refers to the core or most intensely deformed side.

Gravity Spreading

A significantly different and more sophisticated gravity model was partially described by Bucher (1956) and pro-

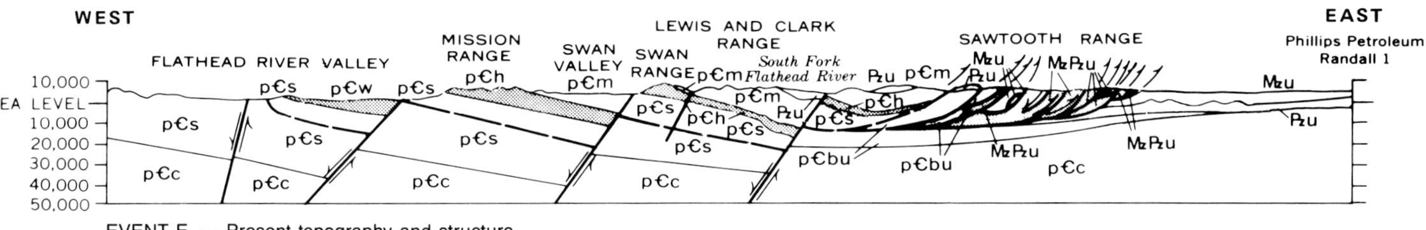

EVENT E. — Present topography and structure.

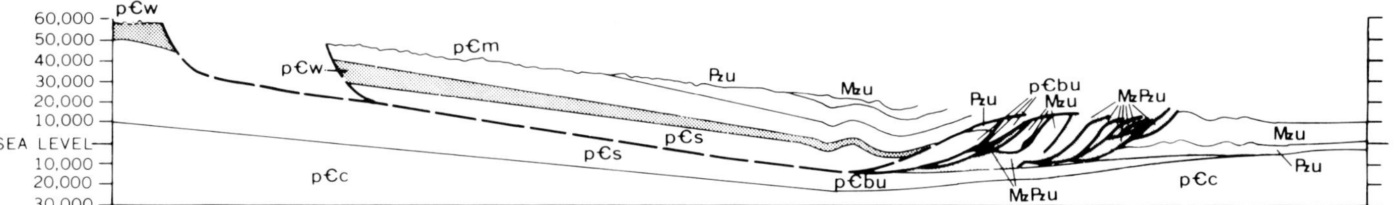

EVENT D. — Gravitational gliding to the east across the small Mesozoic basin, creating folds at the toe and piling one thrust plate upon another. The gap at the pull away zone is strictly diagrammatic. The amount of uplift shown is the maximum that may have occurred during this period.

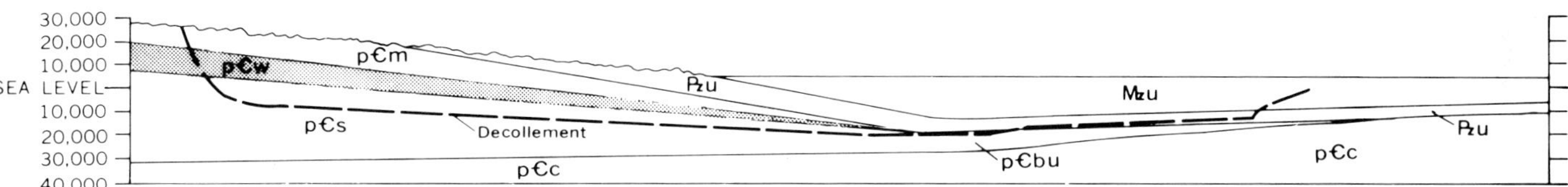

EVENT C. — Renewed uplift of the western area during early Paelocene with continued erosion. Abnormal fluid pressures very likely developed at depths as much as 25,000 feet where the decollement formed.

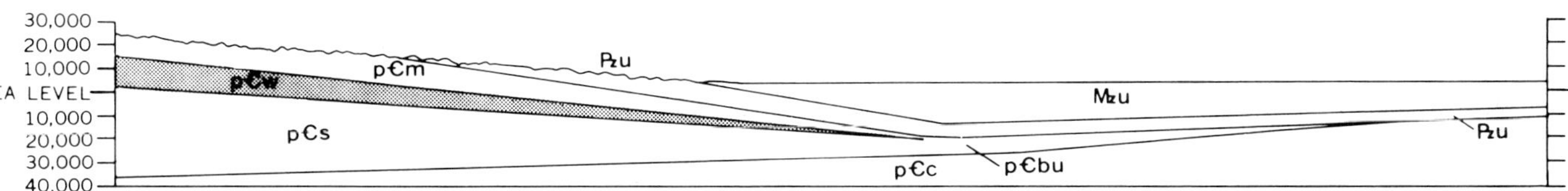

EVENT B. — End of Mesozoic sedimentation. Uplift to the west began during Late Jurassic, and continued periodically with some igneous activity through the Cretaceous. The thickest Mesozoic sediments accumulated in the narrow elongate basin along the east side of the highland.

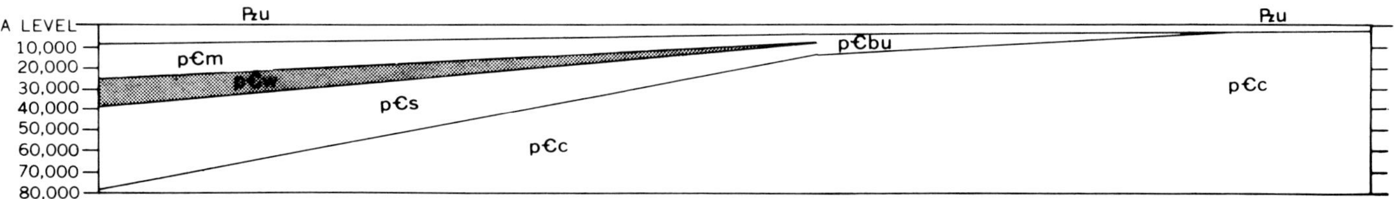

EVENT A. — End of Mississipplian sedimentation. Paleozoic deposition was in a miogeosyncline to the west which was superimposed on the Precambrian geosyncline.

Fig. 6-8 (Mudge, 1970)—Evolution of Montana disturbed belt by gravity sliding. Horizontal and vertical scale equal. Note for Event D section that thrust progression is opposite that known in other belts and that seemingly excessive elevation of the ground surface and slope (8°) on the decollement surface are required to effect gravity sliding up *the west-dipping basement to the east. From Geological Society of America Bulletin. Used with permission.*

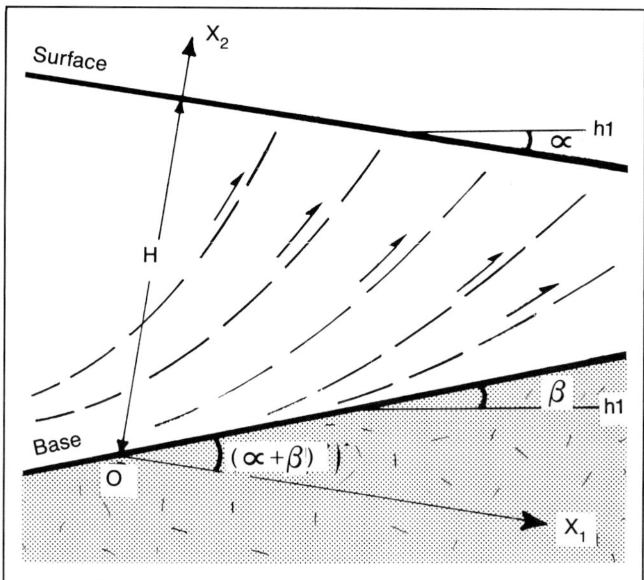

Fig. 6-9 (Elliott, 1976)—Gravity spread model. Note basement slope compared to that required for gravity sliding. From Journal of Geophysical Research. Copyright by the American Geophysical Union. Reprinted by permission.

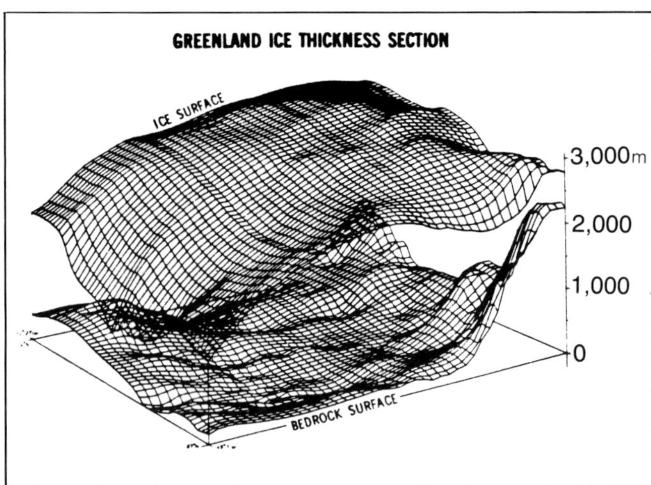

Fig. 6-10 (Geotimes, 1981)—Gravity spread analogy with Greenland ice sheet. Critical gradient for thrusting is ice surface not bedrock surface. Reprinted by permission.

posed by Price (1969, 1973) and Price and Mountjoy (1970) for the Canadian Rockies, and endorsed for even more general conditions by Elliott (1976). In this model surface slopes, not slopes on the actual thrust surfaces, are critical in permitting emplacement of thrust sheets by lateral gravitational spreading (Fig. 6-9). Surface slopes are in turn created by upwelling of an infrastructure of metamorphic and plutonic rocks (including rapidly uplifted magmatic arcs) in the internal part of the orogenic belt. The flow regime is considered analogous to that in large ice sheets, "e.g., Greenland where glacial flow from the central subsea depression to the higher edges of the continent appears to be uphill but is actually controlled by the gradient on top of the ice cap" (Price, 1969; Fig. 6-10). In my opinion the ice cap or glacier analogy is not entirely accurate because the central portion of an ice cap is continually replenished by external material from above, whereas the internal portion of the thrust-fold belt presumably is nourished from below, and not on such a continual basis.

Elliott (1976, p. 961) wondered whether thrusts in the arc-trench gap might not be "driven by the down-surface slope stress, with the regional surface slope being produced by uplift and igneous activity at the magmatic arc." If so, thrusting could not begin until sufficient surface slope had been achieved. It is difficult to understand how uplift and igneous activity at the magmatic arc could have created the surface slope required for thrusting and uplift of trench-deposited turbidites in the trench landward wall, when the latter feature in the case of the Middle America trench, is 150 km (94 mi) from the arc and separated from it by an 80 km-wide (50 mi) unstructured basin (Seely et al., 1974, fig. 12).

Campbell (1973) has questioned the gravitational spreading model for the Canadian Rockies both on the reality of the surface slope and on the basis that the area of the upwelling infrastructure "would seem not to have moved horizontally for a great distance." Another objection can be raised from the Idaho-Wyoming thrust-fold belt where thrusting began before any significant surface slope was attained (Fig. 1-20).

Nevertheless, sedimentary wedges, eventually attaining some degree of surface slope, are present in all thrust-fold belts and it should be noted that the postulated underthrusting model (to follow) and gravity spreading model are not mutually exclusive. Gravity sliding, but only on local gradients, could also modify both models. Underthrusting with associated crustal shortening is regarded here as the primary mechanism for the development of thrust-fold belts and gravity spreading and especially gravity sliding are regarded as distinctly secondary.

Underthrusting

Since the advent of plate tectonics, many geologists, including Coney (1973), Campbell (1973), Burchfiel and Davis (1975), and Dickinson (1976), have appealed to underthrusting for the origin of a variety of detached thrust-fold belts around the world. Inasmuch as underthrusting is ultimately a consequence of lithospheric plate movements, the more modern thrust-fold belts that can be related to plate motion at present are the most readily understood. Thus, Coney (1973, p. 131) was able to describe several Neo-

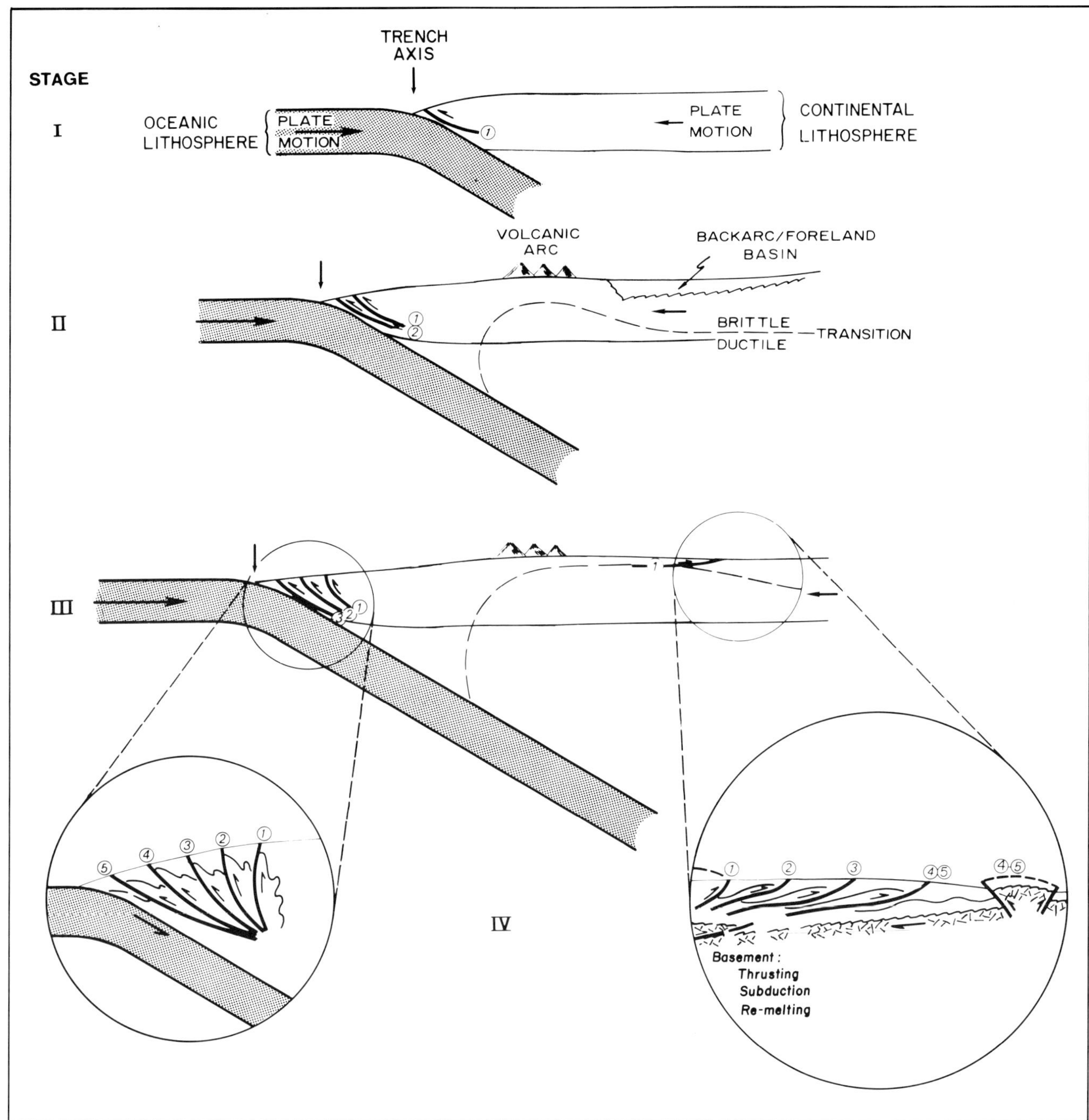

Fig. 6-11 (Modified from Lowell, 1977)—Underthrusting of trench landward wall by oceanic lithosphere and underthrusting of backarc thrust-fold belt by crystalline basement, the two areas separated by volcanic-magmatic arc. Numbers refer to sequence of thrusting, 1- oldest to 5-youngest. Thrust No. 1 in backarc position may be localized and initiated by elevation of critical isotherm which dictates brittle vs. ductile behavior. Thrusts at 5 portray uplift of basement by compression due to underthrusting (cf. Wind River Mountains—east side, fig. 1-20). Backarc belt would be analogous to Argentinian portion of Andes and to the Idaho-Wyoming belt. See text for additional discussion.

gene thrust-fold belts as "mainly due to underthrusting of continental lithosphere on leading edges of actively driving continental plates." These include: the Zagros belt developed by northeastward underthrusting of the Arabian plate and opening of the Red Sea; the Jura Mountains related to southwestward underthrusting of Europe and the late opening of the North Atlantic; the New Guinea thrust-fold belt caused by northeastward underthrusting of the Austra-

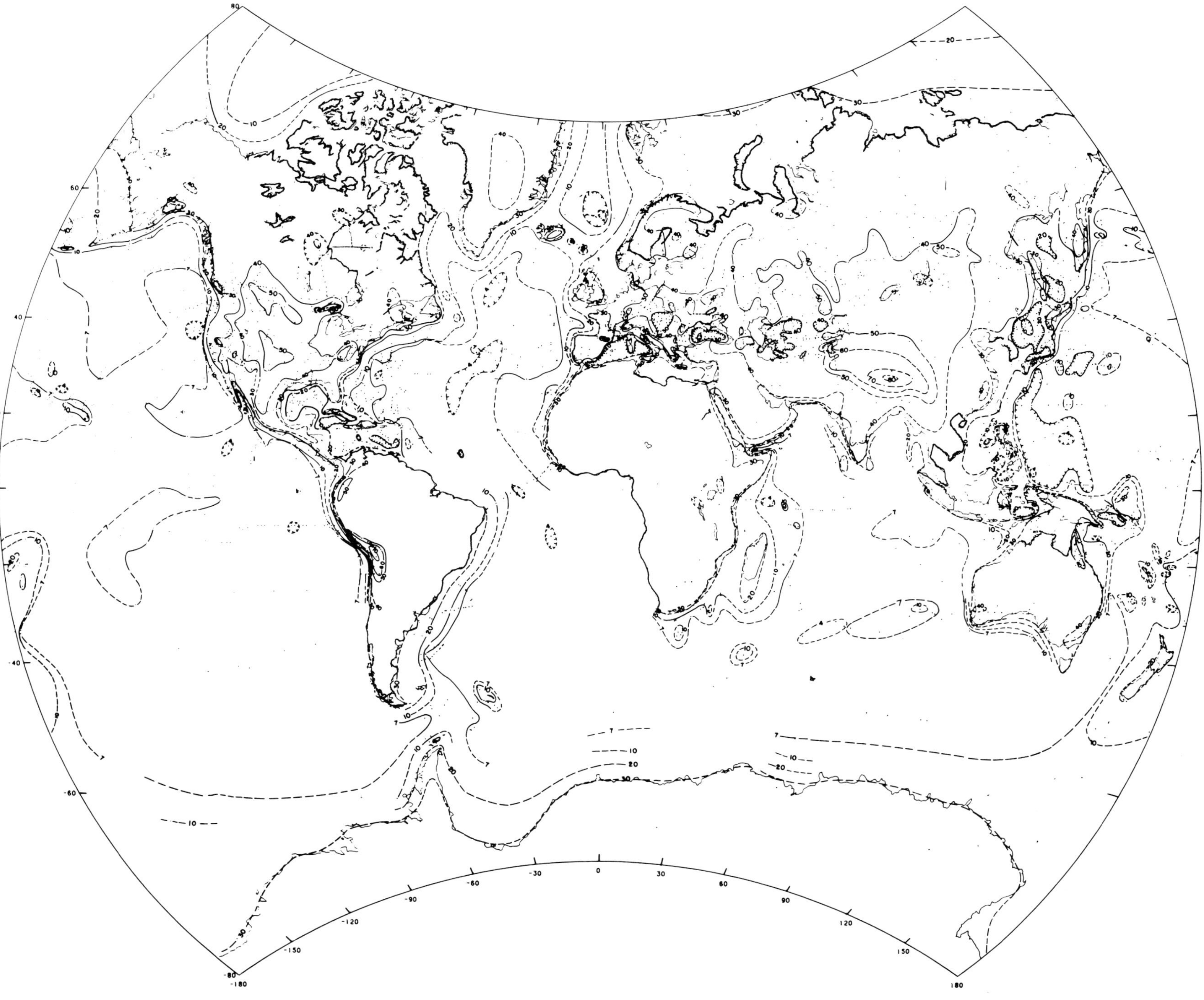

Fig. 6-12 (Soller et al., 1982)—Isopach map of earth's crust or depth to Moho. Dashed lines approximate. C.I.-10 km (6.2 mi). From Tectonics. Copyright by the American Geophysical Union. Reprinted by permission.

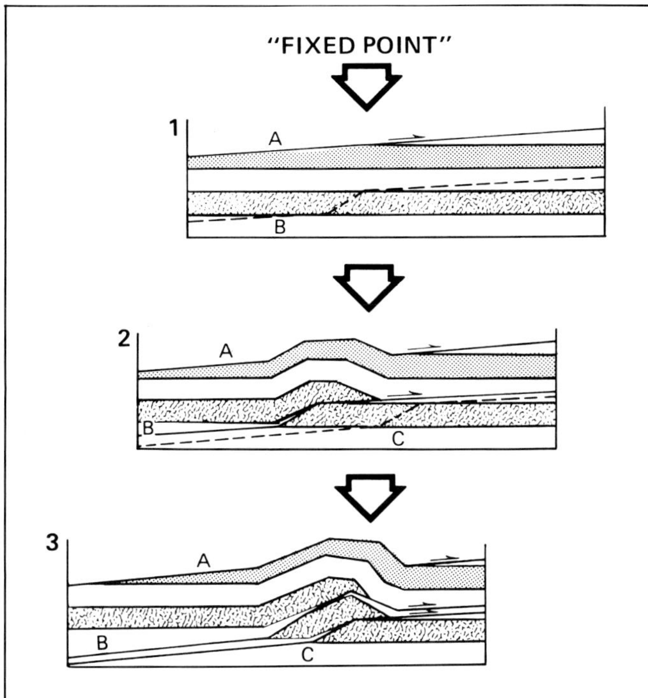

Fig. 6-13 (Modified from Jones, 1971)—Development of folded-fault structure by successive underthrusting of wedges from right to left. Thrust fault progression is toward underthrust plate, A (oldest), B (intermediate), C (youngest). Permission to publish by American Association of Petroleum Geologists.

lian plate as it separated from Antarctica; the Himalayans caused by the northward underthrusting of India and opening of the Indian Ocean; and part of the Andean system created by westward underthrusting of South America and opening of the South Atlantic.

Underthrusting is particularly appealing for the Andean-type system that lacks a direct mechanical drive between subduction of oceanic lithosphere and thrusting in a backarc setting in the superjacent continental lithosphere. Borrowing from the forearc model, underthrusting within the continental lithosphere can be applied to backarc thrust-fold belts (Fig. 6-11). The Idaho-Wyoming thrust-fold belt serves as an example.

In depicting the evolution of the Idaho-Wyoming thrust-fold belt, Royse et al. (1975, p. 48) have postulated that, "crystalline basement is moving west with the sedimentary section above being piled up on thrust faults and folds" (Fig. 1-20). Their model is in excellent accord with that of underthrusting in trench landward walls as shown by the following point-by-point comparison.

1– the sequence of thrusting was from oldest internally (coreward) to youngest externally (cratonward) (Oriel and Armstrong, 1966; Royse et al., 1975) or decreasing in age toward the underthrust plate.

2– the wedging action provided by the insertion of successive thrust sheets from below uplifts and rotates earlier formed thrusts and folds. Royse et al. (1975, p. 48) have stated that "initial uplift of the site of the Bear River and Portneuf ranges (west side of lowest cross-section, Fig. 1-20) is believed to be a result of thrust faulting..." The process of wedging and uplifting the more internal areas of the belt from below by underthrusting insures that the oldest rocks will be exposed in those areas, as they invariably are in most orogenic belts.

3– no deformation occurs to the subhorizontal strata overlying and in contact with the westward moving crystalline basement until they encounter the eastward or cratonward migrating zone of uncoupling, again obviating difficulties of long distance transport of thrust sheets. Furthermore, westward underthrusting appears to be the one mechanism capable of accomodating compressional stress in the basement of the Rocky Mountain foreland east of the Idaho-Wyoming thrust-fold belt, since, as in the case of the latter, stress need not be transmitted long distances and through thermally weakened zones from west to east. Consequently, underthrusting and attendant compression may be partly responsible for the development of foreland compressive basement blocks (Fig. 6-11), e.g., the Wind River Mountains (Fig. 1-20) and the Colorado Front Range, which Boos and Boos (1957) believed had been underthrust from the east during Laramide time. If this is so, however, basement structuring should be prevalent in the forelands of many underthrust-related detached thrust-fold belts; such is not the case, and in addition to basement compression, other mechanisms have been proposed (Lowell 1974a, 1974b, 1983—see Chapter 3).

As underthrusting and compression on one side of a plate proceeds in an orogenic belt, compensating expansion must take place elsewhere, usually on the opposite side of the plate. Thus, convergence or underthrusting on the leading edge of a plate should match in time with divergence on the trailing edge (see previous examples from Coney). Coincidence of timing appears to be very close between the evolution of the central Atlantic Ocean along the trailing edge of the America plate and the Idaho-Wyoming thrust-fold belt along but somewhat behind the leading edge of the same plate. Thrust belt deformation began in Late Jurassic time (Oriel and Armstrong, 1966; Royse et al., 1975) or the age of the oldest oceanic crust on the west and east edges of the central Atlantic Ocean (Pitman et al., 1974). Attenuation of continental lithosphere that precedes conti-

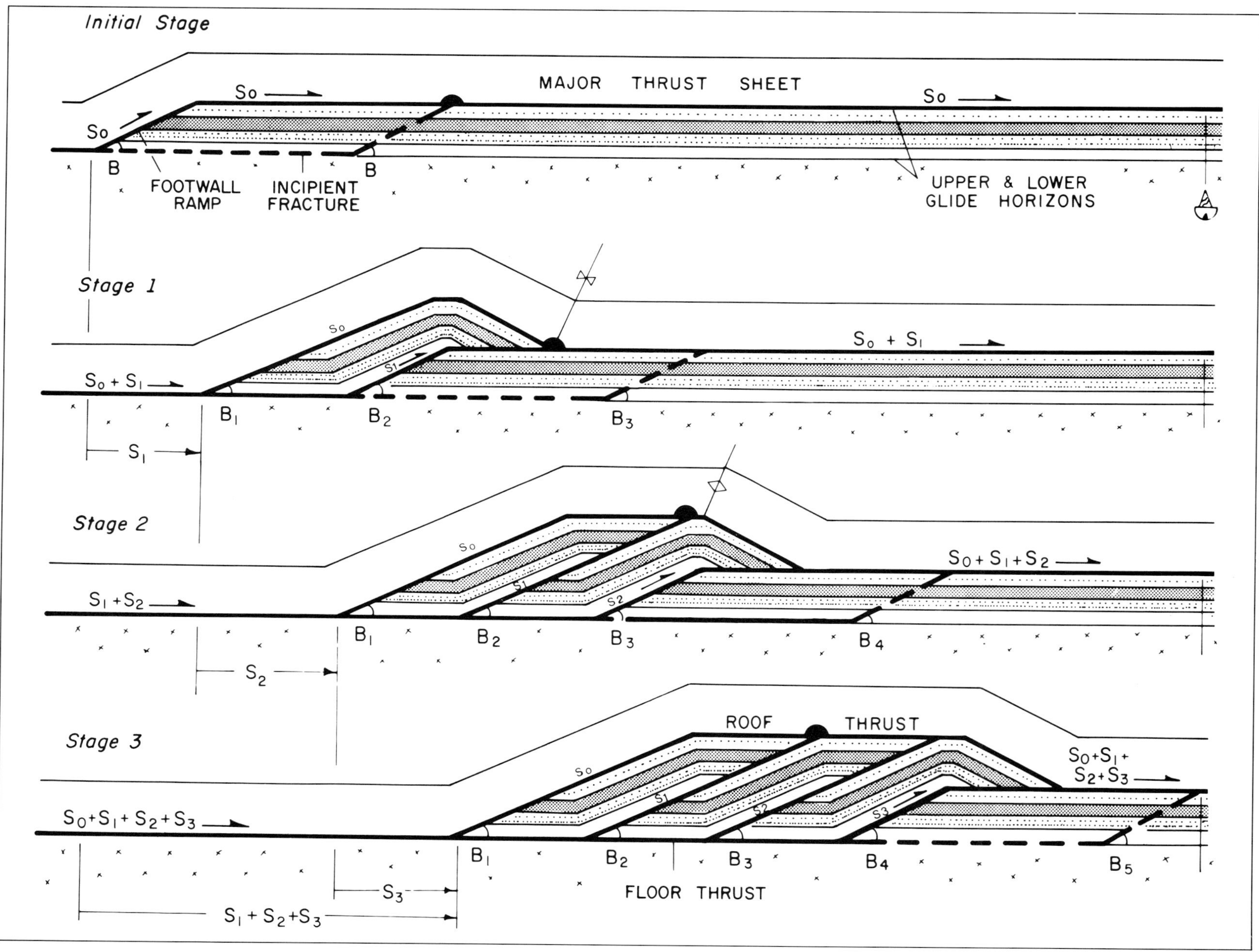

Fig. 6-14 (Boyer and Elliott, 1982)—Development of duplex structure (floor and roof thrusts with connecting imbricates) in measured graphical experiment. Duplex could as easily form by underthrusting as by overthrusting. Permission to publish by American Association of Petroleum Geologists.

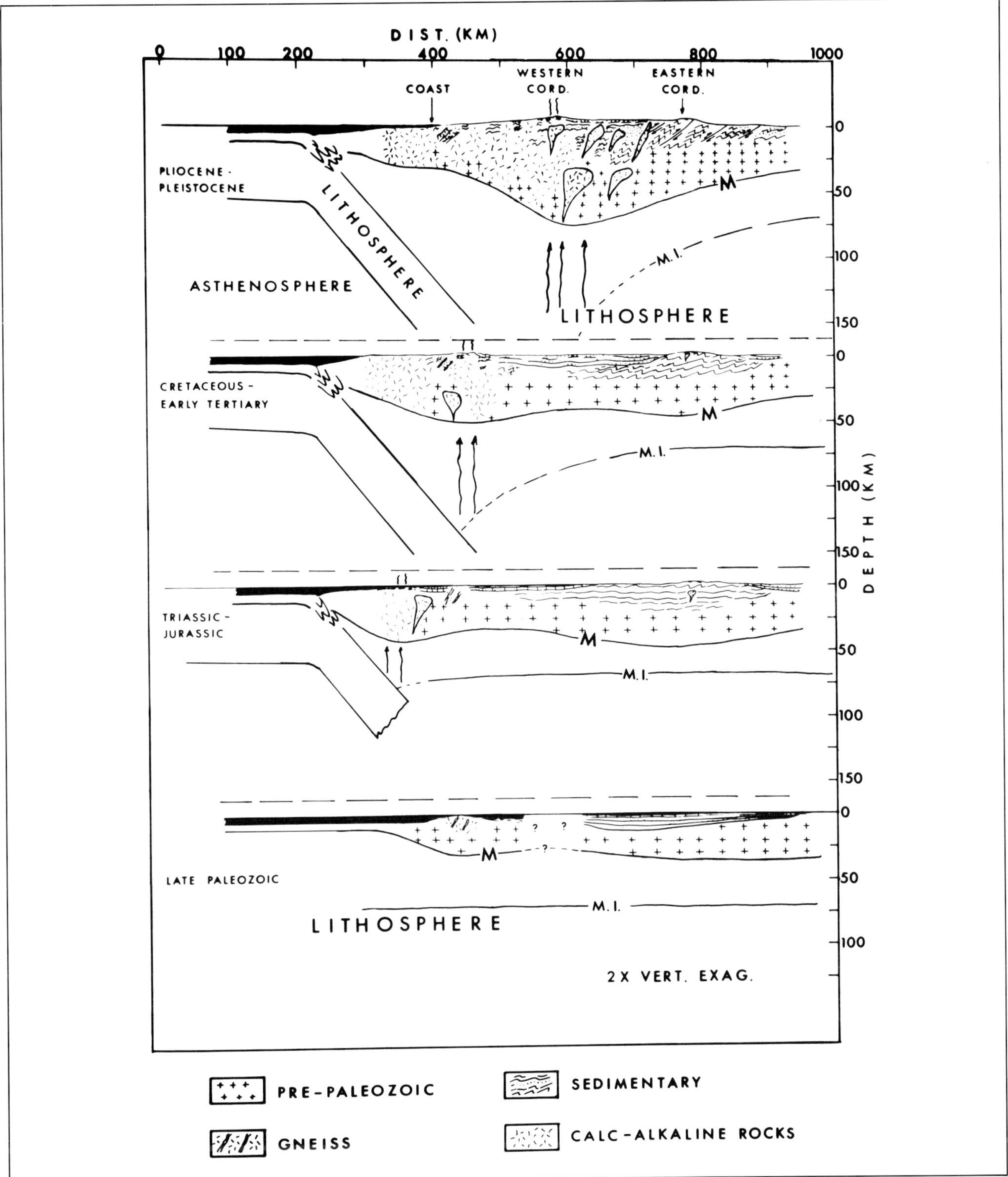

Fig. 6-15 (James, 1971)—Schematic sequence of cross sections illustrating a cordilleran model for the evolution of the central Andes. Letter "M" marks M-discontinuity. "M.I." denotes "magic isotherm" and indicates the presumed temperature at which melting or water release occurs in the lithospheric slab. Arrows denote rising magma. From Geological Society of America Bulletin, V. 82. Used with permission.

Fig. 6-16 (Geotimes, 1978)

nental breakup and sea-floor spreading (Lowell and Genik, 1972) may have been expressed as an earlier period of underthrusting west of the Idaho-Wyoming belt (Misch, 1971). Opening of the Atlantic Ocean has proceeded to the present (Pitman and Talwani, 1972), but termination of thrust belt activity in early Eocene time may be a result of the self-limiting nature of underthrusting described later in this section (Burchfiel and Davis, 1975, p. 385) inasmuch as no major change in plate movements occurred in the western North America-eastern Pacific area at that time (Atwater, 1970).

Shortening of the Idaho-Wyoming belt is generally thought to be on the order of 100 km (62 mi) (Royse et al., 1975, p. 48), which is, of course, far less than the 2600 km (1,622 mi) half-width of central Atlantic spreading. It should be noted, however, that the two figures need not match since the major boundary of the America plate was a subduction zone on its western margin; most of the differential plate motion was taken up at that boundary and the Idaho-Wyoming thrust-fold belt is a backarc intraplate feature.

As underthrusting of rigid sialic basement caused stripping and detachment of the sedimentary cover that amounted to 100 km (62 mi) of shortening in the Idaho-Wyoming belt, westward-moving basement could alleviate space problems by responding in at least three ways. First, basement is known to be thrusted on the internal sides of many thrust-fold belts (Figs. 1-20 and 6-11). Second, basement thrusting can be of sufficient magnitude to effect intra-continental subduction in the original Alpine sense.[3] Subduction and thrusting of basement on the internal sides of orogenic

[3]Although underthrusting is now usually thought to involve convergence between adjacent lithospheric plates, it originally involved downfolding within a single plate (White et al., 1970). Recognizing that there was too much platform section (that originally must have been laid down on continental crust) for the amount of basement present, Alpine geologists invoked the concept of a "Verschluckungs-Zone" or swallowing-up zone to explain the paucity of basement. The amount of shortening in intra-continental subduction is limited because of density considerations. Whereas thousands of kilometers of denser oceanic lithosphere can be subducted beneath lighter continental lithosphere, much less subduction can occur within continental lithosphere where there is no density contrast.

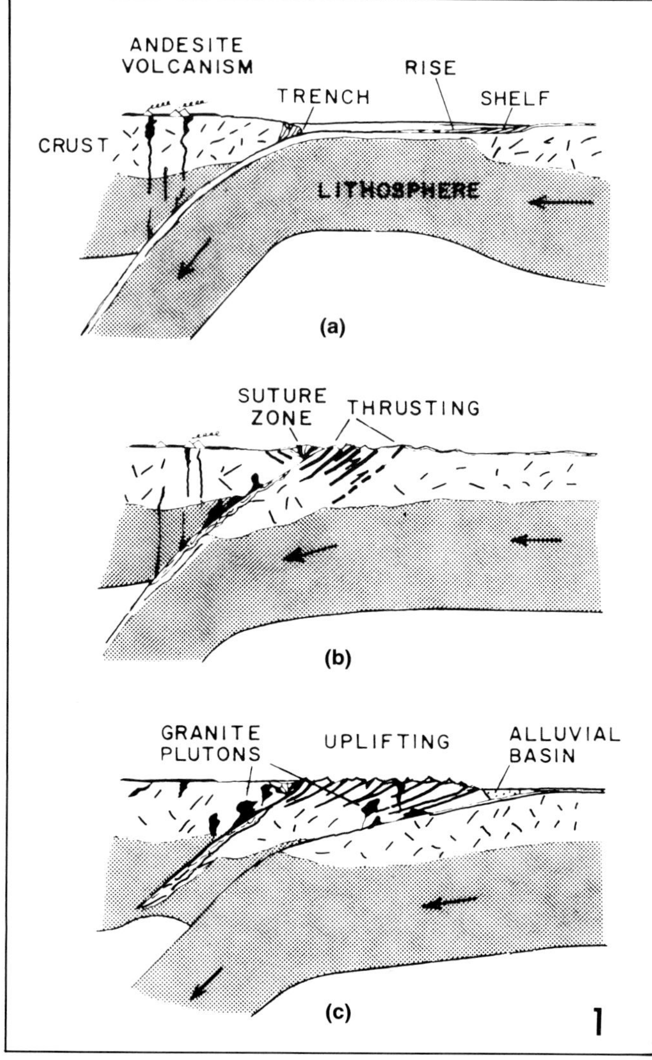

Fig. 6-17—Two versions of continental collision and attendant thrust-fold belt. Note direct mechanical drive, i.e., that sense of thrusting and subduction is the same. (Bird et al., 1975)–1 Three stages in the evolution of a hypothetical continental convergence zone: (a) subduction of the intervening oceanic plate (an Atlantic-type margin approaches a Pacific-type trench with andesitic volcanism produced by underthrusting); (b) subduction of continental rise sediments and then crust of the continental shelf, perhaps with granite magma produced by frictional heat; (c) formation of a new thrust within continental crust and subduction beneath the continental shelf with further granite formation. (Dewey and Bird, 1970)–2 Fragments of oceanic crust are emplaced above continental crust as ophiolites. From Journal of Geophysical Research. Copyright by the American Geophysical Union. Reprinted by permission.

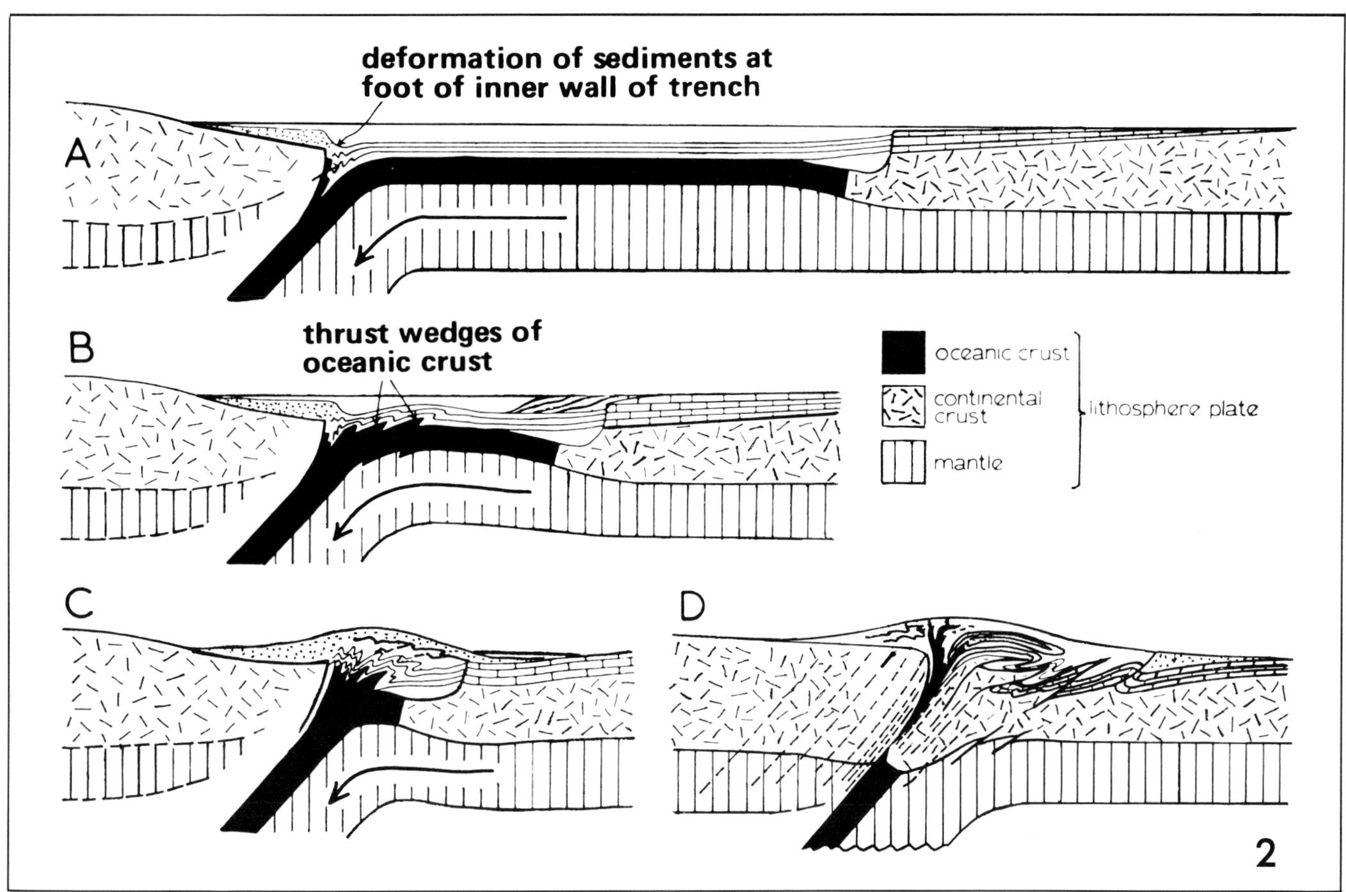

belts should be reflected by large crustal thicknesses there. An isopach map of the earth's crust (Fig. 6-12) shows just this relationship—that crust is thickest beneath orogenic belts, specifically the Himalayas, Alps, Andes, and U.S. Cordillera. Finally, some basement may be heated and partially melted to generate minor plutons.

Folded-fault and duplex structures also accord well with an origin of underthrusting in that they can form from the action of successively inserted thrust wedges from below (Figs. 6-13, 6-14).

Underthrusting Causes

Whereas underthrusting in the landward walls of oceanic trenches is caused by descending oceanic lithospheric slabs (marked by Benioff zones in modern examples) and underthrusting in continental collisions can be attributed to one continent underthrusting another, the cause or causes of underthrusting in barckarc intracontinental or intraplate settings, such as the Idaho-Wyoming thrust-fold belt, are not as easy to understand. Initiation of underthrusting by thermal weakening of the lithosphere caused by rise of a critical isotherm (Fig. 6-11) is a favorite mechanism, but structural anomalies or even lithologic contrasts could also play a part. Burchfiel and Davis (1975, p. 385) envisioned for the Idaho-Wyoming belt westward underthrusting of a lower plate into a ductile zone that was weakened by plutonic activity farther west and therefore required lower critical stress for fracture and development of the first thrust plate:

> As shortening (by underthrusting) continued, rocks east of the first thrust fault to form were brought closer to the zone of crustal yielding... When rocks of the lower plate were brought into the region of critical shear stress a new thrust fault developed in the lower plate (and east) of the first thrust. The process just described is self-limiting...

Dickinson (1976, p. 1272) regarded this type of thrust-fold belt as marking:

> ...an intracontinental locus of relative motion between segments of lithosphere that are partly decoupled. The behavior can be described either as intense intraplate deformation or as partial subduction according to taste. The fold-crumpling of shallow crustal layers and their brittle failure by thrusting occurs in response to shortening accommodated by ductile flowage at deeper crustal layers. The main flowage commonly occurs closer to the arc than the site of the fold-thrust belt, where regional decollement strips the cover off basement being inserted (underthrusted) into the ductile zone.

The model of backarc underthrusting presented here requires convergence of basement in thrust-fold belts. It dif-

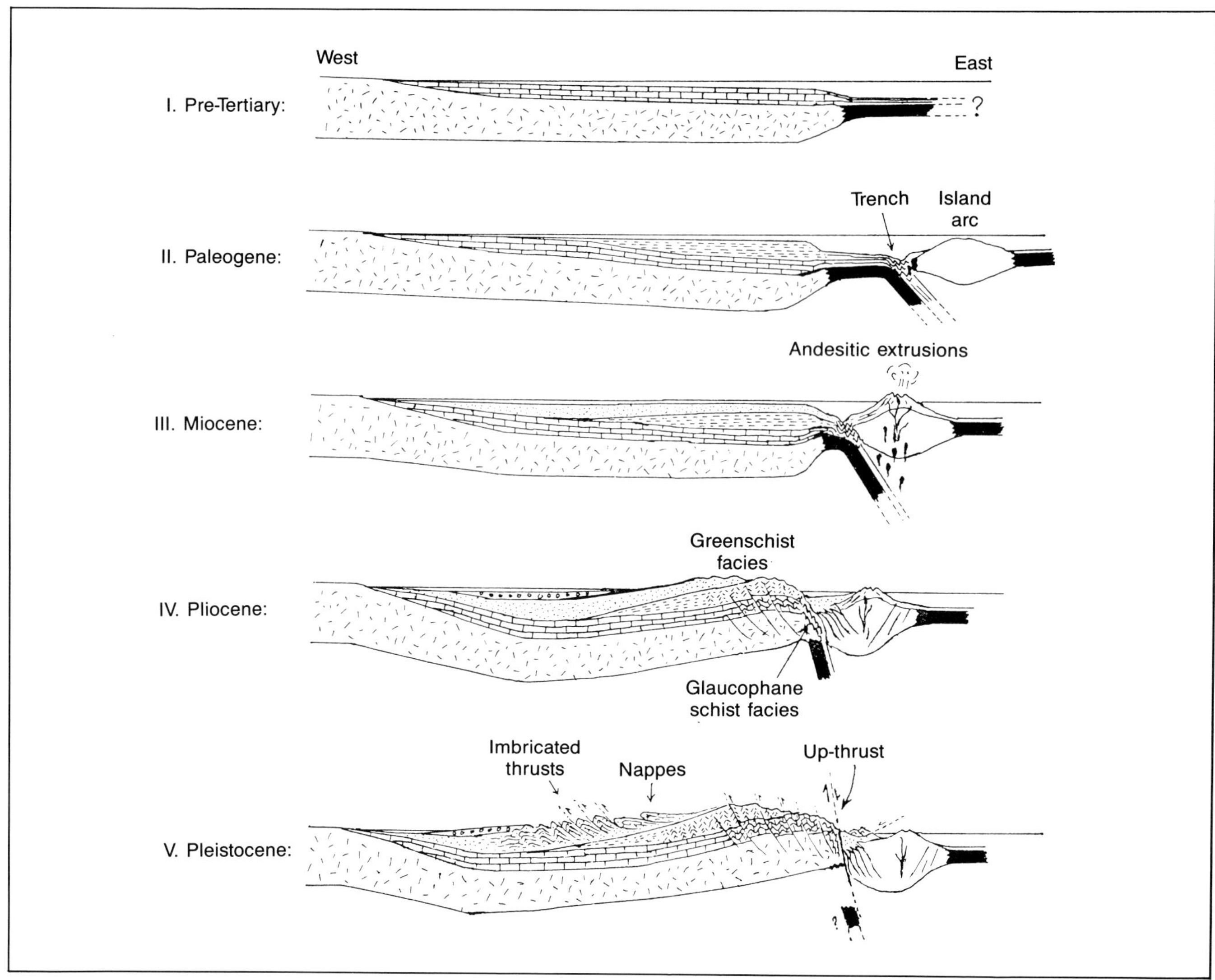

Fig. 6-18 (Chai, 1972)—Schematic sequence of sections illustrating the tectonic evolution of Taiwan by the collision of a continental margin and an island arc. Thrust-fold structure is considerably different from that shown (cf. Fig. 6-55). Circles: Molasse; dotted: Neogene sequence; dashed: Paleogene sequence; brick-pattern: Pre-Tertiary sequence; jackstraw pattern: crystalline basement. From American Journal of Science, V. 272. Used with permission.

fers sharply from more static models such as gravity sliding, gravity spreading, and lateral crowding (Fig. 6-15) that necessitate little to no basement convergence. The dynamic nature of underthrusting is illustrated by a cartoon of a conveyor belt at the University of Hawaii cafeteria (Fig. 6-16). Even this tongue-in-cheek approach shows the salient characteristics of underthrusting: 1– thrust progression toward underthrust plate, 2– uplift mechanism by thrust wedging from below, and 3– no stress transmission and material strength problems.

COLLISIONAL BELTS

Collisional thrust-fold belts are created by the collision of continental plates (Fig. 6-17), an island arc with a continental plate (Fig. 6-18), microcontinents with continental plates or allochthonous terranes (Fig. 6-19), etc. Collisional and backarc belts are not mutually exclusive. The former should have some of the characteristics of the latter when it is preceded by subduction and depending on the width of oceanic crust that is subducted prior to collision. Where relatively narrow basins (100–200 km or 62–124 mi wide) floored by oceanic or greatly thinned continental crust are compressed, however, subduction can only be incipient and an ensuing orogenic belt should have practically only collision-type characteristics.

A collisional belt should have a suture marking the join of the colliding elements. Where exposed, the suture may be delineated at least partially by ophiolites. Defined as gabbros, pillow lavas, and pelagic sediments (in ascending

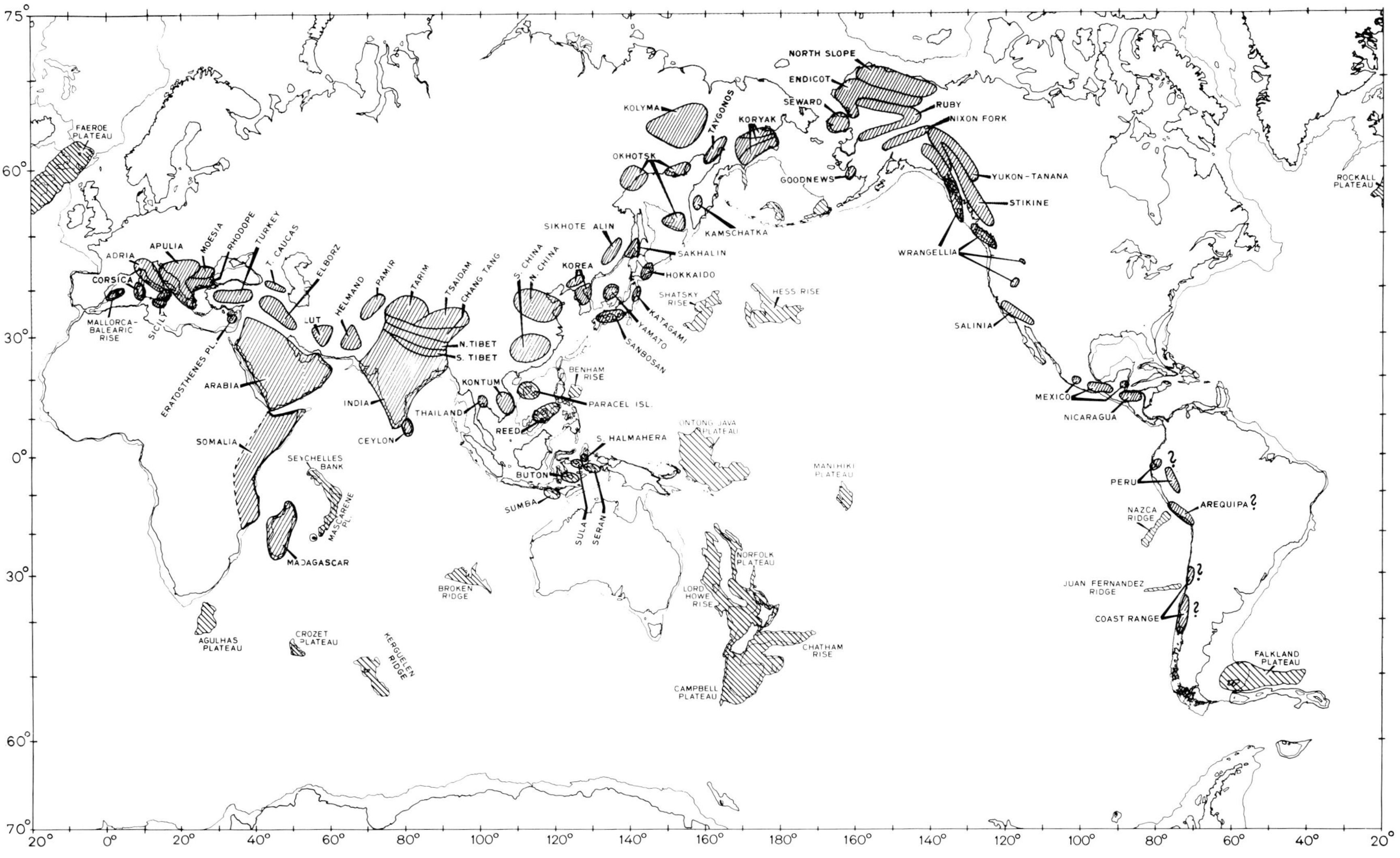

Fig. 6-19 (Nur and Ben-Avraham, 1982)—Sketch of past, present, and future allochthonous terranes, most of which are fragments of continents or have clear continental affinities. Shaded area indicates the Mesozoic-Cenozoic orogenic belts. Allochthonous terranes within the belt are most likely parts of Gondwana which have migrated toward and collided with Europe, SE Asia, and the Pacific margins. Oceanic plateaus, some of which are shown, are moving toward consuming boundaries, possibly to become accreted allochthonous terranes. Several fragments of Africa (e.g., Madagascar, Somalia) will probably collide with the Euroasian plate in the future. Depending on the nature of the edges of both the allochthonous and autochthonous terrances, collisions can create belts of folding and thrusting. Clearly the concentration of allochthonous terranes, or microcontinents, in the shaded area of Mesozoic-Cenozoic orogenic belts is more than fortuitous; collisions have themselves effected part of the orogenic belt deformation. From Journal of Geophysical Research. Copyright by the American Geophysical Union. Reprinted by permission.

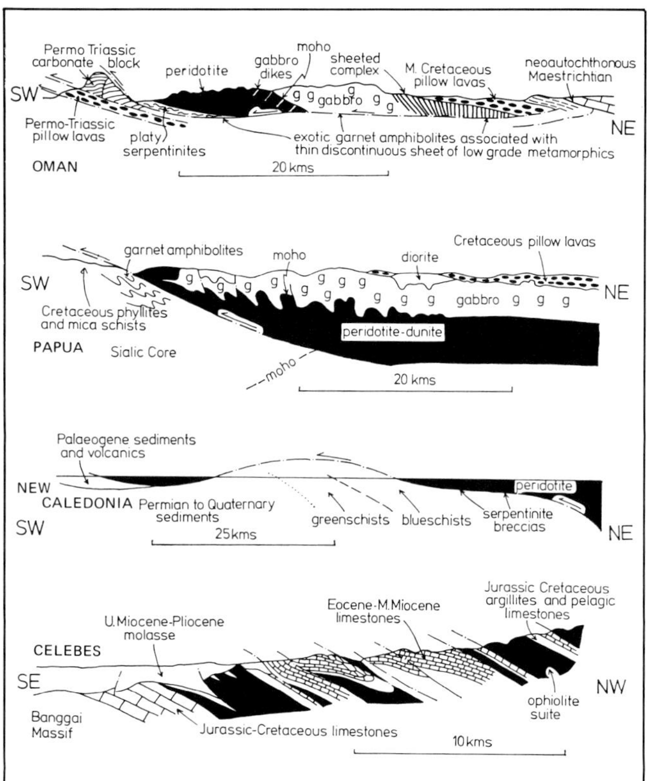

Fig. 6-20 (Dewey and Bird, 1971)—Sections across obducted ophiolite complexes. Semail Nappe, Oman. Papuan ultramafic belt. New Caledonia. Celebes. From Journal of Geophysical Research. Copyright by the American Geophysical Union. Reprinted by permission.

order), ophiolites are now widely believed to be fragments of oceanic crust emplaced by collisional processes on and within continental crust, including the overlying sedimentary rocks (Figs. 6-17, 6-20). Because of the complex deformation of collision, ophiolites are frequently fragmented or dismembered (Fig. 6-20).

As seen from figure 6-17, collision can effect a direct mechanical drive between subduction and thrusting. For this reason collisional fold belts have been termed "synthetic" by Roeder (1973). This direct compressional linkage between deep-seated movement and thrusting can also be viewed as underthrusting. The same characteristics previously described for underthrusting in forearc and backarc settings would apply to collisional belts. Underthrusting in both collisional and backarc settings has created thrust-fold assemblages with potential for oil and gas on the external flanks of orogenic belts.

PLAN VIEW

Regionally, thrust-fold belts often appear festooned as alternating broad salients and somewhat sharper reentrants. Within individual salients a relay pattern, sometimes broken by small tear faults, is dominant (Appendix—Figs. 6A-12 through 6A-14, cf. Fig. 6A-14 with Fig. 4-17).

On a local basis, folds and their associated thrust faults are characteristically parallel (Figs. 6-21, 6-22).

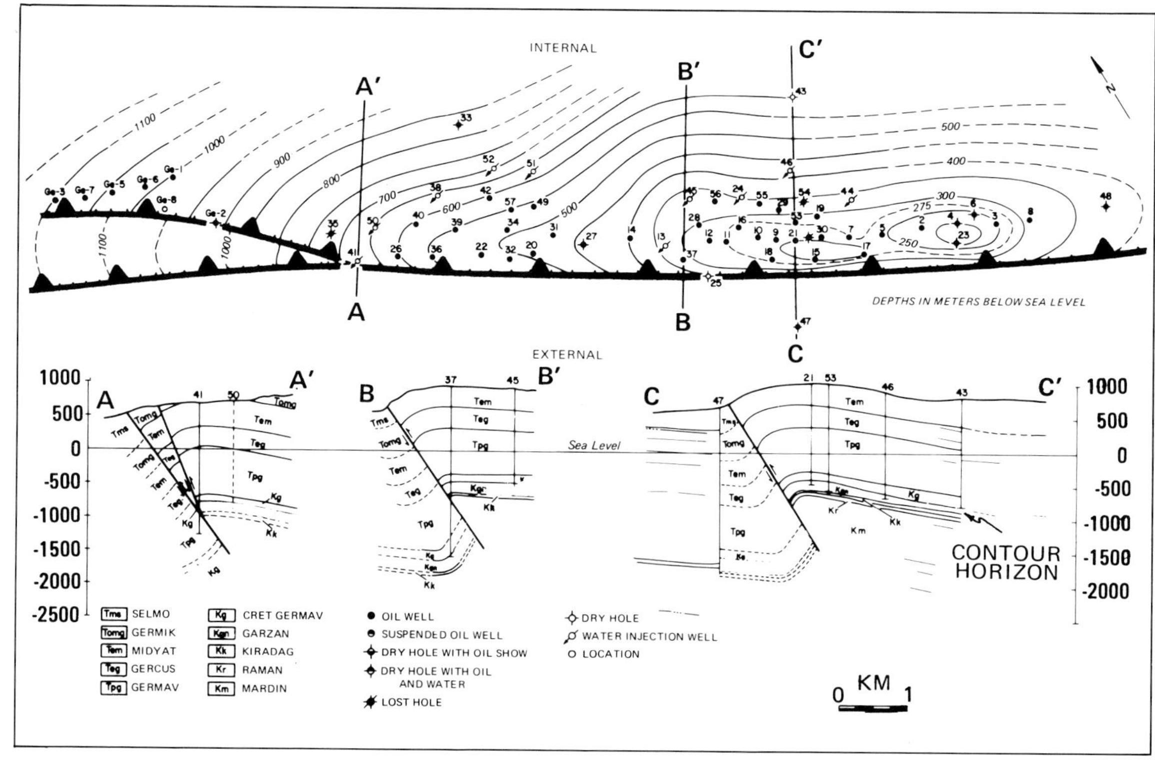

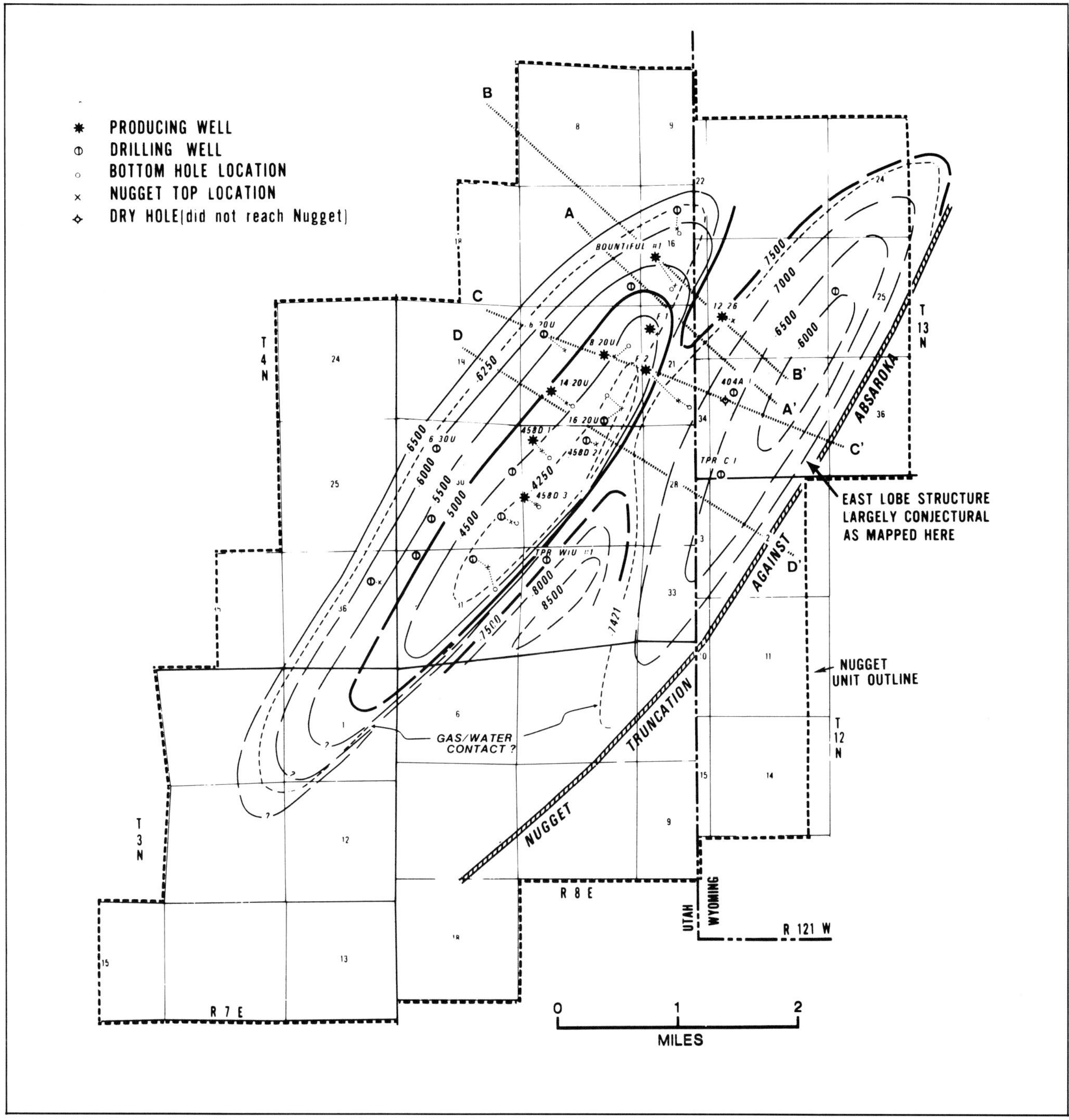

Fig. 6-22 (Lelek, 1982)—Nugget structure map showing west and east lobes in Anschutz Ranch East field. C.I. = 152 m (500 ft). Note parallelism of folds and associated faults. Permission to publish by Rocky Mountain Association of Geologists.

Fig. 6-21 (Sanlav et al., 1963)—Structure contour map on top of Cretaceous Germav shale (just above the productive Garzan limestone) and cross sections through the Garzan-Germik field, Taurus thrust-fold belt, Turkey. Note parallelism of fold and associated fault. Permission to publish by The Institute of Petroleum, London.

SEISMIC EXPRESSION

Some of the characteristics of a regional reflection seismic line through part of the Wyoming thrust-fold belt are shown and discussed in figure 1-19. Regional lines across the Alpine front and Molasse Basin of southern Germany (Fig. 6-23) and the southern Appalachians of the southeastern U. S. (Fig. 6-24) are also presented. Thrust faults are difficult to pick and it is not unusual to be off by two or three cycles, especially where thrusts follow bedding and well control is sparse. Thrusts are frequently marked by divergence between cycles, which most clearly occurs on ramps where hanging-wall events dip more steeply than footwall events (Fig. 6-25). Velocity corrections are always a problem (Figs. 6-24, 6-25).

Figures 6-26 through 6-28 show additional difficulties in thrust-belt seismic interpretation. In the first example, drilling located a thrust not previously interpreted on seismic (Figs. 6-26, 6-27). Next, additional drilling modified the earlier seismic interpretation, especially in defining a high-relief salt feature (Figs. 6-28, 6-29). Salt can also serve as a decollement or glide surface for thrust faults (Fig. 6-30). Finally, good reflectors thought to represent prospective Paleozoic-Mesozoic sedimentary rocks beneath a reflectionless interval believed to represent allochthonous Precambrian turned out instead to emanate from crystalline metamorphic rocks (Fig. 6-31).

On the external sides of many thrust-fold belts, most notably the Canadian Rockies, are zones of backthrusting, also known as delta structures and triangle zones. Such zones are probably localized in external positions because they are at the edge of the original depositional wedge (Fig. 6-32). Backthrusts can combine with foreward-directed thrusts to form delta (Greek capitol letter Δ) or triangle zones that can be prospective for oil and gas. In figure 6-33 the sides of the delta or triangle structure are formed by the NE-dipping backthrust and the SW-dipping blind thrust, and the base is formed by a less steeply NE-dipping blind thrust.

Still other features of some thrust-fold belts are superimposed normal faults. Particularly well known are normal faults that effect sedimentation and preservation of section in half graben in the Canadian Rocky and Idaho-Wyoming thrust-fold belts (Figs. 6-34, 6-35). Early cross sections in these areas projected normal faults to great depth. Subsequent seismic shows events to be continuous beneath the half graben features indicating that associated normal faults must be listric. The position of many normal faults appears to be localized by ramps on underlying thrust faults so that the normal faults sole or go listric on the ramped portions of earlier thrusts. A corollary is that the presence of a half graben with a bounding normal fault at the surface may be a clue to the presence of a ramp in a thrust fault in the subsurface.

Normal faults in the Idaho-Wyoming thrust-fold belt can be viewed as the easternmost expression of Basin-Range extension. The space problem caused by the normal faults being listric is thus alleviated, not by compression at the lower end of the listric fault, but by arching associated with Basin-Range rifting (cf. Fig. 4-2).

Illustration 6-23 begins on next page ⟶

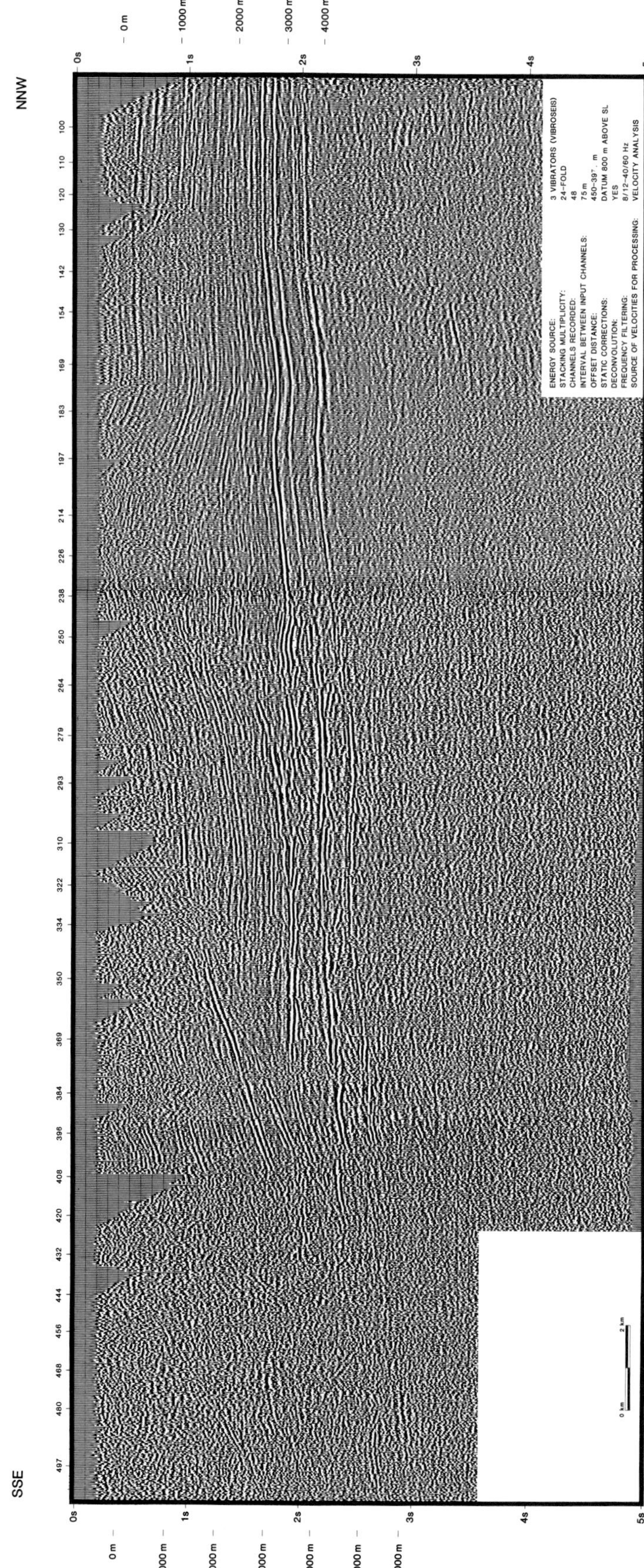
NNW
SSE
0s
1s
2s
3s
4s
5s
0 m
1000 m
2000 m
3000 m
4000 m
5000 m
6000 m
ENERGY SOURCE: 3 VIBRATORS (VIBROSEIS)
STACKING MULTIPLICITY: 24-FOLD
CHANNELS RECORDED: 48
INTERVAL BETWEEN INPUT CHANNELS: 75 m
OFFSET DISTANCE: 450-397 m
STATIC CORRECTIONS: DATUM 800 m ABOVE SL
DECONVOLUTION: YES
FREQUENCY FILTERING: 8/12-40/60 Hz
SOURCE OF VELOCITIES FOR PROCESSING: VELOCITY ANALYSIS
0 km
2 km
100
110
120
130
142
154
169
183
197
214
226
238
250
264
279
293
310
322
334
350
369
384
396
408
420
432
444
456
468
480
497

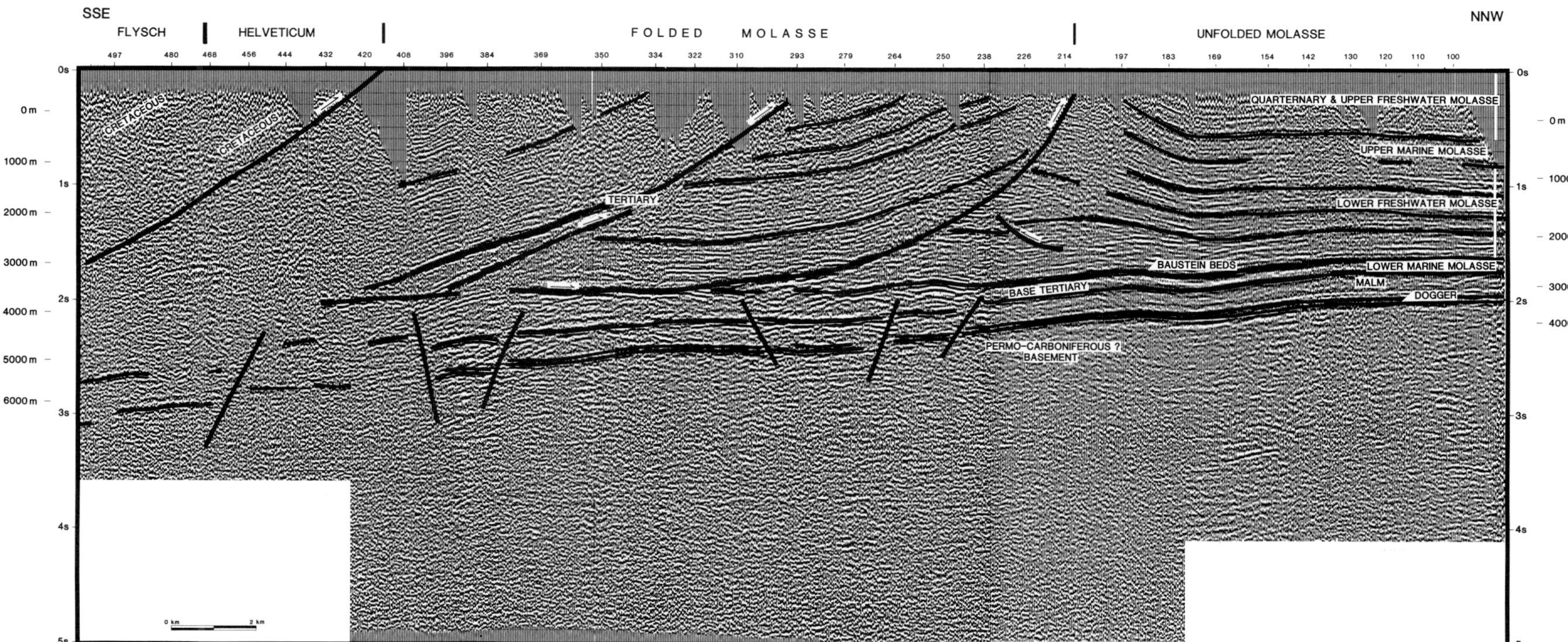

Fig. 6-23 (Bachmann and Koch, 1983)—Seismic line through Alpine front and Molasse Basin approximately 100 km (62 mi) WSW of Munich, Bavaria, showing impingement of thrust sheets on foreland basement, Jurassic, and lower Molasse. Note normal faults of foreland in contrast to compressive blocks (see Chapter 3). Note also listric nature of faults in folded Molasse and backthrust below shotpoint 226 at about 2000 m (6560 ft). Permission to publish by American Association of Petroleum Geologists.

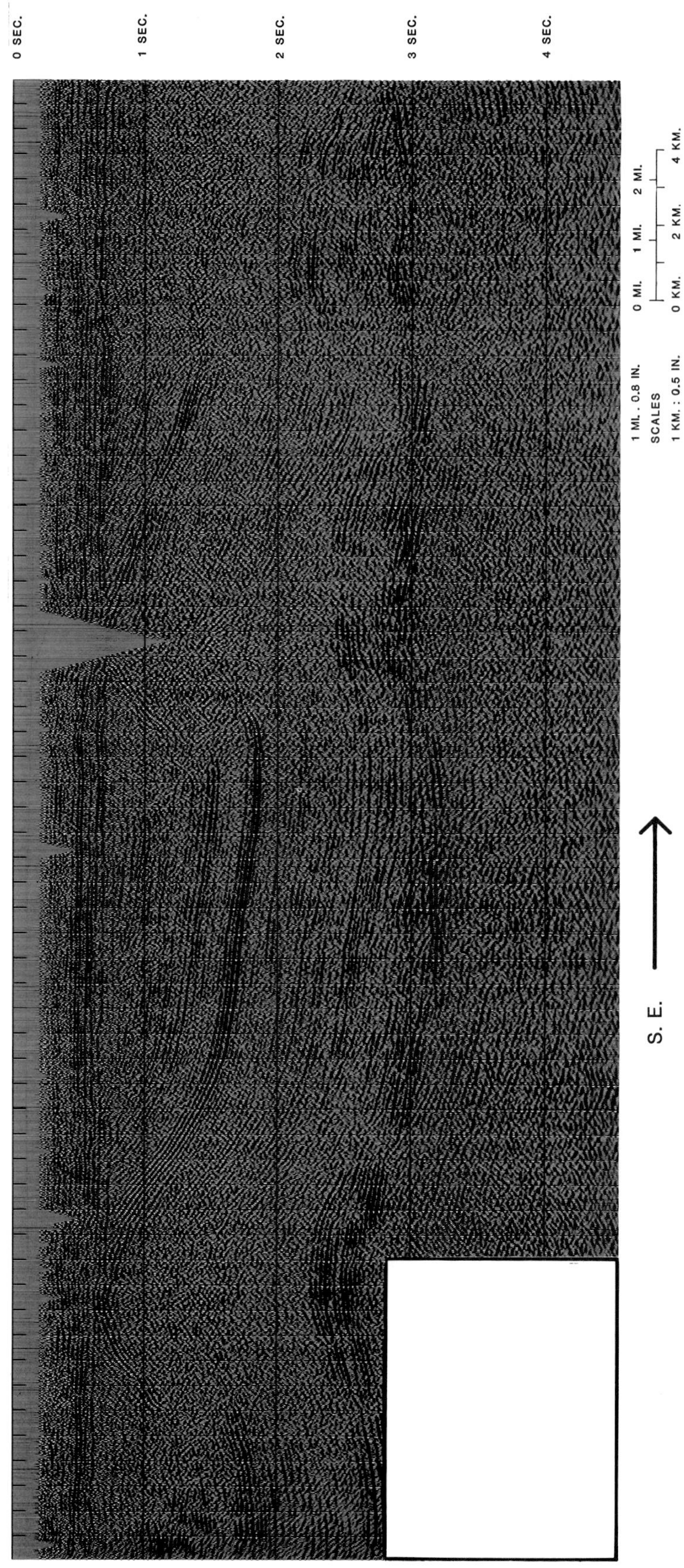
0 SEC.
1 SEC.
2 SEC.
3 SEC.
4 SEC.
0 MI. 1 MI. 2 MI.
0 KM. 2 KM. 4 KM.
1 MI. . 0.8 IN.
SCALES
1 KM. : 0.5 IN.
S. E.

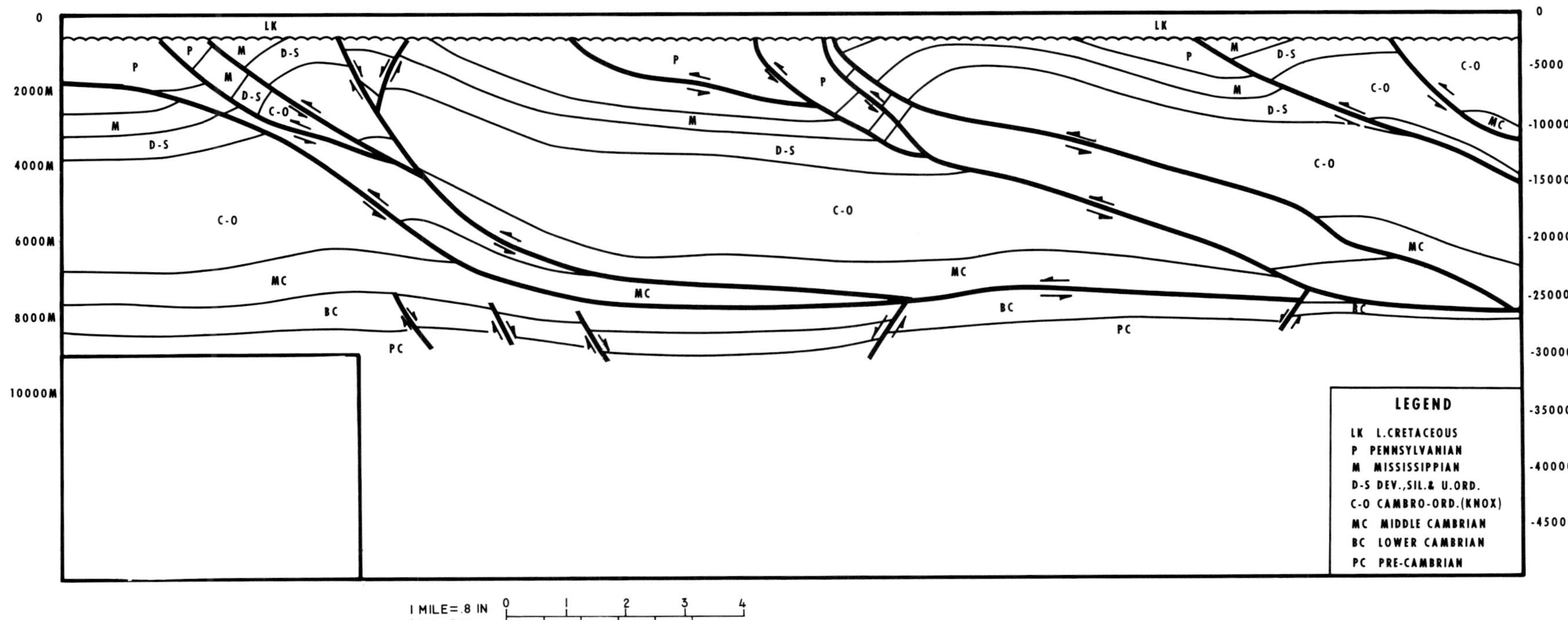

Fig. 6-24 (Sachnik and More, 1983)—Seismic line crossing Mississippi - Alabama state line about 161 km (100 mi) SW of Birmingham, Alabama, showing southern Appalachian thrust-fold belt beneath Lower Cretaceous coastal plain cover. Main detachment is within Lower Cambrian clastics just above basement. Note false velocity pullups on time section beneath the two major hanging wall anticlines and compare with depth section where velocity corrections have been made. Permission to publish by American Association of Petroleum Geologists.

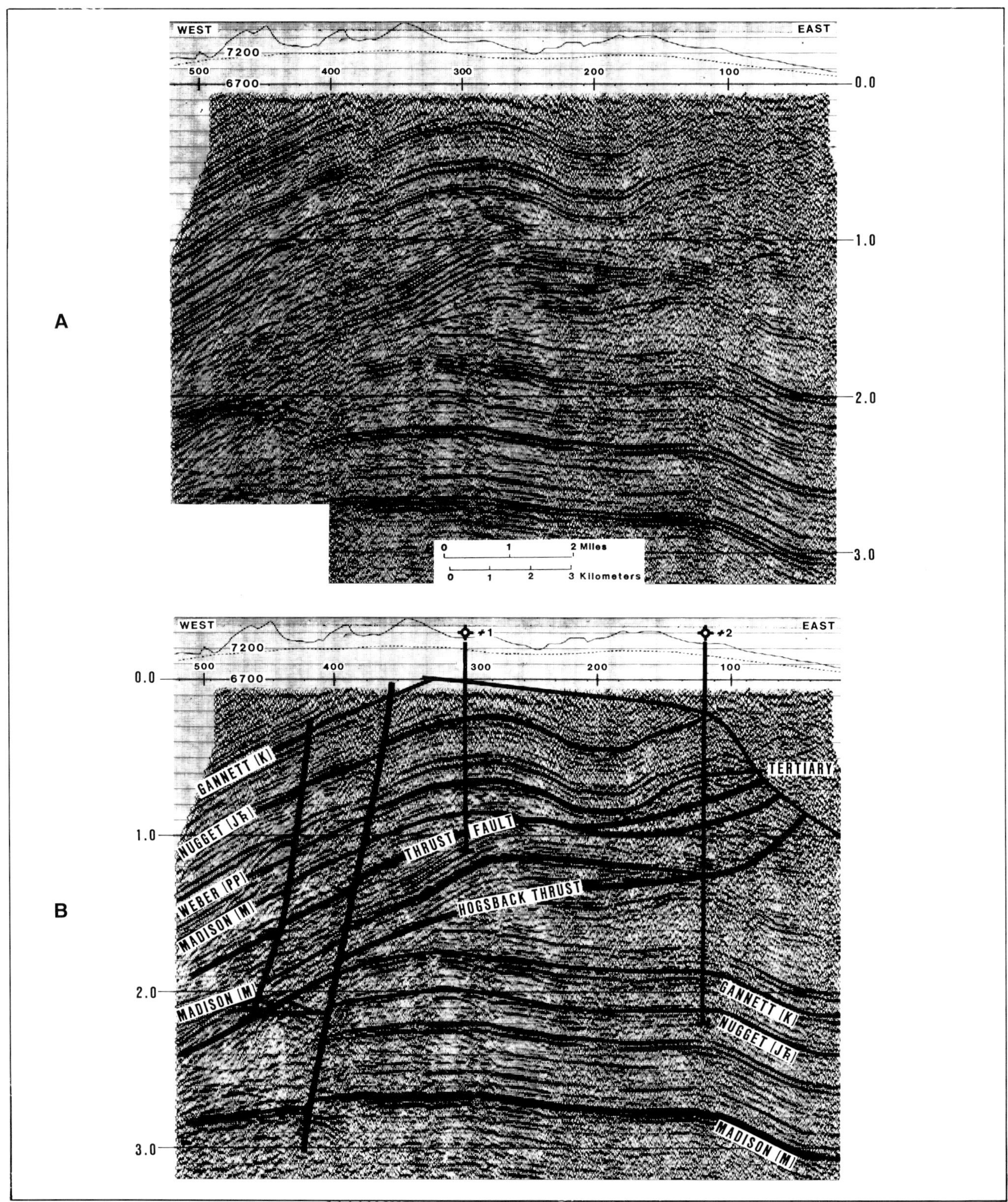

Fig. 6-25 (Norton, 1983)—Seismic line about nine miles (14 km) NE of Kemmerer, Wyoming, showing easternmost thrust (Hogsback) of Wyoming-Utah-Idaho thrust-fold belt. Note divergence of cycles at ramp (west half of section) which also effects hanging-wall fold. Subthrust structure is a velocity anomaly. Permission to publish by American Association of Petroleum Geologists.

Illustration 6-26 begins on next page ⟶

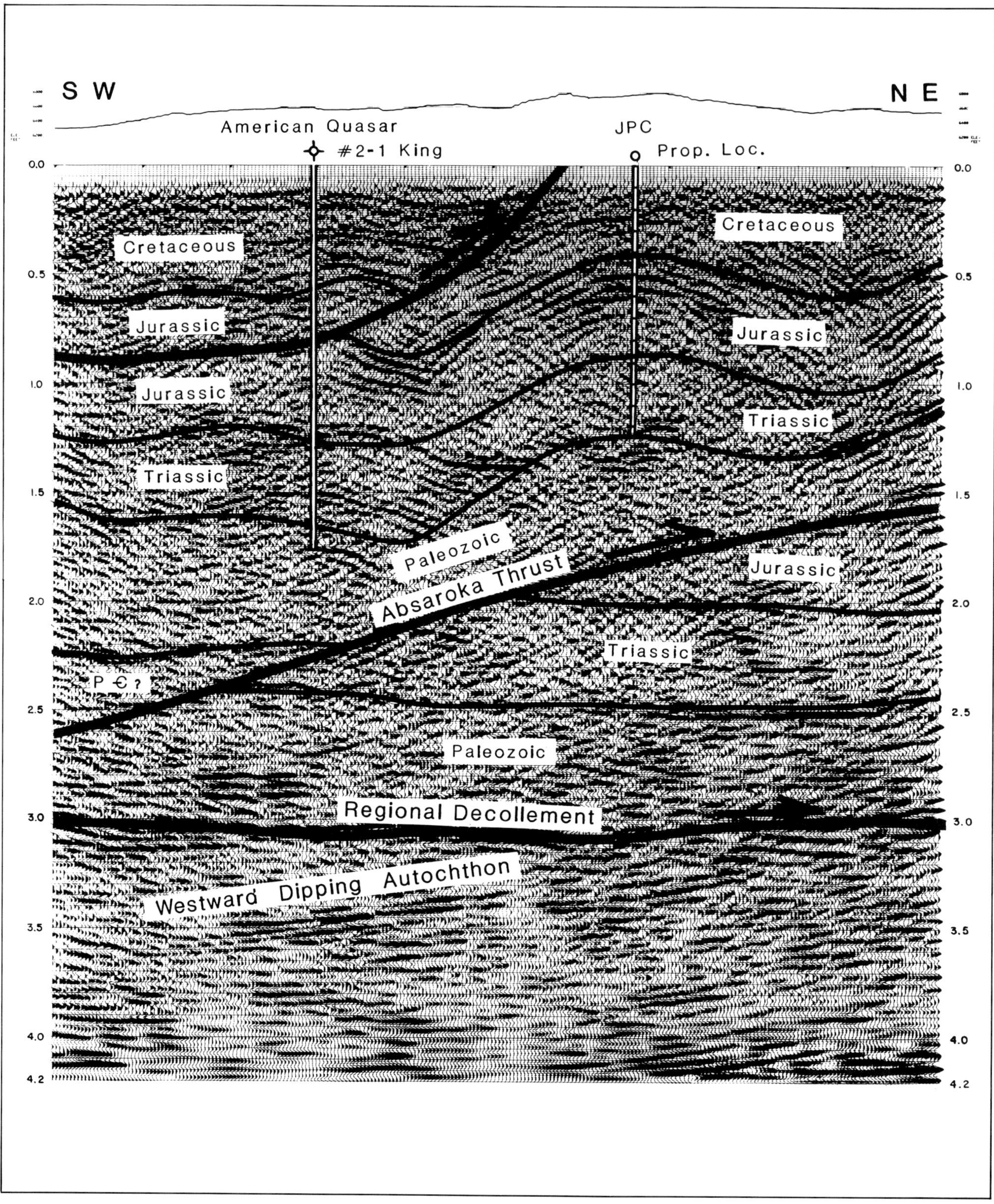

Fig. 6-26 (Butler, 1982)—Seismic line showing a proposed drill location and the interpreted ramp in the Absaroka thrust plate that was thought to bring Paleozoic rocks higher in the proposed location than in the Quasar well, north Caribou prospect, Bonneville County, Idaho. Permission to publish by Rocky Mountain Association of Geologists.

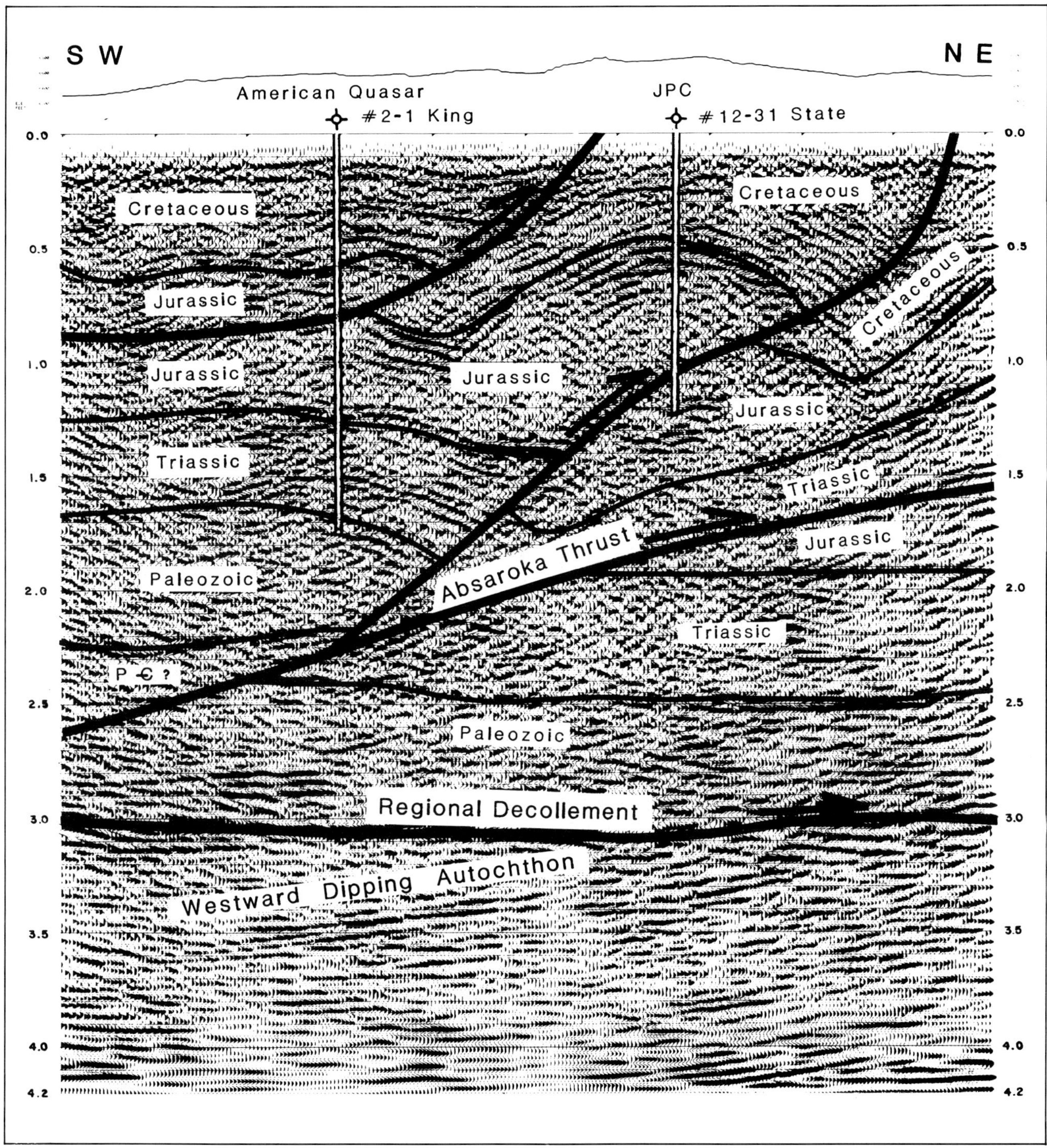

Fig. 6-27 (Butler, 1982)—Interpretation of same seismic line as figure 6-26 after drilling the Juniper Petroleum Corporation No. 12-31 State well. Note thrust fault in lower part of well and missing postulated Paleozoic section. Permission to publish by Rocky Mountain Association of Geologists.

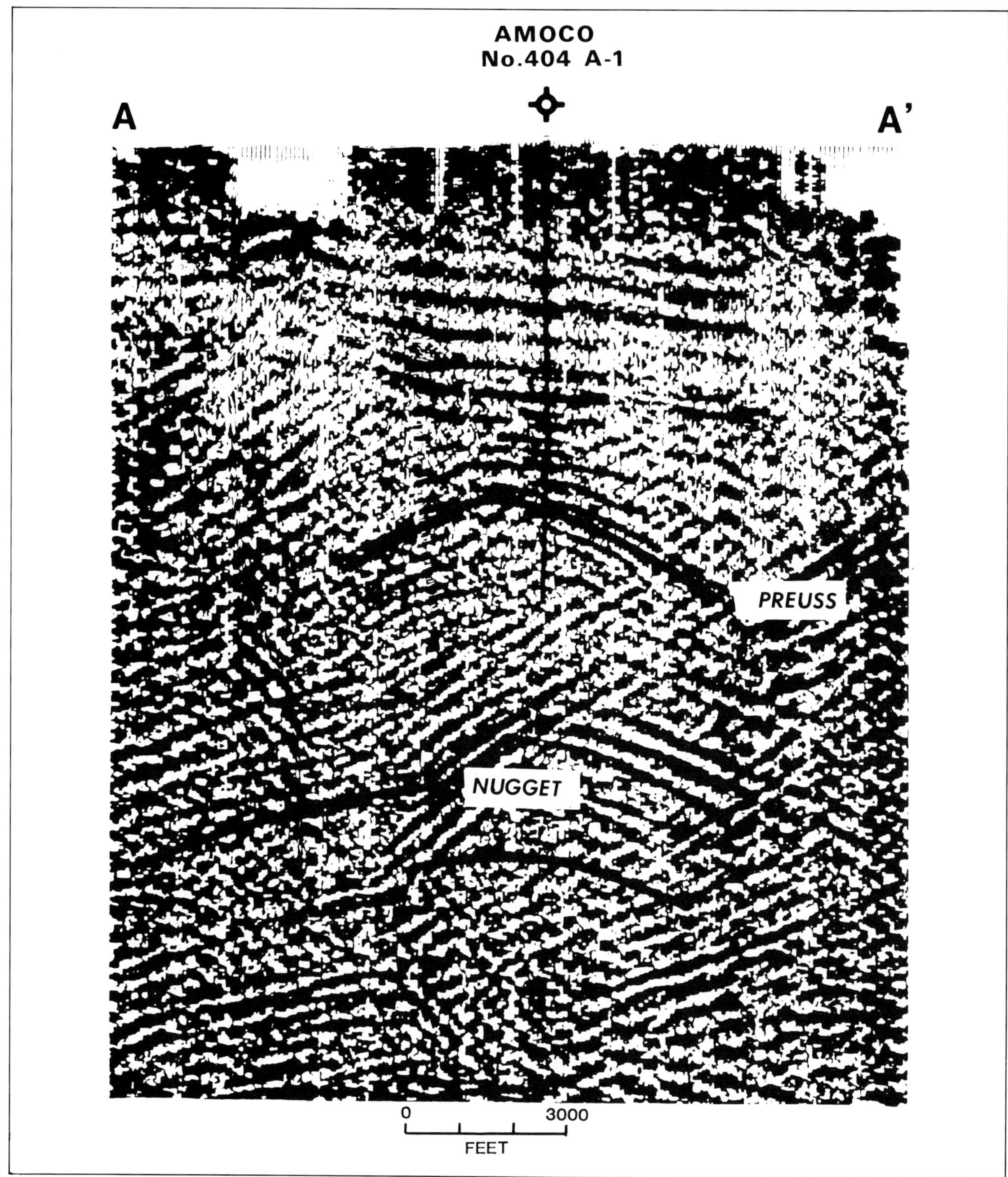

Fig. 6-28 (Lelek, 1982)—Unmigrated seismic section, A-A', showing structure drilled originally by the Champlin 404 Amoco No. 1. Line of section is shown on figure 6-22. Vertical axis is time. Anschutz Ranch East field, NE Utah and SW Wyoming. Permission to publish by Rocky Mountain Association of Geologists.

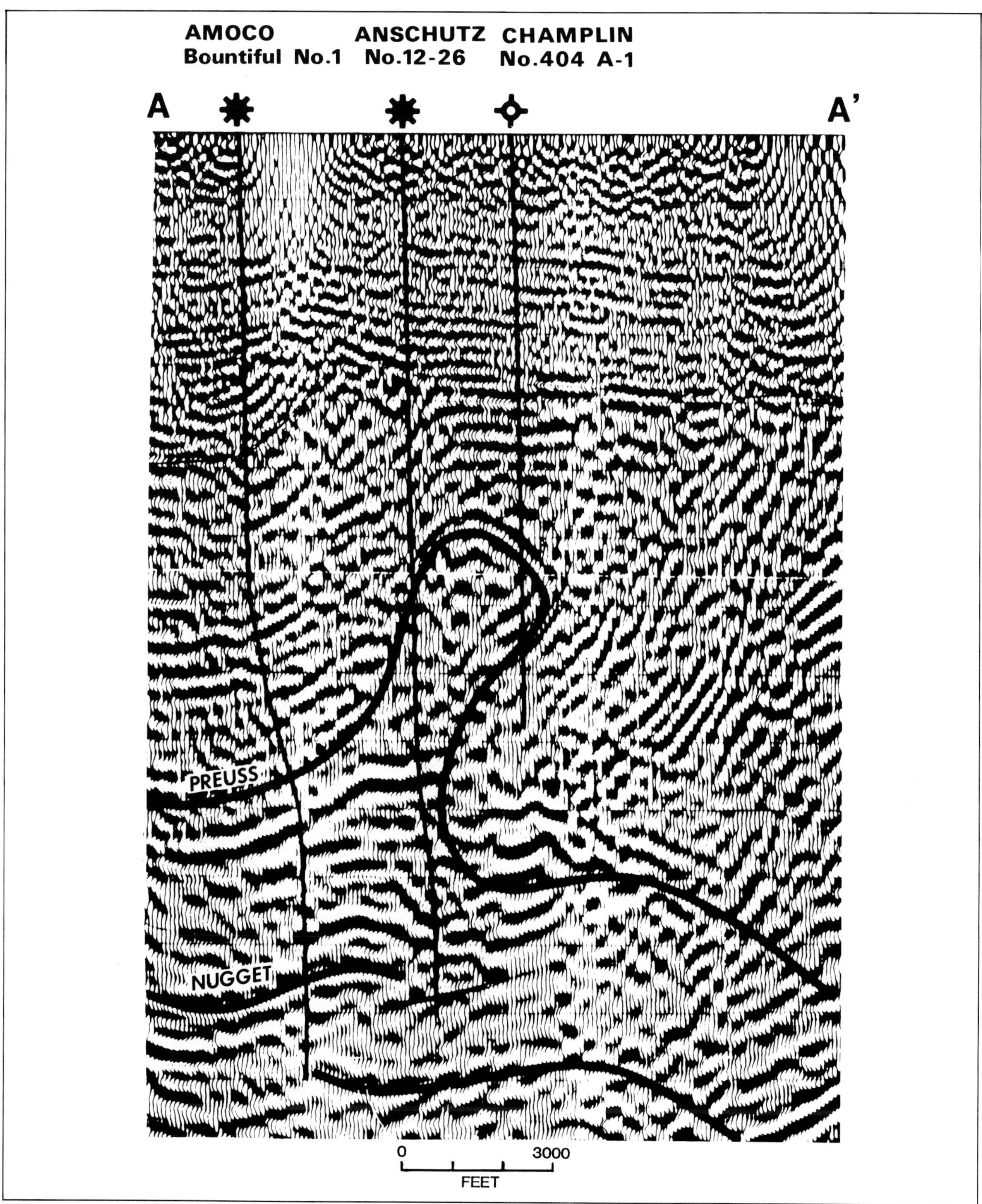

Fig. 6-29 (Lelek, 1982)—Depth-migrated seismic section A-A', generated after additional drilling. Lobe of Preuss is salt. Permission to publish by Rocky Mountain Association of Geologists.

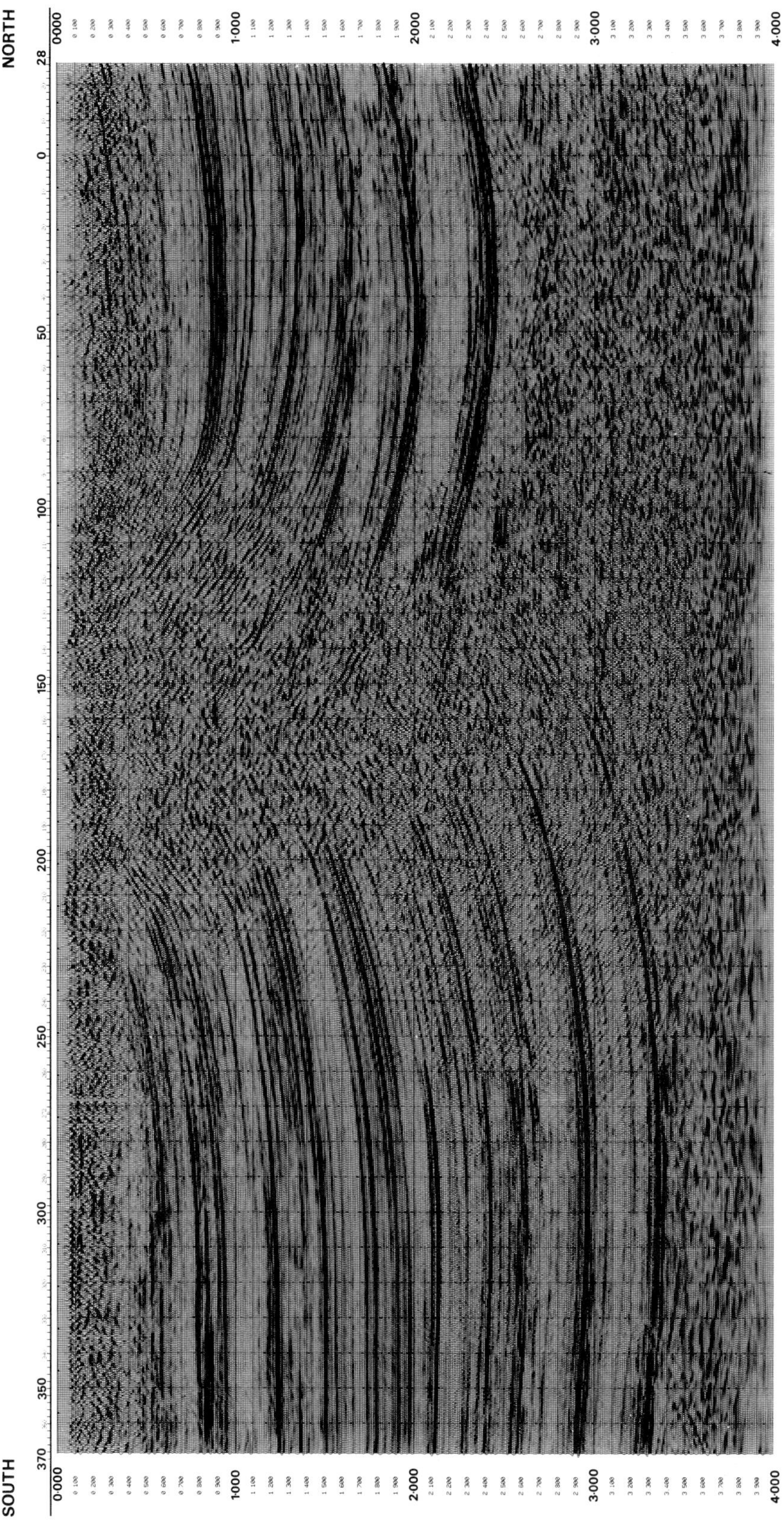
NORTH
SOUTH
28
0
50
100
150
200
250
300
350
370
0·000
1·000
2·000
3·000
4·000

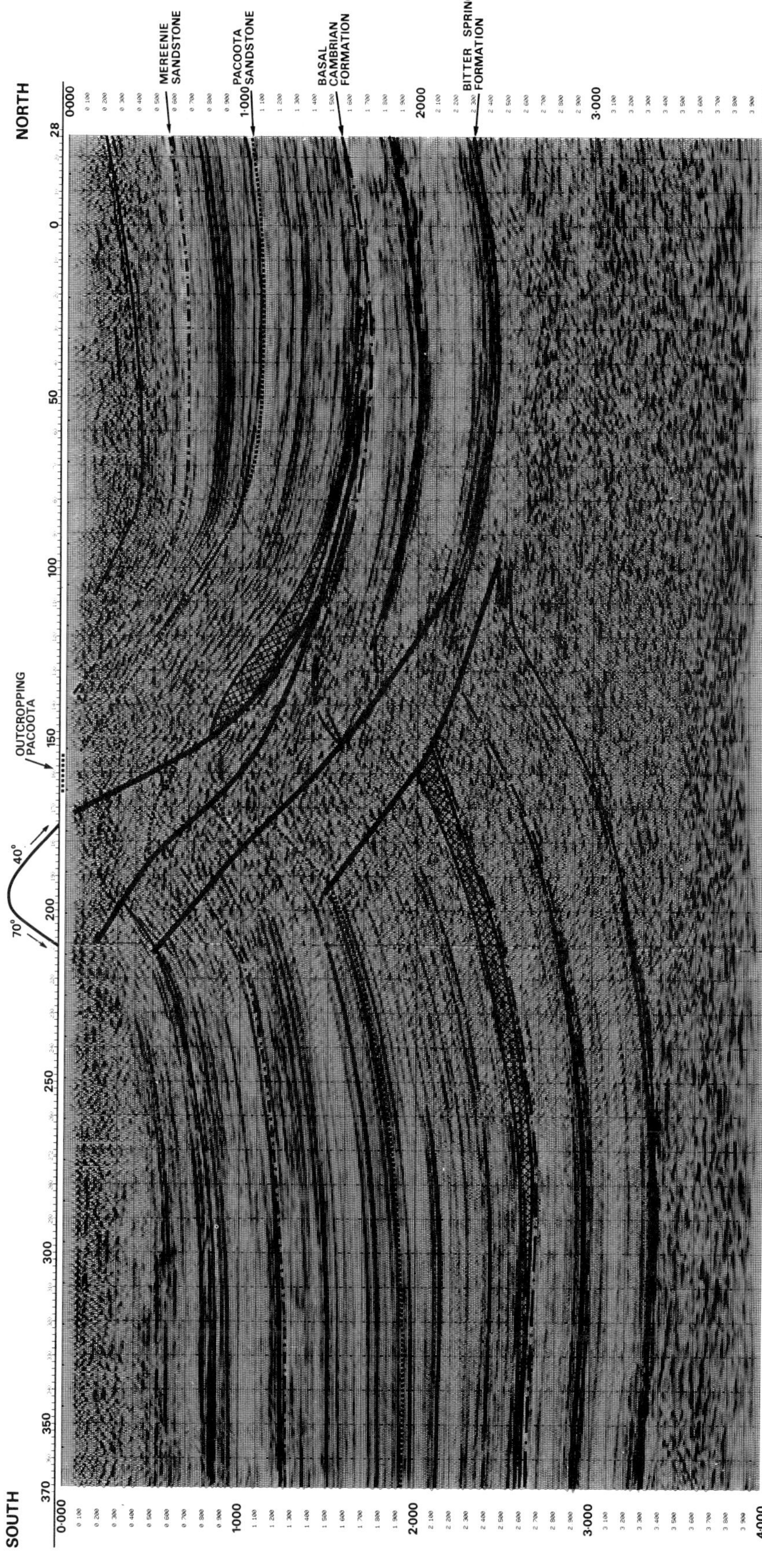

Fig. 6-30 (Schroder, 1983)—Seismic line from the Amadeus Basin of central Australia showing thrusts localized by Lower Cambrian salt (cross-ruled). The Bitter Springs Formation (late Proterozoic) is also evaporite. Note thinning of section, particularly Pacoota to Mereenie interval, on hanging wall of thrust. Time of deformation was Late Devonian (Alice Springs orogeny). Permission to publish by American Association of Petroleum Geologists.

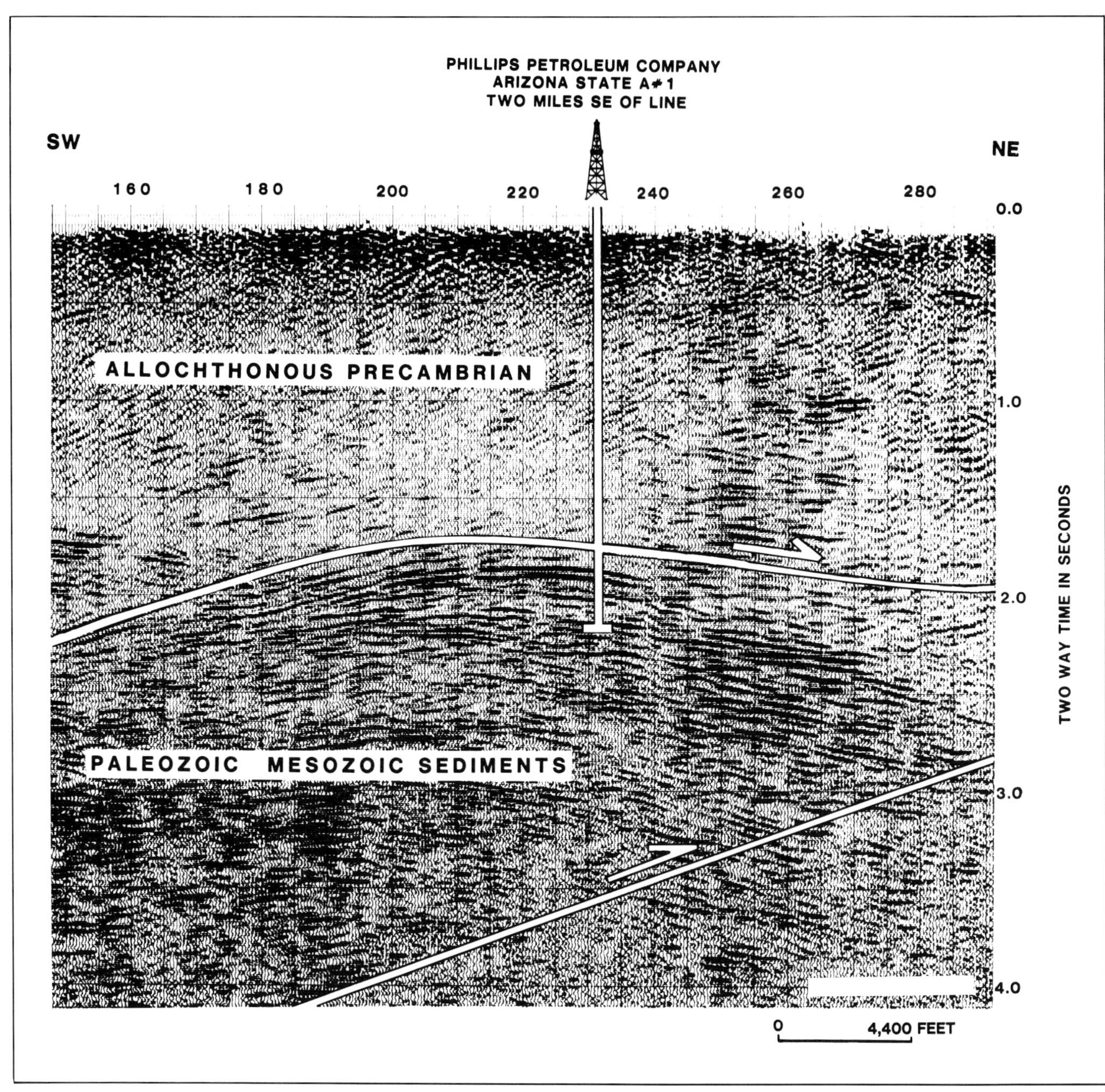

Fig. 6-31 (Robinson, 1982)—Seismic overthrust model of the Brady Wash prospect, Pinal County, Arizona before drilling. Time zero represents a datum of 1700 ft (518 m). Reflections, rather than coming from sediments, appear to be caused by textural differences caused by deformational events in crystalline rocks. Permission to publish by Rocky Mountain Association of Geologists.

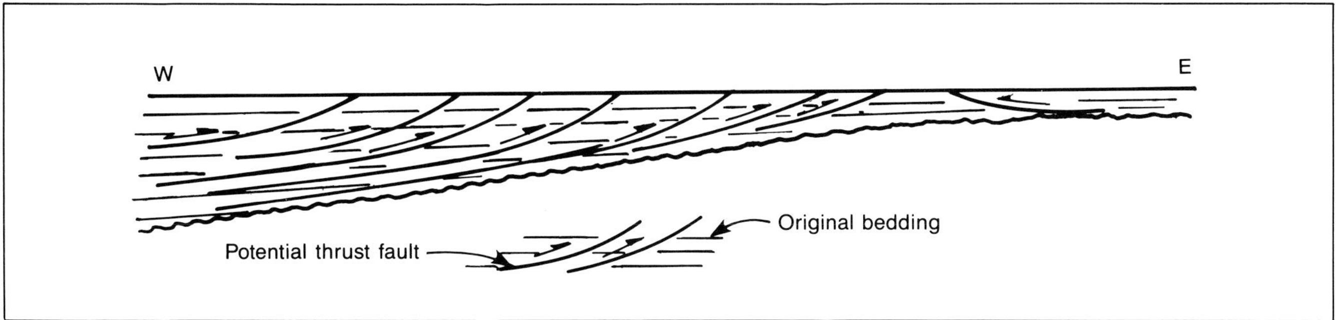

Fig. 6-32—Bedding anisotropy dictates that thrust faults in thick part of wedge be aligned with sedimentary layering; slip is easiest parallel, not across, bedding so that most thrust faults dip to the west even though an equal number should dip east according to shear failure criteria. On thin (east) side of wedge bedding anisotropy is no longer effective and east-dipping thrusts or backthrusts and delta or triangle structures occur.

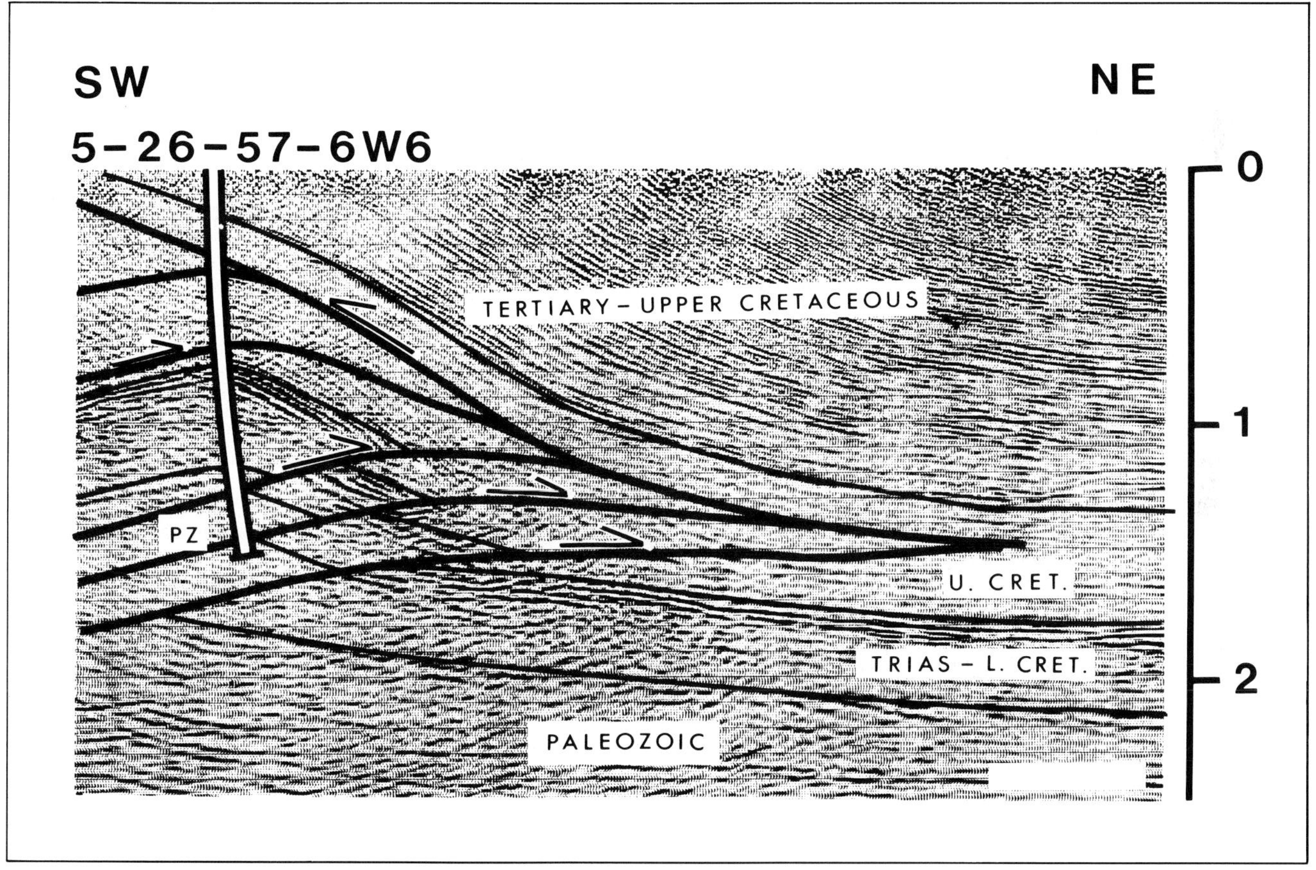

Fig. 6-33 (Jones, 1982)—Seismic line over Findley structure, northern Alberta foothills, Canada. Approximately true scale. Interpretation illustrates backthrust to SW and folded, NE-directed, blind thrusts that merge with the backthrust or upper detachment in a shale unit in the Upper Cretaceous. Structure produces from Mississipian and Triassic carbonates in stacked thrust sheets. The NE-dipping events just NE of well 5-26-57-6W6 commonly occur on thrust belt seismic sections and are sometimes interpreted without thrusting; subsequent well control invariably shows that multiple thrusts are present. Permission to publish by Rocky Mountain Association of Geologists.

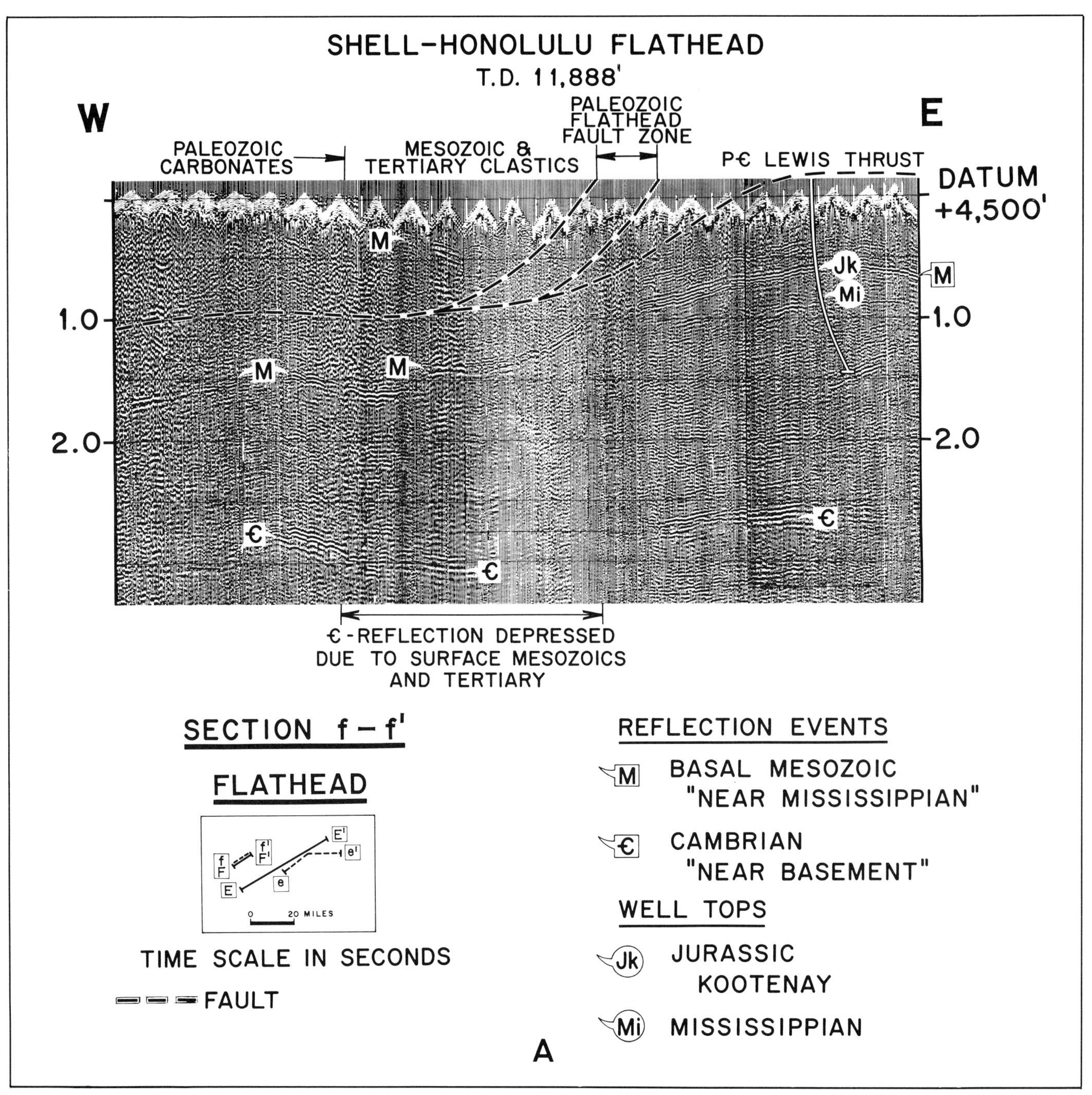

Fig. 6-34 (Bally et al., 1966)—A. Seismic line showing Flathead fault as listric normal fault soleing at or near ramp of Lewis thrust. Note continuity of west-dipping reflectors beneath Flathead fault zone. B. Cross-section interpretation of seismic. Permisson to publish by Canadian Society of Petroleum Geologists.

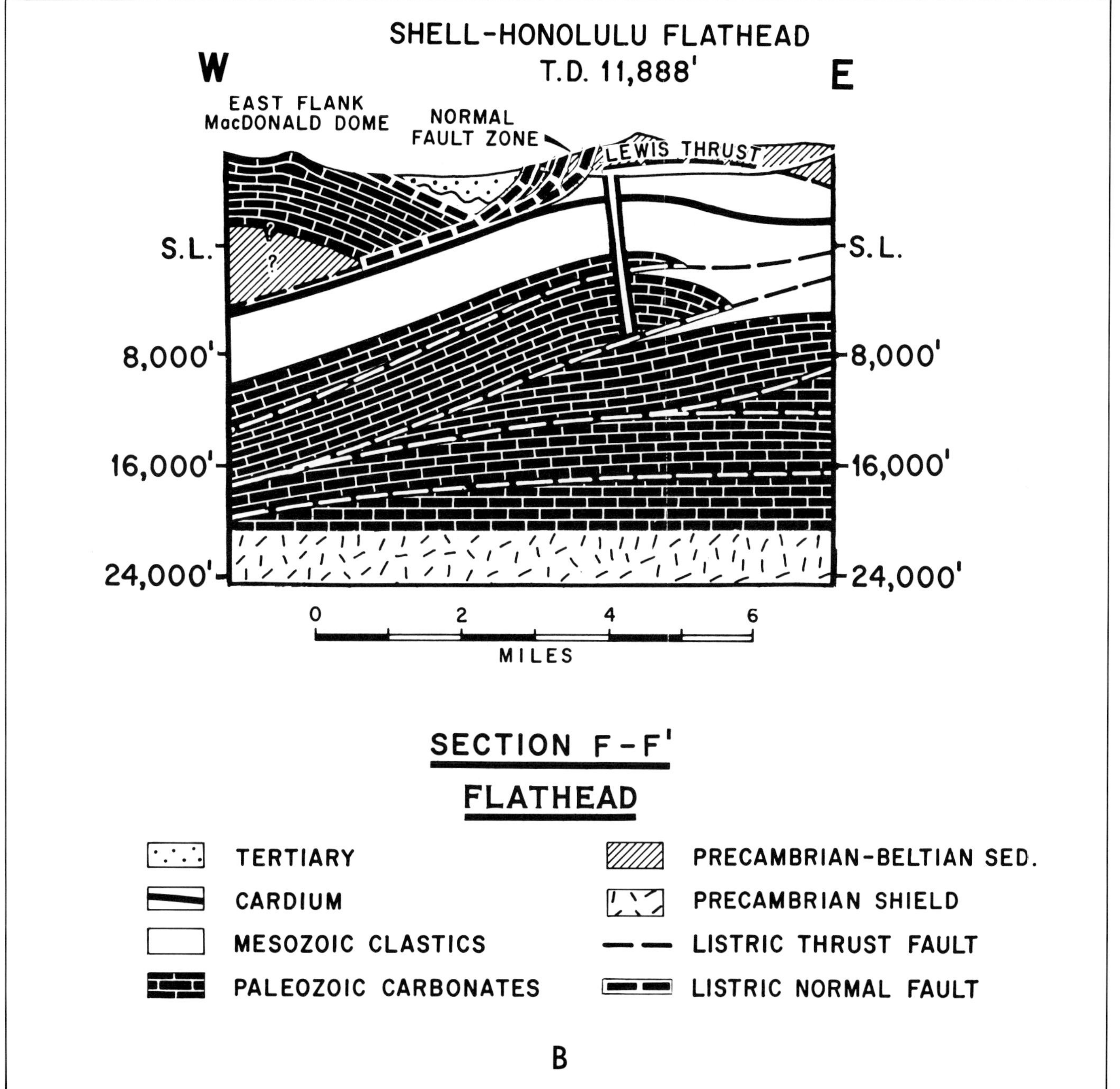

SECTION F-F'

FLATHEAD

B

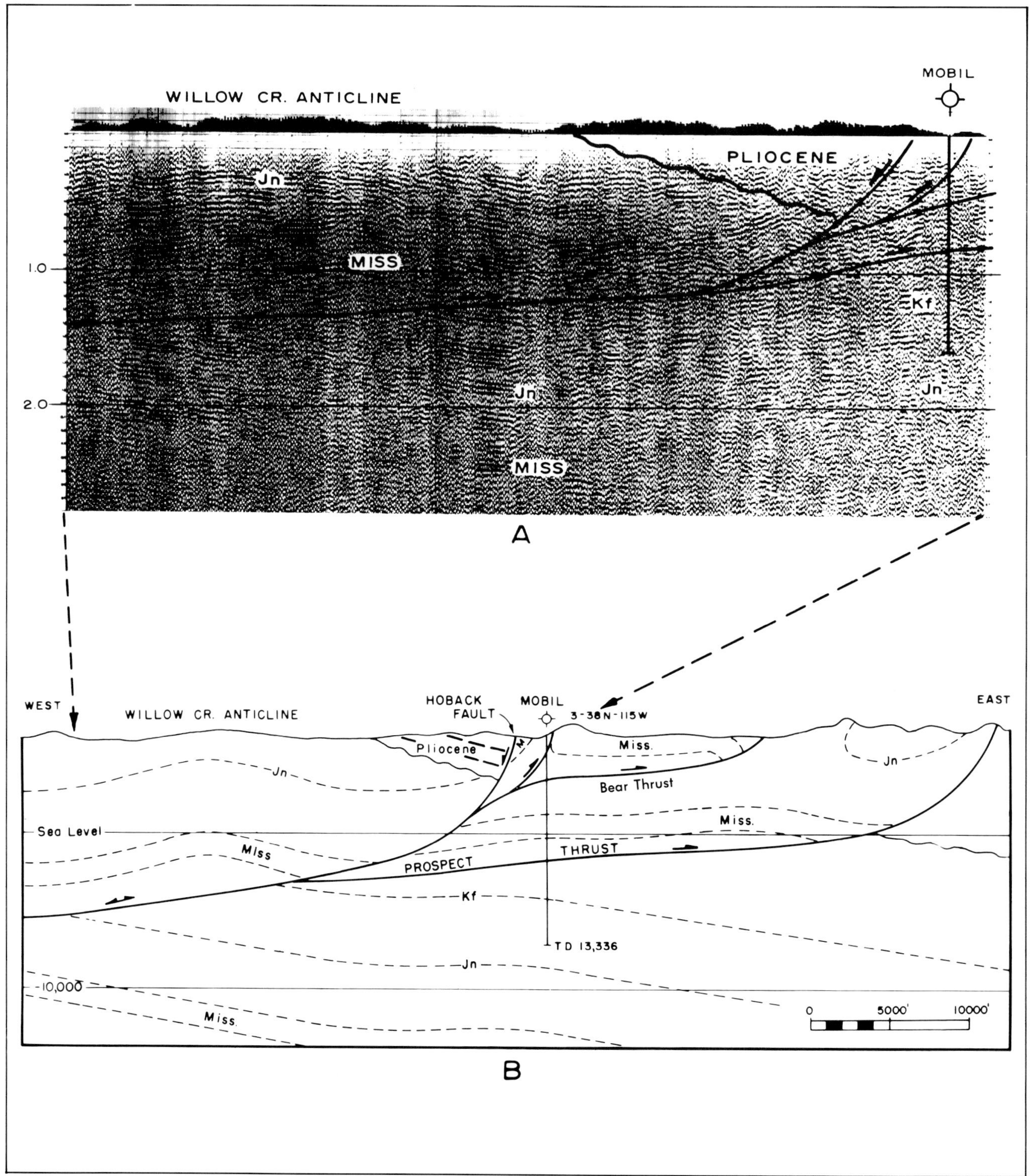

Fig. 6-35 (Royse et al., 1975)—Hoback Canyon area, northwest Wyoming. A. Seismic time section; B. Generalized structural section showing rotation of Pliocene beds (preserved in half graben) into Hoback listric normal fault that soles on ramped portion of Bear thrust. Thus, normal fault is restricted to hanging wall of Prospect thrust. East flank of Willow Creek anticline apparently formed by clockwise rotation along Hoback fault in Pliocene time, well after the termination of thrusting in the Eocene. Permission to publish by Rocky Mountain Association of Geologists.

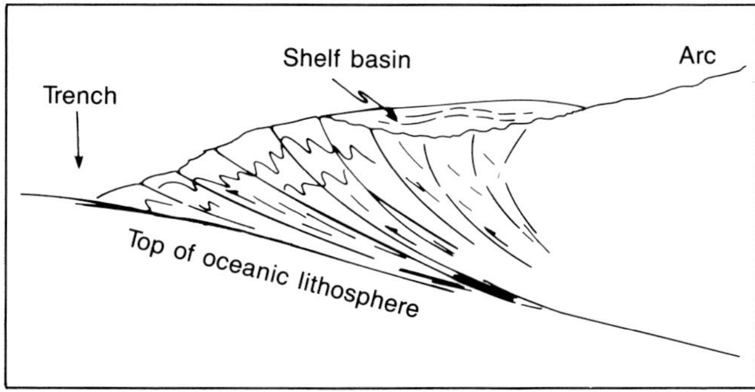

Fig. 6-36—Shelf basin above forearc thrust-fold belt.

PRODUCING EXAMPLES

Forearc

Forearc thrust-fold belts have very low prospectiveness for petroleum for at least four reasons:

1– Complex deformation of intense folding and thrusting, broken formations, and melange (tectonic mixing) has led to an effective reduction, if not destruction, of traps.

2– The same deformation has caused a loss in continuity of potential reservoir rocks. In effect source and reservoir rocks are drilled as a jumble of both rock types. A step-out drilling location from a well with shows will do no better than the original well for conditions are essentially the same in all directions.

3– Potential reservoirs are notorious for having porosity and permeability reduced by plugging by clay altered from volcanic materials which are ubiquitous in forearc settings.

4– Low geothermal gradients associated with downgoing oceanic lithosphere immediately beneath forearc thrust-fold belts have retarded maturation and generation of hydrocarbons in potential source rocks; source rocks are almost invariably immature.

Minor production is known in forearc belts, such as Katalla field in Alaska and from Barbados, but because of the reasons given above their prospectiveness is low.

Prospective basins that are not related to forearc thrust-fold belts do occur in forearc settings. One such basin rests unconformably above the thrust belt in a shelf position (Fig. 6-36). In the Gulf of Guayaquil in Ecuador gas has been found in a structural trap in this type of basin.

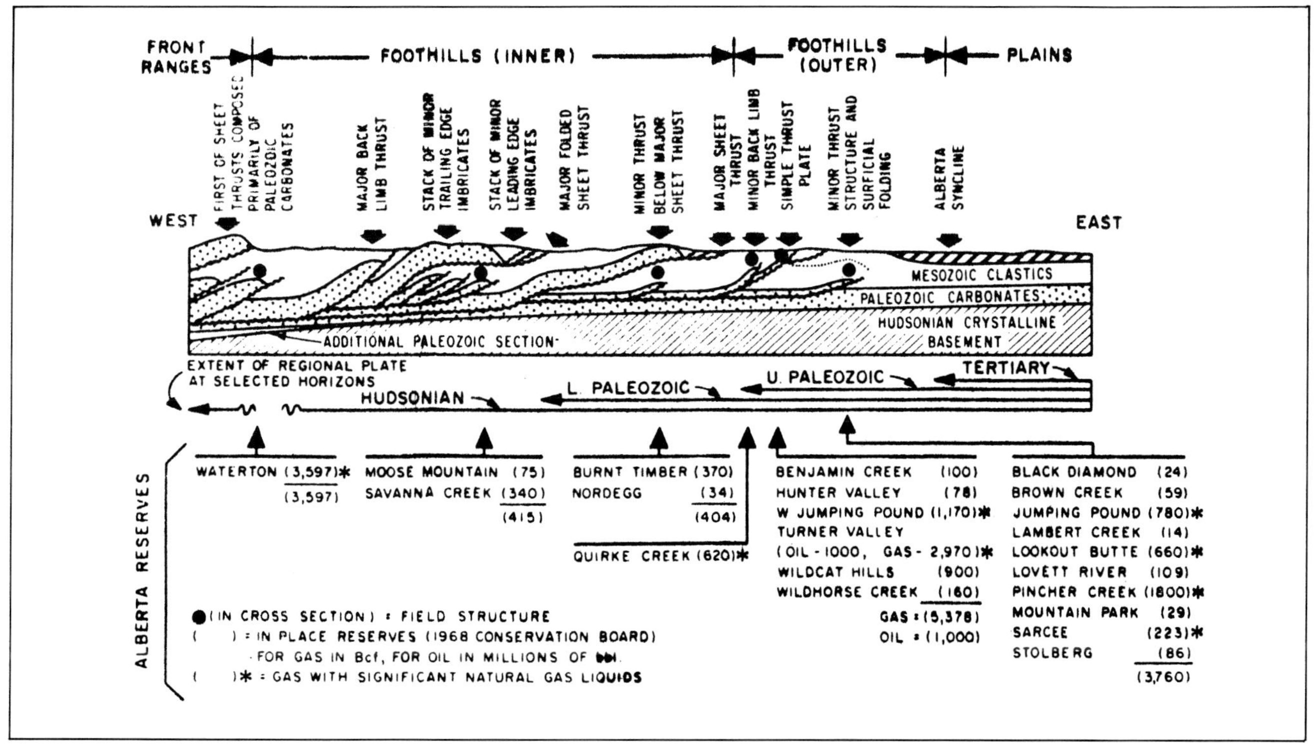

Fig. 6-37 (Dahlstrom, 1970)—Schematic cross section showing zonation of the Canadian Rocky foothills with common structures, field types and reserves. Imbricate thrust zone-foreland detached zone break of figure 6-38 would be placed at the Front Ranges-Inner Foothills boundary. Permission to publish by Canadian Society of Petroleum Geologists.

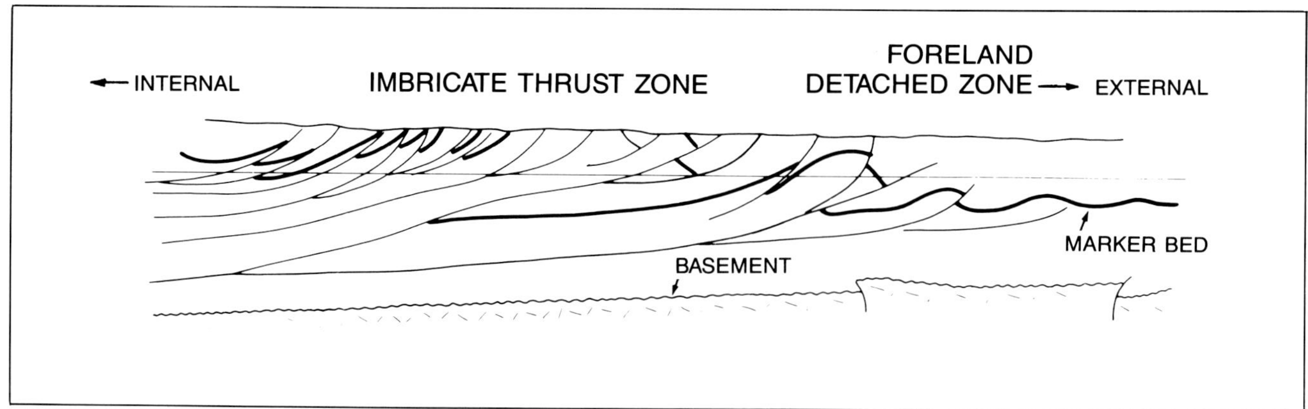

Fig. 6-38 (Lowell, 1973 unpublished)—Idealized thrust-fold belt with imbricate thrust zone and foreland detached zone.

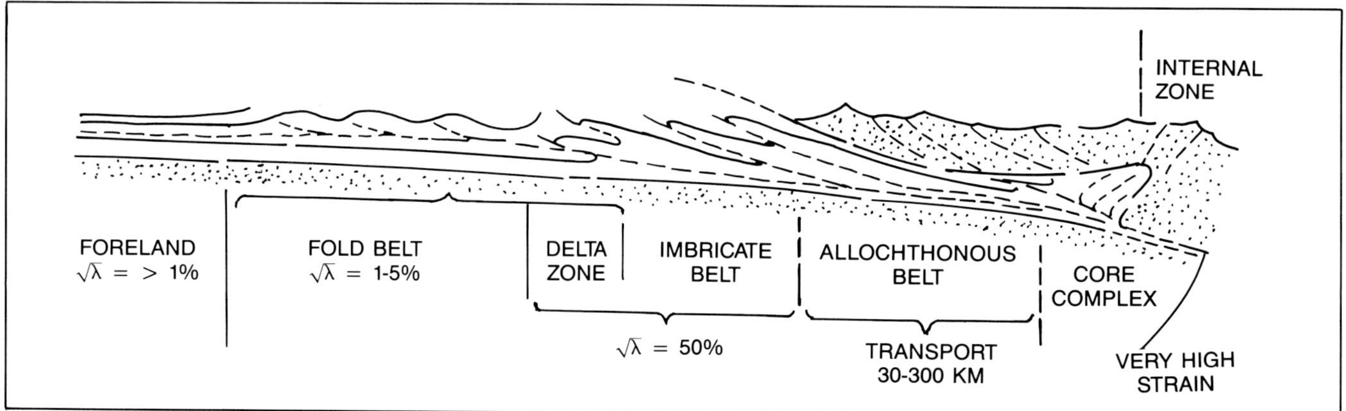

Fig. 6-39 (Roeder, 1983)—Thrust-fold belt zonation with percent strain possible in each zone. Permission to publish by Rocky Mountain Association of Geologists.

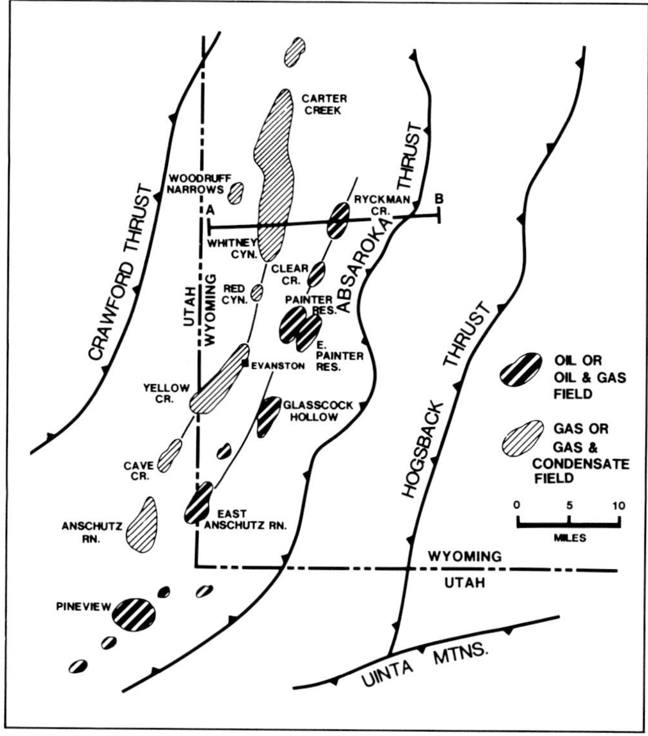

Backarc

Numerous zonations have been made of thrust-fold belts in an attempt to determine their petroleum potential (Figs. 6-37 through 6-39). In all examples the more external portions tend to be most prolific, but the zonations lose much of their value for two reasons. First, any given zonation does not apply to all belts; the Idaho-Wyoming belt, for example, does not have a foreland detached zone. Second, even though most prospects are in external positions, many excellent more internal structures, e.g., Waterton in the Canadian Rockies (Fig. 6-37), might be overlooked if the zonation is rigidly applied.

In general the better prospects in the folds that are only slightly disrupted by thrusts in the external sides of thrust-fold belts are related to good continuity of fold (and therefore reservoir) and drainage paths. Thrust faults can reduce

Fig. 6-40 (Warner, 1982)—Index map of oil and gas fields, southwest Wyoming-northeast Utah. Line of section AB is approximately that of figure 6-41. Permission to publish by Rocky Mountain Association of Geologists.

Illustration 6-41 begins on next page ⟶

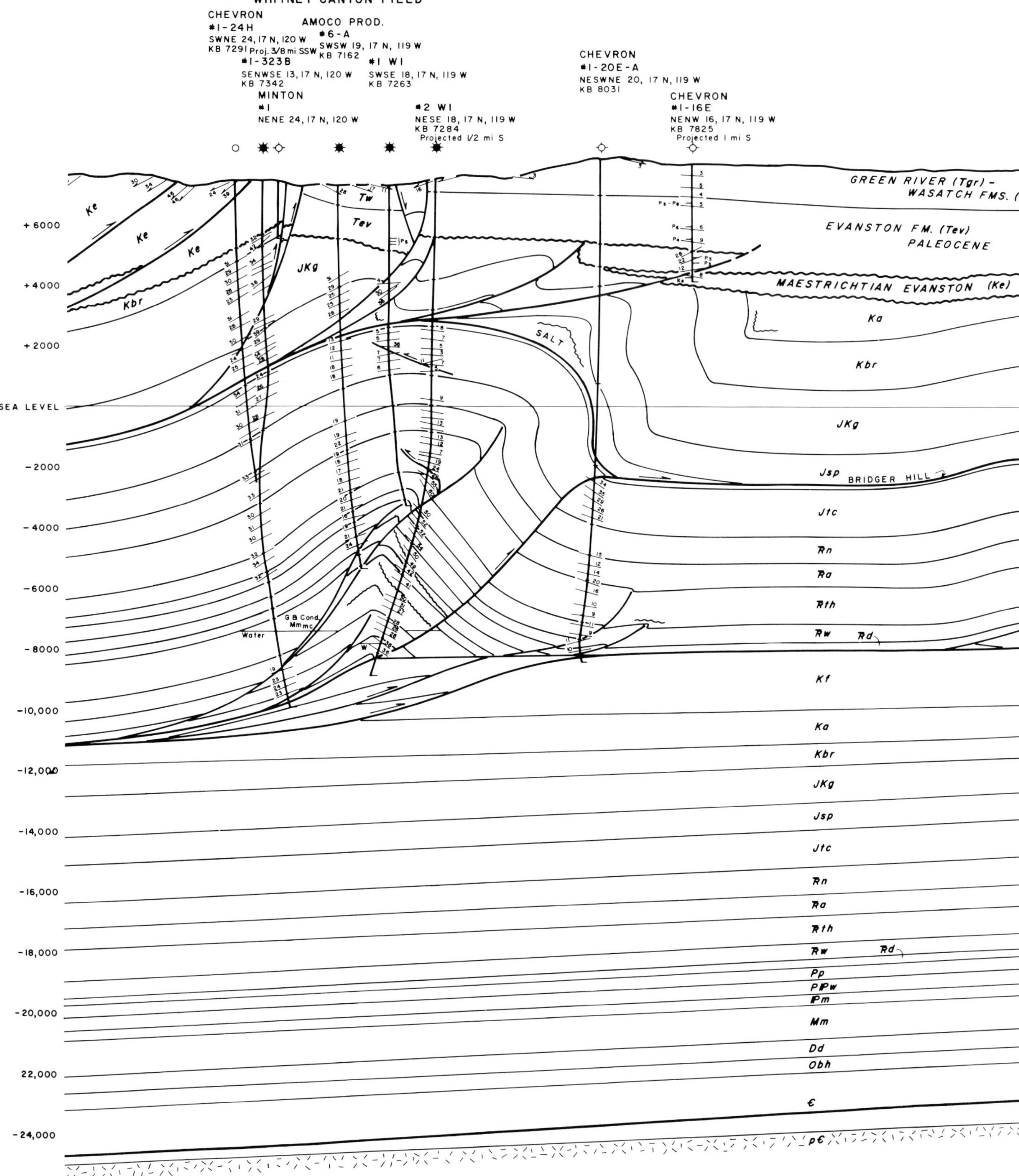

Fig. 6-41 (Lamerson, 1982)—Structural cross section through Whitney Canyon and Ryckman Creek fields. Note long gas column at Whitney Canyon. (See figure 6-40 for approximate line of section.) Permission to publish by Rocky Mountain Association of Geologists.

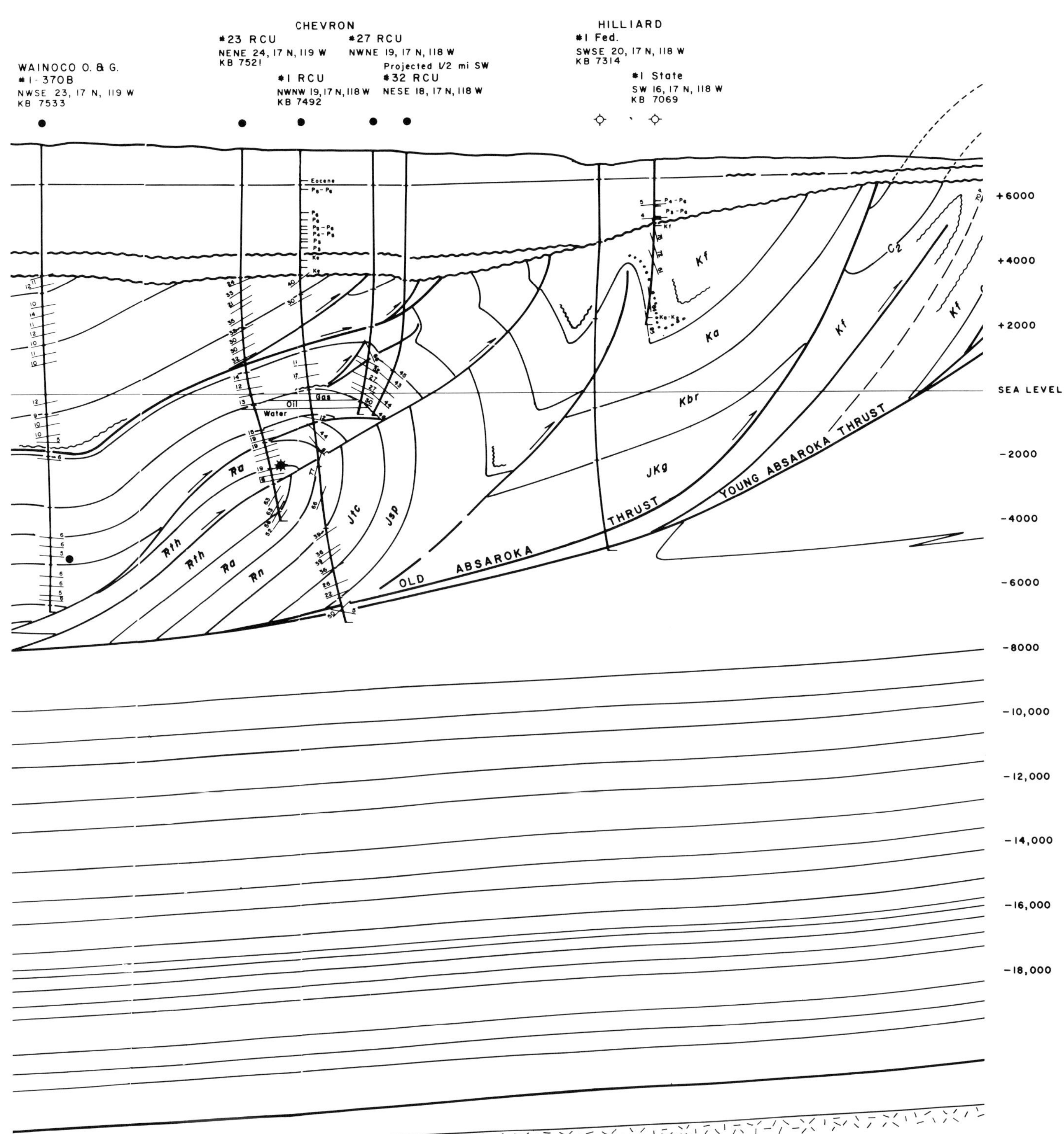

R 119 W
R 118 W
RYCKMAN CREEK FIELD
CHEVRON
#23 RCU
NENE 24, 17 N, 119 W
KB 7521
#27 RCU
NWNE 19, 17 N, 118 W
Projected 1/2 mi SW
#32 RCU
NESE 18, 17 N, 118 W
#1 RCU
NWNW 19, 17 N, 118 W
KB 7492
WAINOCO O. & G.
#1-370B
NWSE 23, 17 N, 119 W
KB 7533
HILLIARD
#1 Fed.
SWSE 20, 17 N, 118 W
KB 7314
#1 State
SW 16, 17 N, 118 W
KB 7069
Eocene
Gas
Oil
Water
Kf
Ka
Kbr
JKg
Jtc
Jsp
Ra
Rth
Rn
OLD ABSAROKA THRUST
YOUNG ABSAROKA THRUST
+6000
+4000
+2000
SEA LEVEL
-2000
-4000
-6000
-8000
-10,000
-12,000
-14,000
-16,000
-18,000

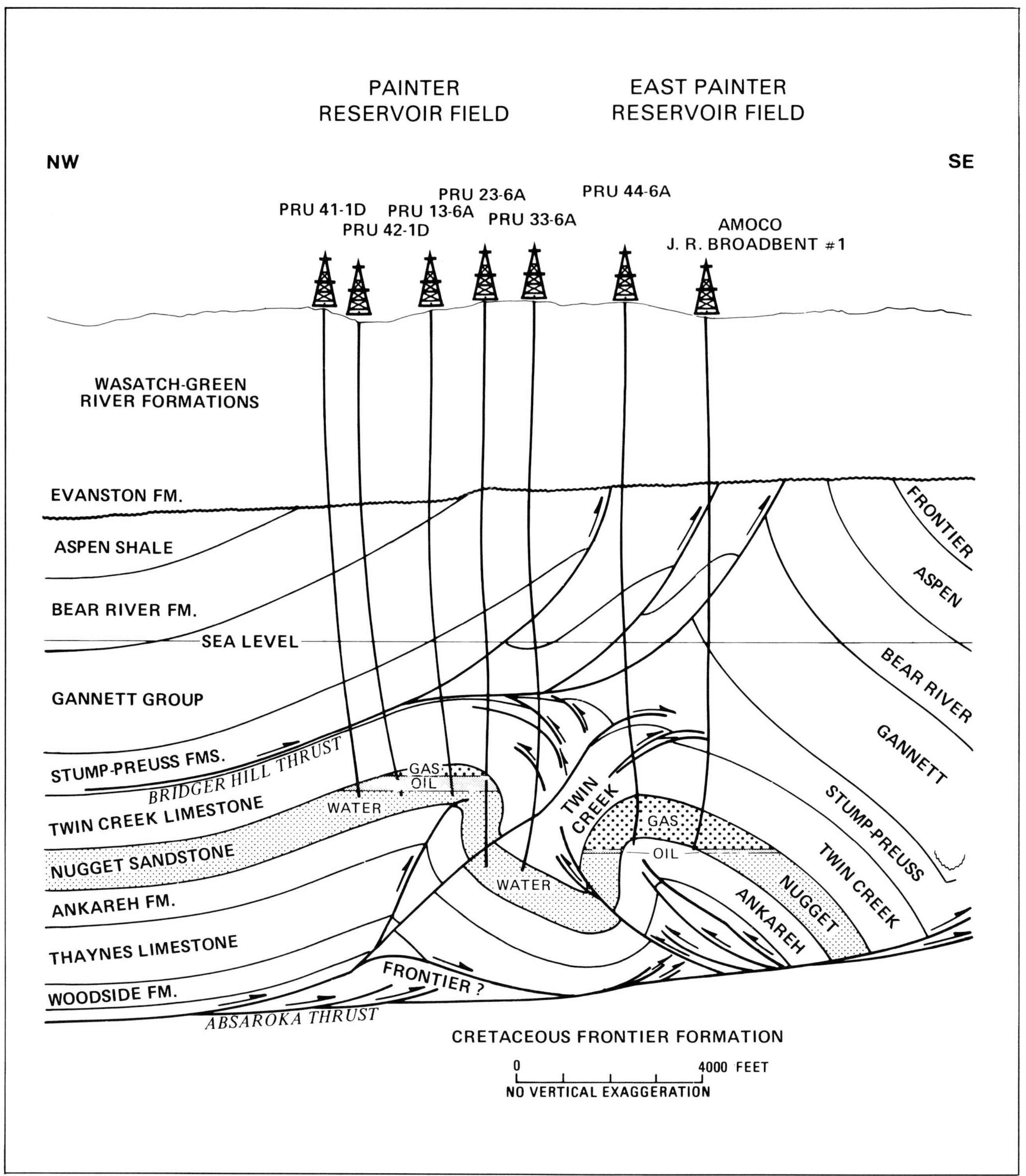

Fig. 6-42 (Frank et al., 1982)—Northwest-southeast structure section through the Painter Reservoir and East Painter Reservoir fields. In these and other Wyoming-Utah thrust-belt fields it is critical that reservoir rocks in the upper plate of the Absaroka thrust be in contact with subthrust Cretaceous rocks that are the source of oil and gas (Warner, 1982). Permission to publish by Rocky Mountain Association of Geologists.

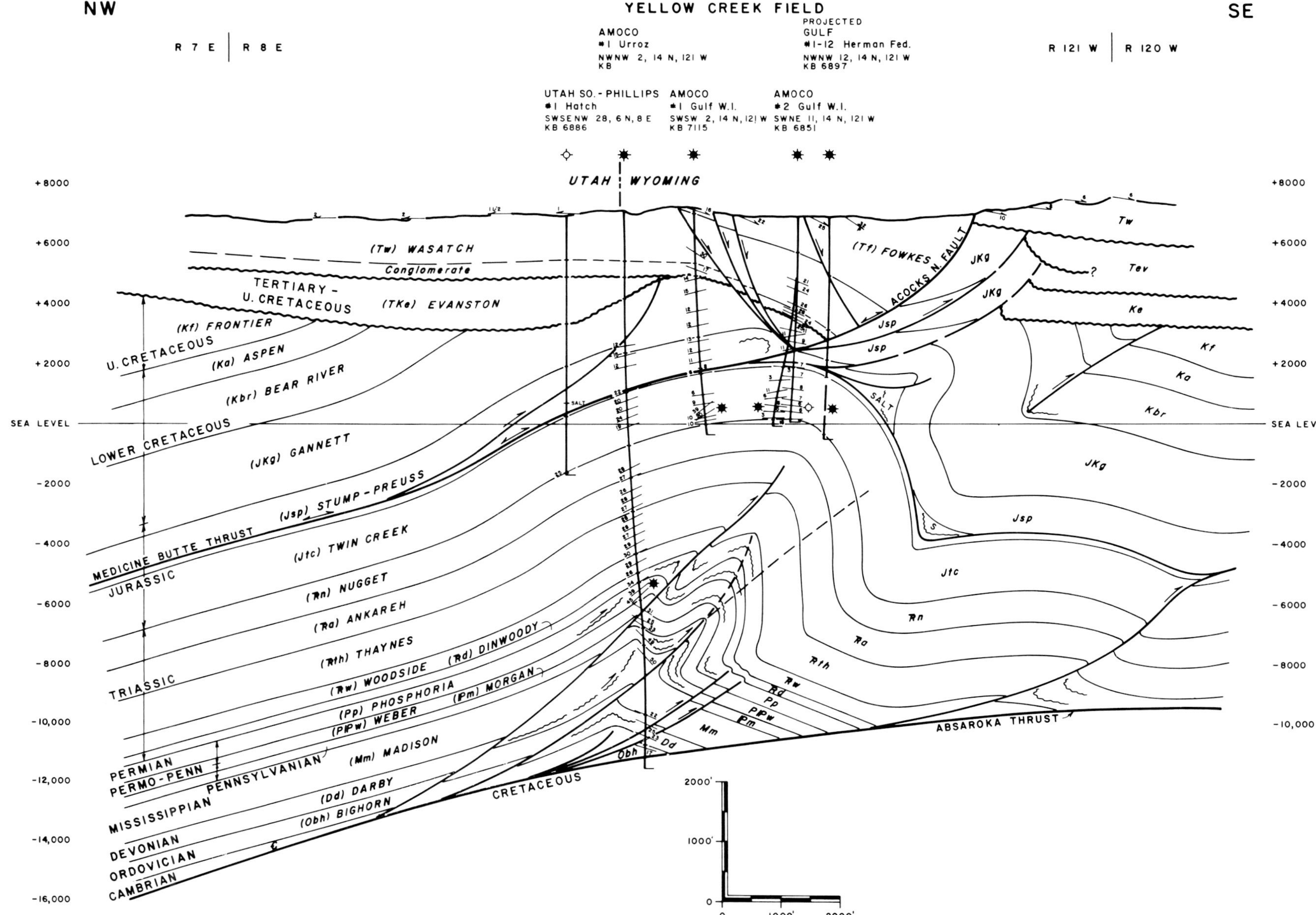

Fig. 6-43 (Lamerson, 1982)—Northwest-southeast structural cross section in Yellow Creek field area. Note open concentric fold at Nugget Sandstone level and complex structure at Paleozoic level as the concentric fold tightens downward. Permission to publish by Rocky Mountain Association of Geologists.

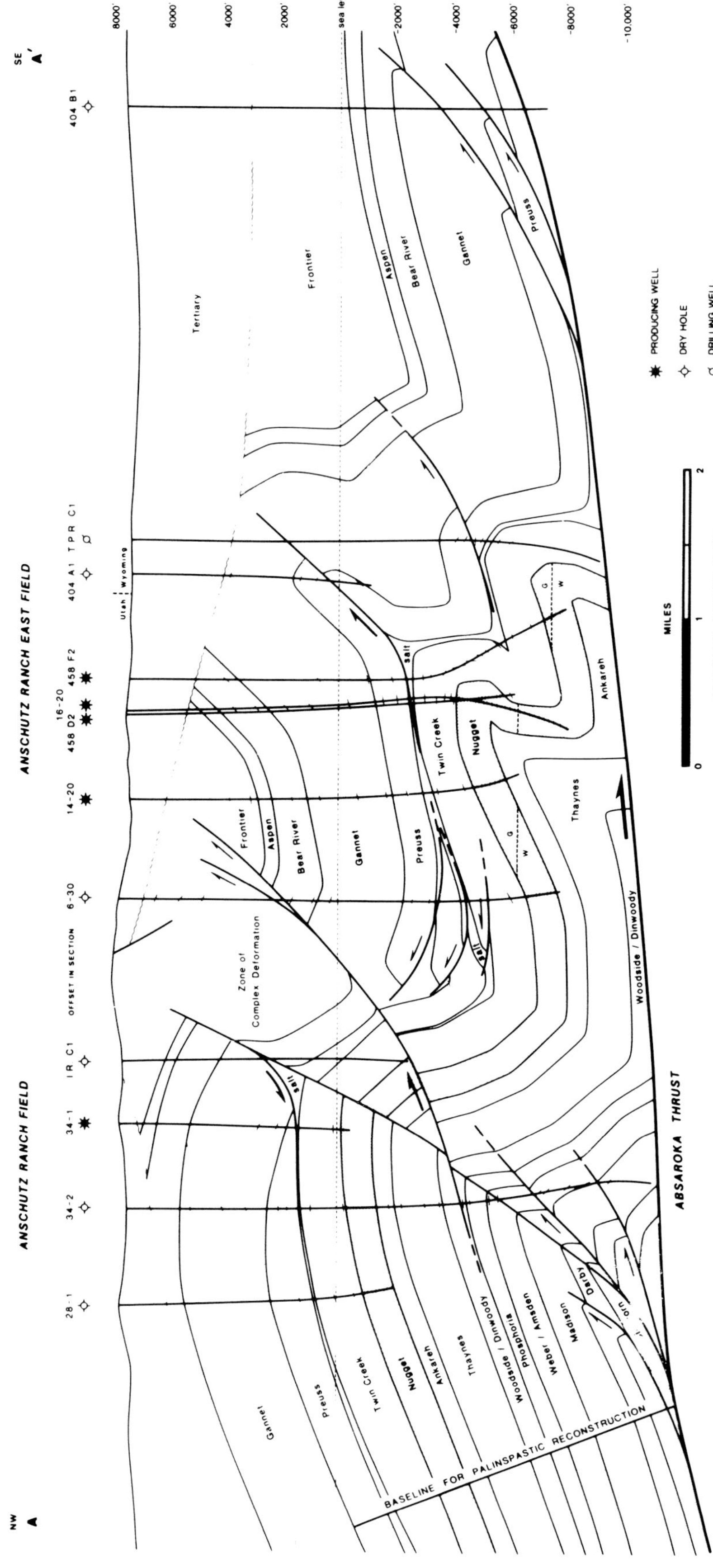

Fig. 6-44 (West and Lewis, 1982)—Structural cross section A-A' through Anschutz Ranch and Anschutz Ranch East fields. Line of section is shown in figure 6-22. Permission to publish by Rocky Mountain Association of Geologists.

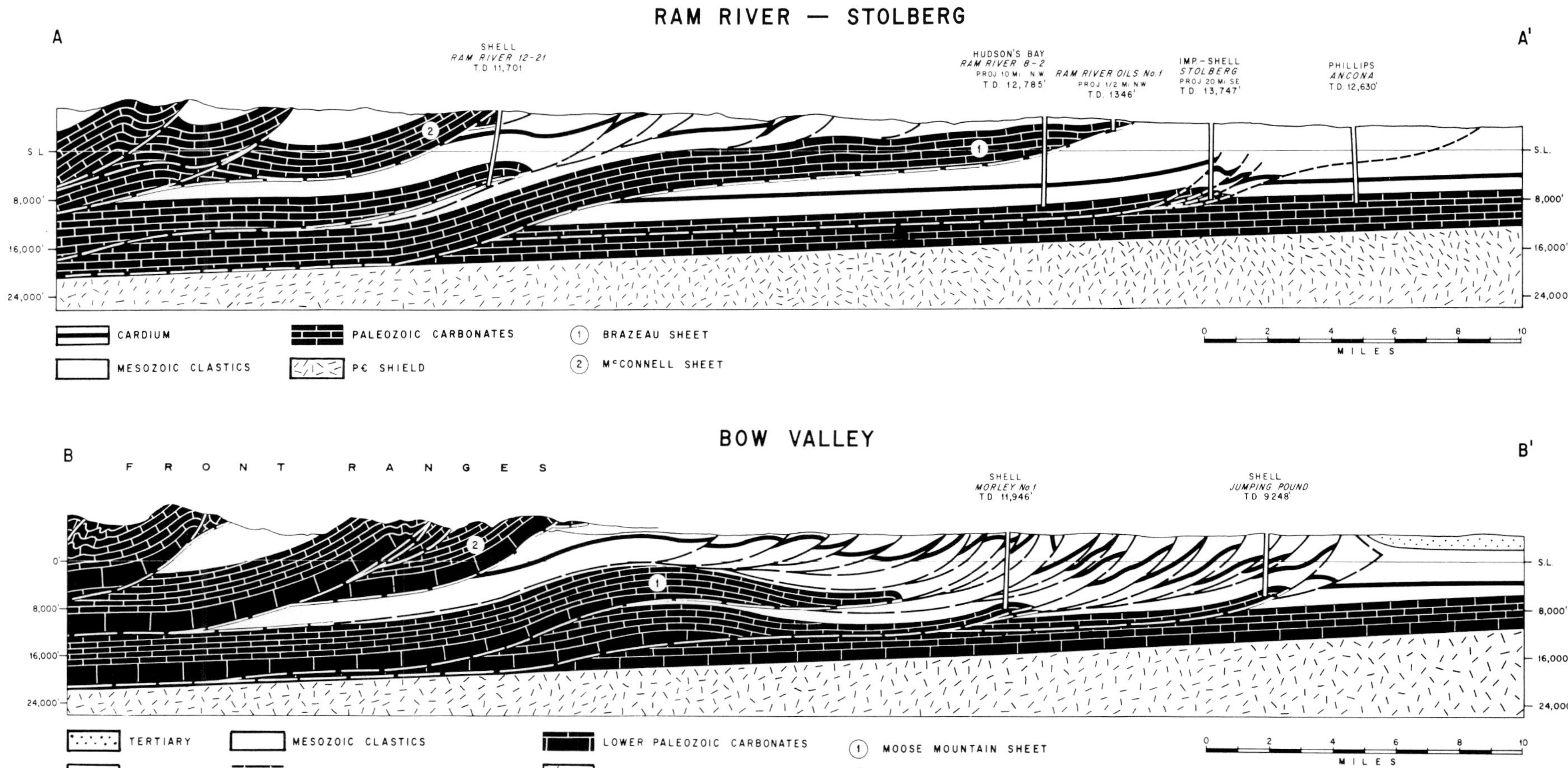

Fig. 6-45 (Bally et al., 1966)—Cross sections through the eastern part of the southern Canadian Rocky Mountains, west of Calgary. Production at Jumping Pound is from leading edge (also termed 'Snake-head' structure) of Upper Paleozoic carbonates (Mississippian Rundle Formation). Note that thrust sheets are buried both on the internal (left) and external (right) sides and thus protected from flushing. Permission to publish by Canadian Society of Petroleum Geologists.

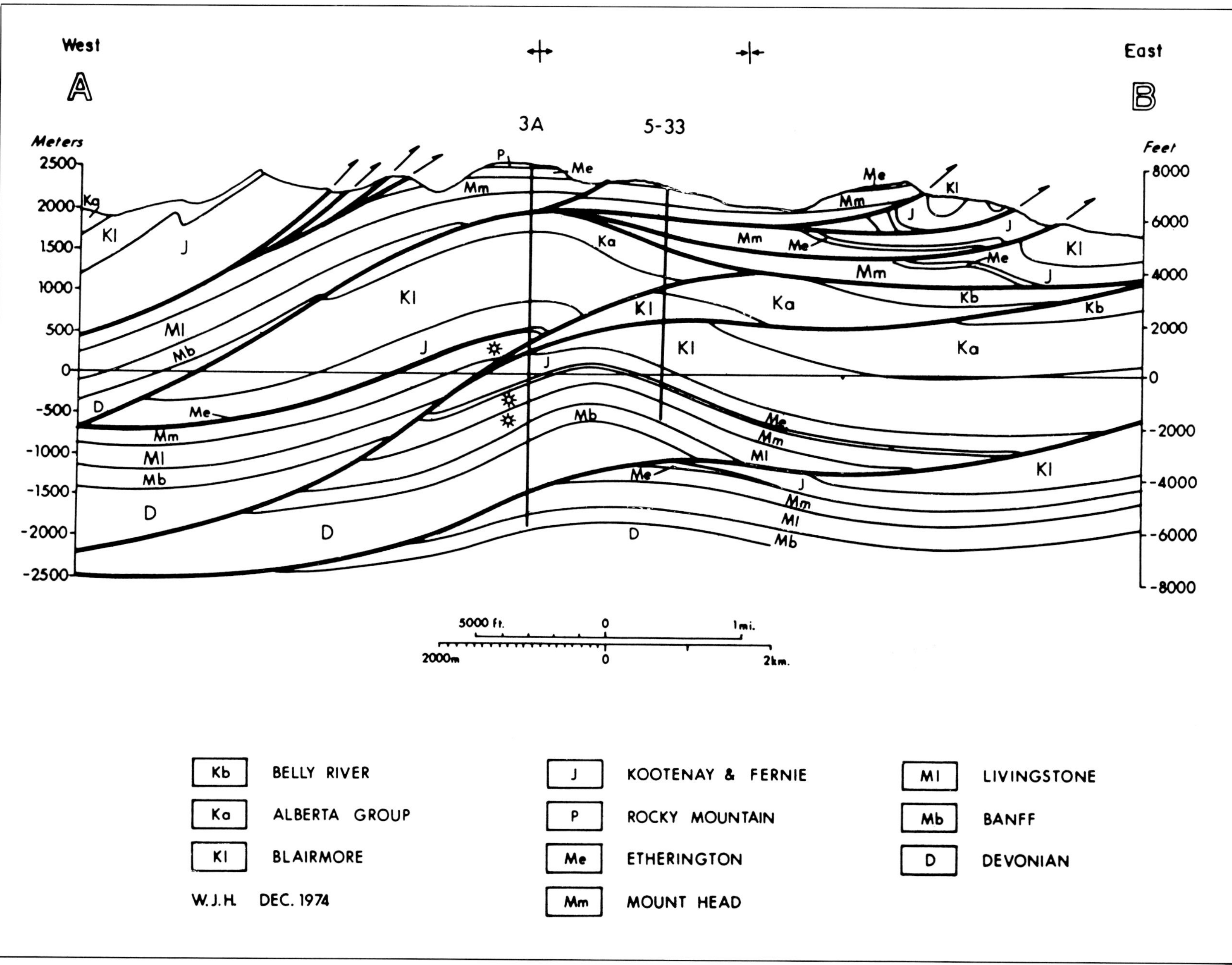

Fig. 6-46 (Hennessey, 1975)—Section across Savanna Creek gas field illustrating folded fault structure. At surface, structure is three miles (4.8 km) wide and 12 mi (19 km) long. Livingston carbonate reservoir has only 3–4% porosity but is highly fractured. Permission to publish by Canadian Society of Petroleum Geologists.

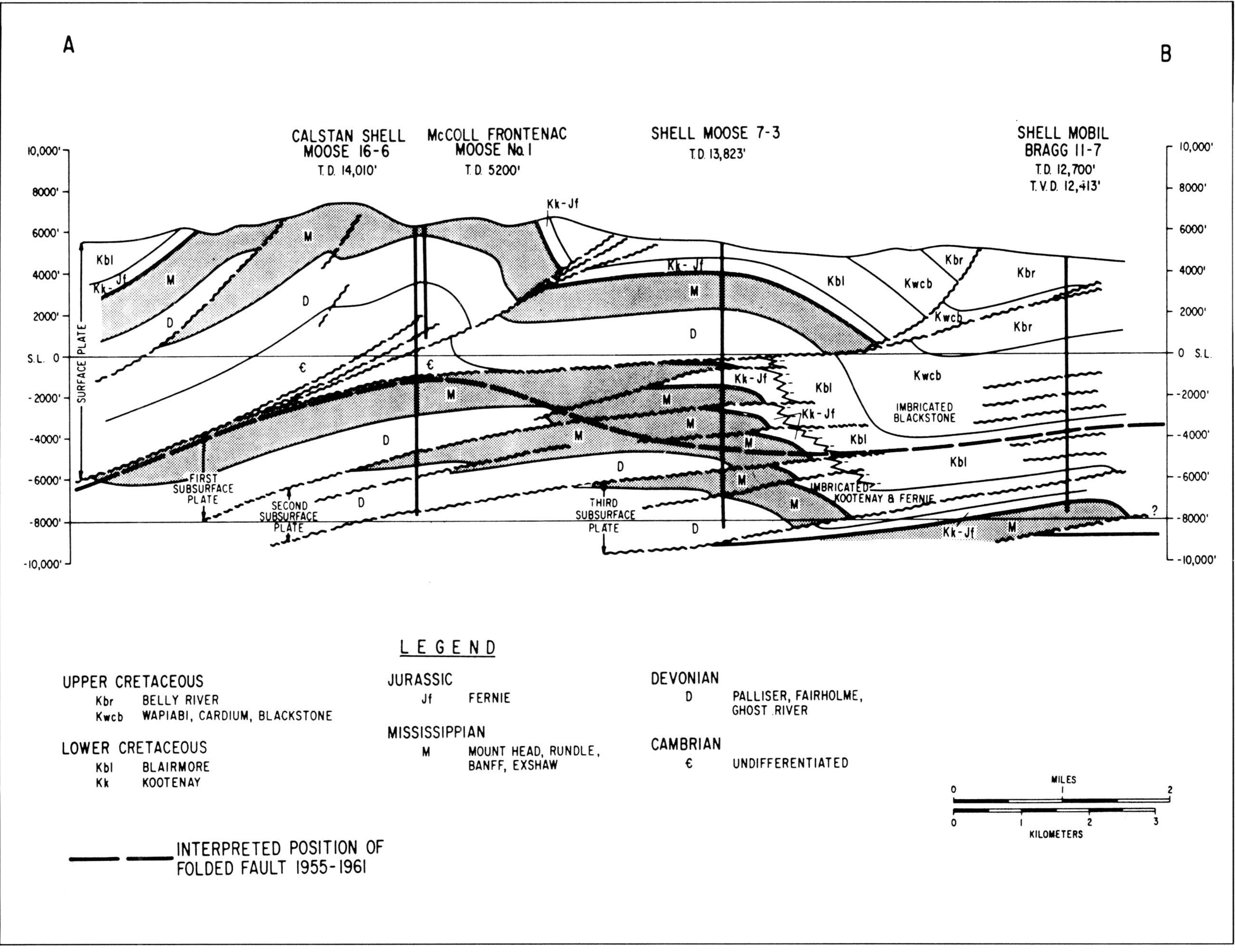

Fig. 6-47 (Ower, 1975)—East-west structural cross section through the Moose Mountain structure showing imbricate slices of Mississippian carbonates rather than a folded-fault configuration as earlier interpreted. Thrust stack in the Shell Moose 7-3 well would probably have seismic expression similar to well 5-26-57-6W6 in figure 6-33. Permission to publish by Canadian Society of Petroleum Geologists.

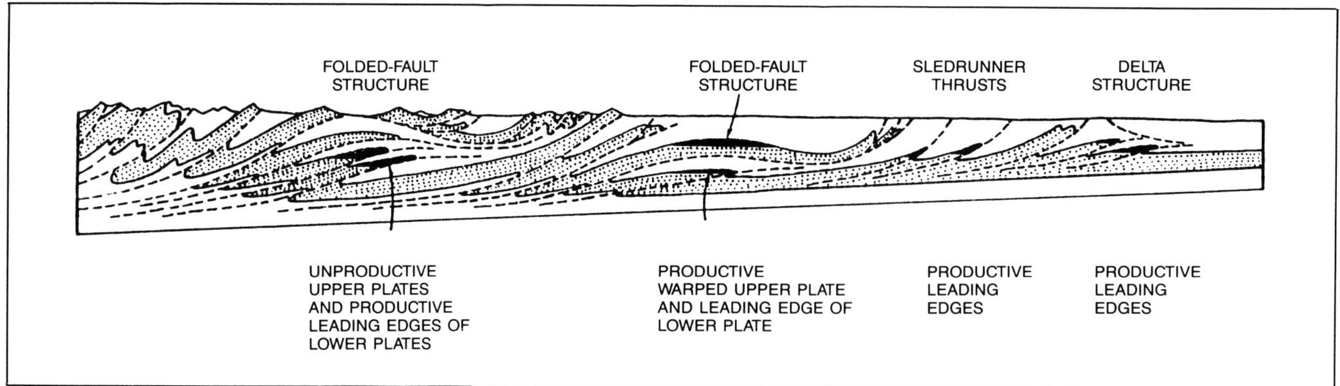

Fig. 6-48 (Roeder and Gilbert, 1977)—Summary of thrust-fold belt petroleum traps based largely on Canadian Rockies. Permission to publish by American Association of Petroleum Geologists.

the thickness of the potentially productive interval and can or cannot juxtapose strata that either provide seals for reservoirs or migration pathways. Neither condition is a problem for fold closures.

The most common trap of the Idaho-Wyoming backarc thrust-fold belt is the hanging wall anticline. Virtually all of the fields shown on figure 6-40 are folds in the hanging wall of the Absaroka thrust. Whitney Canyon, Ryckman Creek, Painter Reservoir-East Painter Reservoir, Yellow Creek, Anschutz Ranch, and Anschutz Ranch East fields are illustrated in figures 6-41 through 6-44. Except for East Painter Reservoir all the fields are asymmetric with the steep limb to the east - in conformance with most thrust-fold belts that, worldwide, are asymmetric with the short limb on the external side. Flexural-slip folding is typically assumed in thrust-fold belts so that, excepting for flexural flow in the Jurassic Preuss salt, all of figures 6-41 through 6-44 are drawn as balanced sections. In this regard, in addition to the discussion of flexural-slip folds in Chapter 3, "Balanced cross sections" by Dahlstrom (1969) is reprinted at the end of this chapter.

In another backarc thrust-fold belt, the Canadian Rockies, trap types differ from the hanging-wall fold. Here, leading edges of massive slabs of thrusted carbonates are a dominant trap (Fig. 6-45). Still another type of trap is the folded-fault structure (Figs. 6-13, 6-14) as exemplified by the Savanna Creek gas field (Fig. 6-46). It is instructive

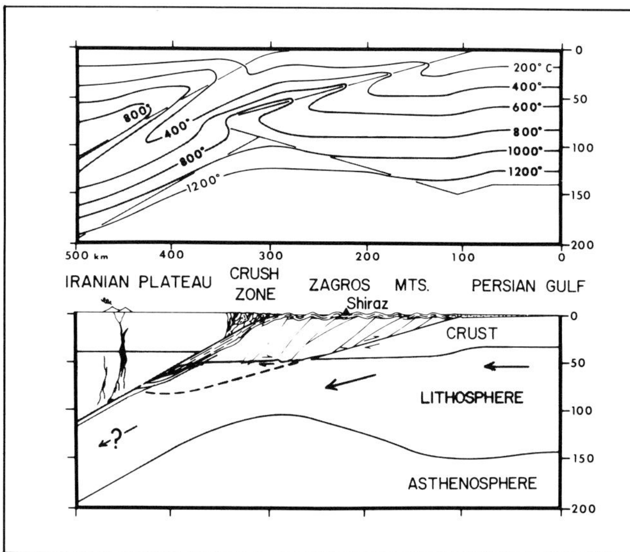

Fig. 6-49 (Bird et al., 1975)—Zagros thrust-fold belt as the result of collision between Iranian block (on left) and underriding Arabian plate (on right). From Journal of Geophysical Research, V. 80. Copyright by the American Geophysical Union. Reprinted by permission.

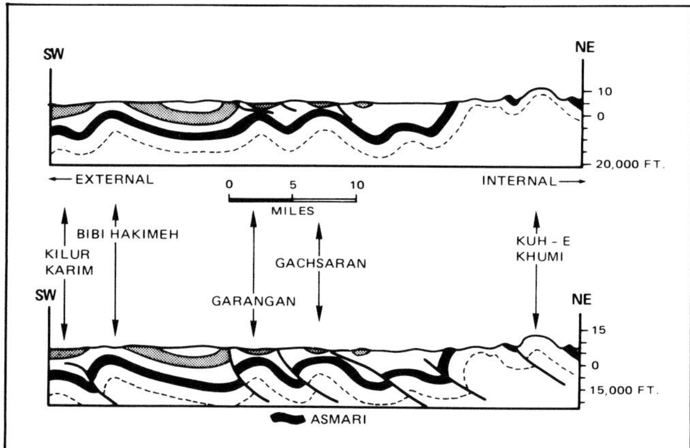

Fig. 6-50 (Hull and Warman, 1970)—Regional cross section through Asmari oil fields, Iran. Note amplitude from crest to trough of some structures is at least 3,049 km (10,000 ft). Note disharmony of structure above Asmari Limestone. Permission to publish by American Association of Petroleum Geologists.

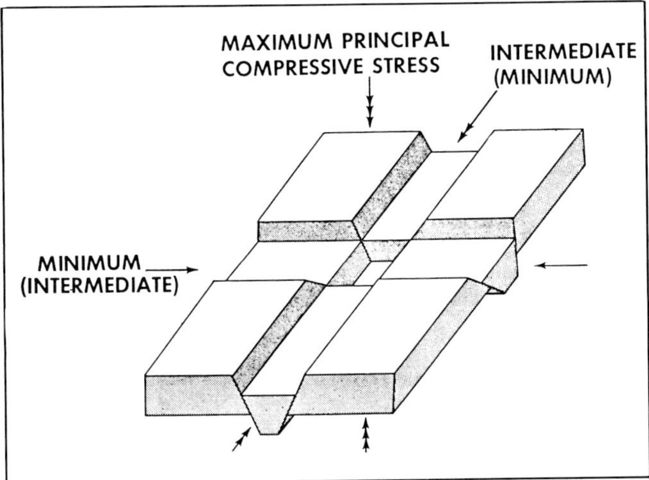

Fig. 6-51—Where maximum principal compressive stress is vertical, as it can be locally in any fold of any structural style, conjugate normal faults and vertical tension fractures form. Some shear may occur on the tension fractures, but the displacement is minor compared to that on the conjugate normal shears. The orientation with respect to structural axes of folds of both normal faults and tension fractures can be longitudinal or transverse or both, depending on the orientation of the least and intermediate principal stresses. Since the absolute values of stress in the least and intermediate directions are often quite close, slight variations in stress will cause the two directions to interchange. Thus, longitudinal and transverse faults, both of the conjugate normal and tension fracture variety, are compatible in the same fold. Longitudinal faults form when intermediate principal stress is horizontal and longitudinal to structure, and least principal stress is horizontal and transverse to structure. Transverse faults form when intermediate and least principal stress directions exchange or flip-flop, i.e., the direction of intermediate principal stress becomes transverse to structure as the least principal stress direction becomes longitudinal to structure. This type of faulting is important in creating fracture porosity and permeability, e.g., as in the case of the Asmari folds.

to note that the Moose Mountain structure was also interpreted as having a folded fault, but later drilling showed this was not correct (Fig. 6-47). Production from delta or triangle structures in the external part of the Canadian Rockies has been discussed previously (Fig. 6-33). A summary of trap types from the Canadian Rockies has been made by Roeder and Gilbert (1977; Fig. 6-48).

Collisional

Zagros

Collisional belts are by far the most prolific of all thrust-fold belts because, prior to 1979, the Zagros collisional belt of Iran (Fig. 6-49) accounted for some 75% of the world's thrust-fold belt production. Curtailed Zagros production and recent Wyoming-Utah discoveries have changed this figure somewhat; nonetheless, Zagros reserves are staggering.

Zagros folds are renown for their great amplitude and large adjacent drainage areas (Fig. 6-50). They can be traced on strike for 160 km (100 mi) and more. These large dimensions of closure together with a thick fractured reservoir (Fig. 6-51) have made it possible for fields such as Gach Saran and Bibi Hakimeh (Fig. 6-50) to have estimated ultimate recoverable reserves of 12 and 8 billion barrels of oil, respectively.

A brief discussion of the stratigraphy of the Zagros thrust-fold belt is essential to understanding its prolific nature. In fact an initial evaluation of the stratigraphy of any thrust-fold belt can be more important than a structural analysis, for these belts are by definition "structured." A stratigraphic reconnaissance possibly can even determine if the belt should be explored and, if so, which parts should be approached first. Moreover, stratigraphic changes often determine or localize structural changes, e.g., position of bedding plane thrusts, behavior of competent and incompetent lithologies, etc.

The main Zagros reservoir, of variable thickness but averaging about 305 m (1000 ft), is the Asmari Formation of Oligocene-Miocene age. In the producing area lithology is dominantly limestone of shallow water, but non-reefal, origin with minor dolomite; matrix porosity and permeability are very low and the unit owes its excellent reservoir performance to extensive fracturing induced by Pliocene-Pleistocene orogenic deformation. Near the head of the Persian Gulf the Asmari limestones interfinger with massive sandstones to become the Ahwaz facies, which was derived from the Arabian shield and has good intergranular porosity. Near the mouth of the Persian Gulf the Asmari is in the Dezac marl facies of unlikely reservoir lithology. Overlying the Asmari are evaporites, usually several thousand feet in thickness, termed the Lower Fars or Gachsaran formation. The presence of this unit, with an impermeable anhydrite at the base, is critical as seal for Asmari production. Erosion of the Lower Fars condemns large areas directly on strike both to the northwest and southeast of the main producing Asmari folds.

In addition to Asmari production, large reserves have been established in the underlying Bangestan carbonate of Cretaceous age. Kent and Warman (1972) have stated that:

> In most of the main producing fields the Asmari reservoirs are full to spill point, and tend to be in fluid connection with the Bangestan, sharing a common oil-water level. In the lesser structures along the southwestern margin there is no connection between the Asmari and Bangestan; the upper reservoirs are poorly supplied with oil which may be heavy and asphaltic; the main body of oil occurs in the underlying Ban-

gestan reservoirs, which have a different pressure regime and are often full to spill point.

Clearly, deeper potential reservoirs should not be overlooked in even the most prolific thrust-fold belts.

Carpathians

Stratigraphic considerations are important in another collisional thrust-fold belt, the Romanian portion of the Carpathians (Fig. 6-52). The following discussion, which may be useful in exploring other thrust-fold belts, is based mainly on an article by Paraschiv and Olteanu (1970) on the well controlled and heavily explored Ploiesti district.

Sandstone reservoirs of Pliocene age contain most of the hydrocarbons in Romania and interbedded fine-grained sediments may have been the source. The Pliocene is confined, however, to the pre-Carpathian foothill and foreland platform areas with the effect that in the imbricate thrust sheets, which extend from the Carpathian bend northward 300 km (187 mi) to the Russian border, production is from older Tertiary rocks and of insignificant amount. Furthermore, although Pliocene rocks are present fringing the Carpathians from the Yugoslavian border on the west to about the Trotus River in east-central Romania, they are in a combination of sufficient thickness and favorable reservoir facies only in the Ploiesti area at the convexity of the Carpathian bend.

In the Ploiesti district the Pliocene is characterized by essentially uninterrupted clastic sedimentation through its four stages, Meotian, Pontian, Dacian, and Levantine, being transgressive at the base and grading upward from brackish with fresh-water intercalations to lacustrine with lignite seams; thus, the Pliocene was deposited in a non-marine basin. It is, however, made up of a sequence of marl and sand in a very favorable proportion for hydrocarbon accumulations. Meotian and Dacian reservoirs have accounted for 1.5 billion of the 1.725 billion barrels produced in the Ploiesti district to January 1, 1968. Alternating sands and marls in the Meotian (porosity 20–25%) and Dacian (porosity 25–35%) effectively act as interbedded

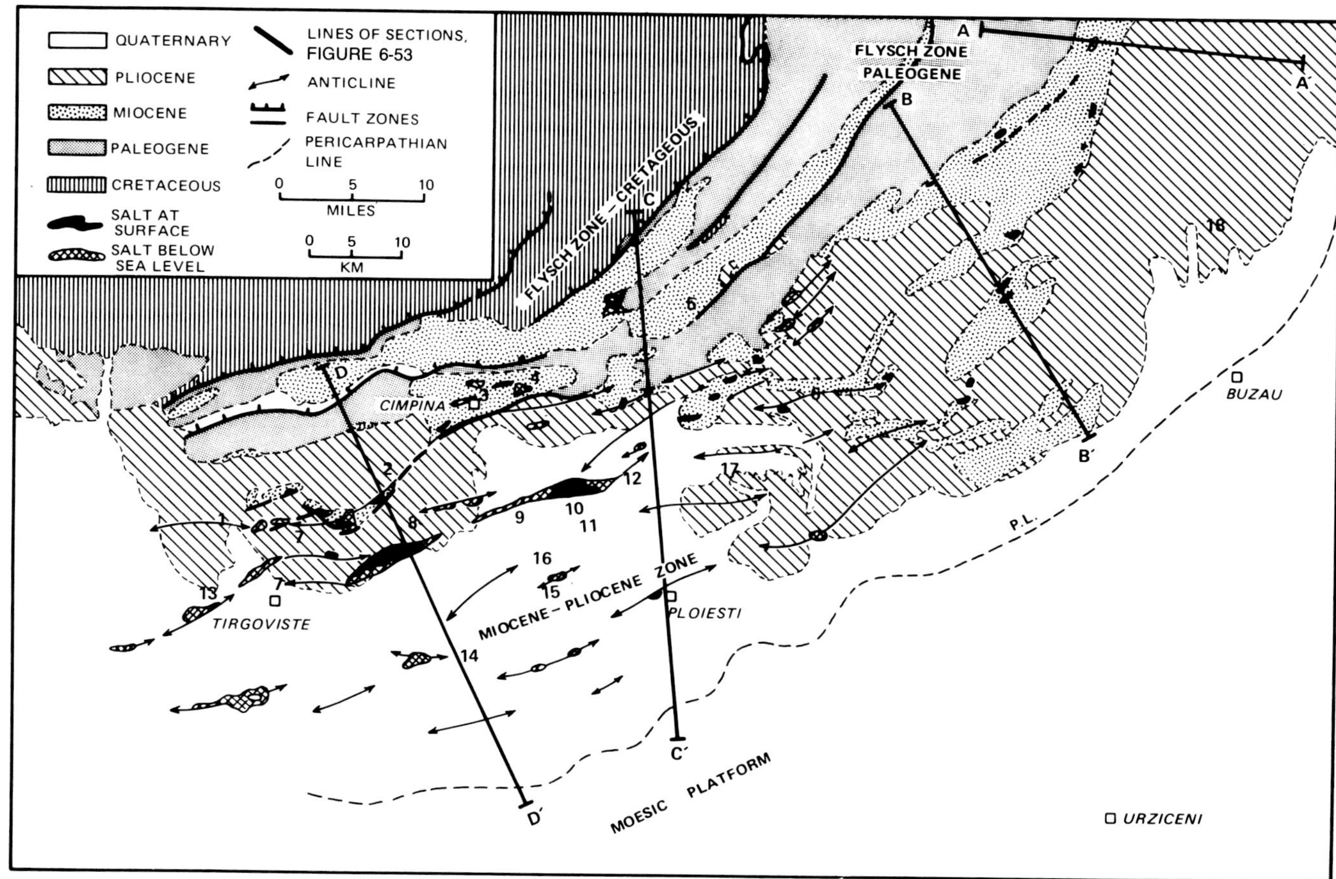

Fig. 6-52 (Paraschiv and Olteanu, 1970)—Geologic-tectonic map of the Ploiesti district, Romania, with lines of section for figure 6-53 and locations of sections for figure 6-54. Permission to publish by American Association of Petroleum Geologists.

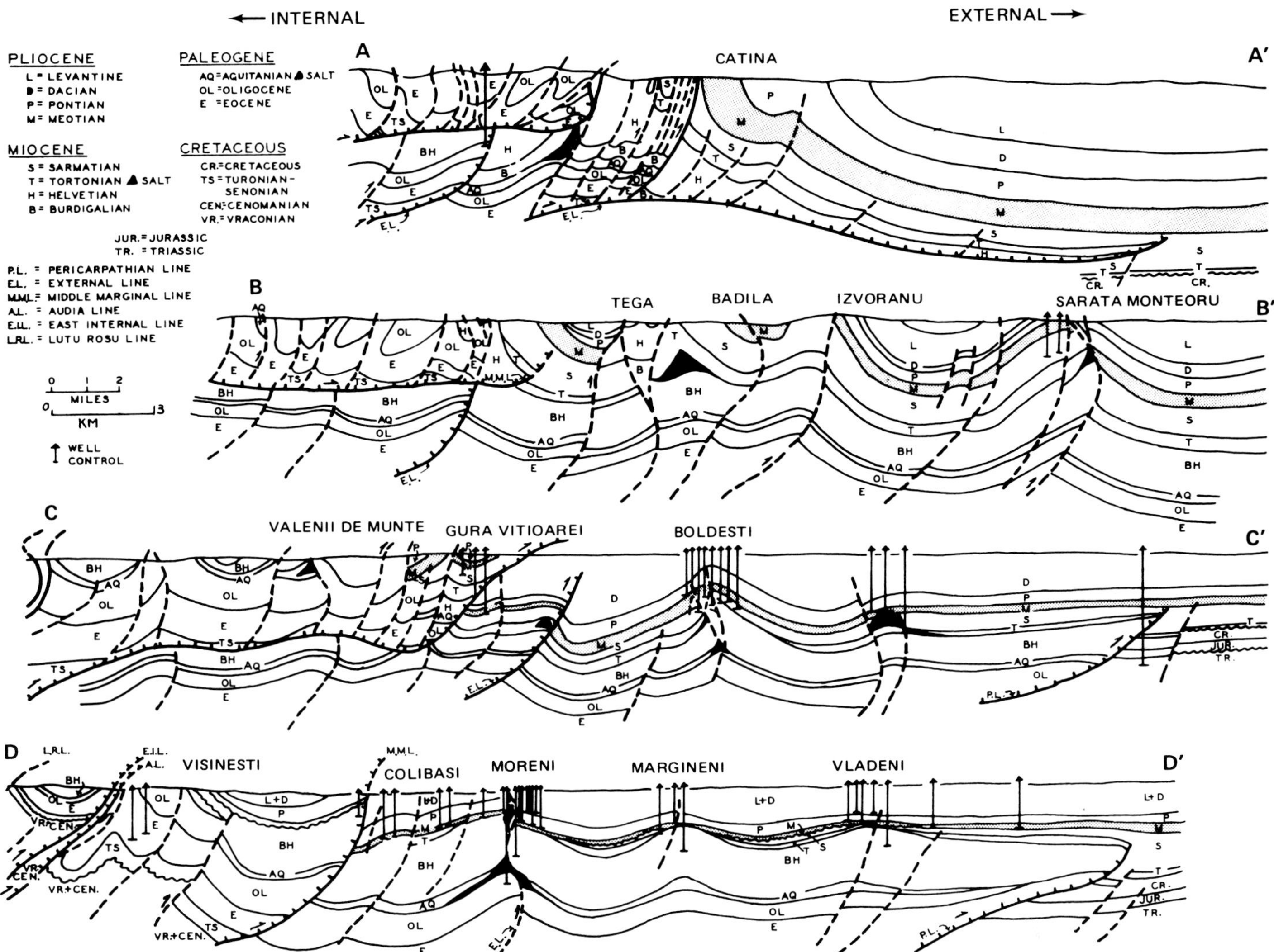

Fig. 6-53 (Paraschiv and Olteanu, 1970)—Regional cross sections in the Ploiesti district. Locations shown on figure 6-52. Best producing structures fundamentally owe their origin to compression derived from the Carpathian orogeny, even though they exhibit only minor thrusting and are south of the main Carpathian thrust sheets. Salt may have been important in localizing earliest structural growth but was tectonically mobilized in the subsequent orogeny. Note that Meotian (shaded) shows good reservoir continuity. Permission to publish by American Association of Petroleum Geologists.

reservoirs (permeability 10–3000 md and higher) and seals, respectively. Salt of Miocene Tortonian and also Aquitanian age played an important dual role in Pliocene accumulation. It served as a locus for earliest structural growth and, having flowed, acted as seal, particularly for Dacian reservoirs.

Though Meotian and Dacian rocks are present both northeast and west from the Ploiesti district, they are unproductive for the following reasons:

1– Both Meotian and Dacian are locally absent or only partially preserved eastward from Ploiesti toward Buzau because of very late erosion (Figs. 6-52, 6-53).

2– The Meotian is too sandy in the east, lacking marl seals and possibly lacking indigenous source.

3– Meotian is thinner in the west and sands play out (Walters, 1946).

4– Dacian sands are discontinuous in the producing area and are even more lenticular beyond.

Both indigenous and older (mainly Oligocene) rocks have been proposed as source for Pliocene oil. A strong case can be made for the former. Geochemical research in Romania has shown organic carbon, bitumen, and asphaltene indices for the fine-grained Meotian and Pontian sediments of brackish and freshwater environments as comparable to those of the more obviously bituminous marine Oligocene.

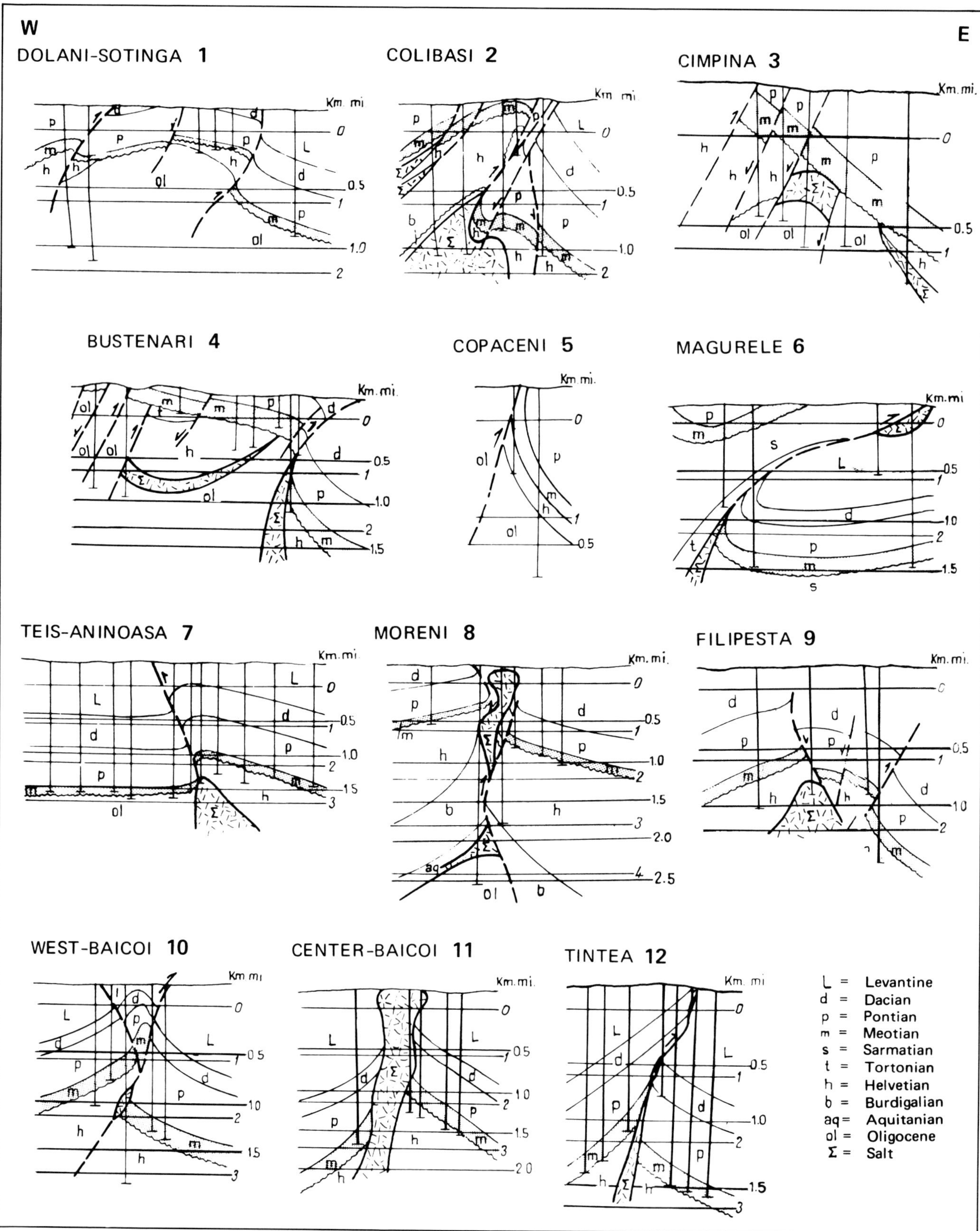
W
E
DOLANI-SOTINGA 1
COLIBASI 2
CIMPINA 3
BUSTENARI 4
COPACENI 5
MAGURELE 6
TEIS-ANINOASA 7
MORENI 8
FILIPESTA 9
WEST-BAICOI 10
CENTER-BAICOI 11
TINTEA 12
Km. mi
L = Levantine
d = Dacian
p = Pontian
m = Meotian
s = Sarmatian
t = Tortonian
h = Helvetian
b = Burdigalian
aq = Aquitanian
ol = Oligocene
Σ = Salt

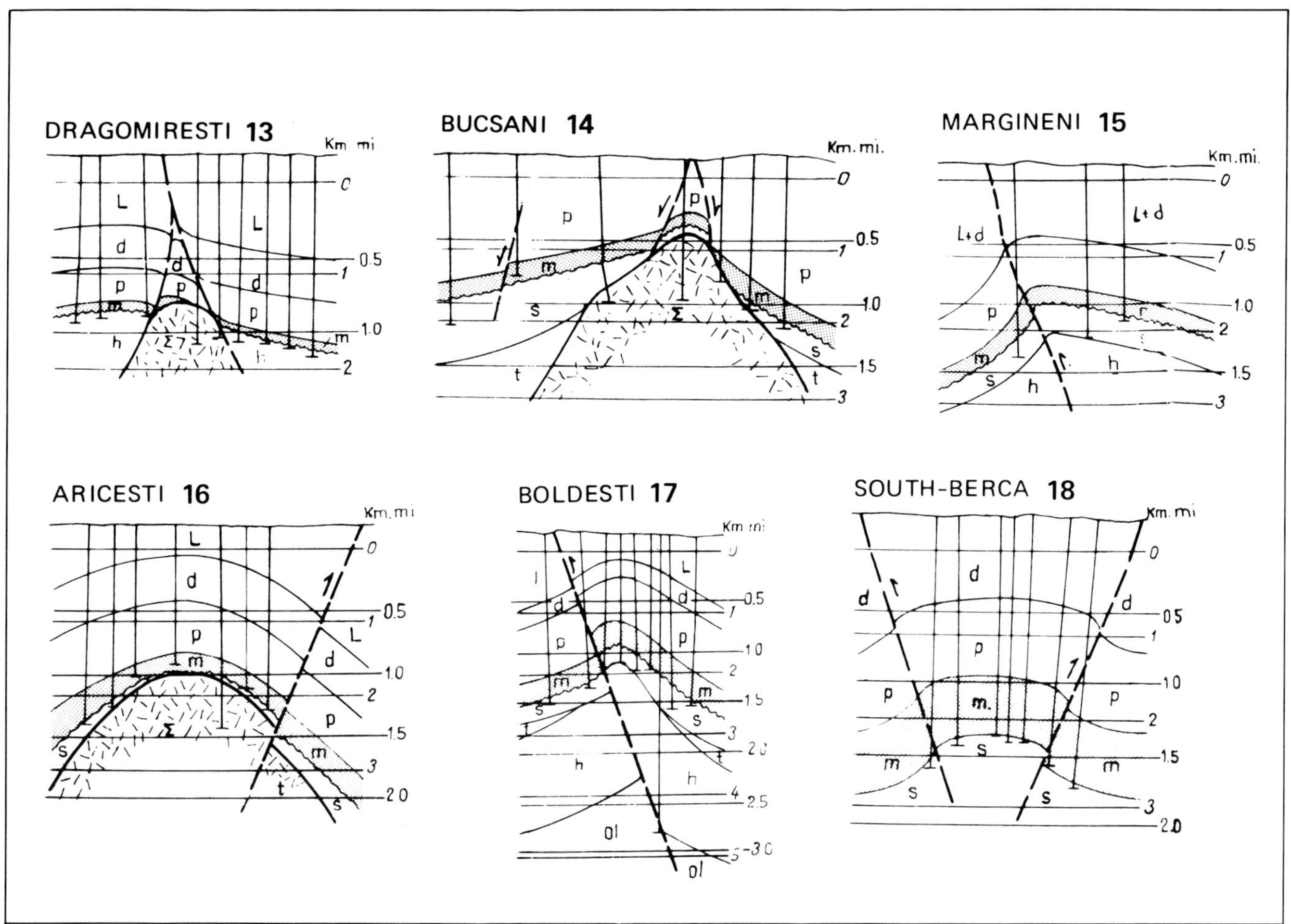

Fig. 6-54 (Paraschiv and Olteanu, 1970)—Local cross sections of producing structures of the Ploiesti district. Locations shown on figure 6-52. Structures arranged in rows west to east with northernmost at top, southernmost at bottom. Northern tier of structures are nearest the Carpathian nappes and have the greatest structural relief; the southern line of structures farthest from the nappes have the least structural relief. Permission to publish by American Association of Petroleum Geologists.

There is no staining of potential Miocene reservoir sands which are situated between the postulated Oligocene source and the ultimate Pliocene destination. Finally Meotian oils are of light, paraffinic variety evidencing greater maturation than Dacian oil, which is heavy and asphaltic. The Meotian appears to be an ideal example of a unit that has sourced its own oil. In the Ploiesti district the Meotian is thickest and shaliest in the south while sands are present to the north. Oil generated from fine-grained sediments in the south is postulated to have migrated into time-equivalent sandstones to the north. The fact that the southern flanks of both the Moreni and Boldesti fields (Figs. 6-53, 6-54) are far more prolific than the northern flanks tends to substantiate the validity of northward migration.

The age of the producing structures in both the Carpathians and Zagros is approximately the same, Pliocene-Pleistocene and probably continuing to the present inasmuch as both areas are seismically active. Both belts demonstrate that late structuring is not detrimental to hydrocarbon accumulation and that no thrust-fold belt is too young to explore.

A final corroboration that no thrust-fold belt is too young comes from Taiwan where an arc-continent collision (Fig. 6-18) has been ultimately responsible for thrusted folds (Fig. 6-55) that are presently growing. In interpreting the growth of Taiwan foothills anticlines by successive imbricate wedging (duplex structure—cf. Fig. 6-14), Suppe (1980) has stated that "correlation of microseismic data... with struc-

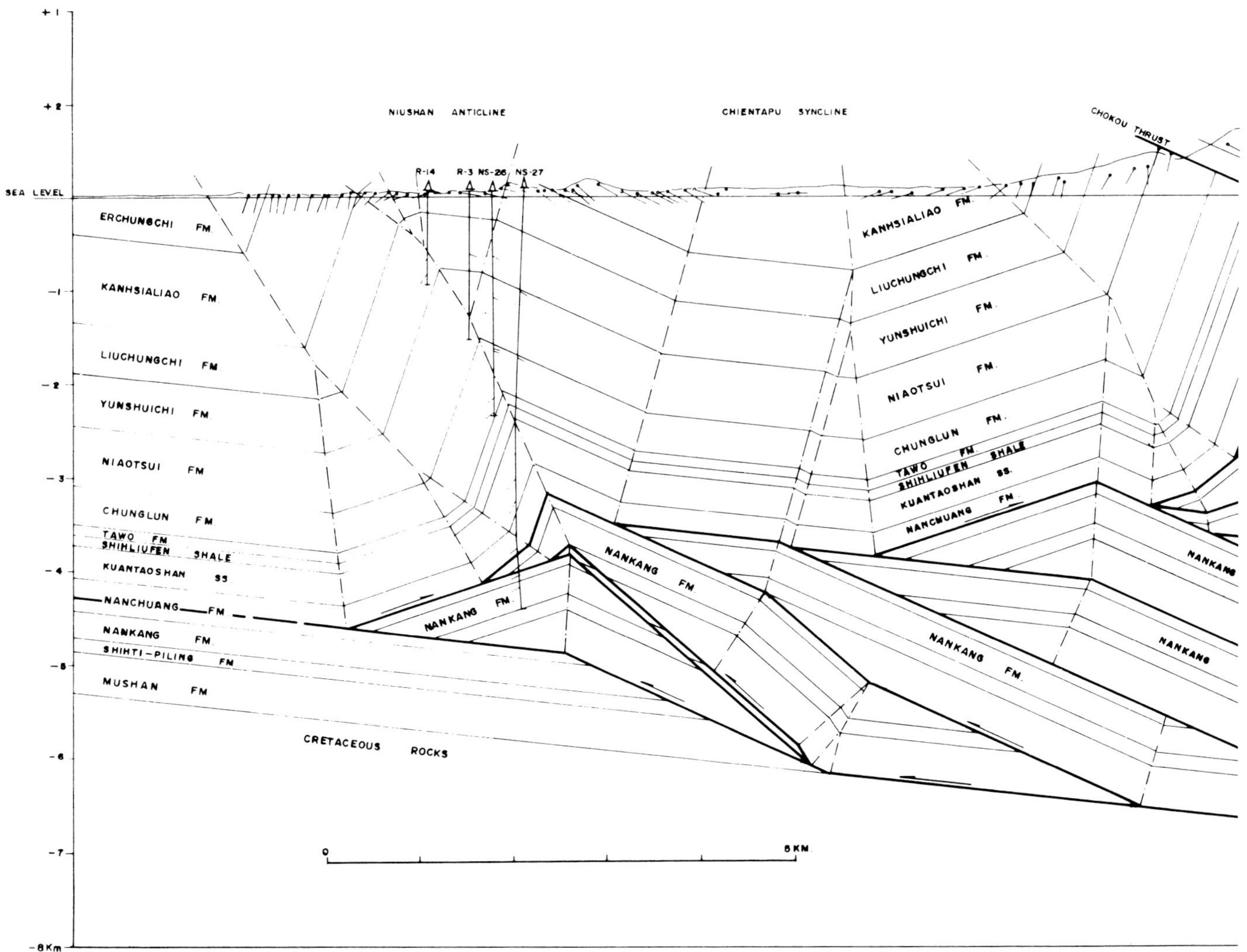

Fig. 6-55 (Suppe, 1980)—Interpretation of Niushan and Nanliao anticlines as duplex structures. Note that technique of imbricate fault-bend folding is used in drawing cross section. Permission to publish by John Suppe.

tural interpretations suggests that the imbricated wedge is currently actively deforming throughout, whereas the rocks above and below are relatively aseismic" (Fig. 6-56).

REFERENCES

Atwater, Tanya, 1970, Implications of plate tectonics for the Cenozoic tectonic evolution of western North America: *Geological Society America Bull.,* V. 81, p. 3513–3536.

Bachmann, G. H., and Koch, K., 1983, Alpine front and Molasse Basin, Bavaria, *in* Bally, A. W., ed., Seismic expression of structural styles: *American Association Petroleum Geologists Studies in Geology #15,* V. 3.

Bally, A. W., Gordy, P. L., and Stewart, G. A., 1966, Structure, seismic data, and orogenic evolution of southern Canadian Rocky Mountains: *Bull. Canadian Petroleum Geology,* V. 14, n. 3, p. 337–381.

Beck, R. H., and Lehner, P., 1974, Oceans, new frontier in exploration: *American Association Petroleum Geologists Bull.,* V. 58, p. 376–395.

Bird, Peter, Toksoz, M. N., and Sleep, N. H., 1975, Thermal and mechanical models of continent–continent convergence zones: *Journal Geophysical Research,* V. 80, n. 12, p. 4405–4416.

Boos, C. M., and Boos, M. F., 1957, Tectonics of eastern flank and foothills of Front Range, Colorado: *American Association Petroleum Geologists Bull.,* V. 41, p. 2603–2676.

Boyer, S. E., and Elliott, D., 1982, Thrust systems: *American Association Petroleum Geologists Bull.,* V. 66, p. 1196–1230.

Bucher, W. H., 1956, Role of gravity in orogenesis: *Geological Society of America Bull.,* V. 67, p. 1295–1318.

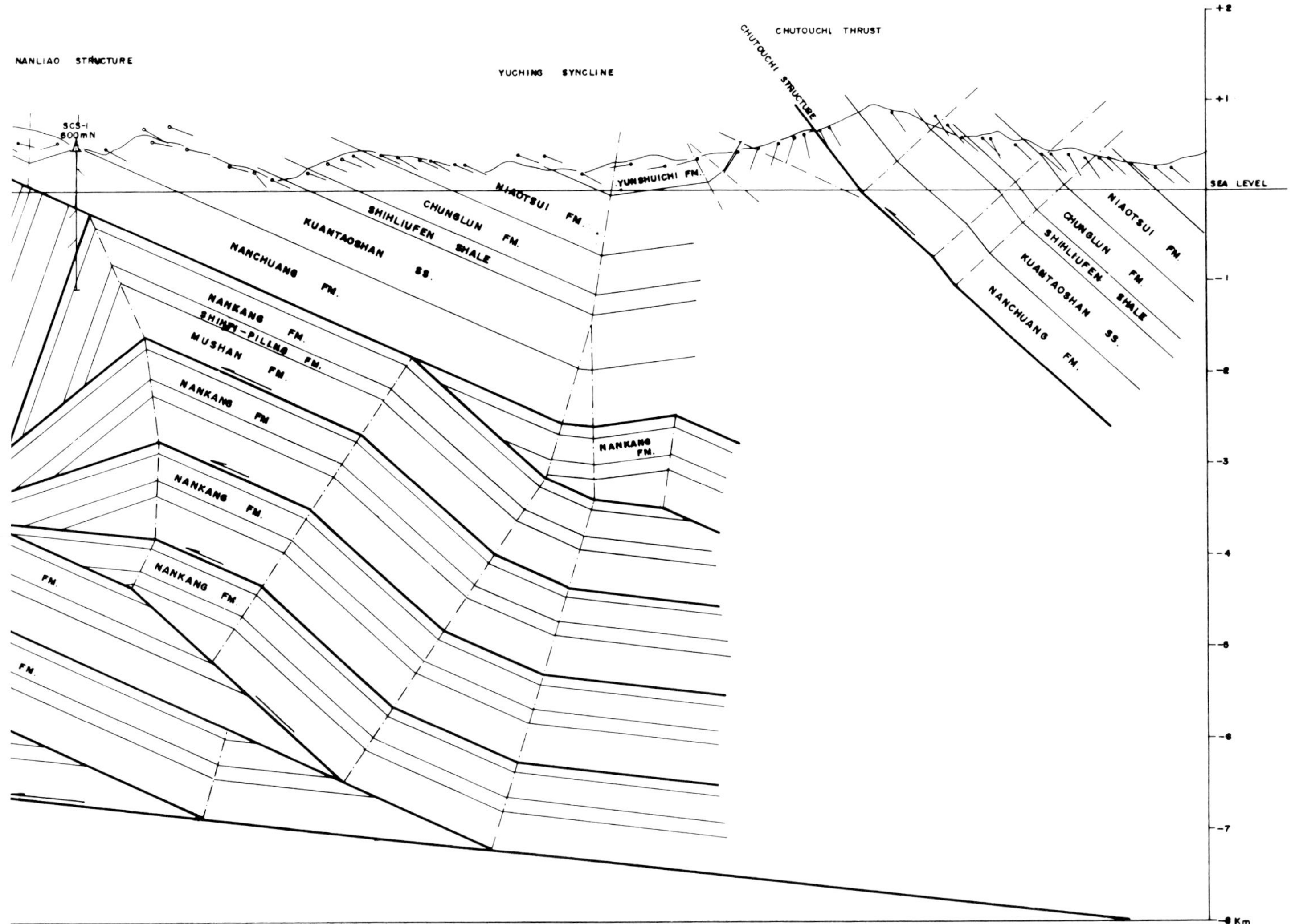

Burchfiel, B. C., and Davis, G. A., 1975, Nature and controls of Cordilleran orogenesis, western United States: extensions of an earlier synthesis: *American Journal Science,* V. 275-A, p. 363–396.

Butler, A. W., III, 1982, Case history of the north Caribou prospect, Bonneville County, Idaho, *in* Powers, R. B., ed., Geologic studies of the Cordilleran thrust belt: Rocky Mountain Association Geologists, p. 649–655.

Campbell, R. B., 1973, Structural cross section and tectonic model of the southeastern Canadian Cordillera: *Canadian Journal Earth Science,* V. 10, p. 1607–1620.

Chai, B. H., 1972, Structure and tectonic evolution of Taiwan: *American Journal Science,* V. 272, p. 389–422.

Coney, P. J., 1973, Plate tectonics of marginal foreland thrust-fold belts: *Geology,* V. 1, p. 131–134.

Dahlstrom, C. D. A., 1969, Balanced cross sections: *Canadian Journal Earth Science,* V. 6, n. 4, p. 743–757.

________, 1970, Structural geology in the eastern margin of the Canadian Rocky Mountains: *Bull. Canadian Petroleum Geology,* V. 18, n. 3, p.

Dewey, J. F., and Bird, J. M., 1970, Mountain belts and the new global tectonics: *Journal Geophysical Research,* V. 75, p. 2625–2647.

Dickinson, W. R., 1976, Sedimentary basins developed during evolution of Mesozoic-Cenozoic arc-trench system in western North America: *Canadian Journal Earth Science,* V. 13, p. 1268–1287.

Elliott, D., 1976, The motion of thrust sheets: *Journal Geophysical Research,* V. 81, p. 949–963.

Fitch, T. J., 1972, Plate convergence, transcurrent faults, and internal deformation adjacent to Southeast Asia and the western Pacific: *Journal Geophysical Research,* V. 77, p. 4432–4460.

Frank, J. R., Cluff, S. and Bauman, J. M., 1982, Painter Reservoir, East Painter Reservoir and Clear Creek fields, Uinta County, Wyoming, *in* Powers, R. B., Geologic studies of the Cordilleran thrust belt: Rocky Mountain Association Geologists, p. 601–611.

Hennessey, W. J., 1975, A brief history of the Savanna Creek gas field, *in* Evers, H. J., and Thorpe, J. E., eds.,

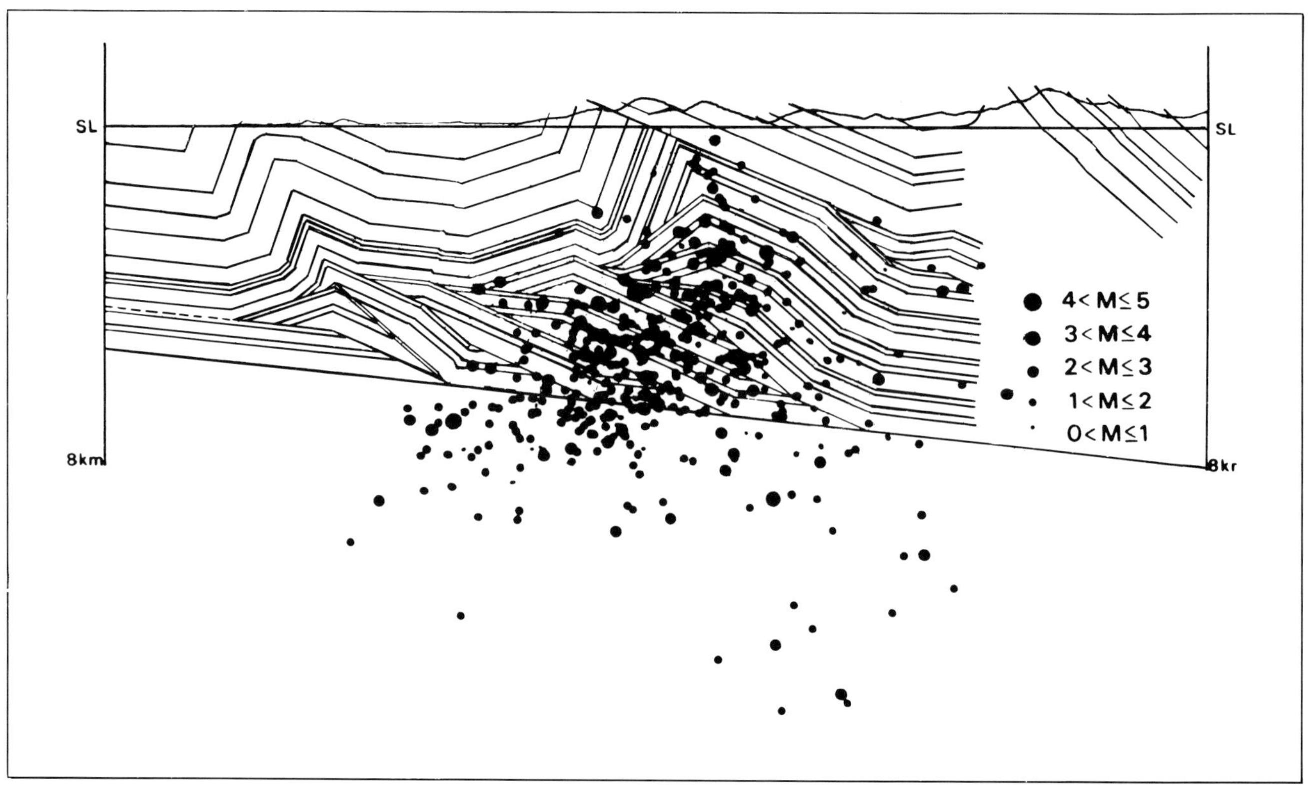

Fig. 6-56 (Suppe, 1980)—Hypocenters of microearthquakes within about 5 km (3 mi) of Niushan cross section (Fig. 6-55). Note that hypocenters are concentrated in duplex wedge. Permission to publish by John Suppe.

Structural geology of the foothills between Savanna Creek and Panther River, S. W. Alberta Canada: Canadian Society Petroleum Geology Guidebook, Exploration Update '75, p. 18–21.

Hobbs, W. H., 1914, Mechanics of formation of arcuate mountains: *Journal Geology,* V. 22, p. 71–90, 166–20.

Hull, C. E., and Warman, H. R., 1970, Asmari oil fields of Iran, *in* Geology of giant petroleum fields: *American Association Petroleum Geologists, Mem. 14,* p. 428–437.

James, D. E., 1971, Plate tectonic model for evolution of the central Andes: *Geological Society America Bull.,* V. 82, p. 3325–3346.

Jones, P. B., 1971, Folded faults and sequence of thrusting in Alberta foothills: *American Association Petroleum Geologists Bull.,* V. 55, p. 292–306.

———, 1982, Oil and gas beneath east-dipping underthrust faults in the Alberta foothills, Canada, *in* Powers, R. B., ed., Geologic studies of the Cordilleran thrust belt: Rocky Mountain Association Geologists, p. 61–74.

Kent, P. E., and Warman, H. R., 1972, An environmental review of the world's richest oil-bearing region—the Middle East: XXIV International Geologic Congress, Proc., sec. 5, p. 142–152.

Lamerson, P. R., 1982, The Fossil Basin area and its relationship to the Absaroka thrust fault system, *in* Powers, R. B., ed., Geologic studies of the Cordilleran thrust belt: Rocky Mountain Association Geologists, p. 279–340.

Lawson, A. C., 1932, Insular arcs, foredeeps, and geosynclinal seas of the Asiatic coast: *Geological Society America Bull.,* V. 43, p. 353–381.

Lelek, J. J., 1982, Anschutz Ranch East field, northeast Utah and southwest Wyoming, *in* Powers, R. B., ed., Geologic studies of the Cordilleran thrust belt: Rocky Mountain Association Geologists, p. 619–631.

Lovering, T. S., 1932, Field evidence to distinguish overthrusting from underthrusting: *Journal Geology,* V. 40, p. 651–663.

Lowell, J. D., 1974a, Plate tectonics and foreland basement deformation: *Geology,* V. 2, p. 275–278.

———, 1974b, Plate tectonics and foreland basement deformation: reply: *Geology,* V. 2, p. 571.

———, 1977, Underthrusting origin for thrust-fold belts with applications to the Idaho-Wyoming belt: *Wyoming Geological Association 29th Ann. Field Conf. Guidebook,* p. 449–455.

———, 1983, Foreland deformation, *in* Lowell, J. D., ed., Rocky Mountain foreland basins and uplifts: Rocky Mountain Association Geologists, p. 1–8.

Lowell, J. D., and Genik, G. J., 1972, Sea-floor spreading and structural evolution of southern Red Sea: *American Association Petroleum Geologists Bull.,* V. 56, p. 247–259.

Misch, P., 1960, Regional structural reconnaissance in central northeast Nevada and some adjacent areas: observations and interpretations: Intermountain Association Petroleum Geology and Eastern Nevada Geologic Society, Guidebook to Geology East Central Nevada, p. 17–42.

———, 1971, Geotectonic implications of Mesozoic decollement thrusting in parts of eastern Great Basin: *Geological Society American Abs.* with programs, V. 3, n. 2, p. 164–166.

Mudge, M. R., 1970, Origin of the disturbed belt in northwestern Montana: *Geological Society America Bull.,* V. 81, p. 377–392.

Norton, M. A., 1983, Kemmerer area, Lincoln County Wyoming, *in* Bally, A. W., ed., Seismic expression of structural styles: *American Association Petroleum Geologists Studies in Geology #15,* V. 3.

Nur, Amos, and Ben-Avraham, Zvi, 1982, Oceanic plateaus, the fragmentation of continents, and mountain building: *Journal Geophysical Research,* V. 87, n. 35, p. 3644–3661.

Oriel, S. S., and Armstrong, F. C., 1966, Times of thrusting in Idaho-Wyoming thrust belt: reply: *American Association Petroleum Geologists Bull.,* V. 50, p. 2614–2621.

Ower, J., 1975, The Moose Mountain structure, birth and death of a folded fault play, *in* Evers, H. J. and Thorpe, J. E., eds., Structural geology of the foothills between Savanna Creek and Panther River, S. W. Alberta, Canada: Canadian Society Petroleum Geology.

Paraschiv, D., and Olteanu, Gh., 1970, Oil fields in Mio-Pliocene zone of eastern Carpathians (District of Ploiesti), *in* Geology of giant petroleum fields: *American Association Petroleum Geologists, Mem. 14,* p. 399–427.

Pitman, W. C., III, and Talwani, Manik, 1972, Sea-floor spreading in the North Atlantic: *Geological Society America Bull.,* V. 83, p. 619–646.

Pitman, W. C., III, Larson, R. L., and Herron, E. M., 1974, The age of the ocean basins: Geological Society America.

Price, R. A., 1969, The southern Canadian Rockies and the role of gravity in low-angle thrusting, foreland folding, and the evolution of migrating foredeeps: *Geological Society America Abs.* with Programs for 1969, Pt. 7, (Ann. Mtg.), p. 284–286.

———, and Mountjoy, E. W., 1970, Geologic structure of the Canadian Rocky Mountains between Bow and Athabasca rivers—a progress report: *Geological Association Canada Spec. Paper n. 6,* p. 25.

Price, R. A., 1973, Large-scale gravitational flow of supracrustal rocks, southern Canadian Rockies, *in* deJong, K., and Scholten, R., eds., Gravity and tectonics, Wiley-Interscience Publications.

Robinson, J. P., 1982, Petroleum exploration in southeastern Arizona: Anatomy of an overthrust play, *in* Powers, R. B., ed., Geologic studies of the Cordilleran thrust belt: Rocky Mountain Association Geologists, p. 665–674.

Roeder, D. H., 1973, Subduction and orogeny: *Journal Geophysical Research,* V. 78, p. 5005–5024.

———, and Gilbert, O. E., 1977, Structure, kinematics, and hydrocarbon prospects of thrust and fold belts: American Association Petroleum Geologists, Structural Geology School lecture notes.

———, 1983, Hydrocarbons and geodynamics of fold-thrust belts: Rocky Mountain Association Geologists, Continuing education short course notes, 216 p.

Royse, F., Jr., Warner, M. A., and Reese, D. C., 1975, Thrust belt structural geometry and related stratigraphic problems, Wyoming-Idaho-northern Utah, *in* Bolyard, D. W., ed., Symposium on deep drilling frontiers in the central Rocky Mountains: Rocky Mountain Association Geologists, p. 41–54.

Sachnik, F. L., and More, R. D., 1983, Southern Appalachian folding and faulting, *in* Bally, A. W., ed., Seismic expression of structural styles: *American Association Petroleum Geologists Studies in Geology #15,* V. 3.

Sanlav, F., Tolgay, M., and Genca, M., 1963, Geology, geophysics, and production history of the Garzan Germik field, Turkey: 6th World Petroleum Congress, Sec. 1, Paper 35, p. 23.

Scholl, D. W., Christensen, M. N., Von Huene, R., and Marlow, M. S., 1970, Peru-Chile trench sediments and seafloor spreading: *Geological Society America Bull.,* V. 81, p. 1339–1360.

Schroder, R. J., Structural interpretaton of Line P81-04, Amadeus Basin, northern Territory, Australia, *in* Bally, A. W., ed., Seismic expression of structural styles: *American Association Petroleum Geologists Studies in Geology #15,* V. 3.

Seely, D. R., Vail, P. R., and Walton, G. G., 1974, Trench slope model, *in* Burk, C. A., and Drake, C. L., eds., The geology of continental margins: Springer-Verlag Publications, p. 249–260.

Soller, D. R., Ray, R. D., and Brown, R. D., 1982, A new global crustal thickness map: *Tectonics,* V. 1, n. 2, p. 125–149.

Suppe, J., 1980, Imbricated structure of western foothills belt southcentral Taiwan: Petroleum Geology Taiwan, n. 17, p. 1–16.

Van Hise, C. R., 1896, Principles of North American Pre-Cambrian geology: United States Geological Survey 16th Ann. Report. 1894–1895, pt. 1, p. 571–843.

Von Huene, R., 1972, Structure of the continental margin and tectonism at the eastern Aleutian trench: *Geological Society America Bull.,* V. 83, p. 3613–3626.

Walters, R. P., 1946, Oil fields of Carpathian region: *American Association Petroleum Geologists Bull.,* V. 30, n. 3, p. 319–336.

Warner, M. A., 1982, Source and time of generation of hydrocarbons in the Fossil Basin, western Wyoming thrust belt, *in* Powers, R. B., ed., Geologic studies of the Cordilleran thrust belt: Rocky Mountain Association Geologists, p. 805–815.

West, J., and Lewis, H., 1982, Structure and palinspastic reconstruction of the Absaroka thrust, Anschutz Ranch area, Utah and Wyoming, *in* Powers, R. B., ed., Geologic studies of the Cordilleran thrust belt: Rocky Mountain Association Geologists, p. 633–639.

White, D. A., Roeder, D. H., Nelson, T. H., and Crowell, J. C., 1970, Subduction: *Geological Society America Bull.,* V. 81, p. 3431–3432.

White, R. S., and Ross, D. A., 1979, Tectonics of the western Gulf of Oman: *Journal Geophysical Research,* V. 84, n. B7, p. 3479–3489.

APPENDIX

The pre-requisite of flexural-slip folds for drawing balanced cross sections has been discussed in Chapter 3. The definitive paper on balanced cross sections by C. D. A. Dahlstrom has been reprinted here. This article assumes that flexural-slip folds are of the concentric type. We have seen thrust-fold belt cross sections however, in which flexural-slip folds have sharper hinges and more planar limbs (Fig. 6-44) than those of concentric folds; this angular geometry seems to be taken to the limit in the "fault-bend fold" cross sections of Suppe, 1980 (Fig. 6-55). Such folds, including chevron folds, have undergone predominantly flexural slip and are also drawn balanced on cross sections.

A summary of empirical guidelines for drawing structural cross sections in thrust-fold belts is provided by Figure 6A-16.

BALANCED CROSS SECTIONS[1]

Post-depositional concentric deformation produces no significant change in rock volume. Since bed thickness remains constant in concentric deformation, the surface area of a bed and its length in a cross-sectional plane must also remain constant. Under these conditions, a simple test of the geometric validity of a cross section is to measure bed lengths at several horizons between reference lines located on the axial planes of major synclines or other areas of no interbed slip. These bed lengths must be consistent unless a discontinuity, like a decollement, intervenes. Consistency of bed length also requires consistency of shortening, whether by folding and (or) faulting, within one cross section and between adjacent cross sections.

The number of possible cross-sectional explanations of a set of data is reduced by the fact that, in a specific geological environment, there is only a limited suite of structures which can exist. This imposes a set of local "ground rules" on interpretation. When these local restrictions are coupled with the geometric restrictions which follow from the law of conservation of volume, it is often possible to produce structural cross sections that have a better-than-normal chance of being right.

The concept of consistency of shortening can be extrapolated to a mountain belt as a whole, thereby indicating the necessity for some kind of transfer mechanism wherein waning faults or folds are compensated by waxing en echelon features. These concepts are illustrated diagrammatically and by examples from the Alberta Foothills.

INTRODUCTION

The purpose of this paper is to describe a method of checking cross sections for geometric acceptability. In the elementary schools of a bygone era children were taught to prove their answers to arithmetic problems by reversing the process: subtraction was checked by addition and division by multiplication. To apply this principle to a geological cross section, one would flatten out the deformed beds and return them to their depositional

[1]Reprinted from Canadian Journal of Earth Sciences, 1969 V. 6, n. 4. Article written by C. D. A. Dahlstrom. Used by permission.

position. If this restoration could be done, one would conclude that the cross section was geometrically possible (although not necessarily true) but, if the beds could not be restored, one would conclude that the cross section was geometrically impossible. Construction of restored cross sections is tedious so they are seldom used to check structural interpretation. However, there are shortcuts which make the principle easy to use.

The method to be discussed is now being used consciously by a few geologists and unconsciously by many more to check their cross sections. The essence of the method has been discussed by Hunt (1957), mentioned by Goguel (1952), and illustrated by the published works of Carey (1962) and Bally et al. (1966), where it is obvious that the method is being consciously applied. However, the rules have not been set forth and discussed so that many geologists have not been exposed to the ideas. This paper is intended to remedy this oversight because the writer believes that a ''balanced'' cross section is a better cross section.

SOURCE OF RESTRICTIONS IN CROSS SECTION CONSTRUCTION

Relatively few geologists still demand for themselves the license of the artist, that is, the right to put on their cross section any interpretation which their imaginations can conceive. Aside from the data itself, most geologists recognize that other restrictions impose boundary conditions within which the imagination must be confined. These restrictions derive from two sources, 1– generalizations derived from observed facts, and 2– geometric principles.

This paper discusses one of the geometric principles and shows how specific rules of interpretation can be derived from that principle. The derivation involves simplifying assumptions which are valid only for specific structural environments. Therefore the same geometric principle will give rise to interpretive rules which vary from one structural environment to another. The writer will not attempt to develop all the possible variants. Discussion is restricted to one simple geological environment, the marginal part of an orogenic belt, and to one example of that environment, the Alberta Foothills. To avoid misunderstanding, let it be emphasized that the interpretive rules derived in this paper are literally applicable only to the Alberta Foothills. With minor variants, they can be used in other marginal belts, but they cannot be applied to another environment, such as a salt dome province. However, the principle and method can be used to develop rules that do apply to those other environments.

GENERALIZATIONS OF OBSERVED FACTS

As a consequence of intensive government and industry search for oil and gas, the Alberta Foothills is well mapped (Bally et al., 1966, and the authors in their bibliography) and sufficient data are available for sound generalizations as to the pattern of structural behavior within this structural province. These generalizations establish the local ground rules for structural interpretation. For example, one of the local ground rules is that folds are ordinarily concentric, rarely chevron but never similar. Recognition of such local interpretive rules is an acknowledgment of the existence of ''familial associations'' of structures. According to this idea a specific geological environment contains a limited suite or ''family'' of structures. A large low-angle thrust, which would be perfectly reasonable in Foothills cross sections, would be as incongruous in a cross section through the plains of Saskatchewan as an elephant on the tundra, because thrusts are not part of the cratonic family of structures. The ''foothills family of structures'' comprise: 1– concentric folds, 2– decollement, 3– thrusts (usually low angle and often folded), 4– tear faults, and 5– late normal faults.

The first restriction then in constructing Foothills structure sections is that one is limited to some variant of these basic structural forms.

The area being used as an example is shown in diagrammatic cross section in figure 6A-1. Of the many generalizations from observed data which can be made about the Alberta Foothills belt, the following are pertinent to the subsequent discussion.

1– The Precambrian basement extends, unbroken, beneath the Foothills structures.

2– The rocks east of the Foothills are essentially undeformed.

3– The Foothills structures were formed in the latter stages of the Laramide orogeny long after most of the rocks now preserved were deposited.

4– There is no thinning or thickening of beds by ''flow'' and therefore folding is concentric.

The through-going basement and the lack of deformation east of the Foothills edge limit the number of solutions which can be postulated for particular geometric problems. In an earlier and simpler day, it was the custom of geologists to draw shallow Foothills cross sections wherein geometric problems were solved by faults which dribbled off the bottom of the cross section. Seismic and drilling data (Bally et al., 1966) have established the basement configuration so that now cross sections like figure 6A-1 are closed systems except at the western end. Now the sophisticated geologist draws deep Foothills cross sections with a through-going basement, wherein geometric problems are solved by faults which are gathered into flat sole faults and hustled westward through the only available exit.

The Laramide orogeny moved progressively from west to east with the Foothills being deformed in the Paleocene Epoch. With the probable exception of the youngest beds in the Alberta syncline, all of the Paleozoic and Mesozoic rocks had been deposited long before the deformation. During the course of their depositional his-

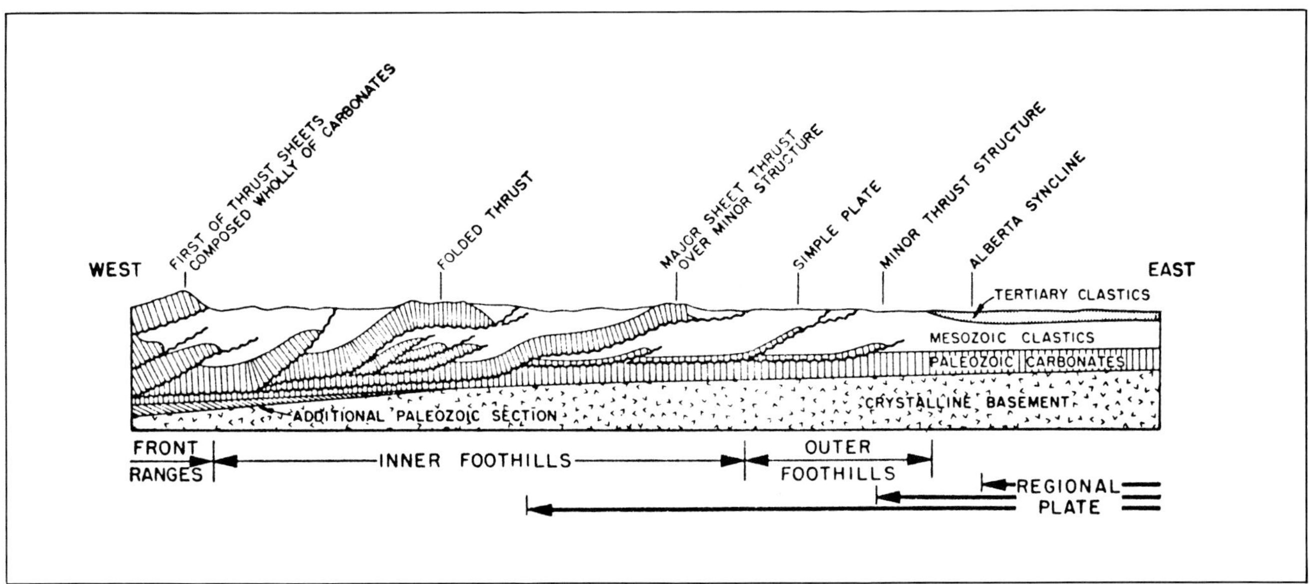

Fig. 6A-1—Schematic cross section of Alberta Foothills (after R. E. Daniel, Chevron Standard). Note that the regional plate extends progressively farther west as older rocks are considered until, at Precambrian level, it extends completely across the section.

tory in the miogeosyncline, these Paleozoic and Mesozoic rocks had indeed been subject to epeirogenic fluctuations and consequent erosion but they had not been orogenically deformed prior to Paleocene time. This single, post-depositional deformation of the Foothills is an important point because it simplifies structural interpretation by eliminating from consideration: 1– pronounced angular unconformites, 2– structures growing during deposition, or 3– substantial amounts of compaction during deformation.

The foregoing terse comments touch only upon those topics which are necessary background for the subsequent geometric discussion. No attempt has been made to summarize Foothills geology.

CONSERVATION OF VOLUME

The law of elementary physics (which Einstein amended) that matter can neither be created nor destroyed is paraphrased in geology as the "law of conservation of volume" (Goguel 1952, p. 147). A rock consists of mineral particles and voids filled with fluids. In the initial stages of clastic deposition, the proportion of voids to mineral grains is large and density is low, but this stage is brief because the compaction rate is very high during the first few hundred feet of burial. Thereafter density increases only slowly with depth. It is worth noting here for future reference that the volume change due to load compaction does reduce the thickness of a bed but it does not alter the areal extent. Volume reduction on account of deformation is thought to be negligible, particularly in the earlier stages of tectonism represented by the marginal part of the Cordillera, because no significant deformation-dependent decrease of porosity (or increase of density) is recognized. For practical purposes, one may assume that the law of volume conservation applies and that rock volume does not change during Foothills type of deformation.

Volume is three dimensional which makes the law of conservation of volume awkward to apply. However, in specific instances it can be made simpler. To start the simplification process one can use the three mutually perpendicular tectonic axes as the three dimensions: "*a*" is the direction of tectonic transport of rock during deformation, "*b*" is the direction of the fold axes, and "*c*" is the third axis which is perpendicular to the "*ab*" plane.

In the Foothills, the fold axes and the fault strikes are parallel, the folds are concentric, and the faults are dip-slip thrusts. Therefore the "action" takes place in the *ac* plane, which is a vertical transverse cross section, and virtually nothing happens in the *b* direction. By ignoring changes in the *b* direction as insignificant, the law of conservation of volume becomes a two-dimensional statement that the cross-sectional area of a bed does not change during deformation. Cross-sectional areas are manageable quantities so the law has been applied in this form to check cross sections (Hunt, 1957) or to calculate depth to detachments (Bucher, 1933).

Even further simplification is possible when folding is concentric because bed thickness does not change during deformation. This has been demonstrated in the Foothills, directly by detailed studies of the anatomy of folds (Price, 1964), and indirectly by many stratigraphic studies, which do not detect any tectonic thinning of units whether they be on gently upright limbs, on vertical limbs, on overturned limbs of folds, or on far-travelled thrust sheets. Because the cross-sectional area of a bed is a function of bed length and bed thickness and because the thickness remains constant, it is possible to eliminate one more dimension and to state the law of conservation of volume in terms of length alone: *In concentric regimes the cross-sectional length of a bed remains constant during deformation.*

CONSISTENCY WITHIN CROSS SECTIONS

In simple deposition the initial areal extent (or length in cross section) of any bed is the same as that of the beds above and below it. Since bed area (or length in cross section) is not appreciably altered by either compaction or by post-depositional concentric deformation, it follows that bed lengths in a cross section must be consistent with one another (Figs. 6A-2a and 6A-2b), unless a discontinuity such as a sole fault or decollement intervenes (Fig. 6A-2c) between the longer and shorter beds.

Using the idea of consistent bed lengths to check a cross section is very simple. The first step is to establish a pair of reference lines at either end of the section in areas of no interbed slip. These reference lines should be the axial planes of major anticlines or synclines or other planes of no slip such as a plane perpendicular to the regional dip of the undisturbed "plains" section at the eastern end of a regional Foothills cross section. The second step is to measure the bed lengths of selected horizons between these reference lines. They should all be the same so that bed lengths "balance". If the cross-sectional lengths do not balance then the cross section must show a valid explanation of why they do not. In regional Foothills cross sections for instance, there is never a balance between the length of the Mesozoic and Paleozoic beds and the length of basement. The "cover" beds are always too long, which is explained in cross section (Fig. 6A-1) by a sole fault along which the upper beds have been moved (shoved? glided?) into the cross section from the west. This is not simply a way of sloughing one's problems into an adjacent area, because the necessary implication of such a section is that the sole fault continues to the west until the bed length anomaly is resolved by shortening of the basement (compressional hypothesis), or extension of the Mesozoic and Paleozoic "cover" rocks (glide hypothesis). One should note that checking Foothills cross sections for balance by measuring bed lengths is purely a geometric test which is quite independent of the genesis of the structures.

The fundamental rule, that the cross-sectional length of a bed remains constant during concentric deformation, has led to the idea that bed lengths in a cross section ought to be consistent with one another. If this is true, then the displacement on thrust faults ought to be consistent as well. In figure 6A-3 the displacement at *B* must be the same as at *A* unless it is postulated that elastic, plastic, or compactional deformation of the rock takes place between A' and B'. None of these things happen in a concentric regime.

Despite this conclusion that thrust displacement ought to be consistent, there are many instances where the displacement can be observed to change along the fault plane. There are only two basic ways in which this paradox (Fig. 6A-4a) can be resolved:

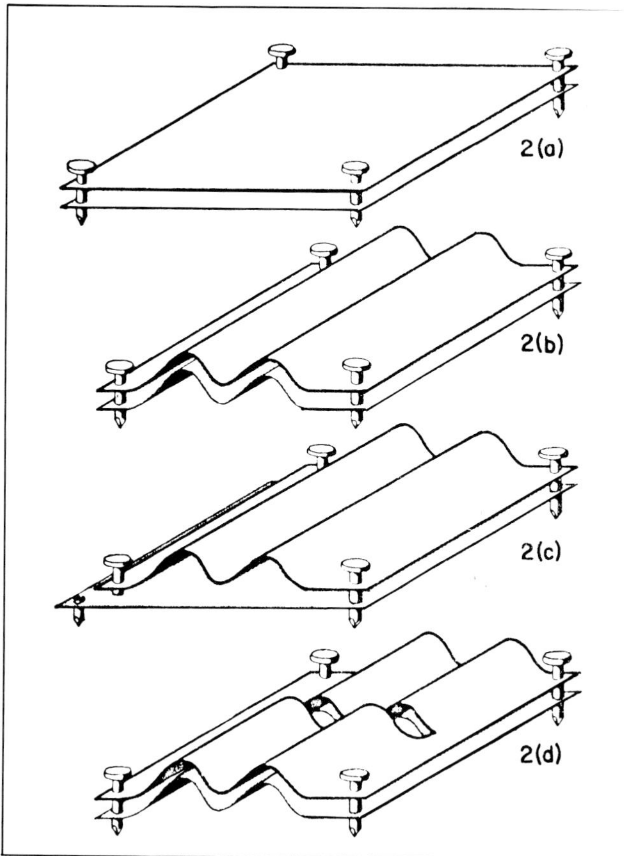

Fig. 6A-2—Consistency of bed length.

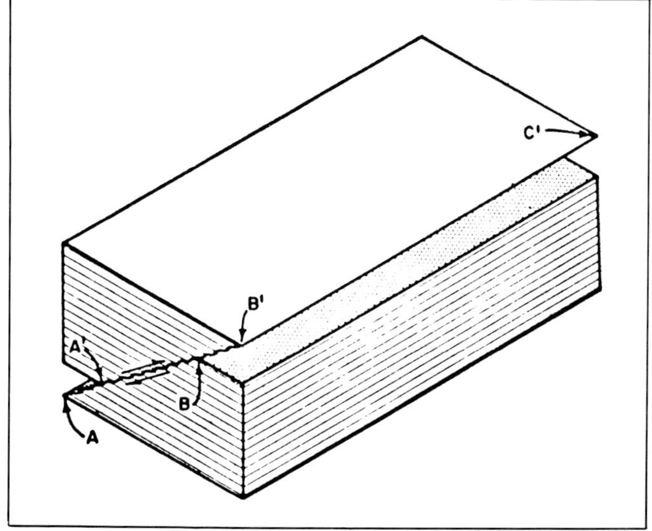

Fig. 6A-3—Consistency of thrust displacement.

1– by interchanging fold shortening and fault displacement (Fig. 6A-4b), or
2– by imbrication (Fig. 6A-4c).

Thrust faulting and folding are both mechanisms for making a packet of rock shorter and thicker than it was originally, so that one could expect the two mechanisms to be interchangeable. Imbrication is the distribution of displacement from one large fault to several minor ones. Figure 6A-4c shows the common imbricate pattern, but imbricates can develop in the footwall rather than the hanging wall of the main thrust, and they can be in the hanging wall but dipping in the opposite direction to the principal fault (''back thrusts'').

In figure 6A-4 the thrust faults were arbitrarily and diagrammatically represented as simple planar features. With planar faults and constant bed length, a substantial amount of interbed slippage is required, which would alter the originally vertical ends of the blocks to the curved shapes shown. From these block diagrams it is fairly evident that:

1– faults with changing displacement are apt to be curved in cross section rather than planar (can be confirmed by observation);
2– interbed slippage is a necessary part of the thrust faulting process just as it is in concentric folding;
3– interbed slippage can contribute to the change of displacement along a fault plane and, in extreme instances, could become a species of imbrication.

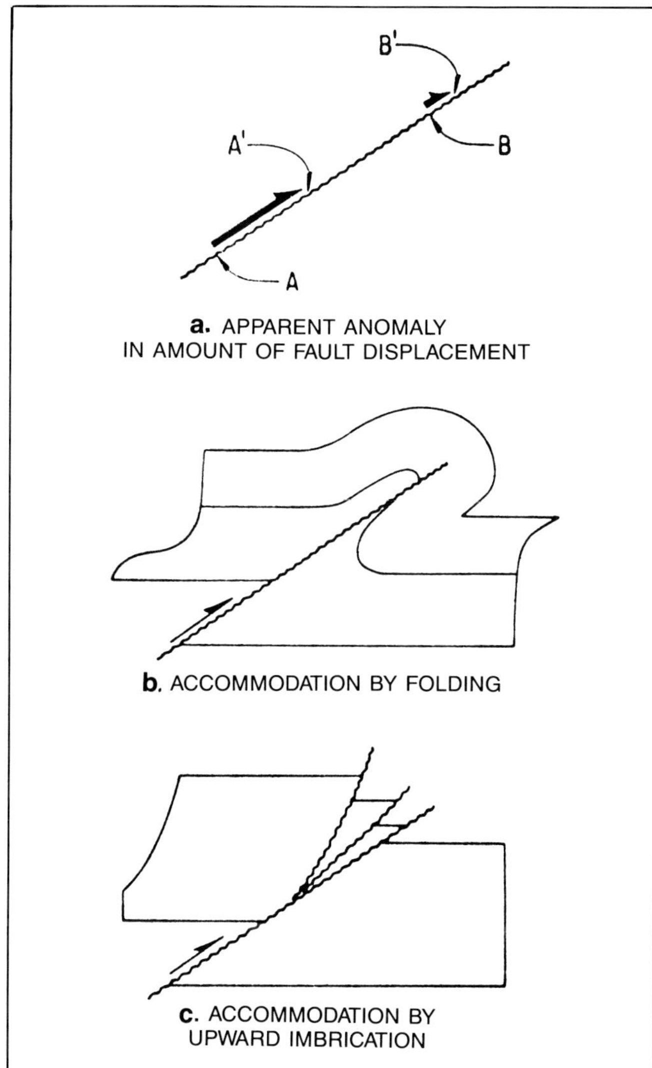

Fig. 6A-4—Changes in thrust displacement. See text for explanation of curved ends on some fault blocks.

The Turner Valley cross section (Fig. 6A-5) is based on good well, seismic, and surface geological control. It shows a remarkable change from more than 3.2 km (2 mi) of fault displacement at depth to virtually none at surface and the accommodation of this change by folding. This is an excellent example of a balanced transition from shortening by faulting to shortening by folding. In checking this cross section for consistency, one would probably use the vertical ends of the cross section as reference lines even though strictly speaking, neither is a proper reference line. The eastern end is in the undisturbed plains section, but the reference line should be perpendicular to the regional dip rather than vertical. The western end of the section does not extend as far as the synclinal axial plane, which would be the proper place for the second reference line. However, at this end of the section, the Rundle member has ''returned to regional'', which provides for practical purposes an acceptable place to draw a reference line, although it too should be perpendicular to the regional dip. (Please observe the behavior of the top of the Paleozoic rocks in the central part of figure 6A-1 for an appreciation of the significance of a horizon's return to its regional elevation.)

EXAMPLES OF THE USE OF BALANCED CROSS SECTIONS

Interpreting a cross section through the culmination of the Panther River anticline posed an interesting problem when a well drilled on the crest of the structure penetrated a thrust of Cambrian over Jurassic rocks, a stratigraphic throw of some 2440 m (8000 ft). This was particularly perplexing because at the surface, the west-dipping fault on the east flank of the structure thrust Jurassic over Lower Cretaceous rocks, a throw of only a few hundred feet. Some well control, surface geology (Fig. 6A-6), and seismic data were available to provide the critical data shown in figure 6A-7. Data in the line of section could be interpreted in two basically different ways (Fig. 6A-8). The interpretation with one major thrust appeared most likely to be correct because it correlated two major thrusts with stratigraphic throws of the same order of magnitude. The interpretation with two major thrusts seemed unreasonable because it linked a major fault in the subsurface with a minor fault at the surface

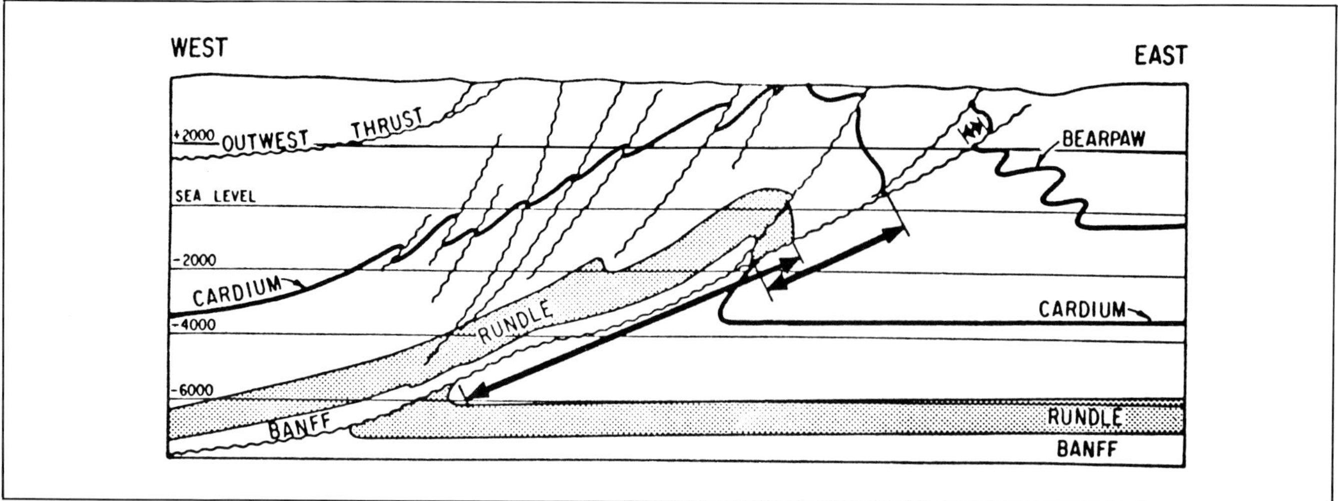

Fig. 6A-5—Changes in thrust displacement in the Turner Valley structure (after Gallup 1951, with the permission of the American Association Petroleum Geologists).

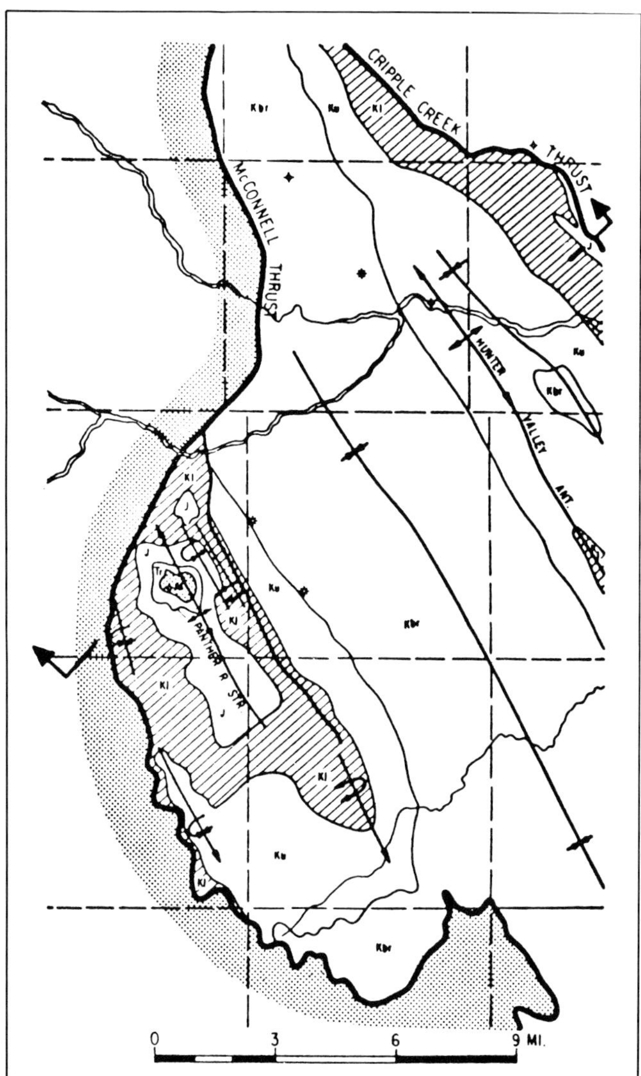

Fig. 6A-6—General geology of the Panther River area.

(one order of magnitude difference in stratigraphic throws). Which alternative one selected had considerable economic significance, because it affected where and at what depth one could expect to find closure in the potentially hydrocarbon-bearing objective horizon at the top of the Paleozoic section.

However, the geological map provided some data off the line of section, which had not been used. R. E. Daniel of Chevron Standard statistically analyzed the fold (Dahlstrom 1954), found it to be cylindrical, and calculated plunge values. The data on the geological map were projected up plunge (Stockwell 1960) into the plane of a vertical cross section passing through the culmination to provide the information shown above ground level at the west end of the section in figure 6A-9. On this cross section it is apparent that the axial planes of the anticline and syncline are converging in depth, which is confirmed by the reduced length of the steep fore limbs in successively lower stratigraphic horizons. Having the fold disappear with depth would produce a geometrically unacceptable discordance in bed lengths unless the fold disappearing downward could be compensated by the fault dying out upward. This would be a repetition of the Turner Valley situation in figure 6A-5. Once this fundamental point was grasped, it was apparent that the superficially unlikely looking cross section in figure 6A-8 with the two major thrusts was the correct one. The cross section which R. E. Daniel constructed according to this concept was subsequently confirmed by the drilling of a well on the deep structure immediately to the east, and by the deepening of the original well.

Hypotheses of regional significance can develop from balancing cross sections. Figure 6A-10 shows two alternative interpretations of seismic, well, and surface data. The data in the two sections are the same, but in Section A, the simply migrated seismic section, the vertical dimension is essentially two-way transit time, whereas B is the normal natural scale structural cross section. The Mesozoic and Paleozoic sections are both cut by thrusts,

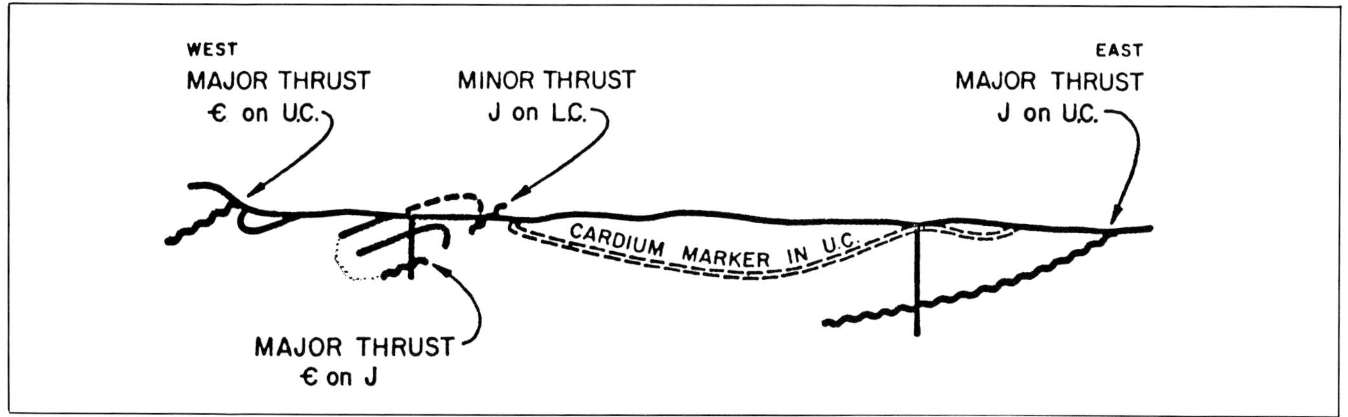

Fig. 6A-7—Basic data available for the original interpretation of the Panther River cross section.

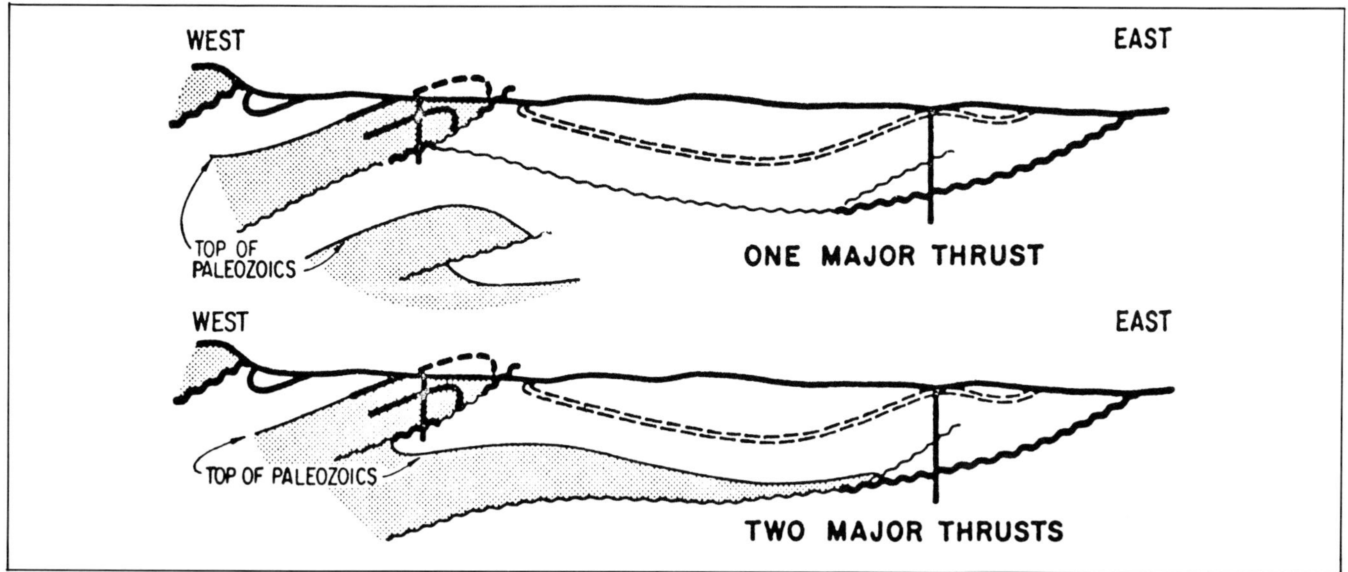

Fig. 6A-8—Two alternate cross sections which would satisfy the data on the original line of section.

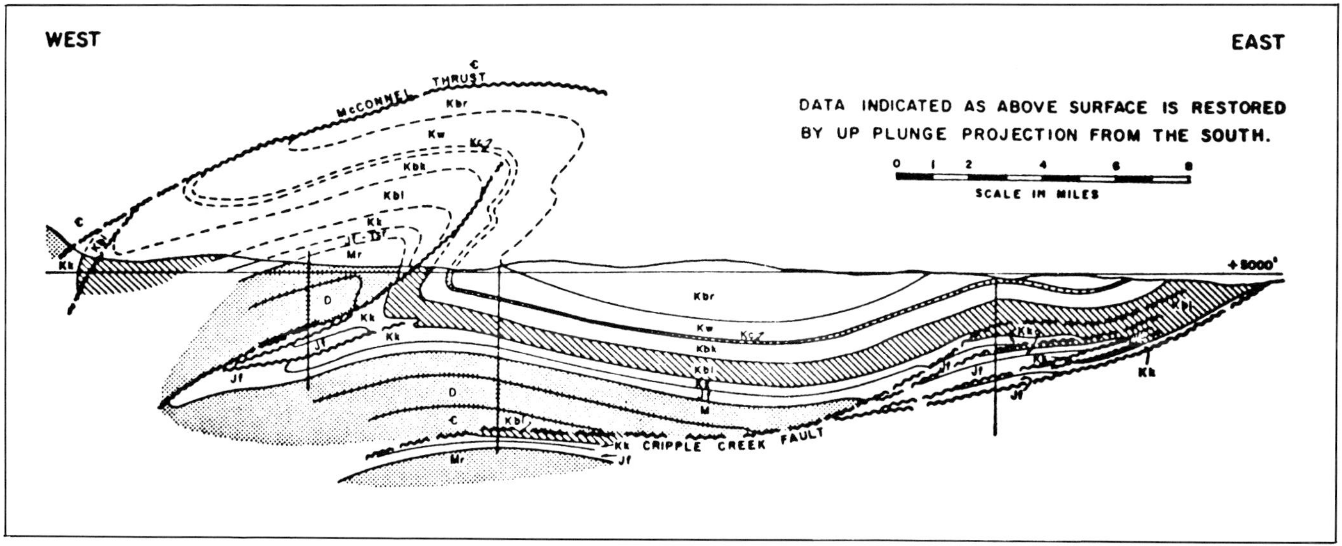

Fig. 6A-9—Panther River cross section as subsequently shown by drilling (after R. E. Daniel, Chevron Standard).

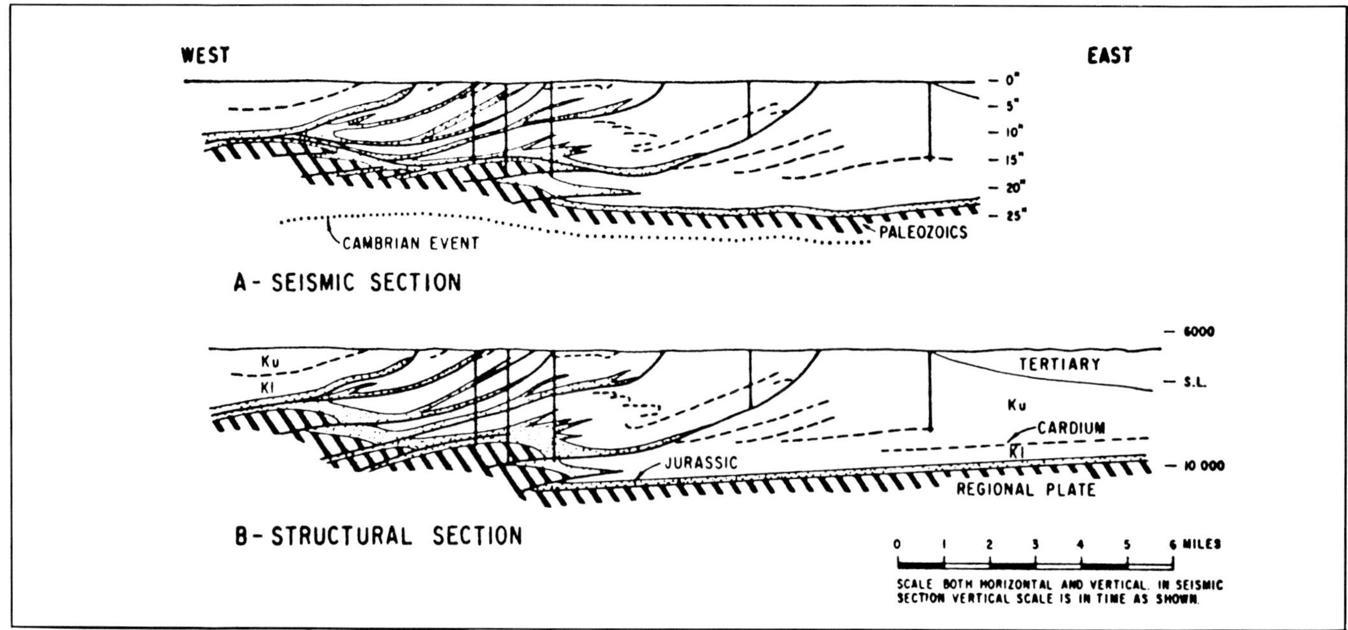

Fig. 6A-10—Two alternate interpretations of surface, well and seismic data in a Waterton area cross section. Minor fault traces have been omitted but their existence is shown by marker bed offsets. Note the inconsistencies of fault displacement in Section B which make this alternate untenable.

but even a casual inspection shows that the shortening in the Mesozoic section is far in excess of that in the Paleozoic. This difference in shortening precludes the kind of interpretation that was attempted in Section B, because any simple fault that one tries to draw will have far more displacement at Mesozoic level than at Paleozoic level. To avoid an unexplained inconsistency it is necessary to postulate a decollement separating the Mesozoic rocks from the Paleozoic rocks. This decollement would be in the basal Jurassic beds just above the top of the Paleozoic section. Since this horizon is itself now deformed by both folding and faulting, it becomes necessary to postulate a two-stage sequence of deformation. According to this hypothesis the Mesozoic rocks were first deformed by folding and thrusting above a sole fault (decollement) in the basal Jurassic beds, and the deformation in the Paleozoic rocks is a subsequent stage of the tectonic cycle. Balancing this cross section shows that the western (upper) thrust system was formed before the eastern (lower) one, and demonstrates (Bally et al., 1966, fig. 6) that, in the regional sense, deformation advanced from west to east.

CONSISTENCY BETWEEN CROSS SECTIONS

The foregoing discussion has been concerned with a one-dimensional check of the internal consistency of an individual cross section by simple measurement and comparison of bed lengths. During the stage of reducing the law of conservation of volume from three dimensions to one, there was a two-dimensional stage wherein one could state that deformation changes the form but does not alter the surface area of an individual bedding plane. This stage is represented in figure 6A-2d, where it is evident that no abrupt change can occur in fold form and (or) location unless there is a discontinuity. Similarly, in the block diagram of figure 6A-3, the fault displacement between B' and C' cannot change abruptly unless a discontinuity intervenes. Such transverse discontinuities (tear faults) do occur in a few places in the Foothills, where they may affect either or both thrusts and folds. From such considerations of the two-dimensional version of the law of conservation of volume one can derive the rule that: *In adjacent cross sections the amount of "shortening" at a specific horizon between comparable reference lines must be nearly the same unless there is a tear fault between them.*

In this context, "shortening" is the difference between actual bed length and the horizontal distance it now occupies. This statement does not deny the possibility of a gradual change in the shortening along an individual structure nor along a mountain belt as a whole, but it does deny that these changes can occur abruptly without tear faulting.

The reader can convince himself of the reasonableness of this rule by taking a long thin strip of light paper and attempting to produce a long parallel system of folds and faults. Paper is a suitable medium for this kind of demonstration because its lack of plasticity obliges one to maintain a constant bed area. The Rocky Mountain belt is 1610 km (1000 mi) long from the Idaho batholith to the Liard River. The deformed width of say 100 miles (160 km) is, perhaps, half of the width of the rocks as originally deposited. These figures provide some ap-

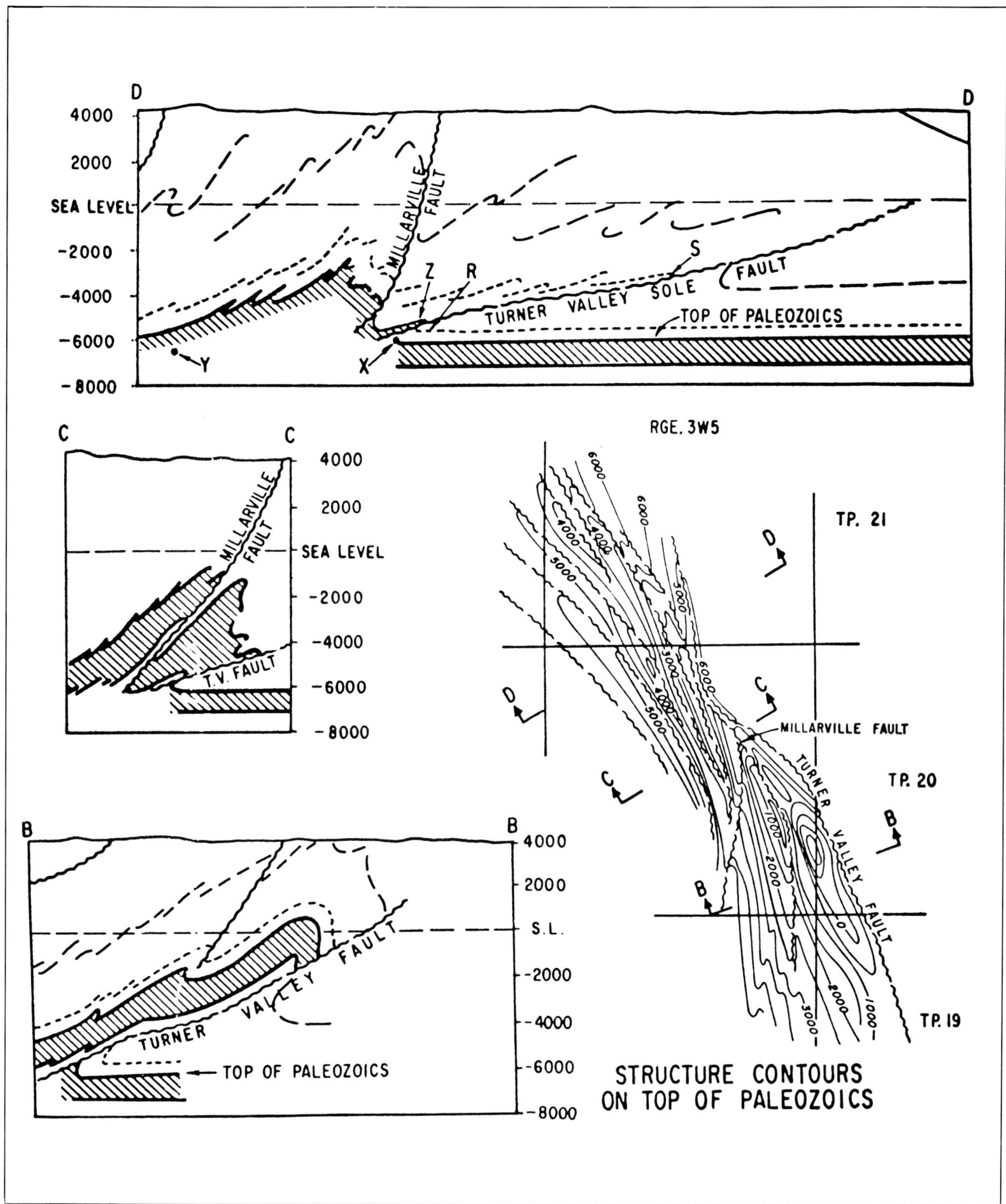

Fig. 6A-11—North end of the Turner Valley structure (after W. B. Gallup, 1951, with the permission of the American Association Petroleum Geologists). Note that the cross-sectional trace of minor faults has been omitted, but their presence is indicated by offsets in the marker horizons. See text for discussion.

proximate dimensions for the paper experiment.

Having performed the experiment actually or mentally, the reader may now have reason to suspect that there should be comparable amounts of shortening in adjacent cross sections. Figure 6A-11 shows the north end of Turner Valley, where there is an abrupt change in structure from the simple faulted anticline of Section B–B to the sheaf of imbricates represented by Section C–C. Despite this change in structural form, the shortening at the top of the Paleozoic level in both Sections B–B and C–C in the published sections is in phenomenally good agreement at a figure of 4571 m (15000 ft). These two sections are consistent. Consider the third section D–D. In Section D–D the well control does not establish how far to the left (west) the regional plate extends under the Turner Valley sole fault. The natural tendency is to put the footwall "cut-off" of the top of the Paleozoic rocks at X. If this is done, the shortening in Section D–D becomes approximately 1830 m (6000 ft), some 2740 m (9000 ft) less than in Section C–C. A 60% reduction in shortening from C–C to D–D would be quite inconsistent, and prompts an interpretation where the top of the Paleozoic beds in the regional plate extends back to Y. This would also produce better internal consistency within section D–D by making the fault displacement on the Home Sand (R–S) equivalent to that on the top of Paleozoic section (Y–Z).

The two southern sections have provided a clear example of consistency of shortening in adjacent sections despite rather drastic changes in structural form, and the third section shows how the rules of consistency can be used to choose between alternate interpretations in the absence of definitive well control.

TRANSFER ZONES

Previously it was stated that shortening on the local scale in individual structures and in the regional scale in a whole mountain belt could change, but that the change would be gradual. Using the same word "gradual" for change at both scales is not really appropriate because the rate of change is substantially greater for individual structures than it is for the mountain belt as a whole. The Lewis Thrust, for instance (Dahlstrom et al., 1962), has a minimum of 37 km (23 mi) of thrust displacement at the United States border and 217.2 airline km (135 mi) to the north, the fault displacement is zero. Over the same distance the overall shortening in the mountain belt as a whole may have diminished, but certainly not by 37 km (23 mi). Since the whole does not change as rapidly as its component parts, it follows that there must be some sort of compensating mechanism at work whereby displacement is "transferred" from one structure to another. Such mechanisms have been observed in the *ac* cross-sectional plane (Fig. 6A-5), so it is not unreasonable to expect that comparable phenomenon would function in the horizontal *ab* plane (as in Fig. 6A-11).

The compensatory mechanism for thrusts is a kind of lap joint wherein the fault whose displacement is diminishing is replaced by an echelon fault whose displacement is increasing. Clearly such a "transfer zone" could not exist unless all of the faults involved in the transfer

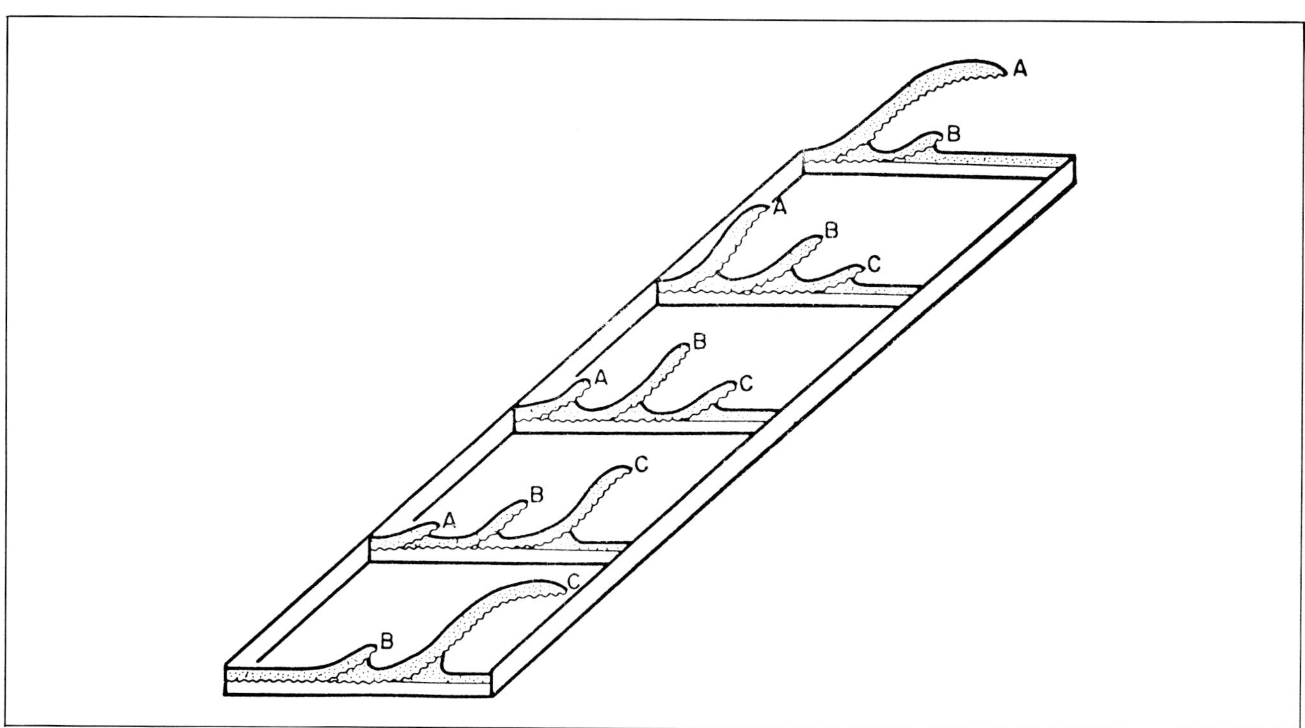

Fig. 6A-12—Shift of displacement between thrusts in a transfer zone.

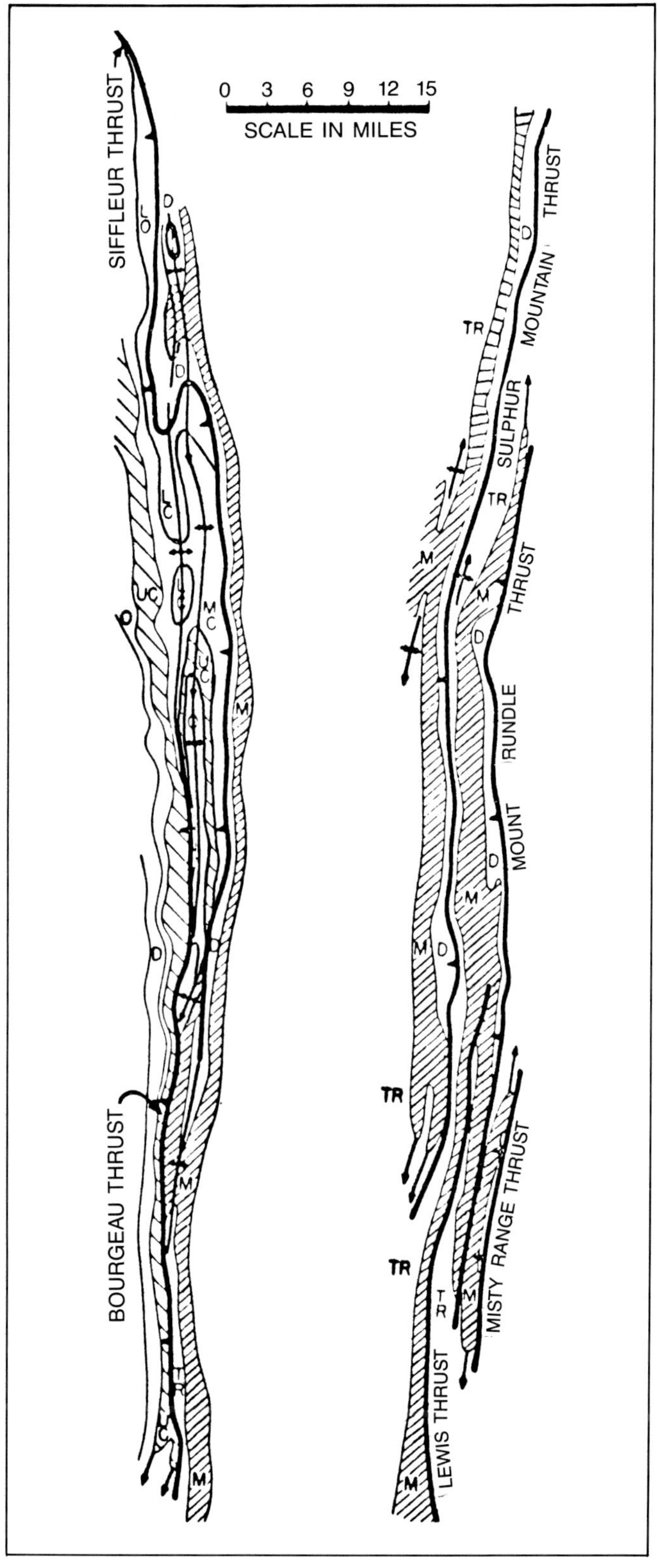

Fig. 6A-13—Two transfer zones in the Front Ranges of Alberta. The abbreviations designate formational ages (i.e., M - Mississippian).

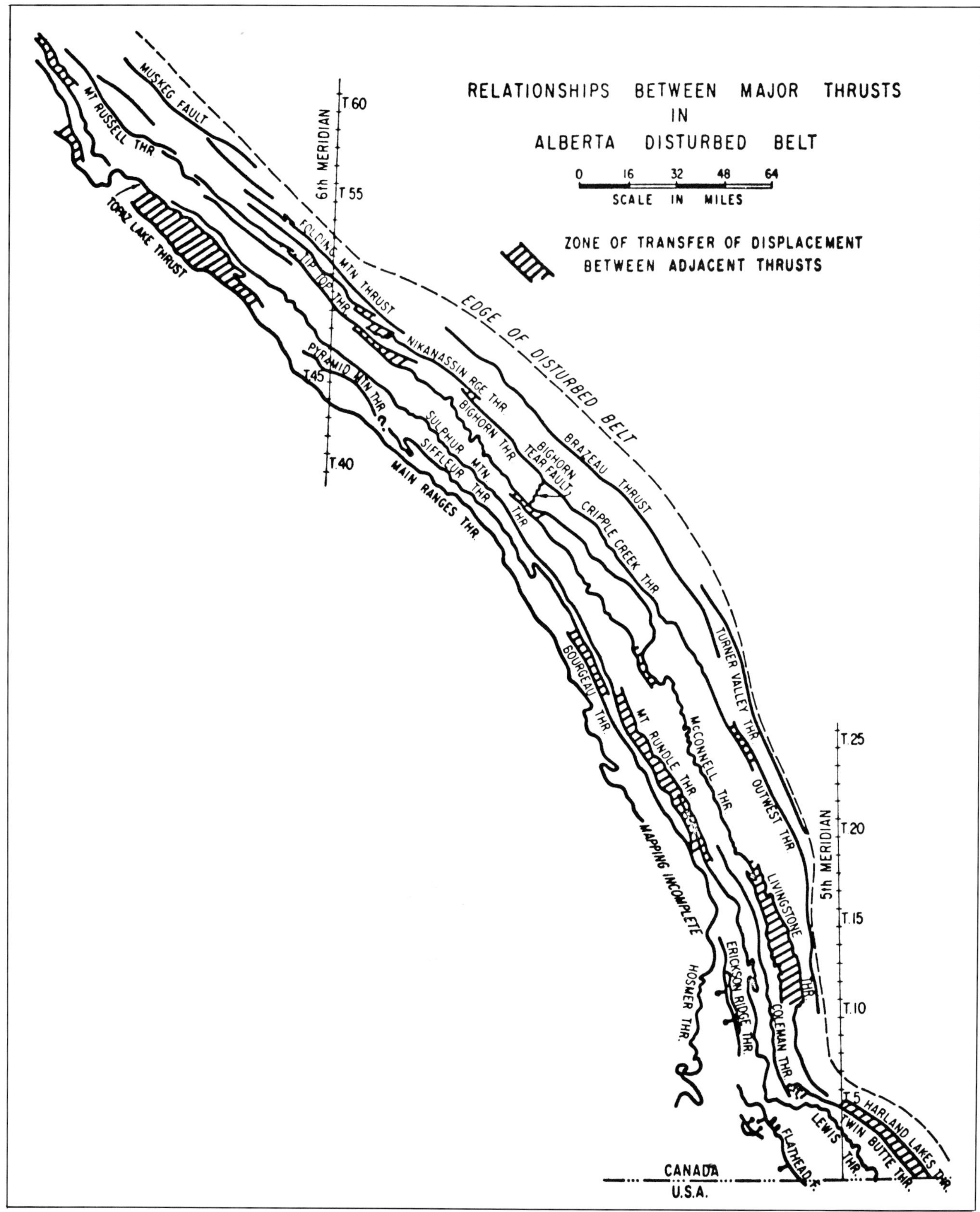

Fig. 6A-14—Use of transfer zones in correlating zones of thrusting. (Thrust faults define a relay pattern–Lowell).

zone are rooted in a common sole fault. Figure 6A-12 shows a relatively simple transfer zone consisting of three faults. In each of the five cross sections the shortening is exactly the same, although at one end virtually all of the displacement is on fault C and, at the other, on fault A. In natural examples, the pattern is often complicated by folds and folded thrust imbricates as shown by the two mapped examples in figure 6A-13.

Recognition of transfer zones enables correlations to be made between thrust faults. Although one fault terminates, its place is taken by another and the zone of thrusting persists. On this basis, a frontal zone and five principal zones of thrusting can be identified within the Foothills and Front Ranges over a distance of some 643.7 km (400 mi) (Fig. 6A-14). At the north end correlation fails because the surficial deformation was primarily folding, and faults are not prominent. The persistence and parallelism of these zones of thrusting over substan-

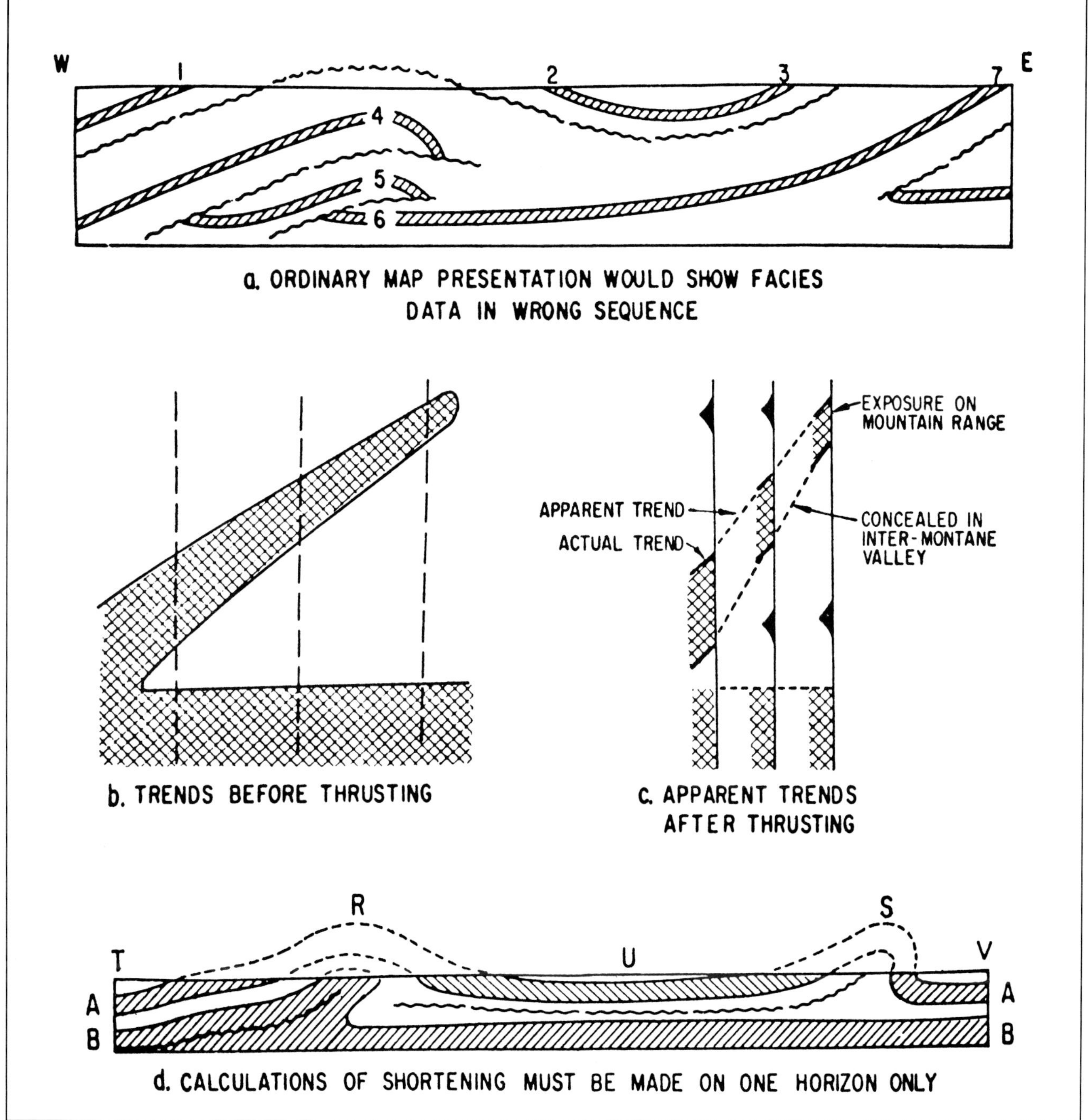

Fig. 6A-15—Elements of palinspastic map construction. See text for discussion.

tial distances lends credence to the suggestion that overall shortening is reasonably consistent along the mountain trend. In one respect, designating six zones of thrusting may be rather arbitrary because at depth all of these zones will join to a common sole fault above the through-going basement. Consequently, in the gross view, one would expect some transfer of displacement between the six thrust zones.

It is not proposed to discuss en echelon[2] folding, although it is a part of the Foothills movement pattern (Fitzgerald 1968) wherein the transfer mechanism operates for folds as it does for faults. En echelon folds are tied to one another by a sole fault (decollement) and maintain consistent shortening by replacing a dying structure with an en echelon growing equivalent (Wilson 1967).

PALINSPASTIC MAPS

The ultimate check of cross sections in deformed terrane is whether or not the process can be reversed and the beds put back into their depositional position without introducing inexplicable bed length anomalies. Since the construction of restored cross sections is a tedious process, the discussion has been concerned with shortcuts which would be adequate for checking the geometric acceptability of individual cross sections and suites of cross sections. However, when stratigraphic studies are being done in deformed terrane, the shortcuts are inadequate, and it becomes necessary to make a restored palinspastic map (Dennison and Woodward 1963), which shows beds in their depositional position.

When stratigraphic data obtained by study of deformed terrane are recorded and interpreted on an ordinary map, it is possible to make serious errors. In figure 6A-15a, seven pieces of surface and subsurface stratigraphic data are shown in cross-sectional view. On the normal geographic presentation this data would appear in the sequence 1, 4, 5, 6, 2, 3, and 7. Obviously, what is needed for stratigraphic work is a restored map where the data are returned to their original depositional sequence of 1, 2, 3, 4, 5, 6, and 7. The normal presentation can also produce misleading stratigraphic trends (Figs. 6A-15b, 6A-15c).

To avoid these pitfalls, one begins a palinspastic map by constructing a suite of cross sections at regular intervals across the study area. These cross sections must be checked for internal consistency and for consistency between contiguous cross sections according to the methods previously discussed. When a consistent suite of sections is available, then shortening is determined for each fault plate and a palinspastic map constructed wherein individual thrust sheets are unfolded and pulled back to their original locations. Some geographic reference points must be maintained on the palinspastic map so that stratigraphic data can be plotted and the interpretations applied to present-day land positions.

The commonest error in palinspastic map construction is using improper values for shortening in the reconstruction. In figure 6A-15d, two horizons A and B are shown where virtually all of the shortening in A takes place at position S, while B is shortened at R. In calculating shortening for palinspastic map construction, it would be very easy to add the shortening at R and the shortening at S together, to arrive at an answer that was twice the proper value. The best method to determine shortening is to do all the measurement on one horizon. If one is forced to change horizons, this can only be done in an area where there is no discontinuity between the reference horizons. In figure 6A-15d, one could change from horizon A to horizon B at positions V or T, but certainly not at position U.

CONCLUSIONS

The quality of structural cross sections can be improved by testing them for geometric validity. In a concentric regime where the structure is post-depositional there is no significant change of rock volume during deformation. Since bed thickness remains constant, it follows that the surface area of any bedding plane remains constant during deformation. Because the area of successive beds in an undeformed depositional sequence is constant, it is necessary that bed lengths in individual cross sections be consistent and that adjacent cross sections have consistent amounts of shortening. Apparent inconsistencies in or between cross sections develop in consequence of discontinuities such as sole thrusts, decollement, or tear faults. Actual inconsistencies are ordinarily due to conceptual, observational, or drafting errors, which might pass undetected were there no tests for geometric validity.

In some lines of geological endeavor, cross sections are significant only as diagrams that are used to convey concepts. In geology applied to oil and mining exploration or to engineering projects, cross sections are used to convey predictions as to rock behavior and they must be conceptually and, more important, geometrically correct. In these areas it is important to the geologist and to his client that there be some way of checking cross-section interpretations prior to drilling. It should be emphasized that a cross section which passes the geometric tests is not necessarily correct, because completely ridiculous cross sections can be drawn which abide by the law of conservation of volume. However, if a cross section passes the geometric tests, it could be correct, and if it has been drawn with due regard for the "local ground rules" it probably is correct. On the other hand, a cross section that does not pass the geometric tests could not possibly be correct.

The rules developed in this paper pertain to a simple, concentrically deformed packet of "layer cake" geology in the Alberta Foothills. The rules themselves cannot be transported to a more complicated area, but the basic concept and method can be used to generate sets of rules which apply to other environments. Thus, it is possible

[2]Actually the pattern is relay, not en echelon—Lowell.

to devise amended rules, which apply to concentric extensional structures and to diapiric structures that grow during deposition or to "similar fold" regimes. Each of these areas requires a different set of rules, which must be derived from the basic geometric principles and observational restrictions. The interpretational rules for similar folding were discussed by Carey (1962). In other areas the geologist will have to develop his own rules, an endeavor that he should find interesting and rewarding.

REFERENCES

Bally, A. W., Gordy, P. L., and Stewart, G. A., 1966, Structure, seismic data, and orogenic evolution of southern Canadian rocky mountains: *Bull. Canadian Petroleum Geologists,* 14, p. 337–381.

Bucher, W. H., 1933, The deformation of the Earth's crust: Princeton University Press, Princeton, N.J., 518 p.

Carey, S. W., 1962, Folding: *Jour. Alberta Society Petroleum Geoogists,* 10, p. 95–144.

Dahlstrom, C. D. A., 1954, Statistical analysis of cylindrical folds: Trans. Canadian Institute Mining Metallurgy, 57, p. 140–145.

———, Daniel, R. E., and Henderson, G. G. L., 1962, The Lewis thrust at Fording mountain: *Jour. Alberta Society Petroleum Geologists,* 10, p. 373–395.

Dennison, J. M. and Woodward, H. P., 1963, Palinspastic maps of central Appalachians: *American Association Petroleum Geologists Bull.,* V. 47, p. 666–680.

Fitzgerald, E. L., 1968. Structure of British Columbia foothills, Canada. *American Association Petroleum Geologists Bull.,* V. 52, p. 641–664.

Gallup, W. B., 1951. Geology of Turner Valley oil and gas field, Alberta, Canada. *American Association Petroleum Geologists Bull.,* V. 35, p. 797–821.

Goguel, J., 1952, *Tectonics* (1962 translation): Freeman and Company, San Francisco, 384 p.

Hunt, C. W., 1957, Planimetric equation: *Jour. Alberta Petroleum Geologists,* V. 5, p. 259–264.

Price, R. A., 1964, Flexural slip folds in the Rocky mountains, southern Alberta and British Columbia, *in* Seminars on tectonics—IV: Dept. Geological Sciences, Queen's Universiy, Kingston, Ont. (Also as Geological Survey Canada, Reprint 78, 16 p.)

Stockwell, C. H. 1960, The use of plunge in the construction of cross sections of folds: Proc. Geological Association Canada, 3, p. 97–121.

Wilson, G., 1967, The geometry of cylindrical and conical folds: Proc. Geological Association, London, 78, p. 179–210.

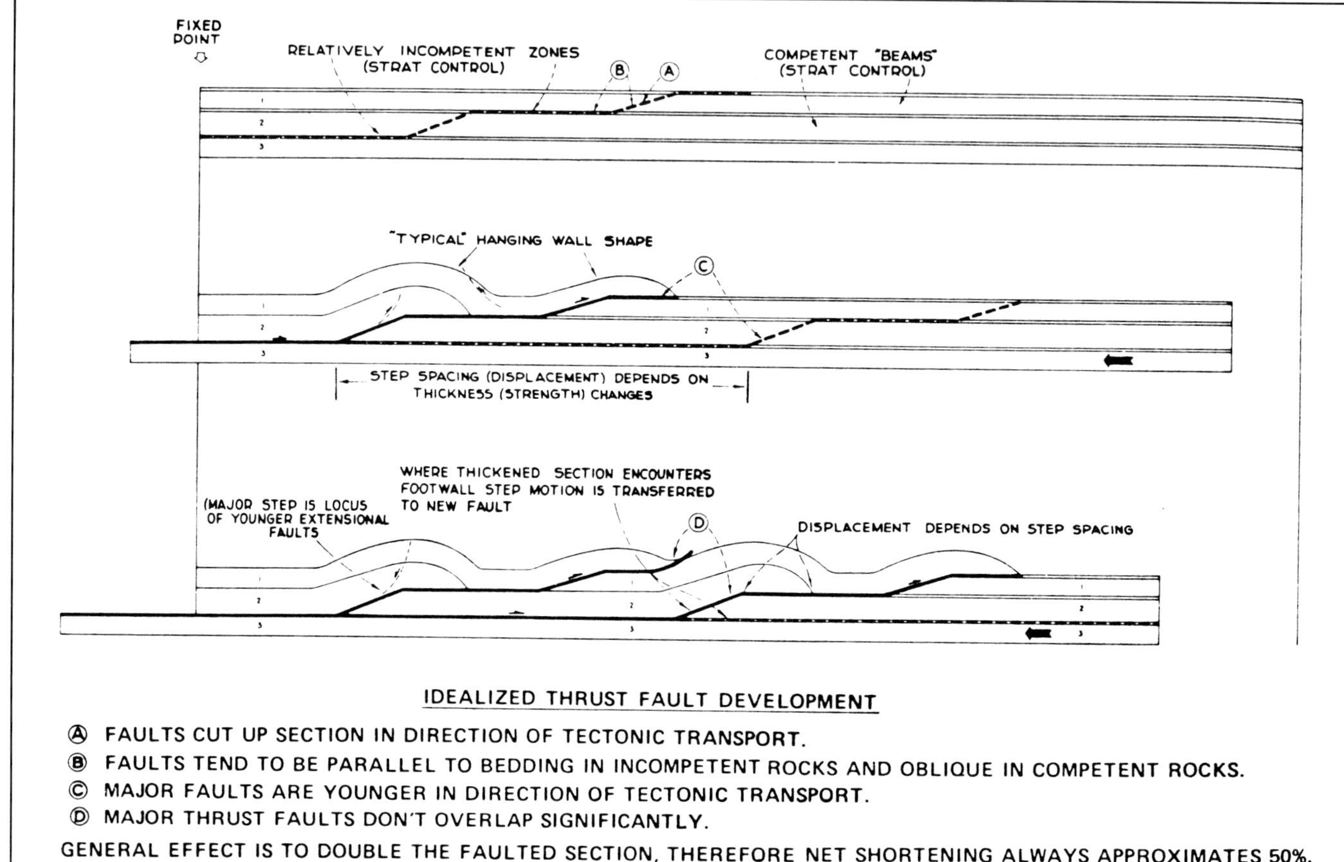

Fig. 6A-16 (Royse et al., 1975)—Schematic diagrams showing empirical guidelines for drawing thrust-fold belt cross sections. Faults cut up section in the direction of relative tectonic transport because they seek the lowest pressure gradients. Ever lower pressures are encountered in ever younger stratigraphic sections so that thrusts eventually reach a free surface. Conversely faults rarely cut down section because in so doing they encounter higher pressures. Major faults are also younger in the direction of relative tectonic transport and do not overlap significantly between the lead edge of one thrust and a major ramp of another. The ramp position of the structurally lower thrust may control the lead edge of the structurally higher thrust, even though the lower thrust is younger than the higher.

7 NORMAL-FAULT ASSEMBLAGE

INTRODUCTION

Normal faults present in sedimentary section and not cutting to basement occur in virtually every structural style (Figs. 1-6; 2-29 through 2-31; 3-31; 4-7; 6-34, 6-35, 6-43, 6-44, and 6-51; 8-6, 8-8, 8-9, 8-11, 8-21, 8-31, 8-32, 8-33, 8-34, 8-37, and 8-40; and 9-2, 9-5, and 9-7). As a rule, however, their amount of displacement is small and they are subordinate to the style in which they occur. These faults, which can take any number of trajectories, are not the subject of this chapter, but rather we are considering normal faults that are distinctly listric and sole at some level above basement (Fig. 1-21). The petroleum industry has been the leader in understanding and refining this style because reflection seismic and drilling have been critical in its recognition.

Mechanically, listric normal faults, interchangeably referred to as extensional glide-plane faults, behave as faults that bound slump blocks. Curved slip surfaces demand a backward and downward rotation (reverse drag) of strata into the upper part of the fault as lateral movement of material proceeds on the more distal, flatter part. This process has already been encountered as part of the compressive block assemblage on the south slope of the Owl Creek Mountains in Wyoming (Fig. 3-31) and in listric normal faulting in thrust-fold belts (Figs. 6-34, 6-35).

Extensional glide-plane faults form a structural style of their own because of their frequent occurrence in unconfined depositional sites, usually in prograding (regressive) stratigraphic sequences. Thus, their preferential habitats are large marine deltas such as the Mississippi, Niger, Orinoco, and Mackenzie and the edges (shelf-break portion) of continental margins where incompetent, sometimes unconsolidated, material is free to creep basinward. Extensional glide plane faults are often called contemporaneous or growth faults because they are active during sedimentation and have thicker sections on their downthrown sides. Down-to-basin fault is synonymous because most displacements as they accommodate thickened sections must be down toward the basin. Backward and downward rotations are expressed as downbend or rollover anticlines which are the prime traps in this structural style.

MECHANICS

In simplest form detached listric normal faulting can be treated as a gravity phenomenon of basinward creep of sediments or sedimentary rocks requiring an upslope area of extension and a compensatory downslope region of compression (Fig. 7-1). Unless there is regional extension (Fig. 4-2), zones of extension must be matched by zones of compression.

Neither salt nor shale need be present for listric normal faults to develop. However, since both lithologies have an affinity for regressive stratigraphic sequences in unconfined depositional sites (often on trailing continental margins), they have usually interacted with and facilitated the faulting. The purely gravitational and extensional motivation of listric normal faults will be discussed first, followed by consideration of the roles of shale or salt, or both, in the sedimentary section.

Detached listric normal faults that comprise the style described in the present chapter are syndepositional and preserve thicker section on their downthrown sides. However, it is instructive in understanding the mechanics of the style to consider experimental models and examples of post-depositional listric normal faults. Of particular interest is a series of clay model experiments that reproduced many of the features that characterize listric normal faults (Cloos,

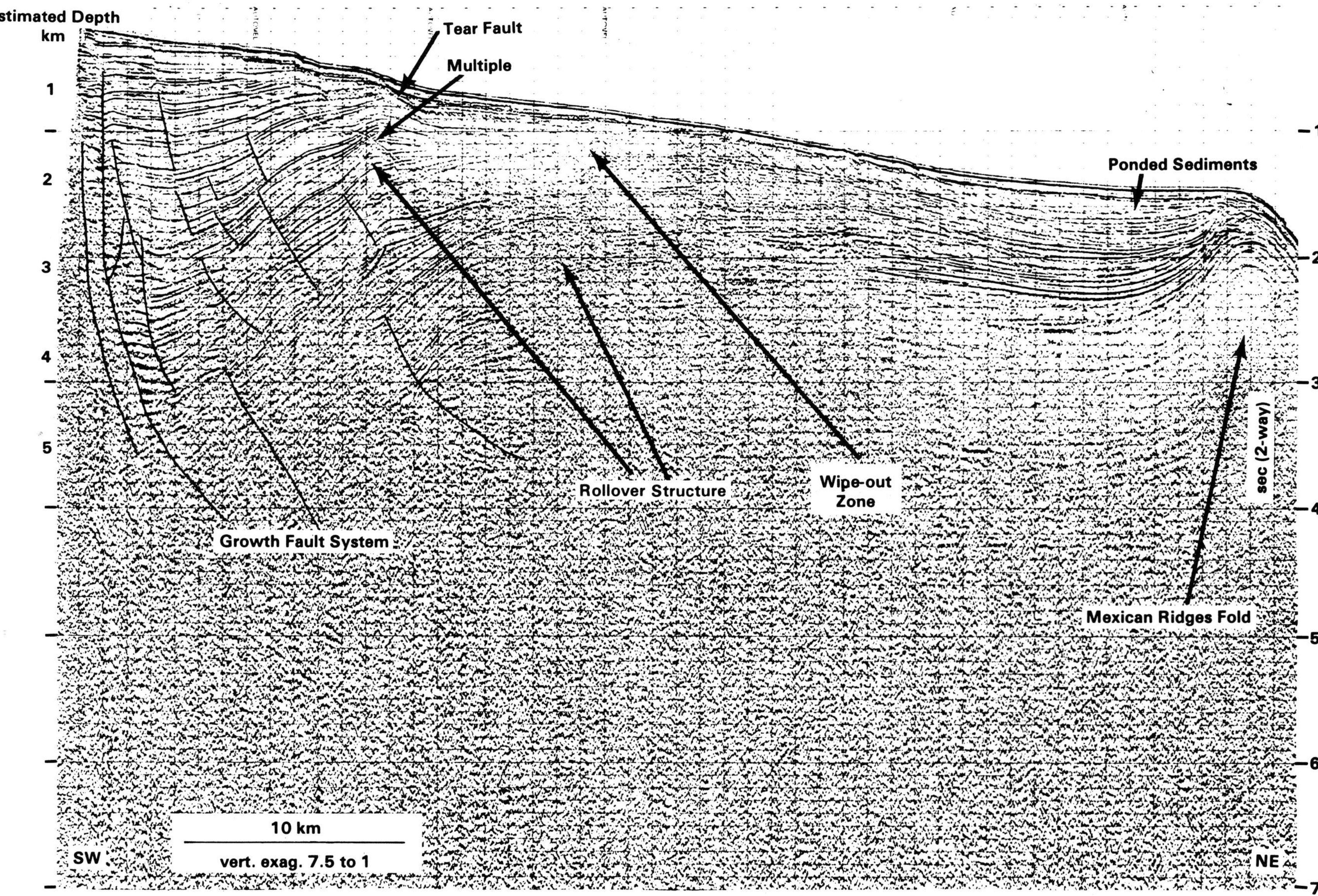

Fig. 7-1 (Shaub, 1983)—Seismic line, offshore Vera Cruz, Mexico, showing normal faults evidencing extension at left, and folds evidencing compensating compression at right.

Fig. 7-2 (Cloos, 1968)—Clay model on overlapping metal sheets before deformation. Left end of upper arrow is at edge of right, upper sheet. Horizontal lines simulate bedding but are scored only on surface of clay for reference during deformation. Permission to publish by American Association of Petroleum Geologists.

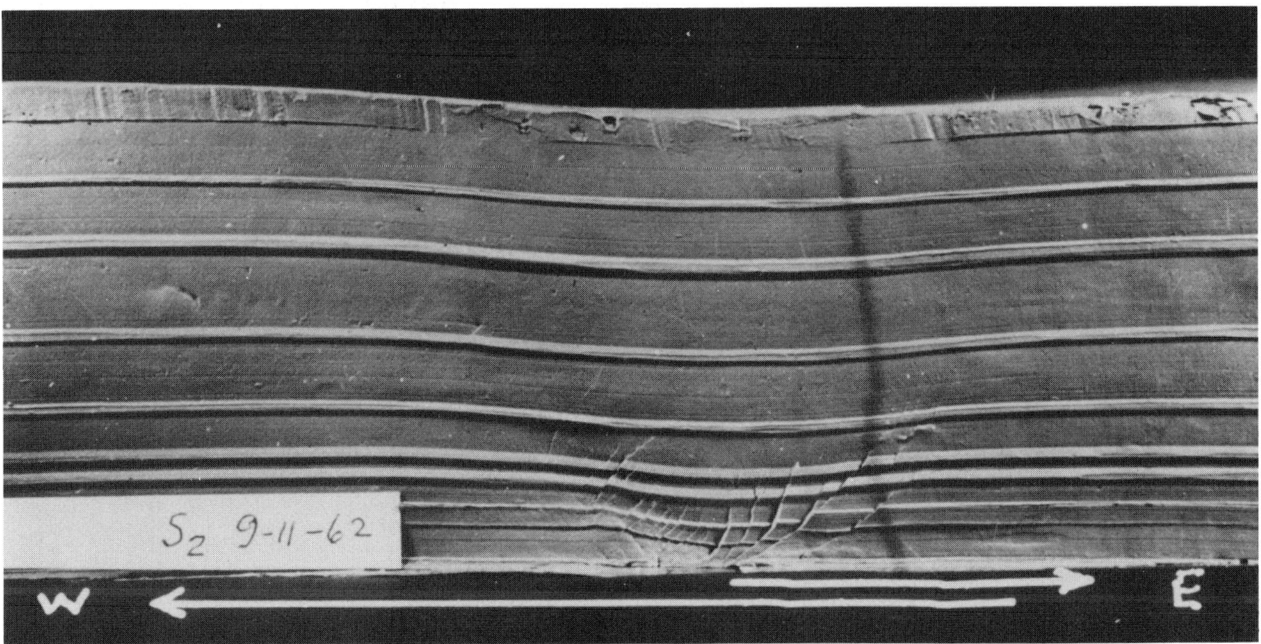

Fig. 7-3 (Cloos, 1968)—Asymmetrical graben after deformation begins. Faults appear first at base and downbending develops in lowest layers. Permission to publish by American Association of Petroleum Geologists.

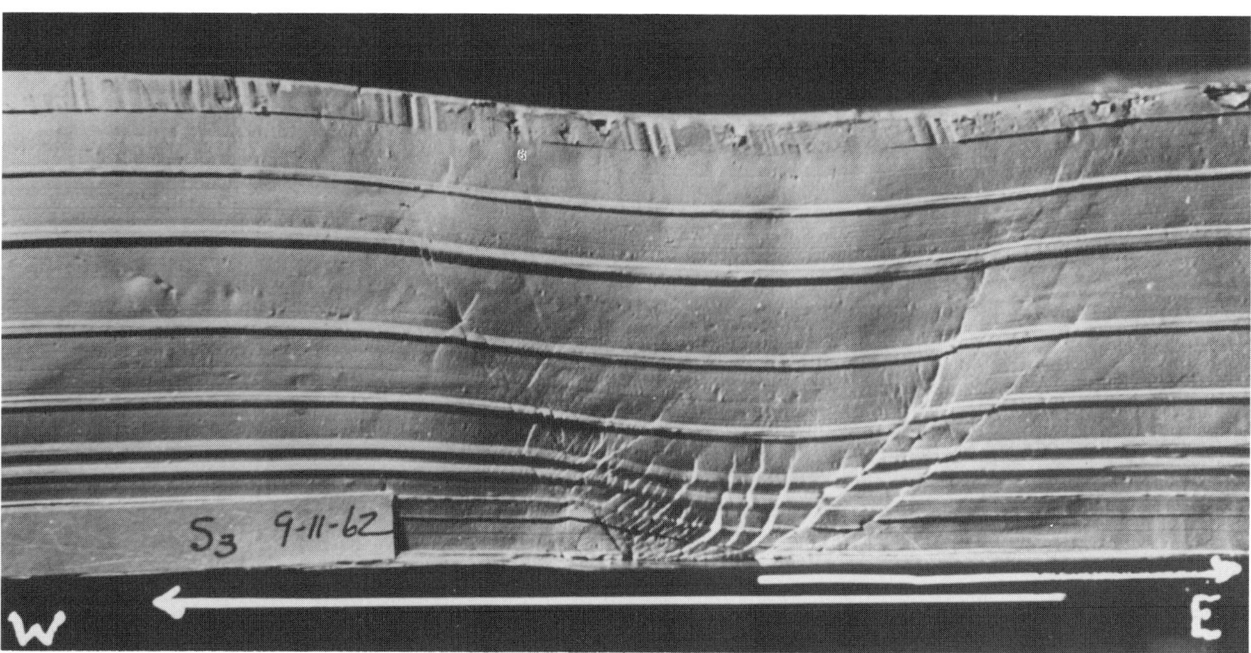

Fig. 7-4 (Cloos, 1968)—Faults at right and fault zone at left have reached surface. Asymmetry grows more pronounced. Downbending grows upward and includes upper layers. In growth faults sagging localizes deposition on low side of fault. Permission to publish by American Association of Petroleum Geologists.

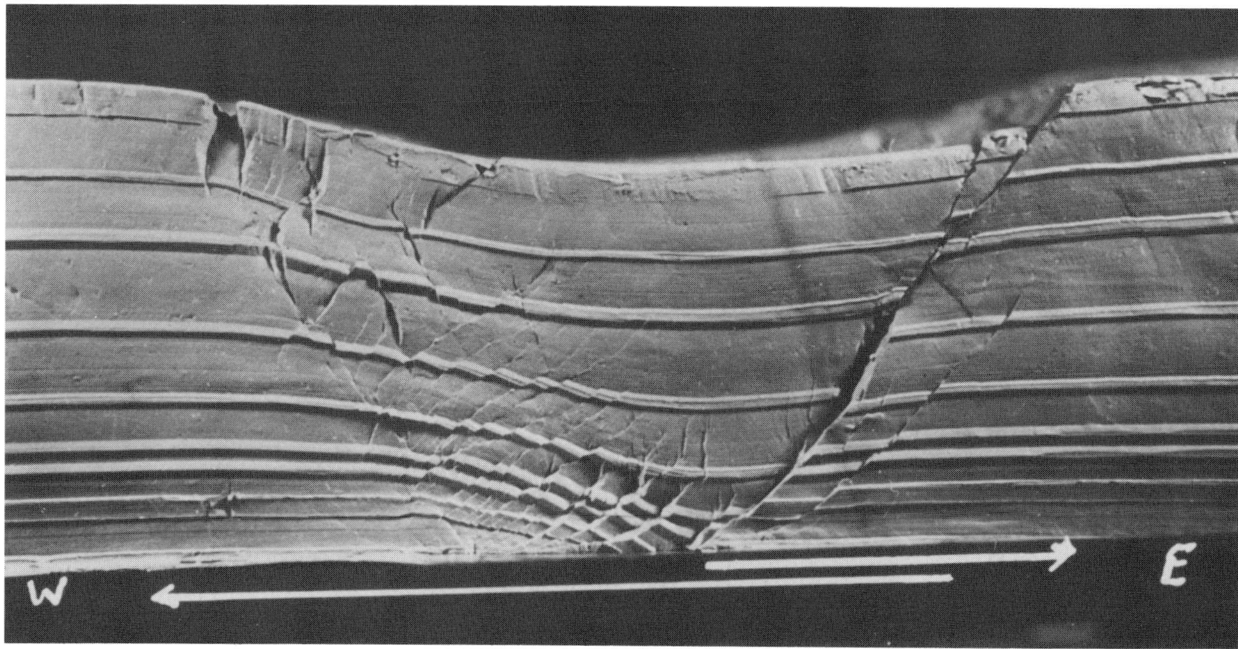

Fig. 7-5 (Cloos, 1968)—Master fault at right separates right block from asymmetrical graben. Left fault zone is wide and accommodates downbending in lower layers, where most individual fault blocks rotate clockwise. Note that if graben was filled with sediments, section would be much thicker on downthrown side of master fault. Permission to publish by American Association of Petroleum Geologists.

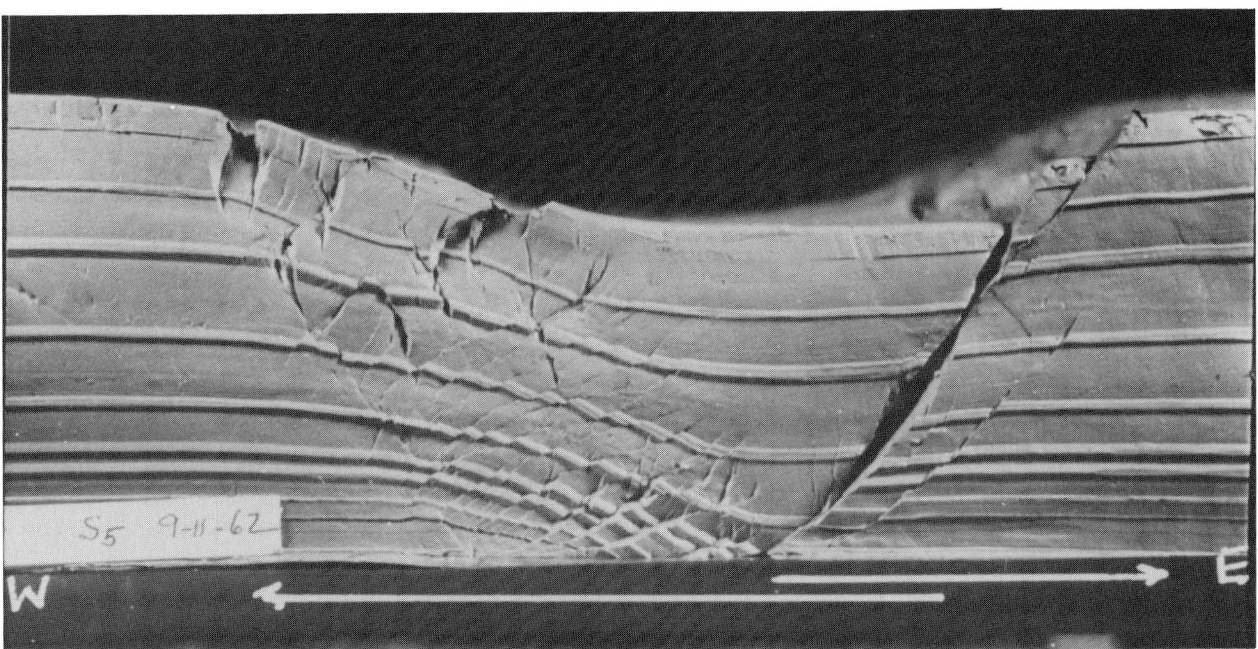

Fig. 7-6 (Cloos, 1968)—Faults antithetic to downbend dip left and blocks rotate in downbend. Master fault is sharp and prominent, flattening downward, until it becomes horizontal at the top of the metal sheets. Gap along the master fault is artificial owing to strength of clay and would not occur in nature; it would be compensated by additional downbending. Permission to publish by American Association of Petroleum Geologists.

Fig. 7-7 (Cloos, 1968)—Oblique photo showing reverse drag, master fault and other faults with striae indicating normal dip-slip movement. Right block moved right; left block remained stationary. Note that slip along the curved fault surface necessitates backward rotation or downbending of beds into fault. Permission to publish by American Association of Petroleum Geologists.

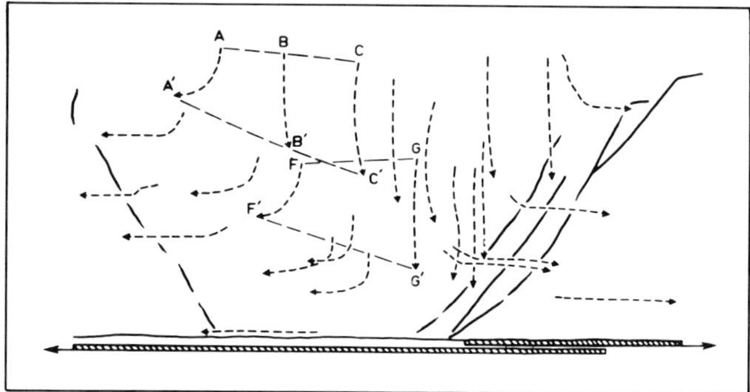

Fig. 7-8 (Cloos, 1968)—Migration of reference points in asymmetrical graben experiment (Figs. 7-2 through 7-7). Line A-C rotates into position A'-C', line F-G becomes F'-G'. Traced from successive photographs. Permission to publish by American Association of Petroleum Geologists.

1968; Figs. 7-2 through 7-8). A master listric normal synthetic fault and other synthetic faults, antithetic faults, and the downbend or rollover anticline (complexly faulted) are all created by extension of clay.

Listric normal faults can occur in consolidated rocks in post-depositional settings such as the Heart Mountain slide in northwest Wyoming (Pierce, 1941), where massive carbonates are downbent on a nearly flat glide plane, and in syndepositional settings, such as the Hoback Canyon area of Wyoming (Fig. 6-35), where a coarsely conglomeratic unit is preserved only on the downthrown side of the Hoback normal fault. Usually, however, listric normal faults occur on basin flanks in the presence of very incompetent sediments, mainly shale or salt, which can significantly modify the faulting process.

Some of the modifiers of faulting caused by slow gravity creep and associated extension are as follows. Differential loading of denser sands and shales on lower-density undercompacted shale has created shale ridges on whose seaward sides faults can sole (Bruce, 1973; Figs. 7-9, 7-10). Flowage within the shale ridge or residual shale mass appears to slightly warp the listric normal fault surface (Figs. 7-10, 7-11).

Shale flowage is most often associated with undercompaction and overpressuring, which can further influence faulting. Overpressuring increases the ductility of shales, which in turn leads to an increased dihedral angle, 2θ. This condition together with the vertical orientation of maximum principal compressive stress that characterizes the extensional glide plane style dictates that faults flatten in zones of high pressure (Fig. 7-12).

Greater ductility of shale in contrast to that of other lithologies also affects fault attitudes. An interpretation of reflection seismic (Fig. 7-13) shows that faults cut pure shale sequences at lower (flatter) angles than interbedded sand and shale sections. Subsequent compaction would be greater in shale than in sand and shale and would also help effect the rotation and flattening of fault surfaces. Differential compaction has also created other types of minor detached faults in the growth fault environment (Fig. 7-14).

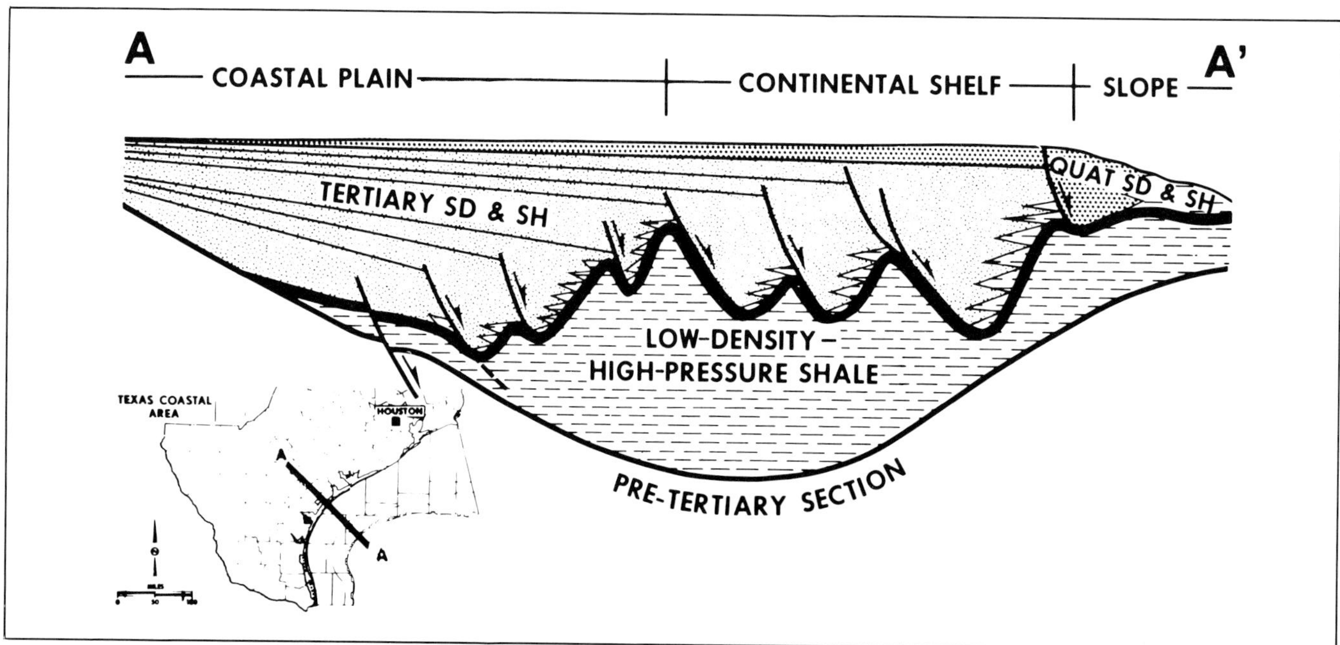

Fig. 7-9 (Bruce, 1973)—Diagrammatic cross section from Texas part of northern Gulf of Mexico Basin illustrating residual masses of pre-Tertiary undercompacted shale created by differential loading and compaction from sand and shale above. Permission to publish by American Association of Petroleum Geologists.

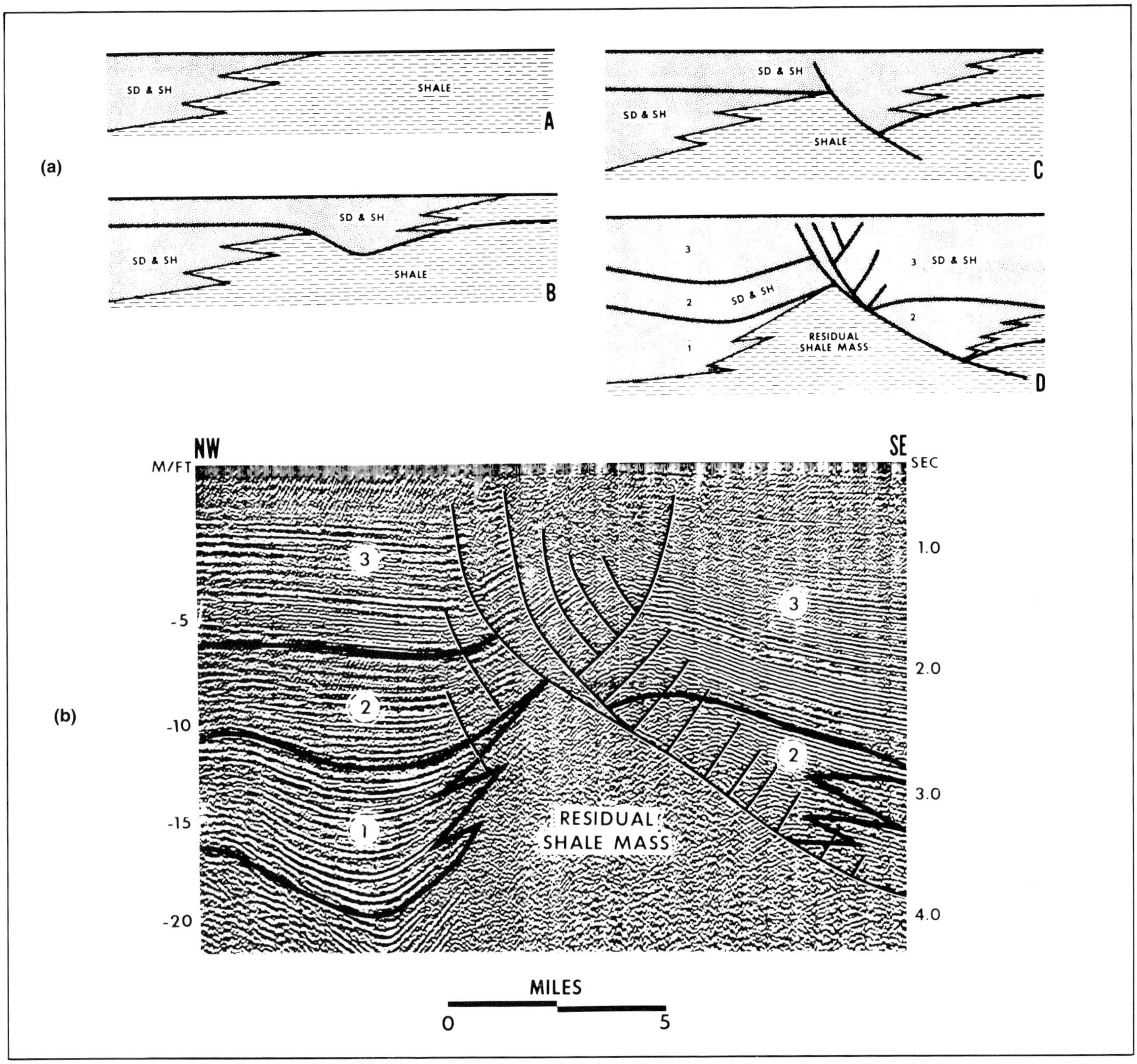

Fig. 7-10 (Bruce, 1973)—a. Diagrammatic illustration of four stages in development of shale mass shown in b.—seismic profile. Numbers of depositional sequences correspond to section D of illustration a. Note slight warp of fault surface on SE side of shale mass. Permission to publish by American Association of Petroleum Geologists.

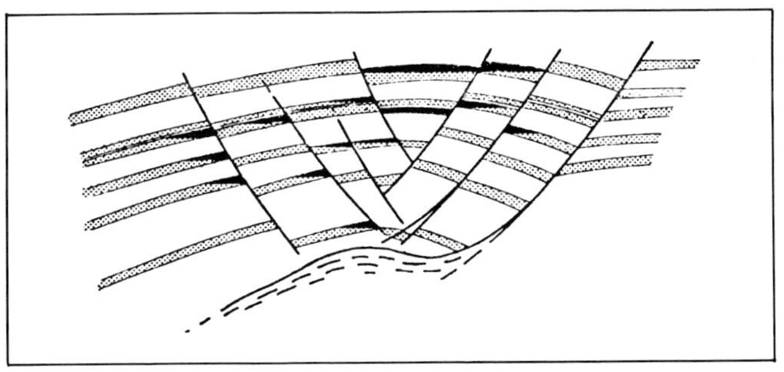

Fig. 7-11 (Weber, 1971)—Example of listric normal fault and rollover anticline with petroleum traps (black) from Niger delta. Note that listric fault, and possibly lower part of rollover anticline, may be warped by shale flowage at depth. Permission to publish by Geologie en Mijnbouw.

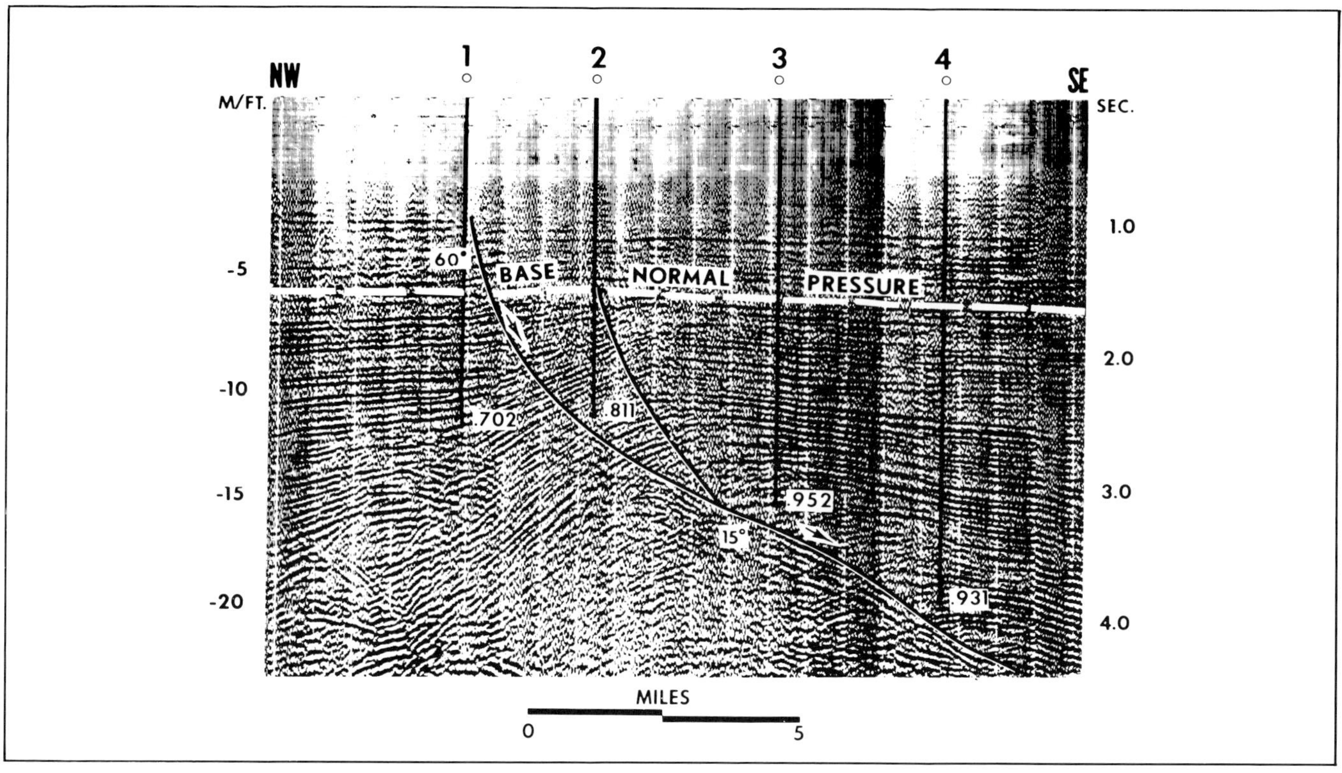

Fig. 7-12 (Bruce, 1973)—Seismic section with well locations showing fluid pressure/overburden ratio at total depth adjacent to fault plane. Below base of normal pressure ductility increases which in turn leads to flattening of fault (see text for additional discussion). Note slight warp of fault trace. Permission to publish by American Association of Petroleum Geologists.

Finally, salt can also modify and effect detachment of listric normal faults (Fig. 1-22). A regional system of detached listric normal faults is known around the upper Gulf Coast of the United States (Fig. 7-15), localized and controlled by flowage and withdrawal of salt at the depositional edge of the Jurassic Louann salt (Figs. 8-3, 8-6). Ocamb (1961) interpreted a detached normal fault system related to salt flowage in a more basinward position (Fig. 7-16). Even more basinward, contra-regional (up-to-basin) faults caused by salt withdrawal are known in the Gulf of Mexico (Fig. 7-17). The faults are of regional dimension and not simply antithetic faults in a down-to-basin system.

HABITAT

Extensional glide-plane faults do not have a dominant plate tectonic habitat as much as they require a prograding or regressive stratigraphic, often deltaic, sequence fronting an unconfined depositional site into which movement can occur (Chapman, 1973). The best known occurrence of detached listric normal growth faults are the Gulf Coast-Mississippi (Fig. 7-15), Niger (Fig. 7-18), Orinoco, and Mackenzie deltas. Such faults are probably often overlooked in fossil delta systems and should be anticipated there.

Evamy et al. (1978) stated that "the overall shape of the (Niger) deltaic wedge, particularly during its initial stages, was (probably) controlled by downfaulting in the basement" (Fig. 7-18). As progradation proceeded, each depositional mega-unit developed in a seaward direction. In a prograding sequence deposition exceeds subsidence and depositional packages become younger seaward (Fig. 7-19A). Locally deposition and subsidence may for a time be the same such that any given package may become very thick (Fig. 7-19B). Instances where deposition is less than subsidence are rare (Fig. 7-19C). In all cases stratigraphic thicknesses are greatest on the downthrown sides of faults as sedimentation is focused in depressions created by the faulting.

PLAN VIEW

In the most regional sense detached listric normal faults follow the outline of the basin (Fig. 7-15) which may in turn dictate the curvature of the upper part of a delta flanking the basin (Fig. 1-25). In extensional experiments Cloos (1968) simulated the fault pattern of the Gulf Coast Basin by pulling a metal sheet from the base of a clay model, this move-

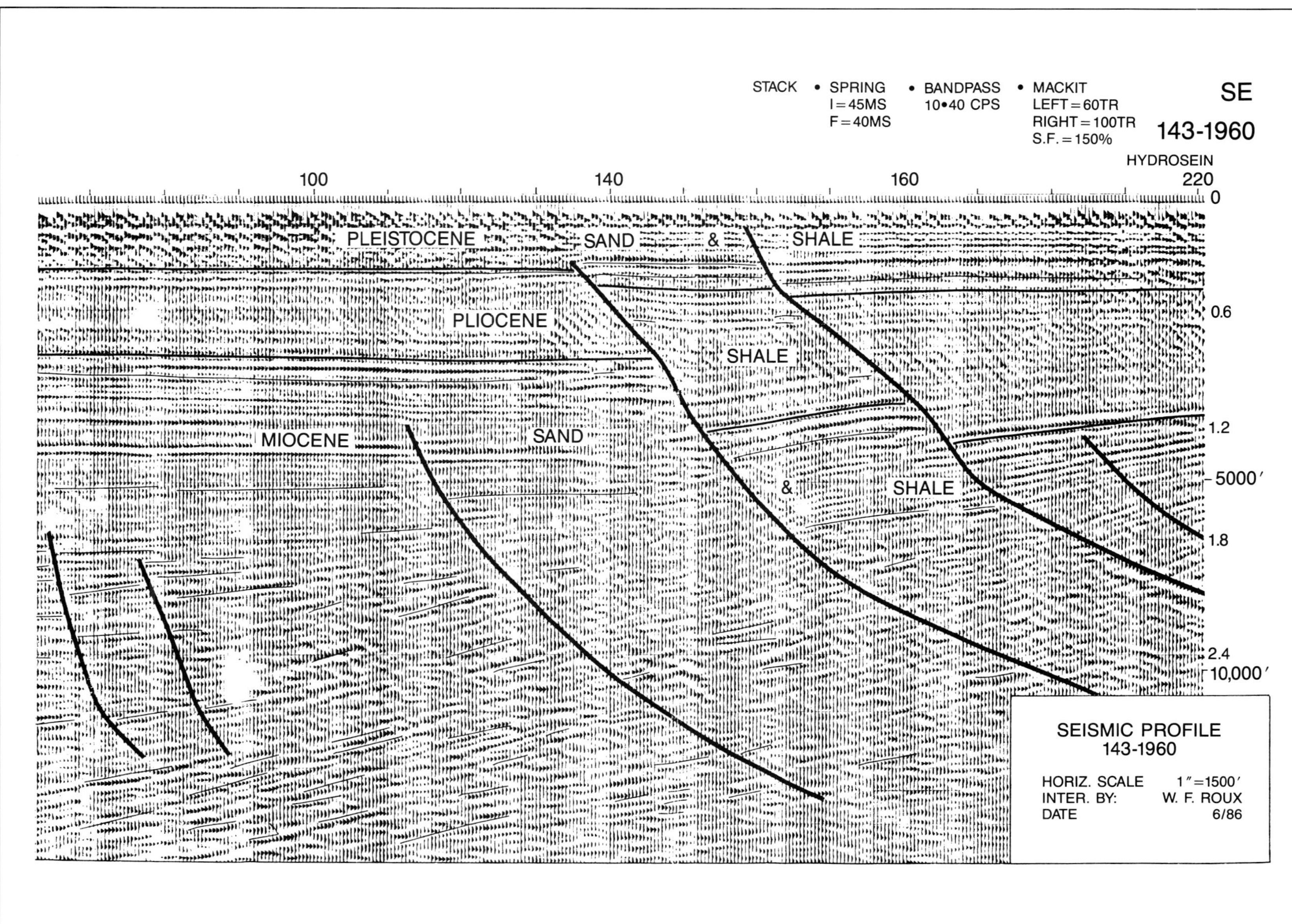

Fig. 7-13 (Roux, 1977)—Seismic profile from U.S. Gulf Coast Basin showing flattening of faults through Pliocene shale section. Conversely structural geology can be viewed as useful in picking stratigraphic boundaries.

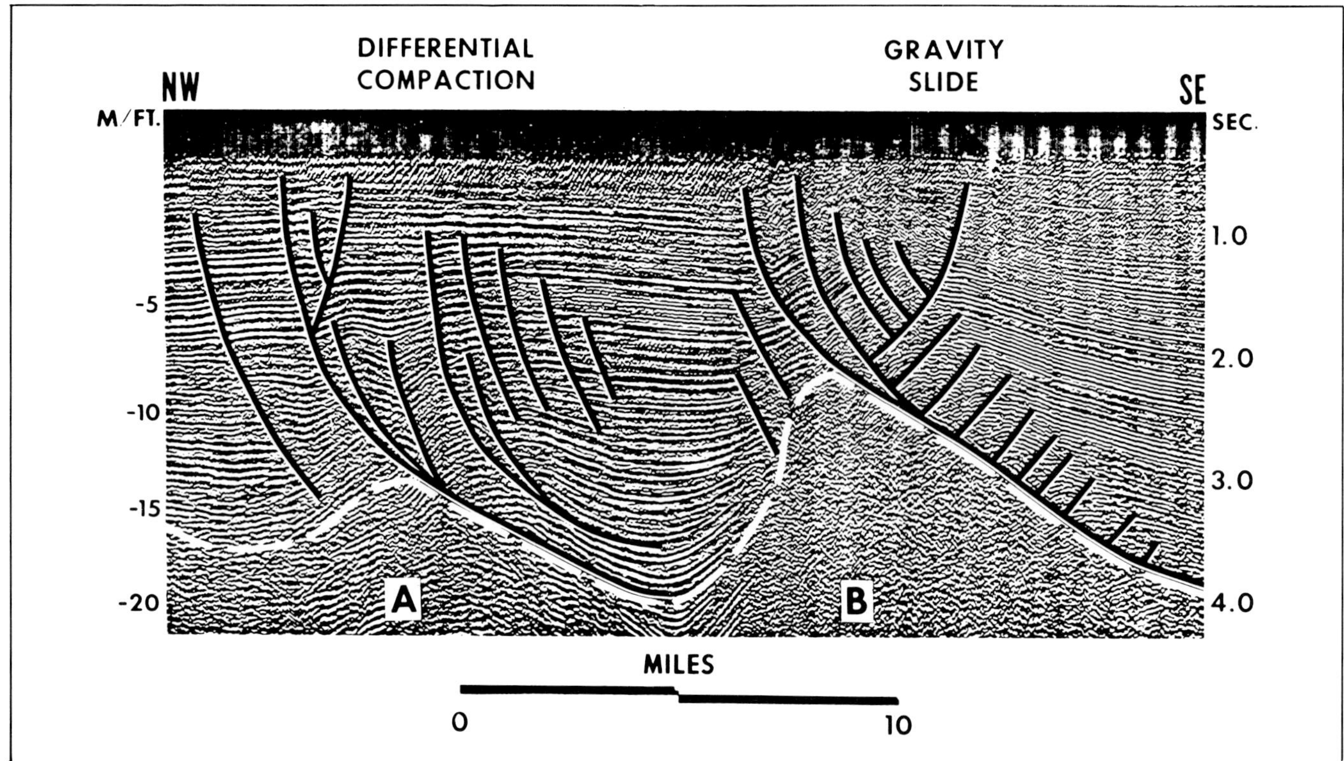

Fig. 7-14 (Bruce, 1973)—Seismic illustration showing differences between fault systems formed by differential compaction and gravity slide. Differential compaction faults usually do not sole, do not effect downbending or growth, and have continuous reflections both above and below them. Note that faults on landward (NW) side of residual shale masses (A and B) terminate abruptly rather than sole. Dashed white line shows configuration of shale masses. Permission to publish by American Association of Petroleum Geologists.

ment creating a curved fault pattern on the clay surface above (Fig. 7-20).

Within the regional pattern faults typically bifurcate and coalesce to form a complex network (Figs. 1-23, 1-25). In detail faults are curved, cuspate, or crescentic almost always in a manner that is concave in a seaward direction (Figs. 1-23, 1-25). Curvature in plan view is probably a consequence of curvature of the fault surface in profile. In at least some examples the central part of the curvature or crescent coincides with the maximum sediment (including sand) accumulation (Fig. 7-21). The axes of rollover anticlines usually parallel their associated and causative faults (Fig. 1-23).

SEISMIC EXPRESSION

Many of the characteristics of the detached listric normal fault style are illustrated on seismic lines, figures 7-22 through 7-25. These include the main listric normal fault (extensional glide-plane fault); rollover or downbend anticline, often complexly faulted and with a crest that migrates rapidly seaward with depth; expanded section with much greater fault throw at depth than shallow; and associated shale ridges and salt features.

PRODUCING EXAMPLES

The rollover or downbend anticline on the downthrown side of its associated and causative listric normal down-to-basin growth fault is by far the most common trap wherever this style is present (Figs. 7-26 through 7-29). Much less frequently production has been established on the upthrown sides of such faults; of course, the upthrown side of one fault can also be the downthrown side of the next adjacent fault. Production is also known from the upthrown side of an antithetic up-to-basin fault (Fig. 7-30).

Fig. 7-15 (Cloos, 1968)—Diagrammatic representation of principal extensional glide plane fault systems in northern Gulf Coast. Note northern system, Balcones, Mexia-Talco, south Arkansas, Pickens, Gilbertown, follows depositional edge of Jurassic Louann salt (cf. Fig. 8-6). Permission to publish by American Association of Petroleum Geologists.

Map views of listric normal faults and their associated downbend anticlines are shown in figures 1-23 and 7-31 through 7-34. As discussed earlier in this chapter, fault patterns can be extremely complex. They are a result of gravity creep, regional extension, differential loading, and movement of shale and salt.

In summary, the fact that listric normal faults and their rollover anticlines grow during sedimentation results in early structures that are always available as traps.

REFERENCES

Bergsma, M., 1983, Upthrown fault closure, Calhoun County, Texas, in Bally, A. W., ed., Seismic expression of structural styles: *American Association Petroleum Geologists Studies in Geology #15,* V. 2.

Bol, A. J., and Van Hoorn, B., 1978, Structural styles in western Sabah offshore: Geological Society Malaysia, 2nd Petroleum Seminar, Kuala Lumpur.

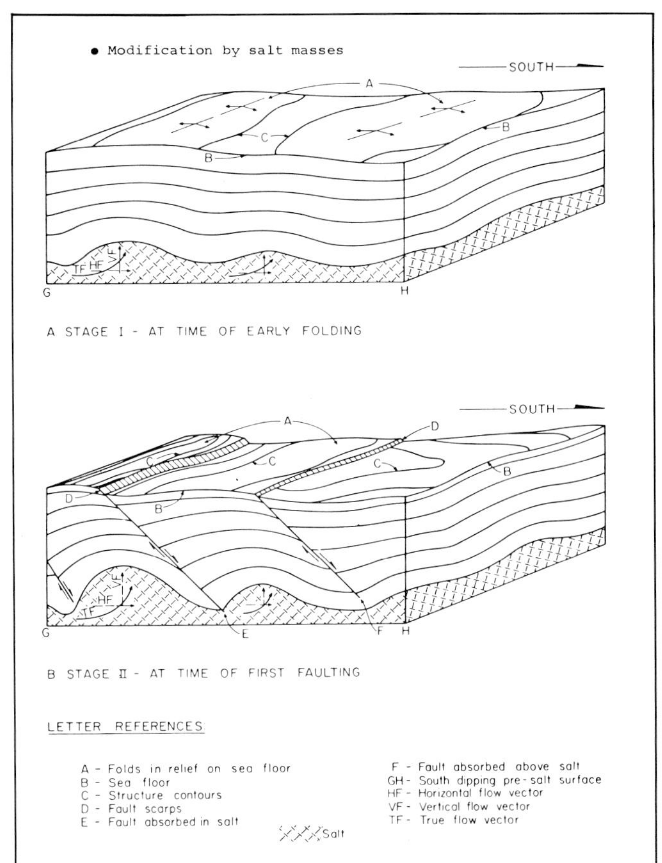

Fig. 7-16 (Ocamb, 1961)—Formation of Eocene trend of detached normal faults as a consequence of salt movement. Permission to publish by Gulf Coast Association of Geological Societies.

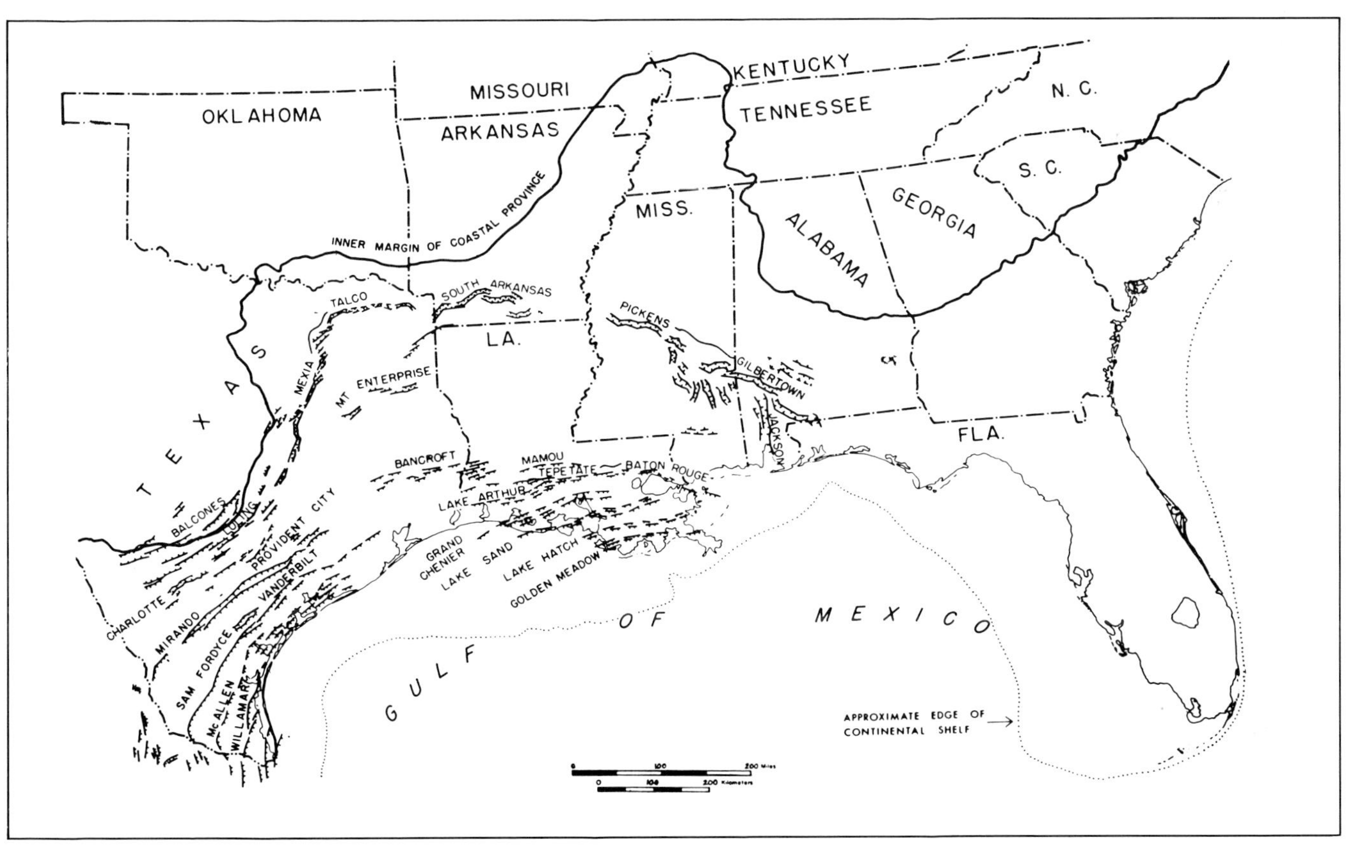

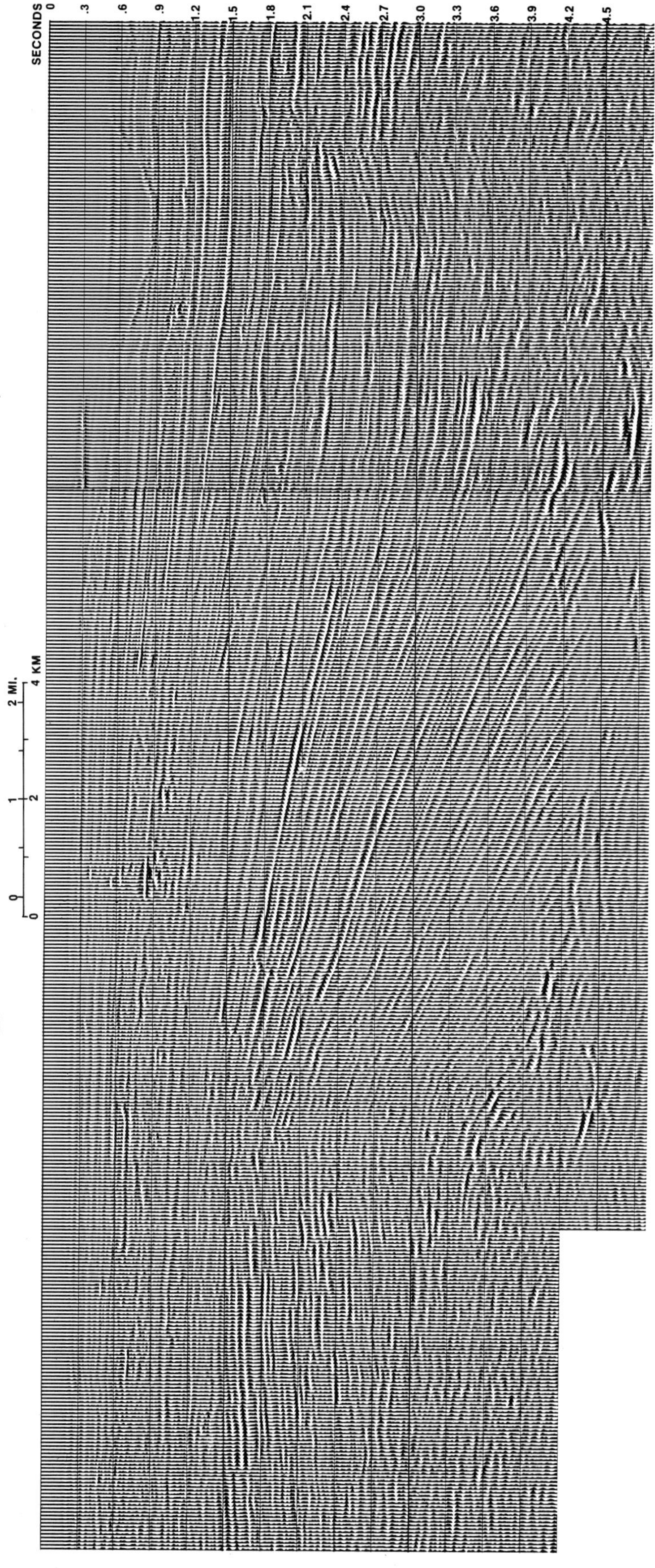
SECONDS
0
.3
.6
.9
1.2
1.5
1.8
2.1
2.4
2.7
3.0
3.3
3.6
3.9
4.2
4.5
0 1 2 MI.
0 2 4 KM

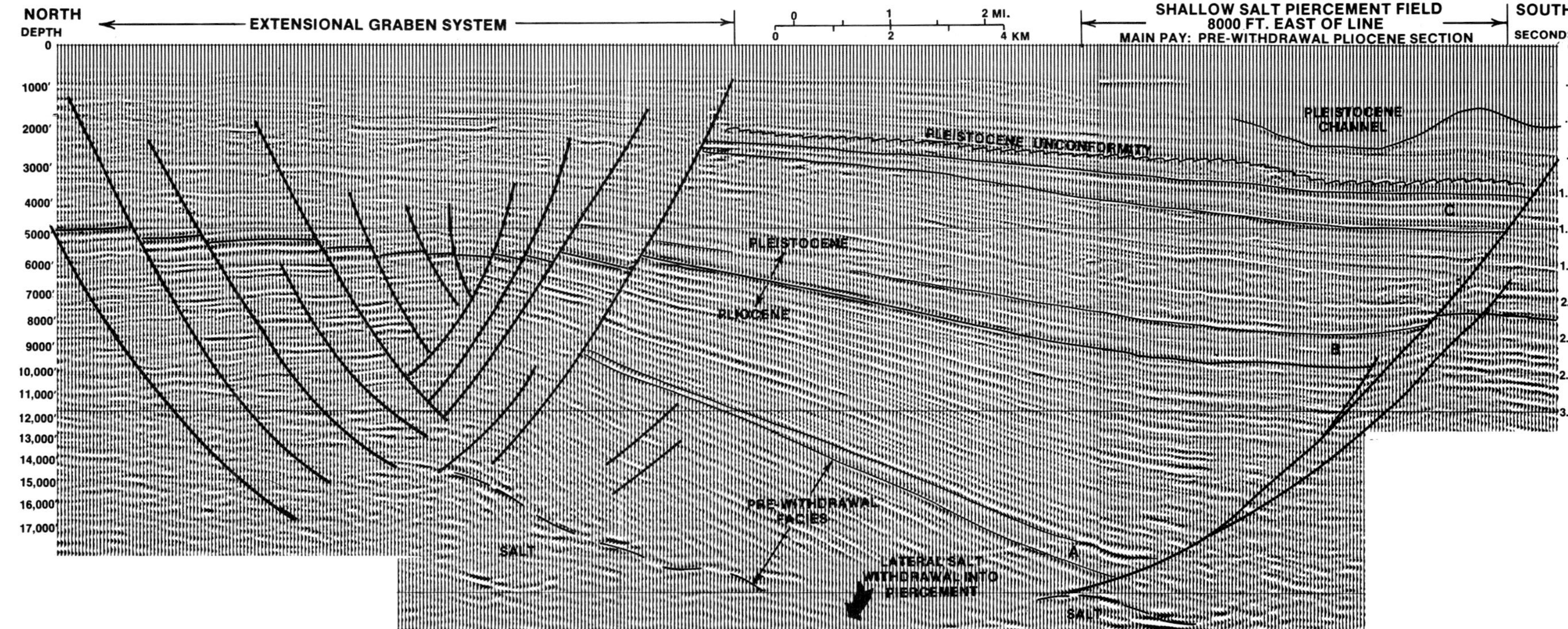

Fig. 7-17 (Larberg, 1983)—Seismic line from offshore Louisiana showing contra-regional or up-to-basin fault caused by lateral flowage of salt from ridge to adjacent diapir. Fault goes listric over remnant salt below 5,183 m (17,000 ft) and reflects periods of salt withdrawal by expanded section (intervals A, B, and C) on downthrown side. Permission to publish by American Association of Petroleum Geologists.

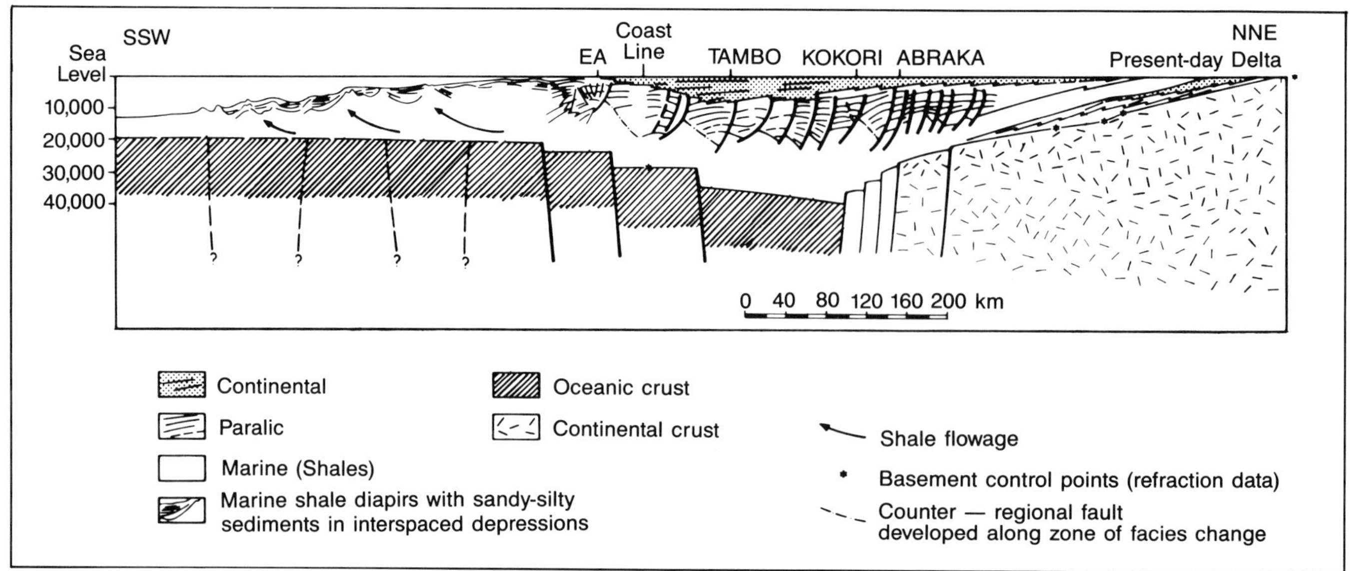

Fig. 7-18 (Evamy et al., 1978)—Tertiary Niger delta prograded over continental-oceanic crustal boundary onto oceanic crust. Note suggestion of compressional structures at downstream (SSW) end of section. Permission to publish by American Association of Petroleum Geologists.

Bruce, C. H., 1973, Pressured shale and related sediment deformation: Mechanism for development of regional contemporaneous faults: *American Association Petroleum Geologists Bull.*, V. 57, p. 878–886.

Chapman, R. E., 1973, Petroleum geology: A concise study: Elsevier, Amsterdam, London, and New York, 304 p.

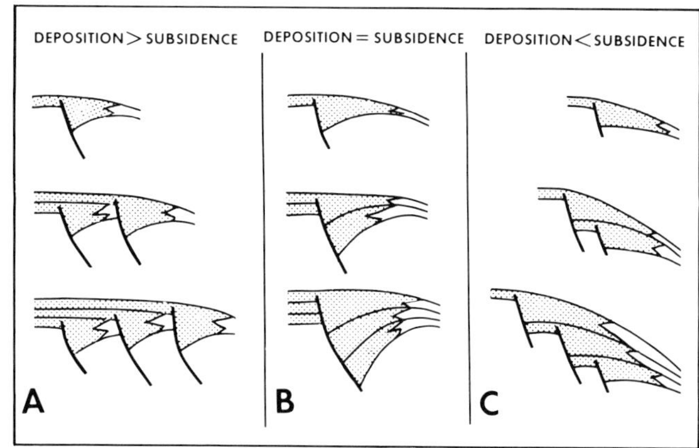

Fig. 7-19 (Bruce, 1973)—Diagrammatic illustration showing development of three types of contemporaneous fault systems as a result of interplay of deposition and subsidence rates. A is by far most frequent. B and C might represent modifications of a single depositional sequence of A. In A, faults are obviously younger in direction of progradation, however locally within a depocenter younger faults can develop landward as in C. Permission to publish by American Association of Petroleum Geologists.

Christensen, A. F., 1983, An example of a major syndepositional listric fault, *in* Bally, A. W., ed., Seismic expression of structural styles: *American Association Petroleum Geologists Studies in Geology #15,* V. 2.

Cloos, E., 1968, Experimental analysis of Gulf Coast fracture patterns: *American Association Petroleum Geologists Bull.,* V. 52, p. 420–444.

Conatser, W., 1983, Lake Hatch field Terrebonne Parish, Louisiana, *in* McCormick, L. L., ed., Oil and gas fields of southeast Louisiana: New Orleans Geological Society, p. 24.

Erxleben, A. W., and Carnahan, G., 1983, Slick Ranch area, Starr County, Texas, *in* Bally, A. W., ed., Seismic expression of structural styles: *American Association Petroleum Geologists Studies in Geology #15,* V. 2.

Evamy, D. D., Haremboare, J., Kamerling, P., Knaap, W. A., Molloy, F. A., and Rowlands, P. H., 1978, Hydrocarbon habitat of Tertiary Niger Delta: *American Association Petroleum Geologists Bull.,* V. 62, p. 1–39.

Halbouty, M. T., and Barber, T. D., 1961, Port Acres and Port Arthur fields, Jefferson County, Texas: *Gulf Coast Association Geological Societies Trans.,* V. 11, p. 225.

Inderwiesen, P. L., 1983, Contemporaneous fault system, Texas Gulf Coast, *in* Bally, A. W., ed., Seismic expression of structural styles: *American Association Petroleum Geologists Studies in Geology #15,* V. 2.

Johnson, D. C., Fifer, R. G., and McQuillan, K. A., 1983, Expansion fault—Gulf of Mexico, *in* Bally, A. W., ed., Seismic expression of structural styles: *American Association Petroleum Geologists Studies in Geology #15,* V. 2.

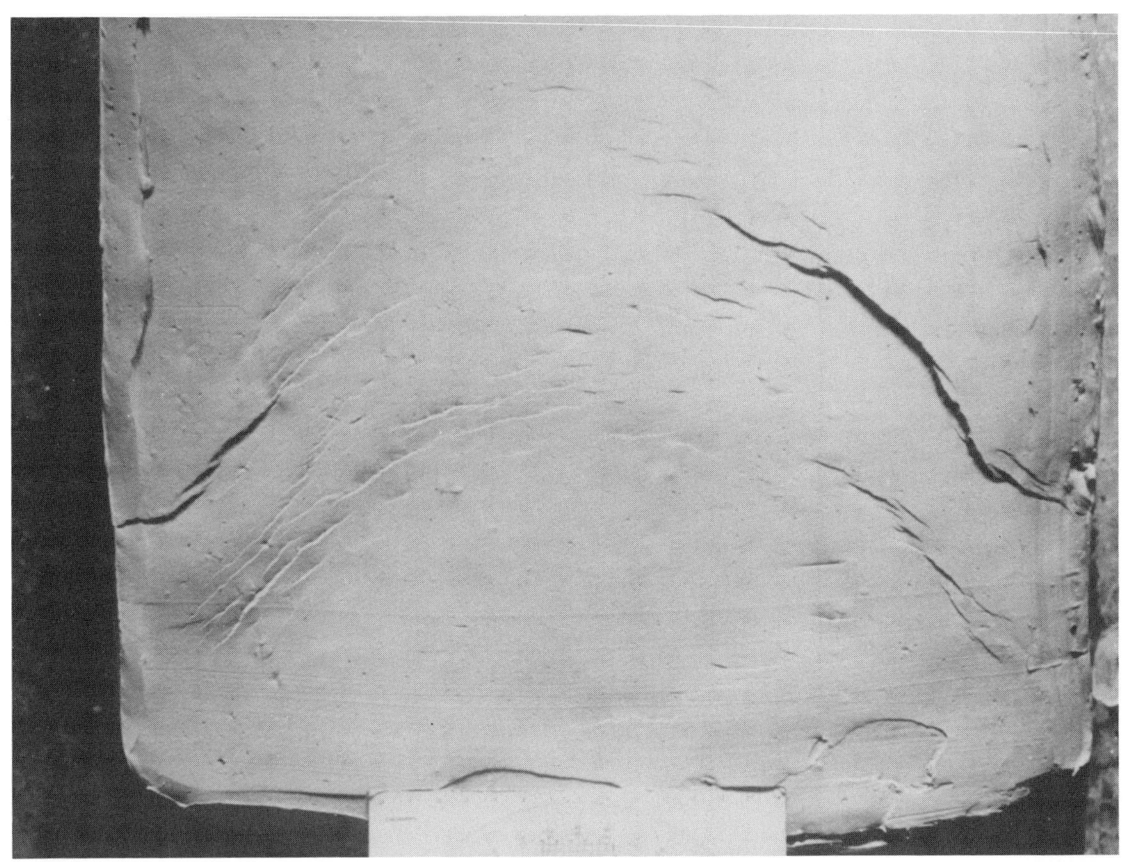

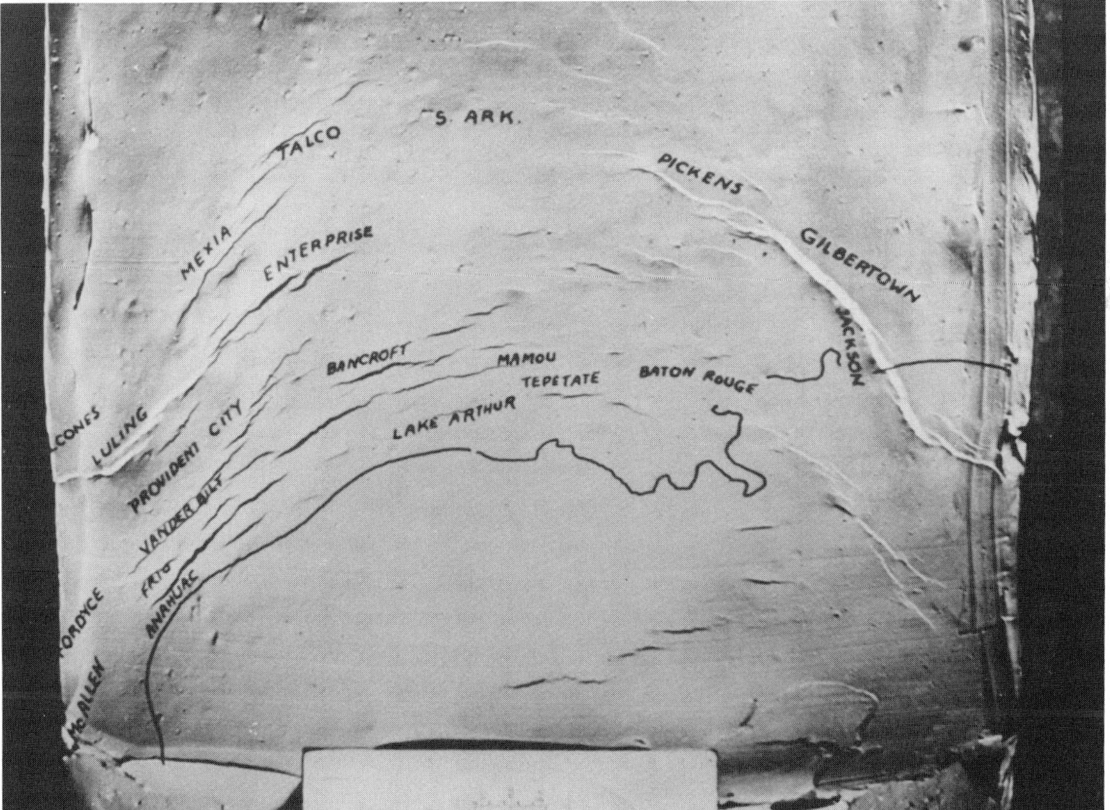

Fig. 7-20 (Cloos, 1968)—A. Fault pattern on surface of clay model illuminated at low angle from bottom of page. Shadows are on north-dipping faults; light fault scarps dip south. Metal sheet at base of clay was pulled to the south; sheet occupied only the "south-central" part of the clay. B. Map of Gulf Coast fault pattern superimposed on model. Permission to publish by American Association of Petroleum Geologists.

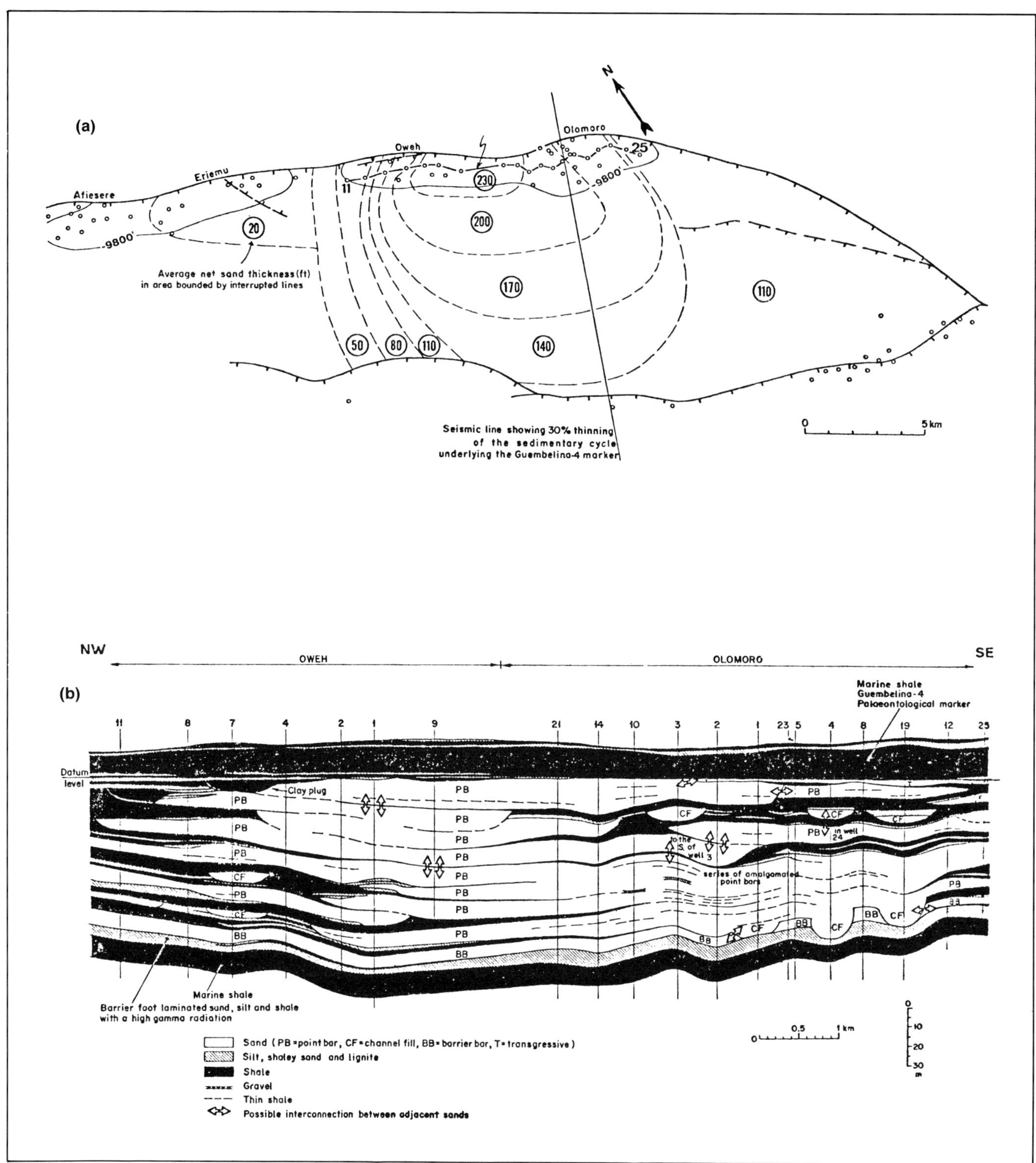

Fig. 7-21 (Weber, 1971)—(a). Fault map of the Nigeria Olomoro-Oweh fields area at the level of the Guembelina-4 marker, together with the measured and estimated net sand thickness in the underlying sedimentary cycle. (b). Cross section from wells 11 through 25. Examples of a thick lens-shaped sedimentary cycle mainly composed of point bar sands. Note maximum sand between wells 9 and 21. Permission to publish by Geologie en Mijnbouw.

Illustration 7-22 begins on next page ⟶

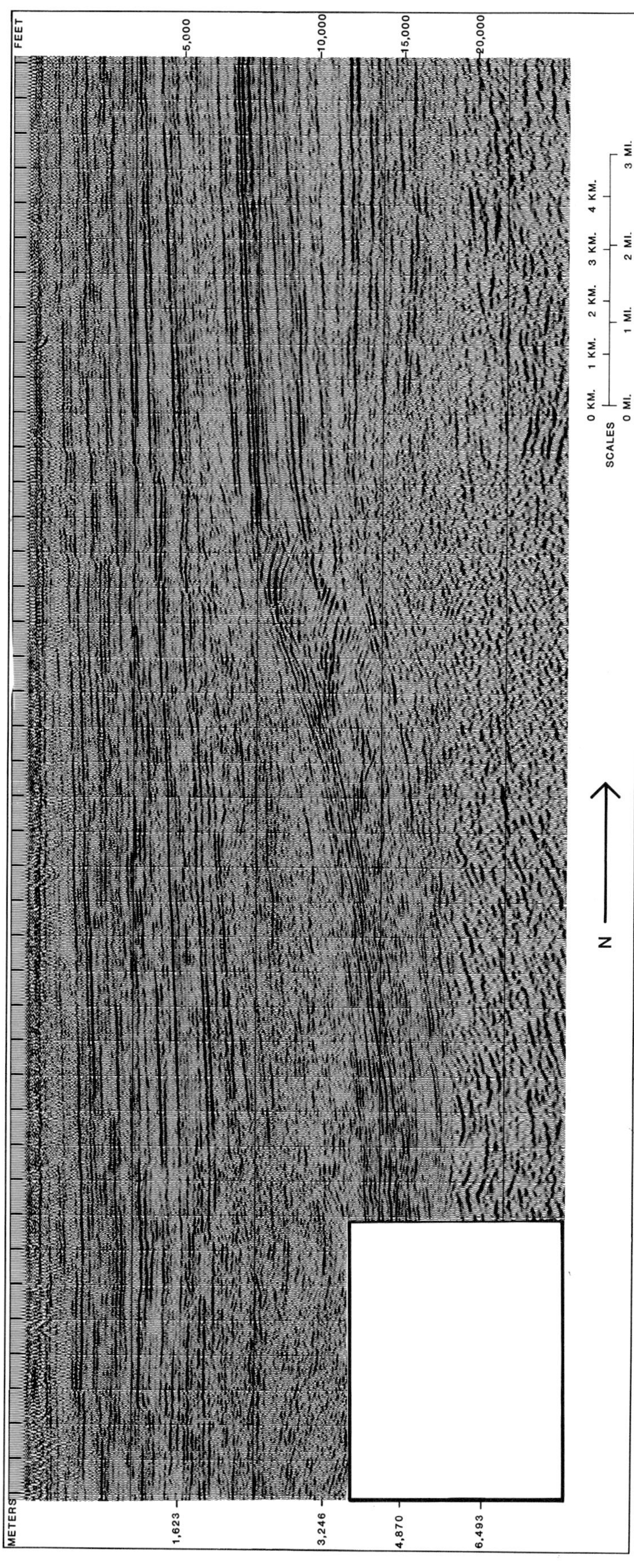

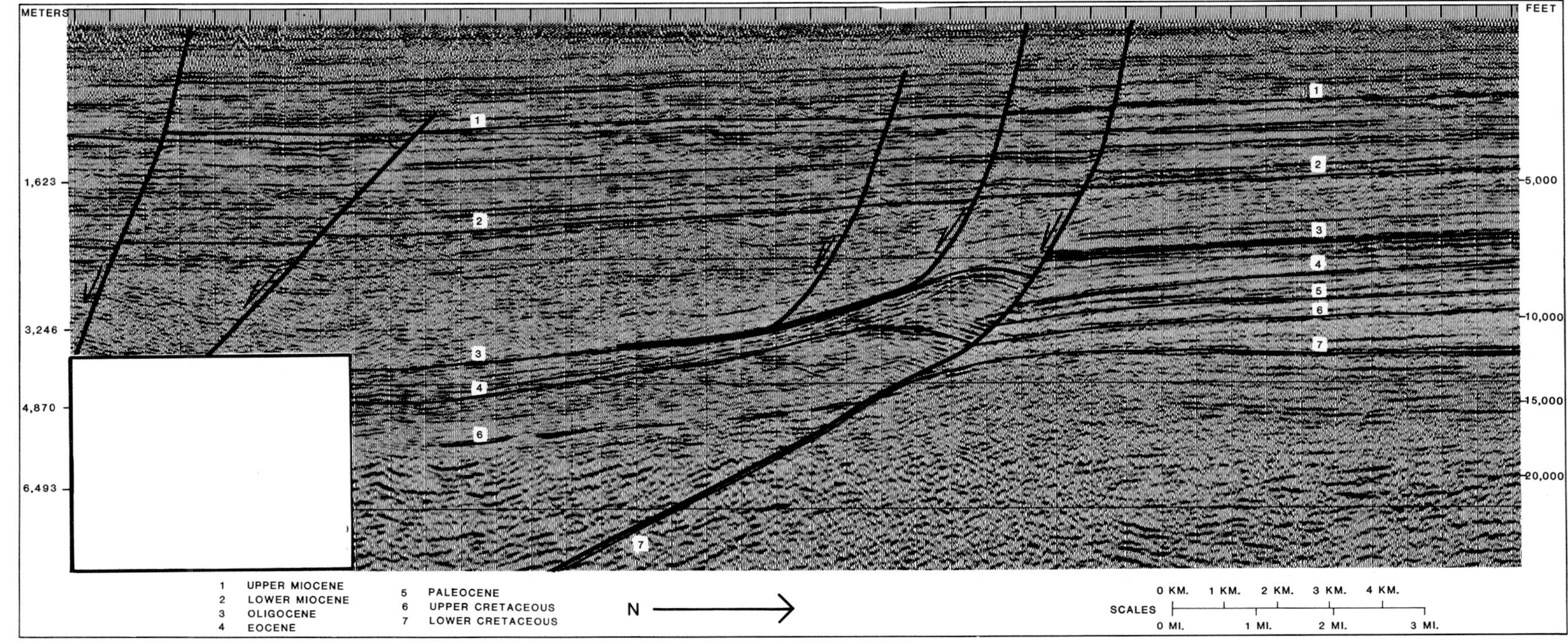

Fig. 7-22 (Reed, 1983)—Seismic line about 135 km (84 mi) SE of New Orleans, offshore Louisiana showing Lower Cretaceous shelf edge which localizes listric normal faults, a rollover anticline, and expanded section above. Permission to publish by American Association of Petroleum Geologists.

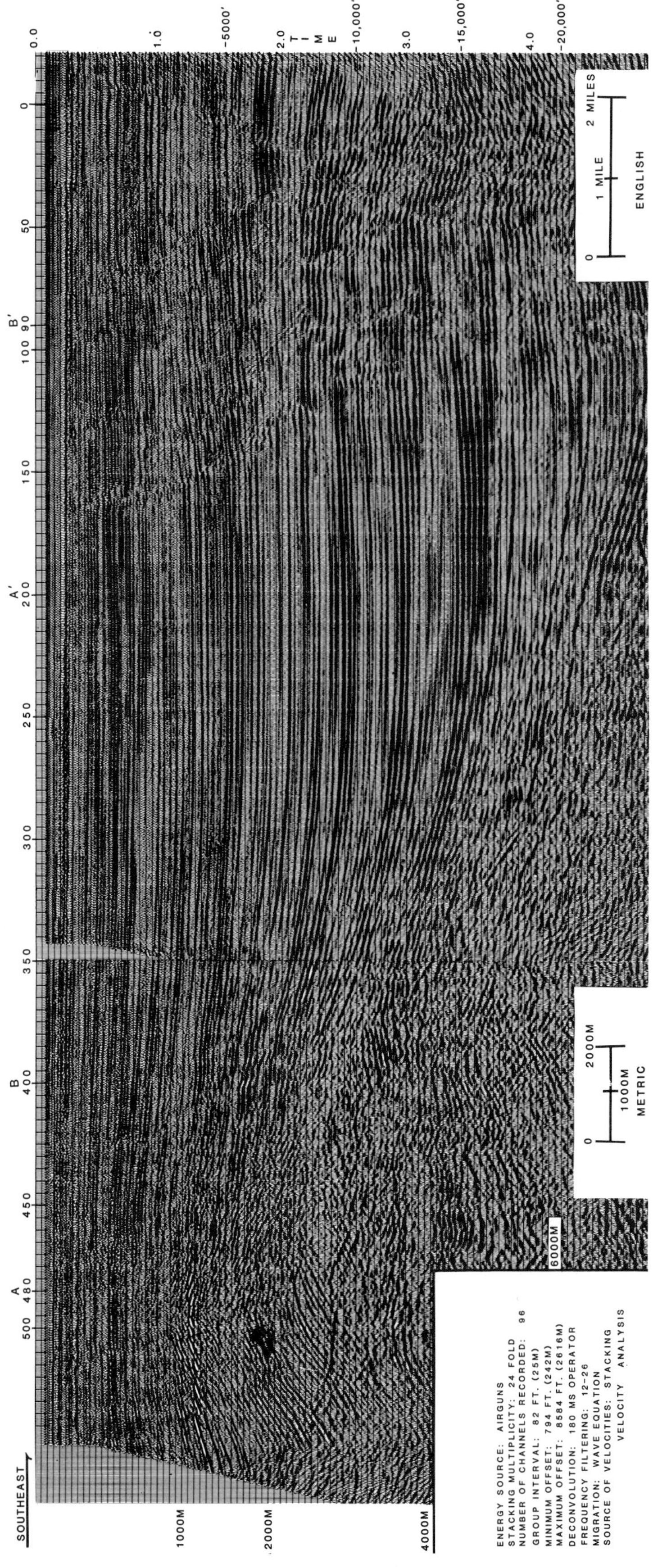

SOUTHEAST
0.0
1.0
-5000'
2.0
TIME
-10,000'
3.0
-15,000'
4.0
-20,000'
0
50
B'
90
100
150
A'
200
250
300
350
B
400
450
A
480
500
1000M
2000M
4000M
6000M
0
1 MILE
2 MILES
ENGLISH
0
1000M
2000M
METRIC
ENERGY SOURCE: AIRGUNS
STACKING MULTIPLICITY: 24 FOLD
NUMBER OF CHANNELS RECORDED: 96
GROUP INTERVAL: 82 FT. (25M)
MINIMUM OFFSET: 794 FT. (242M)
MAXIMUM OFFSET: 8584 FT. (2616M)
DECONVOLUTION: 180 MS OPERATOR
FREQUENCY FILTERING: 12-26
MIGRATION: WAVE EQUATION
SOURCE OF VELOCITIES: STACKING VELOCITY ANALYSIS

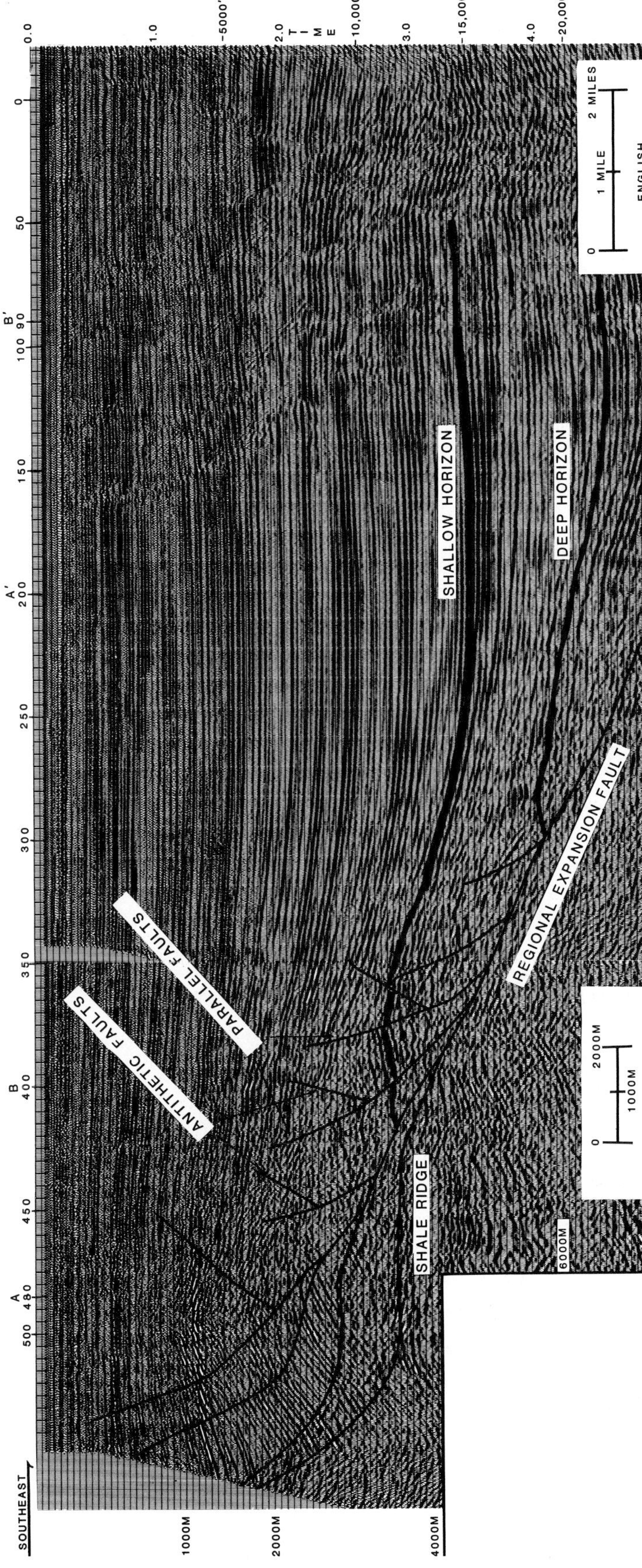

Fig. 7-23 (Johnson et al., 1983)—Seismic line from offshore Texas about midway between Corpus Christi and Galveston showing faulted rollover structure on glide plane fault that soles on seaward side of shale ridge. Authors describe dip changes in fault plane to "gravitational imbalance in conjunction with the shale ridges plastic deformation." Permission to publish by American Association of Petroleum Geologists.

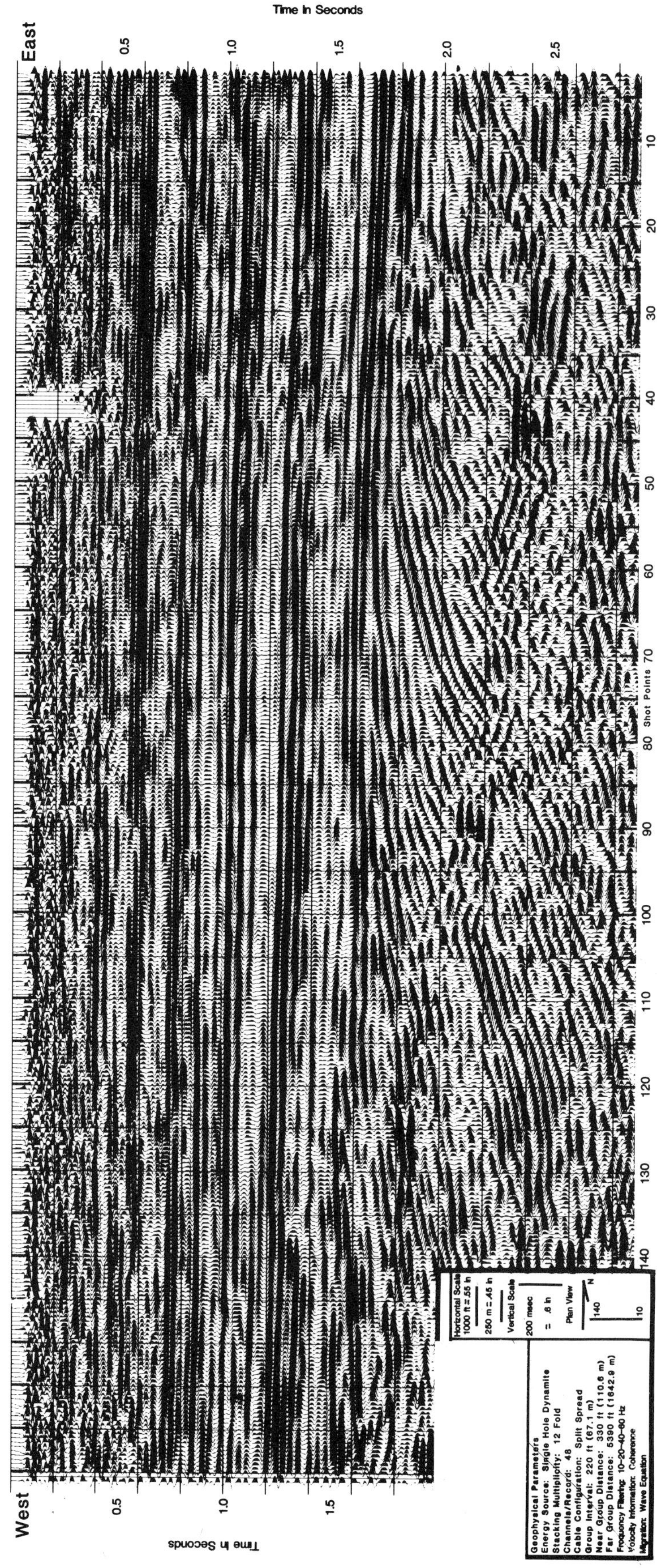
Time In Seconds
East
West
0.5
1.0
1.5
2.0
2.5
Shot Points
Horizontal Scale
1000 ft = .55 in
250 m = .45 in
Vertical Scale
200 msec
= .6 in
Plan View
140
N
10
Geophysical Parameters
Energy Source: Single Hole Dynamite
Stacking Multiplicity: 12 Fold
Channels/Record: 48
Cable Configuration: Split Spread
Group Interval: 220 ft (67.1 m)
Near Group Distance: 330 ft (110.6 m)
Far Group Distance: 5390 ft (1842.9 m)
Frequency Filtering: 10-20-40-60 Hz
Velocity Information: Coherence
Migration: Wave Equation

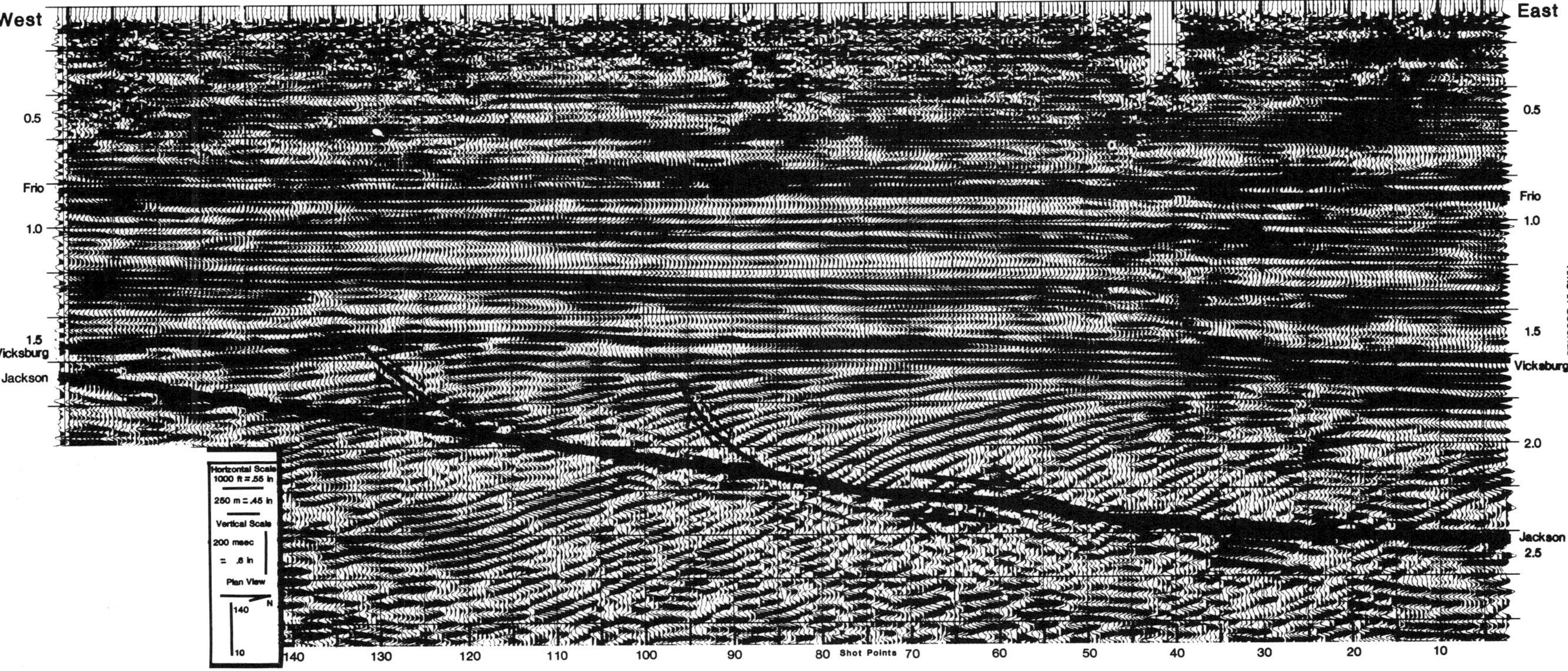

Fig. 7-24 (Inderwiesen, 1983)—Seismic line commencing 1.6 km (1 mi) east of Sam Fordyce-Vanderbilt fault zone, Brooks County, south Texas, showing downbending in Vicksburg (Lower Oligocene) sediments on glide plane on Jackson shale. Fault may be the local slope transport type of Yorston and Weisser (1976). Permission to publish by American Association of Petroleum Geologists.

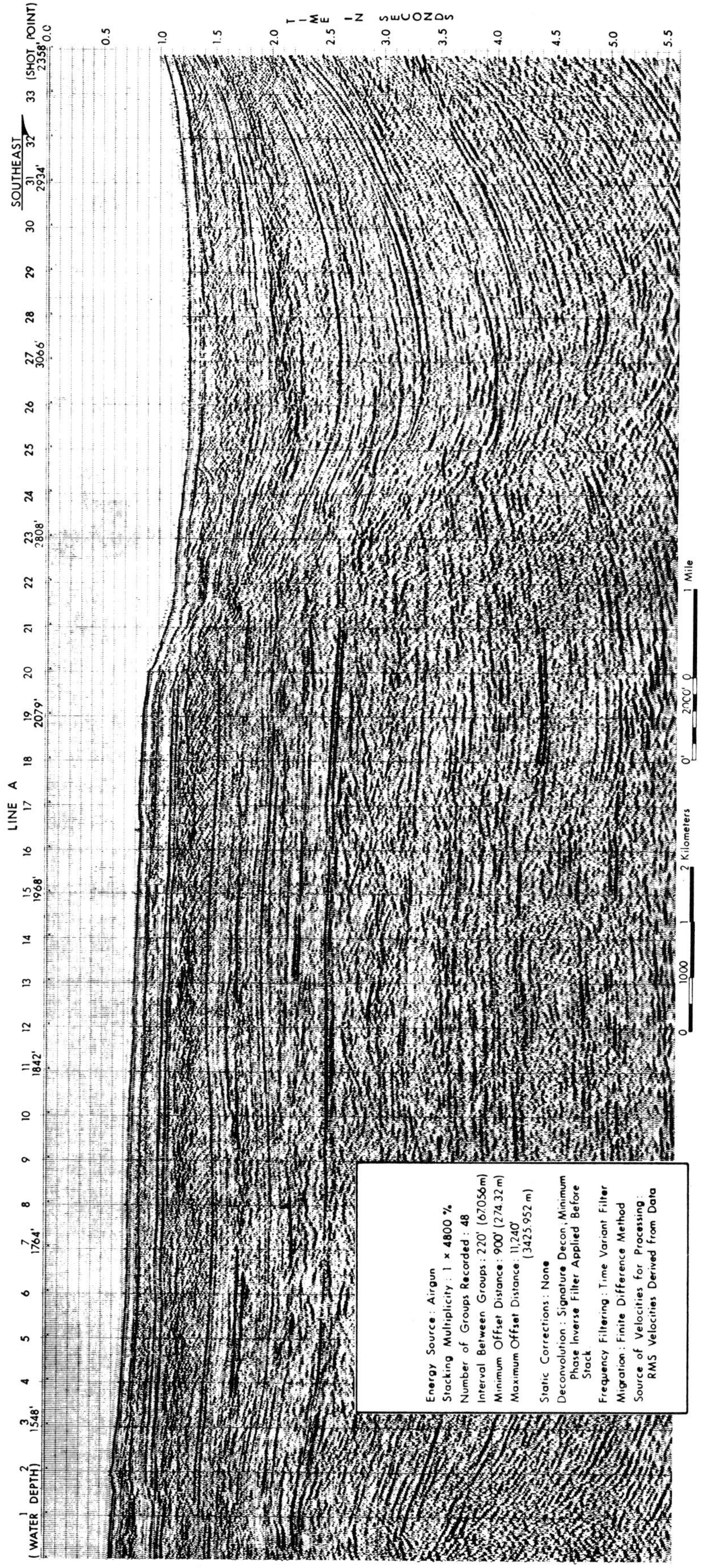
LINE A
SOUTHEAST
(SHOT POINT)
(WATER DEPTH)
1 2 3 4 5 6 7 8 9 10 11 12 13 14 15 16 17 18 19 20 21 22 23 24 25 26 27 28 29 30 31 32 33
2358' 2934' 3066' 2808' 2079' 1968' 1842' 1764' 1548'
TIME IN SECONDS
0.0 0.5 1.0 1.5 2.0 2.5 3.0 3.5 4.0 4.5 5.0 5.5
0 1000 1 2 Kilometers
0' 2000' 0 1 Mile
Energy Source: Airgun
Stacking Multiplicity: 1 × 4800 %
Number of Groups Recorded: 48
Interval Between Groups: 220' (67.056m)
Minimum Offset Distance: 900' (274.32 m)
Maximum Offset Distance: 11,240' (3425.952 m)
Static Corrections: None
Deconvolution: Signature Decon; Minimum Phase Inverse Filter Applied Before Stack
Frequency Filtering: Time Variant Filter
Migration: Finite Difference Method
Source of Velocities for Processing: RMS Velocities Derived From Data

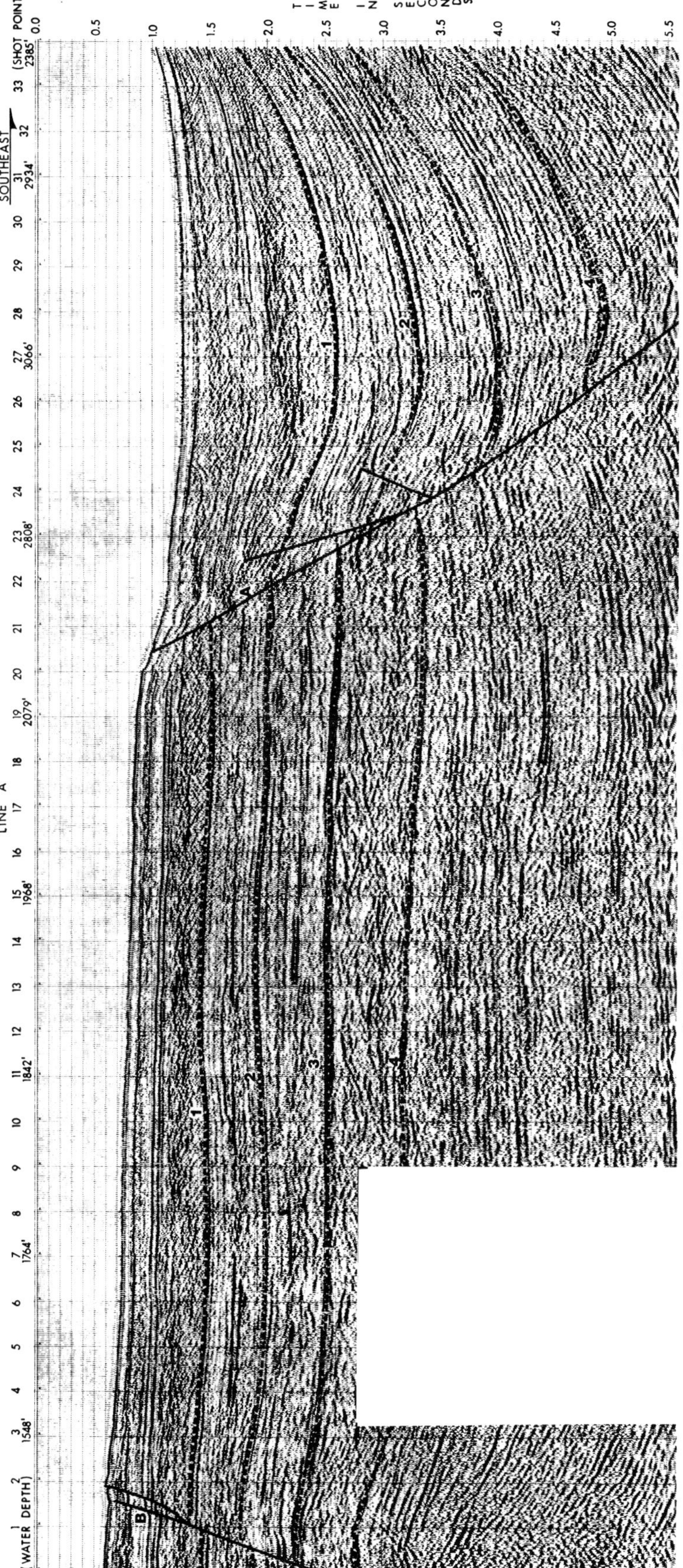

Fig. 7-25 (Wanslow, 1983)—Seismic line from upper continental slope, offshore Louisiana showing a growth fault in the upper slope environment. Throw on horizon 1 is less than 578 m (1700 ft) increasing to almost 1829 m (6000 ft) on horizon 4. Note that fault affects sea bottom. Salt uplift is present to SE (off the section) and strongly influences the dip of beds on downthrown side of fault. Permission to publish by American Association of Petroleum Geologists.

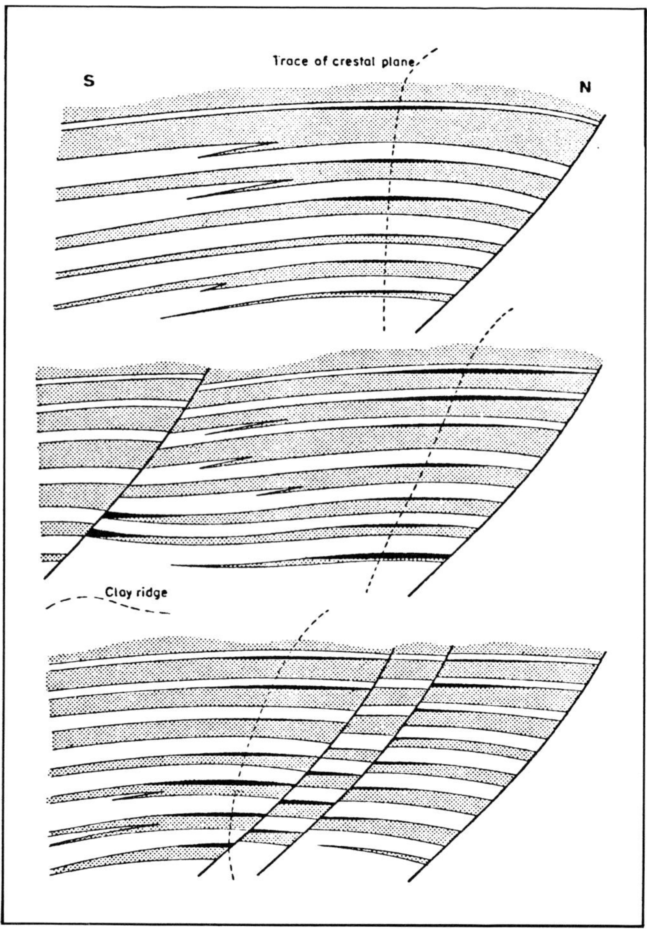

Fig. 7-26 (Weber, 1971)—Principal types of hydrocarbon traps (in black) associated with growth faults in the Niger delta. Note that virtually all are related to downbending. Permission to publish by Geologie en Mijnbouw.

Larberg, G. M. B., 1983, Contra-regional faulting: Salt withdrawnal compensation; offshore Louisiana, Gulf of Mexico, *in* Bally, A. W., ed., Seismic expression of structural styles: *American Association Petroleum Geologists Studies in Geology #15,* V. 2.

Melancon, J. A., 1970, The Bayou Henry field, Iberville Parish, Louisiana, *in* Typical oil and gas fields of southwest Louisiana: Lafayette Geological Society, p. 34.

Ocamb, R. D., 1961, Growth faults of south Louisiana: *Gulf Coast Association Geological Societies Trans.,* V. 11, p. 139–175.

Pierce, W. G., 1941, Heart Mountain and South Fork thrusts, Park County, Wyoming: *American Association Petroleum Geologists Bull.,* V. 25, p. 2021–2045.

Reed, J. M., 1983, Lower Cretaceous shelf edge, offshore Louisiana, Main Pass area, Line "B", *in* Bally, A. W., ed., Seismic expression of structural styles: *American Association Petroleum Geologists Studies in Geology #15,* V. 2.

Roux, W. F., Jr., 1977, The development of growth fault structures: American Association Petroleum Geologists Structural Geology School lecture notes.

Shaub, F. J., 1983, Growth faults on the southwestern margin of the Gulf of Mexico, *in* Bally, A. W., ed., Seismic expression of structural styles: *American Association Petroleum Geologists Studies in Geology #15,* V. 2.

Wanslow, J. B., 1983, Growth fault—upper continental slopes, *in* Bally, A. W., ed., Seismic expression of structural styles: *American Association Petroleum Geologists Studies in Geology #15,* V. 2.

Weber, K. J., 1971, Sedimentalogical aspects of oil fields in the Niger Delta: *Geologie en Mijnbouw,* V. 50, p. 559–576.

Yorston, H. J. and Weisser, G. H., 1976, The seismic analysis of glide plane fault structures; style, mechanisms, and relationship to sedimentation (abs.): Society Exploration Geophysicists, 46th Annual Internat. Mtg. Abs., p. 85.

Illustration 7-27 begins on next page ⟶

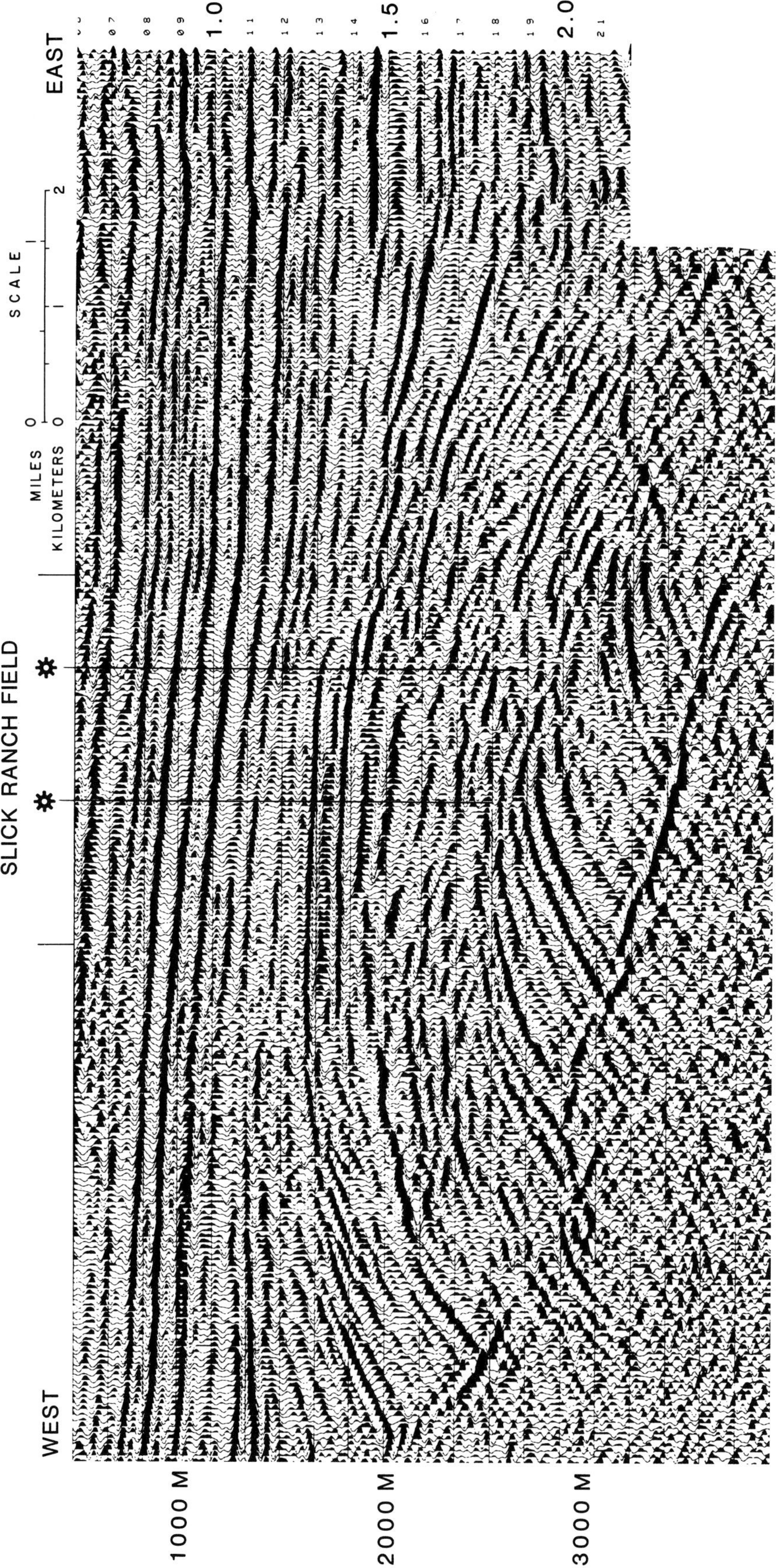
WEST
EAST
SLICK RANCH FIELD
SCALE
MILES
KILOMETERS
0
1
2
1.0
1.5
2.0
1000 M
2000 M
3000 M

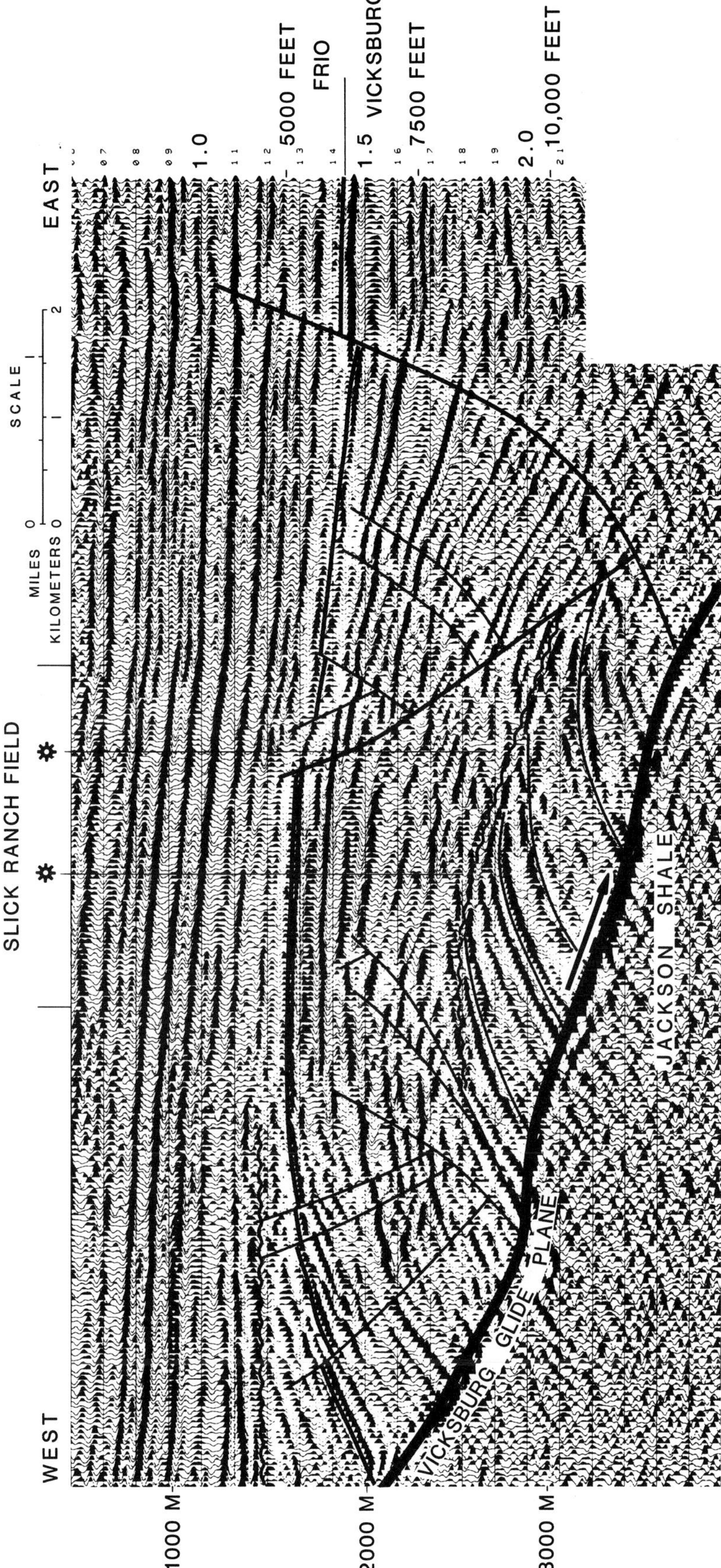

Fig. 7-27 (Erxleben and Carnahan, 1983)—Seismic line from Slick Ranch area, Starr County, south Texas, showing productive rollover anticline in lower Oligocene sediments on a glide plane that soles on older (Eocene-Jackson) shale. Note: unconformity in center of section between 1.8–2.1 sec, antithetic faults, and east dip in upper beds. Production is best beneath the aforementioned unconformity, where the thick wedges of sandstone and shales thin out in the crestal part of the rollover structure. Permission to publish by American Association of Petroleum Geologists.

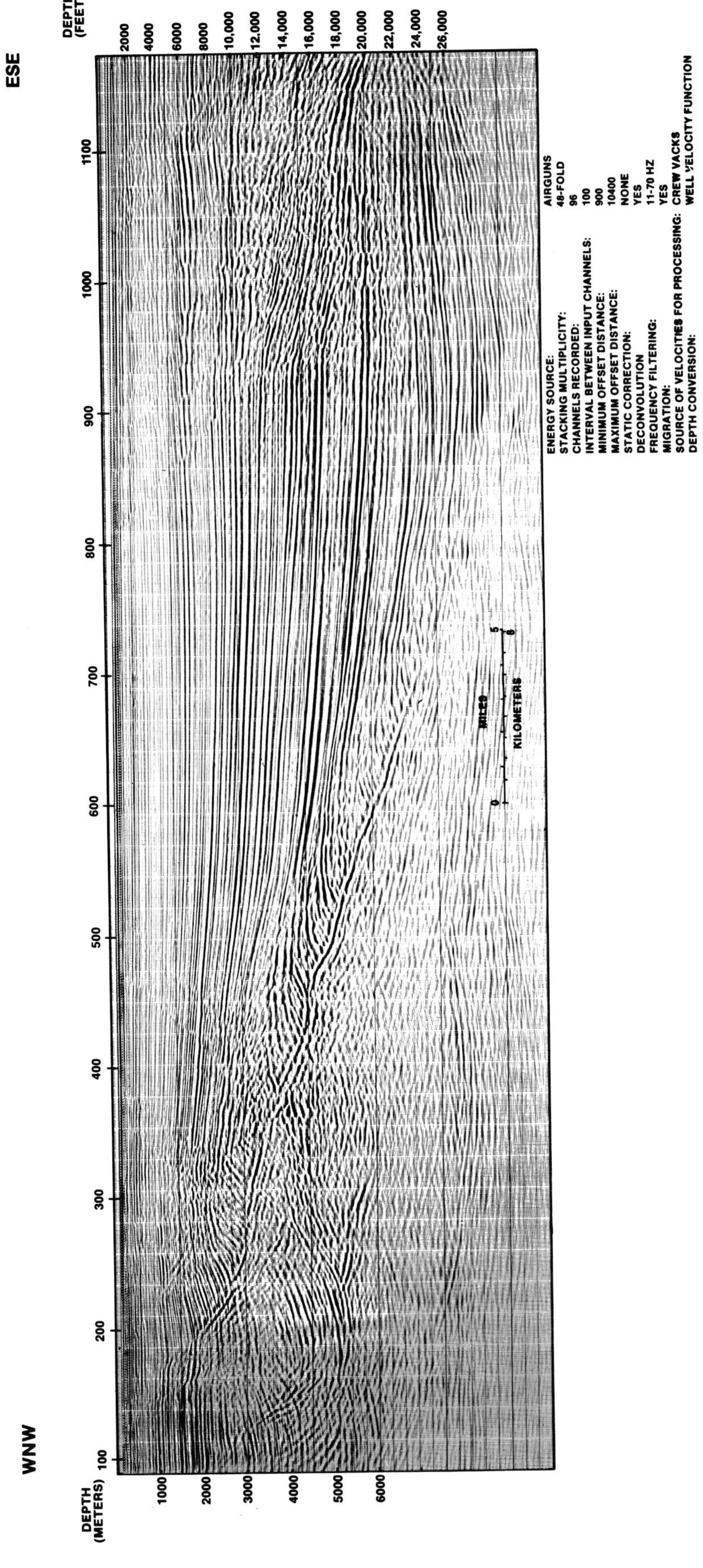
ESE
WNW
DEPTH (FEET)
2000
4000
6000
8000
10,000
12,000
14,000
16,000
18,000
20,000
22,000
24,000
26,000
DEPTH (METERS)
1000
2000
3000
4000
5000
6000
100
200
300
400
500
600
700
800
900
1000
1100
MILES
0
5
KILOMETERS
0
ENERGY SOURCE: AIRGUNS
STACKING MULTIPLICITY: 48-FOLD
CHANNELS RECORDED: 96
INTERVAL BETWEEN INPUT CHANNELS: 100
MINIMUM OFFSET DISTANCE: 900
MAXIMUM OFFSET DISTANCE: 10400
STATIC CORRECTION: NONE
DECONVOLUTION YES
FREQUENCY FILTERING: 11-70 HZ
MIGRATION: YES
SOURCE OF VELOCITIES FOR PROCESSING: CREW VACKS
DEPTH CONVERSION: WELL VELOCITY FUNCTION

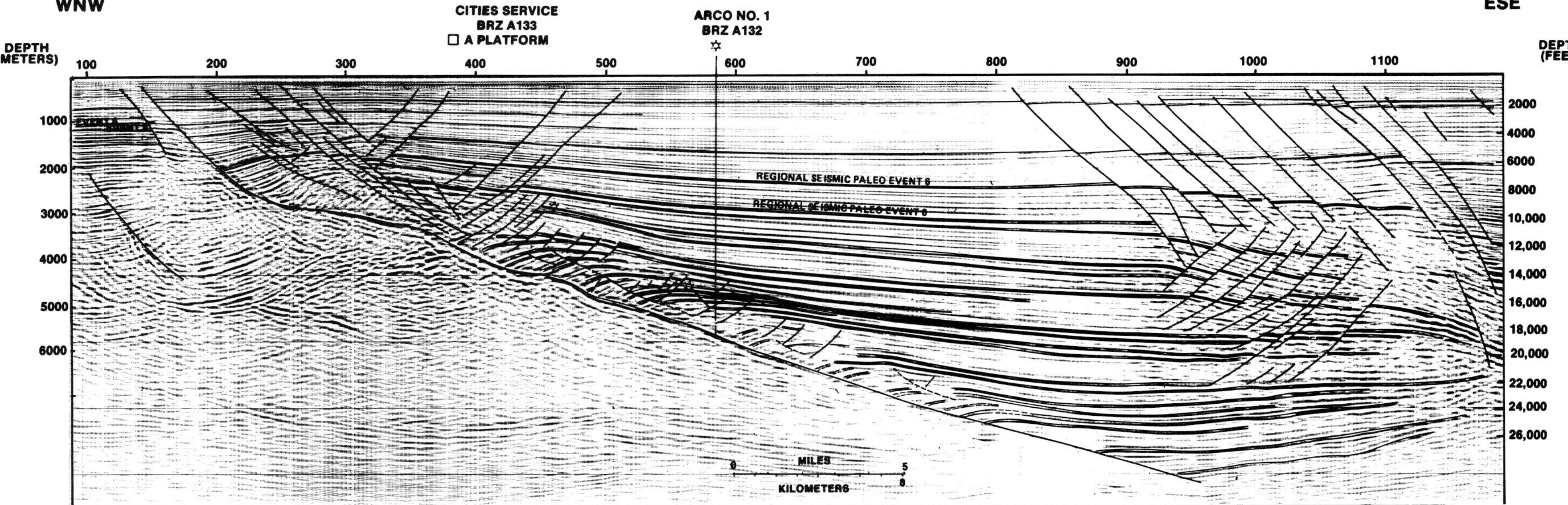

Fig. 7-28 (Christensen, 1983)—Seismic line 30–70 km (20–42 mi) from Matagorda Island, offshore Texas, showing thickening of upper Miocene section across Brazos Ridge fault system. Vertical displacement is more than 5200 m (17,000 ft) and horizontal displacement more than 16.5 km (10 mi) along the fault system. Superb fault plane reflection is created by moderately pressured (0.65) clastics juxtaposed against high-pressured (0.8) shales below.

"Hydrocarbons are found primarily in two types of traps: shallow gas accumulations occur in sandstone reservoirs within the keystone faulted zone and often on the upthrown sides of major antithetic faults; while deeper gas zones occur within regressive wedges of interbedded sandstone and shale that have anticlinal closure associated with rotation into the fault plane.

Downdip of this growth fault system is another interesting trend composed of chevron faulting. The opposing fault directions resulted from an early salt roll with westerly dipping faults on its flank. Then, with withdrawal of salt and collaspse of the younger measures, basinward-dipping faults developed in the upper interval. This withdrawal of salt also provided the space filled by the downthrown sediments of the Brazos Ridge fault and the resulting huge heave."

Permission to publish by American Association of Petroleum Geologists.

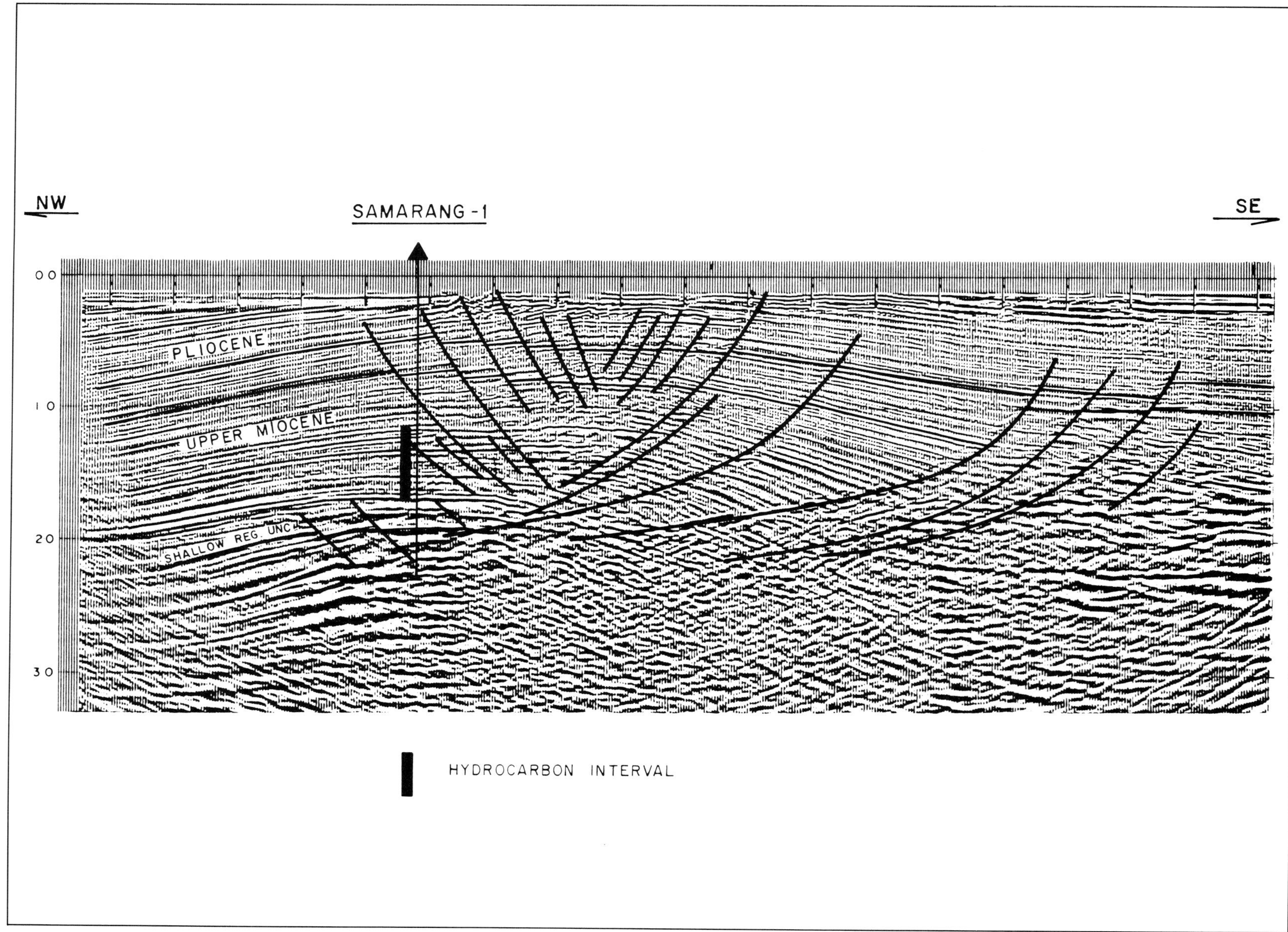

Fig. 7-29 (Bol and Van Hoorn, 1978)—Seismic section across Samarang structure, western offshore Sabah showing trap in downbend anticline. Section is migrated. Scale 1:50,000. (See Fig. 2-28 for line of section). Permission to publish by Geological Society of Malaysia.

Illustration 7-30 begins on next page →

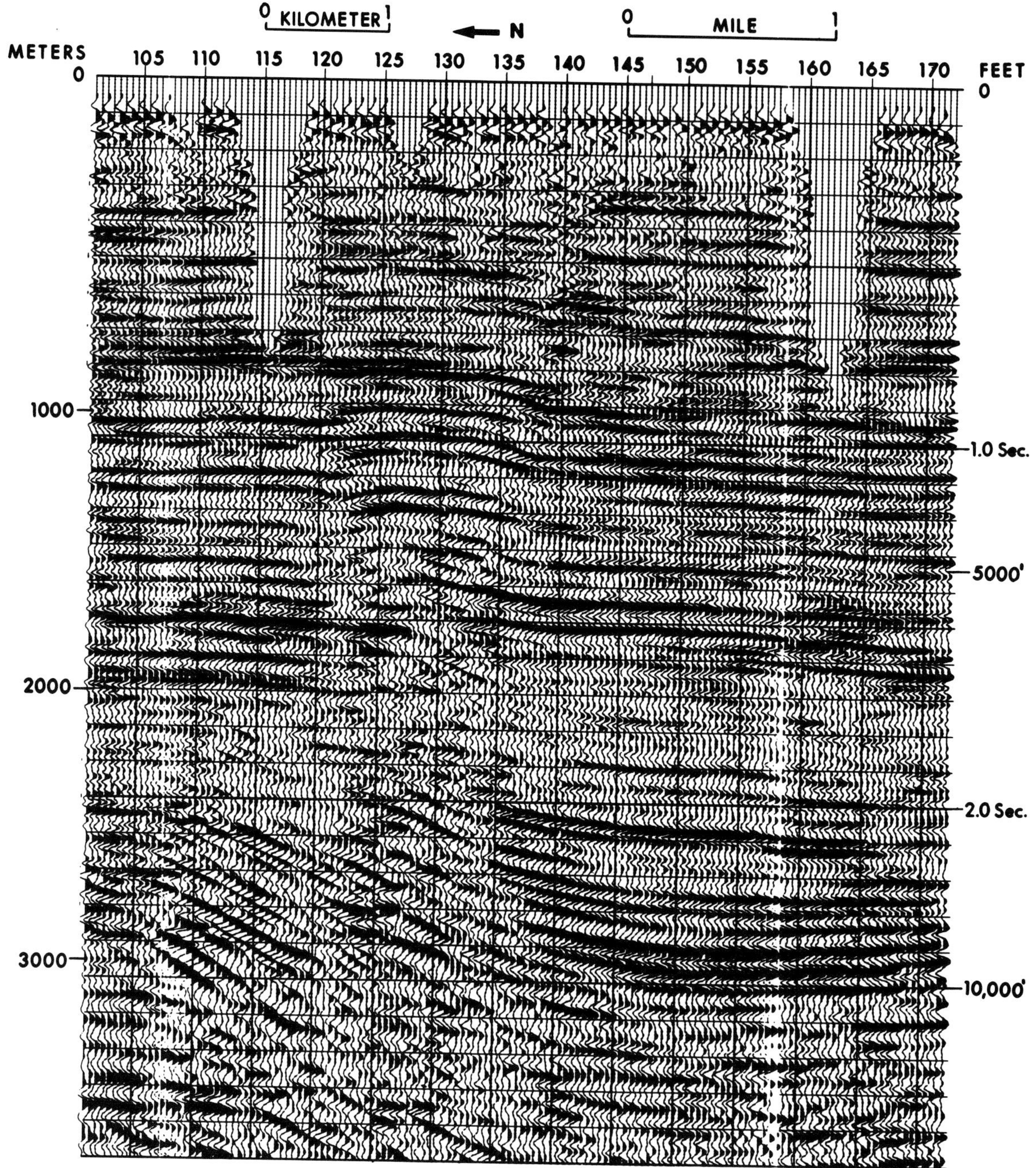
0 KILOMETER 1
N
0 MILE 1
METERS
0
105 110 115 120 125 130 135 140 145 150 155 160 165 170
FEET
0
1000
2000
3000
1.0 Sec.
5000'
2.0 Sec.
10,000'

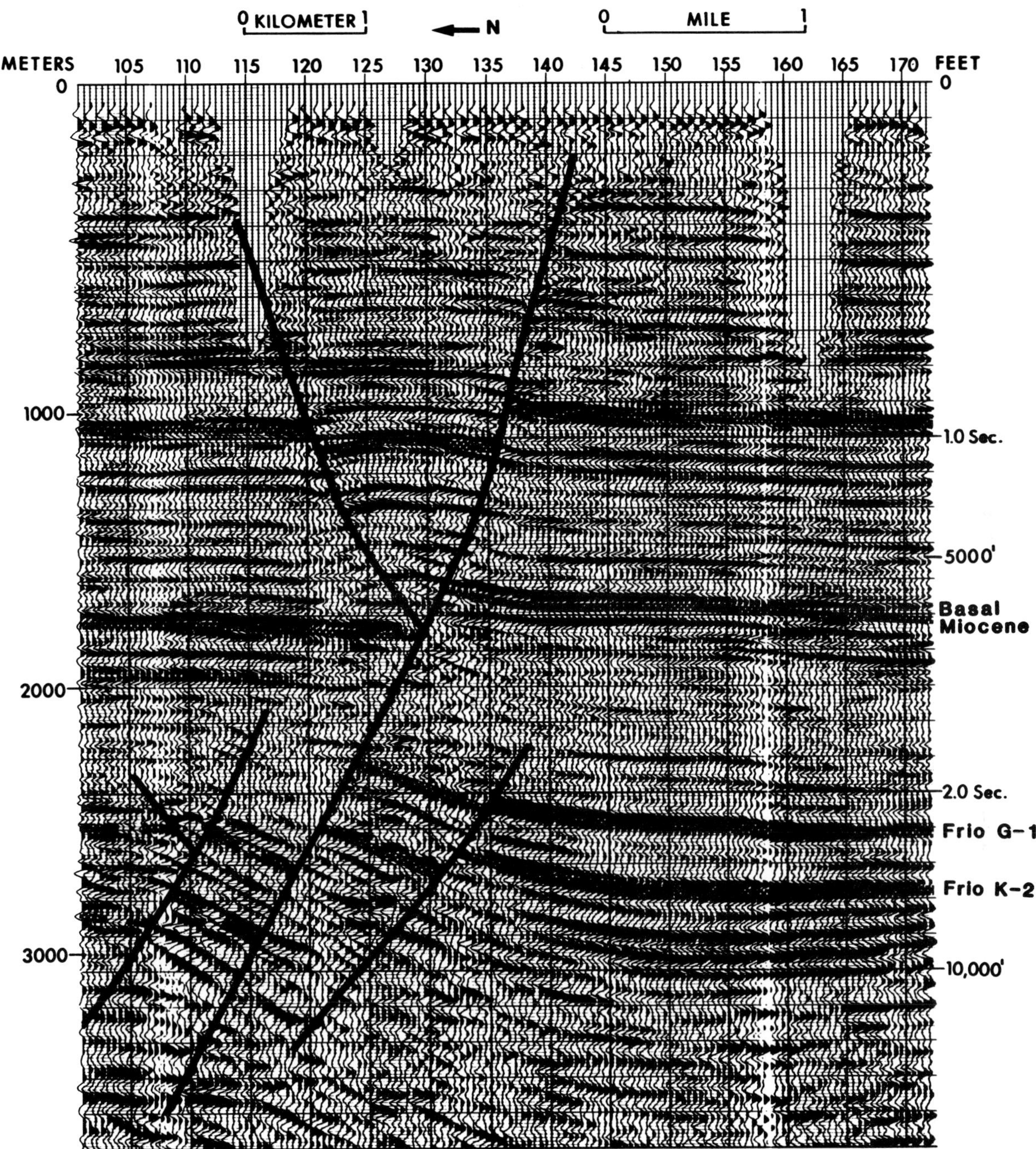

Fig. 7-30 (Bergsma, 1983)—Seismic line from Calhoun County, south Texas, showing counter-regional, up-to-the-basin (down-to-the-NW) fault that has production from the Frio G-1 sand in an upthrown fault closure. A throw of 366 m (1200 ft) juxtaposes productive sandstone with a younger (presumably sealing) shale on the downthrown side. The lowest faults could intersect a listric down-to-basin fault. Permission to publish by American Association of Petroleum Geologists.

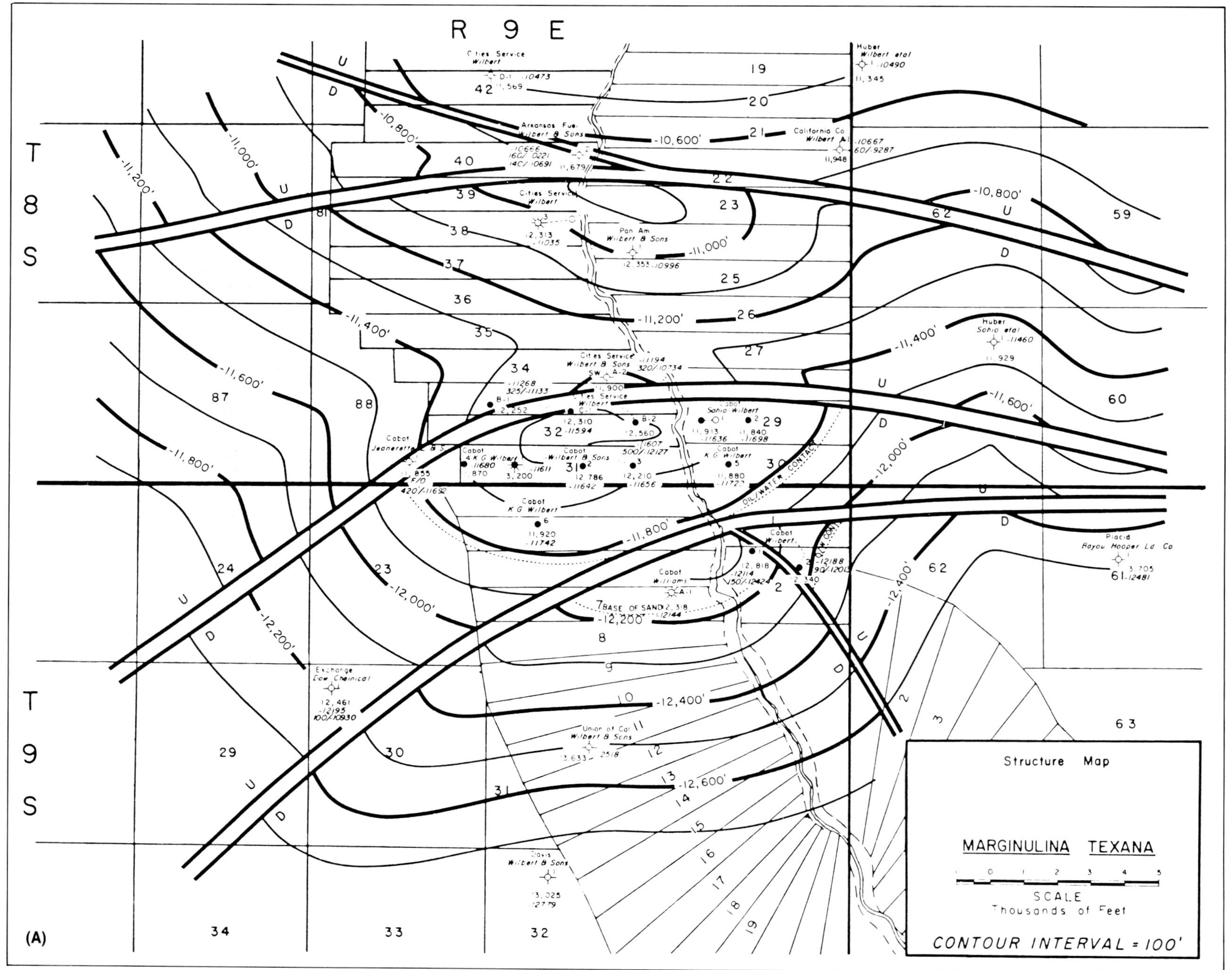
R 9 E
T 8 S
T 9 S
Structure Map
MARGINULINA TEXANA
SCALE
Thousands of Feet
CONTOUR INTERVAL = 100'
Cities Service Wilbert
Arkansas Fuel Wilbert B Sons
Huber Wilbert etal
California Co Wilbert A1
Pan Am Wilbert B Sons
Huber Sohio etal
Cities Service Wilbert B Sons
Sohio Wilbert
Cabot K G Wilbert
Cabot Jeanerette
Cabot Williams
BASE OF SAND
OIL/WATER CONTACT
Placid Bayou Hooper Ld Co
Exchange Cow Chemical
Union of Cal Wilbert B Sons
Davis Wilbert B Sons
-10,600'
-10,800'
-11,000'
-11,200'
-11,400'
-11,600'
-11,800'
-12,000'
-12,200'
-12,400'
-12,600'
(A)

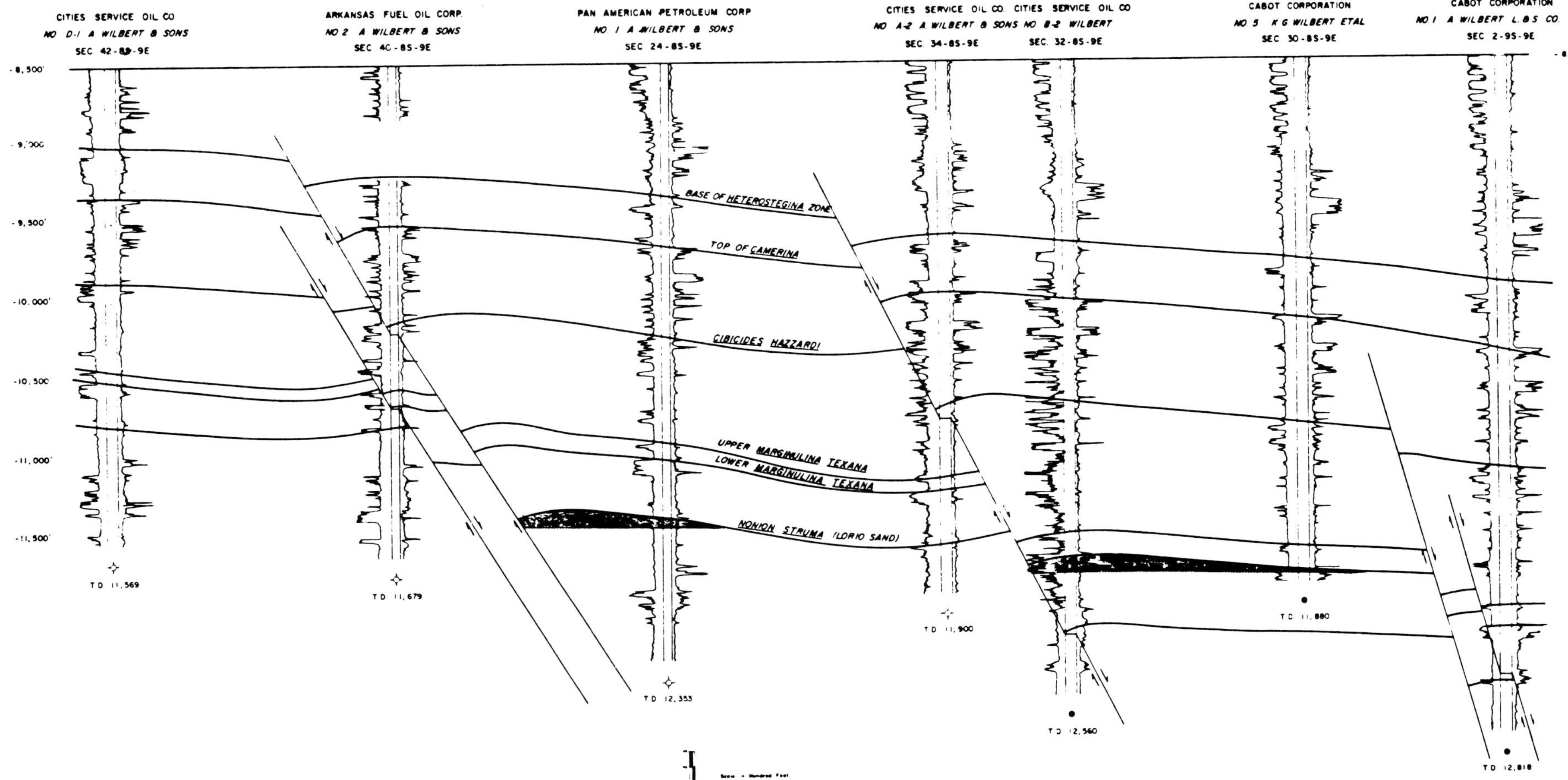

(B)

Fig. 7-31 (Melancon, 1970)—A. Map of Bayou Henry field, Iberville Parish, Louisiana. Note curving faults with closure of rollover anticline occupying center of curvature. B. North-south cross section. Faults no doubt are listric at depth. Permission to publish by the Lafayette Geological Society.

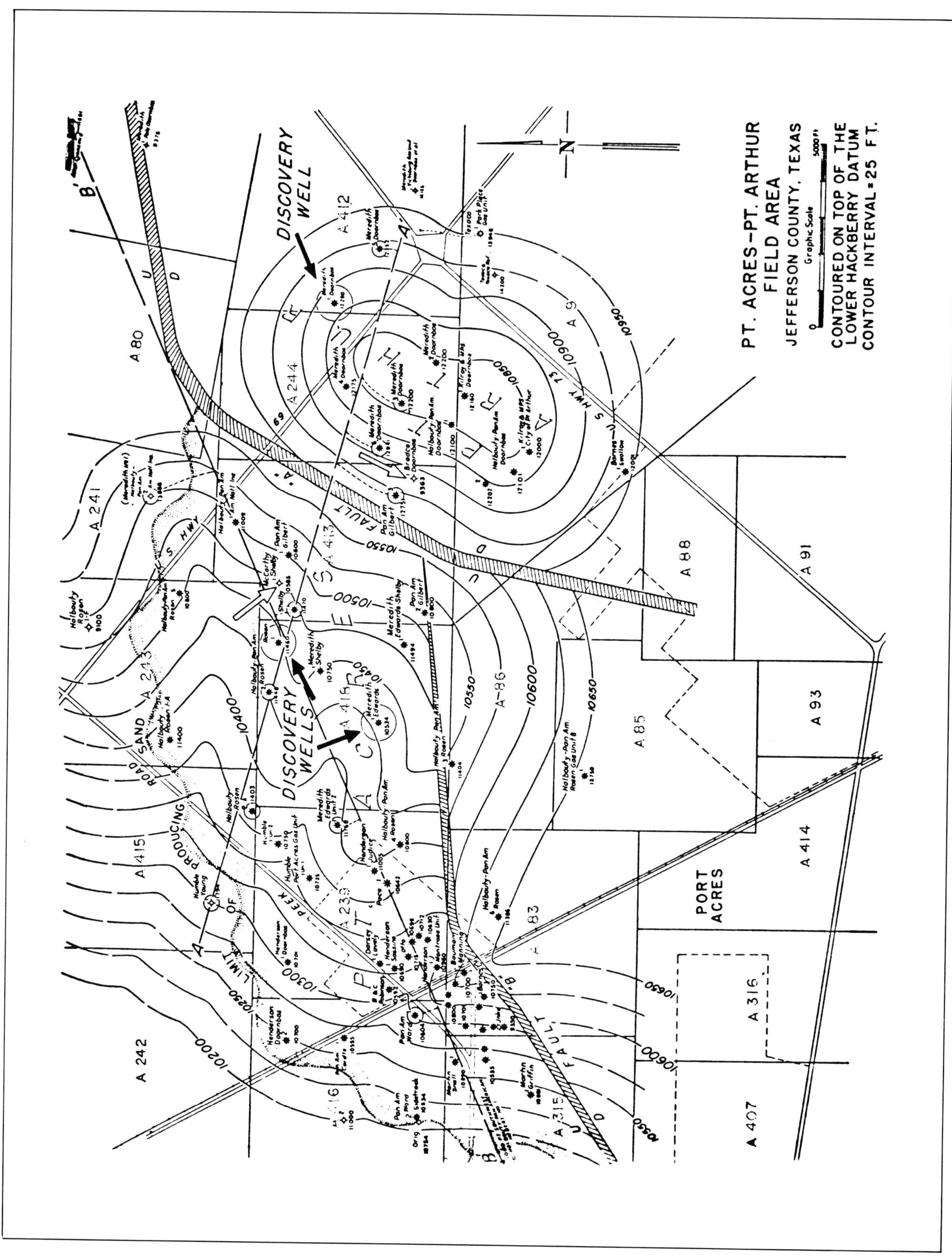
PT. ACRES–PT. ARTHUR
FIELD AREA
JEFFERSON COUNTY, TEXAS
Graphic Scale
CONTOURED ON TOP OF THE
LOWER HACKBERRY DATUM
CONTOUR INTERVAL = 25 FT.
N
DISCOVERY WELL
DISCOVERY WELLS
PORT ACRES
FAULT
LIMIT OF PRODUCING SAND
PT. ACRES
PT. ARTHUR
A
A'
B
B'
U
D
A 80
A 241
A 242
A 243
A 244
A 239
A 412
A 413
A 414
A 415
A 416
A 407
A 316
A 315
A 85
A 86
A 88
A 91
A 93
A 83
A 9
10200
10250
10300
10400
10450
10500
10550
10600
10650
10900
10950

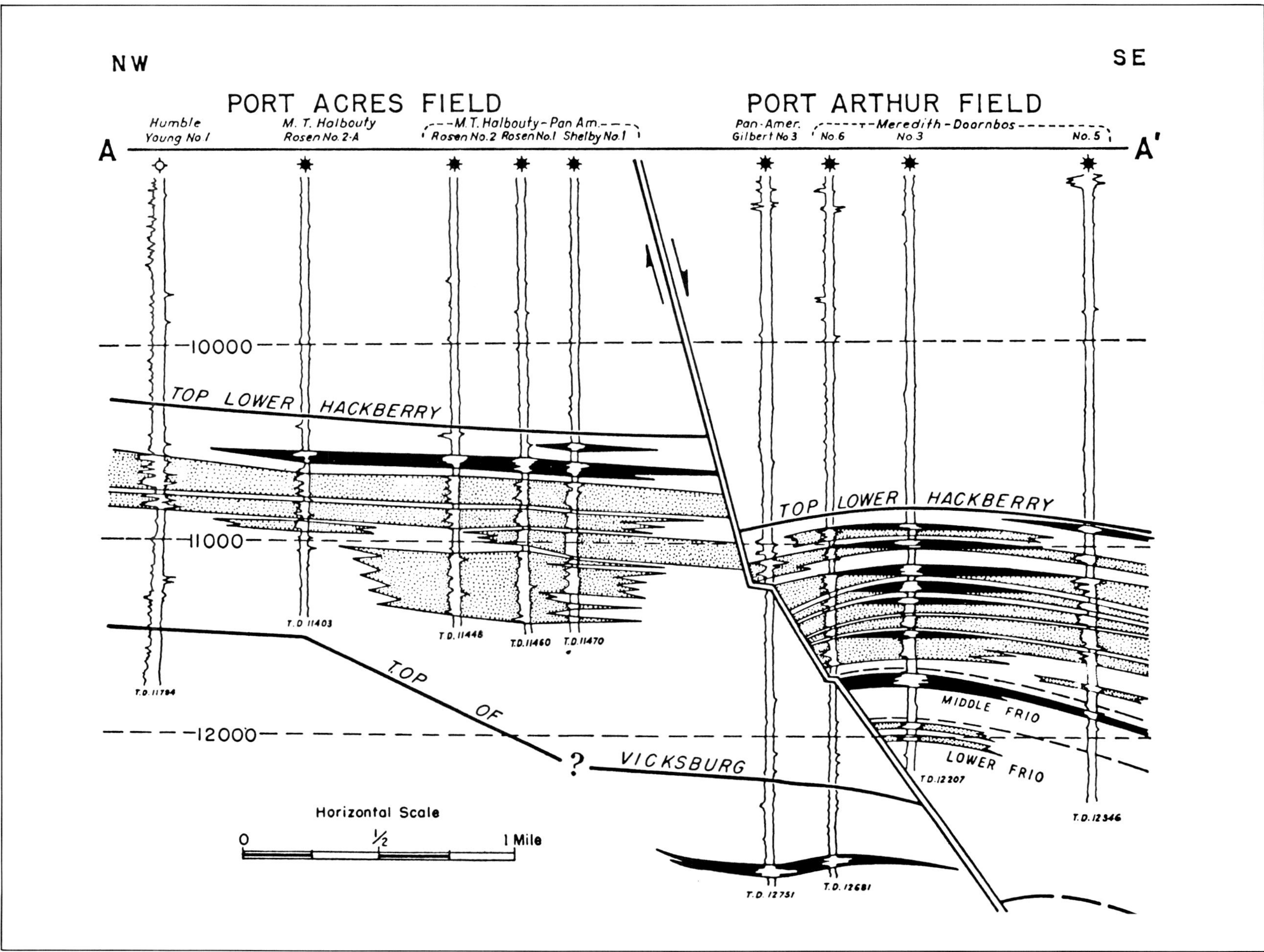

Fig. 7-32 (Halbouty and Barber, 1961)—A. Structural contour map of Port Acres and Port Arthur field area, Jefferson County, Texas. Note parallelism of Port Arthur rollover anticline and associated fault. B. NW-SE structural cross-section A-A' showing rollover anticlinal closure at Port Arthur field. Note deep accumulation of hydrocarbons. Permission to publish by Gulf Coast Association of Geological Societies.

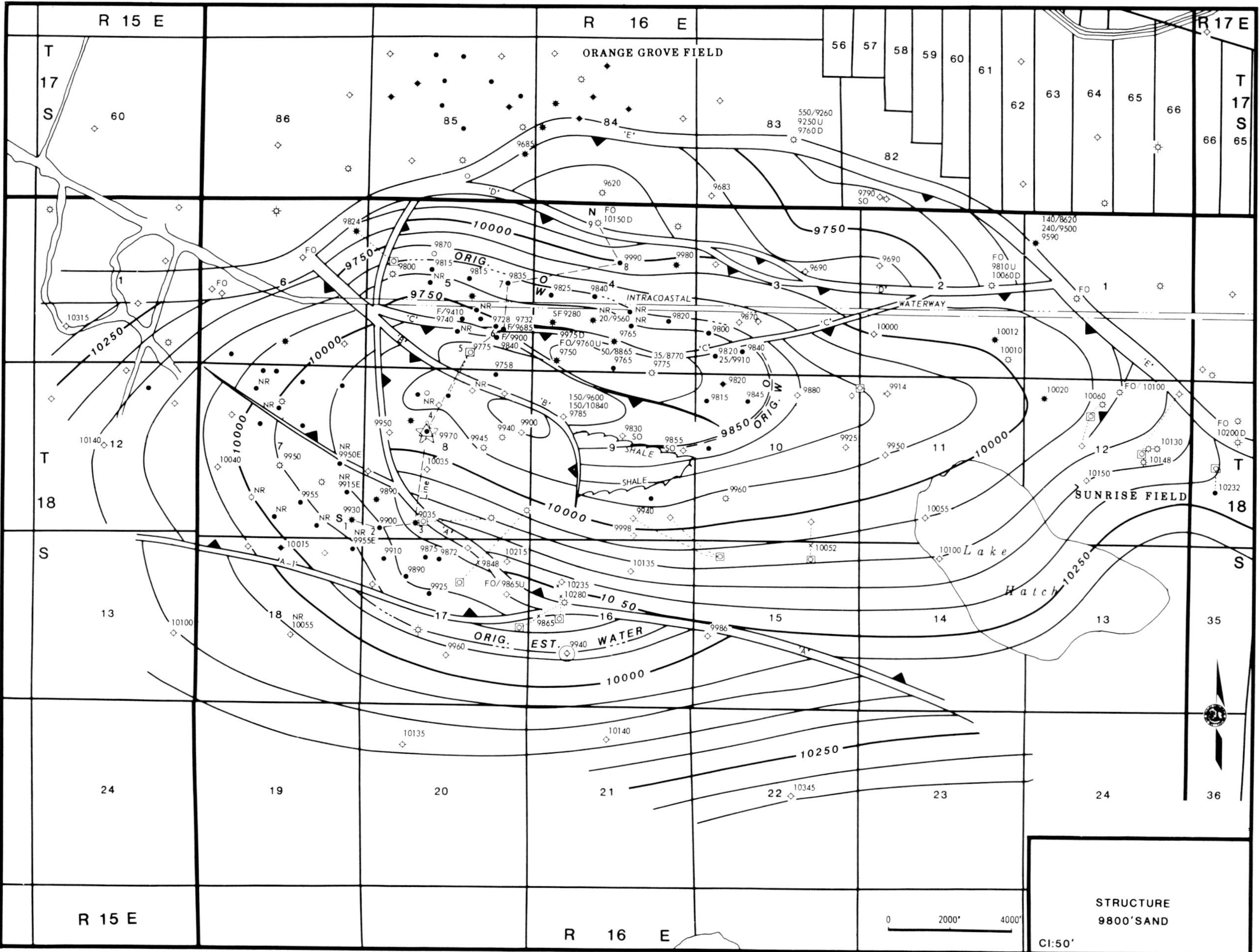

Fig. 7-33 (Conatser, 1983)—A. Structure contour map of Lake Hatch field, Terrebonne Parish, Louisiana, showing anticline on downthrown (south) side of major east-west trending growth fault and modified by additional faults that create discrete traps.

Illustration 7-33B begins on next page ⟶

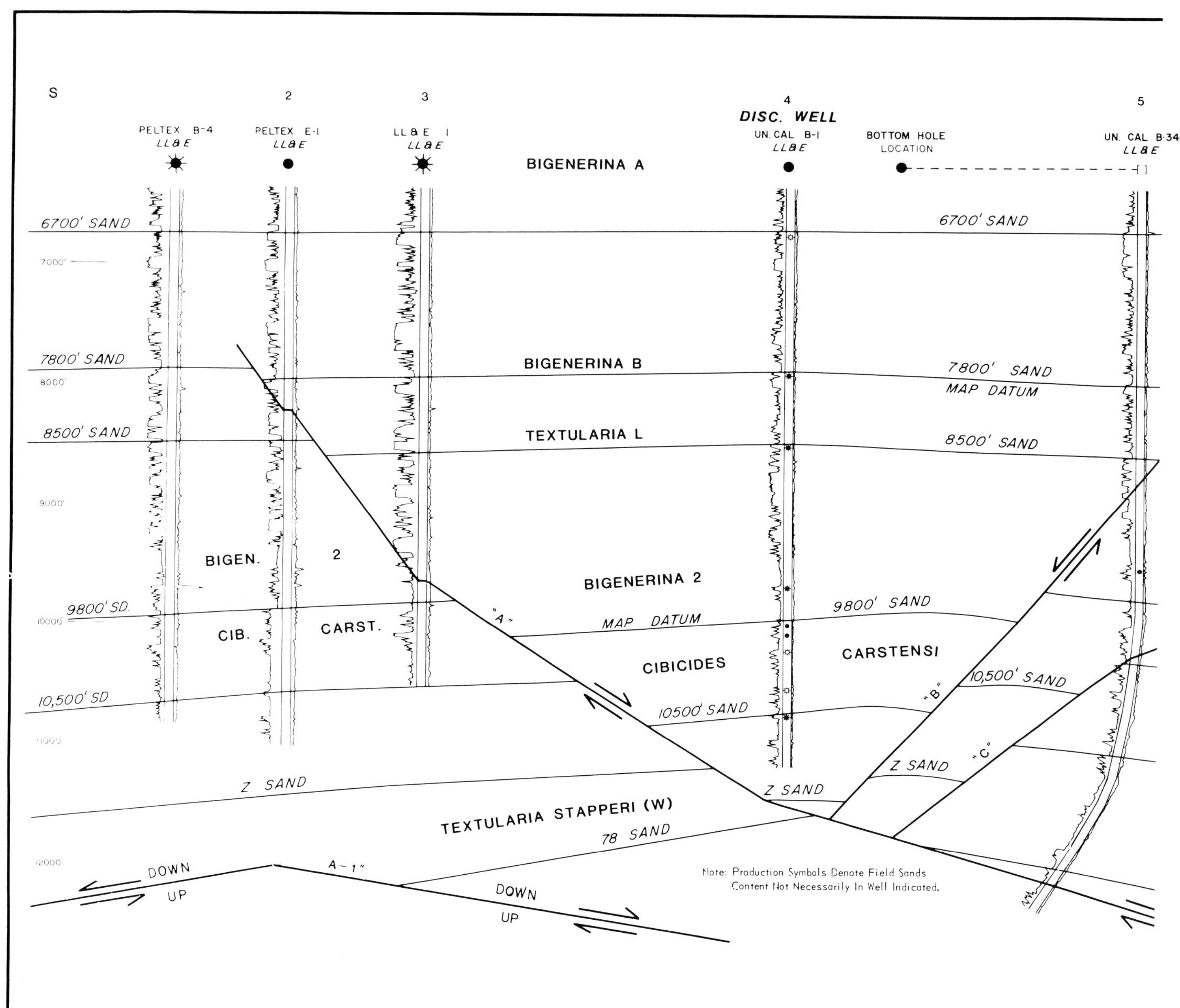

B. Cross section (line of section on map) showing fault complexity. Top of Textularia W zone coincides

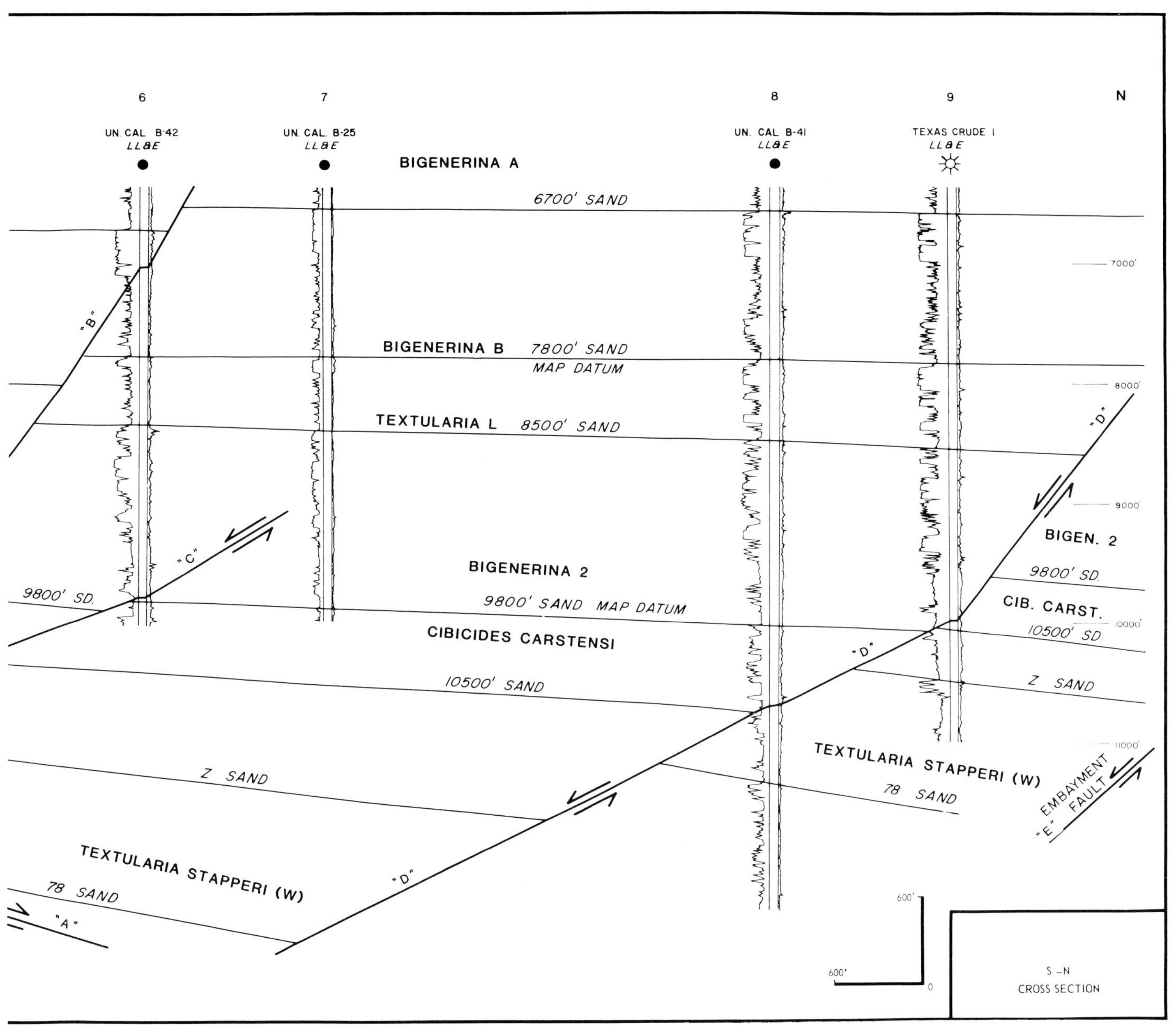

with top of overpressured section. Permission to publish by New Orleans Geological Society.

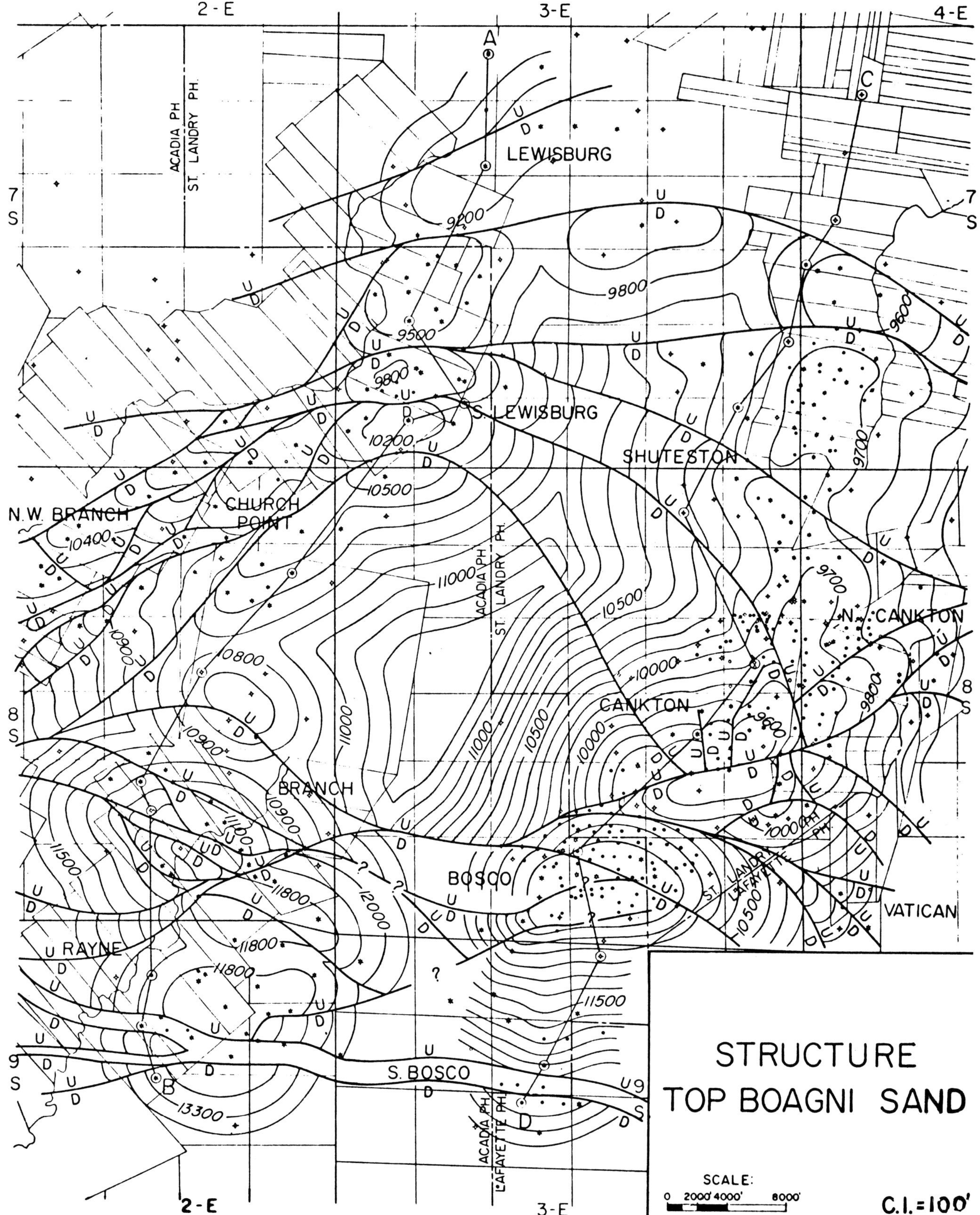

Fig. 7-34 (Ocamb, 1961)—South Louisiana area showing complex network of extensional glide plane faults. Some circular patterns reflect deep-seated salt features. Permission to publish by Gulf Coast Association of Geological Societies.

8 SALT STRUCTURES

OCCURRENCE

It has not been possible to consider other structural styles to this point without already having dealt with salt structures. O'Brien (1968) reported diapiric salt and shale features from more than 40 sedimentary basins around the world (Fig. 8-1, Table 8-I). Since 1968 diapiric salt has been discovered in several other basins, including the Suez in Egypt, southern Mendoza in Argentina, the Grand Banks of Newfoundland, and the North Sea. In addition there are many other basins with salt structures that are concordant to bedding and, hence, not diapiric.

Salt structures have been encountered in virtually every deformational environment, specifically with Red Sea extensional blocks (Figs. 4-7, 4-8), in the Idaho-Wyoming (Fig. 6-29) and Romanian Carpathian (Figs. 1-27 and 6-54) thrust-fold belts, and with United States Gulf Coast detached normal faults (Fig. 7-16). An examination of world-wide occurrences reveals that the preferred plate tectonic habitats of salt are divergent continental margins and aborted rift systems. Narrow, restricted, rifted troughs (particularly in hot, arid, low latitudes), formed by continental breakup, and sometimes having drainage diverted by earlier arching, are a natural setting for thick evaporite accumulations.

Salt when present in and influenced by other structures has been termed tectonic salt. Most early studies of deformed salt either stated or assumed tectonic deformation. The term "diapir" was first used in the Carpathian Mountains of Romania for salt structures that were highly discordant to bedding and obviously affected by tectonic forces that shaped the Carpathian thrust-fold belt (Figs. 1-27 and 6-53). Much later, about 50 years ago, hypotheses of buoyant salt movement were formulated.

Because of its relatively low density (2.2g/cm^3 vs. 2.5–2.6g/cm^3 of consolidated sediments) and high ductility, salt has the capability of moving buoyantly under the influence of gravity in the complete absence of tectonic movement. Buoyant salt movement has sufficient distinguishing characteristics to constitute a style of its own, which is the subject of this chapter. It should be kept in mind, however, that many salt features are initiated by some tectonic perturbation and only after this event does buoyant movement occur and become dominant. Moreover, as O'Brien (1968) has stated, "every conceivable transition between the two types (buoyant and tectonic salt structures) is to be found in the world".

In addition to tectonic triggering of salt flowage the thickness of both overburden and salt also influence salt movement. Trusheim (1960), from experience with the Permian Zechstein salt of Germany, has given an overburden thickness of 1000 m (3280 ft) and a salt thickness of 300 m (984 ft) (presumably assuming pure salt without intercalation of competent lithologies) required to initiate flow. These numbers must be considered as very general for Hughes (1968) in studying the Jurassic Louann salt of the United States Gulf Coast found flowage had taken place in very thin salt beneath only a few hundred feet of overburden; a basin slope of several degrees appears to have prompted salt movement. On the other hand, in the southern Red Sea almost 3049 m (10,000 ft) of salt beneath several thousand feet of overburden has not flowed enough to appreciably disturb that overburden.

ZONATION

In northwest Germany, Trusheim (1960) observed a basinward progression of salt pillows, stocks, and walls (Fig. 8-2). This zonation is related in large part to the increased thickness of salt from basin flank to basin center. A Lower Permian Rotliegend salt joined with the Upper Permian

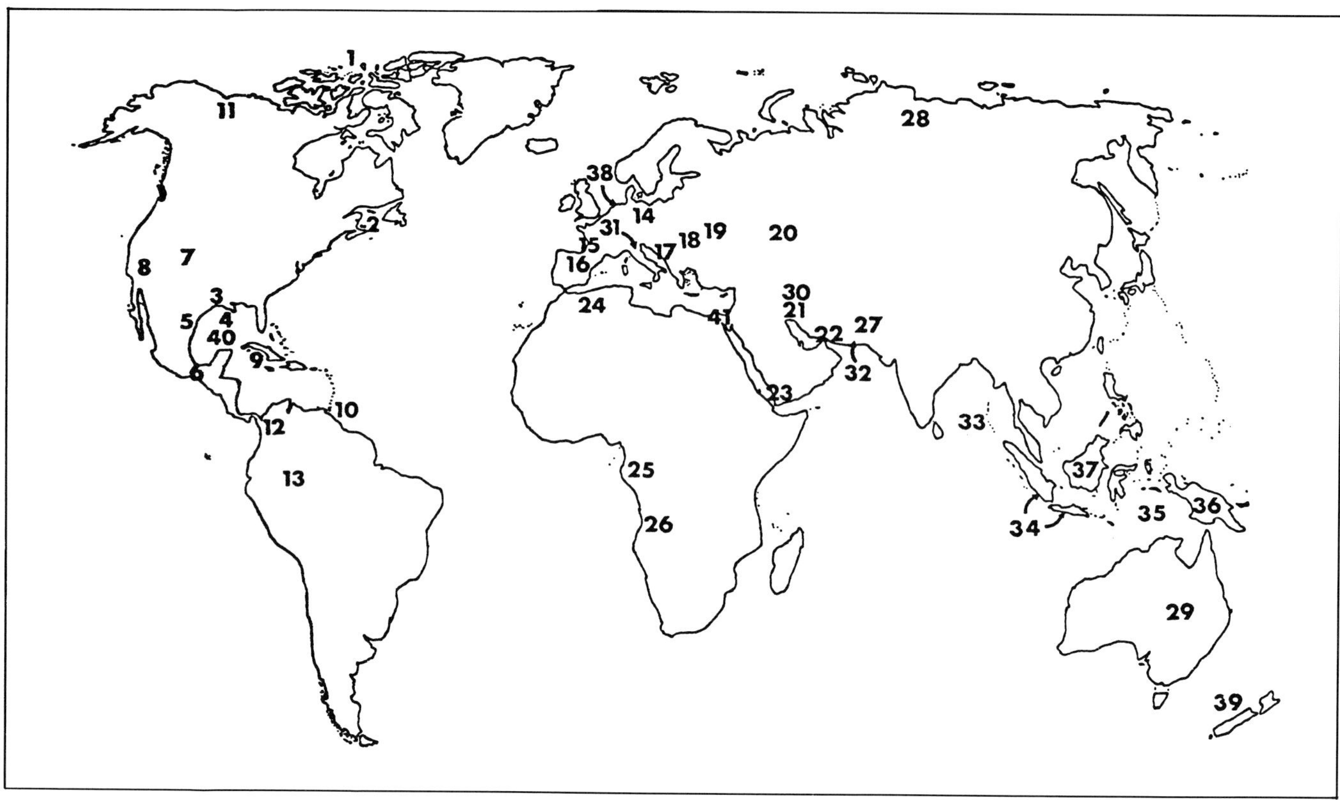

Fig. 8-1 (O'Brien, 1968)—Distribution of diapiric structures in sedimentary basins. Numbers indicate localities listed in Table 8-I. Permission to publish by American Association of Petroleum Geologists.

Zechstein salt to supply truly great volumes of salt in the most basinward zone of salt walls.

In the interior Gulf Coast salt basin of Mississippi, Hughes (1968) made a somewhat different zonation for the Jurassic Louann salt. As shown in figure 8-3, the Louann salt wedges out against the eroded Paleozoic floor to the north, and thickens progressively southward. Salt structures having similar characteristics are developed in areas of similar salt thickness. A zone of "peripheral salt ridges" parallels the wedgeout; adjacent to these is a zone of "low relief salt pillows"; basinward from these is a zone of "intermediate salt anticlines"; and in thicker salt basinward is a zone of "high relief salt anticlines". The mother salt thickens more rapidly basinward on the western side of the area than on the eastern side. Structure contour maps and structural cross sections of a field typical of a structure in each zone are given in figures 8-4 through 8-9.

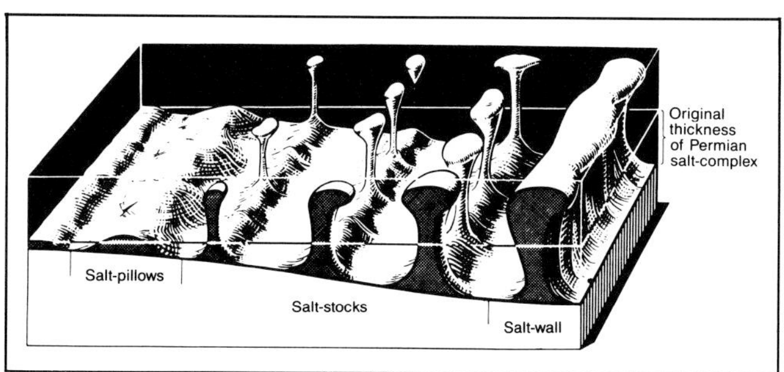

Fig. 8-2 (Trusheim, 1960)—Diagram of different types of salt structures in relation to original thickness of Permian salt complex of northwest Germany. Permission to publish by American Association of Petroleum Geologists.

Every salt basin no doubt has its own zonation differing only slightly or very greatly from another depending on thicknesses of salt and overburden, basin configuration, basin slopes or regional dip, rates of sedimentation, and sediment densities. One characteristic that seems common to most, however, is the "basin-edge effect" or downbend of sedimentary beds above the salt edge that takes place probably because of down-dip salt flowage and possibly because of salt solution (Figs. 4-7, 4-8, and 8-6).

ASSOCIATED ELEMENTS OF BUOYANT SALT STRUCTURES

The region of the Hainesville salt dome in east Texas is remarkable in demonstrating in one small area practically all of the elements associated with buoyant salt structures. Primary and secondary rim synclines, a turtle structure, and a residual high are all present along with a salt dome and

TABLE 8-1

(O'Brien, 1968)—Tabulation of diapiric structures keyed to Figure 8-1. Permission to reprint by American Association of Petroleum Geologists.

Locality (Fig. 8-1)	Geologic Feature	Geologic Setting	Geographic Location	Type of Diapir	Diapiric Material	Age of Diapiric Material	Reference
1	Gypsum domes	Sverdrup basin	Canadian Arctic Islands	Domal	Gypsum	Ordovician	Gould and deMille, this vol.
2	Salt anticlines	Moncton basin	Maritime Provinces, Canada	Anticlinal-domal	Salt	Mississippian	Gussow, 1953
3	Shale domes	U.S. Gulf Coast basin	Onshore and offshore, Gulf of Mexico	Domal	Salt, shale	Jurassic or Triassic; Tertiary	Murray, this vol.
4	Mudlumps	Mouth of Mississippi River	Gulf of Mexico	Domal	"Mud"	Recent	Morgan, this vol.
5	Gypsum anticlines	Coahuila marginal folded belt	Northeastern Mexico	Anticlinal-domal	Gypsum	Jurassic or Triassic	Wall *et al.*, 1967
6	Salt domes	Isthmian Salt basin	Isthmus of Tehuantepec, Mexico	Domal	Salt	Jurassic	Contreras and Castillon, 1960 (trans., this vol.)
7	Salt anticlines	Paradox basin	Western interior, U.S.	Anticlinal	Salt	Pennsylvanian	Elston *et al.*, 1962
8	Serpentine diapirs	Diablo Range	California	Domal	Serpentine	?	Oakeshott, this vol.
9	Salt domes		Camaguey Province, Cuba	Anticlinal-domal	Salt	Jurassic or Triassic	Meyerhoff and Hatten, this vol.
10	Mud flows, diapiric anticlines		Trinidad	Anticlinal-domal	Mud	Oligocene	Suter, 1954
11	Pingoes, ice diapirs	MacKenzie River delta	Northwest Territories, Canada	Domal	Ice	Pleistocene	Müller 1959
12	Mud volcanoes		Northwest Columbia; East Venezuela	Domal	Mud	Miocene	Gansser, 1960
13	Salt domes	East flank of Andes	Peru; Colombia	Domal	Salt	Permian	Benavides, 1962
14	Salt domes	Hannover basin	Northern Germany, North Sea	Domal	Salt	Permian	Sannemann, this vol.
15	Salt diapirs	Aquitaine basin	Southern France	Anticlinal	Salt, marl	Triassic	Dupouy-Camet, 1953
16	Diapirs	Pyrenees, Cantabrian, Andalusian Mtns.	Spain	Anticlinal-domal	Salt, marl	Triassic	Loegters and Brinkman, this vol.
17	Serpentine masses	Inner Dinaric Alps	Yugoslavia	Anticlinal	Serpentine	?	Milovanovic and Karamata, 1960
18	Diapirs	Carpathian Mtns.	Rumania	Anticlinal	Salt	Miocene	Voitesti, 1925
19	Salt domes	Dneiper-Donets basin	Southeastern Russia	Domal	Salt	Devonian	Kityk, 1959
20	Salt domes	Emba region	South-Central Russia	Domal	Salt	Permian	Sanders, 1939
21	Diapiric anticlines	Foothills of Zagros Mtns.	Eastern Iran	Anticlinal	Salt, anhydrite	Miocene	O'Brien, 1957
22	Salt plugs	Persian Gulf area	Southern Iran	Domal	Salt	Cambrian	O'Brien, 1957
23	Salt plugs	Yemen, Farson Islands	Southern Red Sea area	Domal	Salt	Triassic	Wade, 1936
24	Domes, anticlines, and laminae	Atlas Mtns.	North Africa	Anticlinal	Marl, gypsum salt	Triassic	LaCoste, 1934
25	Salt anticlines	Congo basin	Gabon, Africa	Anticlinal	Salt	Cretaceous	Pegand and Reyre, 1959
26	Salt anticlines	Cuanza basin	Angola, Africa	Anticlinal	Salt	Cretaceous	Brognon, Verrier, and Masson, 1959
27	Salt anticlines	Punjab salt range	Pakistan	Anticlinal	Salt	Eocene or Cambrian	Krishnan, 1962
28	Salt dome	Nordvick Bay	North Siberia	Domal	Salt	Silurian	Tolmachoff, 1926
29	Salt diapirs	Amadeus basin	Northern Territory, Australia	Domal	Salt	Precambrian	McNaughton *et al.*, 1962
30	Mud volcanoes	Southeast Caucasus	USSR; Northern Iran	Domal	Mud	Miocene	Goubkin, 1934
31	Clay diapirs	Po basin	Northern Italy	Anticlinal	Clay	Cretaceous	Weidenmayer, 1950
32	Mud volcanoes	Makran Coast range	Southern Iran; Pakistan	Domal-anticlinal	Mud	Miocene-Oligocene	Snead, 1964
33	Mud volcanoes	Burmese-Andaman trend	Burma; islands in Bay of Bengal	Anticlinal	Clay	Oligocene-Miocene	Chibber, 1934
34	Mud volcanoes		Sumatra; Java	Anticlinal	Clay	Miocene	Höfer, 1909
35	Clay diapirs	Molucca, Banda Sea	Indonesia	Domal	Clay	Miocene	Heim, 1940
36	Mud volcanoes		New Guinea	Anticlinal	Clay	Miocene	Höfer, 1909
37	Mud volcanoes		Borneo	Anticlinal	Clay	Miocene	Höfer, 1909
38	Peat diapirs	Flevoland	Netherlands	Domal	Peat	Recent	Paine, this vol.
39	Mud volcanoes		New Zealand	Anticlinal	Clay	Miocene	Höfer, 1909
40	Diapiric structures	Sigsbee Deep, upper continental slopes	Gulf of Mexico	Domal-anticlinal	Salt (?)	Jurassic (?)	Ewing *et al.*, 1966
41	Clay diapir	East edge of Libyan Desert, Palmyrean chain	Egypt; Syria	Anticlinal	Clay, gypsum, and anhydrite, with clay and marl	Jurassic	Omara, 1964

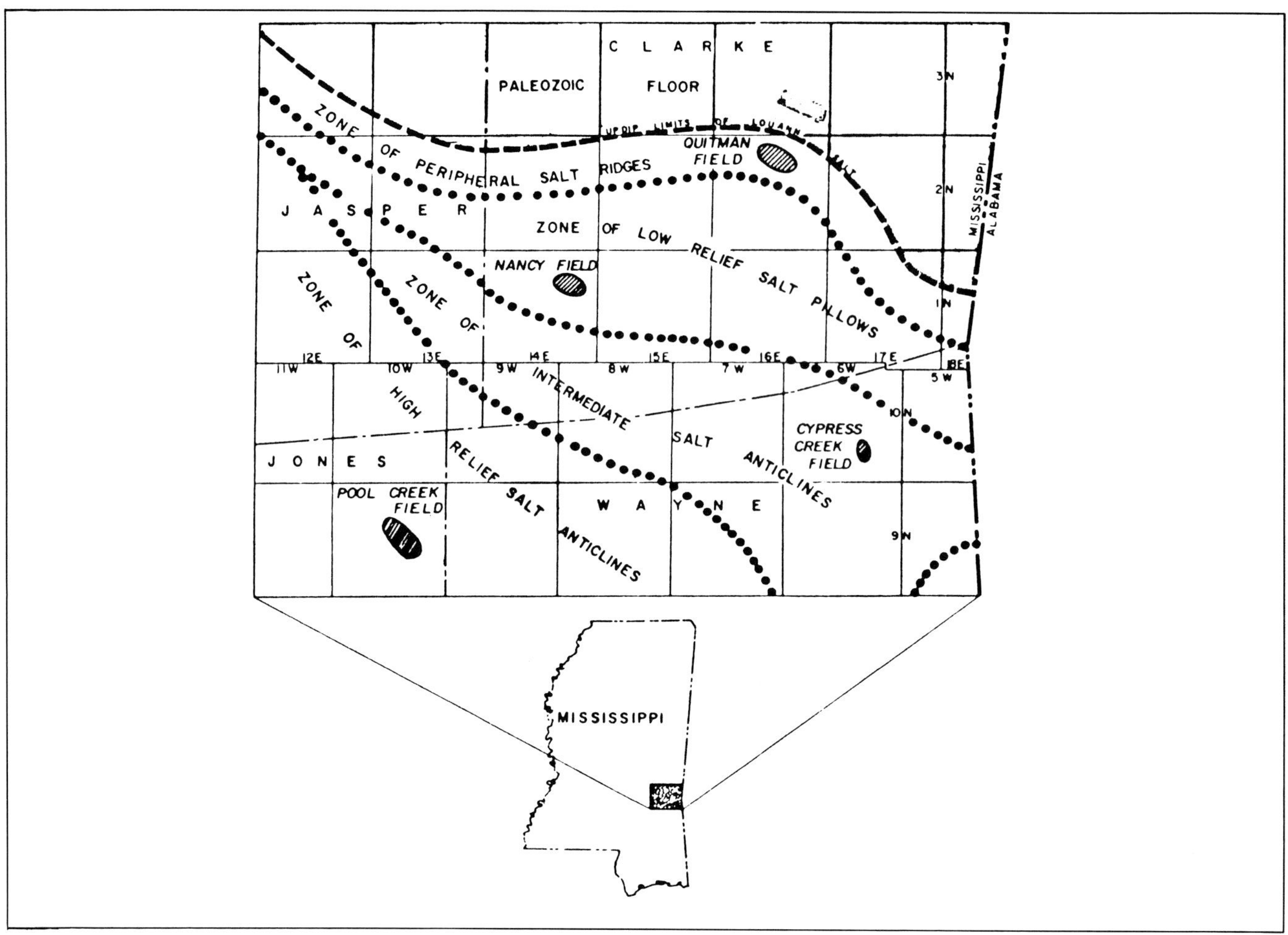

Fig. 8-3 (Hughes, 1968)—Zonation of salt structures in Mississippi salt basin, U.S. Gulf Coast. Permission to publish by Gulf Coast Association of Geological Societies.

a salt anticline on one seismic section (Fig. 1-26).

Rim synclines record the growth history of nearby salt structures by the thinning of sedimentary units over or toward the salt feature from the adjacent synclinal lows (Figs. 8-10 through 8-12). In figure 8-10, no salt movement is shown during deposition of the Buntsandstein (Lower Triassic) and Muschelkalk (Middle Triassic). Initial salt flowage is reflected in the thinning of the Keuper (Upper Triassic). As evidenced by thinning of higher units, salt growth persisted in a concordant manner through the Lower Cretaceous transgression; all thinning up to that time represented salt evacuating a primary rim syncline surrounding the salt feature. Through the Lower Cretaceous and later, salt growth was discordant and secondary rim synclines developed in Lower Cretaceous, Upper Cretaceous, and Tertiary sections. The switch from deposition in primary rim synclines to secondary rim synclines may reflect the change from concordant to discordant salt growth.

Turtle structures are interdomal features that are created by salt flowage (Fig. 8-13). Initial flowage creates a sag with thick sediments. Flowage of salt and consequent withdrawal of support from the flanks of the sag turns the sag inside-out into an anticlinal feature that can be prospective (Fig. 8-14). The Bryan field in Mississippi has been interpreted as an interdomal turtle structure or sediment-cored anticline between the Pool Creek (Fig. 8-9), Eucutta, and Heidelberg salt-cored anticlines (Figs. 8-15, 8-16).

Residual salt highs are areas of original (mother) salt

Fig. 8-5 (Hughes, 1968)—Structure contours on Louann salt are shown for the Cypress Creek field (intermediate salt anticline), and the Pool Creek field (high relief salt anticline). Note change in contour intervals. Figure 8-8 is cross section C-C′ of the Cypress Creek structure. Figure 8-9 is cross section D-D′ of the Pool Creek structure. Permission to publish by Gulf Coast Association of Geological Societies.

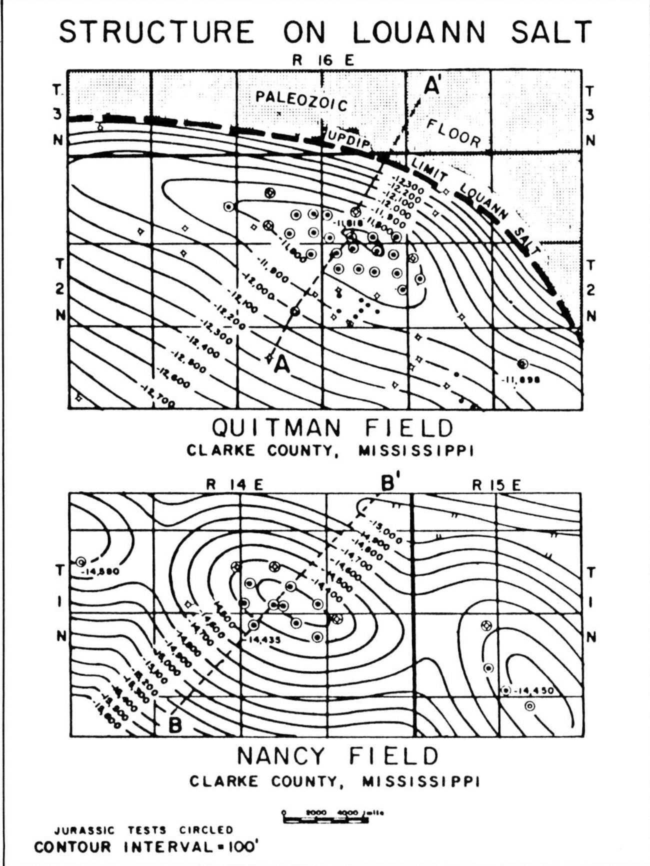

that underwent only partial or no flowage as adjacent salt features were created (Fig. 1-26). Structures can form in beds above residual highs inasmuch as the salt remains positive relative to sedimentation around it. Structure can be accentuated by differential compaction because salt is noncompactable and acts as a solid core. The prolific Katy field near Houston, Texas, is considered a residual salt high (Fig. 8-17).

FAULTS AND FAULT PATTERNS

A great variety of faults accompanies salt structures. All are normal faults with the exception of a rare few that show reverse movement. Reverse faults seem confined to the overhangs on the flanks of salt domes (Figs. 8-18 through 8-20); in essence the rising salt can be viewed as creating an upthrust (of the dip-slip variety) on the flank of a dome.

Normal faults abound in the vicinity of salt structures. They may occur in response to flexing created by the growth of rim synclines (Fig. 1-26), but more often they occur above

Fig. 8-4 (Hughes, 1968)—Structure contours on Louann salt of Quitman field (peripheral salt ridge), and Nancy field (low relief salt pillow). Figure 8-6 is cross section A-A′ of the Quitman structure. Figure 8-7 is cross section B-B′ of the Nancy structure. Permission to publish by Gulf Coast Association of Geological Societies.

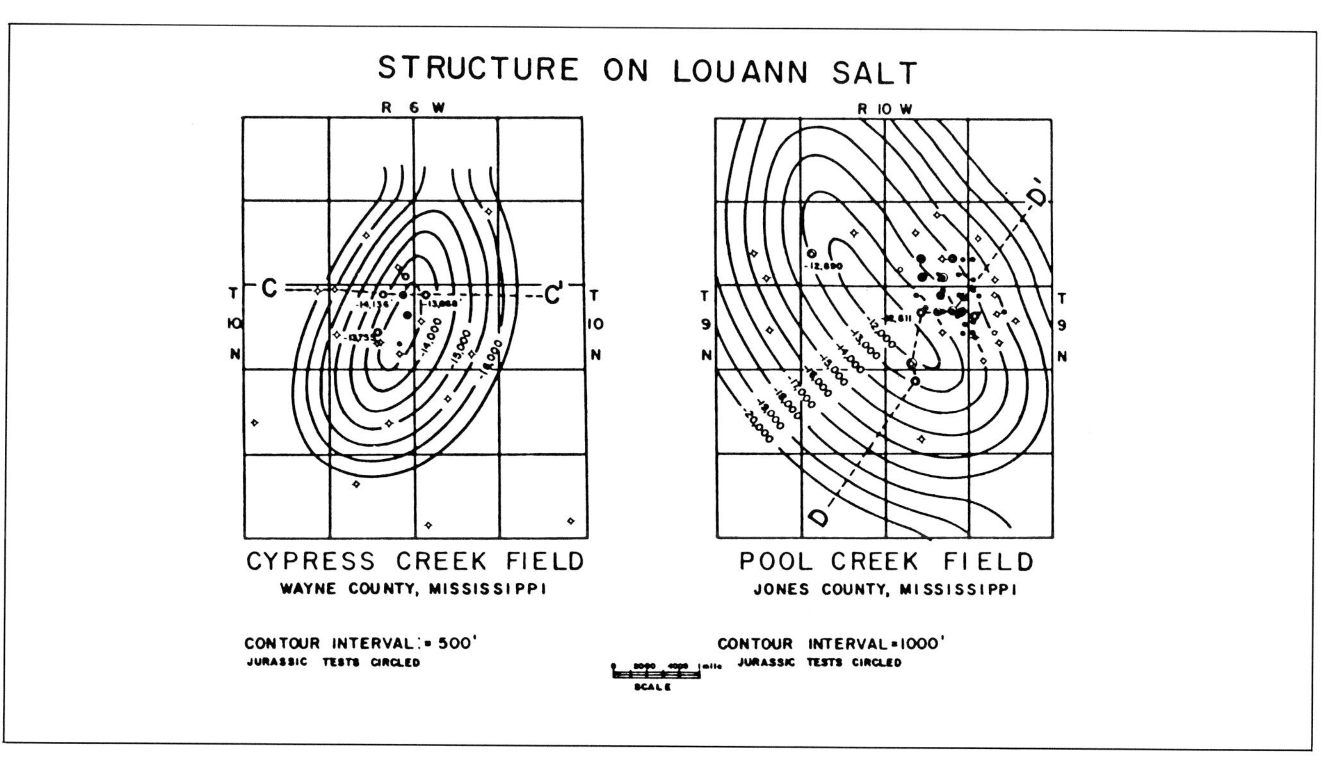

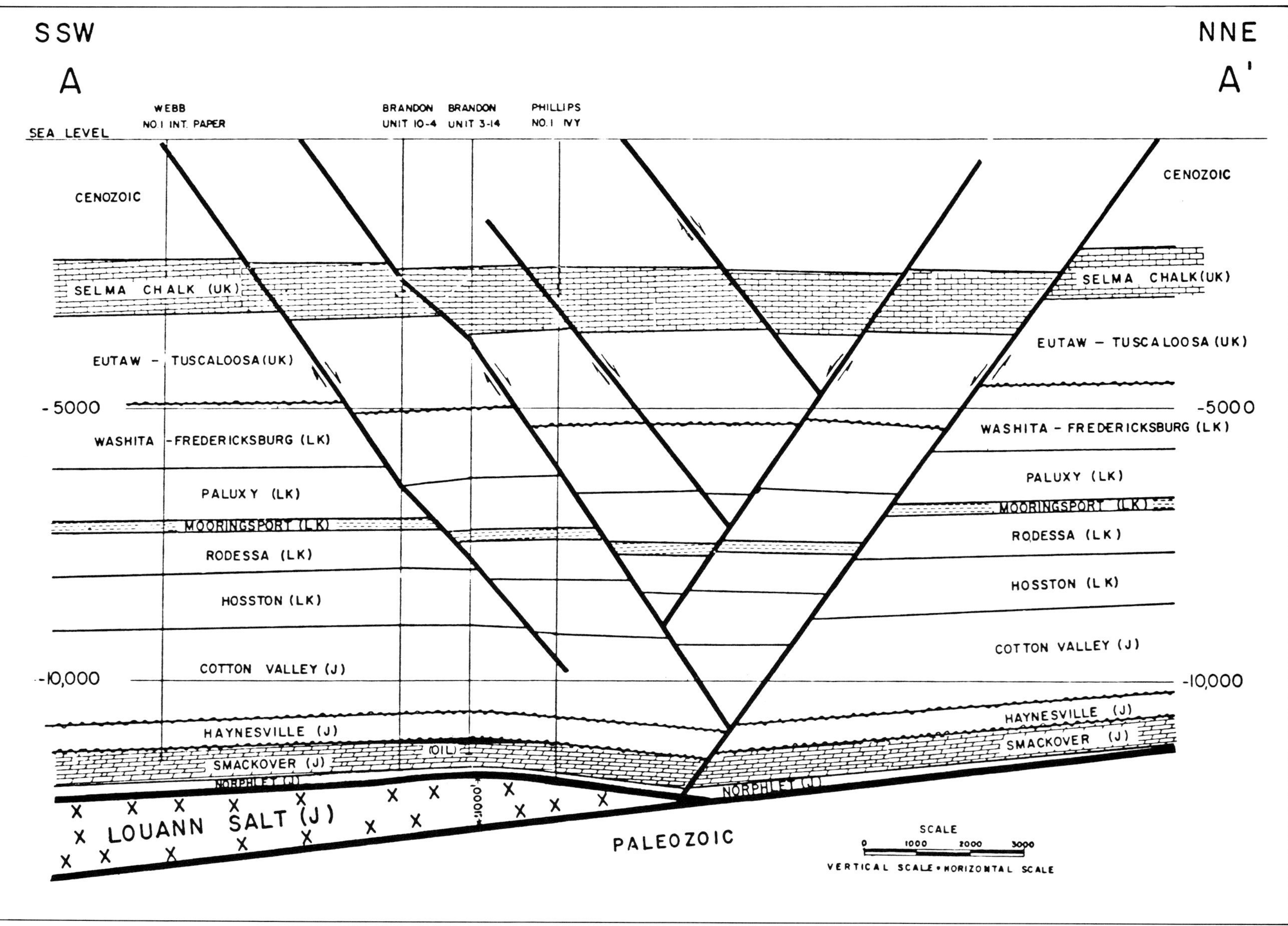

Fig. 8-6 (Hughes, 1968)—Quitman field of peripheral salt ridge zone. Downbending of sedimentary section and fault system above salt edge are probably caused by downslope flowage of salt; solution withdrawal could be a contributing factor. I would draw the south-southwestern most fault to continue downward to the salt edge. Permission to publish by Gulf Coast Association of Geological Societies.

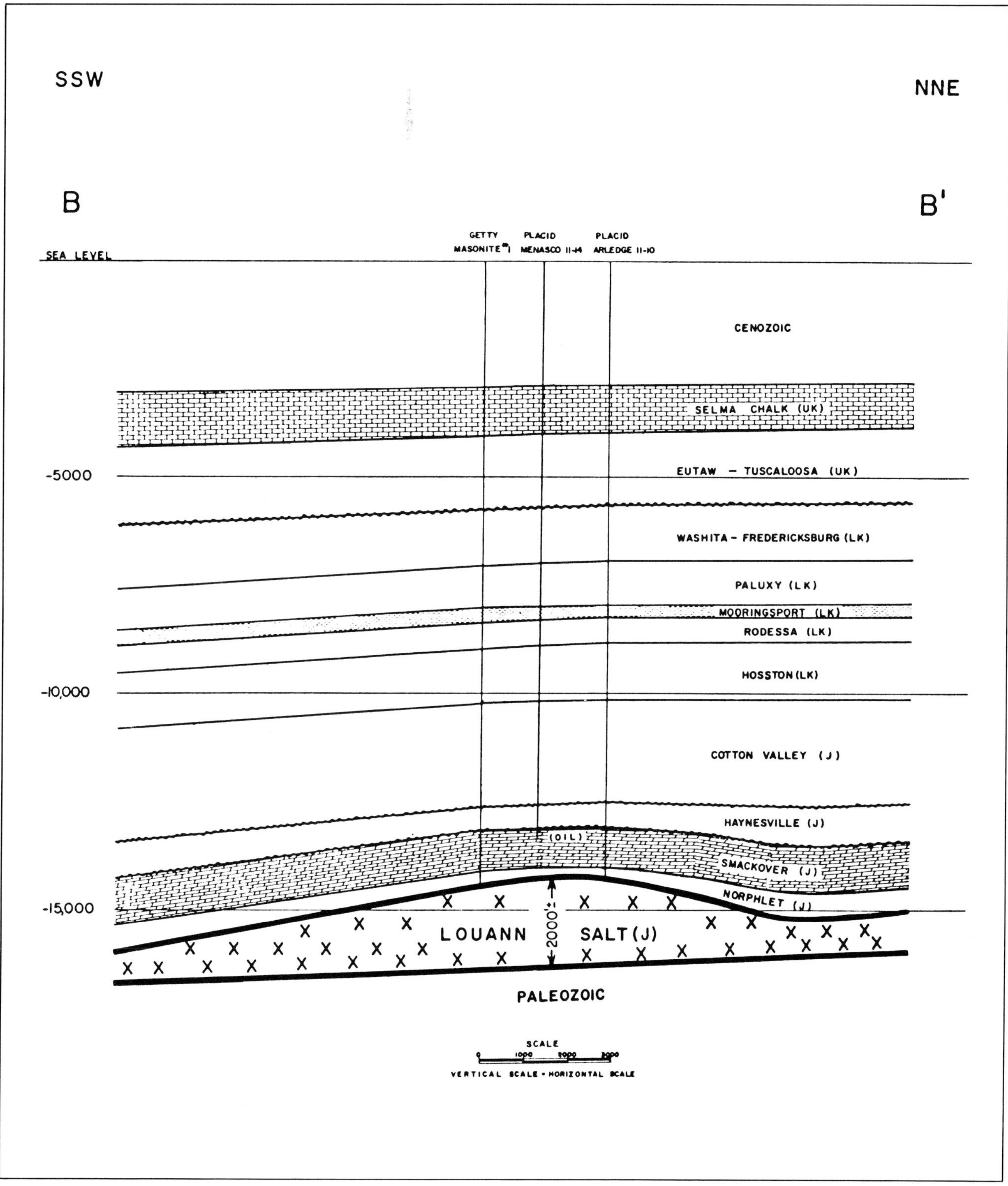

Fig. 8-7 (Hughes, 1968)—Nancy field of low relief salt pillow zone. Very early growth implied by thinning of Norphlet Formation above the salt pillow. Although not shown, normal faults at the Paleozoic-Louann interface could trigger growth of the pillow (cf. Fig. 8-28). Permission to publish by Gulf Coast Association of Geological Societies.

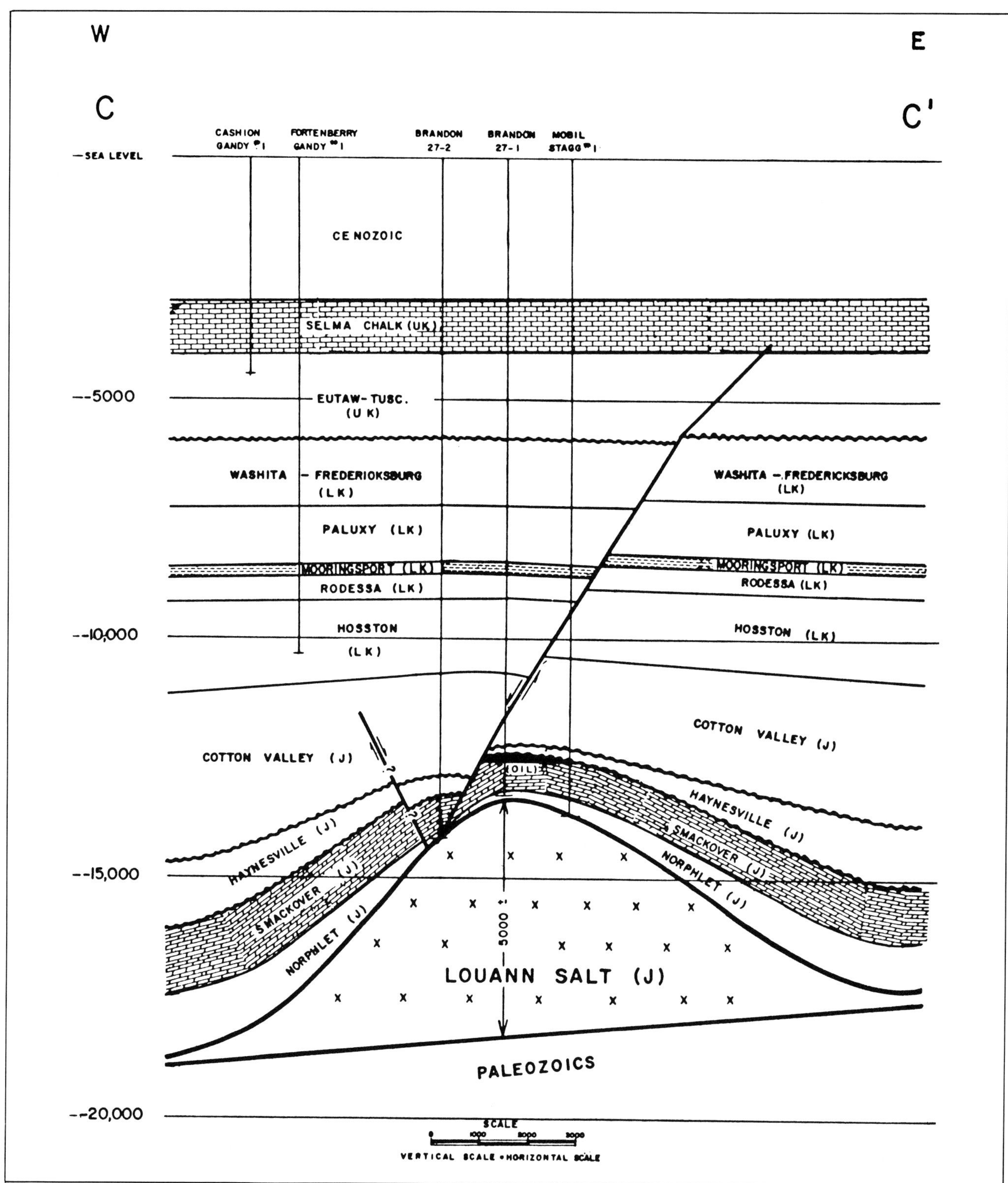

Fig. 8-8 (Hughes, 1968)—Cypress Creek field of intermediate salt anticline zone. Thinning of Norphlet again reflects early growth of salt feature. Note that salt surfaces are concordant to bedding. Sufficient uplift has occurred to create normal faults above the salt. Permission to publish by Gulf Coast Association of Geological Societies.

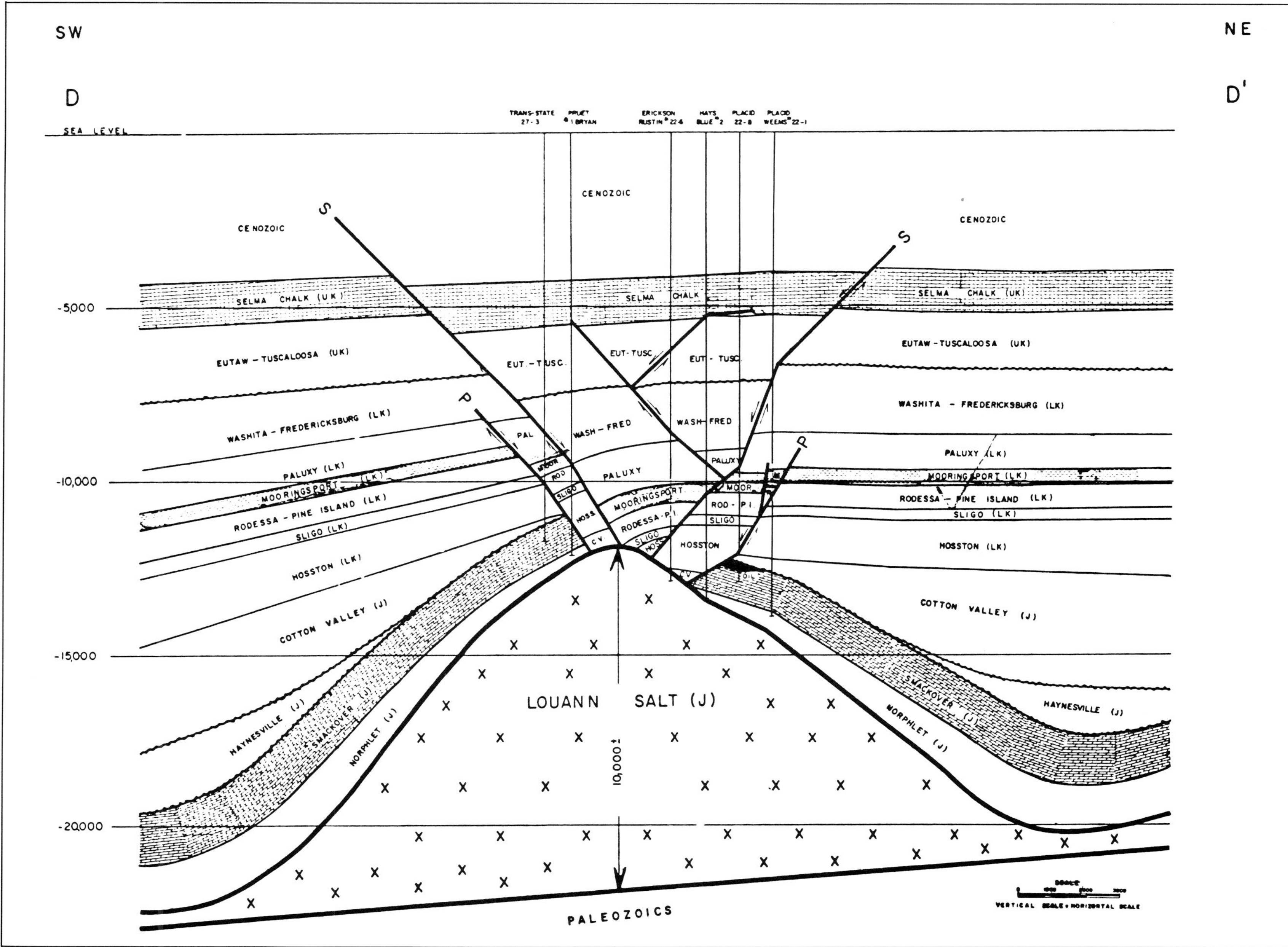

Fig. 8-9 (Hughes, 1968)—Pool Creek field of high relief salt anticline zone. Note: early salt movement; normal faults (P=primary, S=secondary); discordant contact (incipient diapirism) between salt and sediments at top of anticline. Permission to publish by Gulf Coast Association of Geological Societies.

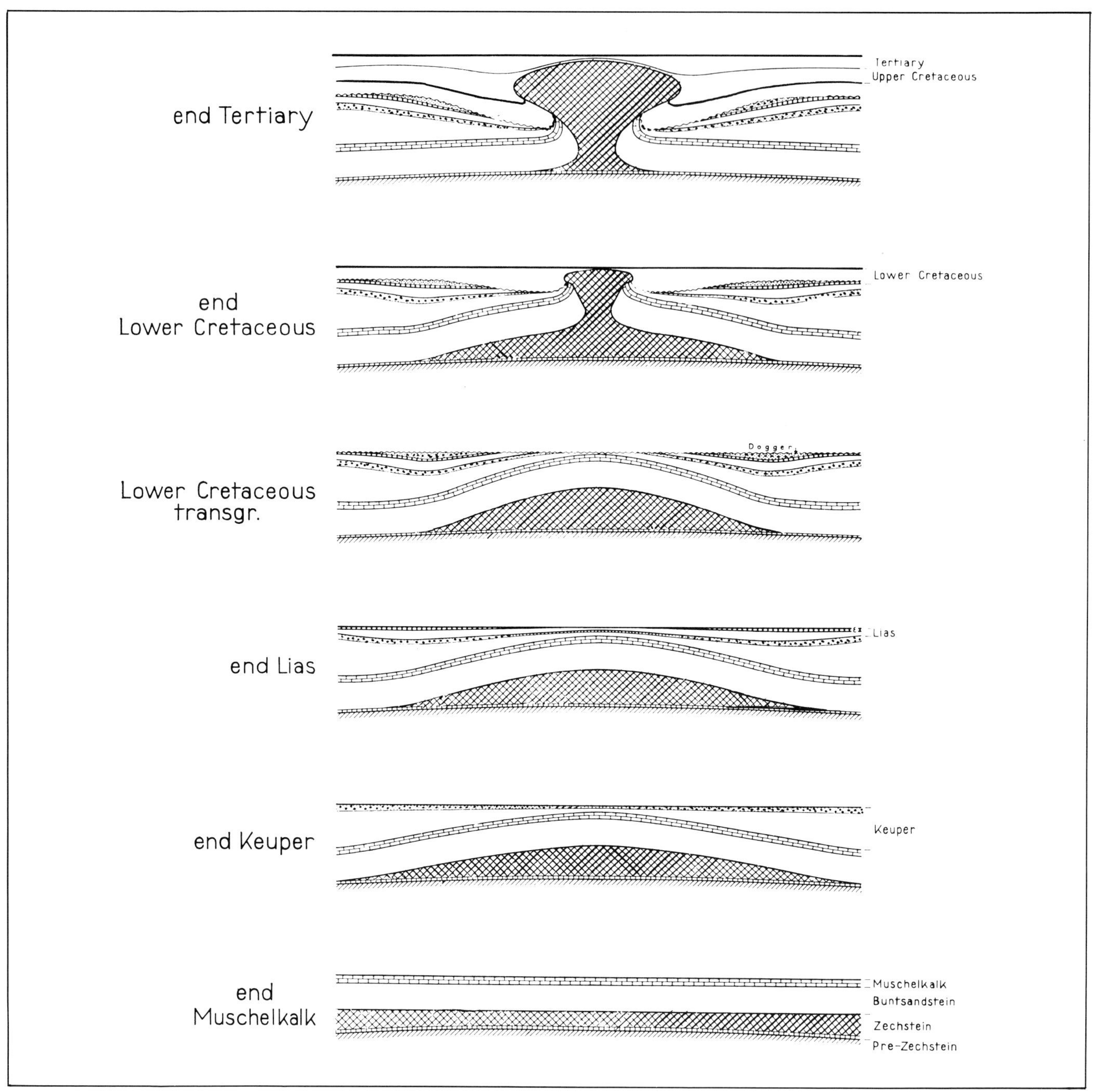

Fig. 8-10 (Trusheim, 1960)—Diagrammatic development of Zechstein salt stock and associated rim synclines based on examples from northwest Germany. Permission to publish by American Association of Petroleum Geologists.

Illustration 8-11 begins on next page ⟶

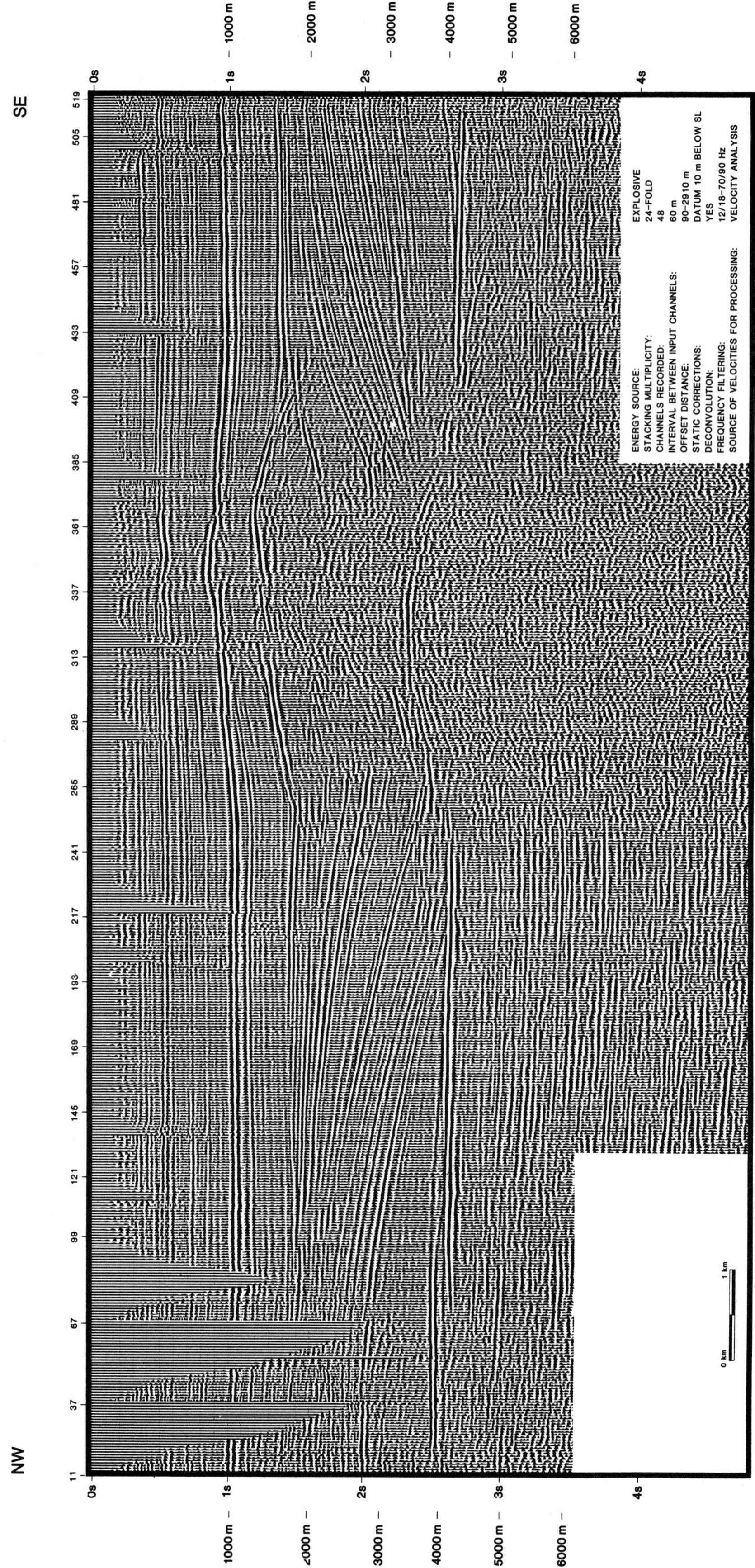
SE
NW
0s
1s
2s
3s
4s
1000 m
2000 m
3000 m
4000 m
5000 m
6000 m
11
37
67
99
121
145
169
193
217
241
265
289
313
337
361
385
409
433
457
481
505
519
ENERGY SOURCE: EXPLOSIVE
STACKING MULTIPLICITY: 24-FOLD
CHANNELS RECORDED: 48
INTERVAL BETWEEN INPUT CHANNELS: 60 m
OFFSET DISTANCE: 90-2910 m
STATIC CORRECTIONS: DATUM 10 m BELOW SL
DECONVOLUTION: YES
FREQUENCY FILTERING: 12/18-70/90 Hz
SOURCE OF VELOCITIES FOR PROCESSING: VELOCITY ANALYSIS
0 km
1 km

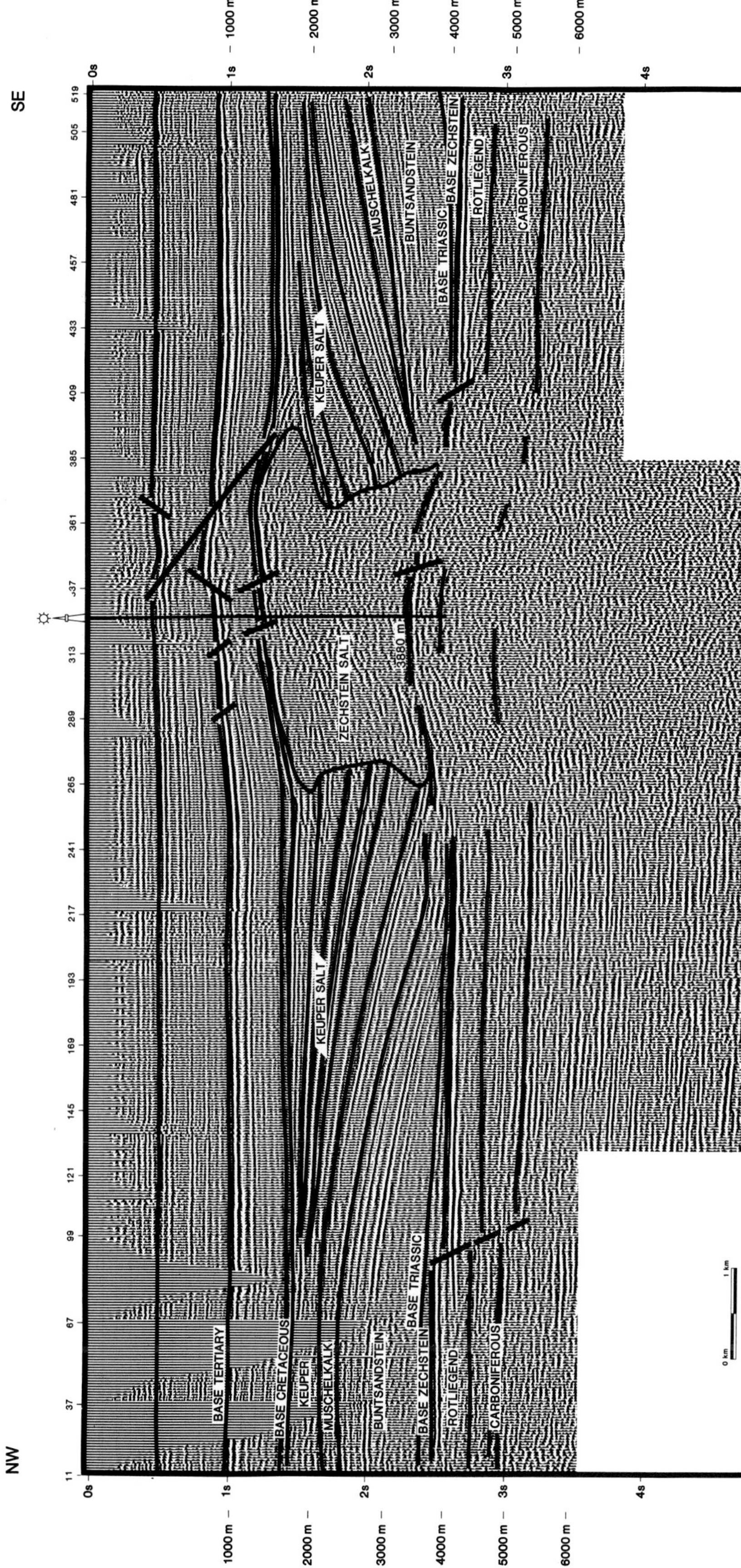

Fig. 8-11 (Lukic et al., 1983)—Seismic line from northwest German Basin illustrating salt plug with associated primary rim syncline in Buntsandstein (NW & SE) and Muschelkalk (SE) and secondary rim syncline in Muschelkalk (NW) and Keuper (NW & SE). Keuper salt is probably dissolved and reprecipitated salt from Zechstein. Gas discovered in Rotliegend red beds beneath Zechstein salt plug. Permission to publish by American Association of Petroleum Geologists.

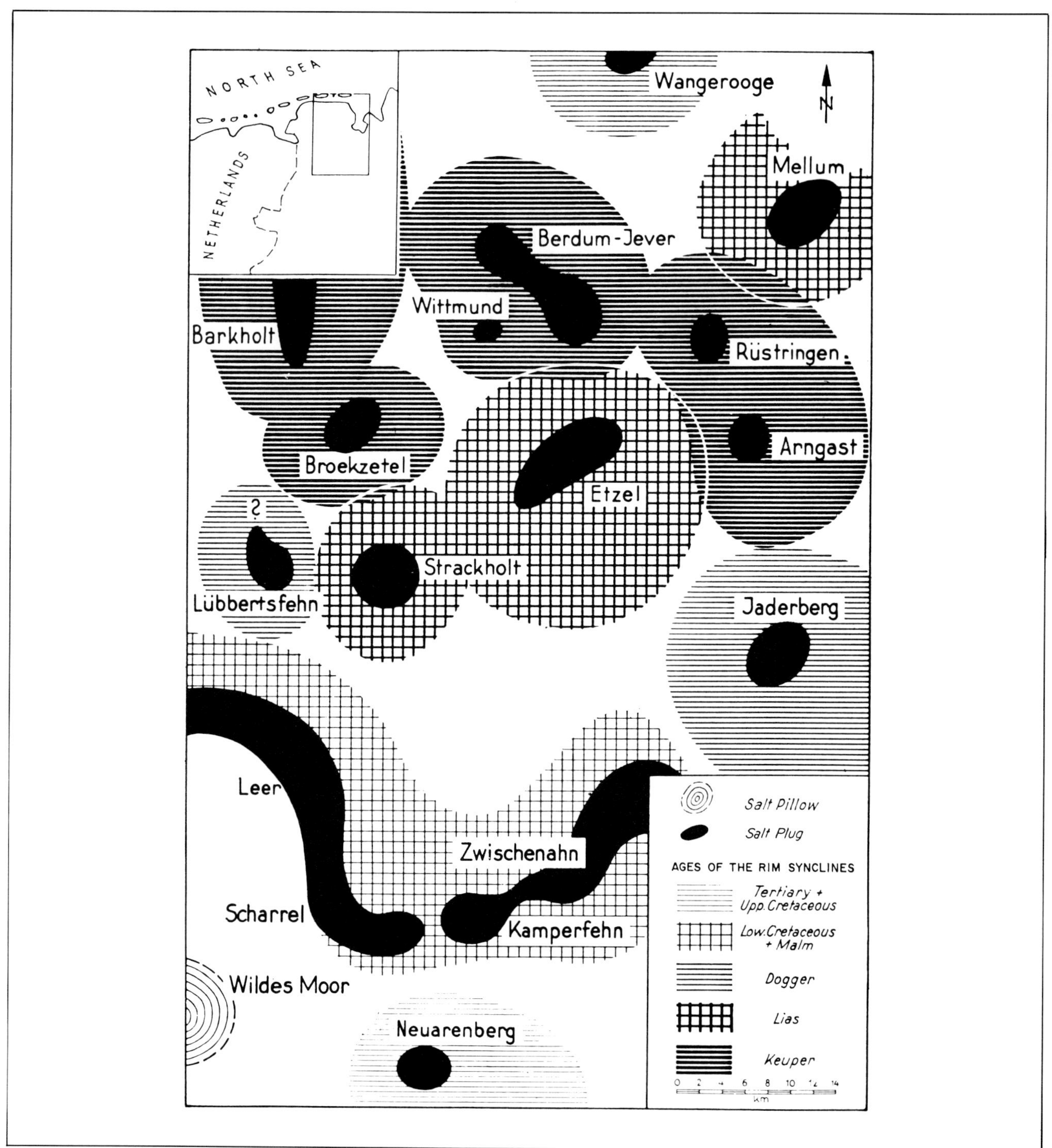

Fig. 8-12 (Trusheim, 1960)—Part of salt dome family in northwest Germany, showing different ages of secondary rim synclines. Oldest salt stocks (secondary rim synclines in Keuper) surrounded by younger diapirs. Permission to publish by American Association of Petroleum Geologists.

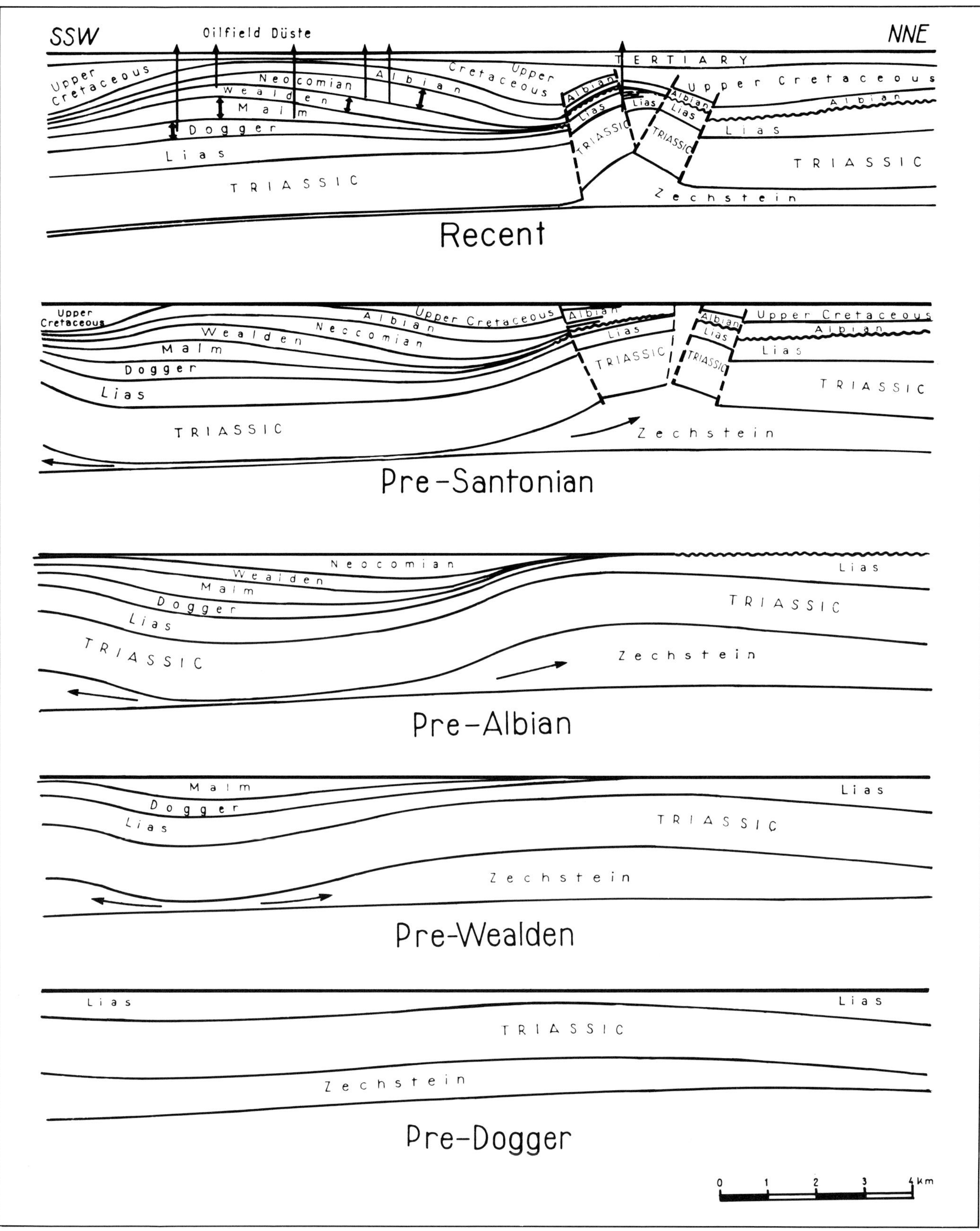

Fig. 8-13 (Trusheim, 1960)—Evolution of turtle structure (upper cross section) Lower Saxonian Basin of Germany. Arrows = direction of salt migration. Double arrows = maximum thicknesses of units in earlier formed sag. Permission to publish by American Association of Petroleum Geologists.

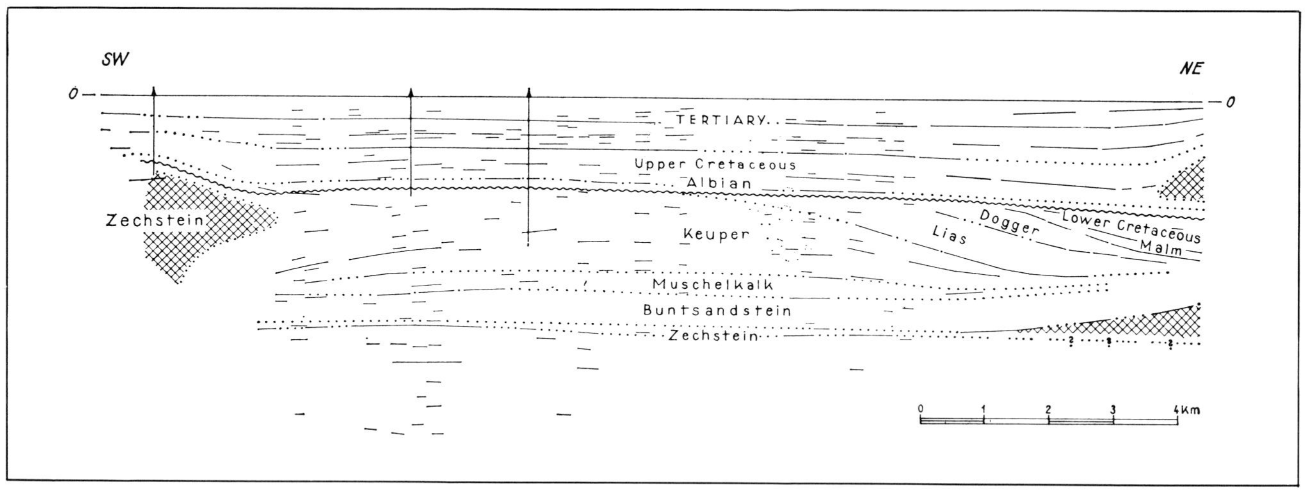

Fig. 8-14 (Trusheim, 1960)—Seismic profile through interdomal turtle-shape structure at Asendorf, northwest Germany. Permission to publish by American Association of Petroleum Geologists.

salt structures as a result of upward movement of the salt generating a vertically oriented maximum principal compressive stress (Fig. 8-21). Currie (1956), simulated normal faults above a salt dome by forcing a piston into a layered section confined by two plexiglass sheets. Sedimentation concurrent with faulting was achieved by successive build up of layers during deformation (Fig. 8-22). Some supradome fault patterns can be exceedingly complex (Fig. 4-34, 4-35).

In plan view, fault patterns theoretically should be radial above circular salt domes. Radial faults, resembling spokes on a wheel, have been created in clay model experiments (Fig. 8-23) and are well known in nature (Figs. 8-24, 8-25). When uplift is elongate rather than domal, or when there is a component of regional extension, a preferential direction of faults ensues (Figs. 8-26, 8-27).

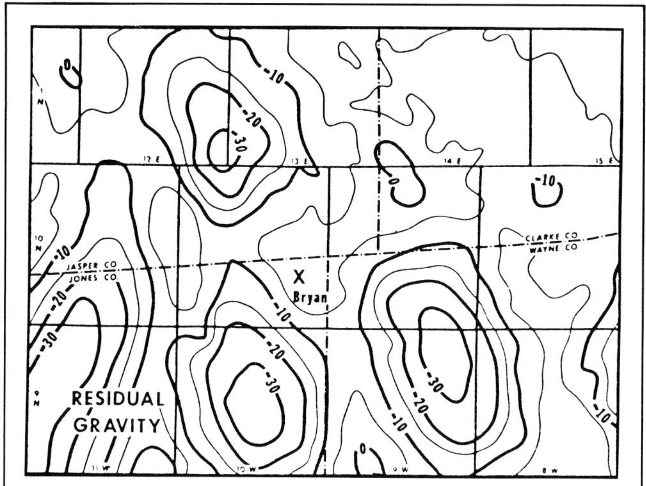

Fig. 8-15 (Oxley and Herlihy, 1972)—Residual gravity map showing relationships between Bryan field and nearby salt features, Mississippi. Permission to publish by The Society of Exploration Geophysicists.

PLAN VIEW

Trends of salt structures are practically limitless. Where strongly influenced by the tectonic forces that shape other structural styles, they assume the trends characteristic of those styles, as in the Carpathian thrust-fold belt of the Ploesti area of Romania (Fig. 1-27). If external tectonic forces are weaker, their influence on the trends of salt structures is more subtle, as in numerous rift basins where underlying normal faults merely trigger salt movement (Fig. 8-28) which then becomes essentially buoyant.

Salt structures that have undergone dominantly buoyant movement theoretically should have less organized trend patterns. This seems to be true for salt structures in the northern North Sea (Fig. 1-28). However, the more elongate salt piercement and pillow structures in the southern North Sea have been at least partly controlled by underlying normal faults. The distribution of primarily buoyant salt structures is related to original basin configuration (Fig. 8-29) and to zonation partly dictated by original salt thickness (Fig. 8-3).

SEISMIC EXPRESSION

Seismic expression of some salt domes, salt anticlines, rim synclines, turtle structures, and residual highs has already been demonstrated by figures 1-26, 4-7, 8-11, and 8-28. It is instructive, however, to consider additional diapiric or discordant structures and concordant structures, some of whose characteristics have not yet been discussed.

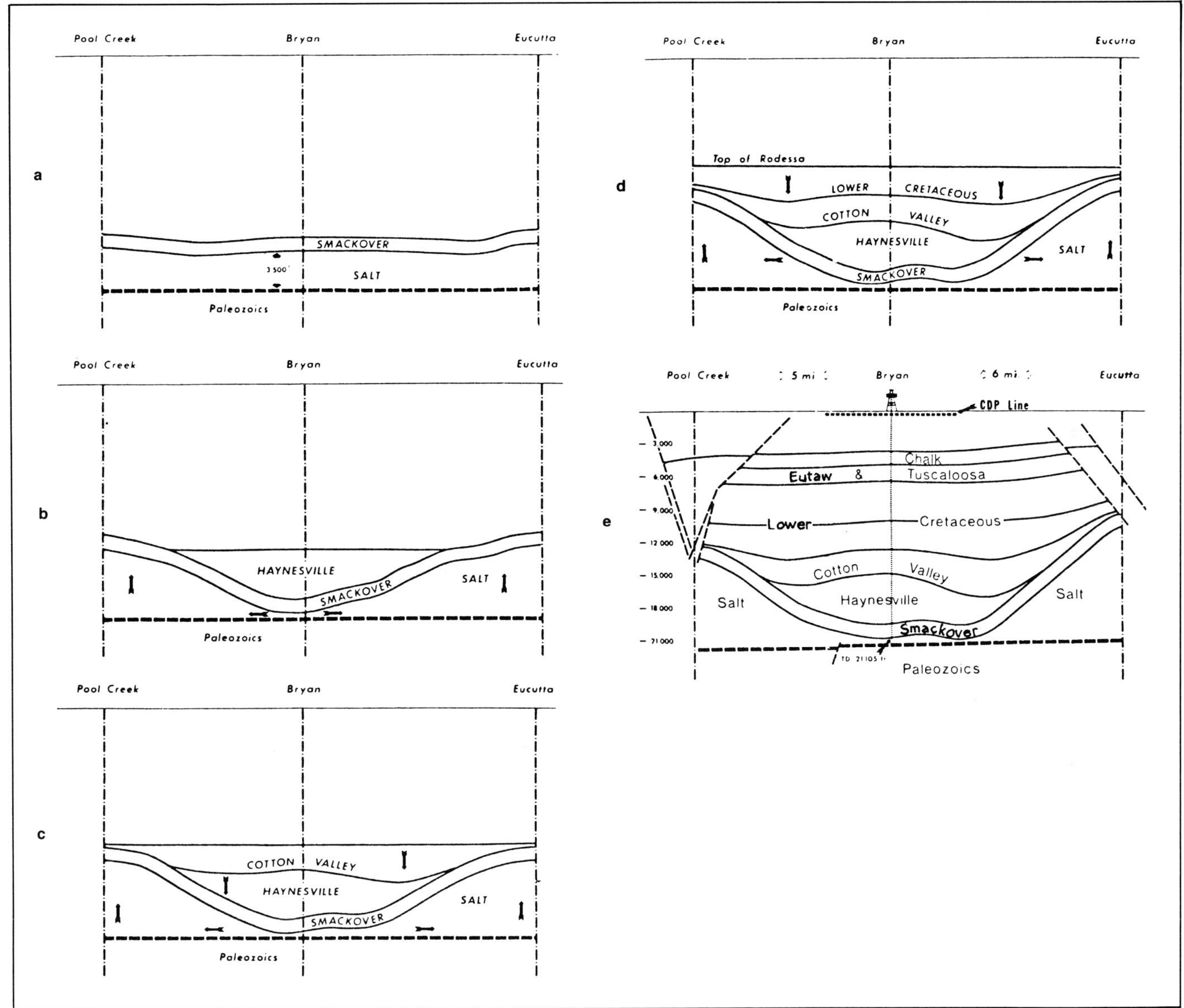

Fig. 8-16 (Oxley and Herlihy, 1972)—Schematic cross sections showing evolution of Bryan field as a turtle structure. Faulting not shown. a. End of Smackover. b. End of Haynesville. c. End of Cotton Valley (top of Jurassic). d. End of Rodessa, Lower Cretaceous (maximum hydrocarbon accumulations). e. Present. Equal volume of sediments maintained in each sequence. Horizontal equals 1.77 × vertical. Permission to publish by The Society of Exploration Geophysicists.

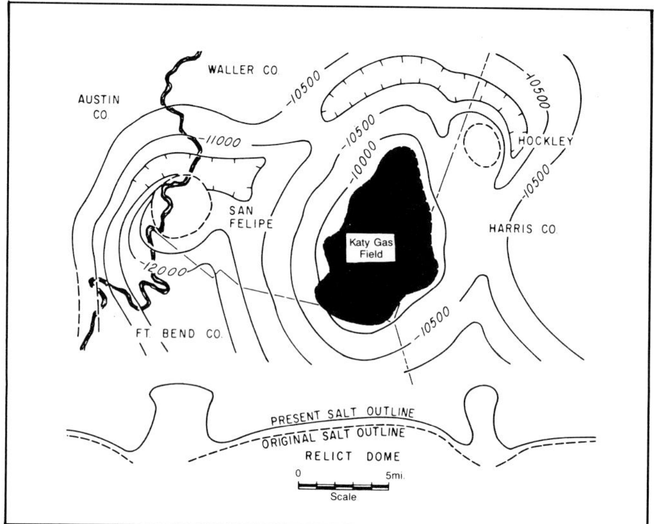

Fig. 8-17 (Halbouty, 1967)—Structure contour map and idealized section of the San Felipe-Katy-Hockley area, Waller and Harris counties, Texas. Katy is postulated as a relict salt mass, a primary uplift in which the locus or path of growth has shifted and ceased to develop into a diapiric structure. From the synclinal area adjacent to Katy, San Felipe and Hockley domes grew rapidly after Katy became stable. The oil and gas at Katy had already migrated and been trapped before Hockley and San Felipe reached their present stage of uplift. This may explain why those shallow piercement domes are virtually barren of oil and gas accumulation. From Salt Domes—Gulf Region, United States and Mexico, Second Edition, by Michel T. Halbouty, copyright © 1979 by Gulf Publishing Company, Houston, Texas. Used with permission. All rights reserved.

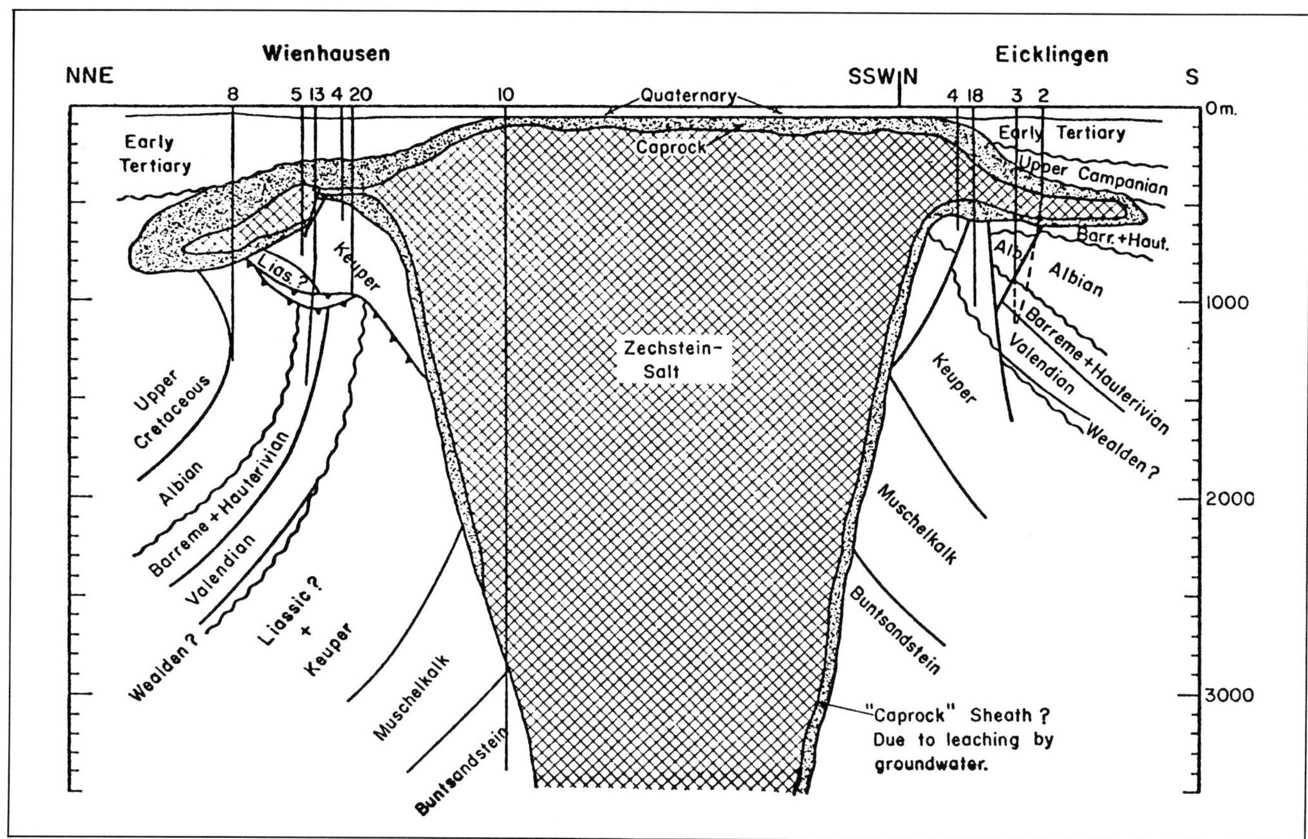

Fig. 8-18 (Gussow, 1968)—Wienhausen salt dome, northwest Germany. Note reverse faults on NNE side. Permission to publish by American Association of Petroleum Geologists.

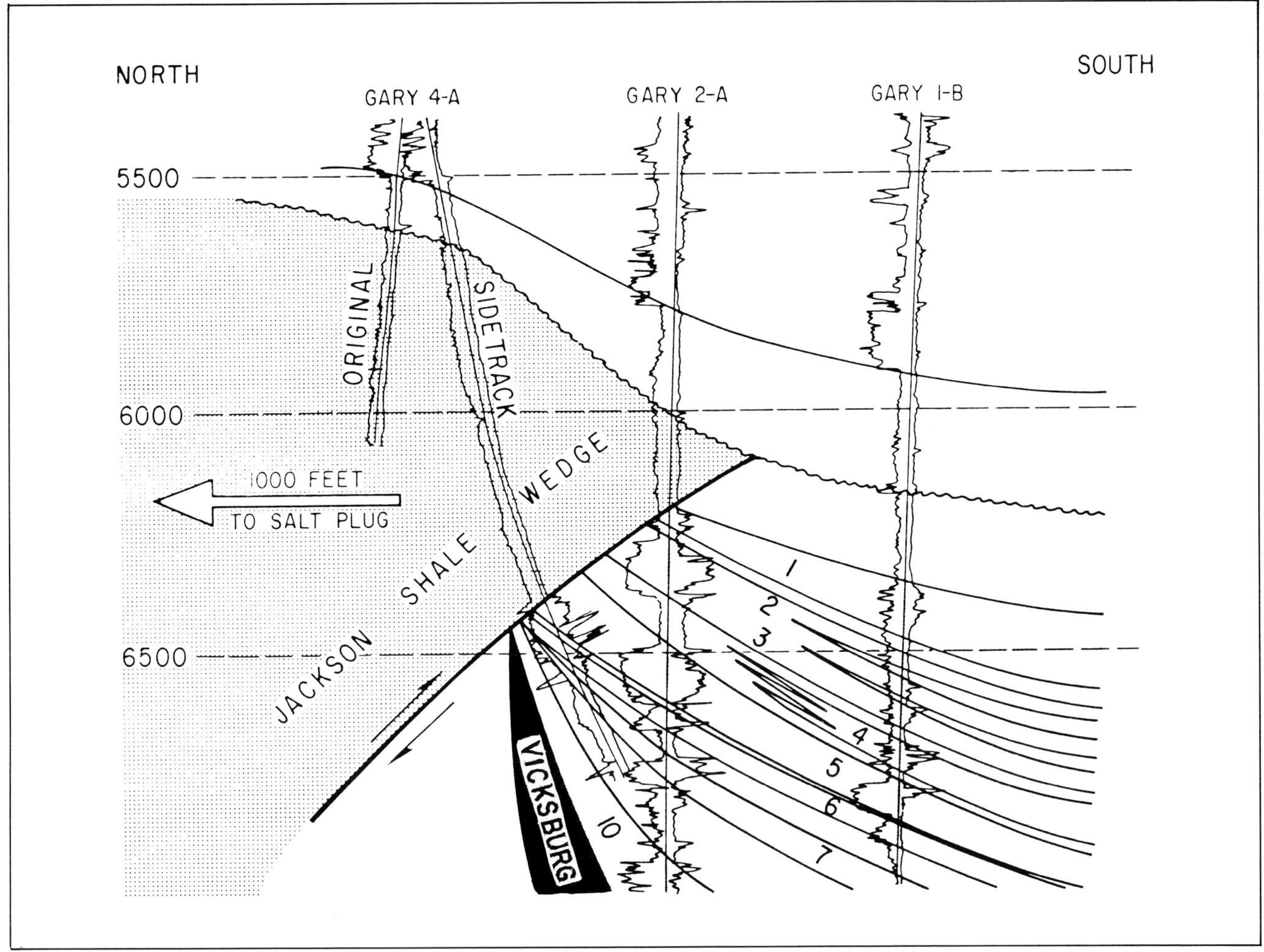

Fig. 8-19 (Halbouty, 1967)—Cross section through South Boling field, U.S. Gulf Coast, showing Jackson shale wedge as upthrust over productive Frio sands (arbitrarily numbered for ease in correlation). Reprinted from "Salt Domes—Gulf Region, United States and Mexico" by Michel T. Halbouty, WORLD OIL, 1967.

Different times of salt growth are illustrated on the same line in the Aquitaine Basin (Fig. 8-30). A salt diapir without associated secondary rim synclines is known from offshore Louisiana (Fig. 8-31). Salt pillows from the southern North Sea (Fig. 8-32) and east Texas (Fig. 8-33) both show clear growth histories and both sub-salt and supra-salt normal fault systems. The effects of downslope movement of salt are illustrated from Cabinda, Angola (Fig. 8-34). Finally, at the base of progradational sedimentary systems that undergo gravity creep, downstream compression can be represented by extrusion or thrusting of salt (Fig. 8-35).

PRODUCING EXAMPLES

By far the most petroleum related to salt structures is in traps concordant to those structures. Ekofisk field in the North Sea (Byrd, 1975) and the Dubai fields in the Persian Gulf are examples of prolific trapping structures formed in sedimentary rocks uplifted by deep-seated salt movement. This type of structure can be relatively simple effecting large continuous traps and reservoirs (cf. Fig. 1-1); but such structures can also be literally shattered by normal faults (Figs. 8-36, 8-37).

Discordant traps formed by abutment of beds and stratigraphic pinchouts against the sides of salt structures (Fig. 1-1 and 8-38) are important, but are usually smaller and less prolific than concordant traps. They, too, can be complexly faulted and their size thus reduced (Figs. 8-39, 8-40).

Salt flowage is frequently triggered by sub-salt normal faulting as shown by figures 8-11, 8-28, 8-30, 8-32, and 8-33. Consequently an extensional fault block play may exist beneath many salt structures.

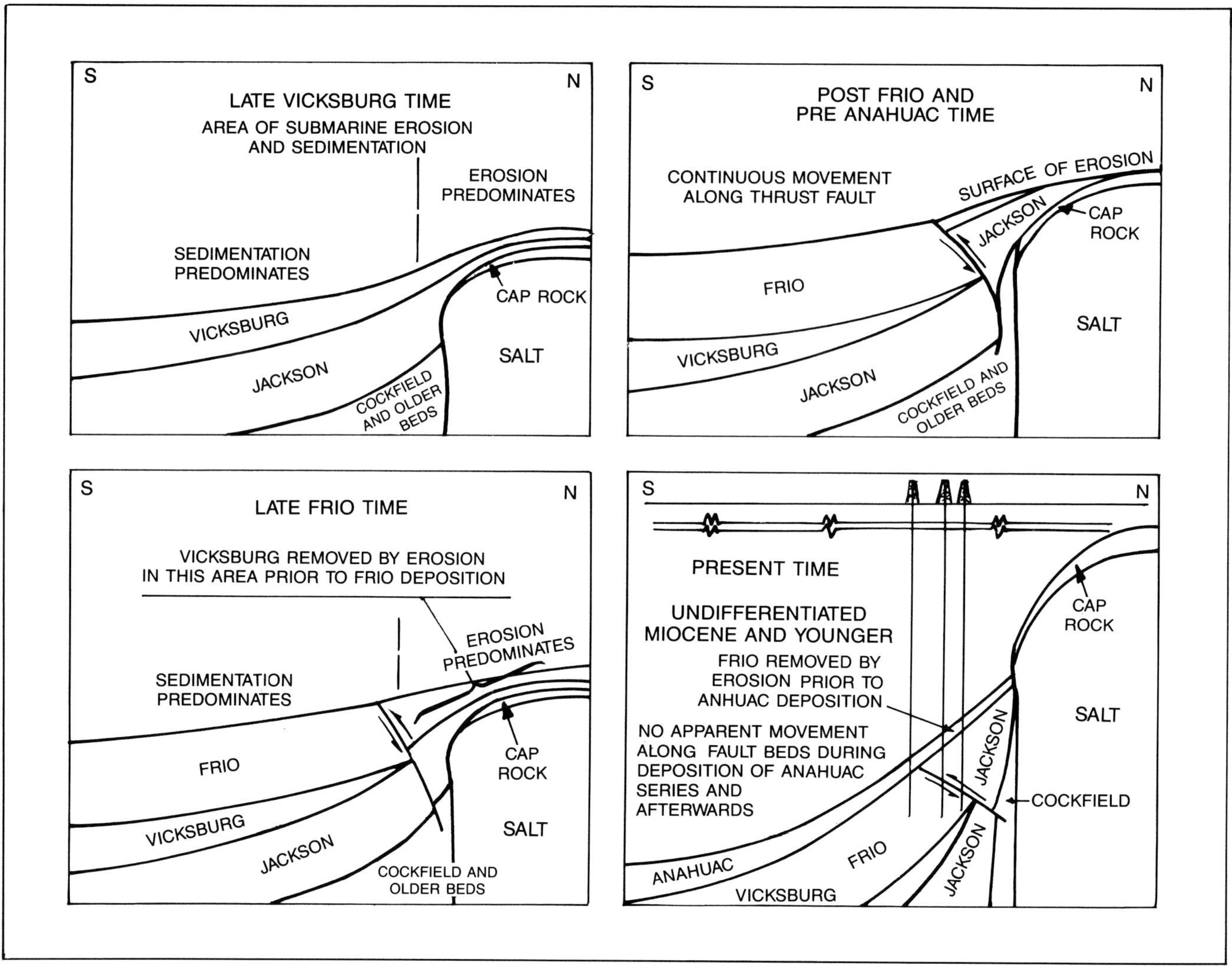

Fig. 8-20 (Halbouty, 1967)—Sequence of cross sections showing the concept of formation of South Boling upthrust in connection with growth of Boling salt dome. Reprinted from "Salt Domes—Gulf Region, United States and Mexico" by Michel T. Halbouty, WORLD OIL, 1967.

REFERENCES

Amery, G. B., 1969, Structure of Sigsbee Scarp, Gulf of Mexico: *American Association Petroleum Geologists Bull.,* V. 58, p. 2480–2482.

Brice, S. E., Cochran, M. D., Pardo, Georges, and Edwards, A. D., 1982, Tectonics and sedimentation of the south Atlantic rift sequence: Cabinda, Angola, *in* Watkins, J. S., and Drake, C. L., eds., Studies in continental margin geology: *American Association Petroleum Geologists Mem. 34,* p. 5–18.

Byrd, W. D., 1975, Geology of the Ekofisk field, offshore Norway, *in* Woodland, A. W., ed., Petroleum and continental shelf of northwest Europe: V. 1, Geology, Halstead Press, p. 439–444.

Bullard, T., 1973, Eugene Island block 126 field, offshore St. Mary Parish, Louisiana: Offshore Louisiana Oil and gas fields, Lafayette Geological Society, p. 111.

Curnelle, R., and Marco, R., 1983, Reflection profiles across the Aquitaine Basin (salt tectonics), *in* Bally A. W., ed., Seismic expression of structural styles: *American Association of Petroleum Geologists Studies in Geology # 15,* V. 2.

Currie, J. B., 1956, Role of concurrent deposition and deformation of sediments in development of salt-dome graben structures: *American Association Petroleum Geologists Bull.,* V. 40, p. 1–16.

Galloway, W. E., Ewing, T. E., Garrett, C. M., Tyler, N. and Bebout, D. G., 1983, Atlas of major Texas reservoirs: Bureau Economic Geology, University of Texas at Austin, 139 p.

Gullett, M. K., 1983, Good Hope field, St. Charles Parish,

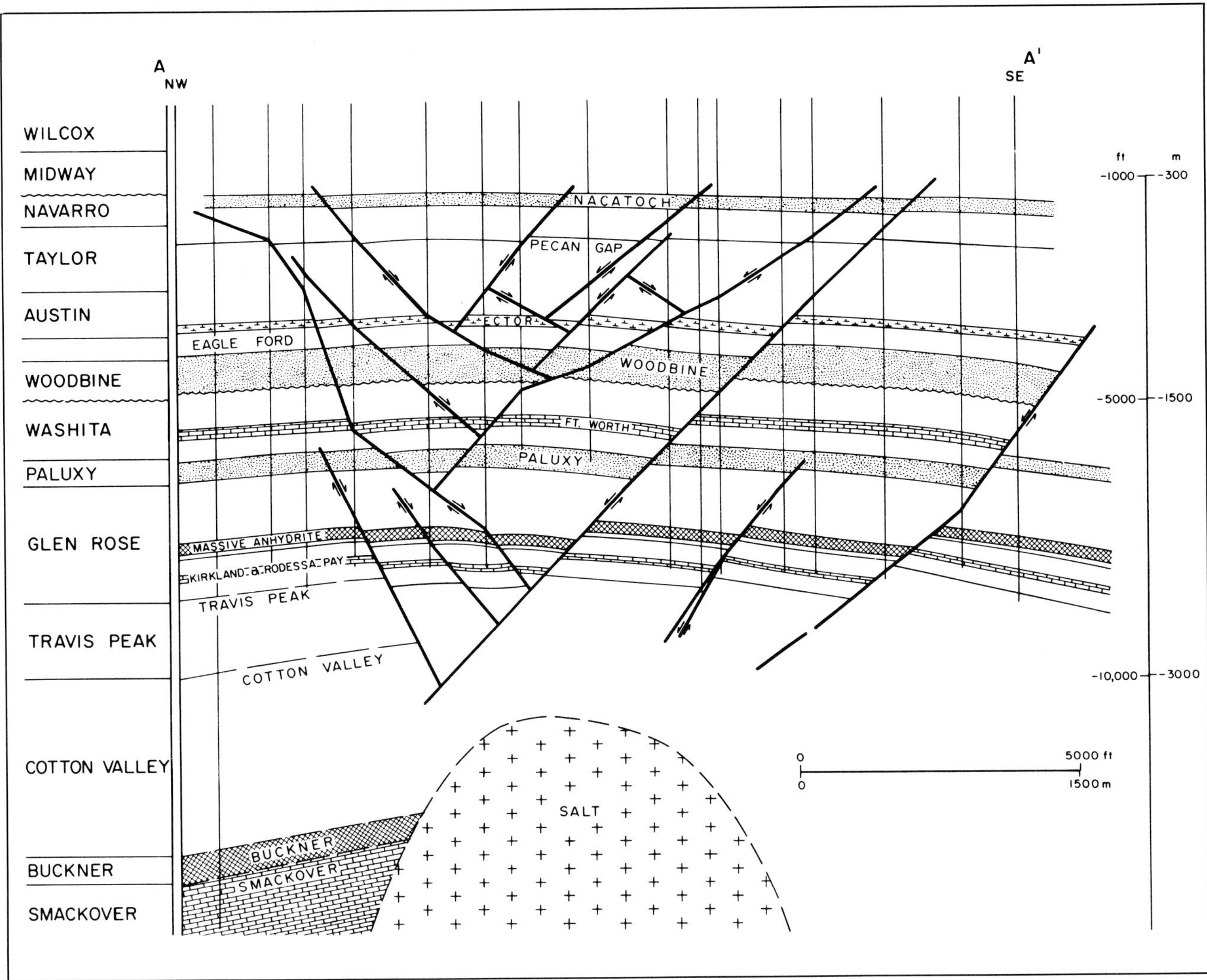

Fig. 8-21 (Galloway et al., 1983)—Cross section through Quitman field, Wood County, Texas, showing complex normal fault systems in sedimentary section in response to vertical uplift of salt below. Permission to publish by American Association of Petroleum Geologists.

Louisiana, *in* McCormick, C. C., ed., Oil and gas fields of southeast Louisiana: New Orleans Geological Society, p. 20.

Gussow, W. C., 1968, Salt diapirism: Importance of temperature, and energy source of emplacement: *American Association Petroleum Geologists Mem. 8,* p. 16–52.

Halbouty, M. T., 1967, Salt domes—Gulf region, United States and Mexico: Gulf Publishing Company, Houston, 425 p.

Hughes, D. J., 1968, Salt tectonics as related to several Smackover fields along the northeast rim of the Gulf of Mexico basin: *Gulf Coast Association Geological Societies Trans.,* V. XVIII, p. 320–330.

Inderwiesen, P. L., 1983, Salt anticline—east Texas basin, *in* Bally, A. W., ed., Seismic expression of structural styles: *American Association of Petroleum Geologists Studies in Geology #15,* V. 2.

Lafayette Geological Society, 1970, Typical oil and gas fields of southwest Louisiana.

Lowell, J. D., Genik, G. J., Nelson, T. H., and Tucker, P. M., 1975, Petroleum and plate tectonics of the southern Red Sea, *in* Fisher, A. G., and Judson, S., eds., Petroleum and global tectonics: Princeton University Press, p. 129–153.

Lukic, P., Maselek, W., and Bachmann, G. H., 1983, Salt plug Siegelsum in the northwest-German Basin, *in* Bally, A. W., ed., Seismic expression of structural styles: *American Association Petroleum Geologists Studies in Geology #15,* V. 2.

Martin, R. G., 1980, Distribution of salt structures in the Gulf of Mexico: United States Geological Survey Map MF-1213.

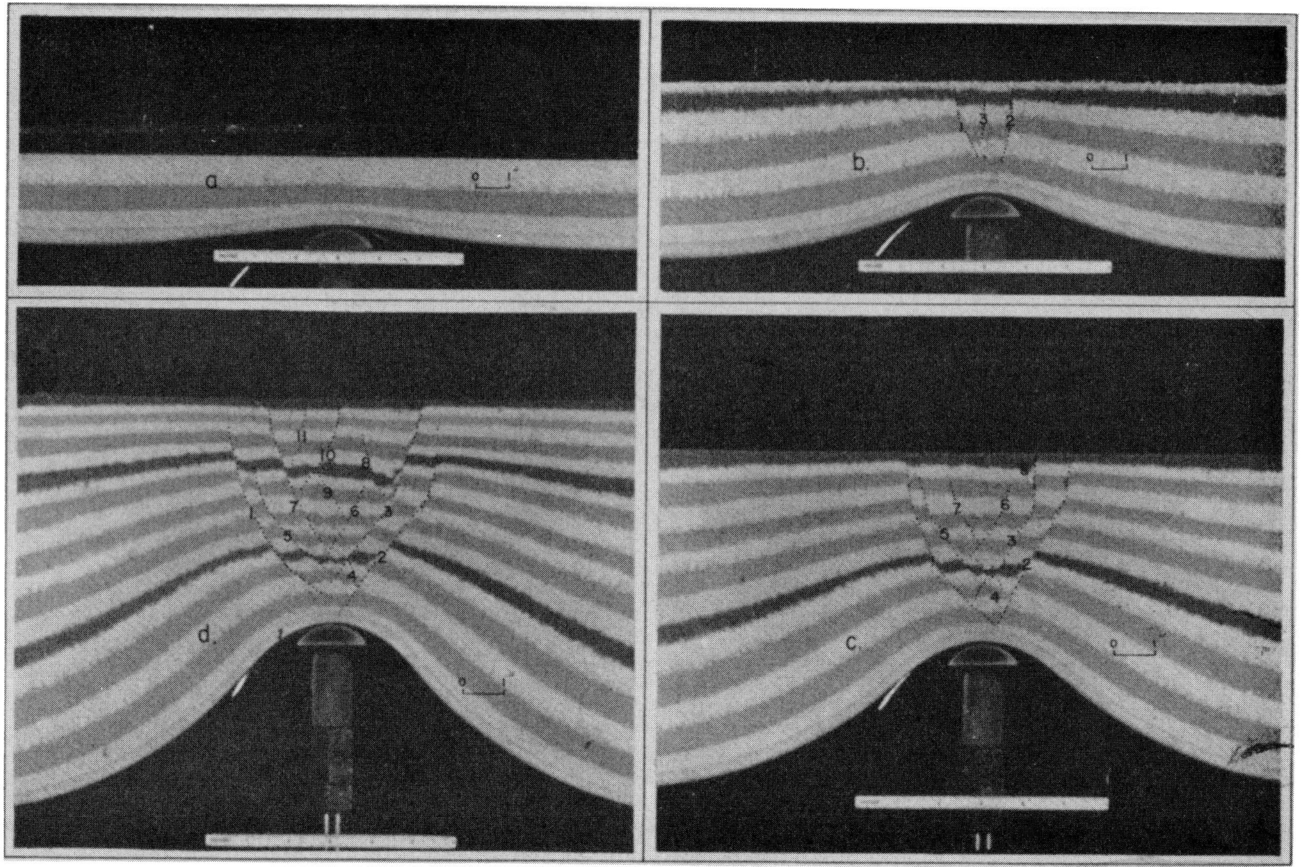

Fig. 8-22 (Currie, 1956)—Model simulation of normal faults above a salt dome. Note sequence of faulting; youngest faults are always in center. Permission to publish by American Association of Petroleum Geologists.

O'Brien, G. D., 1968, Survey of diapirs and diapirism: *American Association Petroleum Geologists Mem. 8,* p. 1–9.

Owen, P. F., and Taylor, N. G., 1983, A salt pillow structure in the southern North Sea, *in* Bally A. W., ed., Seismic expression of structural styles: *American Association Petroleum Geologists Studies in Geology #15,* V. 2.

Oxley, M. L., and Herlihy, D. E., 1972, The Bryan field—a sedimentary anticline: *Geophysics,* V. 37, p. 59–67.

Robinson, E. C., 1970, Gueydan field, Acadia and Vermillion parishes, Louisiana: Typical oil and gas fields of southwest Louisiana, Lafayette Geological Society, p. 12.

Sunwall, M. T., McQuillan, K. A., and Nick, C. J. 1983, Salt diapir—Gulf of Mexico, *in* Bally A. W., ed., Seismic expression of structural styles: *American Association Petroleum Geologists Studies in Geology #15,* V. 2.

Trusheim, F., 1960, Mechanism of salt migration in northern Germany: *American Association Petroleum Geologists Bull.,* V. 44, p. 1519–1540.

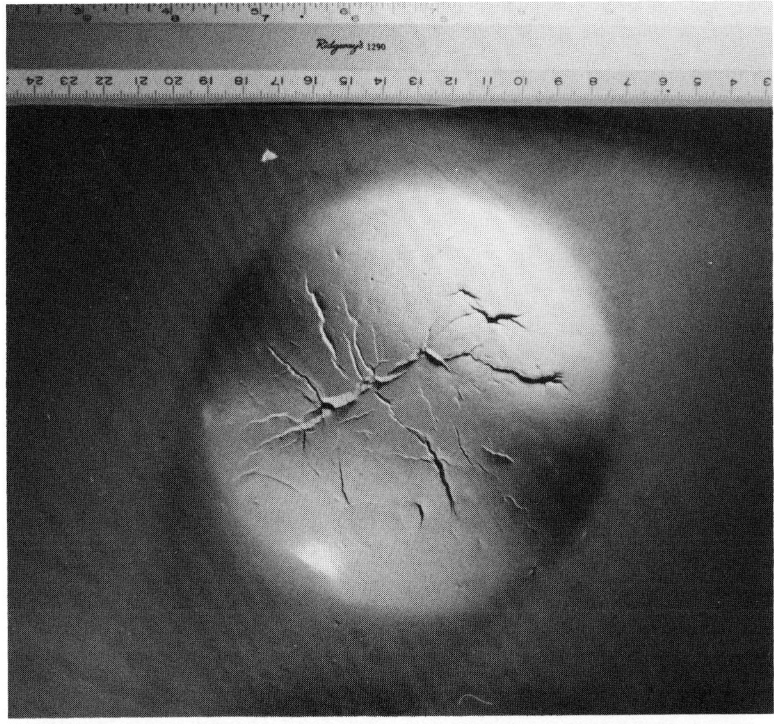

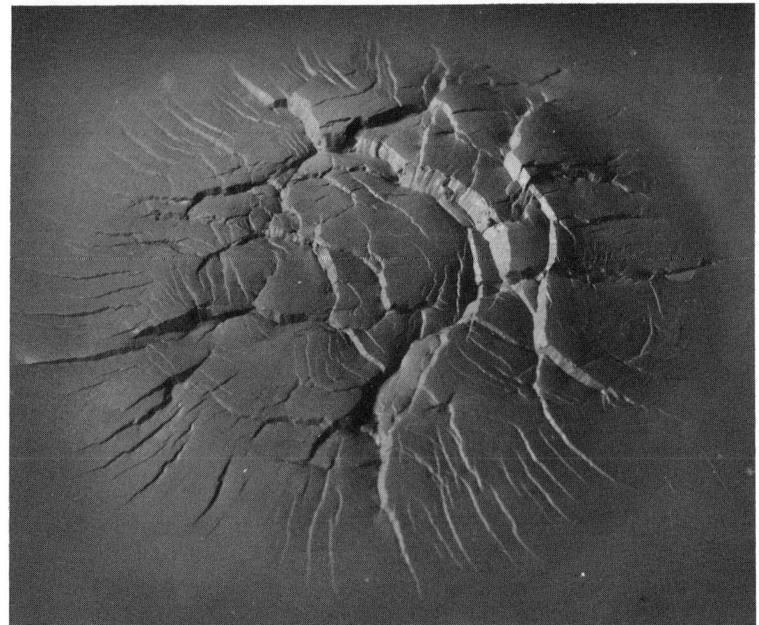

Fig. 8-23—Clay models illustrating radial fault patterns. Perfectly circular or domal uplift is achieved by inflating rubber balloon through a circular cutout in a metal template beneath the clay.

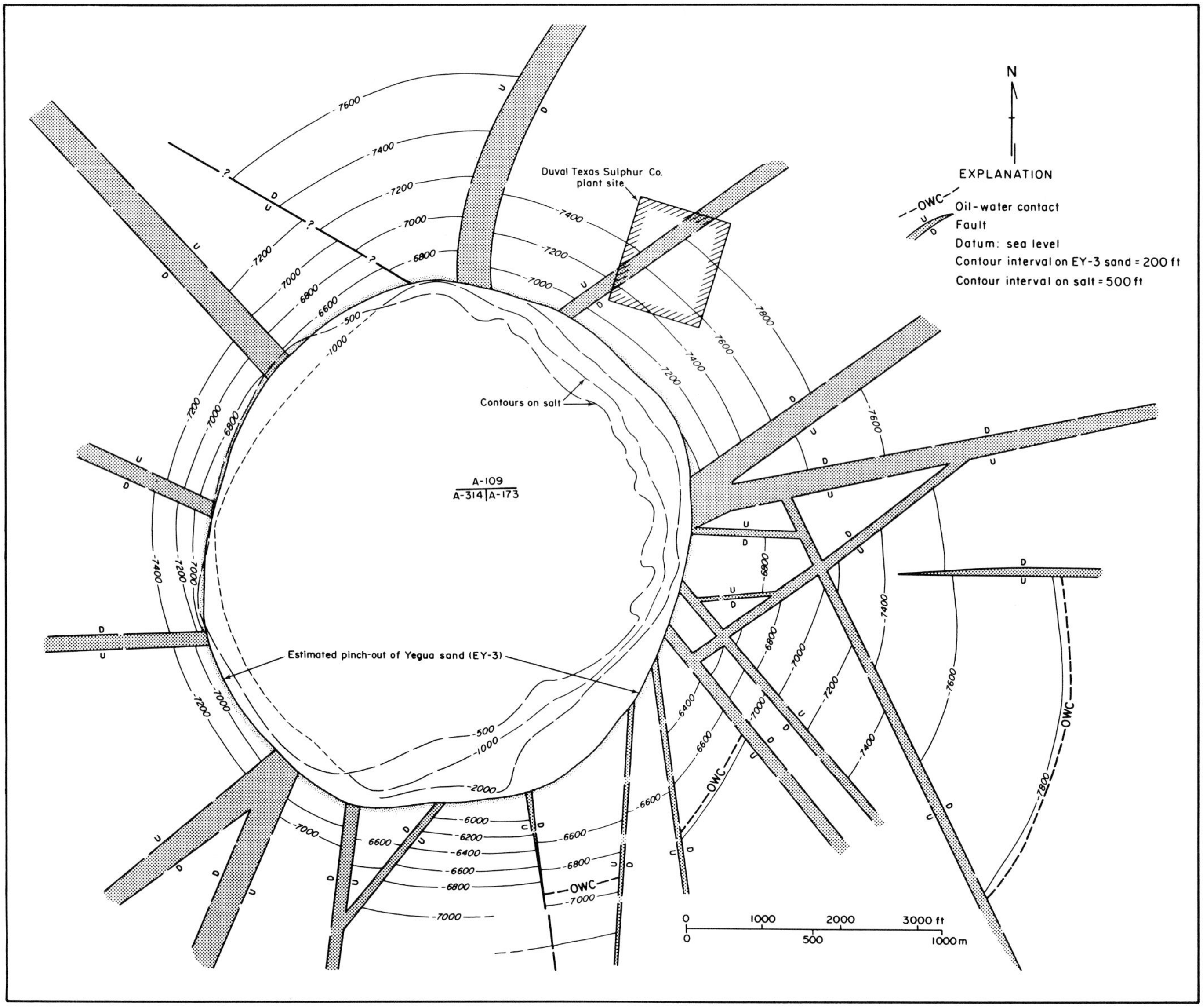

Fig. 8-24 (Galloway et al., 1983) (above)—Structure contour map (on top of salt-dashed and top of EY-3 Yegua Eocene sand-solid) of Moore's Orchard field, Fort Bend County, Texas, showing radial fault pattern as a response to circular shallow doming of salt. Permission to publish by Bureau of Economic Geology, University of Texas, Austin.

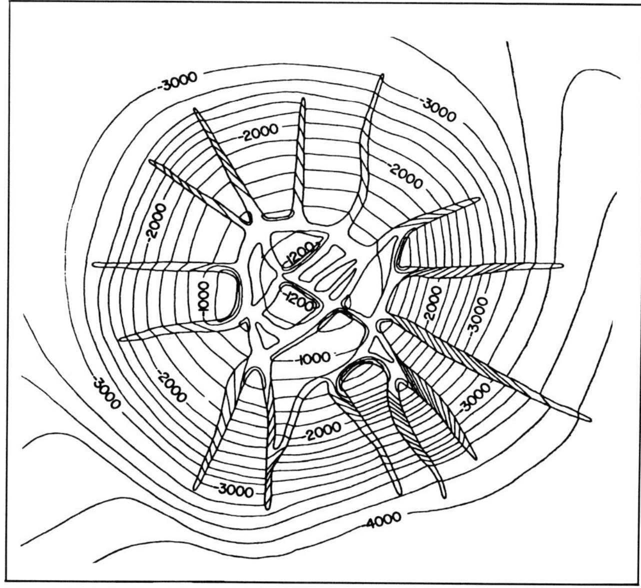

Fig. 8-25 (Gussow, 1968)—Wilcox structure map, Clay Creek dome, Washington County, Texas, showing pattern of radial fractures and central graben above a salt plug. Datum is top of Wilcox. Contour interval is 61 m (200 ft). Permission to publish by American Association of Petroleum Geologists.

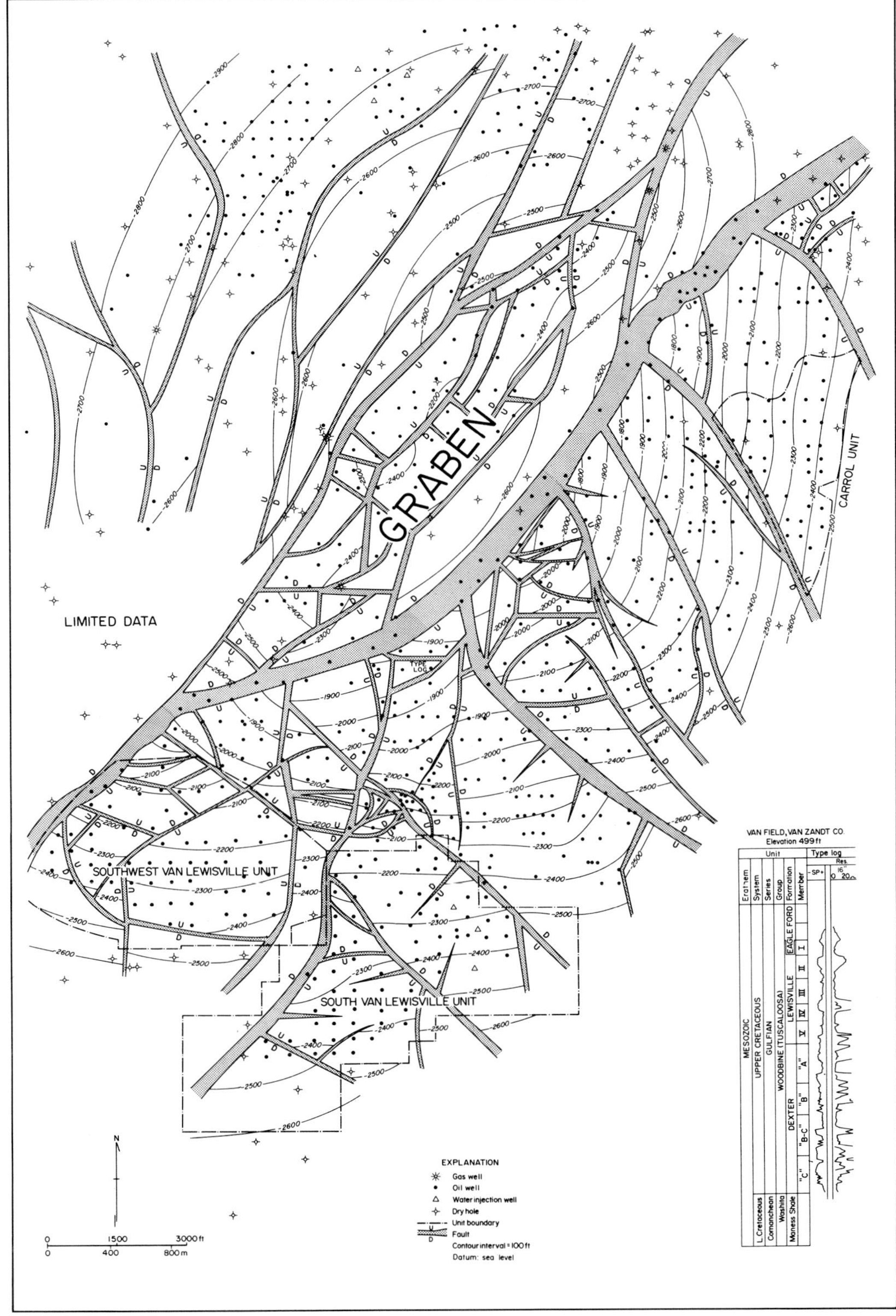

Fig. 8-26 (Galloway et al., 1983)— Structure contour map (on top of Woodbine) of Van field, Van Zandt County, Texas, showing fault pattern from elongated uplift by deep-seated salt movement. Permission to publish by Bureau of Economic Geology, University of Texas, Austin.

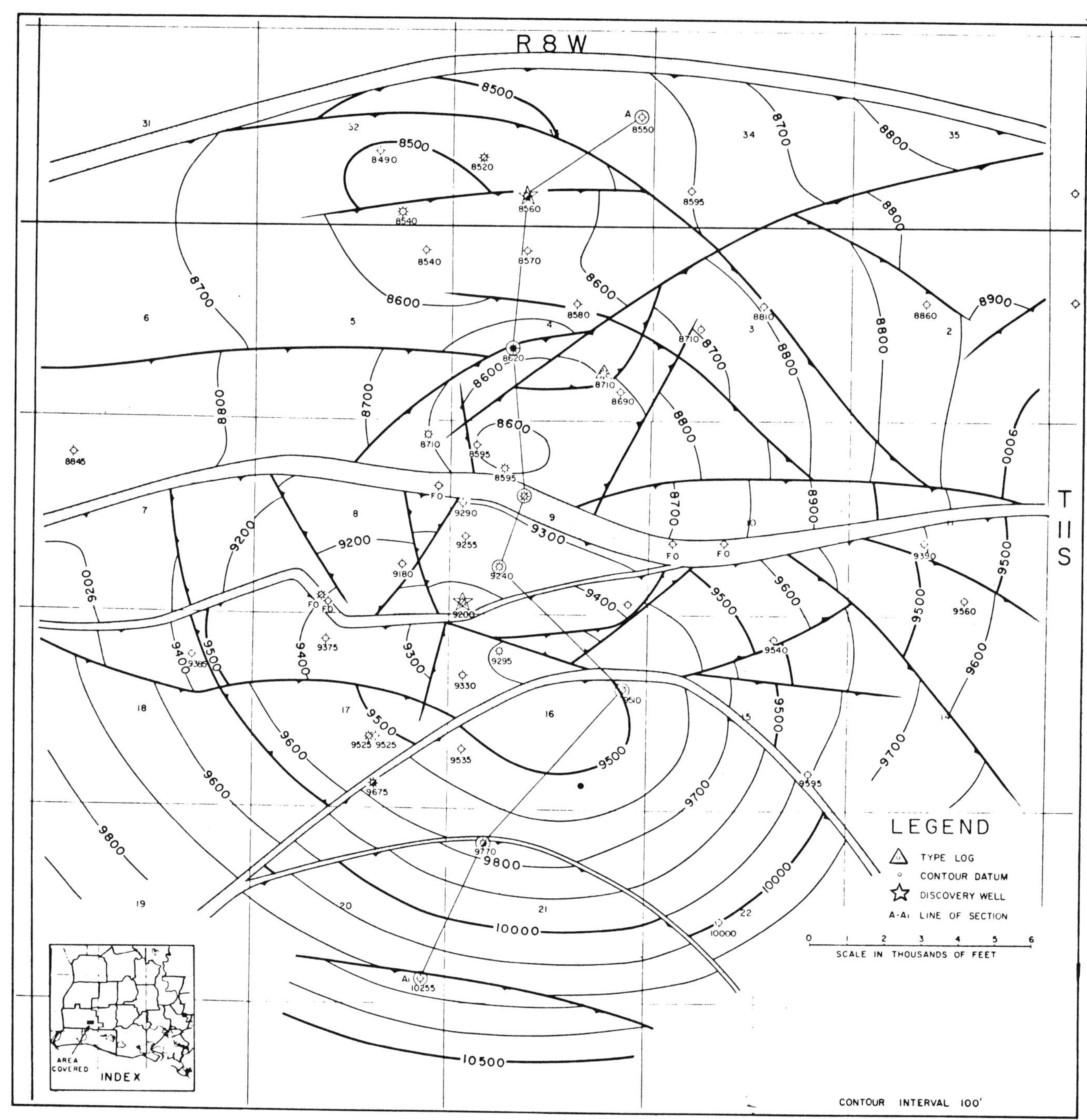

Fig. 8-27 (Lafayette Geological Society, 1970)—Structure contour map (on Bolivina perca lime) of South Lake Charles and Coulee Hippolyte fields, Calcasieu Parish, Louisiana, showing interference of concave seaward faults of detached normal fault style with deep-seated salt features. Permission to publish by Lafayette Geological Society.

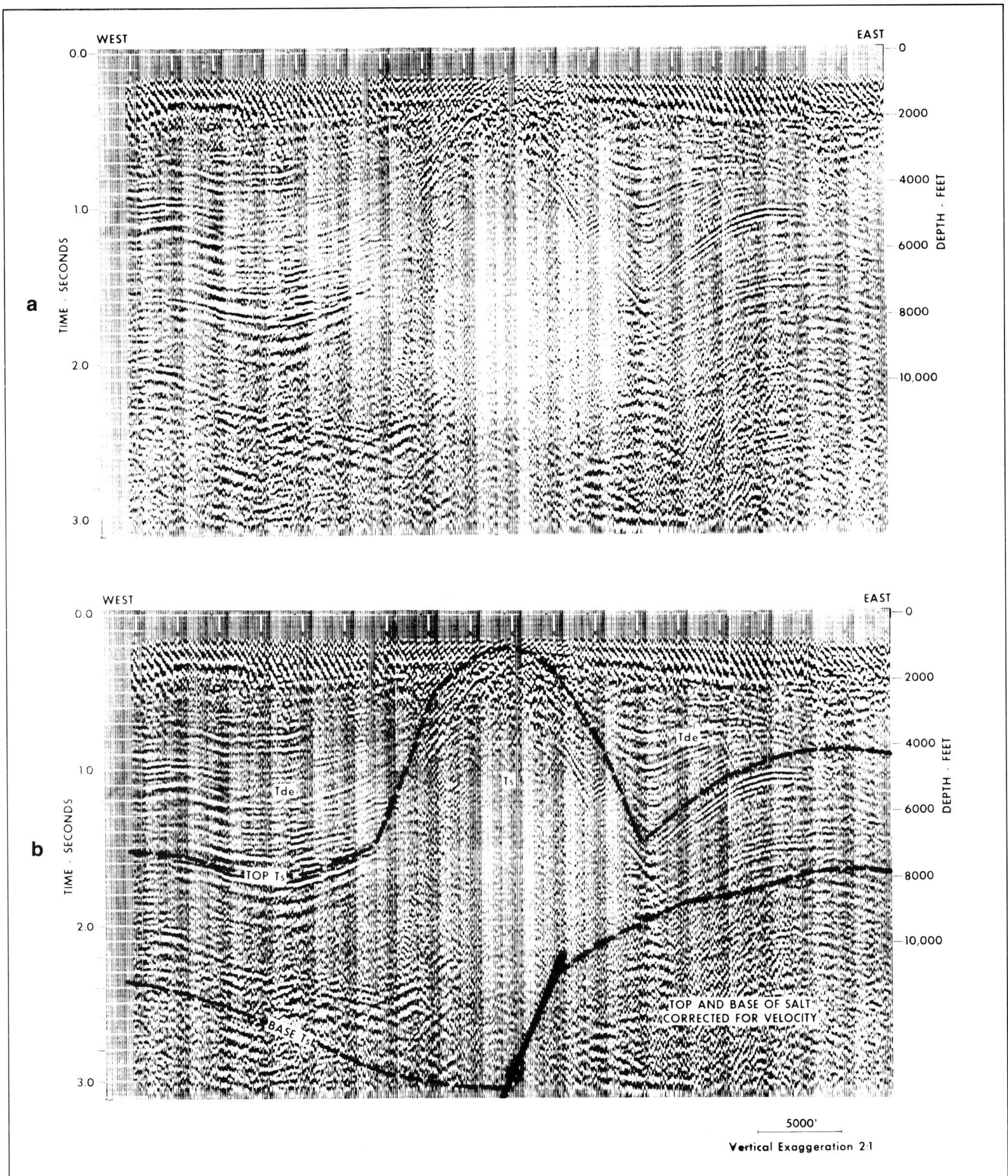

Fig. 8-28 (Lowell et al., 1975)—Reflection seismic section in Ethiopian waters of the southern Red Sea showing piercement salt dome with normal fault below. Ts = salt (Miocene). Tde = Desset series (Pliocene). From Petroleum and Global Tectonics, copyright © Princeton University Press. Used with permission.

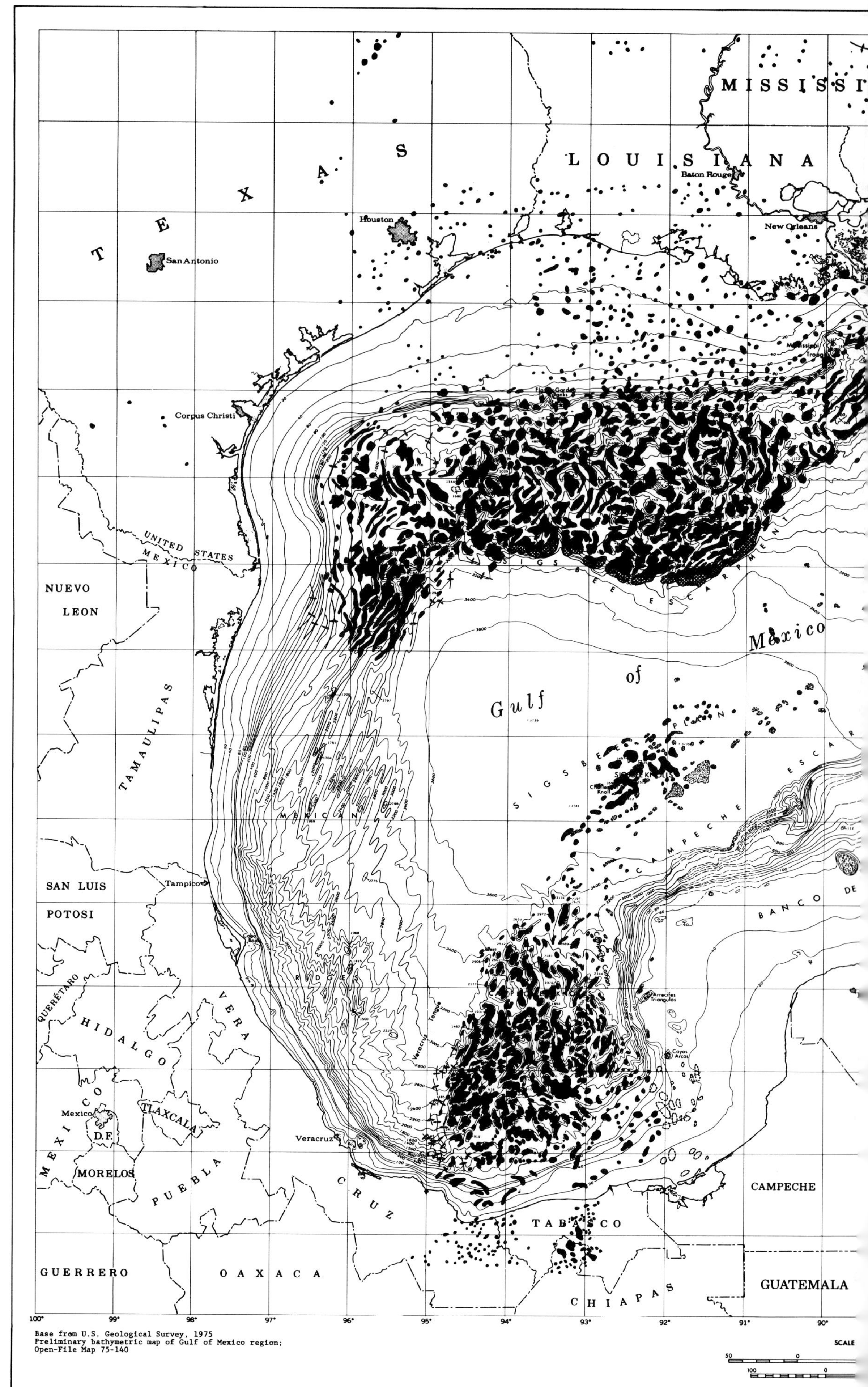

Fig. 8-29 (Martin, 1980)—Distribution of salt structures in the Gulf of Mexico and environs.

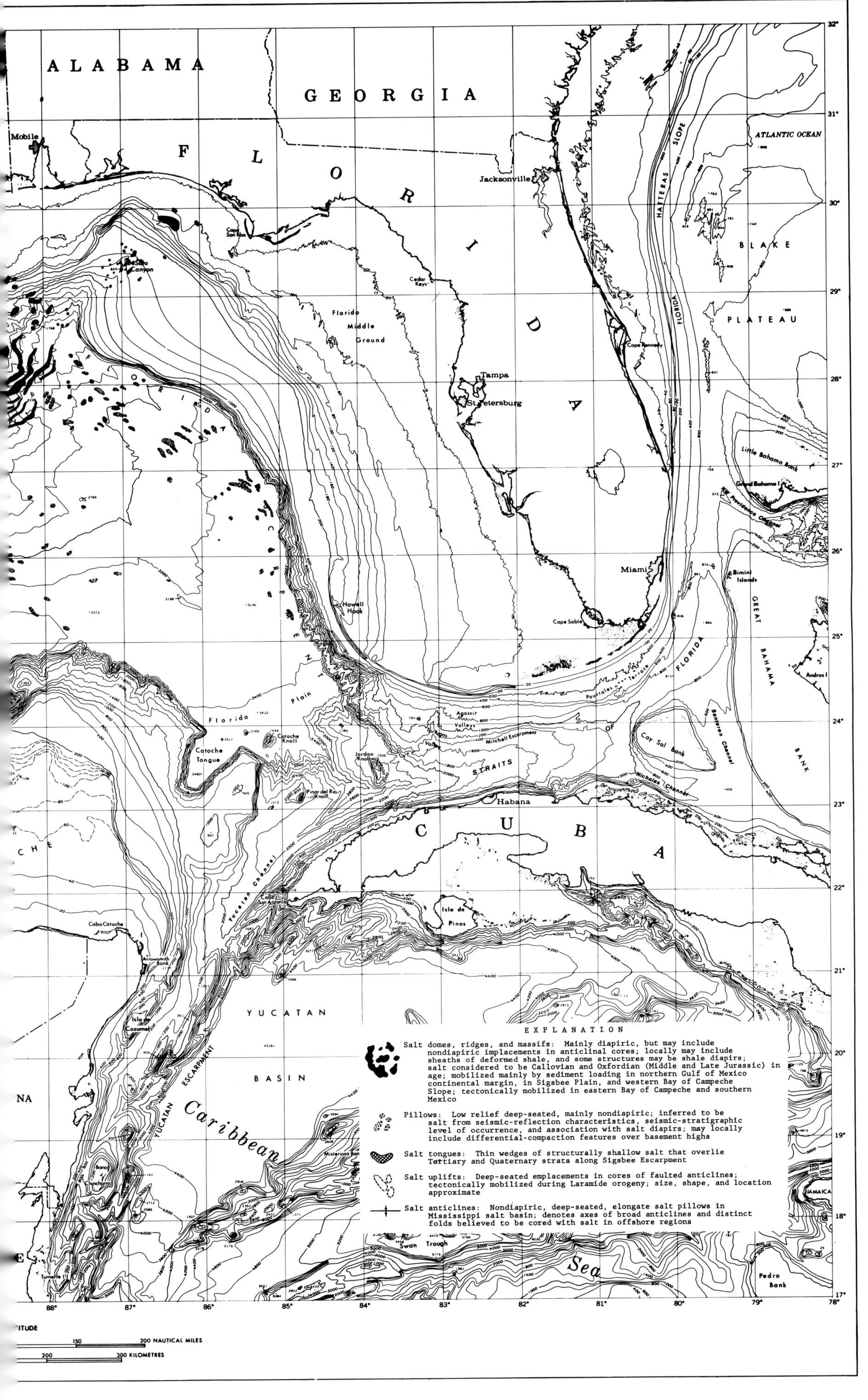
ALABAMA
GEORGIA
FLORIDA
ATLANTIC OCEAN
Mobile
Jacksonville
Cape San Blas
DeSoto Canyon
Cedar Keys
Florida Middle Ground
Tampa
St. Petersburg
Cape Kennedy
HATTERAS SLOPE
FLORIDA SLOPE
BLAKE PLATEAU
Little Bahama Bank
Grand Bahama I.
NW Providence Channel
Miami
Bimini Islands
GREAT BAHAMA BANK
Andros I.
Cape Sable
Howell Hook
FLORIDA ESCARPMENT
Florida Plain
Catoche Knoll
Catoche Tongue
Jordan Knoll
Pinar del Rio Knoll
Agassiz Valleys
Mitchell Escarpment
Pourtales Terrace
FLORIDA
STRAITS OF FLORIDA
Cay Sal Bank
Santaren Channel
Nicholas Channel
Habana
CUBA
Yucatan Channel
Cabo San Antonio
Isla de Pinos
Cabo Catoche
Arrowsmith Bank
Isla de Cozumel
YUCATAN BASIN
YUCATAN ESCARPMENT
Caribbean Sea
Misteriosa Bank
Swan Trough
JAMAICA
Pedro Bank
Turneffe I.
32°
31°
30°
29°
28°
27°
26°
25°
24°
23°
22°
21°
20°
19°
18°
17°
88°
87°
86°
85°
84°
83°
82°
81°
80°
79°
78°
150
200 NAUTICAL MILES
200
300 KILOMETRES
EXPLANATION
Salt domes, ridges, and massifs: Mainly diapiric, but may include nondiapiric implacements in anticlinal cores; locally may include sheaths of deformed shale, and some structures may be shale diapirs; salt considered to be Callovian and Oxfordian (Middle and Late Jurassic) in age; mobilized mainly by sediment loading in northern Gulf of Mexico continental margin, in Sigsbee Plain, and western Bay of Campeche Slope; tectonically mobilized in eastern Bay of Campeche and southern Mexico
Pillows: Low relief deep-seated, mainly nondiapiric; inferred to be salt from seismic-reflection characteristics, seismic-stratigraphic level of occurrence, and association with salt diapirs; may locally include differential-compaction features over basement highs
Salt tongues: Thin wedges of structurally shallow salt that overlie Tertiary and Quaternary strata along Sigsbee Escarpment
Salt uplifts: Deep-seated emplacements in cores of faulted anticlines; tectonically mobilized during Laramide orogeny; size, shape, and location approximate
Salt anticlines: Nondiapiric, deep-seated, elongate salt pillows in Mississippi salt basin; denotes axes of broad anticlines and distinct folds believed to be cored with salt in offshore regions

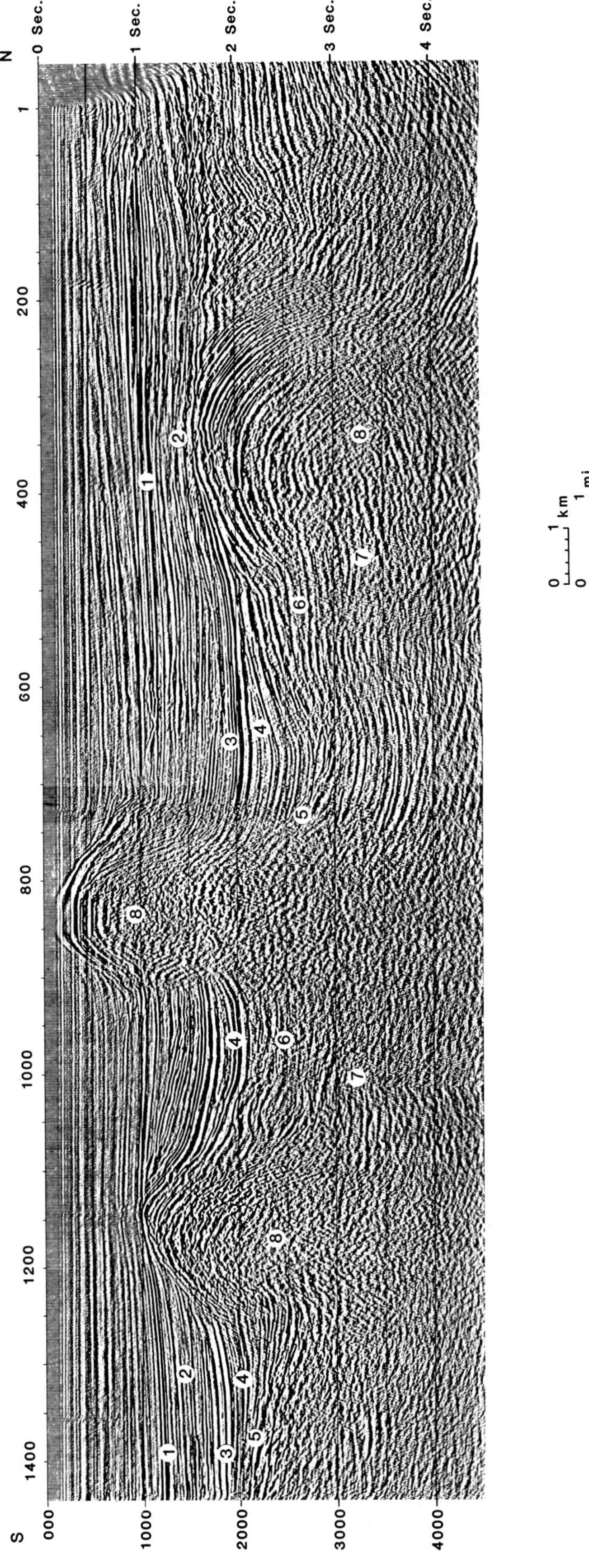
N
S
0 Sec.
1 Sec.
2 Sec.
3 Sec.
4 Sec.
1
200
400
600
800
1000
1200
1400
000
1000
2000
3000
4000
0 1 km
0 1 mi

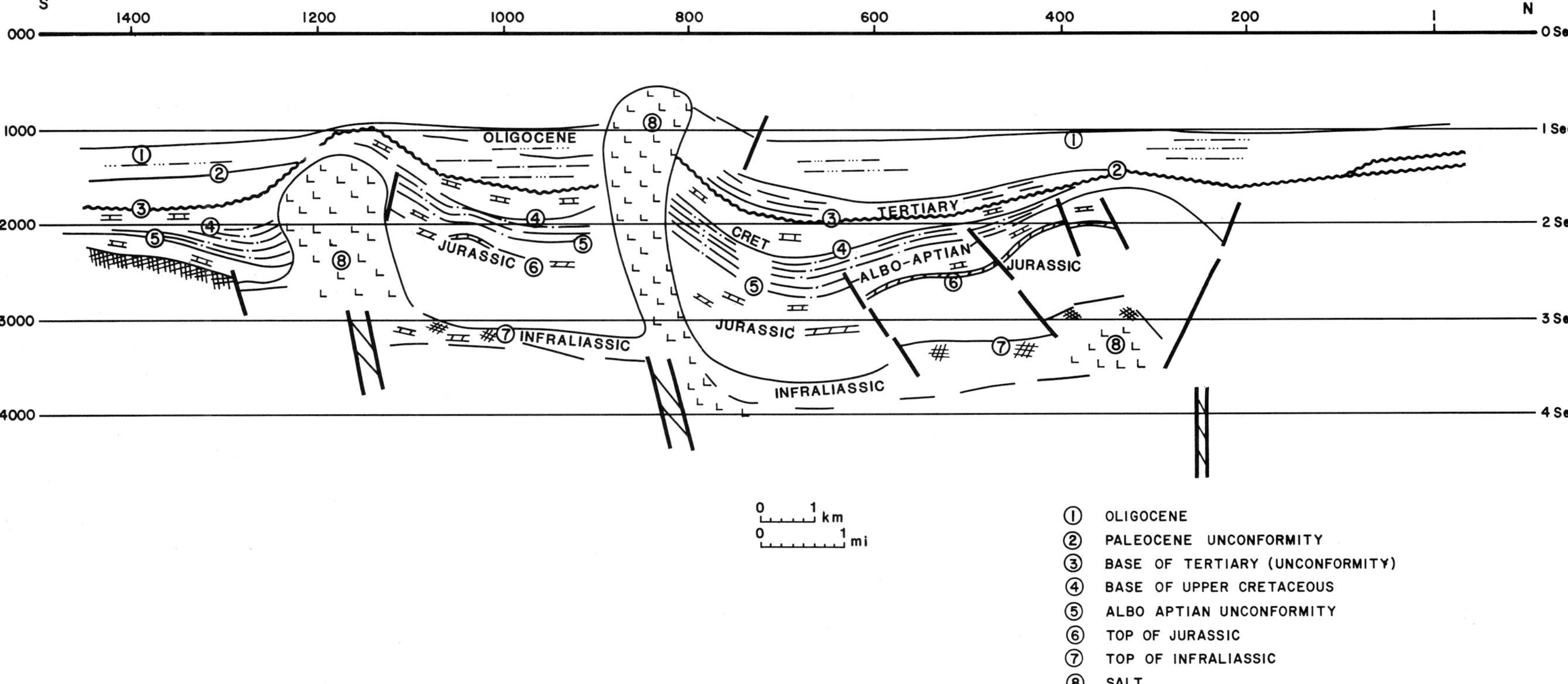

Fig. 8-30 (Curnelle and Marco, 1983)—Seismic line and geologic interpretation from offshore Aquitaine Basin showing Cretaceous growth and early Tertiary rejuvenation on northernmost salt structure (concordant), very late (young Tertiary) growth on center diapir, and Late Cretaceous and early Tertiary growth on southernmost piercement structure. Note suggested triggering of salt structures by underlying normal faults. Permission to publish by American Association of Petroleum Geologists.

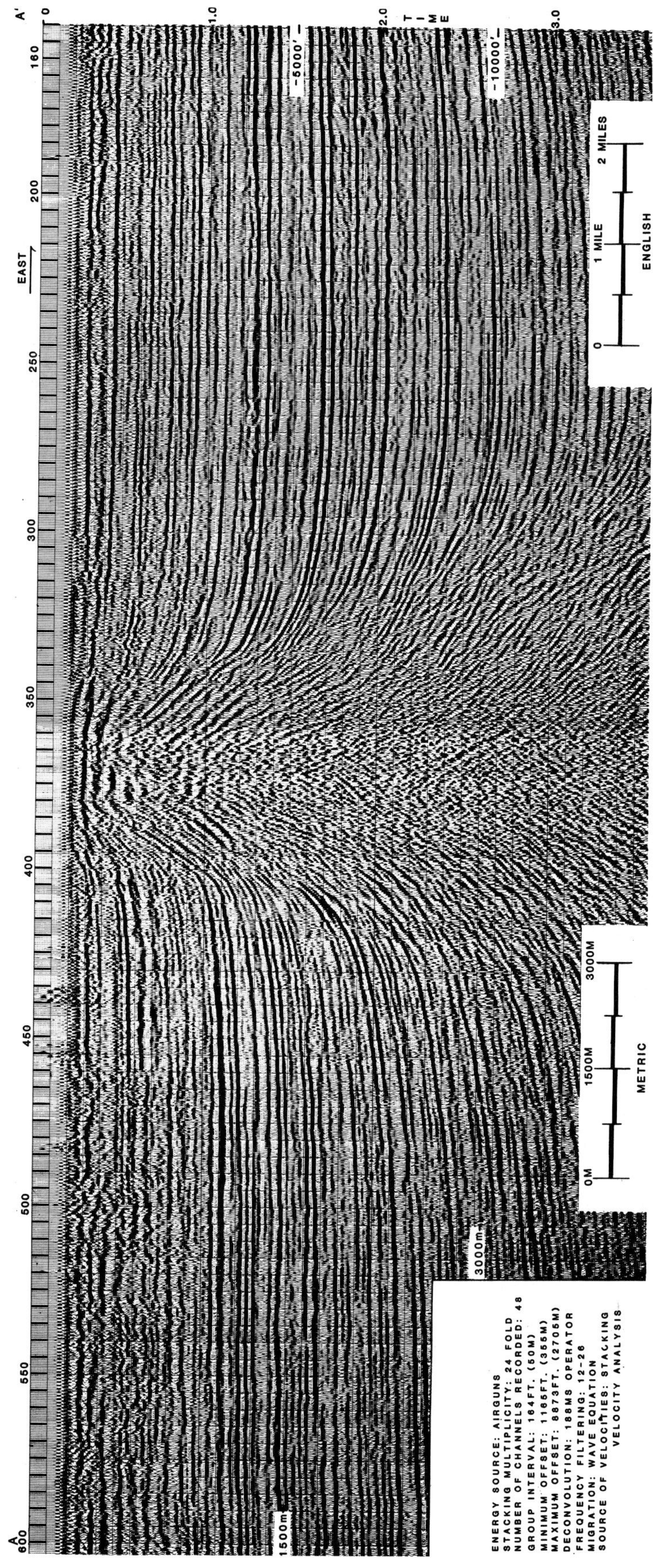
A'
A
600
550
500
450
400
350
300
250
200
160
EAST
0
1.0
2.0
3.0
TIME
-5000'
-10000'
1500m
3000m
0
1 MILE
2 MILES
ENGLISH
0M
1500M
3000M
METRIC
ENERGY SOURCE: AIRGUNS
STACKING MULTIPLICITY: 24 FOLD
NUMBER OF CHANNELS RECORDED: 48
GROUP INTERVAL: 164FT. (50M)
MINIMUM OFFSET: 1165FT. (355M)
MAXIMUM OFFSET: 8873FT. (2705M)
DECONVOLUTION: 188MS OPERATOR
FREQUENCY FILTERING: 12-26
MIGRATION: WAVE EQUATION
SOURCE OF VELOCITIES: STACKING VELOCITY ANALYSIS

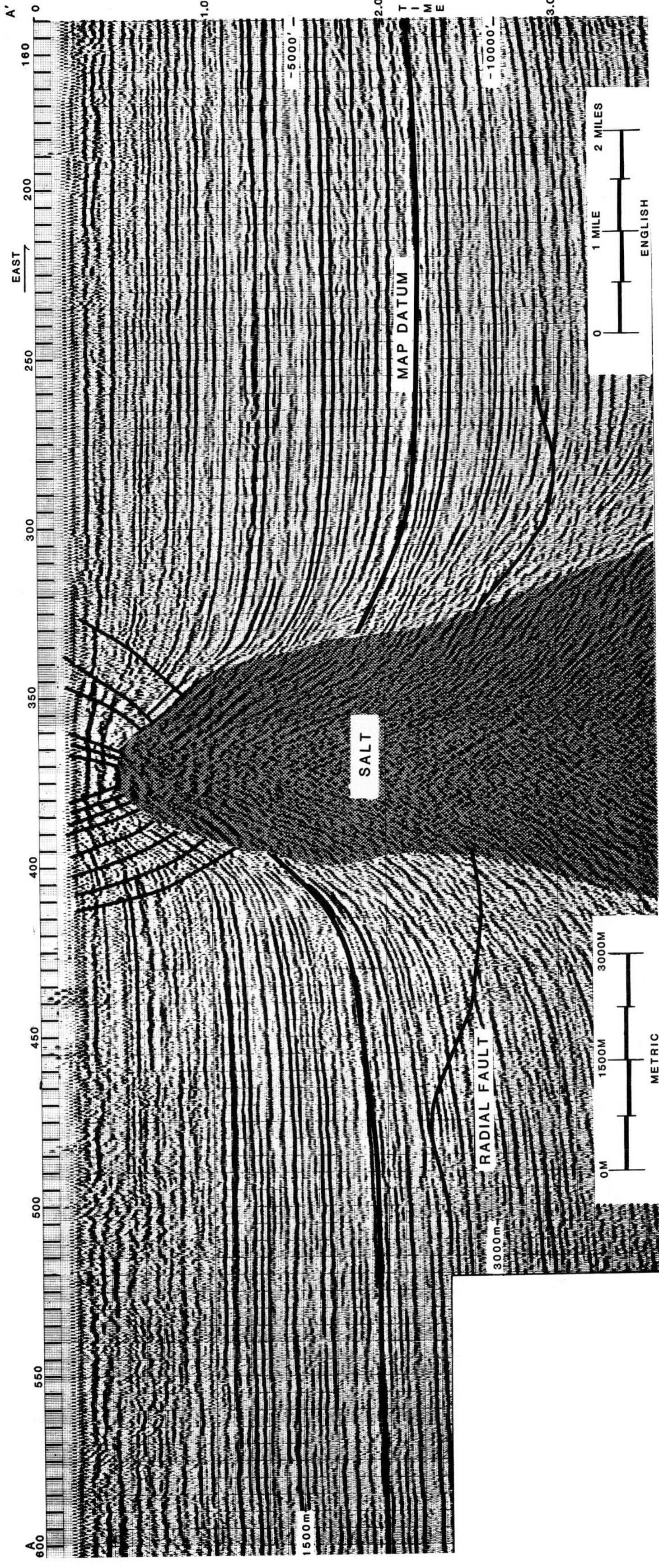

Fig. 8-31 (Sunwall et al., 1983)—Seismic line from offshore Louisiana, U. S. Gulf Coast, showing young salt diapir (characterized by reflection cut-outs) with superjacent normal faults and lower radial faults which strike parallel to seismic line. Salt configuration based on well control, gravity, and reflection and refraction seismic. Note that stratigraphic thinning toward diapir begins early (below 2 sec), but secondary rim synclines are not developed. Permission to publish by American Association of Petroleum Geologists.

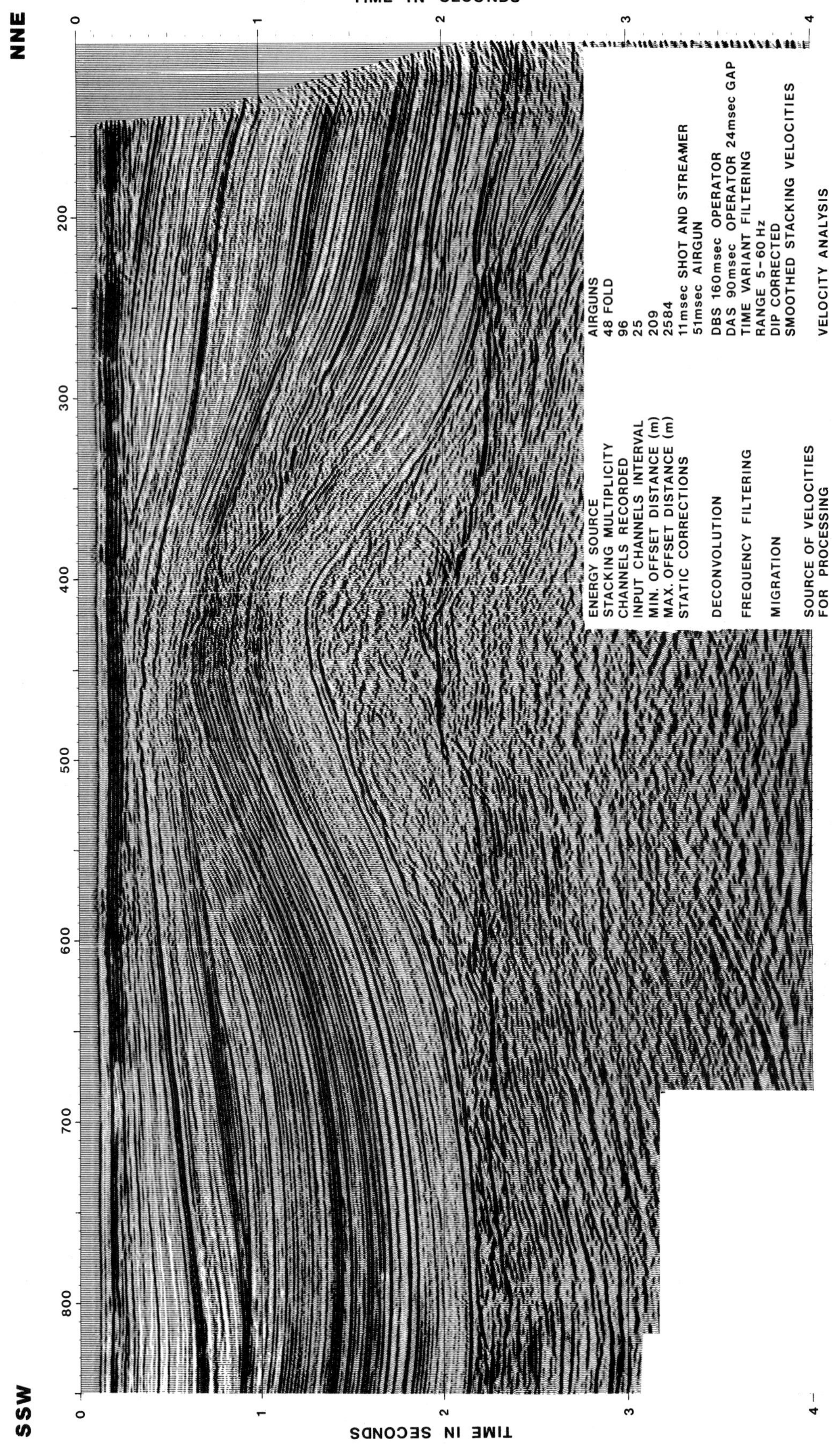
TIME IN SECONDS
NNE
SSW
0
1
2
3
4
200
300
400
500
600
700
800
ENERGY SOURCE AIRGUNS
STACKING MULTIPLICITY 48 FOLD
CHANNELS RECORDED 96
INPUT CHANNELS INTERVAL 25
MIN. OFFSET DISTANCE (m) 209
MAX. OFFSET DISTANCE (m) 2584
STATIC CORRECTIONS 11msec SHOT AND STREAMER
51msec AIRGUN
DECONVOLUTION DBS 160msec OPERATOR
DAS 90msec OPERATOR 24msec GAP
FREQUENCY FILTERING TIME VARIANT FILTERING
RANGE 5-60 Hz
MIGRATION DIP CORRECTED
SMOOTHED STACKING VELOCITIES
SOURCE OF VELOCITIES FOR PROCESSING VELOCITY ANALYSIS

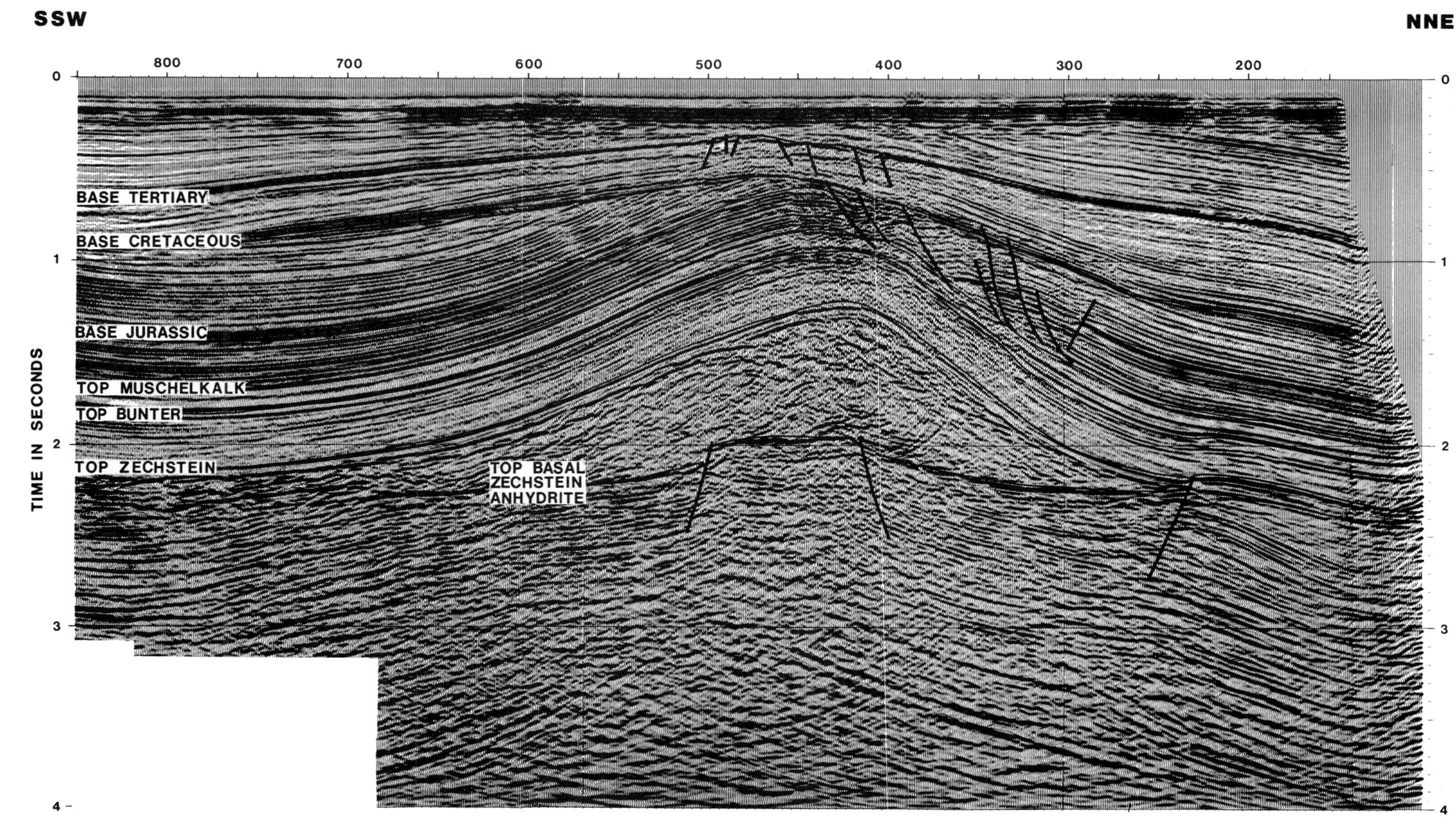

Fig. 8-32 (Owen and Taylor, 1983)—Seismic line from southern North Sea (U.K. waters) showing salt pillow with pronounced growth in Late Jurassic and again in late Tertiary. Listric faults on NNE side may be more a result of regional tectonic activity than of salt uplift. Asymmetry of pillow caused by change in direction of salt sourcing. Note sub-salt horst could have triggered salt movement. Permission to publish by American Association of Petroleum Geologists.

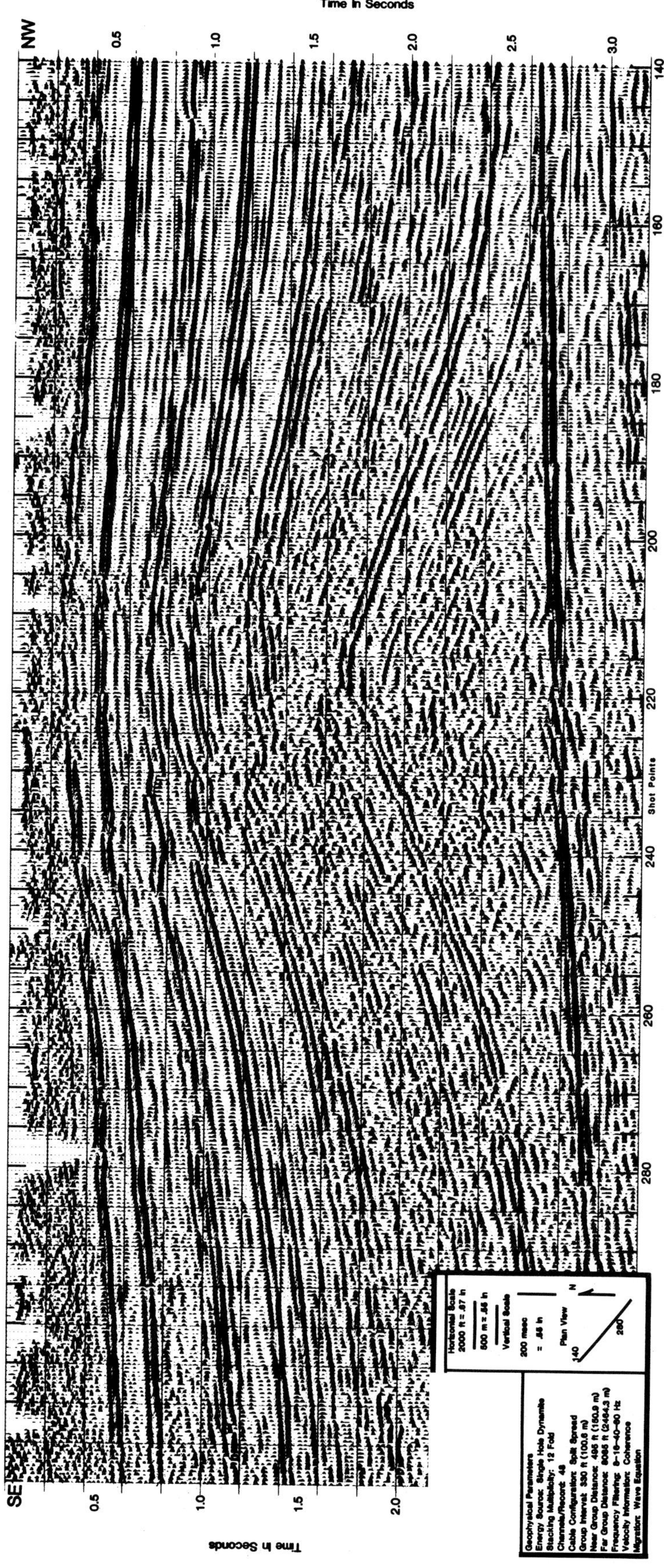

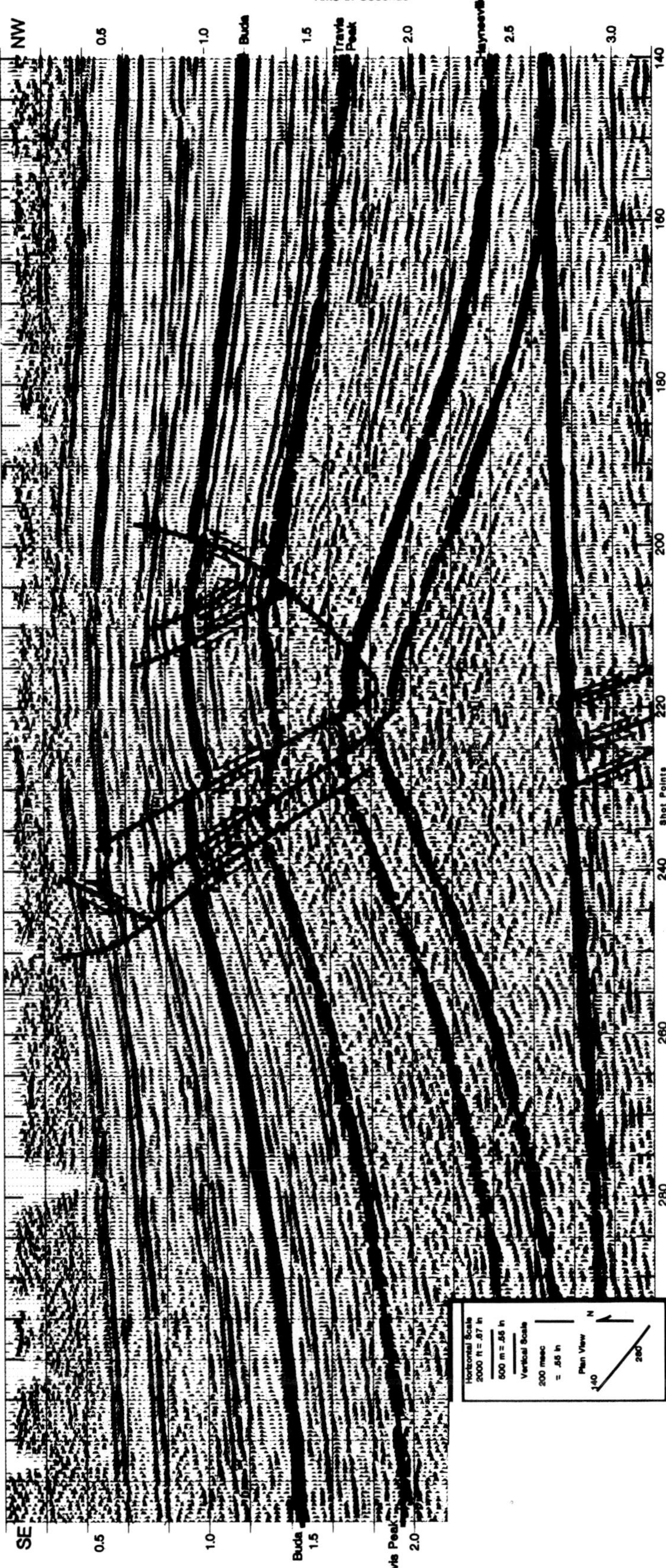

Fig. 8-33 (Inderwiesen, 1983)—Seismic line from Van Zandt County, east Texas Basin showing salt pillow structure with probable triggering normal faults below and an extensional system above triggered in turn by salt movement. Maximum growth in Haynesville-Travis Peak interval. Note more complete salt withdrawal from NW than from SE side of pillow. Permission to publish by American Association of Petroleum Geologists.

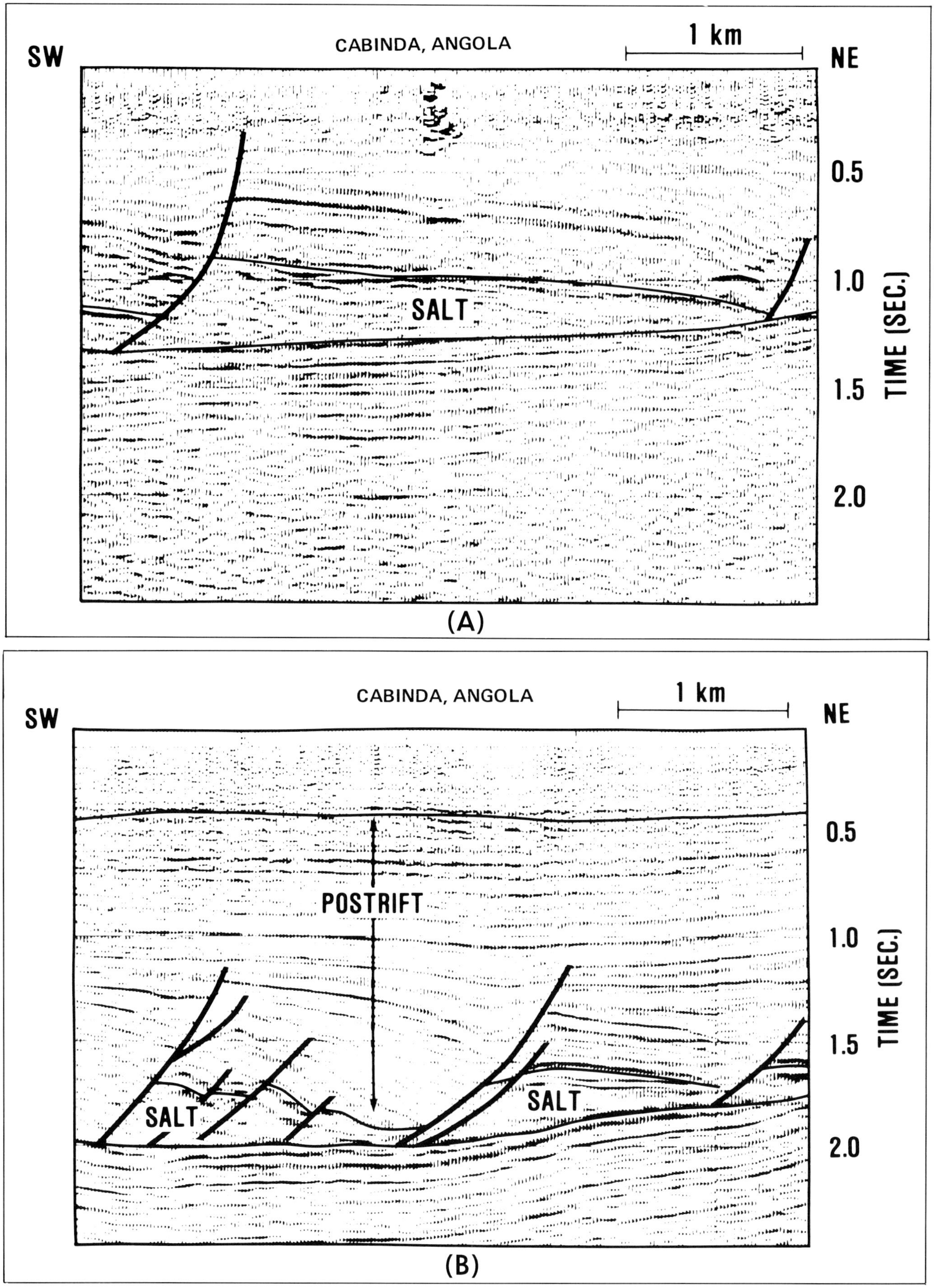

Fig. 8-34 (Brice et al., 1982)—Seismic lines from Cabinda, Angola, showing (A) downbending of section above salt and (B) growth fault system affecting salt and carbonates above. Both effects may be caused by down-slope movement of salt. Permission to publish by American Association of Petroleum Geologists.

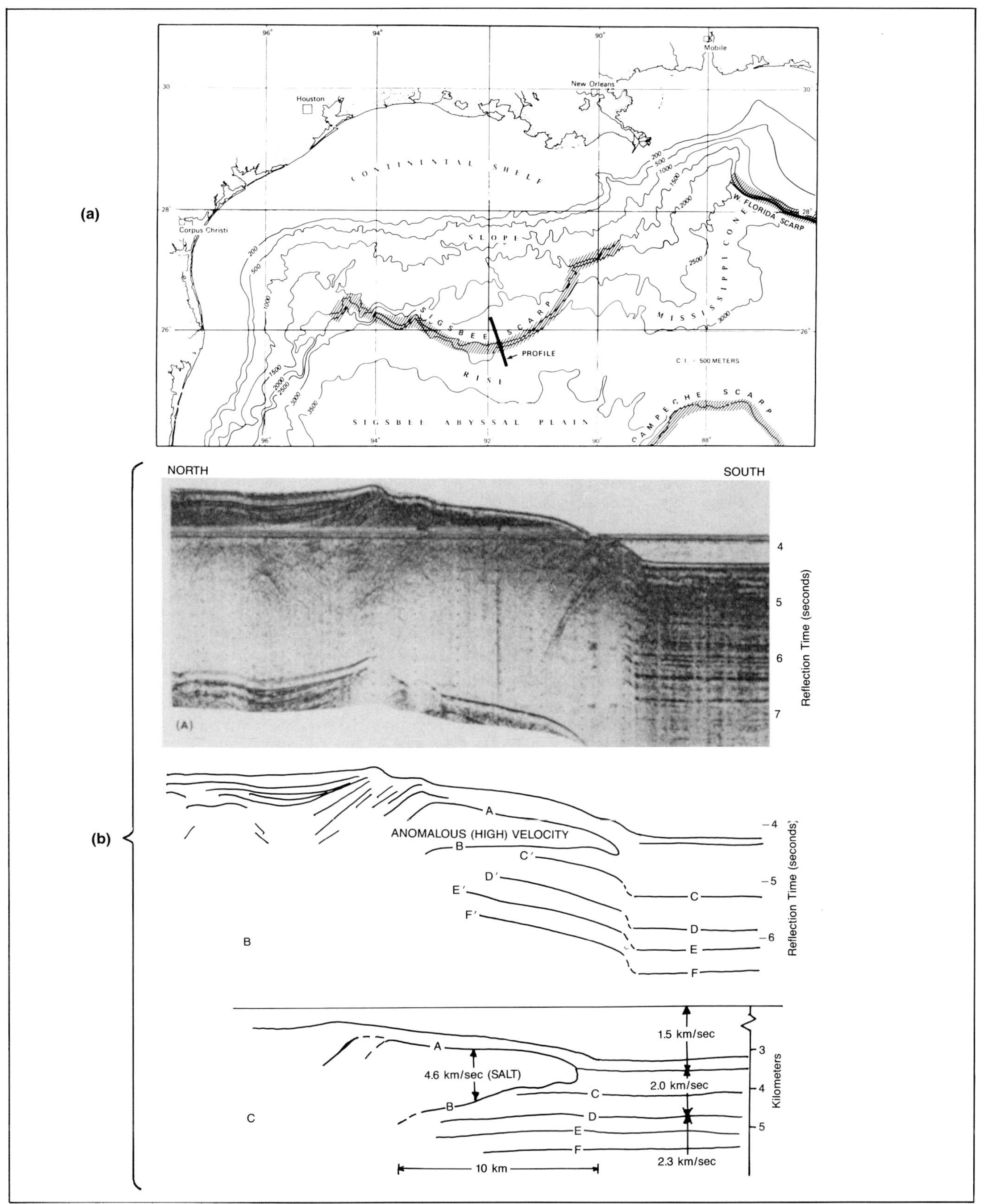

Fig. 8-35 (Amery, 1969)—(a). Physiography of northern Gulf of Mexico and location of (b). Sparker profile across Sigsbee scarp showing wedge of salt extruded over flat continental-rise sediments. A. Photograph of sparker profile across Sigsbee scarp showing deep reflections north of scarp and flat, continuous reflections south of scarp. B. Line drawing from A, indicating wedge of material with anomalously high velocity and reflection continuity beneath scarp. C. Cross section constructed from B, showing velocity values used to convert reflection time to depth. Prism AB is presumed to be a salt extrusion. Vertical exaggeration 2:1. Permission to publish by American Association of Petroleum Geologists.

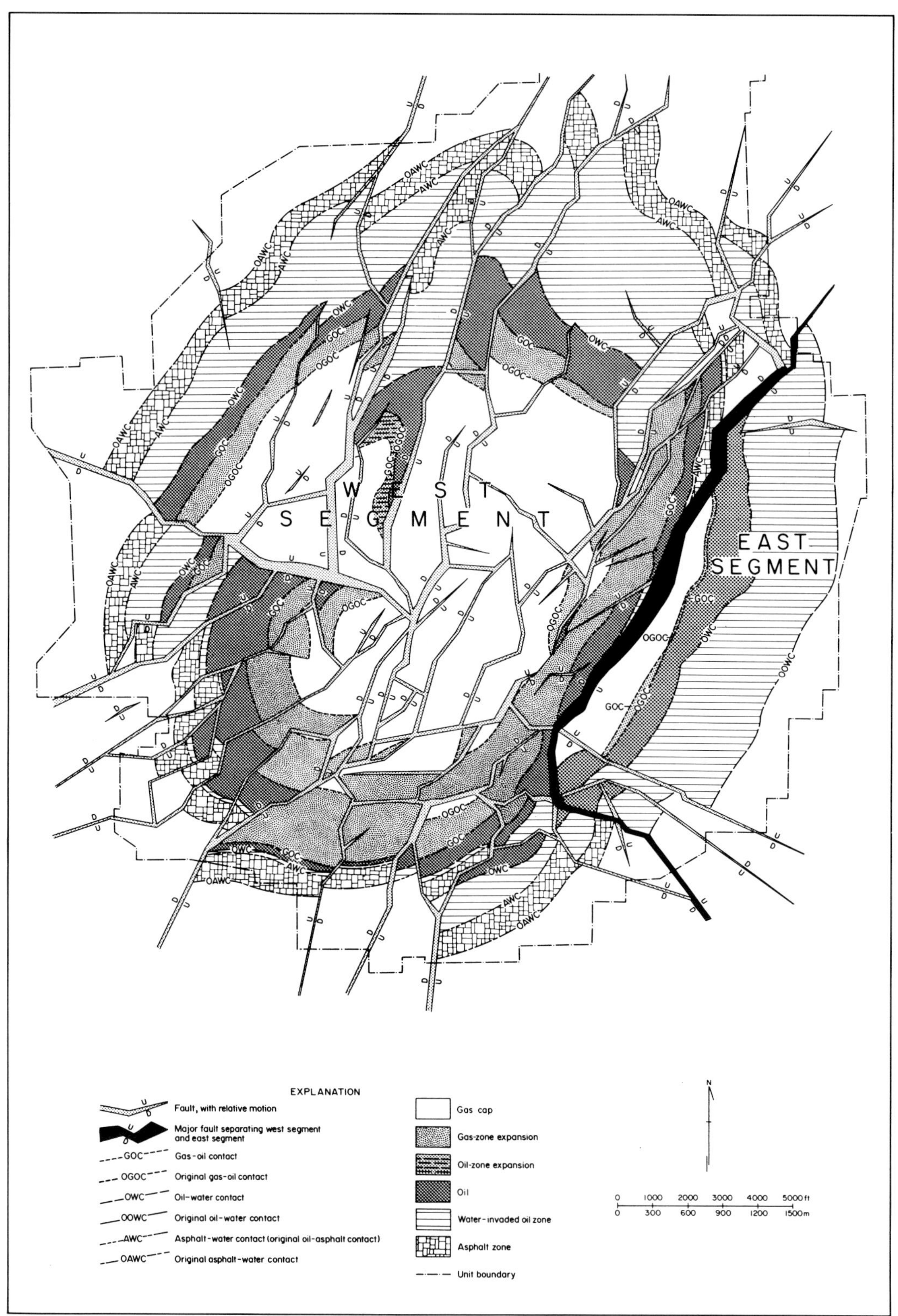

Fig. 8-36 (Galloway et al., 1983)—Reservoir map of Hawkins field, Wood County, Texas, showing complicated fault system. Inasmuch as most of the faults are nonsealing, however, sandstones are in pressure communication. Permission to publish by Bureau of Economic Geology, University of Texas, Austin.

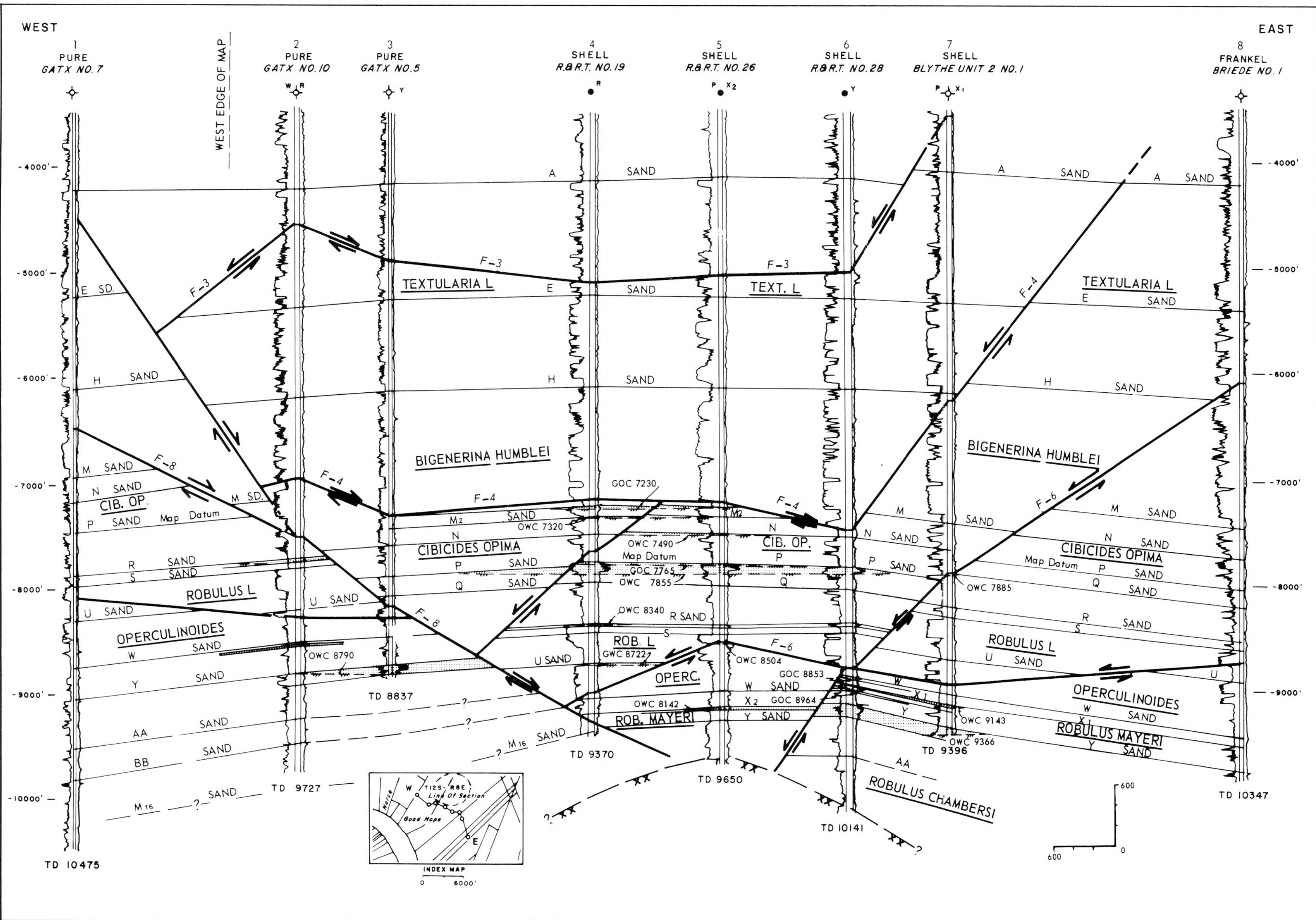

Fig. 8-37 (Gullett, 1983)—Cross section of Good Hope field, St. Charles Parish, Louisiana, showing complex pattern of normal faults cutting sedimentary rocks above a salt feature. Permission to publish by the New Orleans Geological Society.

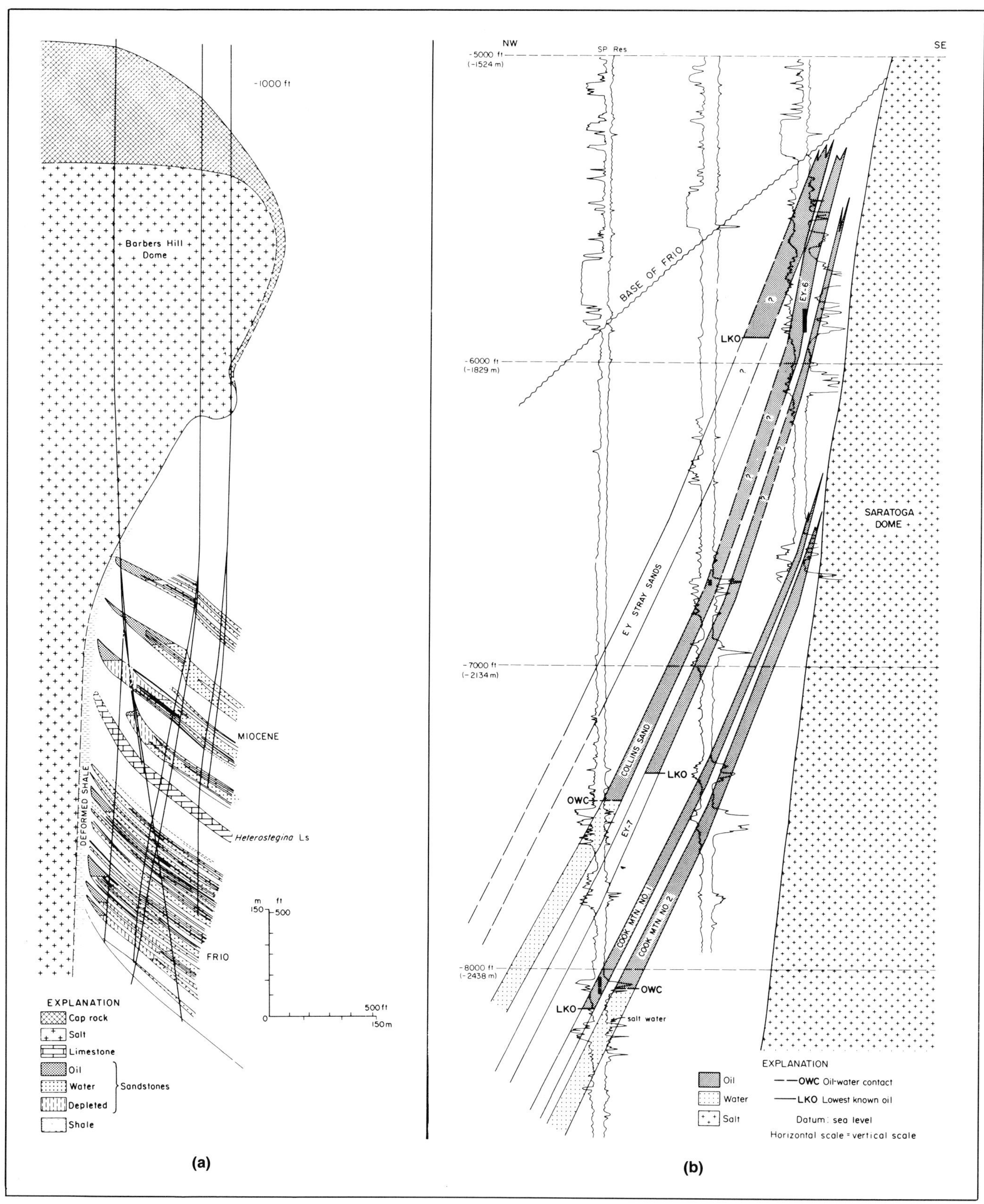

Fig. 8-38 (Galloway et al., 1983)—Cross sections of (a). east flank of Barbers Hill dome, Chambers County, Texas and (b). West Saratoga field, Harden County, Texas. Both sections have good well control showing oil accumulations related to stratigraphic pinch outs and to abutment of sands against diapiric deep water marine shale (Barbers Hill) at the edge of salt dome. Permission to publish by Bureau of Economic Geology, University of Texas, Austin.

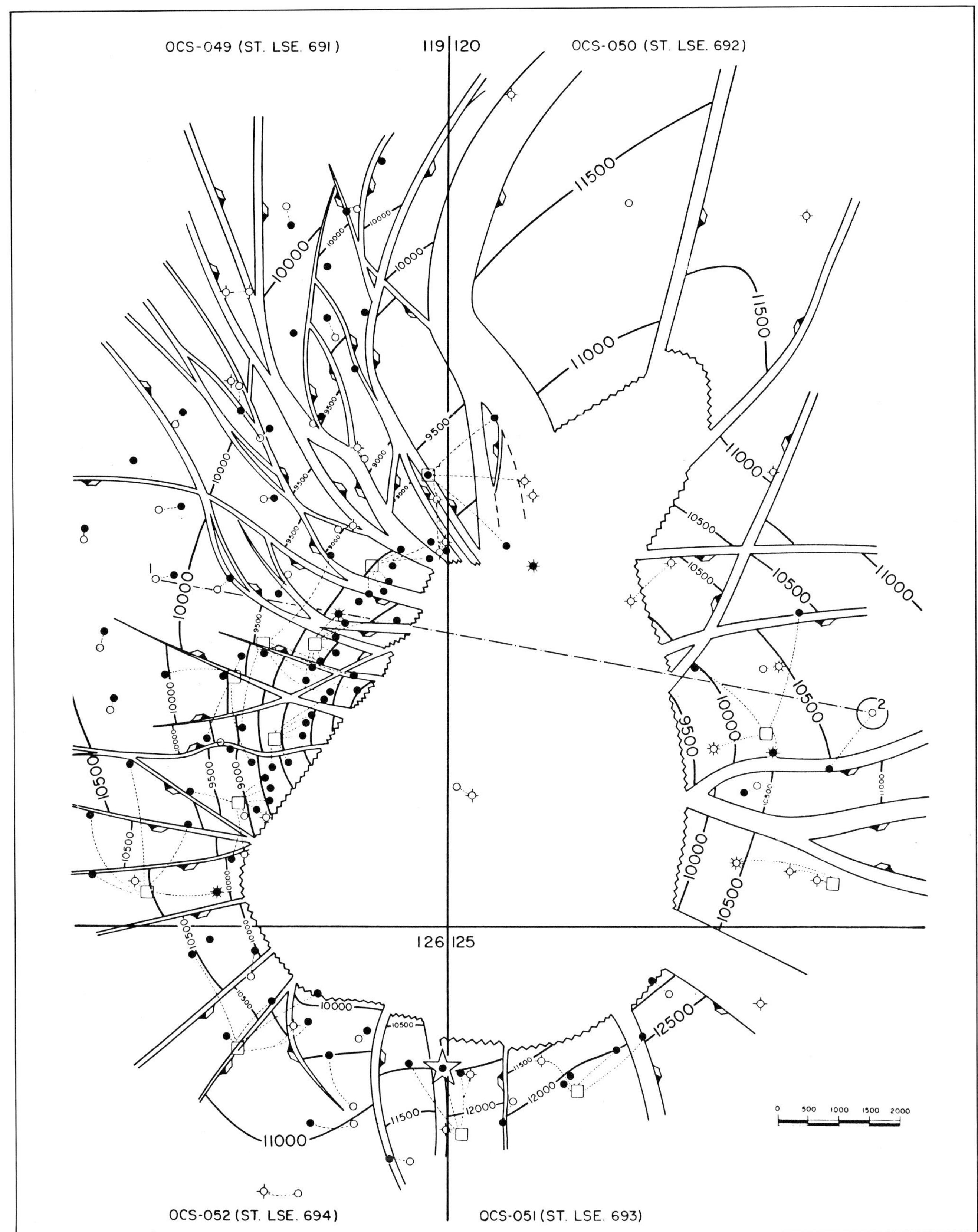

Fig. 8-39 (Bullard, 1973)—Structure contour map (on Bigenerina A horizon) of Eugene Island Block 126 field, offshore Louisiana, showing complex radial fault system affecting accummulation of hydrocarbons. Permission to publish by Lafayette Geological Society.

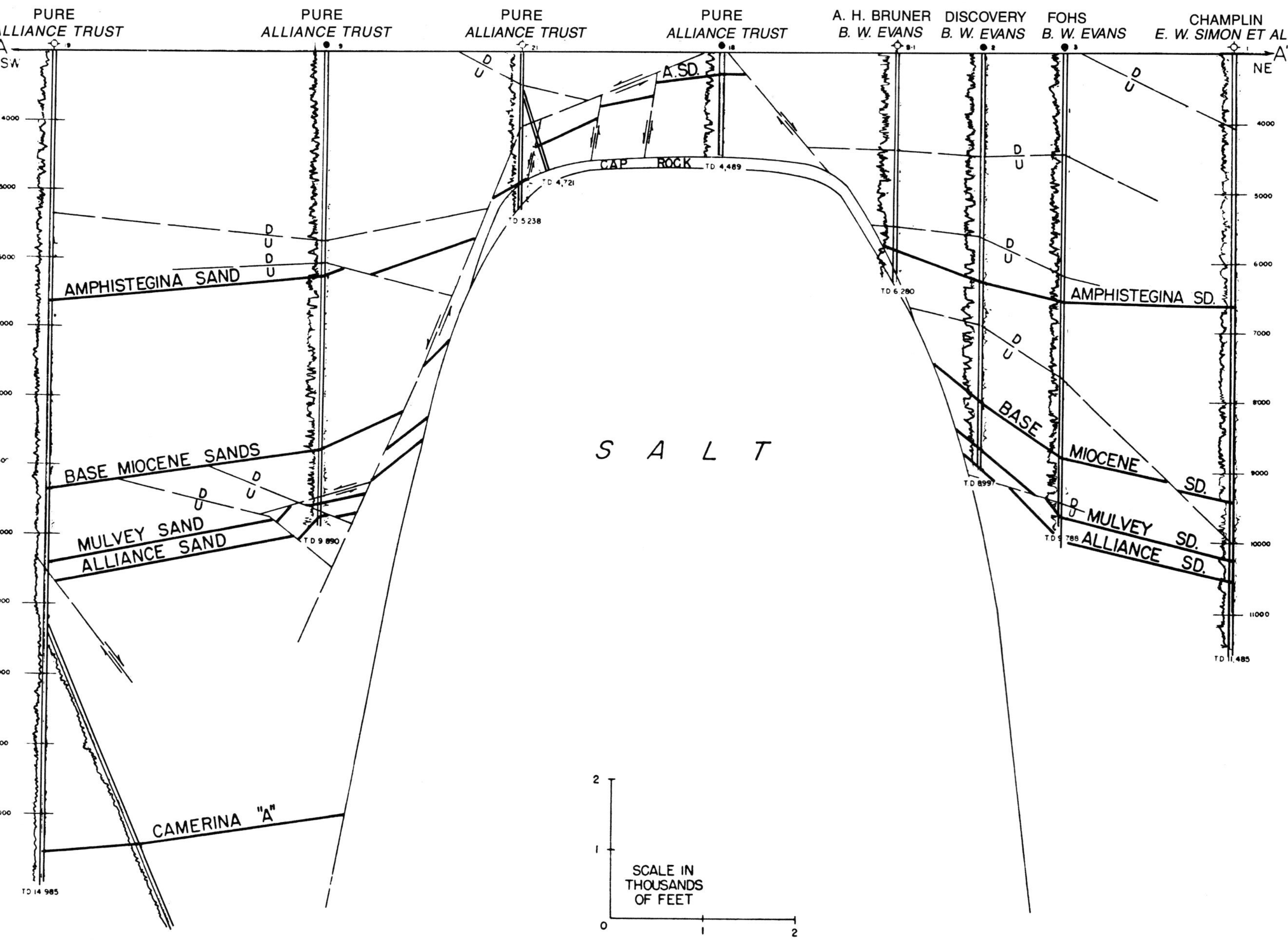

Fig. 8-40 (Robinson, 1970)—Cross section through Gueydan field, Acadia and Vermillion parishes, Louisiana, showing that prospective sands are compartmentalized around salt dome by complex normal faults. Permission to publish by Lafayette Geological Society.

9 SHALE STRUCTURES

Shale structures have both motivated and modified detached listric normal faults and hence already have been partly covered in Chapter 7. Moreover, since some aspects of salt deformation also apply to shale, parts of Chapter 8 are also applicable to shale structures. As seen from figure 8-1 and table 8-1, shale structures span a broad range of plate tectonic environments and are mainly dependent on the presence of thick overpressured shales in the sedimentary section.

Shales sometimes play a virtually interchangeable role with salt, as in a compound domal mass of salt and shale (Fig. 9-1). Salt rises buoyantly because its constant density is often less than surrounding and shallower sediments, whereas shale's buoyant rise is caused by high fluid pressure that can be present in some shales but not others.

Diapiric shale and salt both typically respond as blank zones on reflection seismic. They cannot always be distinguished by gravity work since their densities may be the same. If interval velocities can be determined, shale has a much slower velocity than salt, which is constant at 4500 m/sec (14,760 ft/sec). However, frequently the top of a diapiric feature can be mapped, but not the bottom. The Susu structure in Indonesia is an excellent example of a diapir with both a mappable top and bottom (Fig. 9-2). It is cored by shale as suggested on the time section by a velocity low at its base and confirmed by drilling.

Besides the distinction of shale and salt by interval velocity, the presence or absence of rim synclines may be another criterion for differentiation. Table 9-1 shows that the very flowage characteristics that are exhibited by salt are the same ones that would tend to form rim synclines.

	FLOWAGE:	
	Salt	Shale
Time or Duration:	Very long (geologic periods)	Shorter
Rate:	Very slow ($<$ mm/yr)	Faster (Sometimes explosive with entrained gases)
Distance:	Very great ($\approx$6 mi/10 km)	Shorter

TABLE 9-1
Relative Flowage Characteristics of Salt vs. Shale

Primary rim synclines are often several miles from a salt dome that has evolved very slowly over a period of tens of millions of years. Shale flowage can also be slow over a great duration for long distances, but not as slow or for as great a duration or for as long a distance as salt. A rationale for this difference is that shales cannot remain in an overpressured state indefinitely. Tectonic activity or erosion will eventually break the pressure seal so that flowage will cease. Further, it is of interest that shale can go from diapiric (overpressured) to bedded in a very short distance (Fig. 9-1) whereas some salt layers are present over an entire basin.

Seismic examples of shale behavior, in addition to those shown in figures 1-22, 1-29, 7-10, 7-12, 7-13, 7-14, 7-23, 7-27, and 7-28, are shown in the figures 9-3, 9-5 and 9-7.

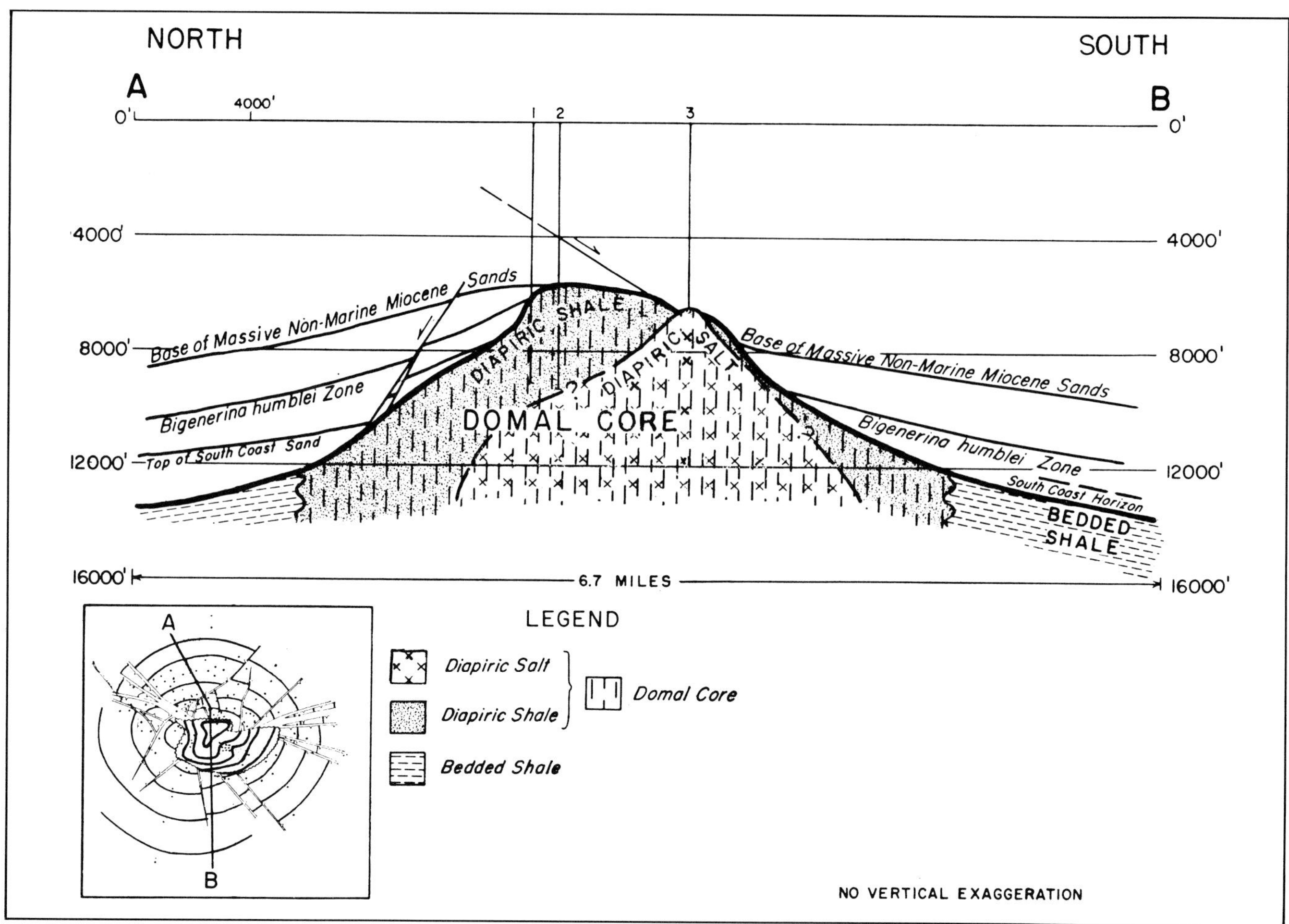

Fig. 9-1 (Atwater and Forman, 1959)—Structural cross section of Valentine salt dome, Lafourche Parish, Louisiana, illustrating compound domal mass of salt and shale which forms a diapiric core. Permission to publish by American Association of Petroleum Geologists.

REFERENCES

Atwater, G. I., and Forman, M. J., 1959, Nature of growth of southern Louisiana salt domes and its effect on petroleum accumulation: *American Association Petroleum Geologists Bull.*, V. 43, p. 2592–2622.

Buffler, R. T., 1983, Structure of the Mexican ridges foldbelt, southwest Gulf of Mexico, *in* Bally, A. W., ed., Seismic expression of structural styles: *American Association Petroleum Geologists Studies in Geology #15*, V. 2.

Mulhadiono and Marinoadi, 1977, Notes on hydrocarbon trapping mechanism in the Arun area, north Sumatra: *Indonesian Petroleum Association Proc. 6th Ann. Conf.*, p. 95–115.

Rao, Y. S. N., 1980, Geology of the eastern margin of the Indian subcontinent: *Southeast Asia Petroleum Exploration Society Proc.*, V. V, p. 161–178.

Whittle, A. P., and Short, G. A., 1978, The petroleum geology of the Tembungo field, East Malaysia: Southeast Asian Petroleum Exploration Society Offshore Southeast Asia Conf.

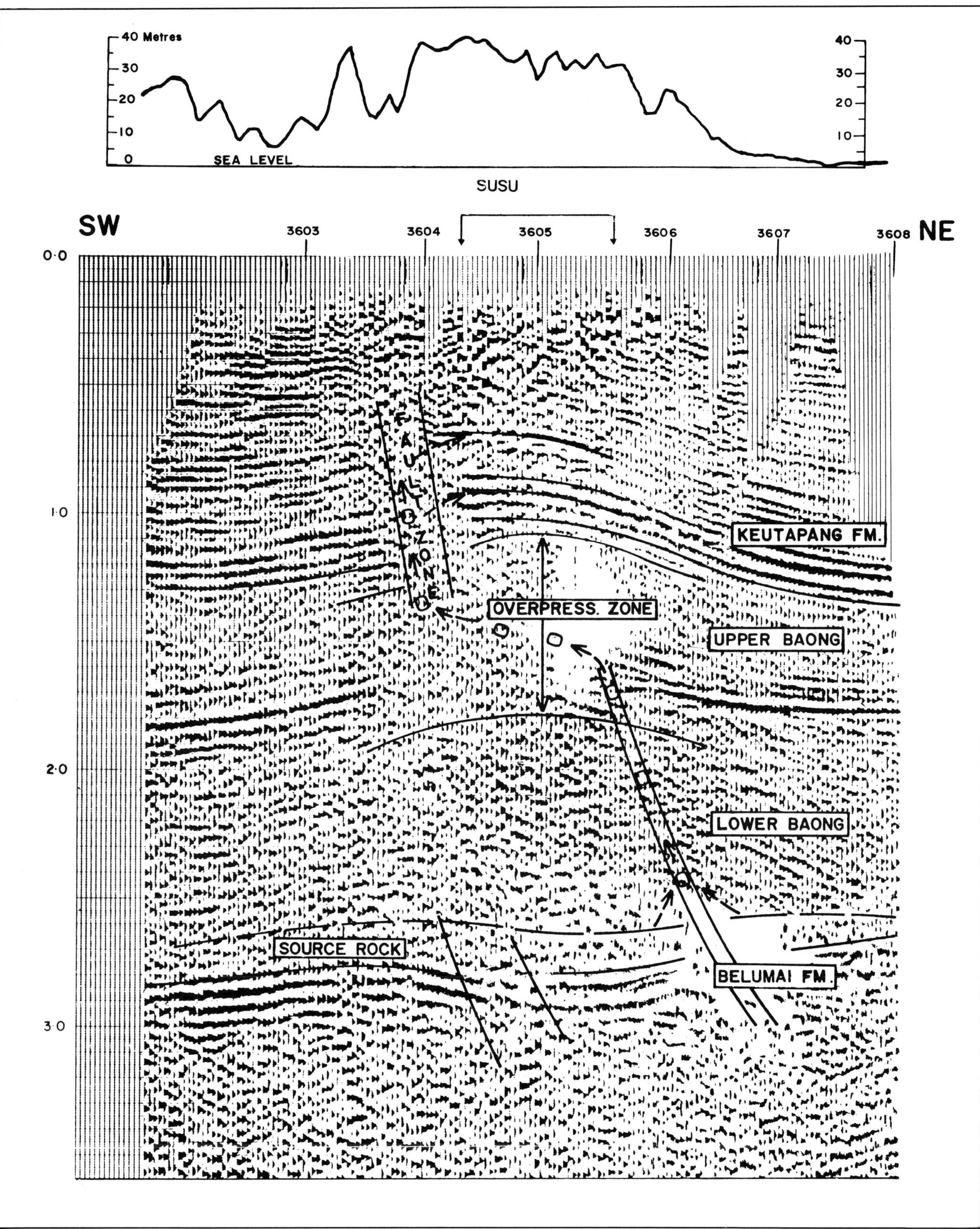

Fig. 9-2 (Mulhadiono and Marinoadi, 1977)—Seismic section across Susu field, north Sumatra, Indonesia. Note overpressured zone and reflection cut-outs coincident with mobile shale and a velocity low beneath shale. Permission to publish by Indonesian Petroleum Association.

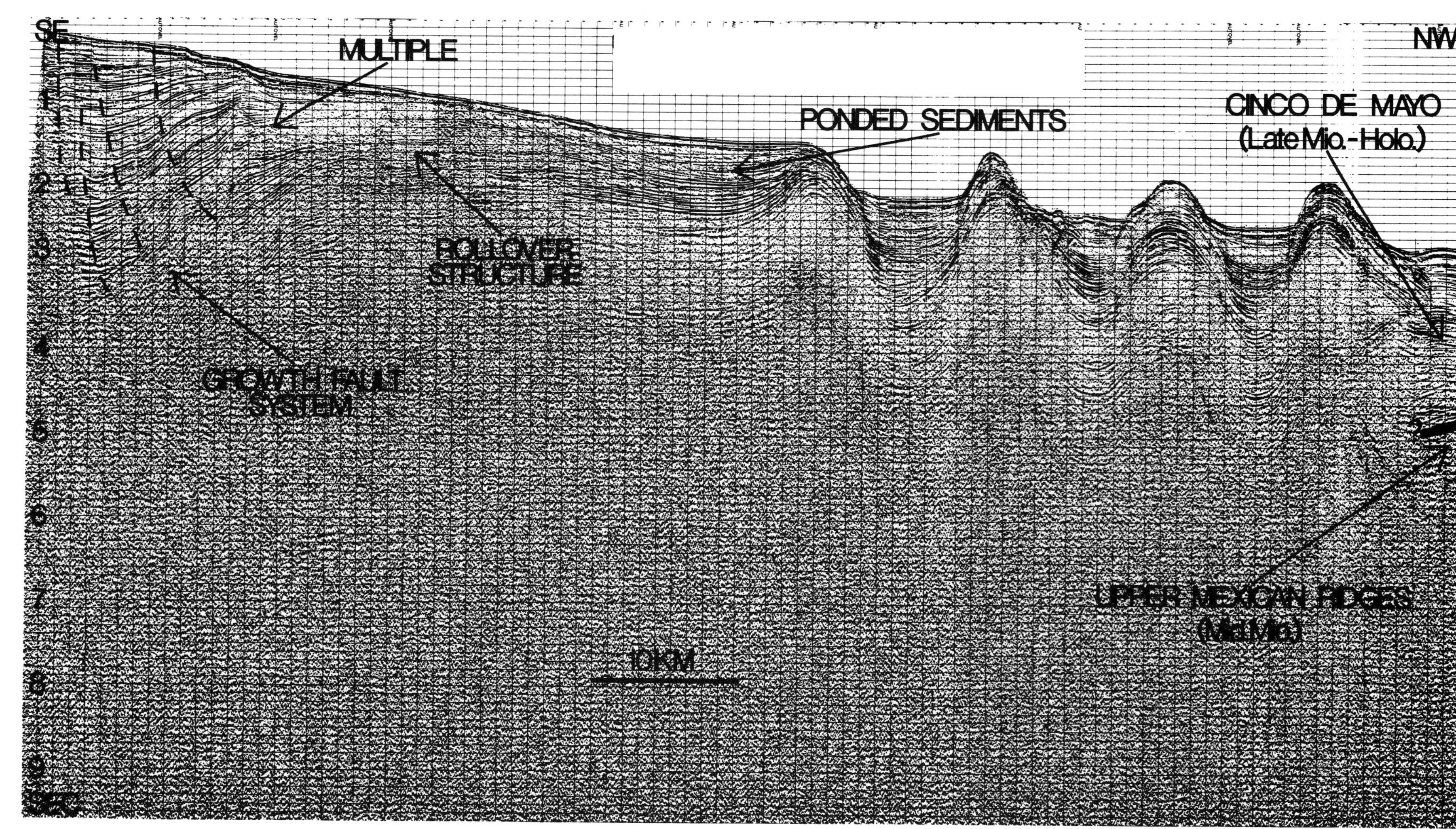
SE
NW
MULTIPLE
PONDED SEDIMENTS
CINCO DE MAYO
(Late Mio.-Holo.)
ROLLOVER
STRUCTURE
GROWTH FAULT
SYSTEM
UPPER MEXICAN RIDGES
(Mid.Mio.)
10KM
1
2
3
4
5
6
7
8
9
SEC

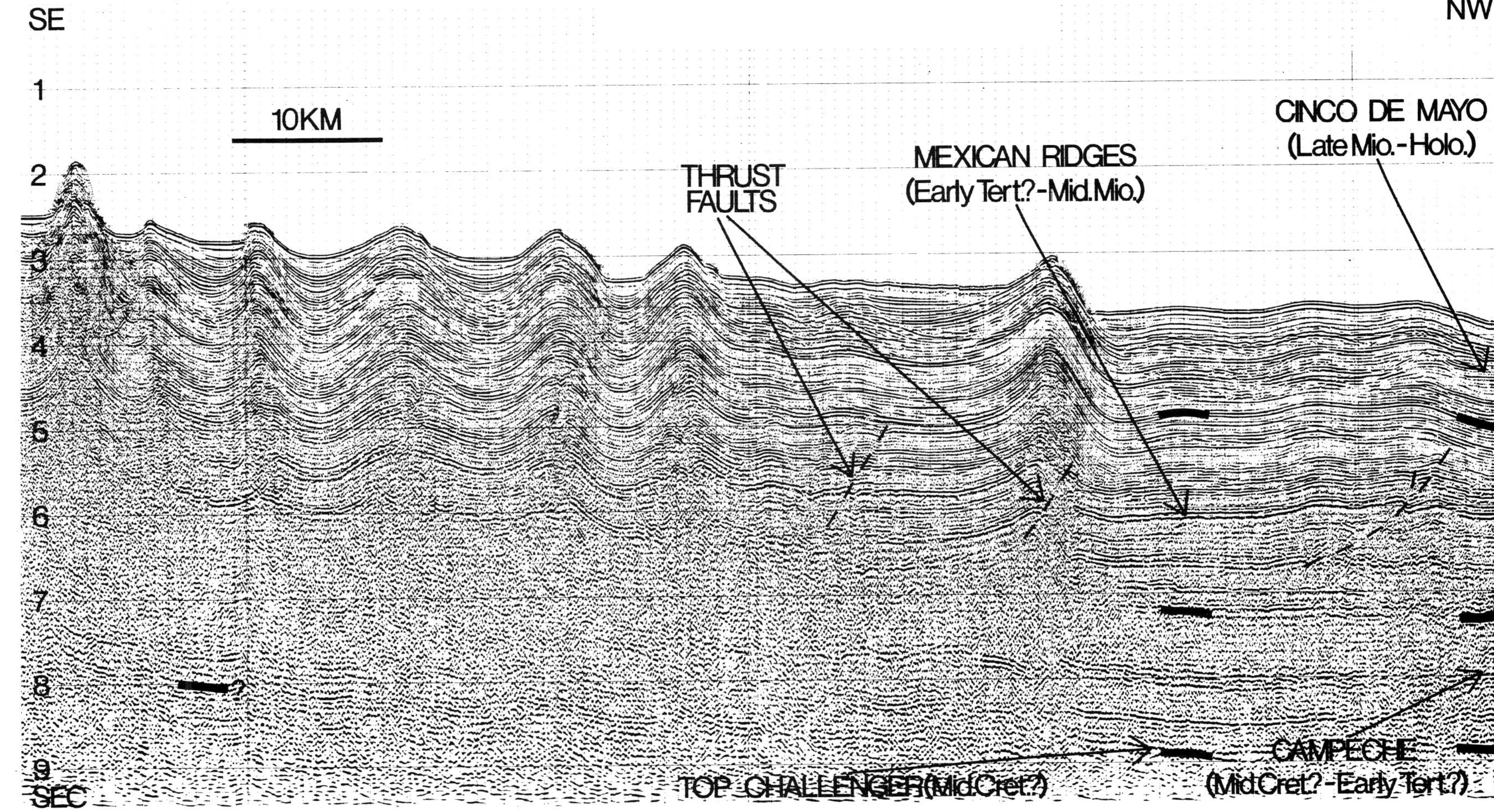

Fig. 9-3 (Buffler, 1983)—Seismic line offshore Veracruz, Mexico, showing growth fault system with downstream Mexican ridges foldbelt as "probably a huge late-Tertiary gravity-slide feature detached from flat-lying older Tertiary strata below along a decollement or deformed zone within weak, possibly geopressured, slope shales". Lines are more or less continuous with about 20 km (12.5 mi) of overlap. Permission to publish by American Association of Petroleum Geologists.

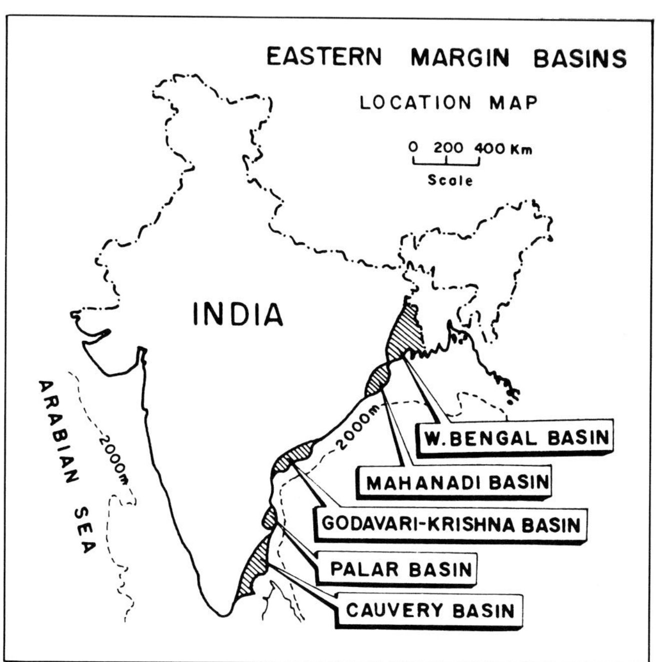

Fig. 9-4 (Rao, 1980)—Location of Godavari-Krishna Basin on east Indian coast. Reprinted with permission from SEAPEX.

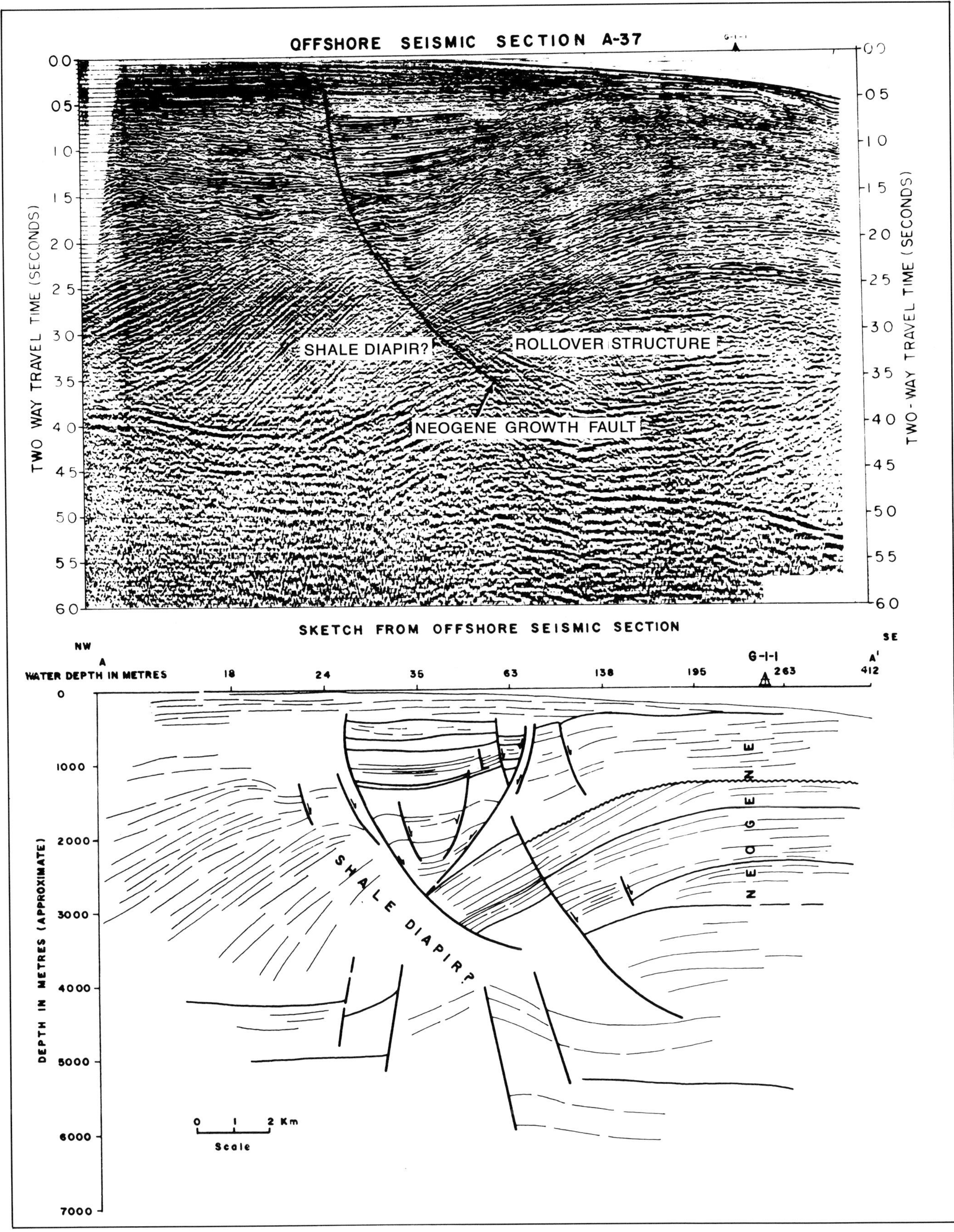

Fig. 9-5 (Rao, 1980)—Seismic section and interpretation showing shale diapirism in Godavari-Krishna Basin. Oil and gas are present in Neogene sands at about 2150 m (7052 ft) in rollover structure. Reprinted with permission from SEAPEX.

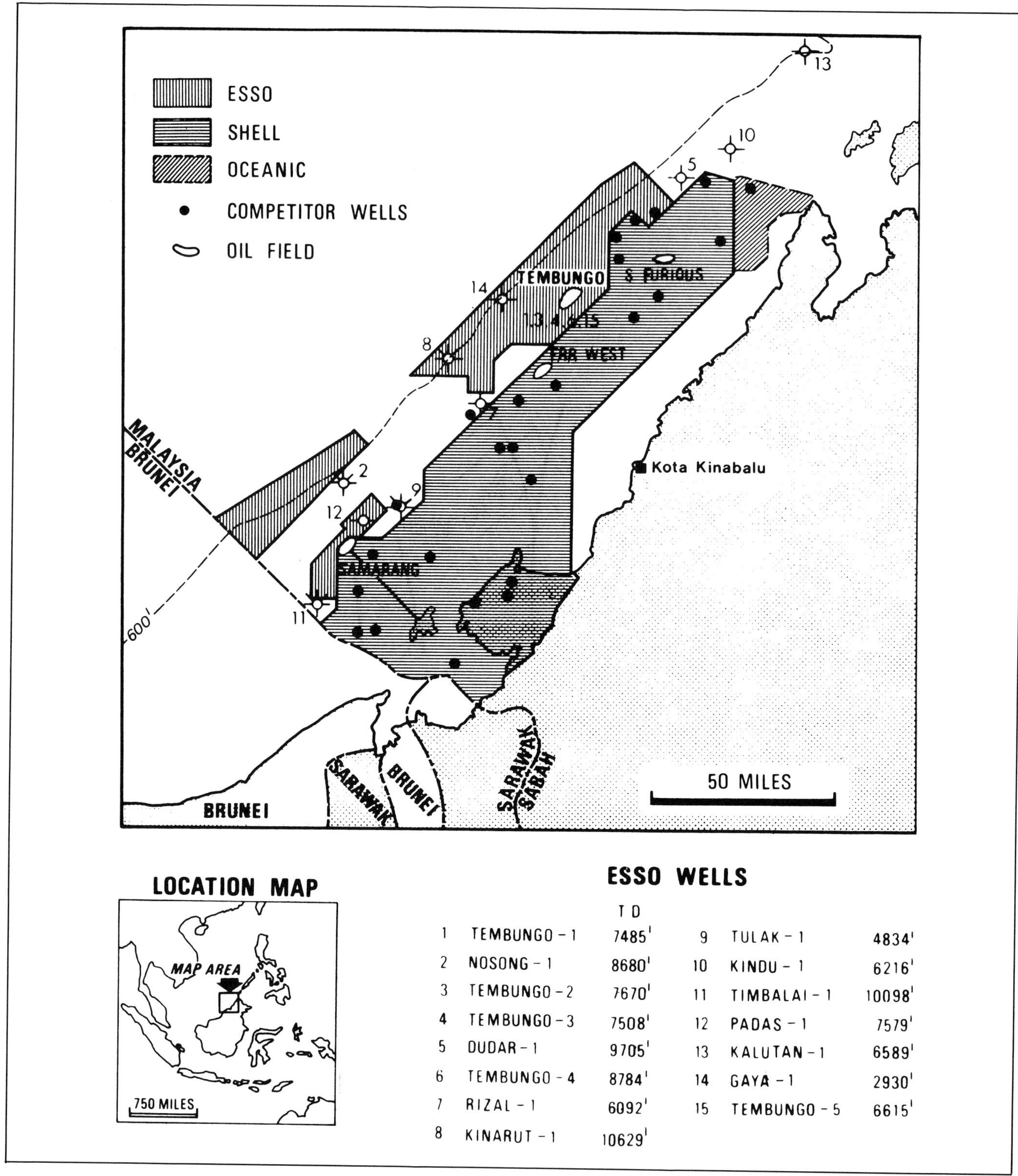

		TD			
1	TEMBUNGO - 1	7485'	9	TULAK - 1	4834'
2	NOSONG - 1	8680'	10	KINDU - 1	6216'
3	TEMBUNGO - 2	7670'	11	TIMBALAI - 1	10098'
4	TEMBUNGO - 3	7508'	12	PADAS - 1	7579'
5	DUDAR - 1	9705'	13	KALUTAN - 1	6589'
6	TEMBUNGO - 4	8784'	14	GAYA - 1	2930'
7	RIZAL - 1	6092'	15	TEMBUNGO - 5	6615'
8	KINARUT - 1	10629'			

Fig. 9-6 (Whittle and Short, 1978)—Index map and wells, to July, 1977, offshore, West Sabah, Malaysia, with location of Tembungo structure. Reprinted with permission from SEAPEX.

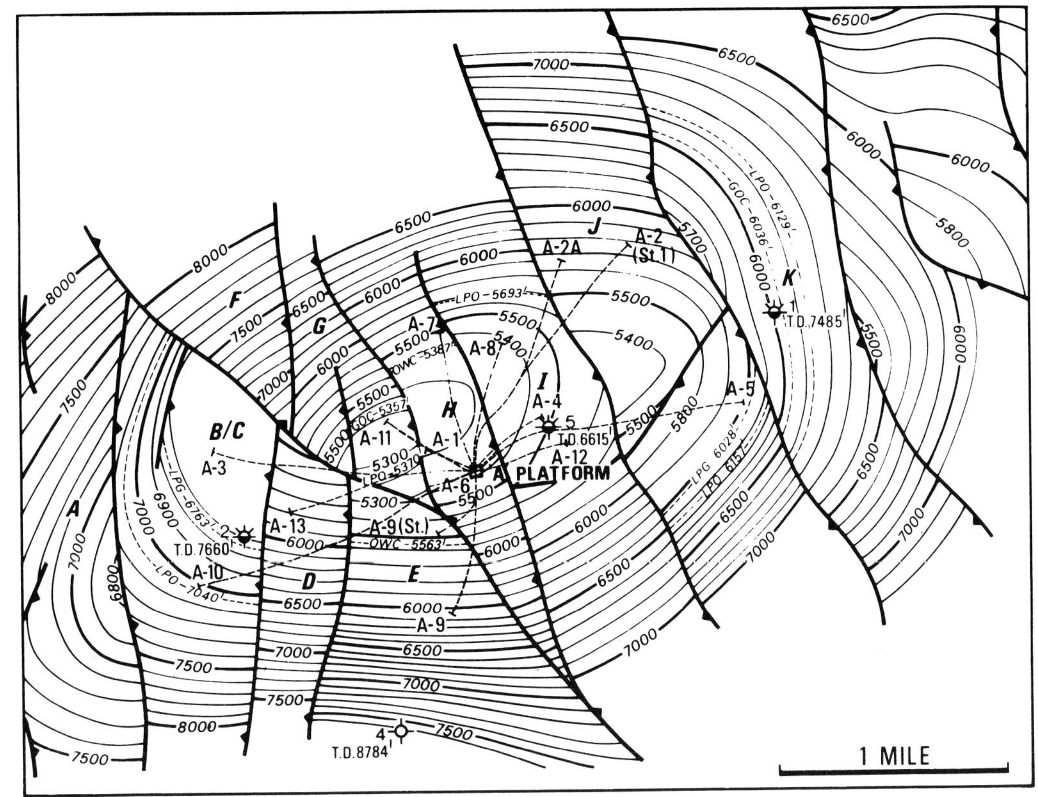

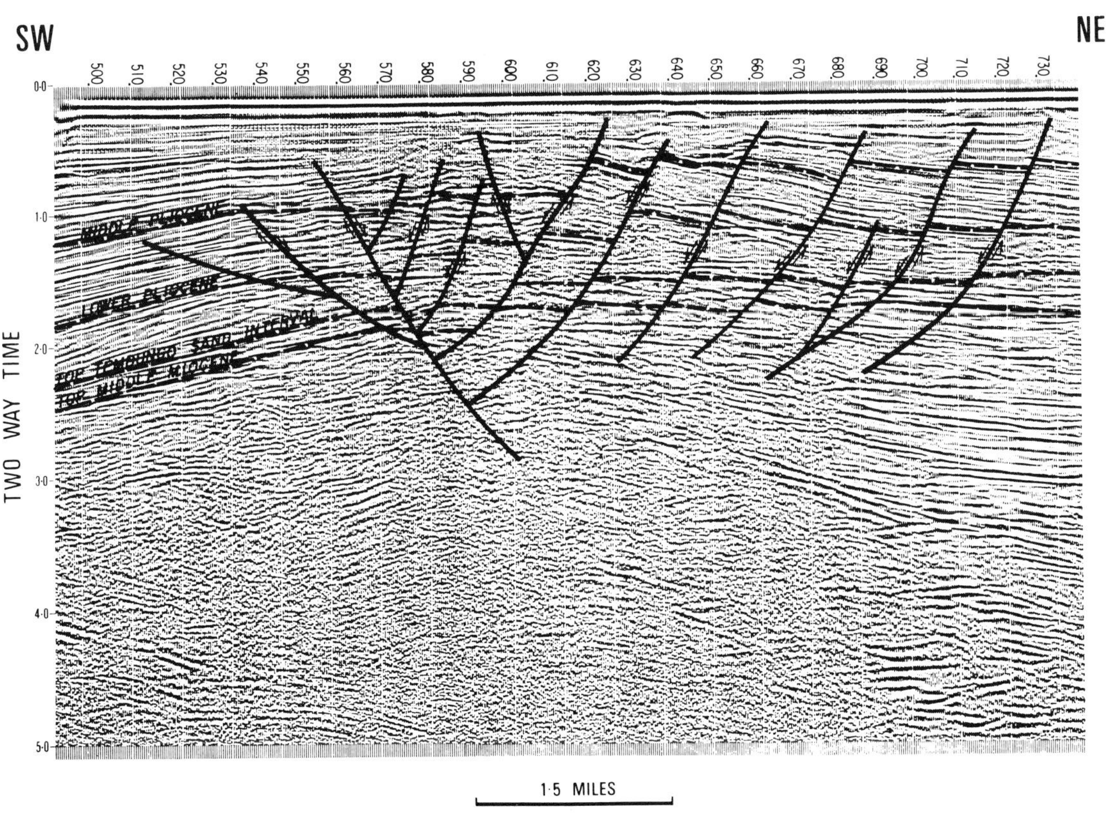

Fig. 9-7 (Whittle and Short, 1978)—Tembungo field, A. structure (top of upper Miocene) contour map and B. seismic section (through about middle of field) showing complexly normal-faulted structure interpreted to be induced by shale flowage. Additional drilling revealed a much more complicated fault pattern than originally mapped. Reprinted with permission from SEAPEX.

10 SUPERPOSED STYLES

INTRODUCTION

The intent of the previous chapters has been to identify and describe each structural style in its purest form with minimal regard to variation and diversity. It is essential that each style be understood before more complex style combinations and superimpositions can be unraveled. Even with specific intent, however, the "end member" approach to analysis of styles has not been entirely possible. For example, Chapters 7, 8, and 9 have shown the extremely close interrelationship between detached listric normal faults, salt, and shale and the impossibility of discussing them all separately. The close association between salt and extensional blocks has been noted in Chapter 4, and the occurrence of evaporites in thrust-fold belts noted in Chapter 6. The presence of detached structures amidst basement-involved compressive blocks, as discussed in Chapter 3, further demonstrates the interdependence of styles.

Variations in style and diversity of style occurrence are, of course, important for they often affect the styles of sedimentary basins that are explored; some discussion of these factors is given in Chapter 1. Combinations (as discussed above) and superpositions of styles are also critical to exploration as many petroleum provinces have more than one structural style. Of the instances of superposed or overprinted styles, perhaps none are more significant than structural inversion, a phenomenon that has become widely recognized.

STRUCTURAL INVERSION

Structural inversion is a reversal of deformational processes: most commonly a depressed region is uplifted. Deposition occurs in the presence of normal faults such that a thick section is developed on the downthrown sides, e.g., half graben or small rifted basin. Low areas are then turned inside out into highs, usually by a combination of compression and wrenching. The final product is an anticline or structural high into which sedimentary units thicken. Structural inversion can strongly affect the petroleum prospectiveness of an area in which it occurs.

Inversion may have been first recognized, or at least well publicized, in northwest Europe including the North Sea and England, but it has also occurred in many other parts of the world, several of which are discussed here.

North Sea Area

Regions of structural inversion in the North Sea and environs are shown in figure 10-1. A typical inversion structure is the Broad Fourteens Basin, offshore Netherlands (Fig. 10-2), where thick Triassic and Jurassic sediments accumulated in a normal-faulted trough and were subsequently uplifted and partially eroded (Fig. 10-3). A comparable feature is known from the Wessex Basin of southern England (Fig. 10-2) where thicker section is present on the downthrown side of the Purbeck normal fault zone, which was subsequently reversed to effect uplift of the basin (Fig. 10-4). In both examples the Upper Cretaceous chalk is not present across the inversion structures. In early seismic mapping of the North Sea (circa 1964) defining areas of missing chalk first led to the recognition of inversion structures.

Structural inversion is well documented but its cause or causes are puzzling. Voight (1962) attributed structural inversion in continental Europe to isostatic rebound of the depressed lighter sediments. This verticalist interpretation seems too restrictive and Ziegler (1975, 1981) believed that compressional and wrench deformation in the foreland of and associated with the Alpine orogeny was the cause of

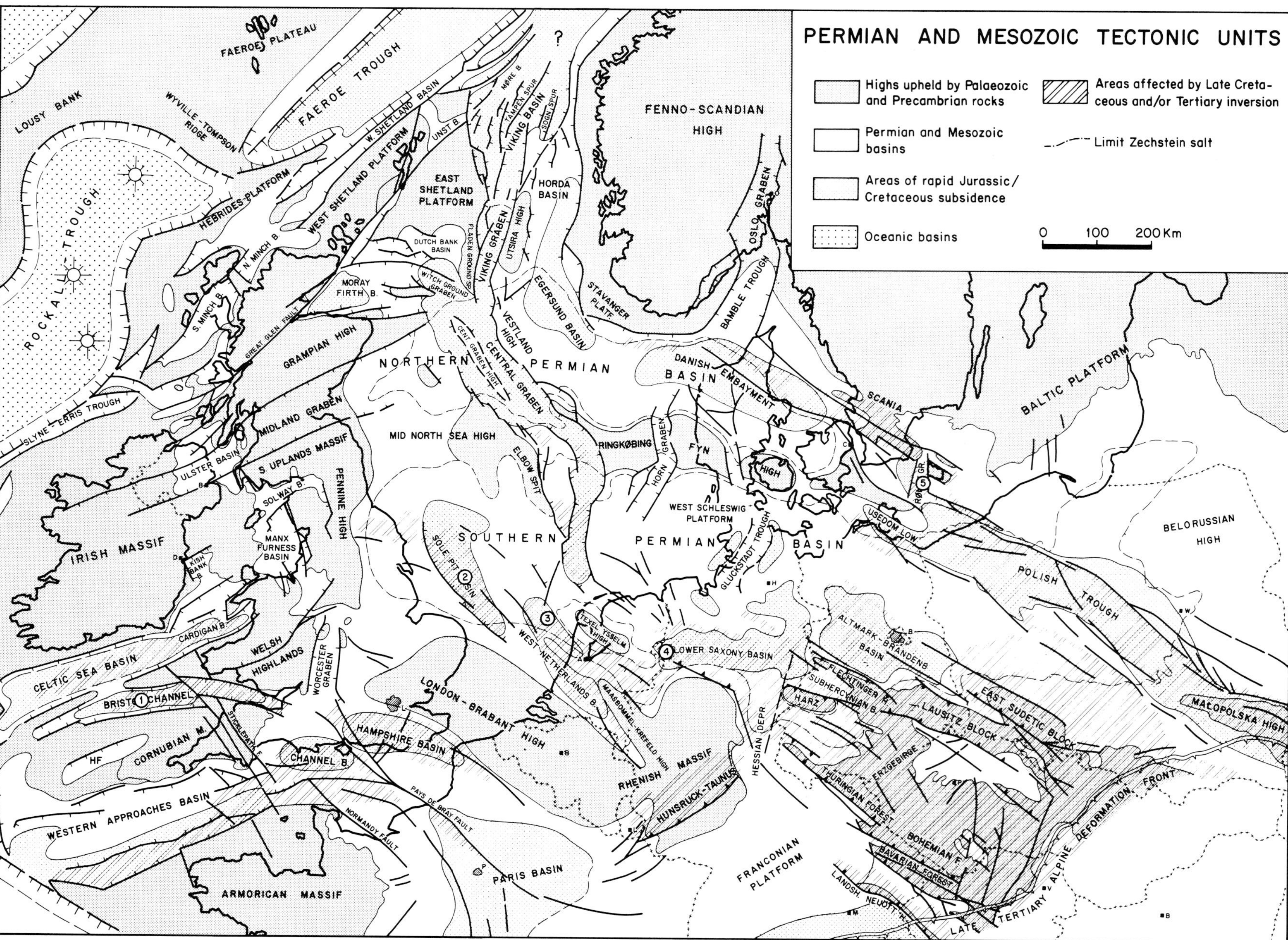

Fig. 10-1 (Ziegler, 1975)—Axes of inverted basins of northwest Europe with Upper Cretaceous paleogeography; Upper Cretaceous has been eroded from inversion structures. Inversion is also known west of U.K. and north of S. German high and Bohemian massif. Permission to publish by Elsevier Applied Science Publishers Ltd.

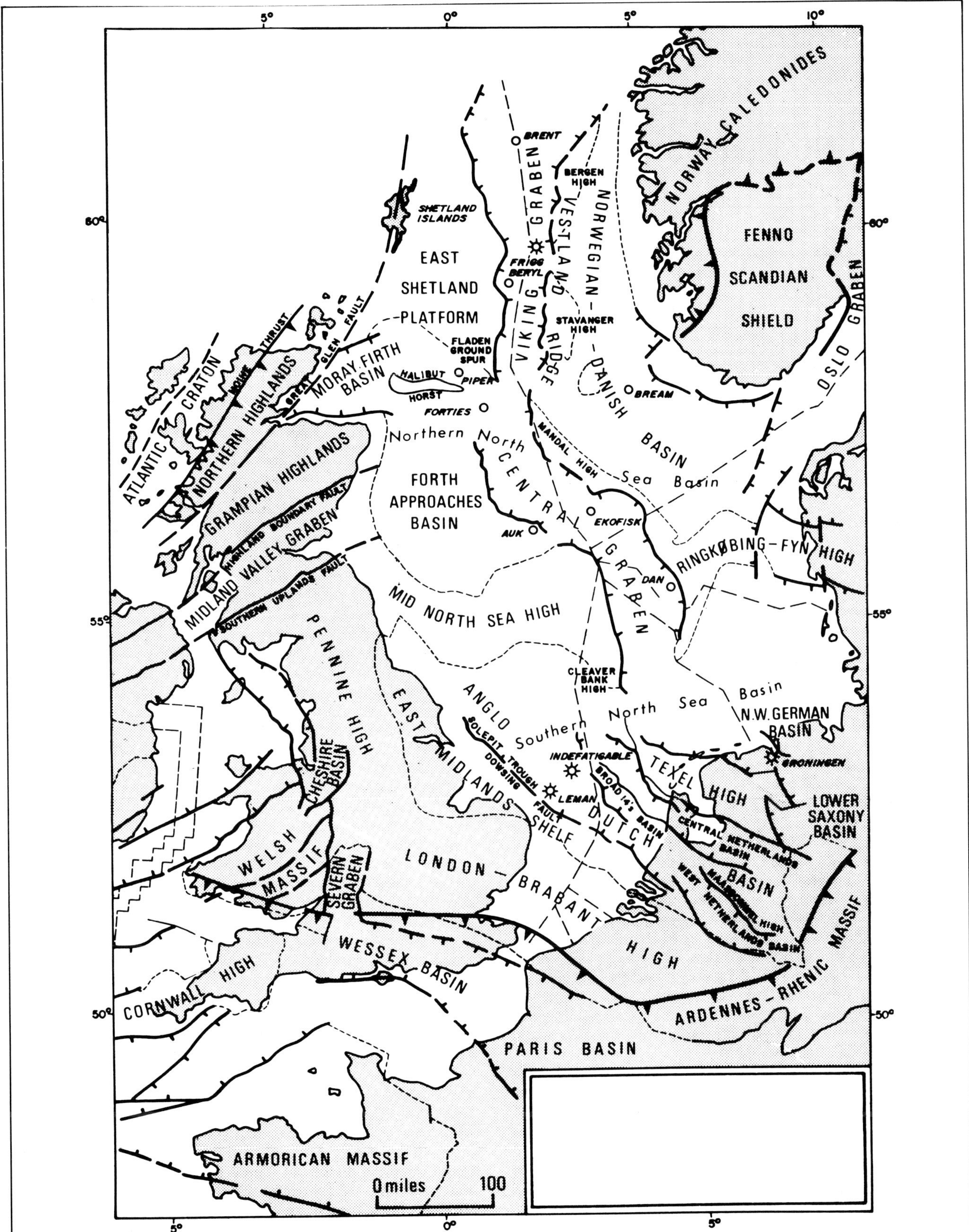

Fig. 10-2 (Blair, 1975)—Index and tectonic map for North Sea. Permission to publish by Elsevier Applied Science Publishers Ltd.

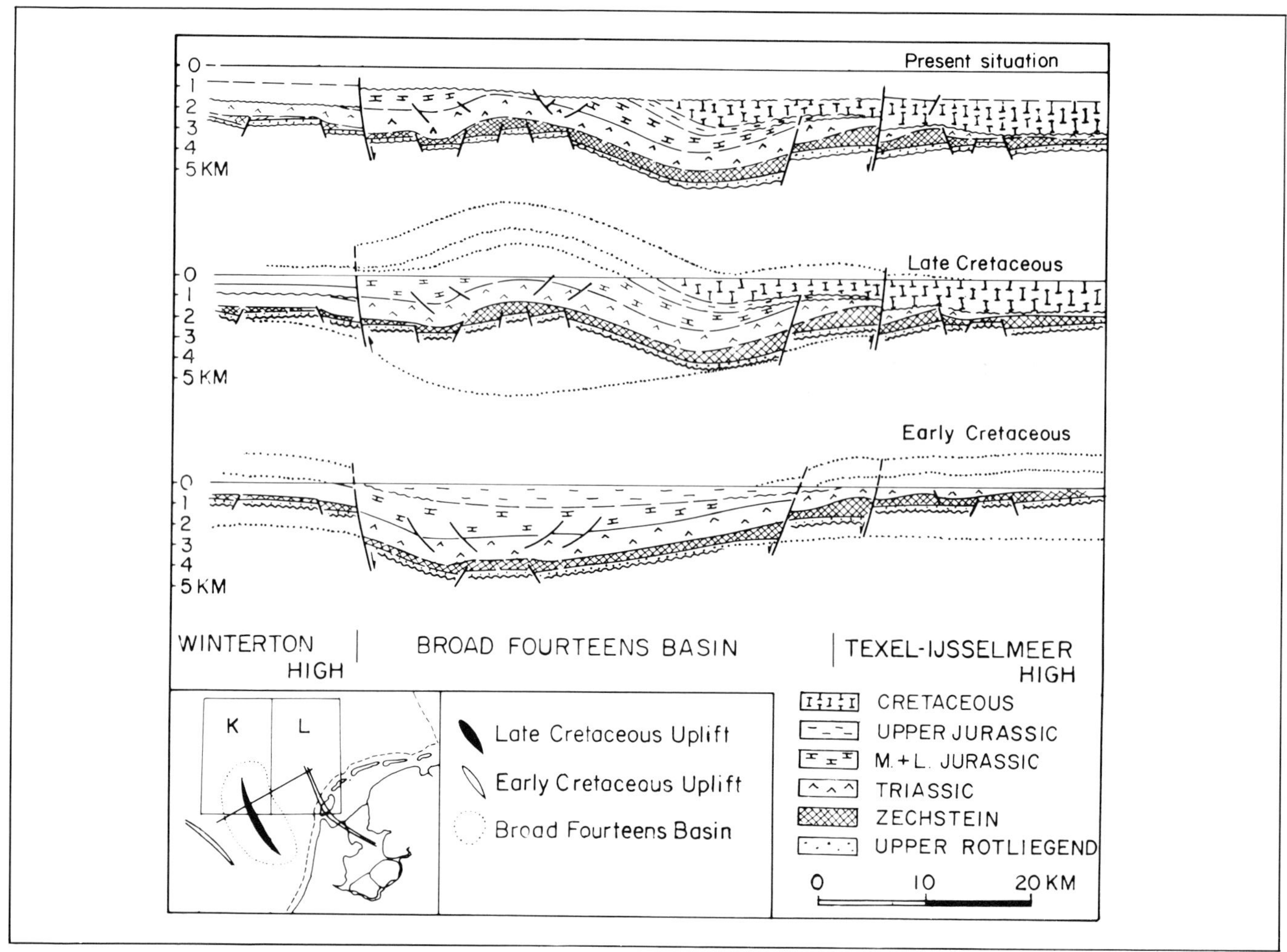

Fig. 10-3 (Oele et al., 1981)—Interpretation of Broad Fourteens Basin (offshore Netherlands) as inversion structure. Permission to publish by The Institute of Petroleum, London.

inversion. The distance of more than 1000 km (620 mi) between some of the inverted structures to the Alpine orogenic front (Fig. 10-1) appears too great for compression from that source to be effective in inversion. The process of southeastward underthrusting of the northwest European plate during Alpine deformation could, however, provide an explanation. Compression would work progressively northwestward in the European plate[1] during underthrusting and upon "feeling" the normal-faulted basins would shorten and thereby uplift or invert them. Compression and associated shortening were selective and acted only on those normal-faulted basins because they were areas in which the lithosphere was previously thinned and thus weakened by incipient rifting.

[1]It is pertinent to the hypothesis of underthrusting that present day in situ stresses in the northern Alpine foreland (from east of the Rhine graben to the Atlantic coast and including the Paris Basin) have been measured as consistently compressive, horizontal, and directed north-northwest (Griener and Lohr, 1979; Froidevaux et al., 1979).

Where such basins were at some angle other than perpendicular to regional compression, a wrench component of predictable displacement would be introduced. For the inversion features immediately north of the south German high (Fig. 10-1), for example, that component should be right lateral in theory. Independent evidence of wrenching can be found in the North Sea (Figs. 10-5, 10-6). At the northeast edge of the inverted Sole Pit trough (Fig. 10-2), faults with horizontal slickensides, upthrust profile (flower structure), repeated section, and both normal and reverse senses of displacement, all features characteristic of wrenching, have been reported from the Indefatigable field. None of these features accord with a vertical uplift hypothesis of inversion.

A test of the underthrusting hypothesis would be possible by analyzing the timing of structural inversion. The onset of inversion in Cretaceous time is consistent with the first Alpine orogenic movements. Inversion should then be progressively less intense (Ziegler, 1981) and younger to the

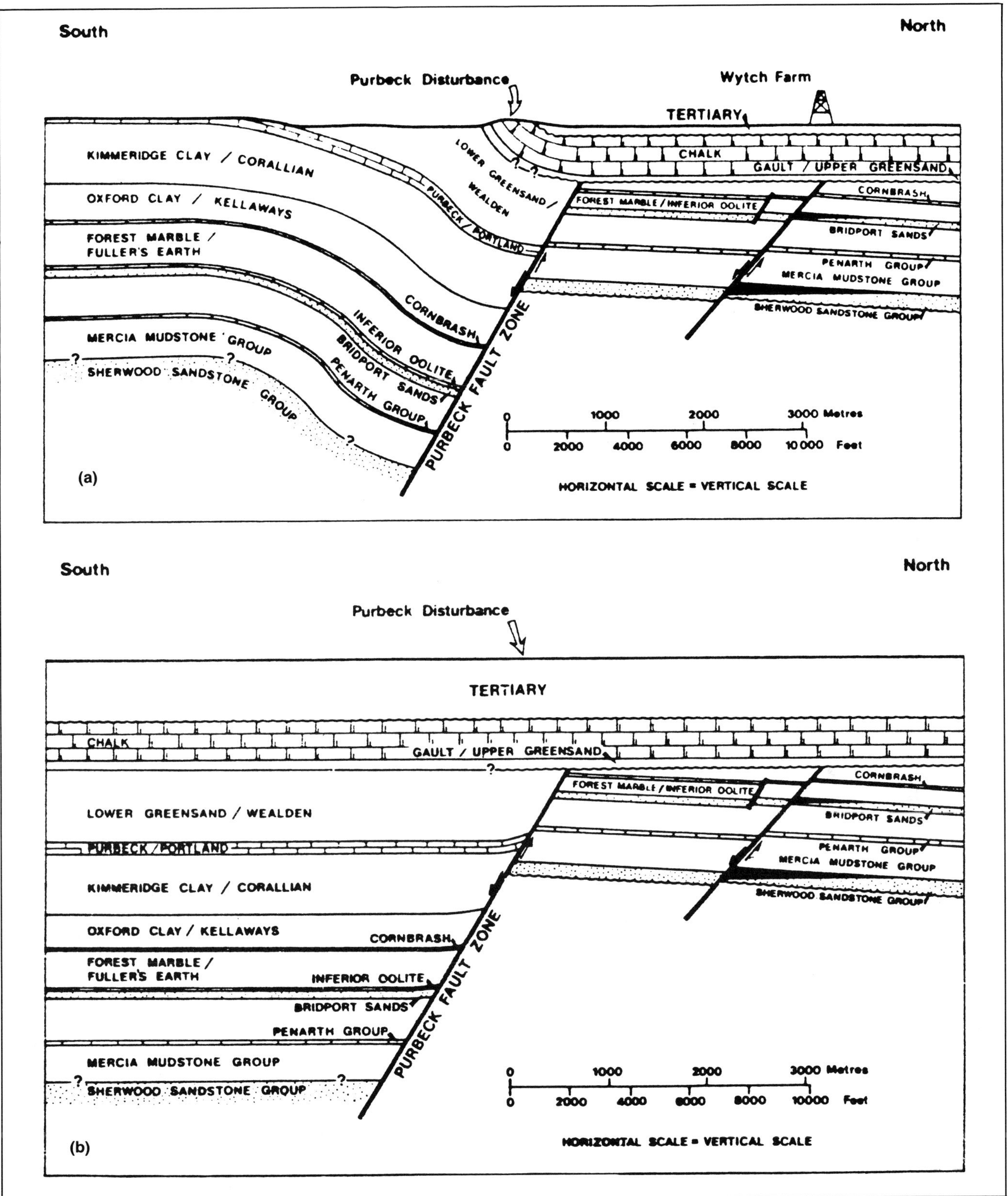

Fig. 10-4 (Colter and Harvard, 1981)—Interpreted structural sections through Wytch Farm field and Purbeck fault, southern England, illustrating structural inversion; (a). present day; (b). Tertiary, before Alpine earth movements. Permission to publish by The Institute of Petroleum, London.

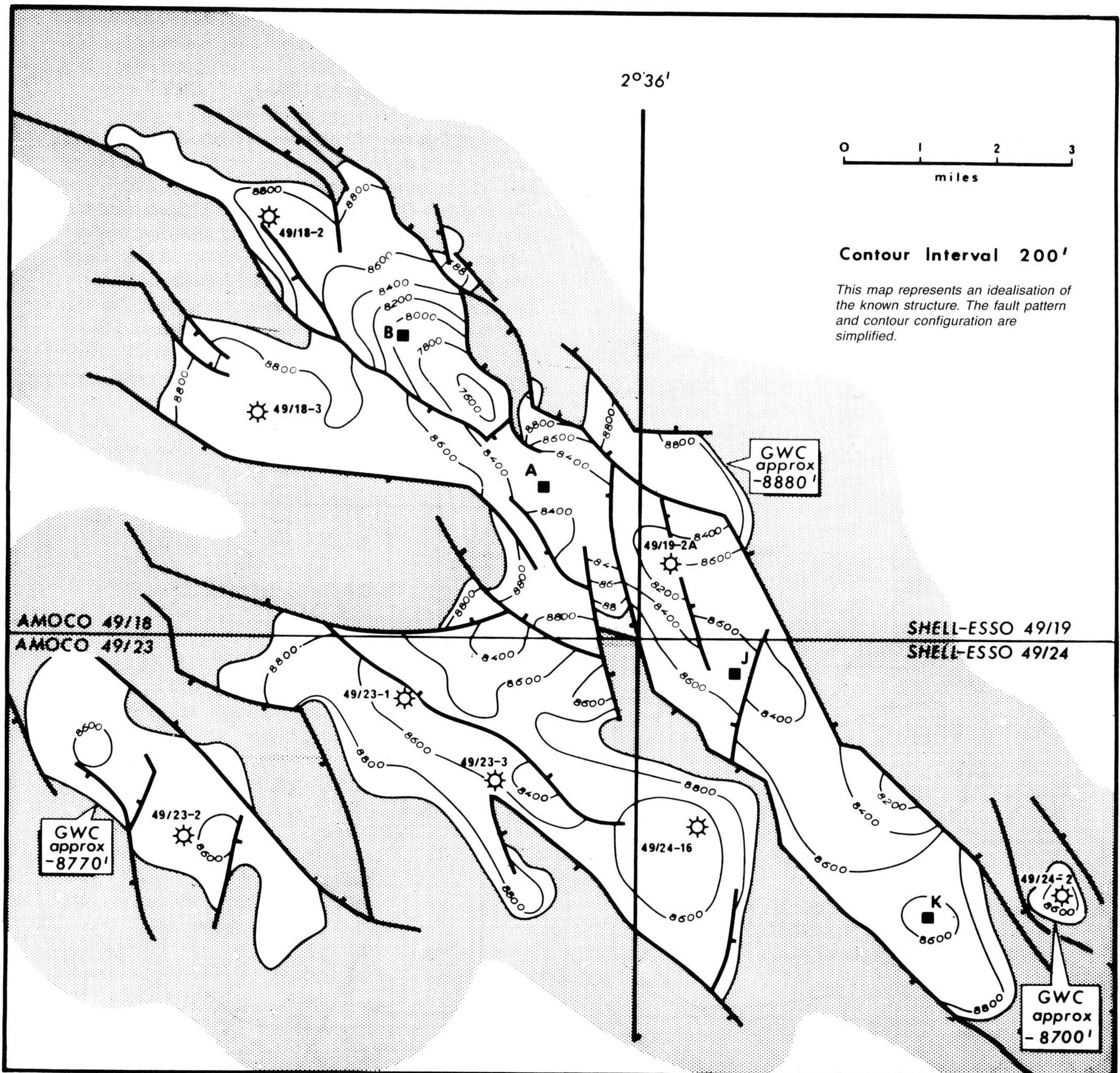

Fig. 10-5 (France, 1975)—Structure contour map on Lower Permian Rotliegendes Formation in Indefatigable field. Fault pattern is typical of rifting, but wrench motion could be concealed (see model studies -this chapter). Permission to publish by Elsevier Applied Science Publishers Ltd.

northwest. Inversion could occur as late as slightly later than the youngest Alpine movements. It is curious that the southwest side of the Sole Pit trough was not uplifted until latest Cretaceous time, whereas the northeast side experienced Early Cretaceous uplift (Glennie and Boegner, 1981). If inversion progressed as a northwestward-advancing wave, first the southeast, then the northwest ends of both flanks should have been uplifted at about the same time. Uplift would have been slightly earlier on the southeast end of the southwest flank. Possibly the earlier uplift of part of the northeast side was related to salt flowage.

Structural inversion has influenced the petroleum potential of many basins in the North Sea and elsewhere by affecting traps, reservoirs, source rocks, rock velocities, and even hydrocarbon accumulations themselves. Compression and wrenching that combine to uplift former structural lows

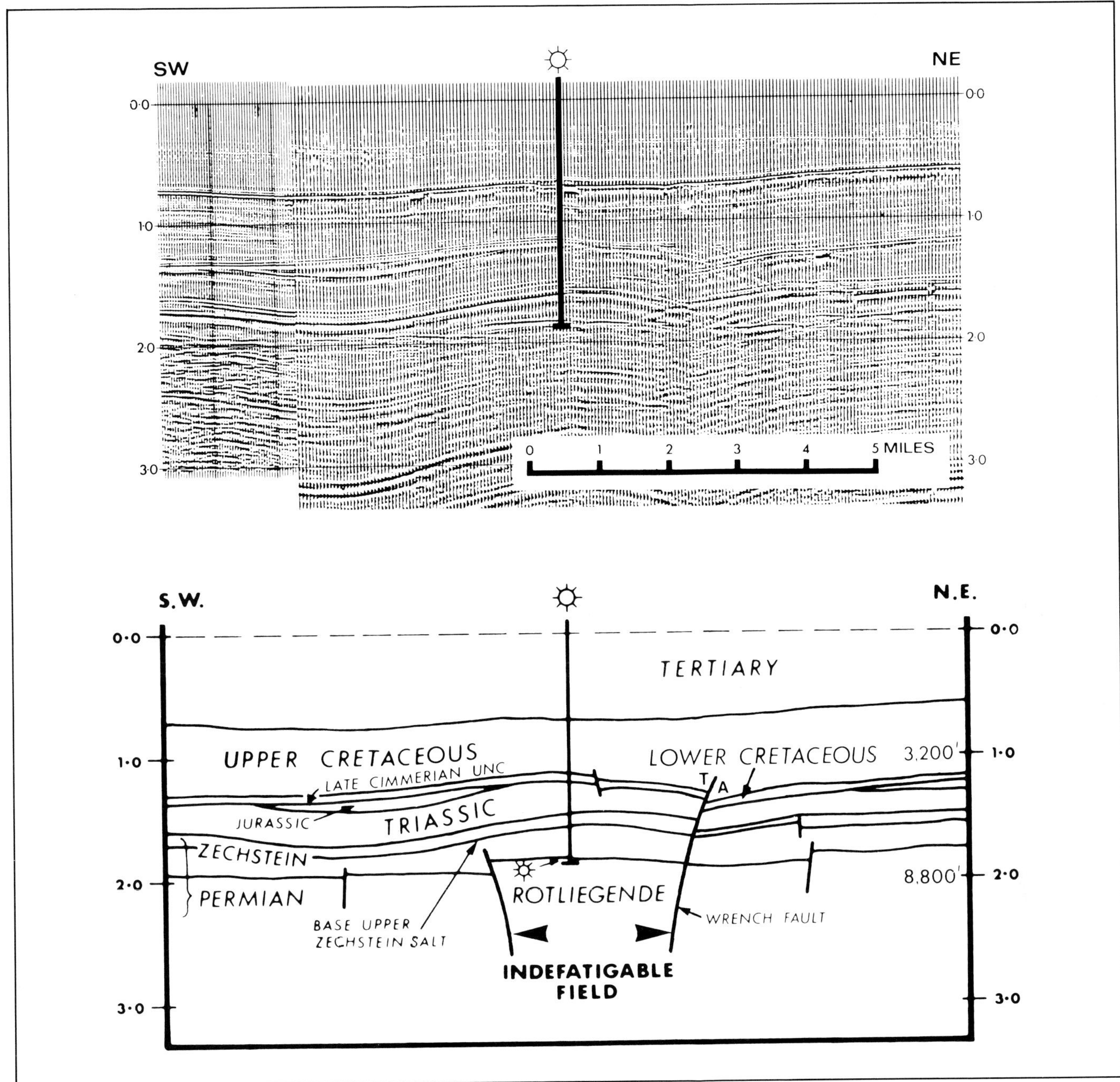

Fig. 10-6 (Blair, 1975)—Seismic line with geologic interpretation showing Indefatigable field as related to wrenching. Permission to publish by Elsevier Applied Science Publishers Ltd.

can also form potential traps. In the North Celtic Sea off southeastern Ireland, the Kinsale Head gas field is a large anticline that was formed during inversion along the axis of a graben (Colley et al., 1981). Anticlinal traps were also created during inversion of the southwest Netherlands Basin (Bodenhausen and Ott, 1981).

Two basic aspects of inversion, early deep burial and later compression, can reduce porosity and permeability of potential reservoirs below values that would be anticipated for rocks at their present (post-inversion) depths. Glennie and Boegner (1981) have specifically cited destruction of porosity and permeability in the Rotliegendes sandstone of the Sole Pit area, as have Oele et al. (1981) in the Netherlands offshore. Potential source rocks on the other hand might benefit from early deep burial and associated higher temperatures that prompt maturation of petroleum. This includes both "conventional" fine-grained source rocks and coals that are susceptible to gasification (Oele et al., 1981).

Both the processes of deep burial and subsequent uplift by compression have increased velocities of the sedimentary rocks involved in inversion. By comparing velocities of the Triassic Bunter shale in the Sole Pit trough against those of regions in which the Bunter had undergone more normal continuous subsidence, Glennie and Boegner (1981) estimated that "the Bunter shale of parts of the Sole Pit High was once buried more than 1500 m (5000 ft) deeper than at present". Oele et al. (1981) also reported abnormally high velocities for lower Bunter Shale at its present depth of burial in the offshore Netherlands. The amount of velocity increase attributed to deep burial on the one hand and later horizontal compression on the other, however, could not be determined.

In the northern part of the southwest Netherlands Basin Bodenhausen and Ott (1982) have interpreted that hydrocarbons were destroyed by erosion during inversion.

Argentina

Structural inversion that leaves little doubt as to its cause has occurred in Salta province in northern Argentina. Reflection seismic clearly shows rift-fill section inverted by reverse movement on reactivated normal faults (Fig. 10-7). The Salta occurrence is particularly valuable in elucidating the origins of structural inversion. Inversion took place in the Andean foreland along a rifted trough that strikes ENE-WSW into the N-S striking Andes. Part of the trough is on strike with foreland compressive blocks and lies about 100 km (62 mi) from the front of the Andean thrust-fold belt. The geographic position of the inverted area together with the late time of inversion being coincident with the young time of Andean deformation show conclusively that structural inversion was caused by Andean compression, or, more specifically, compression related to underthrusting. The angle of the inversion axis to the deformation front requires that right-lateral shear was also important, if not dominant, in the inversion process.

Brazil

Structural inversion comparable to that in Argentina has also occurred in the Acre Basin of westernmost Brazil. Here, a few hundred kilometers east of the Andes, a thick Triassic-Jurassic section has been inverted along a reverse fault that cuts Tertiary and younger units (Fig. 10-8).

Central and South Sumatra

A particular type of inverted fold in central Sumatra has been termed a "Sunda" fold by Eubank and Makki (1981). Figure 10-9 shows Sunda folds typically to be fault-bounded shallow anticlines that are synclinal or half graben at depth so that depositional units thicken into the fold. Sunda folds are attributed by Eubank and Makki (1981) to wrench deformation of an earlier normal-faulted section; however, given their plate tectonic and regional structural setting in the foreland of the Barisan Mountains (Figs. 1-2C and 2-70), compression in addition to wrenching has also probably occurred. Sunda folds are not restricted to central Sumatra but occur elsewhere on the Sunda shelf.

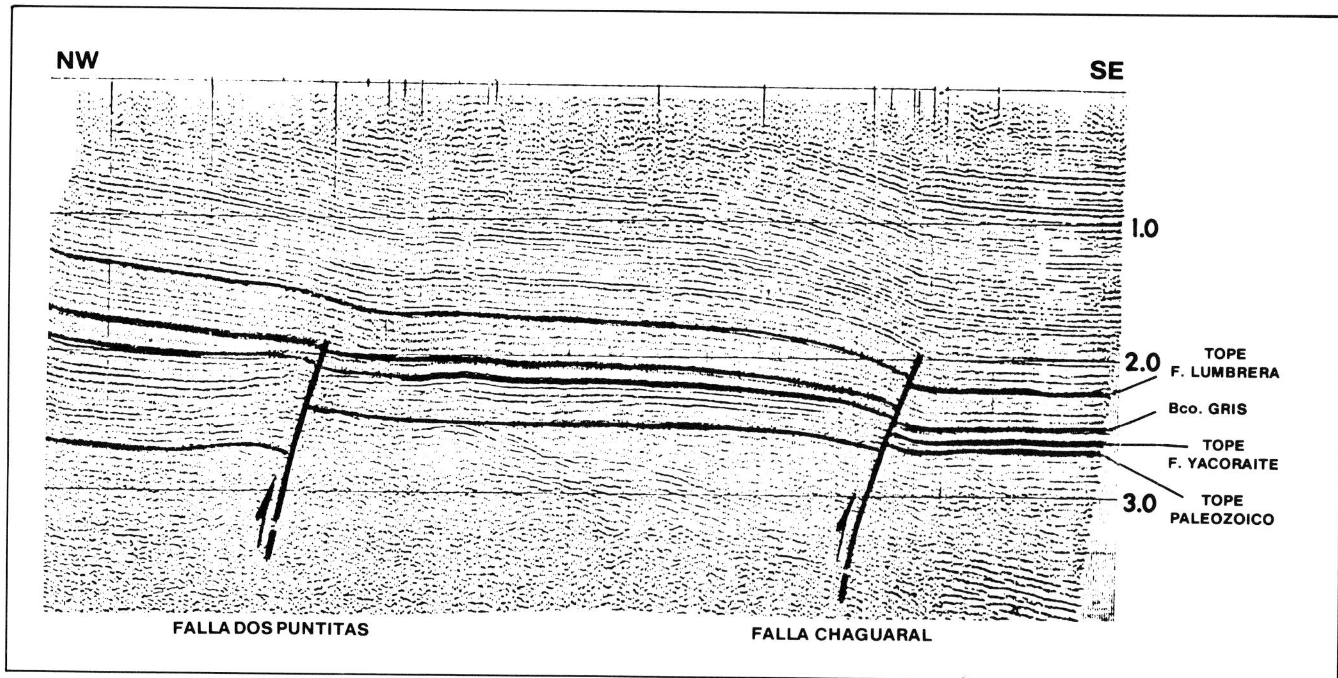

Fig. 10-7 (Bianucci et al., 1982)—Seismic line from Chaguaral area of Salta province, northern Argentina showing late structural inversion of Upper Cretaceous rocks (from top Paleozoic to top F. Lumbrera).

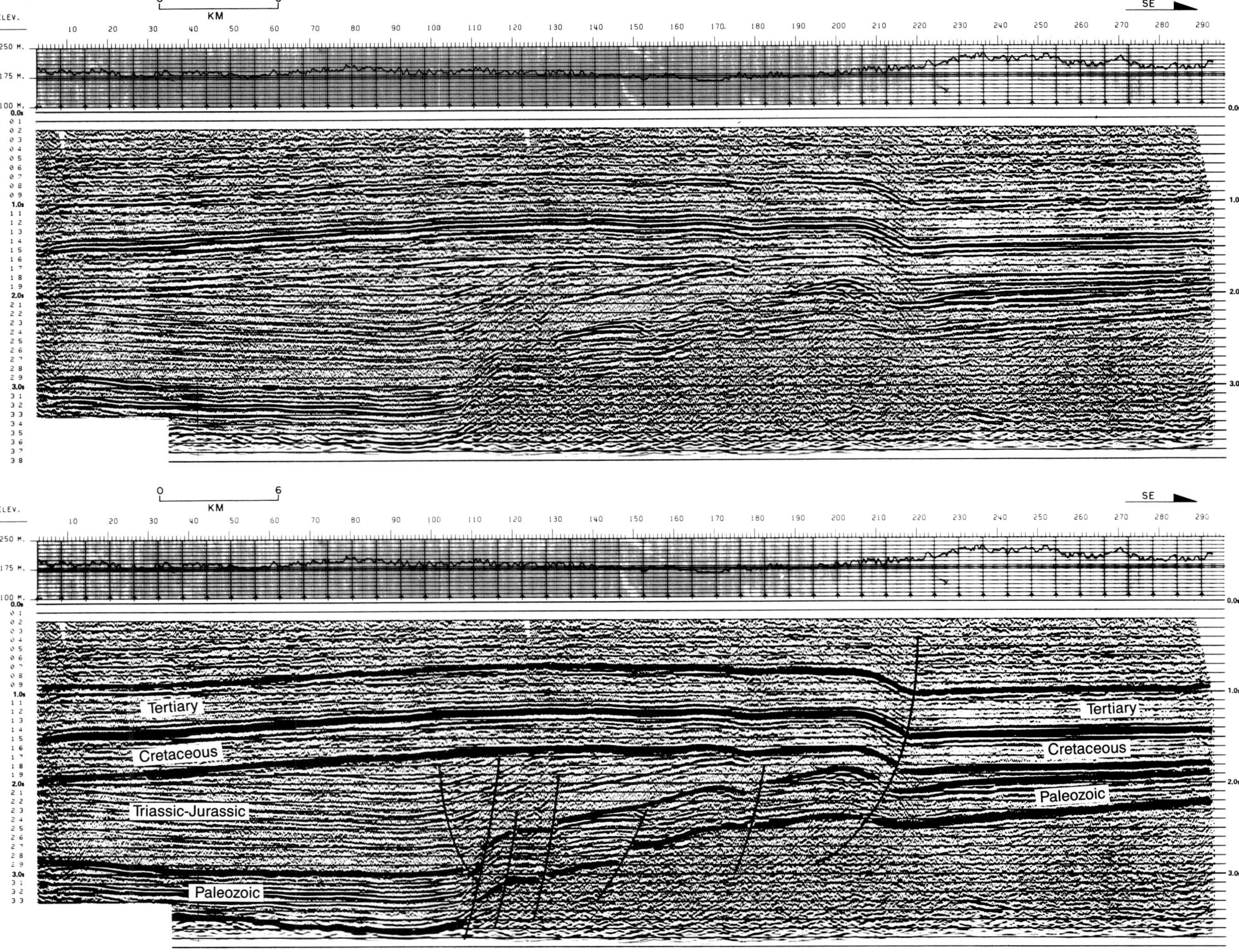

Fig. 10-8 (Petrobras, 1983)—Seismic line from Acre Basin, westernmost Brazil, showing inversion of Paleozoic-Lower Mesozoic basin. Permission to publish by American Association of Petroleum Geologists.

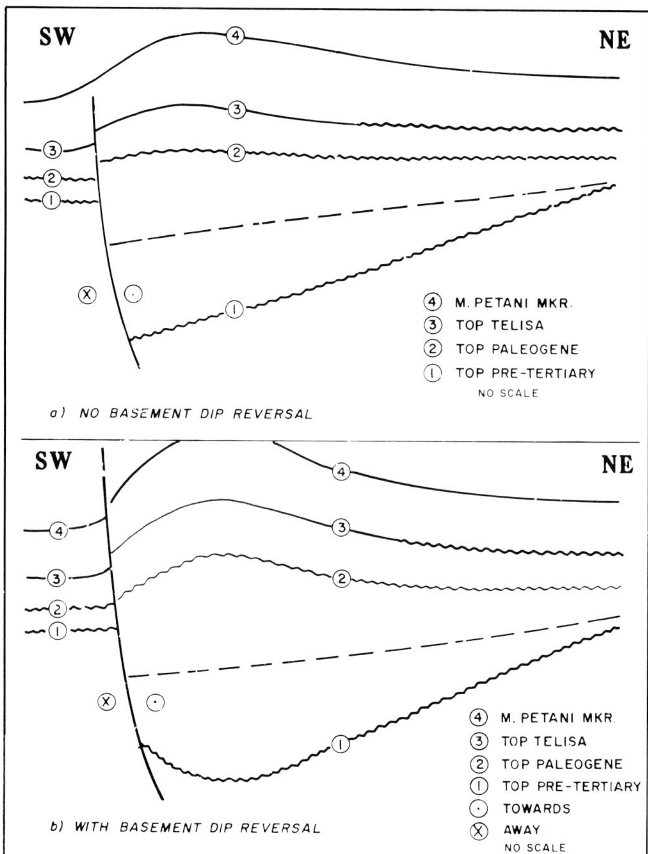

Fig. 10-9 (Eubank and Makki, 1981)—Diagrammatic cross sections of "Sunda" folds as inversion structures in the central Sumatra back-arc basin. Permission to publish by Indonesian Petroleum Association.

Inversion has occurred on a larger scale in south Sumatra where part of a basin has been inverted (Harding, 1983). Miocene to Oligocene sedimentary rocks are about three times thicker beneath a subsequently inverted basinal area than they are on an adjacent platform that is at present structurally low to the former basin on the top of Miocene surface (Fig. 10-10). Harding (1983) attributed this inversion dominantly to compressionally driven dip-slip on earlier normal faults rather than to wrenching because: 1– the map pattern suggests compression and 2– the listric faults do not resemble wrench faults.

Southeastern Australia

Another instance of large scale, basin inversion comes from the Gippsland Basin, offshore southeastern Australia (Fig. 10-11). Extensional deformation marked by expanded sections on the downthrown sides of normal faults dominated the basin from Early Cretaceous through early Eocene (upper part of Latrobe Group) time. Compression related to right-lateral wrench movement is interpreted to have inverted the basin by reactivation of older, normal faults in late Eocene and early Oligocene time.

Central Montana

Inversion has occurred along the Sand Creek fault zone in central Montana where post-Mississippian (Charles) through Jurassic (Morrison) section is noticeably thicker on the present uplifted side of the fault whose earlier downthrow was reversed by Laramide (post-Eagle) deformation (Plawman, 1983; Fig. 10-12). Multiple inversion may have occurred for the Cambrian (Flathead) through Charles interval appears slightly thicker on the present downthrown side of the fault. Wrenching appears to have been important in the inversion because the Sand Creek fault is steep and bifurcates at the top. It is even possible that strike-slip could have partially juxtaposed thick and thin units across the fault and that some of the inversion is more apparent than real.

MODEL STUDIES

Clay model experiments have been performed to simulate superposed deformations of normal faulting followed by wrenching (Figs. 10-13 through 10-15). Clay was first extended (Fig. 10-13) then subjected to right-lateral wrench motion (Figs. 10-14, 10-15). The earlier extension of the clay imparted a normal-fault fabric that controlled the later wrench deformation. Subsequent wrench motion took place entirely on the earlier formed normal faults without creating any of the diagnostic signs of wrenching. Apparently from a least work principle it was easier for wrench movement to be accommodated on the pre-existing dip-slip normal faults than to create any of the characteristic structures of the wrench assemblage. Model studies suggest, then, that significant wrench motion occurring after an earlier episode of rifting could be overlooked from simply geometric considerations. Piercing points would be required to determine actual slip.

CORE COMPLEXES

Some 26 core complexes are known in the North American Cordillera from southern British Columbia through the western United States to Sonora province in Mexico (Fig. 10-16). The essential components of a core complex are shown on figure 10-17. Such features are, indeed, complex and still imperfectly understood. The recent Geological Society of America Memoir 151 (1980) emphasized their late (Tertiary) history of extension and metamorphism. This deformation, however, seems to have been superposed on an earlier compressional terrane.

Core complexes are interpreted here as having been initiated during pre-Late Jurassic metamorphism and atten-

Illustration 10-10 begins on next page ⟶

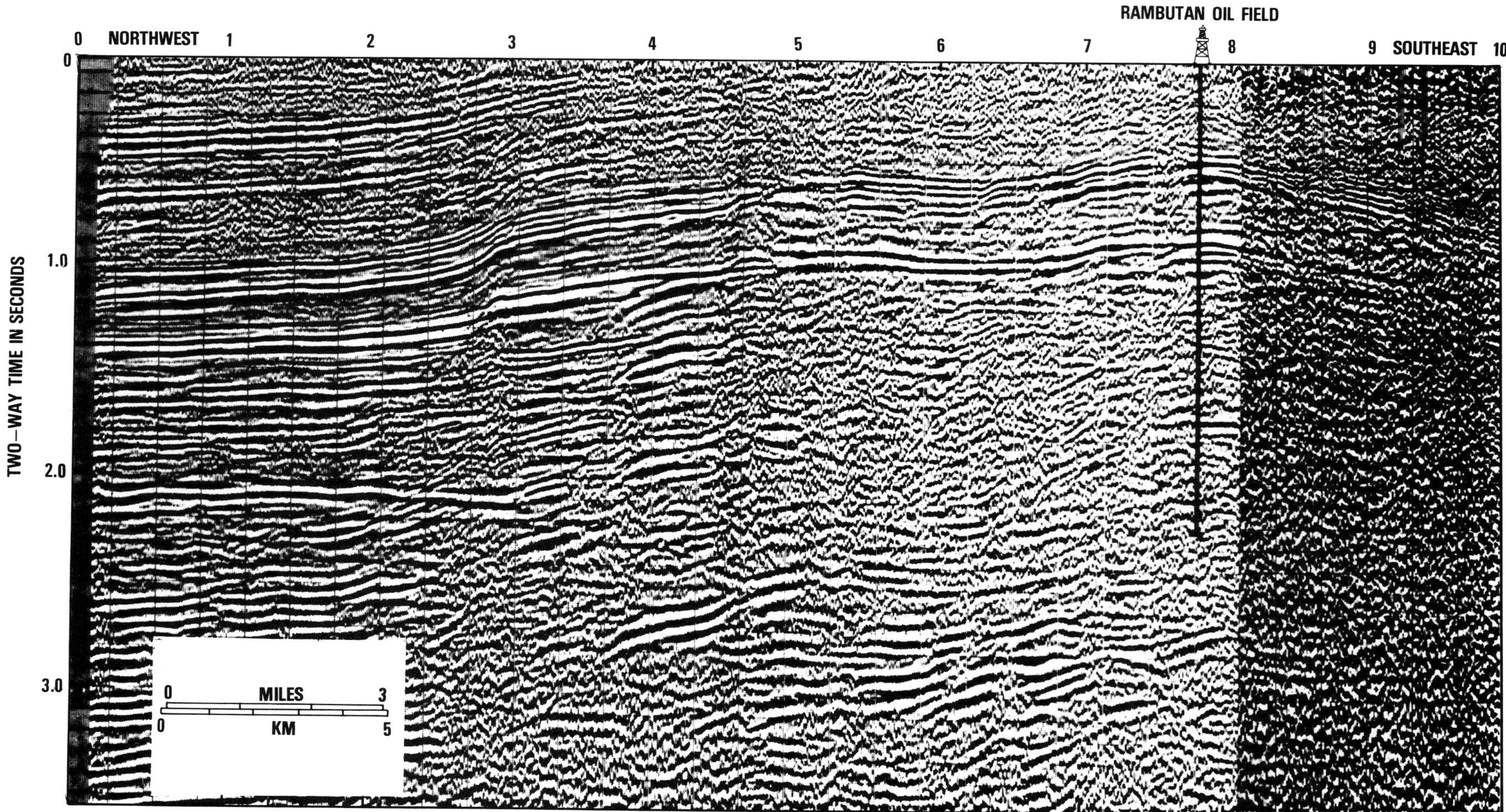
RAMBUTAN OIL FIELD
0 NORTHWEST 1 2 3 4 5 6 7 8 9 SOUTHEAST 10
TWO—WAY TIME IN SECONDS
0
1.0
2.0
3.0
0 MILES 3
0 KM 5

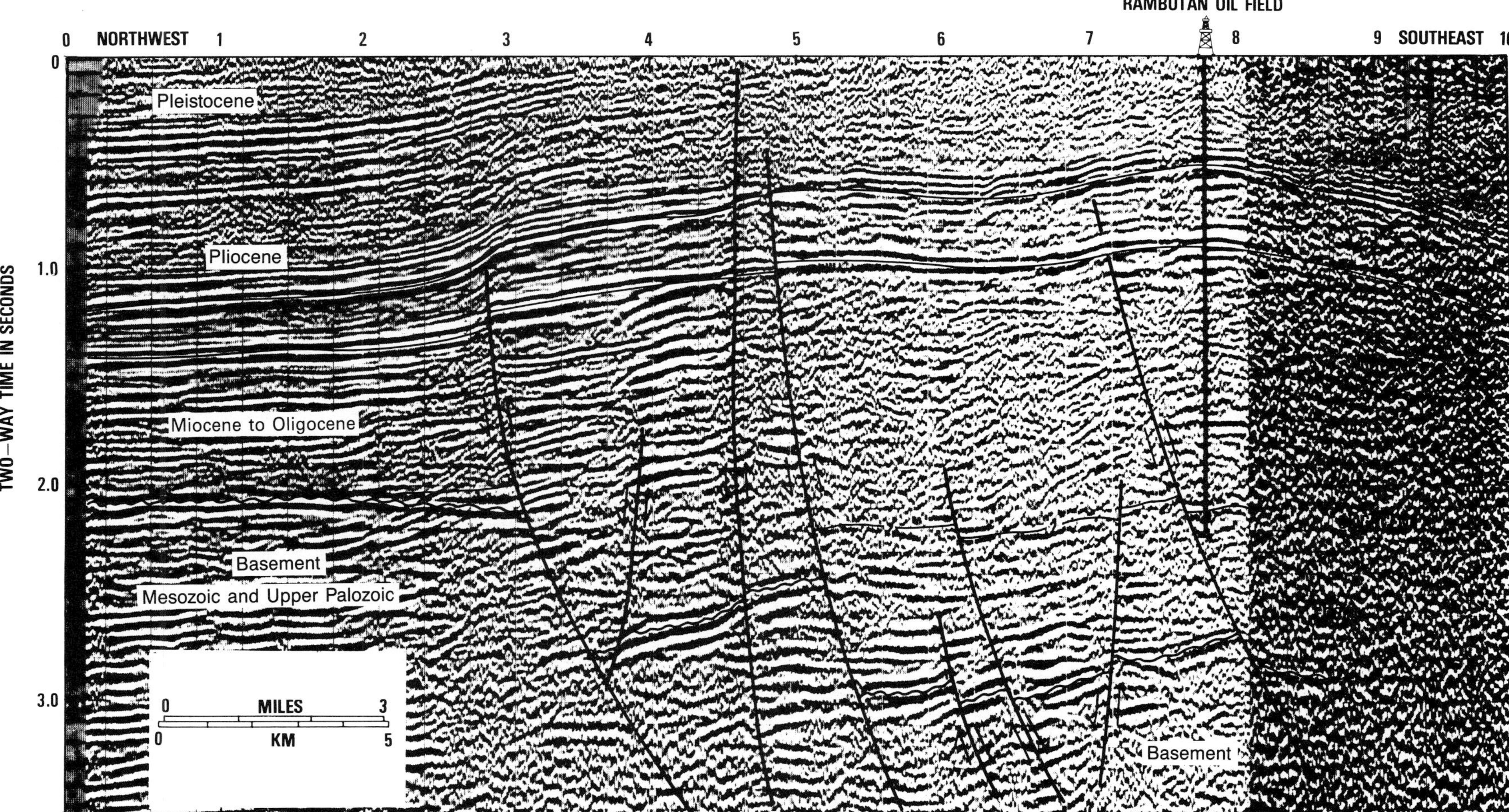

Fig. 10-10 (Harding, 1983)—Seismic section from south Sumatra showing basin inversion of mainly Miocene to Oligocene sections of Benakat gulley. Note that inversion is very late, affecting Pleistocene sediments. Compare shallow monoclines caused by reverse-sense displacement on earlier normal faults to those of figures 10-4, 10-6, 10-7, 10-8, and 10-11. Rambutan field has minor oil production from lower Pliocene and upper Miocene sandstones that are structurally high because of late inversion. Permission to publish by American Association of Petroleum Geologists.

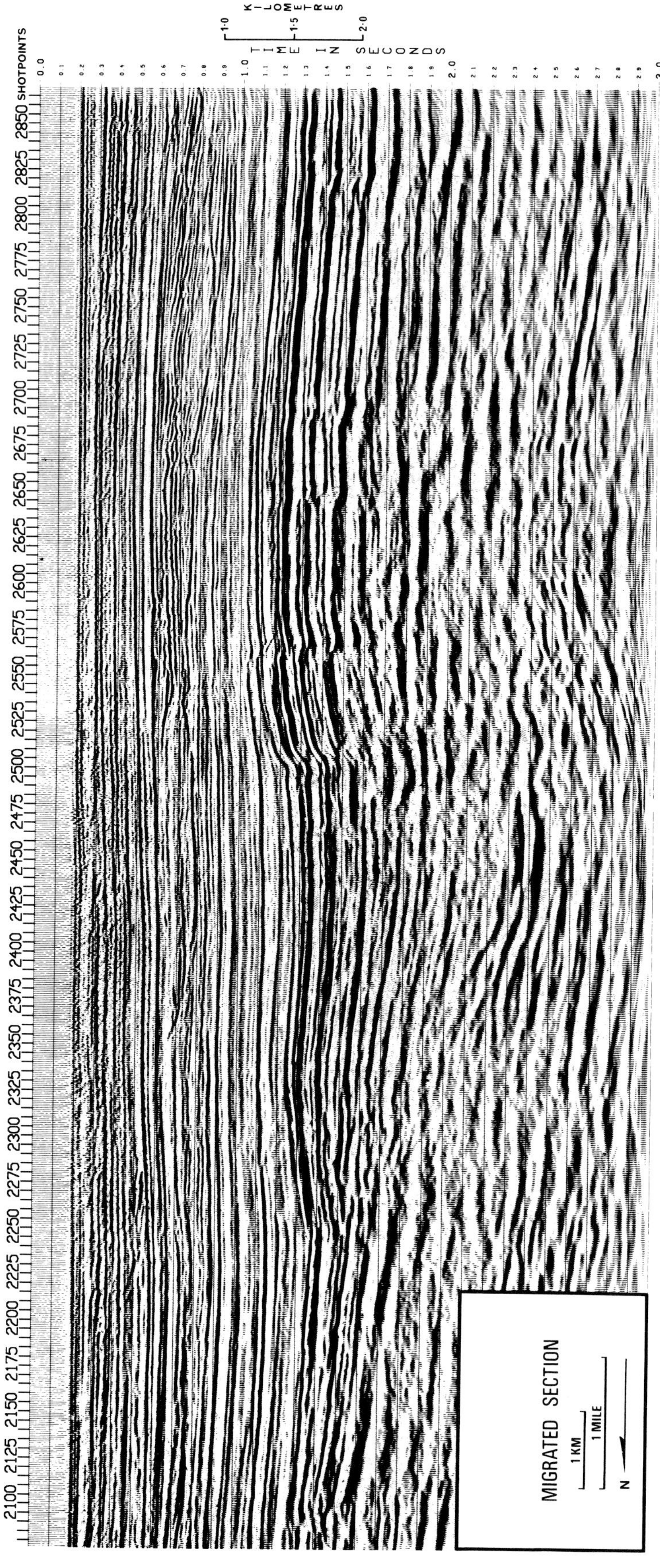
SHOTPOINTS
2100 2125 2150 2175 2200 2225 2250 2275 2300 2325 2350 2375 2400 2425 2450 2475 2500 2525 2550 2575 2600 2625 2650 2675 2700 2725 2750 2775 2800 2825 2850
TIME IN SECONDS
KILOMETRES
1·0 1·5 2·0
MIGRATED SECTION
1 KM
1 MILE
N

Fig. 10-11 (Davis, 1983)—Seismic line from offshore Gippsland Basin, southeastern Australia, showing inverted Latrobe Group (Upper Cretaceous-upper Eocene). Type 1 faults effected rift structure and sedimentation. Type 2 faults effected inversion. Permission to publish by American Association of Petroleum Geologists.

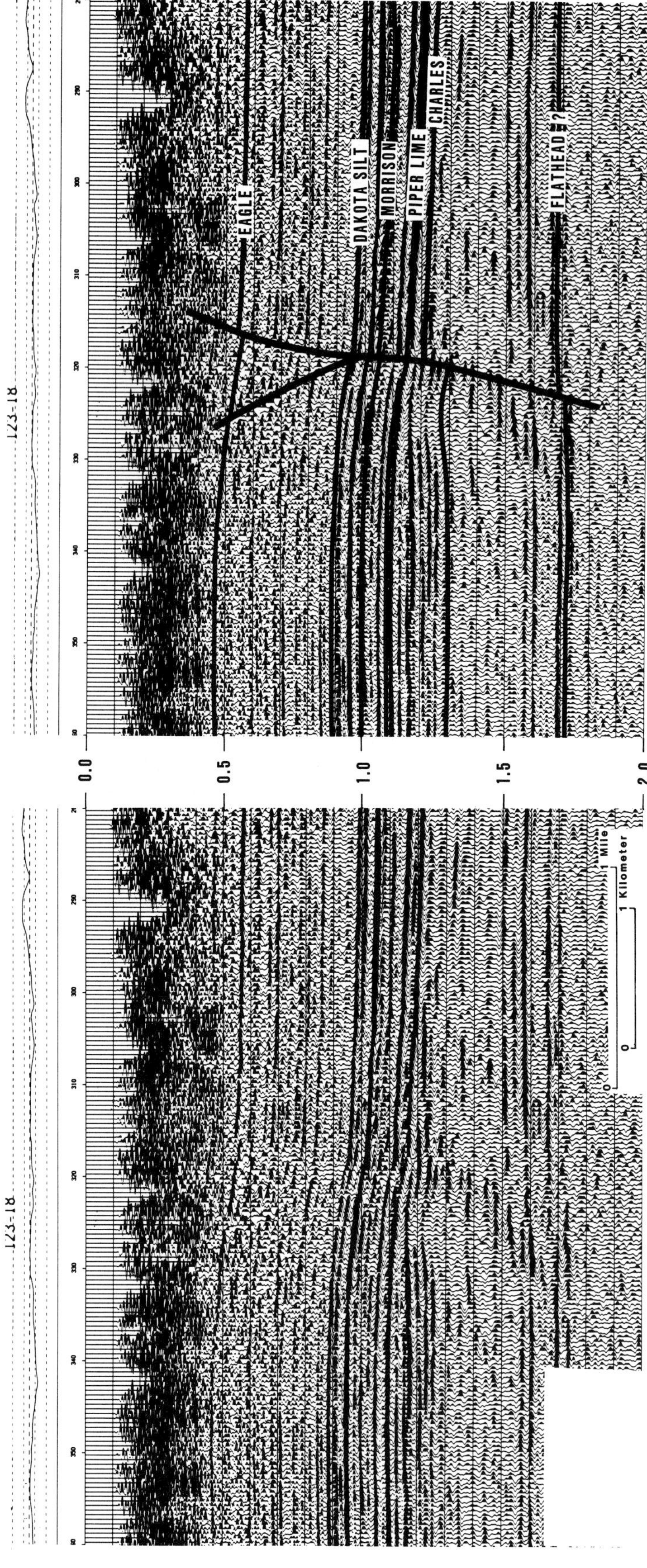

Fig. 10-12 (Plawman, 1983)—Seismic line across Sand Creek fault, central Montana, showing structural inversion. See text for additional discussion. Permission to publish by American Association of Petroleum Geologists.

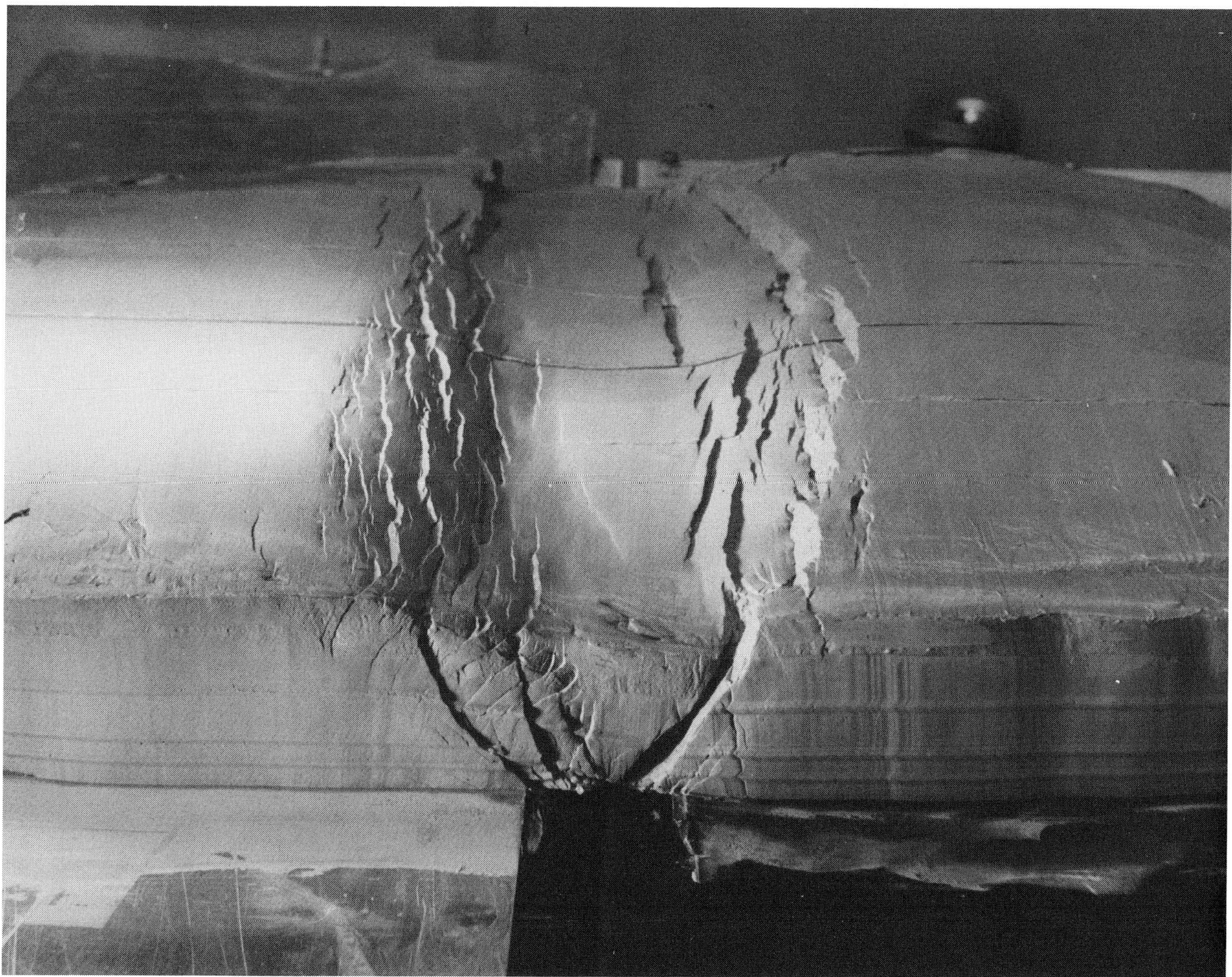

Fig. 10-13—Oblique view of clay model showing rifted normal fault pattern resulting from pure extension at base of clay. Note tin sheet beneath clay on left side.

dant underthrusting related to the Sevier orogeny; their Tertiary history was related to the lingering effects of the underthrust plate. As continental lithosphere was thickened by thrust and intraplate subduction telescoping, melting of the thickest portions of lithosphere created core complexes, some of which reached the near surface and were exposed by later erosion or unroofing. The tectonic occurrence of core complexes, which is enigmatic when related to thicknesses and depths of burial of Paleozoic section or to pre-existing Precambrian trends (Coney, 1980), would be related instead to the amount of uplift attendant upon lithospheric thickening by underthrusting. A prediction of the present hypothesis is that core complexes are part of a continuous subsurface trend. Their present surface exposures are simply a function of height of uplift and depth of erosion. Basically, they are present continuously along the North American Cordillera, although the regions of profound later extension are restricted to the most elevated portions. Needless to say core complexes should not be drilled for oil and gas.

SUCCESSOR BASINS

Successor basins are younger basins superposed on older terrain that was usually marked by a different structural style. Figures 6-34 and 6-35 show younger sedimentary rocks preserved in half graben effected by listric normal faults superposed on areas of earlier thrust faults.

An example of a successor basin with production from both the younger basin and the rocks on which it was superposed comes from the Vienna Basin of Austria (Fig. 10-18).

CONCLUSION

The foregoing examples of superposed styles by no means exhaust the possibilities of superpositions. Lowell (1971,

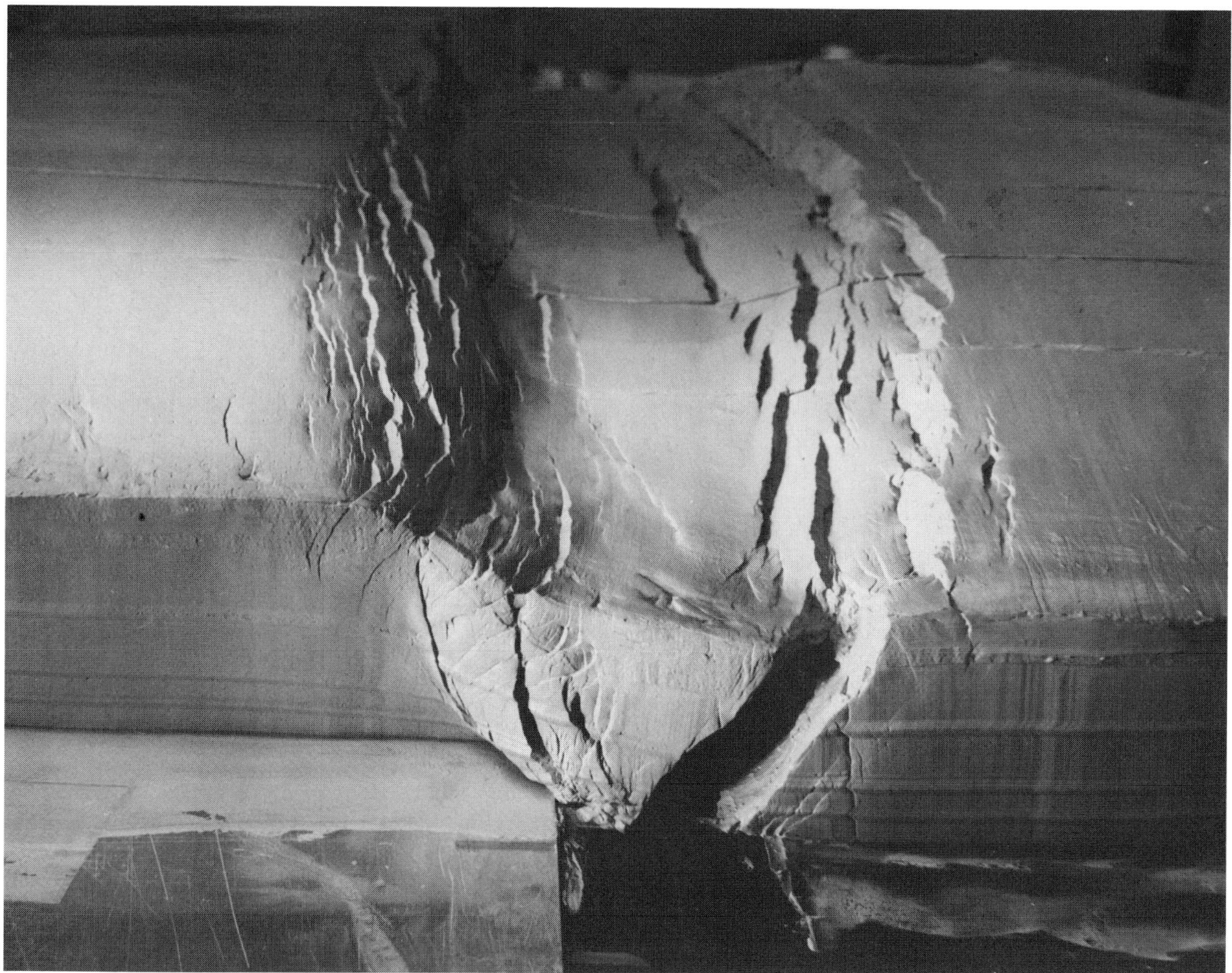

Fig. 10-14—Oblique view of clay model showing negligible effects of wrenching on pre-existing normal faults. As tin sheet (refer to Fig. 10-13) is pulled away from observer, imparting right-lateral slip to clay, wrench motion is absorbed on earlier normal faults.

unpub.), Helwig (1976), and Cohen (1982) suggested later reverse movement on earlier normal faults in thrust-fold belt and foreland settings (Fig. 10-19). One advantage of these interpretations is that the trends of foreland compressive blocks could then be inherited from those of earlier extensional blocks; as we have already seen the trends of both styles have much in common. Clearly numerous other possibilities of superposed styles also exist.

In addition to the concept of a simple "end member" style being complicated by combinations and superpositions of style, what might be termed "hybrid deformation" can also modify a particular style. Hybrid deformation is here considered a combination of motions that lead to oblique slip on associated faults; it has been anticipated by the kinematics of convergent and divergent wrenching (Chapter 2). The opening of the Red Sea illustrates hybrid deformation in that the direction of opening was not perpendicular to the trends of the main bounding faults (Fig. 10-20) so that slip must have been oblique. No doubt most rifting has occurred in a similar fashion. However the obliquity of opening rarely seems to mask the essential characteristics and patterns of the rifting process.

REFERENCES

Bianucci, Hugo, Homovc, J. F., and Acevedo, O.M. 1982, Inversion tectonica y plegamientos resultantes en la comarca Puesto Guardian-Dos Puntitos, Dpto. Oran, Provincia de Salta: 5 Congreso Latino-americano de Geologia, p. 23–30.

Blair, D. G., 1975, Structural styles in North Sea oil and gas fields, *in* Woodland, A. W., ed., Petroleum and the continental shelf of north-west Europe, V. 1 Geology: Halstead Press, p. 327–338.

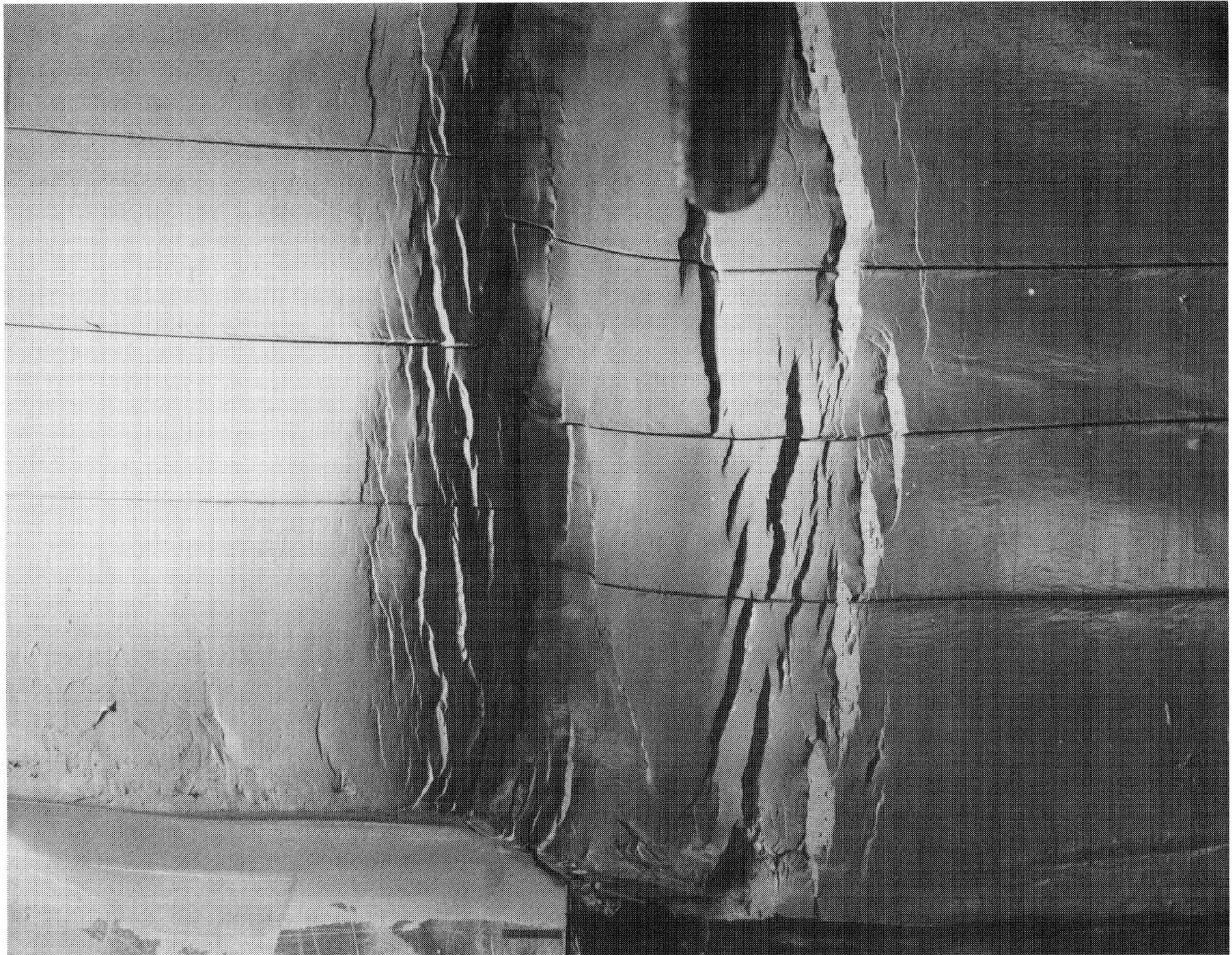

Fig. 10-15—Plan view of clay model showing right-lateral slip (offset lines) superposed on pre-existing, modified relay pattern of dip-slip normal faults. Effects of strike slip are negligible.

Bodenhausen, J. W. A., and Ott, W. F., 1981, Habitat of the Rijswijk oil province, onshore, The Netherlands, *in* Illing, L. V., and Hobson, G. D., eds., Petroleum geology of the continental slope of north-west Europe: The Institute of Petroleum, London, Heyden & Son Ltd., p. 301–309.

Cloos, H., 1939, Hebung-Spaltung-Vulkanismus: Elemente einer geometrischen Analyse irdischer Gross-formen: *Geologische Rundschau,* Bd. 30, Heft 4A, p. 405–527.

Cohen, C. R., 1982, Model for a passive to active continental margin transition: Implications for hydrocarbon exploration: *American Association Petroleum Geologists Bull.,* V. 66, p. 708–718.

Colley, M. G., McWilliams, A. S. F., and Myers, R. C., 1981, Geology of the Kinsale Head gas field, Celtic Sea, Ireland, *in* Illing, L. V., and Hobson, G. D., eds., Petroleum geology of the continental slope of north-west Europe: The Institute of Petroleum, London, Heyden & Son Ltd., p. 504–510.

Colter, V. S., and Harvard D. J., 1981, The Wytch Farm oil field, Dorset, *in* Illing, L. V., and Hobson, G. D., eds., Petroleum geology of the continental slope of north-west Europe: The Institute of Petroleum London, Heyden & Son Ltd., p. 494–503.

Coney, P. J., 1980, Cordilleran metamorphic core complexes: An overview: *Geological Society America Mem. 153,* p. 7–31.

Davis, P. N., 1983, Gippsland Basin, Southeastern Australia, *in* Bally, A. W., ed., Seismic expression of structural styles: *American Association Petroleum Geologists Studies in Geology #15,* V. 3.

Eubank, R. T., and Makki, A. C., 1981, Structural geology of the central Sumatra back-arc basin: Indonesian Petroleum Association, 10th Ann Conv.

France, D. S., 1975, The geology of the Indefatigable gas-field, *in* Woodland, A. W., ed., Petroleum and the continental shelf of north-west Europe, V. 1 Geology: Halstead Press, p. 233–239.

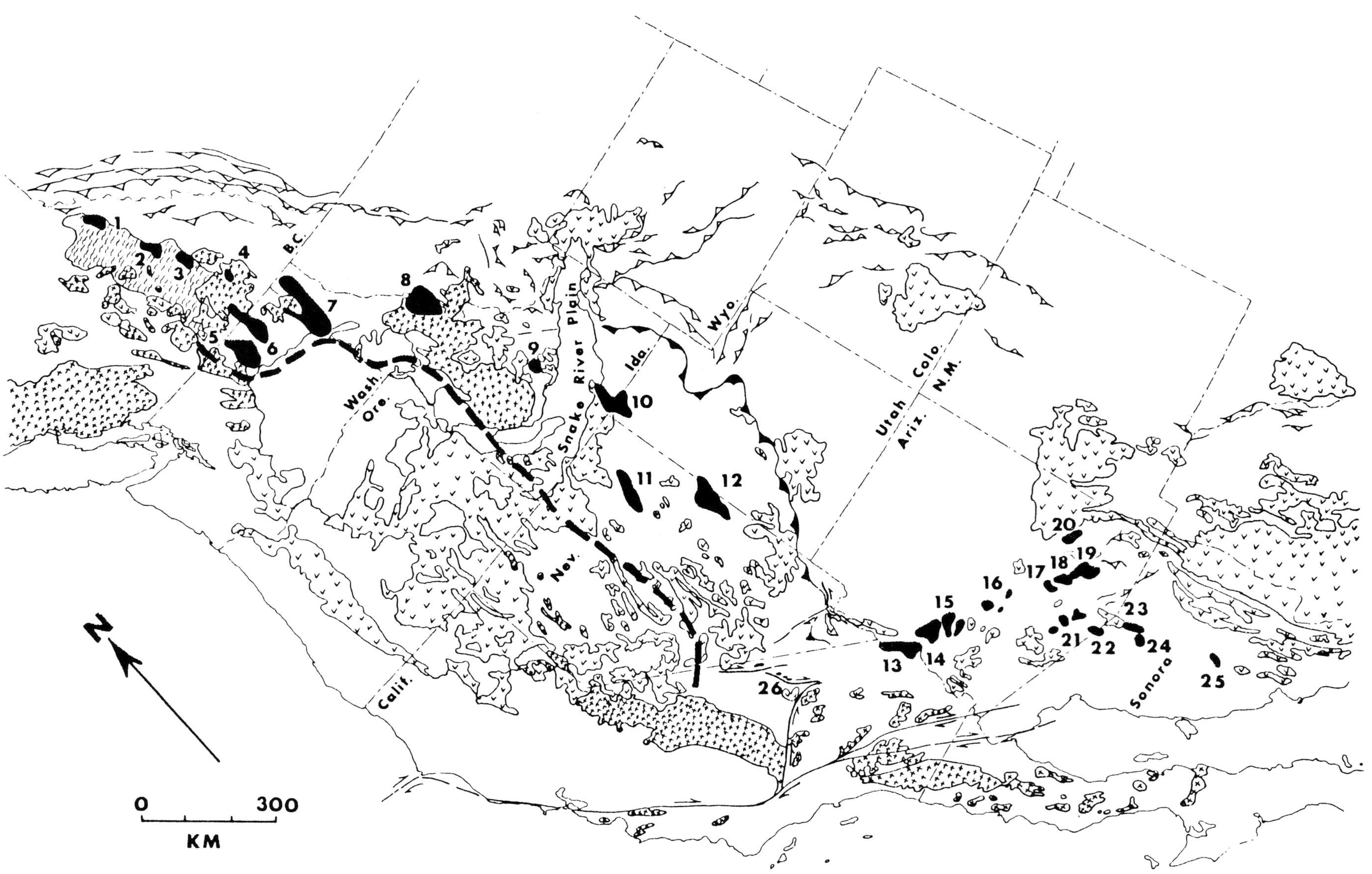

Fig. 10-16 (Coney, 1980)—Distribution and general regional tectonic setting of Cordilleran metamorphic core complexes, numbered as follows: 1– Frenchman's Cap; 2– Thor-Odin; 3– Pinnacles; 4– Valhalla; 5– Okanogan; 6– Kettle; 7– Selkirk; 8– Bitterroot (Idaho batholith); 9– Pioneer; 10– Albion-Raft River-Grouse Creek; 11– Ruby; 12– Snake Range; 13– Whipple; 14– Harcuvar; 15 Harquahalla; 16– South Mountains– White Tank; 17– Picacho; 18– Tortolita; 19– Catalina-Rincon; 20– Santa Teresa-Pinaleno; 21– Comobabi-Coyote; 22 Pozo Verde; 23– Magdalena; 24– Madera; 25– Mazatan; 26– Death Valley turtlebacks. From Geological Society of America Memoir 153, p. 7–31. Used with permission.

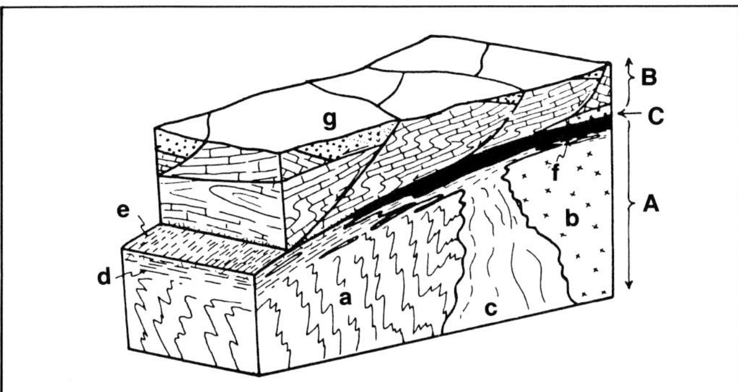

Fig. 10-17 (Coney, 1980)—Schematic structural block diagram of Cordilleran metamorphic core complex; A. basement terrane; B. cover terrane; C. decollement zone; a. older metasedimentary rocks; b. older pluton; c. younger pluton (early to middle Tertiary); d. mylonitic foliation; e. mylonitic lineation; f. marble tectonite (black); g. lower to middle Tertiary sedimentary and volcanic rocks. From Geological Society of America Memoir 153, p. 7–31. Used with permission.

Froidevaux, C., Paquin, C, and Souriau, M., 1979, Tectonic stresses in France: *American Geophysical Transactions,* V. 60, n. 32, p. 607.

Glennie, K. W., and Boegner, P. L. E., 1981, Sole Pit inversion tectonics, *in* Illing, L. V., and Hobson, G. D., eds., Petroleum geology of the continental slope of northwest Europe: The Institute of Petroleum, London, Heyden & Son Ltd., p. 110–120.

Griener, G., and Lohr, J., 1979, Tectonic stresses in the northern foreland of the Alpine system—measurements and interpretation: *American Geophysical Union Transactions,* V. 60, n. 32, p. 607.

Harding, T. P., 1983, Structural inversion at Rambutan oil field, south Sumatra Basin, *in* Bally, A. W., ed., Seismic expression of structural styles: *American Association Petroleum Geologists Studies in Geology #15,* V. 3.

Helwig, J., 1976, Shortening of continental crust in orogenic belts and plate tectonics: *Nature,* V. 260, p. 760–770.

Kroll, A., and Wessely, G., 1973, Neue Ergebnisse beim Tiefenaufschluss im Wiener Becken: *Erdoel-Erdgas-Zeitschrift,* 89. Jg., p. 400–413.

Lowell, J. D., and Genik, G. J., 1972, Sea-floor spreading and structural evolution of southern Red Sea: *American Association Petroleum Geologists Bull.,* V. 56, p. 247–259.

Oele, J. A., Hol, A.C.P., and Tiemens, J., 1981, some Rotliegend gas fields of the K and L blocks, Netherlands offshore (1968–1978)—a case history, *in* Illing, L. V., and Hobson, G. D., eds., Petroleum geology of the continental slope of north-west Europe: The Institute of Petroleum, London, Heyden & Son Ltd., p. 289–300.

Petrobras, 1983, Acre and Upper Amazon Basin, Brazil, *in* Bally, A. W., ed., Seismic expression of structural styles: *American Association Petroleum Geologists Studies in Geology Ser. #15.* V. 2.

Plawman, T. L., 1983, Fault with reversal of displacement, central Montana, *in* Bally, A. W., ed., Seismic expression of structural styles: *American Association Petroleum Geologists Studies in Geology #15,* V. 3.

Voigt, E. (1962) Uber Randtroge vor Schollenrander und ihre Bedeutung im Gebiet der Mitteleuropaischen Senke und angrenzender Gebiete: *Zeitschrift Deutsche Geologische Gesellschaft,* V. 114, 378–418.

Ziegler, P. A., 1975, North Sea basin history in the tectonic framework of northwestern Europe, *in* Woodland, A. W., ed., Petroleum and the continental shelf of northwest Europe, V. 1, Geology: Halstead Press, p. 131–148.

———, 1981, Evolution of sedimentary basins in North-West Europe, *in* Illing, L. V., and Hobson, G. D., eds., Petroleum geology of the continental slope of north-west Europe: The Institute of Petroleum, London, Heyden & Son Ltd., p. 3–39.

———, 1983, Inverted basins in the Alpine foreland, in Bally, A. W., ed., Seismic expression of structural styles: *American Association Petroleum Geologists Studies in Geology, Ser. #15,* v. 3.

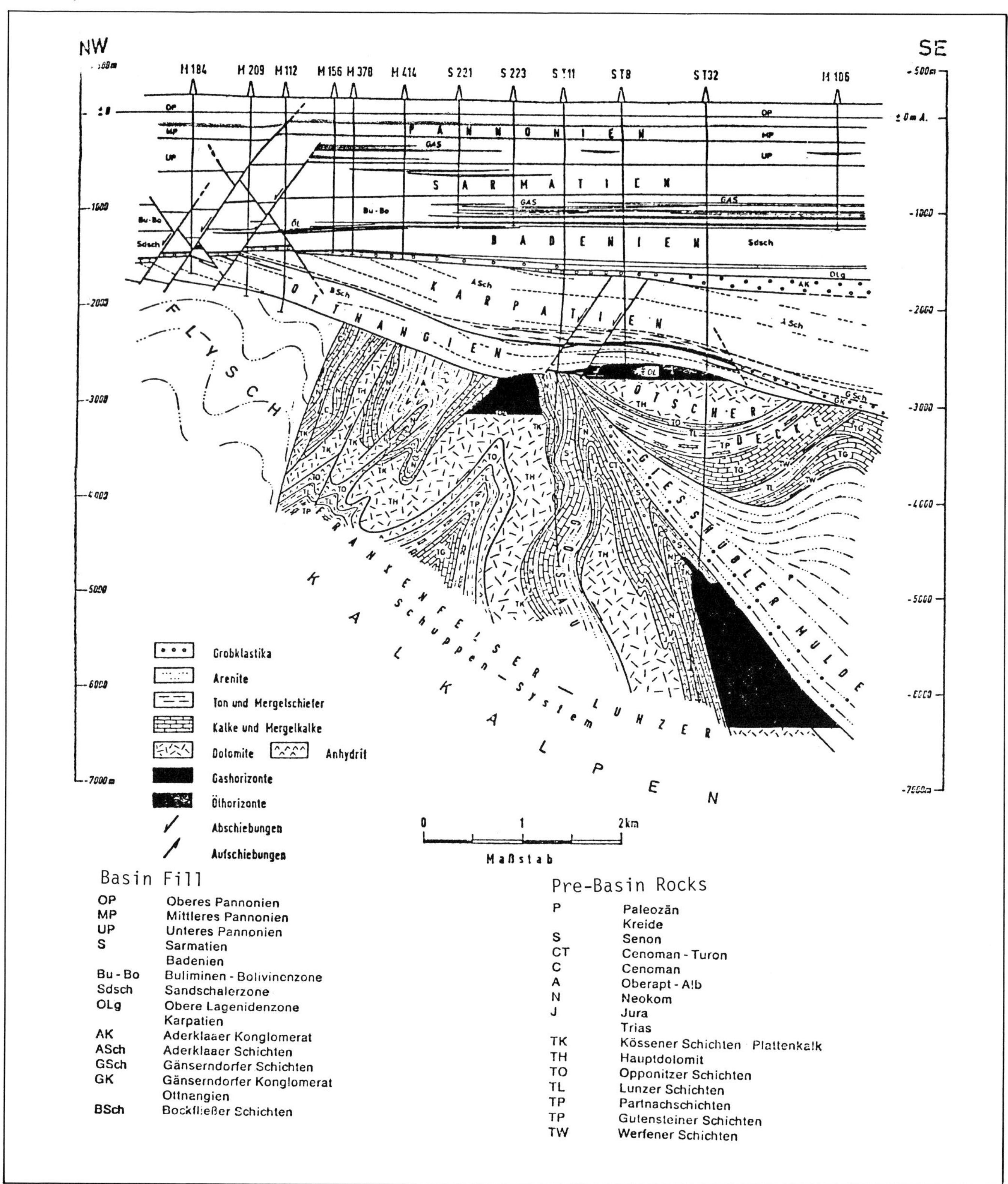

Fig. 10-18 (Kroll and Wessely, 1973)—Cross section in normal-faulted Tertiary Vienna Basin and its older basement of thrust sheets of the calcareous Alps. Hydrocarbons are found in both style settings. Reprinted by permission of OMV Aktiengesellschaft.

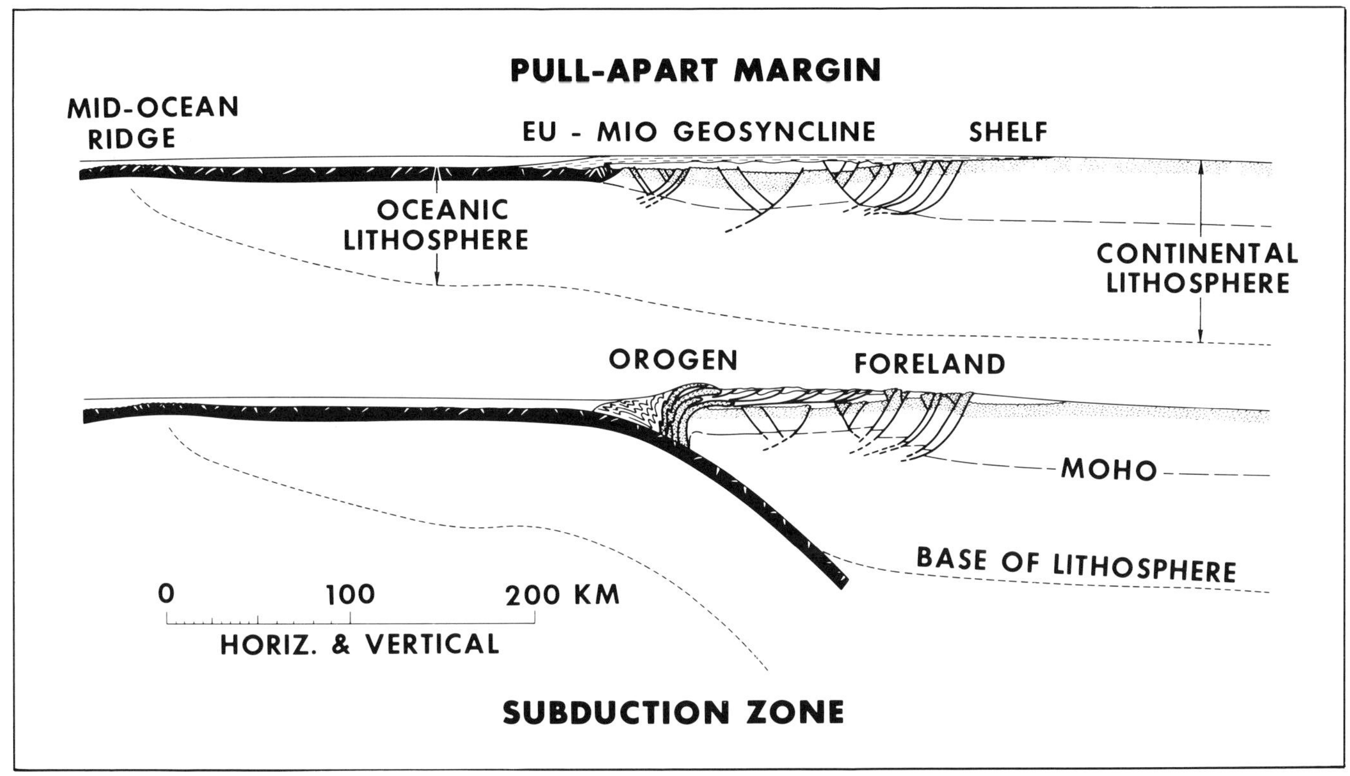

Fig. 10-19 (Lowell, 1971 unpub.)—Diagrammatic sections showing subduction and the conversion of a geosyncline and shelf of a pull-apart or extensional continental margin into an orogen and foreland, respectively. Reverse movement creates compressive blocks along the earlier normal faults that developed during the pull-apart phase.

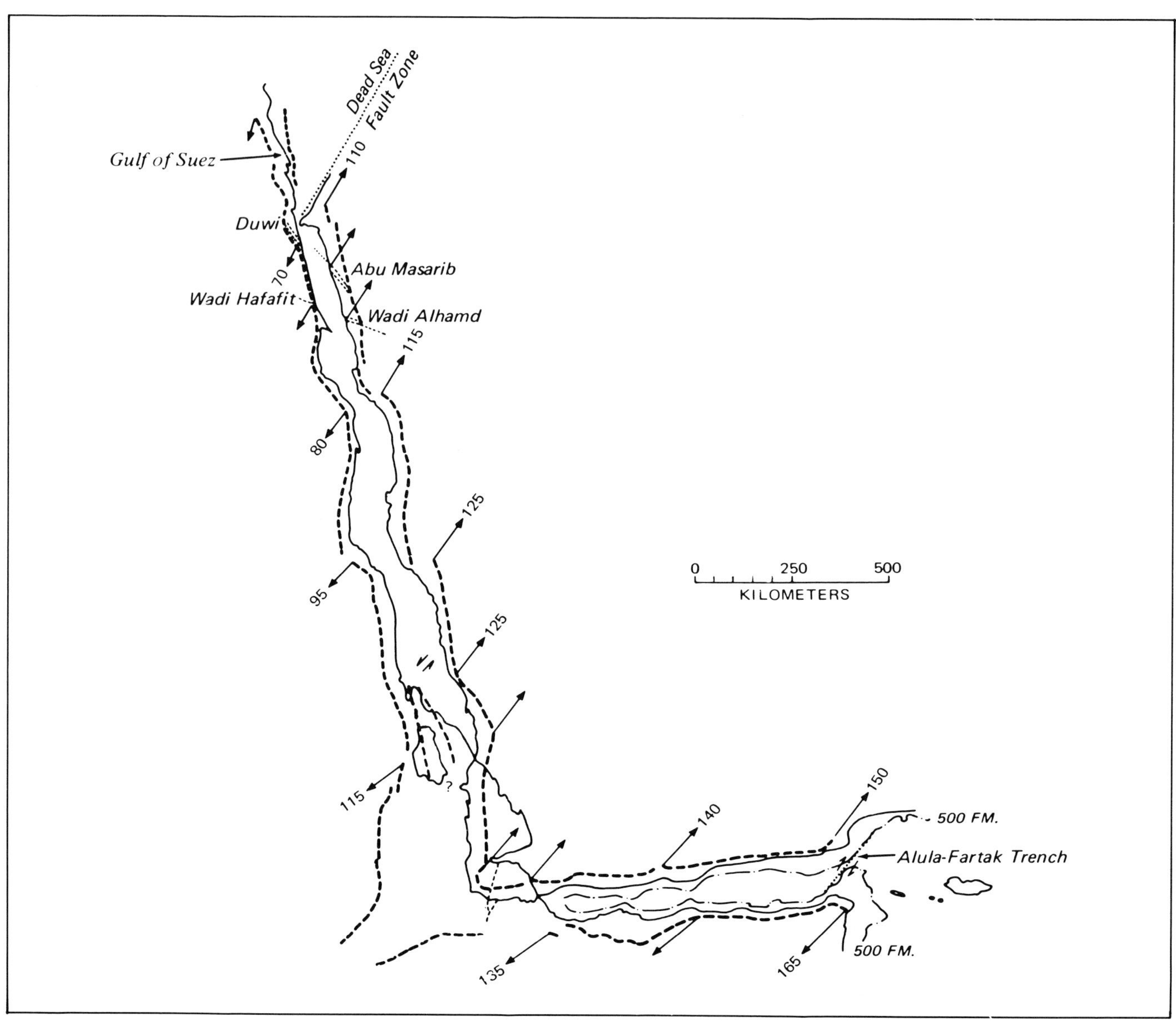

Fig. 10-20 (Lowell and Genik, 1972)—Prerift restoration of Red Sea and Gulf of Aden. Heavy dashed lines define approximate limits of rifting. Relative directions and amounts of separation shown by arrows with accompanying values in kilometers. Normal faults bounding Red Sea are parallel with coastlines rather than perpendicular to direction of separation. They are interpreted to have received this trend initially during arching stage, wherein faulting was longitudinal to broad regional arch (H. Cloos, 1939; E. Cloos, pers. com.); subsequent motion on normal faults from separation of lithospheric plates should have component of oblique slip. Permission to publish by American Association of Petroleum Geologists.

Subject Index

Geographical-Geological Location Index

Author Index